Liaison Functions 2017

Core Programming Area at the 2017 AIChE Annual Meeting

Minneapolis, Minnesota, USA
29 October – 3 November 2017

ISBN: 978-1-5108-5805-3

Printed from e-media with permission by:

Curran Associates, Inc.
57 Morehouse Lane
Red Hook, NY 12571

Some format issues inherent in the e-media version may also appear in this print version.

Printed by Curran Associates, Inc. (2018)

For permission requests, please contact AIChE
at the address below.

AIChE
120 Wall Street, FL 23
New York, NY 10005-4020

Phone: (800) 242-4363
Fax: (203) 775-5177

www.aiche.org

Additional copies of this publication are available from:

Curran Associates, Inc.
57 Morehouse Lane
Red Hook, NY 12571 USA
Phone: 845-758-0400
Fax: 845-758-2633
Email: curran@proceedings.com
Web: www.proceedings.com

TABLE OF CONTENTS

Some abstracts in this book refer to images that were not made available by the publisher

505003 Introductory Remarks by Nada Anid

Abstract:

PAIC aims to identify, respond to, and communicate public-policy issues at local, national, and international levels. It also aids AIChE members in learning how to communicate better with government and the public. In light of this mission, the goal of the Town Hall meeting is 1) to share the findings of three PAIC Task Forces and get the membersâ€™ input on each and on other matters of importance to the AIChE membership and 2) to consider the best means of communication to enhance the standing of the AIChE community and express its stand on policy matters.

504000 Food-Energy-Water Issues

Abstract:

The overlapping needs and technologies for food, energy, and water call for contributions from many people around the world, but chemical engineers bring especially valuable expertise. This briefing will summarize the issues and preview the topical conference that is taking place this week.

504002 Advanced Manufacturing

Abstract:

New â€œAdvanced Manufacturingâ€ approaches are expected to impact the production of chemical products significantly. Recent advances include process intensification, manufacturing for sustainability, new sensors and actuators, data science, and high-level simulations. We will discuss what advances and further developments need to be communicated, why they should be adopted, and how they can be communicated effectively.

504001 Climate-Change Review and Adaptation

Abstract:

The PAIC is leading an AIChE review of climate change and our policy on it. AIChE members are participating in a series of online Engage discussions, each introduced by an article posted on ChEnected. By drawing from the knowledge, dedication, and passion of AIChE members, AIChE can shape a Climate-Change Policy that captures membersâ€™ concerns while helping our profession to grapple with this issue.

505002 PAIC Town Hall

Abstract:

In this Town-Hall-style, moderated discussion, the emphasis will be on identifying other critical public-policy issues where AIChE and chemical engineers can have impact. Attendees are encouraged to submit questions beforehand to the organizers by email or at the session. The goal is to shape action paths and find participants who are interested in engaging with each other and the public on these issues.

509411 Predicting Temperature-mediated Solid Form Transformations in Small Molecule Crystals with Molecular Dynamics

Abstract:

Many pharmaceutical compounds can crystallize into more than one solid form, and the overall solubility and bioavailability of a therapeutic compound can vary markedly between different structures. Selecting a stable and soluble solid form is therefore an important step in the drug development process, and an unintended late-stage transformation into a new solid form can severely delay a therapeutic reaching the market. Computational models of a compoundâ€™s solid form landscape can reveal previously unobserved crystal structures and assist in selecting the appropriate form for commercial manufacturing. However, the current approaches for crystal structure predictions (CSP) often neglect the effects of temperature on solid form stability. Such approaches cannot capture temperature-mediated transformations that can occur during product manufacturing and storage.

In this work, we present a method to compute the temperature-dependent stability of small molecule crystals and capture temperature-mediated solid form transformations. All crystal structures of interest are heated from zero Kelvin up to ambient conditions using molecular dynamics simulations. The sets of simulations are combined to construct temperature-dependent free energy profiles for each structure. We demonstrate this technique on multiple polymorphic small molecules of pharmaceutical interest, and show that current point-charge potentials are capable of capturing the large entropy differences between forms which facilitates crystal transformations at high temperature.

In addition, we show that molecular dynamics simulations can yield visual insights into when crystalline systems undergo critical property changes as a result of a change in temperature. These changes include solid-solid transformations, order-disorder transitions, and lattice minima interconversions.

509412 Polymer D-255 Cement Fluid Loss Additive Control Optimization

Abstract:

Incorporated and optimized new designs with D-255 FLAC technology to meet the same performance given by previous additive D-167. The project had the main goal to improve control of water loss from cement to formation by reducing the permeability of the cement filter cake, and increasing viscosity of the aqueous phase. The testing results concluded new technology to have better stability and compressive strength development. The replacement of new technology with a PC/PR of 0.27 will save $81K to Schlumberger with a 50% sales profit increase in five years.

Modeling Paste Transport Systems for FGD Water Disposal

Trent Rogers

Research & Development

SOUTHERN COMPANY

What is Paste?

- Fly ash encapsulated FGD wastewater
- Non-Newtonian, cementitious fluid pumped into landfills
- FGD wastewater disposal technique that is zero liquid discharge (ZLD) compliant

Uncured paste sample vs cured paste sample

Project Overview

- Model the approximate cost of operating a paste system using a rheometer
- Two dominant cost factors:
  1. Energy/capital costs of pumping (head loss)
  2. Raw material cost of paste (mix design)
- No reliable model for rheopectic pseudo-plastics in mining industry

Mining paste backflow system

Rheometer Testing

- Used to characterize non-Newtonian flow
- Two rheometer measurements
  1. Shear rate: angular velocity of spindle
  2. Shear stress: torque of motor
- These measurements can be used to simulate pipe wall interaction

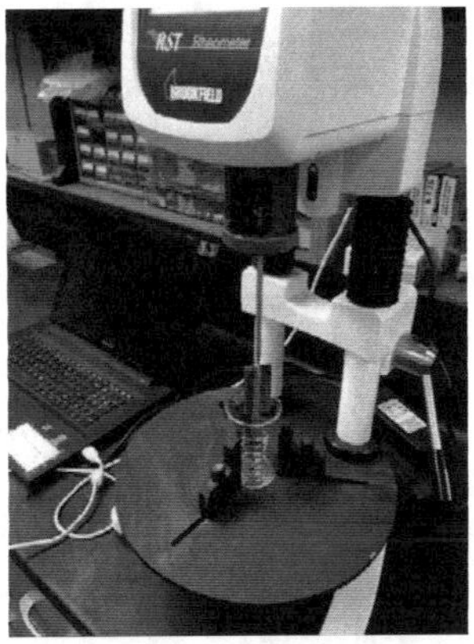

Rheometer Setup

Estimating Head Loss

- Modeling head loss using a rheometer is fairly common; time dependency is often neglected
- Paste is an "apparent rheopectic" fluid due to rapid cementitious reaction
- Modified cement slurry model (Haimoni and Hannent) for rheopectic behavior

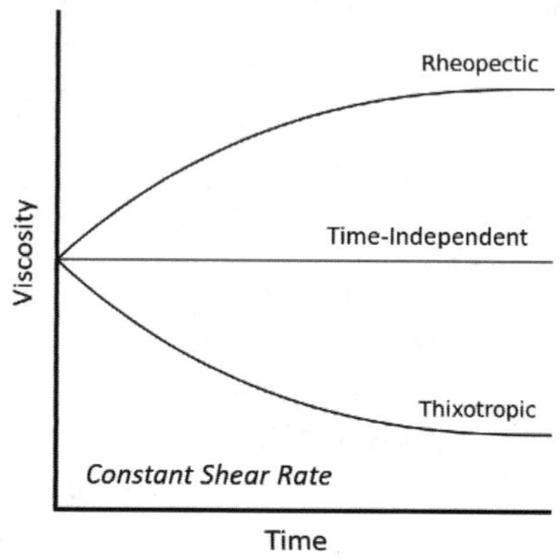

Mix Design Costs

- Chemical additives greatly affect rheology
- Balancing material cost and energy cost is essential
- Mix design can also impact capital costs if a different pump is required

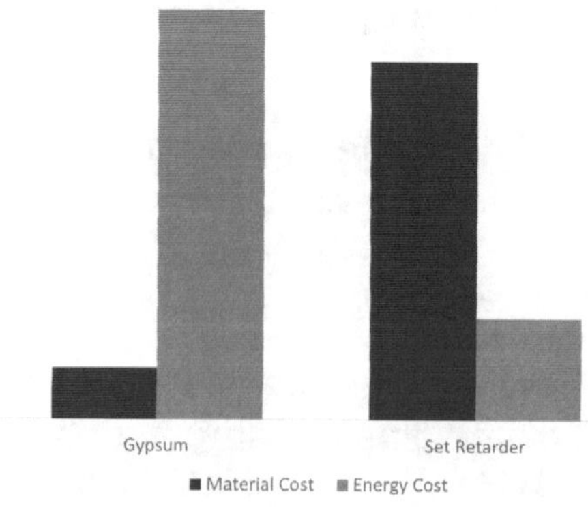

Summary of Modeling Process

1. *Rheometer*
   - Use a designed velocity and rheometer geometry to determine a controlled shear rate
2. *Head Loss Calculation*
   - Use the measured shear stress to simulate the changing pressure loss at the pipe walls
3. *Energy Cost Analysis*
   - Use the pressure loss to determine the energy cost of pumping material at the designed piping distance

Rheometer Results

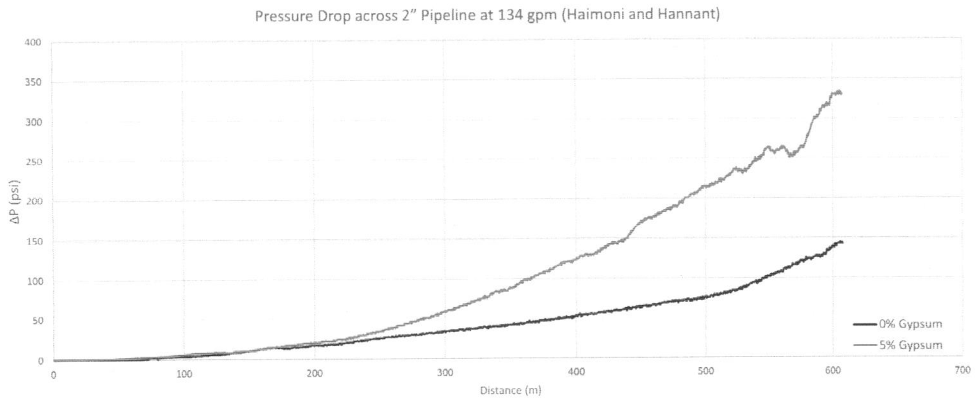

Rheometer Results

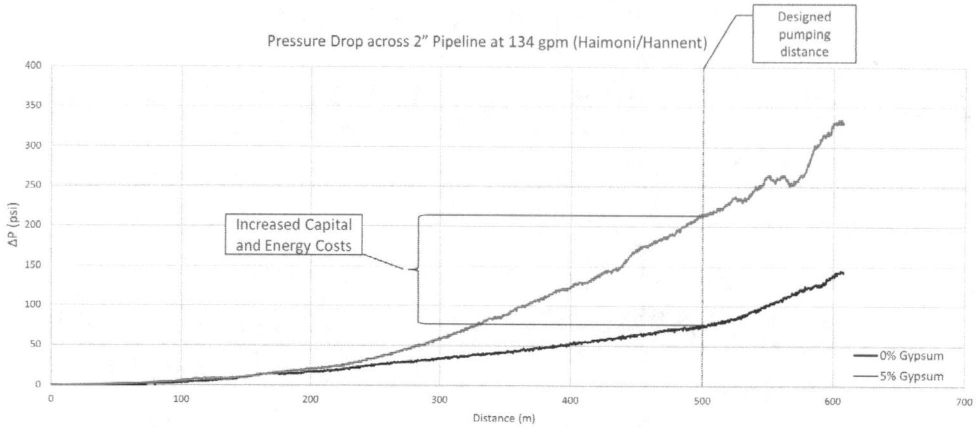

Simulated Case Study: Replacing 5% Gypsum

- Material cost is a constant, but the energy cost increases as time increases

- If pumping beyond the red line, the energy savings outweighs the material savings

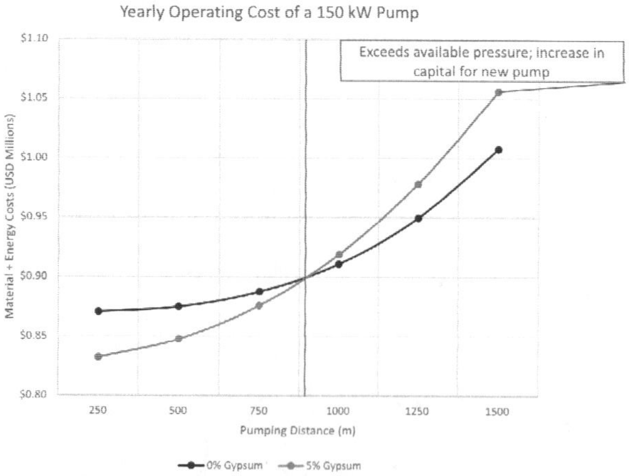

Conclusion

- Next step is confirming these results on a full-scale system
- Simple test that can yield massive amounts of data
- This method can be applied to any variety of time-dependent non-Newtonian fluid

Uncured paste flowing out of mixer

Questions

Research & Development

SOUTHERN COMPANY

511600 Production of Nimesulide Nano-particles via Liquid Anti-solvent Precipitation using Spinning Disc Reactor

Abstract:

This summer I was offered a research fellowship by Indian Academy of Sciences (IAS) to conduct a project on biochemical applications of nanotechnology in Institute of Chemical Technology, Mumbai. According to a study, nearly 40% of the pharmaceutical drugs are poorly water soluble due to which bioavailability of these pharmaceutical drugs is insubstantial and, high dosages of these drugs have to be administered.

My project focused on finding the optimum parameters for producing stable drug nanoparticles in order to increase its dissolution rate. For my study, I chose Nimesulide drug, a non-steroidal anti-inflammatory drug and produced its nanoparticles in a spinning disc reactor using the principles of Liquid-Antisolvent precipitation. The size of the raw drug was found to be 18 microns. The project required a comprehensive knowledge of chemical engineering principles such as transport phenomena, biochemical engineering, reaction engineering and mass transfer. Using these principles, supersaturation of drug was achieved. Spinning disc reactor employed applications of momentum and mass transfer for the production of coated nano-drug. The thickness of boundary layer was also attempted to study.

During my project, I monitored influences of operating parameters such as flow rate of solvent-antisolvent, solvent-antisolvent ratio, nimesulide concentration, disc characteristics and rotating disc speed on the nanoparticles size. At the optimum parameters, particles as small as 190 nm were obtained with a substantial increase in dissolution rate. Samples were analyzed using Dynamic Light Scattering and UV-Vis spectroscopy.

This project is a gateway to finding an efficient method of drug delivery. Using this method, the bioavailability of the drugs could potentially increase which would not only reduce the dosage of the drug administered but also make the drug economically feasible. Production of nano-drug and development of generic pathways for drug delivery could hold the potential cure for epidemic diseases like AIDS, cancer, and tuberculosis.

506141 Into Hot Water: Utilizing Thermal Distributed Energy Resources to Improve Grid Reliability

Abstract:

Combined heat and power (CHP) technology along with grid-integrated water heaters (GIWH) are underutilized thermal distributed energy resources. They can be implemented on a large scale to provide ancillary services for the grid, including load shifting, frequency regulation, and improved grid reliability. The market is ripe for CHP and GIWH to be rapidly implemented on a large scale. This paper discusses the legislative and technical background of CHP and GIWH, the way these technologies provide benefits to a rapidly modernizing smart grid, and concludes with potential policies on the state and federal level that would support greater use of thermal distributed resources. Three policies are discussed in detail. The first of these is to expand the Department of Energyâ€™s CHP Technical Assistance Partnership program to facilitate knowledge and implementation of a shared ownership model for CHP systems between utilities and an institution that will utilize the energy produced. Next, states should encourage utilities to create incentives for consumers to install GIWH to be used in utility programs; these incentives should focus in particular on assisting low-income households to purchase GIWHs. Last, states should set goals to have all utilities that function on retail markets to adopt time-based pricing structures for electricity.

506142 Turning the Tide: Policies to Advance Saltwater Desalination in the United States

Abstract:

Although water remains one of the worldâ€™s most abundant resources, most of the water resides in the ocean. With an already limited supply of freshwater, the increasing global population, changing climate, and worsening droughts are placing additional strains on the worldâ€™s water supply. Currently, several regions of the U.S. are experiencing some level of water shortage, and while the majority of the affected states reside in the West, the effects are felt nationwide. Although both water conservation and reuse remain key components of the national water portfolio, these practices alone will not sufficiently accommodate future water demands. Desalination, which removes salts and other suspended solids from seawater and brackish water, is a technique widely used around the world to convert high-salinity water into water suitable for human use. The unique benefits of seawater desalination remain to be recognized in the U.S., and since the options for augmenting the current water supply are limited, seawater desalination should hold a distinct place in the U.S. water portfolio.

506143 Microgrids for the Macrogrid: Advancing Community Microgrids for Grid Modernization

Abstract:

Global trends - the digital age, prospect of cyberwarfare, climate change, consumer preferences â€“ are irreversibly changing the supply and demand for electricity as well as the risks that threaten access to electricity. The existing grid was designed to serve a different kind of society and is ill-suited for meeting new challenges. A modern electricity system will not only have more actors, but will need to be fundamentally different in how it dispatches resources, communicates information, responds to grid disturbances, and plans for the future. A modern grid is decentralized, flexible, adaptable, and smart (i.e. automated, digitized, self-healing). By these characteristics, electricity access becomes reliable, efficient, and resilient. Microgrids add both resilience and flexibility to the system and policy initiatives can encourage them through technology-agnostic market rules and economic forces.

510175 Small Scale Shock-Proof Biogas Digesters

Abstract:

Affordable, easily maintained biogas digesters have the potential to greatly increase the standard of living in several key communities around the world. From food scraps and animal wastes, these reactors can output high-quality fertilizer in addition to natural gas that can be used for heating, cooking, etc. Designing an assembly guide with the key features of a safe biogas reactor is key to developing this technology in the communities in question; particularly communities with limited access to education or manufacturing capabilities. Development of this assembly guide requires implementation and development of a small â€œtestâ€ reactor, including various mechanical systems, embedded electronics, and other features that would allow a more comprehensive and thorough understanding on the feasibility and overall maximum output possible by this kind of reactor. Secondary research would strengthen this knowledge base and further increase the ability for these reactors to be beneficial to their intended users, this research includes lab-scale experimentation on the growth rates of critical bacteria, determining decomposition rates of certain safety features, and detailing how other physical stimuli may result in differing output rates of the reactor. The presence of a simple, customizable biogas reactor in an impoverished community could potentially raise the standard of living for the surrounding populace by eliminating excess food waste, producing high-quality fertilizer for crops, and supplying a steady stream of clean natural gas for heating or cooking.

503908 Networking for Nerds: How to Land (or Create) Your Dream Job and Keep Your Career Moving Forward!

Abstract:

Wanna land your dream job? Get ready to network! Most jobs and other game-changing career opportunities are not advertised, and even if they are, there is usually a short-list of candidates already in mind. So how do you find out about and access the 90% of jobs and other opportunities that are "hidden"? In this workshop, we will focus on proven networking strategies and tactics to identify new opportunities, locate decision-makers within organizations, solidify your reputation and brand in the minds of those who hire, and gain access to hidden jobs and game-changing opportunities. Discover how networking and self-promotion can enable you to land or even create your dream job from scratch!

504195 Sustaining Innovation / Innovating Sustainably

Abstract:

Optimism is at the heart of both innovation and sustainability. Research creates discoveries that lead to inventions, creating innovations that make the world better. R&D starts with the optimistic view that you can create something better. It is an investment in the future, where resources remain for future generations. At Dow, weâ€™ve earned the ability to continue innovating based on our past performance. R&D isnâ€™t left to chance, we take steps to stack the deck by:

- Rigorously manage our R&D portfolio based on quantifiable metrics
- Focus on R&D productivity and timely delivery of technology
- Collaborating to extend access to new inventions and to create customer intimacy
- Application of technical excellence in both product and process design

Portfolio management, built on a well-established foundation, guides our efforts and incorporates sustainability and other metrics to influence decisions. We use the life-cycle impact of our products to guide our efforts. Far too often, hype has surpassed reality, boundary conditions have been neglected, and actions taken that do not truly lead to a sustainable future. Looking across the life-cycle compels decisions to reduce raw material use, energy use and to limit toxics use. Innovation will create a sustainable future.

504196 Creating Chemistry for a Sustainable Future

Abstract:

Imagine a future where we can use our creativity and innovation power to help solve some of the largest global sustainability challenges of our times. Are we doing enough to enable this?

Investing in businesses that seek positive economic, ecological and social impact is becoming imperative. We will be judged by our communities, by our work-force, by the talent pool of Millennials, by investors. We will be judged in the courts of social media and public opinion. We have to act now; authentically, and for the right reasons

Chemical companies are positioned uniquely to solve the global sustainability challenges. But we have to take a holistic look at these challenges; going beyond our typical value chains and including the larger eco-system of stakeholders. That is, going beyond current business models to co-creating solutions.

For example, Chemistry plays an active role in improving energy efficiency of homes. There are many opportunities to co-create solutions to improve this situation. BASF therefore, is partnering with other stakeholders to change state laws and make it easier to update residential building codes. This challenged us to work with new stakeholders: industry partners, architects, city planners, builders, citizens, governments and many more. This is what we mean by co-creation.

BASF is convinced of the importance of the UN Sustainability Development goals to our long-term success. This is not solely about recycling or bio-based raw materials; those are of course important. Moving forward, however, requires us to build upon existing platforms and address critical issues such as transparency, education, diversity and inclusion â€“ to just name a few.

No single company can address all the seventeen UN sustainability development goals on their own. Co-creation is the way forward to address both the overlaps and the gaps.

504198 Continued Delivery on Impactful Sustainable and Innovative Business and R&D Strategies

Abstract:

Expanding our impact through sustainability is a core value at Owens Corning. From the products we make to the way that we make them, sustainability is an essential element of our business. We design innovative solutions that meet expectations of customers/markets and contribute toward addressing significant global issues such as energy conservation, environmental impact and overall quality of life. We consistently leverage our technological expertise to deliver sustainability-driven innovations to the Insulation, Roofing and Composites markets. Our insulation products enhance energy efficiency; cool roof shingles help reduce air-conditioning and power consumption; reinforced lightweight glass-composites deliver superior mechanical properties with lower environmental impact than traditional materials. Our business strategies promote innovative ways to achieve high-performing, greener and healthier products via high-efficiency manufacturing processes with reduced energy and water consumption. The discussion in this panel will highlight Owens Corningâ€™s innovative strides throughout its core businesses, its continued commitment to robust sustainability progress in its handprint and footprint, and the companyâ€™s significant safety improvement.

506295 Densification of Biomass By Using Natural and Synthetic Binder

Abstract:

Declination of fossil fuels at an accelerated rate along with their significant environmental problems have led to an increasing demand for renewable energy resources. Biomass with the high percentage of domestic availability is the best choice for producing energy. This work primarily focuses on the densification of biomass to form pellets with high heating value by using natural binder and synthetic binder. Pellets of different sizes have been produced by using natural and synthetic binder and then their heating value had been checked for natural and synthetic binder. Conclusions for the concentration of binder, binder Application, pellets size and heating value have been compared against different experimental approaches.

510079 Effects of Selenium on Human Glioblastoma Multiforme and Human Dermal Fibroblast Cell Lines

Abstract:

Selenium is a trace element that is naturally found in human diets. While it can be toxic, the health benefits associated with supplemental selenium for treating numerous illnesses has made it a topic of interest in recent years. Selenium has been found to function as an anticancer therapy that effectively inhibits carcinogenesis at multiple stages. The increased surface area exposure of selenium nanoparticles (SeNPs), relative to micron scale selenium, optimizes the likelihood of cell surface reactions. Combining this effect with the increased efficacy and reduced toxicity that nanotechnology offers, SeNPs could potentially serve as an optimal cancer treatment. SeNPs are also more likely to naturally diffuse across the cell membrane due to their smaller size. In the current study, the efficacy of SeNPs against glioblastoma multiforme (GBM), a grade IV astrocytoma, was investigated. While the mechanism enabling selenoenzymes or selenium ions to inhibit cancer cell growth is not fully understood, these particles were employed to examine effects at certain concentrations on healthy tissue cells. After fabricating SeNPs, their size distribution, typically at or below 100nm, was confirmed using transmission electron microscopy. These particles were then used at varying concentrations in both Human dermal fibroblasts and GBM cell lines to reveal what concentration is most effective at treating the disease while not harming the healthy cells. We have treated cells over a three day period and examined the results via MTS assays. This project then aims uncover the genetic effects that selenium has on cell lines through Polymerase chain reaction testing.

506298 Antibody Adsorption on Fluid-Fluid Interface

Abstract:

Many proteins will adsorb on fluid-fluid interfaces and dynamically unfold. This tendency causes multiple problems in the formulations steps for biomacromolecules in the pharmaceutical industry, which is actively working on developing protein-based drugs. To resolve this issue, it is necessary to study the rate of antibody adsorption on the interface, using air-water interface as an example. We propose to examine Immunoglobulin 1 (IgG1) as a model antibody, and we will examine Bovine Serum Albumin (BSA) in physiologically relevant solvents as a 'control' protein. Circular Dichroism spectrometer will be used to test BSA's stability at different temperatures ranging from room temperature to 90°C, using water bath. The heating and cooling curves will let us determine the BSA's melting point and help to predict the reversibility nature of the protein, if any. Another set of experiments using pendant drop tensiometry technique helps to determine BSA's surface tension and critical micelle concentration using water, NaCl, and surfactants as excipients. This will provide us with enough information to make conclusions about BSA's adsorption properties on the interface. Similar set of experiments would be done using IgG1-excipient mixture to determine its adsorption properties using CD spectroscopy and pendant drop tensiometer. The above studies will allow us to predict the physical stability of the protein antibodies and to define manufacturing conditions biomacromolecules involved in autoimmune diseases and cancer treatments.

506299 Liposome Production and Concomitant Loading of Drug Simulants By Microfluidic Hydrodynamic Focusing

Abstract:

Liposomes are spherical vesicles enclosed by phospholipid bilayers. Nanoscale liposomes with diameters around 100nm are widely employed for drug delivery in the pharmaceutical industry. Compared to conventional batch production of drug-loaded liposomes, liposome fabrication and concomitant drug loading by microfluidics potentially allows better control of liposome size, requires fewer toxic solvents, allows for more efficient use of drug molecules, and enables tunable loading efficiency. In this research project, nanoscale liposomes were fabricated using the microfluidic hydrodynamic focusing (MHF) approach, and the effects of flow rate ratio (FRR) on liposome size and drug loading efficiency were studied. The widths of microfluidic channels were within 100Î¼m and FRR was varied from 10 to 50. For analysis purposes, fluorescein isothiocyanate was used as a hydrophilic drug simulant and Nile red was used as a hydrophobic drug simulant. The drug simulant concentrations were analyzed by a microplate reader and the liposome size was measured through dynamic light scattering. The experiment results showed that hydrophilic drug simulant loading efficiency increased as FRR increased and eventually plateaued. The hydrophobic drug simulant loading efficiency and FRR had a positive linear correlation. It was also confirmed that liposome size decreased as FRR was increased, and that liposomes were larger when loaded with the hydrophobic drug simulant compared to unloaded liposomes and liposomes loaded with the hydrophilic drug simulant alone. The results presented the effect of FRR on liposome size and drug loading efficiency, and suggested that liposome size and drug loading efficiency are tunable with the MHF technique. This provided evidence to encourage further studies of microfluidic liposome fabrication in the pharmaceutical industry.

506537 Tuning Size and Charge of a Multivalent Polymer Library for Enhanced Drug Delivery to Cartilage

Abstract:

Efficient transport of therapeutics through dense biological tissues has proven to be a complex challenge, especially for osteoarthritis, a degenerative joint disease characterized by gradual cartilage degradation. In fact, the avascular, densely packed, and negatively charged cartilage extracellular matrix has stood a barrier to the transport and, thus, therapeutic effect of even locally administered disease modifying osteoarthritis drugs. Such a challenge stems from the fact that achieving cartilage protection by maintaining chondrocyte viability, inhibiting matrix degradation, and stimulating matrix biosynthesis requires that therapeutics be delivered to the middle and deep zones of cartilage. In efforts to transform cartilage from a drug barrier into a drug reservoir, we screened a library of cationic dendritic polymers for cartilage uptake capability and cytotoxicity. Facile control over size and charge of the carriers was achieved through modification of the polymerâ€™s periphery with poly-ethylene glycol (PEG) molecules of various lengths. Fluorescence intensity measurements of nanoparticles incubated in approximately human-thick bovine cartilage explants revealed a significant influence of size and charge on uptake capabilities of the polymers. Larger constructs with higher degrees of charge shielding maintained excellent chondrocyte viability but displayed poor uptake into cartilage. Smaller, more cationic nanoparticles yielded superior tissue binding but exhibited cytotoxic behavior at physiologically-relevant doses. Formulations demonstrating an optimal balance of uptake and cell viability were conjugated with a cartilage-repairing growth factor (GF), and were superior to free GF in transporting it through cartilage.

506538 Building Brains: Marrying Engineering & Medicine in the Fight Against Alzheimer Disease

Abstract:

Neurodegenerative diseases such as Alzheimer disease incur high caregiver costs and significantly diminish economic output. Treating them is an urgent priority. Unfortunately, there are no treatments. Drug discovery has been stymied by the absence of reliable pre-clinical screens for identifying promising leads, which implies that results from pre-clinical testing do not translate well to the clinic. There is a clear requirement to develop better pre-clinical screens to test lead compounds in a brain- specific context. To this end, we are developing a process to manufacture engineered brain tissue whose microenvironment closely mimics the environment within degenerated brains. We refer to healthy brain tissues produced via this method as â€˜cerebroidsâ€™, and assembly of these tissues is achieved in a 2-stage process that employs an innovative method for stem cell differentiation and cellular assembly. The cerebroids are manipulated in the third stage to elicit neurodegeneration. Herein, we are investigating two complementary approaches that facilitate seamless scale-up. Our work combines concepts and insights from stem cell bioengineering, tissue biomanufacturing, and biomedical instrumentation. The manufactured tissue incorporates all clinically relevant cell types and sub-tissues, and we are partnering with a Vancouver-based company to commercialize the manufacturing platform. The cerebroids platform could prove to be a game changer for pharmaceutical testing and molecular medicine.

508958 Smart Manufacturing in the Automobile Industry

Abstract:

General Motors is transforming as customers and the industry move towards new mobility solutions, including electrification, autonomous vehicles and ride and car sharing. These changes drive innovations not only in the vehicles that General Motors makes, but also how it makes them. The companyâ€™s goal is a higher level of data-driven productivity from its global manufacturing network, including active monitoring, analytics and detailed system integration. This presentation will address how GMâ€™s manufacturing process is evolving in the digital era.

508959 Smart Manufacturing in Chemical Industries

Abstract:

Smart manufacturing provides a revolutionary ways for the next generation of production in the chemical manufacturing industries. One of the unique advantages of smart manufacturing is the creation of a knowledge base to reduce the operating and capital costs of chemical processes. The Smart Manufacturing Leadership Coalition (SMLC) has collectively developed and is pursuing a comprehensive technology and business roadmap to develop and implement Smart Manufacturing capabilities that will enable next-generation economic, energy, sustainability and EH&S manufacturing performance and competitiveness. (SMLC, 2011). We viewed Corningâ€™s concept of smart manufacturing as a combination of multiple capabilities (integrating measurements, controls, computing platforms, modeling and analytics) to improve process/product development, manufacturing performance, and reduce manufacturing cost within a plant and across the supply chain as very encompassing. (Michelle Pastel, Corning 2016). The challenges and management considerations in smart manufacturing will be presented.

508960 Innovations in Chemical Engineering: Automation of the Factories of the Future and the Impact of Internet of Things (IoT) from the Control of the machinery in factories to Home Appliances

Abstract:

The industrial world is primed for digitization and General Electric is one of he companies leading the application of Internet of Things to the manufacturing environment. Advanced software is available that provides access to secure industrial strength cloud environment for different machine data to be oaded in a secure environment for analytics to be applied for efficiency in factories of the future. The industrial Internet (HOT) will also apply artificial intelligence, virtual reality, and augumented reality in new manufacturing as well as other uses like the control of home appliances. This paper will focus on the opportunities for chemical engineers to lead these innovations from process control to novel IoT sensors development, and the need for advanced materials to enable transformational changes in the Internet of Things platform.

508961 Reversing the Tide in Science, Engineering, Technology, and Science (STEM): Academically Gifted African American Students in Historically Black Colleges and Universities (HBCU)

Abstract:

What started off as a national concern has escalated to the point of crisis (Bonner, Alfred, Lewis, Nave & Frizell, 2009; Chubin & Malcolm, 2008; Hrabowski, Summers & Hrabowski, 2006). The current state of affairs regarding African American participation in STEM disciplines has become one of the thorniest issues of contemporary focus. This presentation highlights data collected for a NSF Education Research grant. The focus of this HBCU-UP project was to uncover factors contributing to the success of academically gifted (high-achieving) African American students enrolled in engineering programs at the 12 ABET accredited four-year institutions in the U.S. Specifically, this research highlights qualitative case study data including key categories and themes identified as impacting the academic success of this high-achieving cohort across each of the 12 institutions included in this study. Based on the findings of this study, recommendations are provided to increase the success of African American gifted students in STEM fields at HBCUs.

508962 Blacks in Science, Engineering and Medicine: Struggles that Continue, Struggles that are Growing, and Possible Solutions

Abstract:

Blacks in many ways face a crisis in regards to numbers of individuals pursuing medicine, engineering and science. For Instance numbers of Black men matriculating into medical school are now lower than they were almost 40 years ago. The causes are manifold. Solutions are demanded.
To quote Baldwin 'Not everything that is faced can be changed, but nothing can be changed until it is faced.'

The American Institute of Chemical Engineers and its Minority Affairs Committee should take a leadership role in advancing solutions to address what is a growing crisis. Next steps are outlined.

497394 Role of Electrokinetics in Glomerular Capillary Filtration: Toward an Artificial-Kidney

Abstract:

According to the U.S. Department of Health and Human Services, 1 in 10 people have some form of chronic kidney disease (CKD) worldwide. CKD has no cure, kidney transplants are accompanied by long wait times and the possibility of tissue rejection, and dialysis fails to provide a good quality of life. An artificial kidney is an ideal solution to combat suffering due to kidney disease and failure. As a path to an artificial kidney, the micro-rheological properties of blood in the glomerular capillary are analyzed at the microscopic scale based on continuum mechanic approaches. Once the proper understanding of the microscopic behavior is attained (with the possible impact on the macroscopic scale), the system will be downscaled to a size ideal for potentially designing an artificial kidney. As the system is downscaled the importance of electrokinetics comes into question in a more detailed manner. In fact, the role of electrokinetics in kidney function (as well as its malfunction) has come into consideration recently. A glomerular capillary has three main layers of which have various characteristics that determine their filtering specificity due to size and charge. As a preliminary step toward a more detailed understanding, in this presentation, a single capillary of cylindrical geometry will be used as a potential domain for a â€œnephronâ€- the most crucial element of the kidney filtration. Principles of EKHD (see Pascal et al., 2016) will be used in modeling the glomerular capillary domain to understand the role of electrokinetics in the filtration efficiency.Â The role of electrokinetics in influencing the separation performance of the capillary-nephron domain under a variety of operating conditions will be considered in the development of an asymptotic solution for separation (filtration) efficiency.

496314 Patterning Various Commercial Nanofiltration and Reverse Osmosis Membranes

Abstract:

Progressive decline in permeability due to fouling is one of the largest costs associated with membrane processes in water treatment. Innovations in membrane surface modification that limit fouling would have significant economic and societal impacts. Prior studies in this area largely have focused on chemical modifications to the membrane surface. Based on considerable literature and experience, we know that chemical modification strategies alone can be effective but not sufficient for controlling fouling. The discovery that patterning of a membrane surface can improve its fouling resistance provides an orthogonal membrane design parameter. Our hypothesis is that combining chemical modification and patterning will yield membrane surfaces that are more effective at fouling control than either method alone. This strategy is unique in its use of both chemical coating and patterning to combat fouling.

This contribution describes our method to apply a wide variety of patterns to commercial nanofiltration and reverse osmosis membranes produced using various chemistries and coatings. The goal is to better understand what factor(s) contribute to differences in patterning of thin-film nanocomposite membranes, leading to a set of heuristics that can be used by the membrane community for patterning, which is becoming a hot topic in the field. Direct patterning of 15 commercial thin-film nanocomposite membranes was done by thermal embossing using nano- and micropatterned stamps. Patterning studies began by embossing all membranes at one set of fixed conditions followed by AFM characterization. These initial studies uncovered the strong membrane chemistry dependence on pattern replication that sparked a more comprehensive study using controlled conditions of force and time with varying temperatures. AFM, SEM, and laser scanning microscopy (LEXT) were used to determine the extents of patterning. AFM and LEXT were used to determine the local strain as a function of embossing conditions. Membrane performance was evaluated by measuring flux and salt rejection using a magnesium sulfate solution. Flux and salt rejection properties were compared to local strain values and Youngâ€™s Modulus values of the patterned membranes to establish correlations.

493270 Preparation of ZIF-8 Membranes Supported on Polymer Hollow Fibers Using Microwave-Assisted Seeding and Secondary Growth Method

Abstract:

Preparation of ZIF-8 Membranes supported on Polymer Hollow Fibers using Microwave-assisted Seeding and Secondary Growth Method

Moon Joo Lee[1], Mohamad Rezi Abdul Hamid[1], Jongmyeong Lee[3], Ju Sung Kim[3], Young Moo Lee[3] and Hae-Kwon Jeong*[1,2]

[1]Artie McFerrin Department of Chemical Engineering and [2]Department of Materials Science and Engineering, Texas A&M University, College Station, TX 77843-3122

[3]Department of Energy Engineering, College of Engineering, Hanyang University, Seoul 133-791, Republic of Korea

* Corresponding author: hjeong7@tamu.edu

Separation of olefin/paraffin mixtures, e.g. propylene/propane, is one of the most challenging separations due to the similarity of their physicochemical properties. Currently, an energy intensive cryo-distillation process is employed. Though energy-efficient membrane-based separation processes are drawing significant interest, no membranes are commercially available for olefin/paraffin separations mainly due to the limitations of current membrane materials and processing methods to fabricate their membranes with large membrane areas.[1-3]

Zeolitic-imidazolate frameworks (ZIFs), a sub-class of metal-organic frameworks (MOFs) with zeolite topology, are highly crystalline nanoporous materials with aperture sizes of typically less than 5 Å. ZIFs composed of metal ions (typically Zn^{2+} and Co^{2+}) and imidazole-derived organic linkers,[4] have been extensively studied for gas separation applications. Specifically, ZIF-8 is shown very promising as a new membrane material for propylene/propane separation because the effective pore size is in between the size of propylene and propane. While several ZIF-8 membranes showed high propylene/propane separation factors, however, there are critical issues for their commercial applications: 1) high cost of membranes and 2) low productivities. To increase the productivity and to reduce the cost of ZIF-8 membranes, polymeric hollow fibers have been utilized as substrates, but only a few of the resulting ZIF-8 membranes showed promising propylene/propane separation performances.[6-9]

Herein, we would like to present a microwave-based method to prepare well inter-grown ZIF-8 membranes on polymeric hollow fibers readily. The method is modified based on our previously reported microwave-assisted seeding and secondary growth technique.[10] Densely-packed nano-sized ZIF-8 seed crystals were rapidly prepared on either inner or outer surface of hollow fibers under microwave heating. Subsequently, seed crystals were secondarily grown into continuous ZIF-8 films on polyimide hollow fibers. The thickness of the resulting ZIF-8 membranes was much smaller than any ZIF-8 membranes grown on polymer hollow fibers reported, which are several micrometers in thickness. The effects of synthesis parameters and fiber morphologies on the microstructures of ZIF-8 membranes will be discussed. Finally, the gas separation performances of ZIF-8 membranes will be presented.

References and Notes

1 Koros, William J. and Fleming, G.K. (1993) "Membrane-based gas separation", *Journal of Membrane Science*, 83, pp 1-39

2 Baker RW., (2002) "Future Directions of Membrane Gas Separation Technology" *Industrial & Engineering Chemistry Research.*, 41 (6), pp 1393-1411

3 Tanaka, K., Taguchi, A., Hao, JQ, Kita, H., and Okamoto, K., (1996) "Permeation and separation properties of polyimide membranes to olefins and paraffins", *Journal of Membrane Science*, 121, pp 197-207

4 Park KS, Ni Z, Côté AP, et al., (2006) "Exceptional chemical and thermal stability of zeolitic imidazolate frameworks", *Proceedings of the National Academy of Sciences,* 103 (27), pp 10186-10191

5 Zhang C, Lively RP, Zhang K, Johnson JR, Karvan O, Koros WJ., (2012) "Unexpected Molecular Sieving Properties of Zeolitic Imidazolate Framework-8", *The Journal of Physical Chemistry Letters*., 3 (16), pp 2130-2134.

6 Andrew J. Brown, Nicholas A. Brunelli, Kiwon Eum, Fereshteh Rashidi, J. R. Johnson, William J. Koros, Christopher W. Jones, Sankar Nair, (2014) "Interfacial microfluidic processing of metal-organic framework hollow fiber membranes", *Science*, 345, 6192, pp 72-75

7 FernandoCacho-Bailo, SilviaCatalán-Aguirre, MirenEtxeberría-Benavides, Oguz Karvan, VictorSebastian, CarlosTéllez, JoaquínCoronas (2015) "Metal-organic framework membranes on the inner-side of a polymeric hollow fiber by microfluidic synthesis", *Journal of membrane science*, 476, pp 277-285

8 Kiwon Eum, Chen Ma, Ali Rownaghi, Christopher W. Jones, and Sankar Nair (2016) "ZIF-8 Membranes via Interfacial Microfluidic Processing in Polymeric Hollow Fibers: Efficient Propylene Separation at Elevated Pressures", *ACS Applied materials and interfaces*, 8, pp 25337 – 25342

9 Anne M. Marti, Wasala Wickramanayake, Ganpat Dahe, Ali Sekizkardes, Tracy L. Bank, David P. Hopkinson, and Surendar R. Venna (2017) "Continuous Flow Processing of ZIF-8 Membranes on Polymeric Porous Hollow Fiber Supports for CO2 Capture", *ACS Applied materials and interfaces,* 9, pp 5678-5682

10 Kwon HT, Jeong H-K., (2013), "Highly propylene-selective supported zeolite-imidazolate framework (ZIF-8) membranes synthesized by rapid microwave-assisted seeding and secondary growth", *Chemical Communications,* 49, pp 3854-3856

495664 A Zeolitic Imidazolate Framework (ZIF-8) Film for H_2/CO_2 Separation

Abstract:

A Zeolitic Imidazolate Framework-8 (ZIF-8) Film for H_2/CO_2 separation

Eunhee Jang[1], Jungkyu Choi[1*]

[1] *Department of Chemical & Biological Engineering, Korea University, Seoul 02841, Republic of Korea*

ehphoenix@korea.ac.kr, jungkyu_choi@korea.ac.kr

Zeolitic imidazolate framework-8 (ZIF-8) membranes are highly suitable as H_2 selective walls in membranes reactors used for water gas shift reactions because of their efficient molecular sieving properties that favor H_2 transport, coupled with their high thermal stability and processability. In this study, we adopted an *in-situ* counter diffusion method (CD method) to fabricate a ZIF-8 membrane; Zn sources, already placed inside a porous support, were allowed to diffuse out and react with the 2-methylimidazole (mim) ligands in the bulk phase. Because the reaction rates between the Zn source and the mim ligand were very high, their diffusion rates played a key role in determining the final properties of the membranes. To exploit the effect of diffusion rate on preparation of ZIF-8 membrane via the CD method, a hierarchically structured support, i.e., a gamma alumina layer-coated alpha alumina disc (gamma/alpha alumina disc), was used in addition to an intact alpha alumina disc. Existence of ZIF-8 membrane was demonstrated with SEM, EDX of Zn atom, XRD, and contact angle. The position of ZIF-8 grains in the supports was estimated based on the direct measurement of N_2 physisorption. ZIF-8 membranes in the alpha alumina disc (membrane ZIF-8_A) were primarily formed on top similar to a conventional supported-membrane, whereas those in the gamma/alpha alumina disc (membrane ZIF-8_GA) were embedded inside the modified support. From the N_2 physisorption analysis, it was found that the ZIF-8 grains in membrane ZIF-8_GA was pre-dominantly formed inside the alpha alumina disc, while filling ~70% of the mesopores in the gamma alumina layer. As desired, the embedded membrane ZIF-8_GA showed marked H_2 separation performance with a maximum H_2/CO_2 separation factor (SF) of ~9.9 plus-minus 1.2 at 250 deg C (vs. a maximum H_2/CO_2 SF of ~7.5 plus-minus 0.2 for membrane ZIF-8_A). Although both type membranes persisted at 200 and 250 deg C for up to 72 h, at a higher temperature of 300 deg C, the membrane performance started deteriorating after ~2 h and ~10 h for membranes ZIF-8_A and ZIF-8_GA, respectively. This indicates that the gamma alumina layer served as a protective layer for preserving the performance of the embedded ZIF-8 membrane. Their performance was completely degraded due to the eventual conversion of ZIF-8 into ZnO phases.

Figure. 1. H_2/CO_2 separation performance test of ZIF-8-A and ZIF-8-GA at 300 ¡ÆC for 72 h. Here, performance test result obtained for 48 h was expressed. Schematic images for ZIF-8-A and ZIF-8-GA was depicted at the bottom of the performance test.

489268 Scale-up of Electrochemical Carbon Dioxide Separation Using Membrane Electrode Assemblies

Abstract:

Scale-up of multi-cell membrane electrode assembly (MEA) systems require reproducible fabrication of reliable electrodes and membrane materials. Optimization of electrode support and catalyst microstructures ensures a high rate of CO_2 separation through the anion exchange membrane. Separation of CO_2 from a 5–7 mol% CO_2, high-humidity gas mixture, representative of an exhaled breath, was demonstrated by a sub-scale system. Screen-printed Pt and IrO_2 electrodes were fabricated and evaluated as single MEAs to measure baseline CO_2 removal rates and electrochemical behavior. Catalyst microstructures were characterized using scanning electron microscope (SEM) imaging. Electrode mass and charge transfer resistances were quantified using electrochemical impedance spectroscopy (EIS). The electrode efficiency and CO_2 flux were measured over a range of equivalent full-scale breathing flow rates (40 L/min–120 L/min) for 3, 7, 11, and 22 cell stacks at variable current densities. The multi-cell system displayed performance linearity and uniformity with increasing number of cells. For a full-scale flow rate of 80 L/min, the multi-cell system achieved a CO_2 flux of 1.57 L/m^2·min with an electrode efficiency of 75%. Flow field design and cell hydrodynamics were addressed for scale-up of the electrochemical system.

494146 Iron/Palladium Nanoparticle Functionalized Membrane for Chlorinated Contaminates Treatment

Abstract:

Functionalized polyvinylidene fluoride (PVDF) membrane platforms were developed for environmentally benign in-situ nanostructured Fe/Pd synthesis and remediation of chlorinated contaminates (PCB and TCE). To prevent leaching and aggregation, nanoparticle catalysts were integrated into membrane domains functionalized with poly (acrylic acid). Furthermore, nanoparticles of 16-19 nm were observed inside the membrane pores. The quantification of nanoparticles properties (size, distribution and composition) versus depth underneath the membrane surface was also investigated by cross-sectioning the membrane samples with focused ion beam (FIB). The Fe/Pd nanoparticles immobilized membrane showed excellent performance in the degradation of chlorinated organics and also the repeated degradation experiments were conducted to demonstrate the reusability of Fe/Pd immobilized membranes.

To increase the accuracy of kinetic analysis and further simulation of PCB dechlorination, the correlation between nanoparticle properties and depth inside membrane pores was studied by using FIB. Particle size was uniform inside membrane pores at different depths (particle size: 16.7±0.7 nm) but slightly smaller than those nanoparticles located on the surface (19.4±3.2 nm). Furthermore, the fairly similar surface coverage values (0.169±0.011) demonstrated the evenly distributed nanoparticles inside the membrane pores. Besides, the element composition of particles as well as the functionalized membrane domain were analyzed in line-scan mode by using energy dispersive spectroscopy (EDS).

Over 96% degradation of 3,3',4,4',5-pentachlorobiphenyl was achieved at a residence time of 14.7 seconds in the membrane pores. The ksa was calculated (using LFR model) to be 0.171 L/(m^2h) in convective flow mode, which is 2.5 times the rate obtained in batch mode. The effects of temperature and pH values were also investigated.

In XRD analysis, the Fe/Pd particle samples (which were deliberately oxidized and then reduced) exhibited the same crystalline patterns as the original samples. The membrane was tested for reactivity after four degradation cycles with regeneration between each cycle. The increase of surface particle size of 22% resulted in a decrease of 9.7% PCB conversion for the 4 hr reaction time. [1]

This research is supported by the NIEHS-SRP grant P42ES007380.

Reference:

[1] Wan, Hongyi, L. Ormsbee, and D. Bhattacharyya. "Pore functionalized PVDF membranes with in-situ synthesized metal nanoparticles: Material characterization, and toxic organic degradation." Journal of Membrane Science 530 (2017): 147-157.

494584 *In situ* Growth of MOF Membranes Assisted By Electro-Deposition

Abstract:

Metal–organic frameworks (MOFs) have great potential for use a membranes in separation processes because their pore size can be tuned and they have highly designable structures and adsorption properties.[1] The preparation of supported MOF membranes has attracted increasing attention in separation technology.[2] We develop a facile and efficient electro-deposition method to modify inexpensive porous stainless-steel nets for use as substrates in the in situ growth of metal–organic framework membranes, such as
ZIF-8, ZIF-67 and HKUST-1. Using this method, different metal precursors can be electro-deposited depending on the central metals required in the target metal–organic frameworks. The inorganic modifiers prepared by this approach are sufficiently reactive for the one-step growth of continuous metal–organic framework membrane.

We chose stainless-steel nets (SSNs) as the substrate as they are inexpensive
and easily welded. The preparation concept is shown schematically in Fig. 1. A metal oxide/hydroxide with a morphology exactly corresponding to the metal centres of the targeted MOFs is introduced onto the SSN via electro-deposition. For example, ZnO nanorods, $Co(OH)_2$ nanosheets and Cu_2O nanocubes can be electrodeposited onto substrates quickly and provide active sites for the crystallization of Zn–ZIF-8, Co–ZIF-67 and Cu–HKUST-1,
followed by in situ growth to form continues MOF membranes.

The metal precursors of ZnO nanorods, $Co(OH)_2$ nanosheets and Cu_2O nanocubes are electro-deposited onto SSN respectively as inorganic modifiers. As a representative, the microstructure images of ZnO nanorods are shown in Fig. 2a. These reactivity of these modifiers is excellent because electro-deposition is operated at low temperature and also some intrinsic defect exists. The following growth of three different MOF membranes can be finished in one step without any pre-treatment. The microstructure images of ZIF-8 membranes are shown in Fig. 2b, 2c. The pure phase is confirmed by XRD analysis (Fig 2d). This strategy is proved to be general and all the membranes are tested for gas separation performance (shown in Table 1).

An efficient electro-deposition strategy is proposed for the introduction of diverse metal clusters into inexpensive SSN substrates. These metal clusters universally act as inorganic modifiers for the subsequent facile in situ growth of the MOF
membranes. The fast and mild process of modification assisted by electro-deposition is promising for scaled-up production as well as low-cost manufacturing. The excellent reactivity of these metal oxide/hydroxide precursors arising from electro-deposition plays a vital part in the one-step in situ synthesis of various MOF membranes. The proposed electro-deposition-assisted strategy simplifies the procedure, exhibits excellent generality and can be applied to the preparation of different kinds of metal-centered MOF membranes.

495767 Pd/Ta Composite Metallic Membranes for High Purity Hydrogen Separation: Permeability and Durability

Abstract:

Hydrogen is considered as one of the clean energy carrier, but still, more than 90% of hydrogen is produced from hydrocarbon reforming processes involving a hydrogen purification step. Dense metallic membranes are drawing attention as a high purity hydrogen separation technique either as a stand-alone purification unit or to be used with catalyst as a membrane reactor to replace a series of end processes. Palladium based thin metallic membranes have been intensively developed due to their stable nature, but they are costly and permeability is significantly lower than that of body centered cubic (BCC) metal based membranes . In this study, as an effort to overcome this problem, Tantalum/Palladium (Ta/Pd) composite membrane has been fabricated using electroless deposition of Pd on a chemically treated Ta tube. The amount of palladium is significantly reduced while maintaining high permeability of Tantalum. The fabricated membrane had permeability value higher than that of the best Pd based metallic membrane reported, with hydrogen purity of 99.999% under mixed gas operations including N2, CO2, CO, H2O. It was possible to operate the membrane in a desired temperature range of 400 °C ~500 °C with no failure due to hydrogen embrittlement. Durability behavior of the membrane has been evaluated as a function of operating temperature and hydrogen flux, and optimal operating conditions are suggested as a function of Pd thickness.

501077 Fabrication and Characterization of Silicalite Membranes Subject to Knudsen and Surface Diffusion Transportation Regimes

Abstract:

Inorganic membranes composed of zeolitic crystals inside ceramic membrane supports show particular promise for gas separation applications due their high permeability, chemical stability, and low operating costs. The fabrication of membranes of this type is however very challenging due to the significance of even a small number of defects on membrane performance. An elegant characterization model has therefore been developed which can be used to quantify the physical characteristics of defects in novel membranes based on a limited number of pure gas permeation tests, and evaluate novel membranes. Additional refinements to standard fabrication and testing conditions have also been developed which allow greater control of experimental variables, and are presented in this work.

To demonstrate this characterization model, several batches of silicalite membranes have been hydrothermally synthesized using ceramic membrane supports with active layers whose pore sizes are 0.45 and 0.8 μm. Pure gas permeation experiments were then conducted using these membranes with He, N_2, and CH_4 gases in turn. The transport of these gases through membranes of this type proceed according to the regimes of Knudsen diffusion through defects, and surface diffusion through silicalite crystals (excepting He, which is sufficiently small that it is predominantly transported by Knudsen diffusion through silicalite crystals at standard conditions). The results of these permeation experiments, in combination with previously reported diffusion data and adsorption isotherms have then been used to determine the effective silicalite channel path length, and defect size to length ratio that are unique to each membrane. Through the application of the proposed model, effective silicalite channel lengths and defect ratios for membranes that were fabricated in different batches at similar conditions have been found to vary up to 20% and 700% respectively.

Given these results, characteristic values such as permselectivity can be determined that are normalized to exclude the defect transport of gases, and are therefore expected to be more comparable. Furthermore, this model can be applied to many different inorganic type membranes, and enable more reliable simulations of scaled up membrane processes.

Effects of Cyanuric Chloride and Its Derivatives on Gas Separation Properties of Polyurethane Membranes

A. A. Shamsabadi[1], M. Sadeghi[2], M. Dinari[3], A. Ronasi[2], M. Soroush[1]

[1] Chemical and Biological Engineering, Drexel University, Philadelphia, USA
[2] Department of Chemical Engineering, Isfahan University of Technology, Iran
[3] Department of Chemistry, Isfahan University of Technology, Iran

CO_2>400 ppm | Membrane Technology | Mixed Matrix Membranes

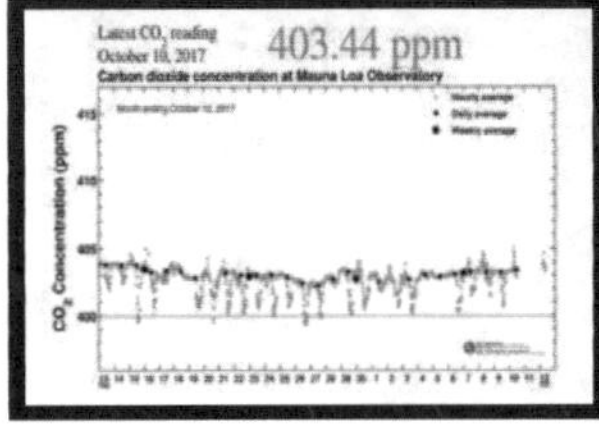

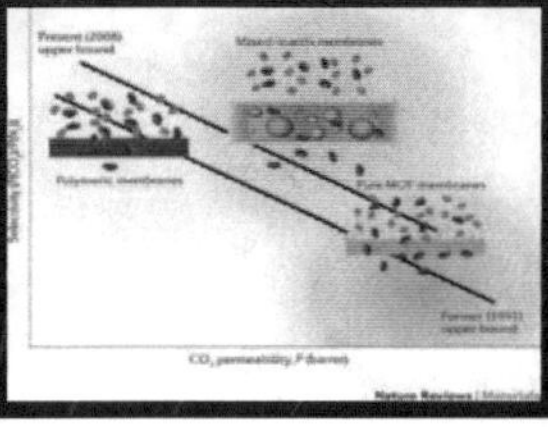

Membranes Fabrication and Characterization

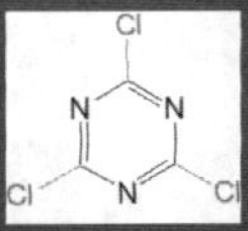

Cyanuric chloride (CC)

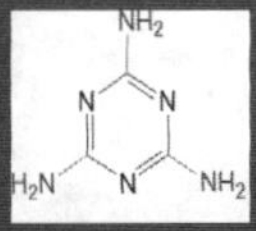

Melamine (M)

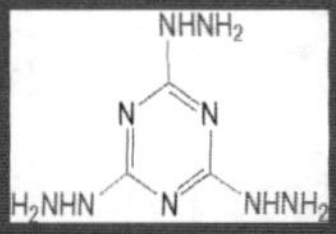

2,4,6-Trihydazino-1,3,5-triazine (THDT)

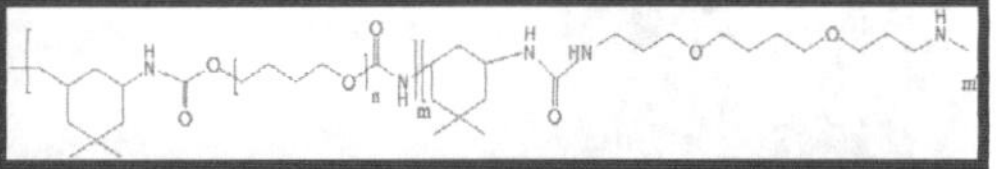

Polyurethane (PU)
Polytetramethylene glycol (PTMG)
+
Isophorone diisocyanate (IPDI)

$$\text{Hydrogen bonding index (HBI)} = \frac{A_{C=O,bonded}}{A_{C=O,free}}$$

FTIR: (a) PU-CC ,(b) PU-M , and (c) PU-THDT

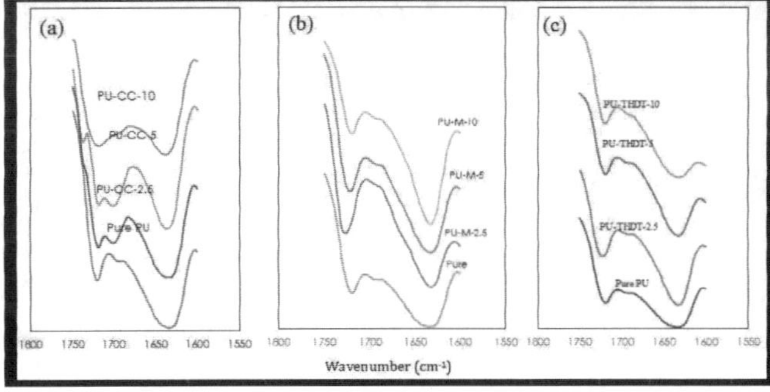

Sample	HBI
PU	1.23
PU-CC-2.5	1.12
PU-CC-5	1.08
PU-CC-10	1.03
PU-M-2.5	1.35
PU-M-5	1.38
PU-M-10	1.86
PU-THDT-2.5	1.44
PU-THDT-5	1.54
PU-THDT-10	1.62

2

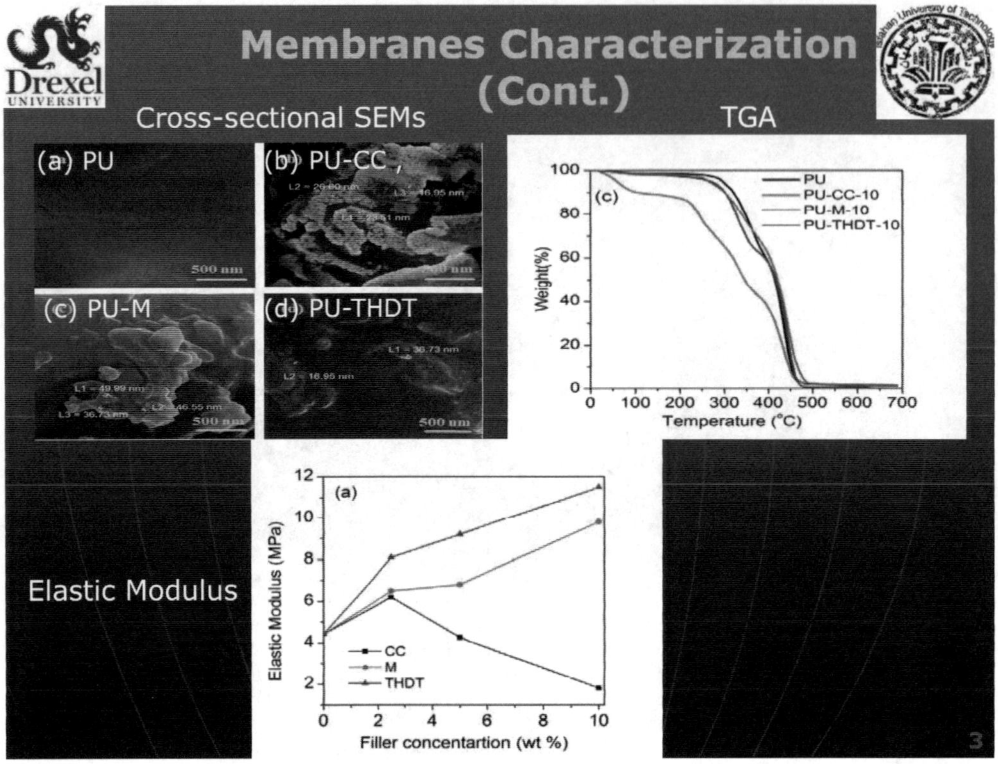
Membranes Characterization (Cont.)
Cross-sectional SEMs
TGA
(a) PU
(b) PU-CC
(c) PU-M
(d) PU-THDT
500 nm
L1 = 49.99 nm
L2 = 46.55 nm
L3 = 36.73 nm
L1 = 36.73 nm
L2 = 16.95 nm
L3 = 16.95 nm
(c)
PU
PU-CC-10
PU-M-10
PU-THDT-10
Weight(%)
Temperature (°C)
0
100
200
300
400
500
600
700
20
40
60
80
Elastic Modulus
(a)
Elastic Modulus (MPa)
Filler concentartion (wt %)
CC
M
THDT
2
4
6
8
10
12
3

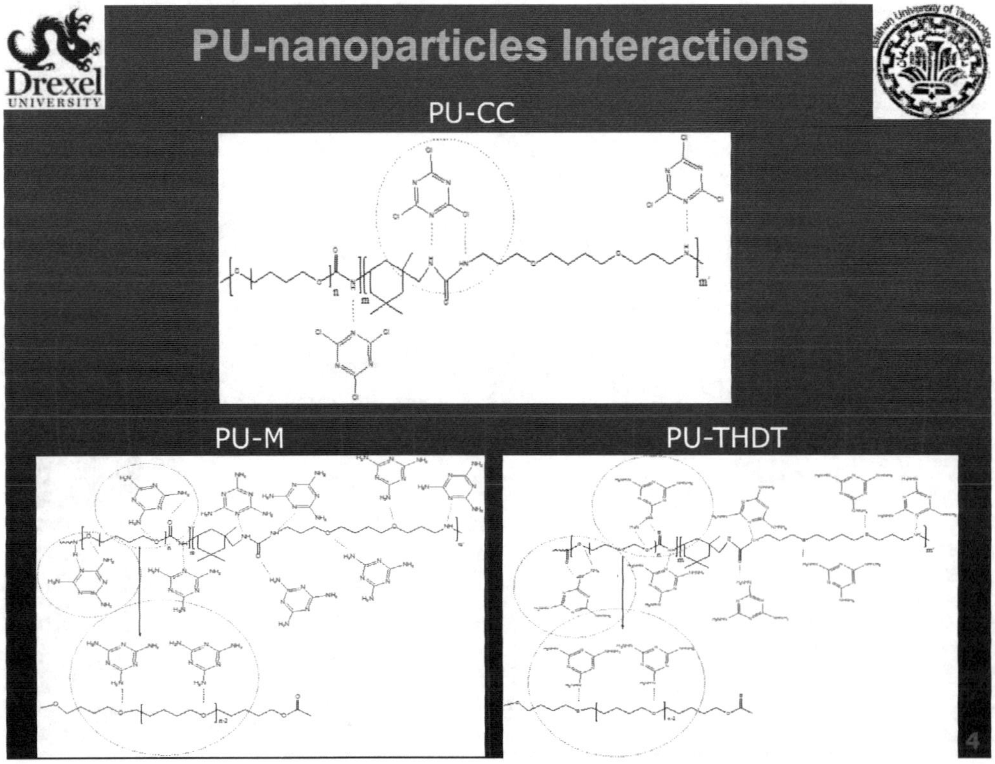
PU-nanoparticles Interactions
PU-CC
PU-M
PU-THDT
4

Membranes Performance

Sample	Permeability (barrer)				Selectivity		
	CO_2	CH_4	O_2	N_2	CO_2/CH_4	CO_2/N_2	O_2/N_2
PU	83.44	11.71	7.26	3.48	7.12	24.01	2.09
PU-CC-2.5	97.21	12.15	8.66	3.6	8	27.02	2.41
PU-CC-5	107.06	12.36	9.13	3.74	8.67	28.66	2.44
PU-CC-10	132.68	14.24	10.73	4.19	9.31	31.65	2.56
PU-M-2.5	90.5	10.55	6.64	3.07	8.58	29.5	2.17
PU-M-5	85.33	8.75	5.82	2.55	9.75	33.49	2.28
PU-M-10	73	7.31	4.92	2.11	9.98	34.57	2.33
PU-THDT-2.5	97.22	10.56	6.68	3.17	9.21	30.7	2.11
PU-THDT-5	104.27	9.58	5.73	2.72	10.89	38.35	2.11
PU-THDT-10	87.9	7.05	4.58	1.91	12.46	46.14	2.4

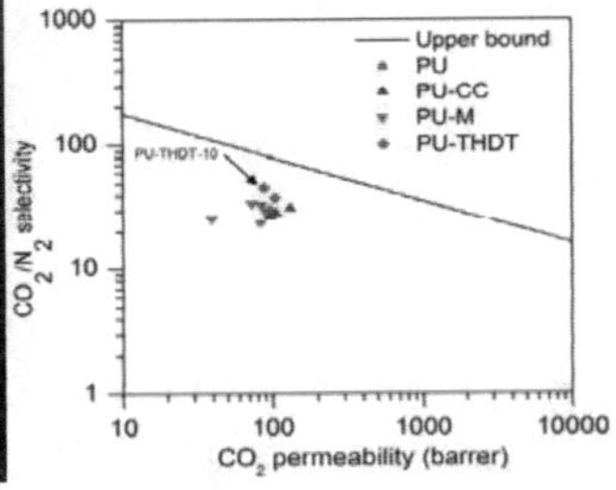

5

Concluding Remarks

- Cyanuric chloride nanoparticles are dispersed in the hard segments, but melamine and THDT particles in the soft segments
- Considerable enhancement of selectivity (75% for CO_2/CH_4 and 92% for CO_2/N_2) with higher CO_2 permeability via embedding 10% THDT into PU matrix
- THDT nanoparticles provide strong H-bonding and fixed facilitated transport for CO_2 removal
- Cyanuric chloride nanoparticles improve the gas permeability and selectivity simultaneously

6

Acknowledgment

Dr. Ali Pournaghshband

7

498772 The Growth of Glycidyl Methacrylate on Ultrafiltration Membrane: Spatial Control on Surface Initiated Aget-ATRP with Chain End Potential Functionalities

Abstract:

The growth of glycidyl methacrylate on Ultrafiltration Membrane: Spatial control on surface initiated AGET-ATRP with chain end potential functionalities

Arijit Sengupta[1], Blaine M. Carter[1], Xianghong Qian[2], S. Ranil Wickramasinghe[1]*

[1]Ralph E Martin Department of Chemical Engineering, [2]Department of Biomedical Engineering, University of Arkansas, Fayettteville, AR 72701

Corresponding author: swickram@uark.edu

Abstract

Surface modification is frequently used to tune membrane properties [1]. Atom Transfer Radical Polymerization (ATRP) was found to be one of the most convincing methods for controlled polymer growth on the outer surface and inside the pore of the membrane depending upon the associated chemistry and pore size [2]. ATRP has been shown to provide great control over the grafted nano structure [3]. Though the ATRP showed very attractive features, the most challenging step of the ATRP is to keep air free or oxidant free environment to sustain the modification [4]. The less stable Cu in +1 oxidation states is mainly responsible for the sustaining of ATRP reaction. The surface initiated Activator generated electron transfer (AGET) ATRP provides the advantages of ATRP removing the stringent condition of traces of oxidant free environment [5]. Ultrafiltration membranes are having large pore size (1-100 nm) frequently being used for size exclusion, pressure driven membrane based separation for protein concentration and buffer exchange. Though the controlled surface modification is the remedy protein fouling of UF membrane; but no spatial control on the modification was so far reported involving AGET-ATRP on UF membrane which is very important because of the large pore size provides possibilities of outside surface or inside pore modification [6].

An attempt was made to have spatial control of the modification on external versus internal pore surface of regenerated cellulose (RC) ultra-filtration (UF) membrane by glycidil methachrylate (GMA) polymers using activator generated electron transfer (AGET) atom transfer radical polymerization (ATRP). This controlled polymerization was optimized for GMA on RC UF resulting epoxy functionalized polymer brush end which on subsequent ring opening generates variety of chain end functionalities. The optimization of the experimental parameters of the surface initiated AGET ATRP revealed that the increase in relative concentration of the reductant i.e. ascorbic acid to the Cu makes a constant ratio of the two oxidation states of Cu for sustaining the AGET-ATRP, while the increased ATRP time can lead to the formation of longer polymer chain on the membrane. With increase in the initiator concentration during initiator immobilization step, its density either on the outside surface of the membrane or on the inside surface of the pore enhanced resulting decrease in the water flux. 100 mM initiator concentration with 1 min of immobilization followed by AGET ATRP for 2 hours with 2M monomer concentration with ascorbic acid to copper ratio 0.4 was found to be the optimized experimental condition. The appearance of carbonyl peak in Infra-Red spectroscopy confirmed the initiator immobilization step on the RC UF membrane while the peak due to epoxy ring confirmed the AGET-ATRP. The opening of epoxy ring by the treatment of sodium sulphonete and monoethanol amine was signature by X-ray Photoelectron Spectroscopic analysis. The performance of the membrane in terms of change in water flux and rejection of specified protein also confirmed the tuning of modification on the outer surface of the membrane vis a vis internal pore surface.

The interfacial modification using different pore filling solvents were found to bring the surface selective modification of the RC-UF membrane compared to the non-selective solution phase modification. The nature of the pore filling solvents, their viscosity coefficient, the interaction of initiator molecules with the pore filling solvents and the solvent of reaction played significant role in tuning the spatial selectivity. Depending upon the interaction of the initiator with these solvents the spatial selectivity of the plugging of initiator and subsequent polymer growth will be determined. The initiator needs to be partitioned between the pore filling solvent and the solvent of reaction and followed by diffusion to the pore to be immobilized. More the partitioning of the initiator into the pore filling solvent more will be the chance of immobilization inside the pore. Lesser the viscosity of the pore filling solvent lesser will be the probability for the initiator to diffused through it and plugged on the surface of the pore. The ratio of flux of ATRP to the flux of initiator immobilization was found to follow the trend: glycerol > L 64 > PEG 400 > ethanol > acetonitrile ~ solution. Since acetonitrile is the solvent used for initiator immobilization, using it as pore filling solvent showed the similar scenario of solution phase immobilization and hence no spatial selectivity was observed. For glycerol, the ratio was found to be 0.9 indicated the selective plugging of the initiator on the external surface of the membrane not on the surface of the pore. High immiscibility of glycerol and acetonitrile makes the partitioning of BIB into the glycerol difficult and again the higher viscosity coefficient of the glycerol makes the diffusion of the BIB even more difficult to reach to the surface of the pore. Therefore, probably, only external surface got modified in case of glycerol as pore filling solvent. In case of other solvent, depending upon their miscibility with acetonitrile and viscosity coefficient, there will be selective initiator immobilization on the external surface over internal surface of the pore as indicated by the flux ratio measurement. The maximum selectivity was found to be achieved by using glycerol as pore filling solvent due to its highly viscous nature as well as immiscibility with reaction medium.

References

1. M. Ulbricht, Advanced functional polymer membranes, Polymer, 47 (7) (2006) 2217
2. Z. Ma, X. Lu, C. Wu, Q. Gao, L. Zhao, H. Zhang, Z. Liu, Functional surface modification of PVDF membrane for chemical pulse cleaning, J. Memb. Sci, 524 (2017) 389
3. K. Matyjaszewski, P. J. Miller, N. Shukla, B. Immaraporn, A. Gelman, B. B. Luokala, T. M. Siclovan, G. Kickelbick, T. Vallant, H. Hoffmann, T. Pakula, Polymers at Interfaces: Using Atom Transfer Radical Polymerization in the Controlled Growth of Homopolymers and Block Copolymers from Silicon Surfaces in the Absence of Untethered Sacrificial Initiator, *Macromolecules*, *32* (26) (1999) 8716
4. J.-Q. Meng, C.-L. Chen, L.-P. Huang, Q.-Y. Du, Y.-F. Zhang, Surface modification of PVDF membrane via AGET ATRP directly from the membrane surface, Applied Surface Science 257 (2011) 6282
5. W. Jakubowski and K. Matyjaszewski, Activator generated by electron transfer for atom transfer radical polymerization, Macromolecules, 2005, 38, 4139.
6. N. Singh, Z. Chen, N. Tomar, S. R. Wickramasinghe, Modification of regenerated cellulose ultrafiltration membranes by surface â€"initiated atom transfer radical polymerization, J Memb Sci, 2008, 311, 225

503250 Carbon Molecular Sieves for Binary Permeation of N_2/CH_4 and CO_2/CH_4Â Gas Pairs

Abstract:

Carbon Molecular Sieve Membrane (CMSM) is derived from the pyrolysis of polymeric precursors. It possesses good mechanical strength, inert surface chemistry, and above all, high permeability and selectivity. It is one of the ideal candidates for the purification of natural gas (NG). The research aims at the removal of N_2 out of NG stream by using membrane technology. It compares favorably to the traditional method of cryogenic distillation at about -180 ^{0}C, which is energy intensive and costly. For N_2/CH_4 gas pair, CMSM 700 at 20 ^{0}C showed a good perm-selectivity of ~ 16.35 with N_2 permeability of 2.370 Barrer and CH_4 permeability of 0.145 Barrer at 250 KPa. The membrane is working more on sieving mechanism than surface diffusion mechanism for N_2/CH_4 gas pair. For CO_2/CH_4 gas pairs, CMSM 600 showed very good selectivity of 129.30 and CO_2 and CH_4 permeability of 342.64 and 2.65 Barrer, respectively at the permeation temperature of 20 ^{0}C.

509054 Solve this! Fundamental Approach to Problem Solving in Industrial Processes I (Invited Talks)

Abstract:

Solving complex plant problems is an everyday exercise that affects all projects and budgets, and it is very rarely introduced to students at any level. In fact, by the time young engineers are few years into working in industry, most have lost the engineering skills that they developed in school. We hope to encourage future engineers to retain those skills and learn the importance of plant problem-solving.

498677 Road Map for Embedding Ethics into ChE Undergraduate Curricula

Abstract:

The author will outline a number of methodologies and a suggested subject sequencing that can be effectively used to further develop an embedded ethics curricula. This curricula will consist of definitions, examples and suggestions on where to embed ethics into any existing ABET-approved, chemical engineering undergraduate course set. The author has lectured extensively at universities on ethics and a number of ethics related subjects.

504018 Views on Ethics in Undergraduate Education

Abstract:

Engineering programs are expected to graduate students who understand their professional ethical responsibilities. A studentâ€™s experience with ethics often begins with academic integrity. Research on student attitudes and behavior regarding academic integrity will be reviewed with suggestions to help faculty nurture a strong ethical foundation in their students.

495849 Ethical Reasoning in the Engineering Curriculum

Abstract:

Although the provision of ethics in engineering curricula has increased in recent years, the teaching of engineering ethics remains patchy and sporadic in a number of countries. Ethics in research, on the other hand, appears to be at a more mature stage. At the same time, the number of ethical codes, codes of practice and research concordats has flourished, both nationally and internationally. The questions remain: are those codes â€œliveâ€? Have those concordats resulted in an improvement of professional integrity? Is the teaching of ethics different and/or disconnected from ethics in research and in the workplace? Is ethics in chemical engineering different from ethics in other engineering disciplines? The paper analyses those questions in relation to an ever changing landscape, influenced by local and international drivers.

RAPID

Transforming Process Industries

Breakthrough Innovation to Transform the Process Industries

Karen Fletcher
CEO - RAPID

Jim Bielenberg
CTO - RAPID

October 30, 2017

ManufacturingUSA

Manufacturing USA Overview

Vision

U.S. Global leadership in Advanced Manufacturing

Mission

Connecting people, ideas, and technology to solve industry-relevant advanced manufacturing challenges, thereby enhancing industrial competitiveness and economic growth, and strengthening our national security.

Goals

1. Increase competitiveness
2. Facilitate technology transition
3. Accelerate the manufacturing workforce
4. Ensure stable and sustainable infrastructure

www.ManufacturingUSA.com

2

RAPID
Transforming Process Industries

Manufacturing USA

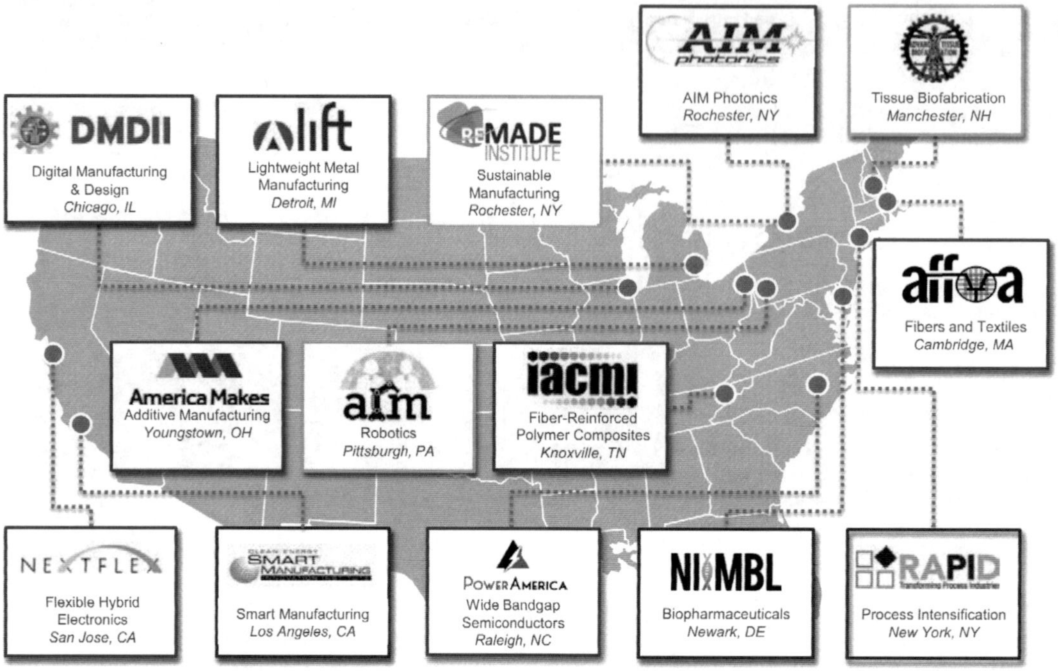

3

RAPID
Transforming Process Industries

RAPID's Industry-Led Vision

A dynamic network of partners who collectively build a sustainable ***ecosystem*** *that:*

... researches, develops and advances new technologies for modular chemical process intensification

... builds a strong portfolio of R&D projects & educational tools

... delivers dramatic reductions in energy, environmental footprint, capital and operating cost

... makes U.S. Manufacturing and our workforce more competitive

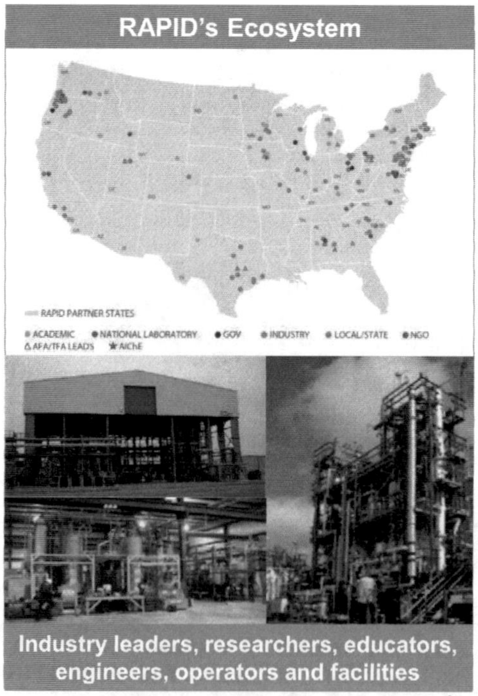

4

RAPID
Transforming Process Industries

Our Mandate

- Lead a national effort to research, develop and demonstrate high-impact modular chemical process intensification solutions for U.S. Manufacturing.
- Actively build RAPID membership through an inclusive and attractive value proposition.
- Leverage $70 million of federal funding with cost share from members.
- Operate the Institute to benefit a wide range of stakeholders.
- Establish an infrastructure that enables access to process intensification resources, tools, expertise, and facilities.
- Bring together private and public entities to co-invest in R&D, commercialization, and deployment of innovative technologies.
- Establish a technical education and workforce development program.

"The goal for these Institutes is to revitalize American manufacturing and support domestic manufacturing competitiveness."
— U.S. DOE

RAPID
Transforming Process Industries

DOE Performance Metrics

- Energy Efficiency
- Energy Productivity improvement
- Intensification in Individual Process Modules
- Cost-Effective Manufacturing of Modules
- Cost Effective Deployment
- Enabling R&D Portfolio
- Industrial Partnerships
- Pathway to Self-Sustainment
- Train the Trainers
- Educate Students
- Annual Planning Process
- Industrial Roadmap
- Emerging Supply Chain
- Diversity of Firms and Individuals in the Eco-System

6

RAPID Transforming Process Industries

52 Members and growing

RAPID Transforming Process Industries

Technical Focus Areas

Chemical Commodity Processes

- Develop guidelines for integration of novel reaction and separation modules
- Validate design tools for process intensification

Renewable Bio Products

- Prototype and scale novel bio-conversion processes
- Improve energy & capital efficiency of existing and emerging processes

Natural Gas Upgrading

- Maximize impact through transfer of technologies and learnings across industries
- Mature and demonstrate transformational and enabling technologies for gas utilization

8

RAPID Transforming Process Industries

Technical Focus Areas

Modeling and Simulation

- Develop methods & tools for design, optimization, and intensification across multiple length and time scales
- Establish methods for intensification of dynamic/periodic operations

Intensified Process Fundamentals

- Advance inherently energy efficient separation processes & reaction platforms
- Develop fundamentals for multifunctional modules such as hybrid separation/ reaction schemes

Module Manufacturing

- Standardize modules and components to drive demand and capital investment within the supply chain
- Lower the cost of PI equipment using advanced manufacturing technology

9

RAPID Transforming Process Industries

2017 Accomplishments

- Completed DOE negotiations & all required documentation
- Stood up the Institute, with strong support/services from AIChE
- Established membership base
- Identified, negotiated and launched 4 Jump Start projects
- Formed Education & WFD Committee and completed a roadmap
- Completed focus area roadmaps with 100+ participants
- Executed call for pre-proposals and call for proposals
- Designed and launched a new website www.aiche.org/rapid
- Hired CTO (Mar), Technology Manager (Oct), Comms Leader (Nov)

10

A Public-Private Partnership Between U.S.DOE EERE & AIChE®

RAPID
Transforming Process Industries

Technology Gaps

RAPID Roadmap

Chemical & Commodity Processing

- **Rxn/Separation**
  Novel reaction/separation schemes that are scalable and drive process efficiencies (e.g., membrane- or sorption enhanced reactors). Applications include processing light paraffins to olefins, increase p-xylene yield vs conventional processes, hydrogen production, managing oxygen supply to reactions, etc.
- **Non-Thermal Drivers**
  Use of alternative, non-thermal driving forces to activate chemical systems at the appropriate (atomic/molecular) scales.
- **Batch Systems**
  Intensification schemes for batch systems. Transferring concepts largely developed for continuous processes to the batch realm could result in increased productivity/lower cost for specialty/fine chemicals.

Natural Gas Upgrading

- **Selective Conversion**
  Concepts that show dramatic increases in desired product yields via fundamental improvements in catalysis, heat and mass transfer, and process concepts. This could include include alternative energy inputs and/or the use of novel reaction systems.
- **Separations**
  Energy efficient separations technology to purify the reaction product mix, condition the feed in preparation for conversion, and to generate co-reactants for participation in natural gas conversion gas.
- **Process Consolidation**
  Process consolidation and modularity to reduce total installed cost by reducing the total number of unit operations and by reducing the amount of field fabrication.

Renewable Bioproducts

- **Primary Separation**
  Technologies to reduce energy demand in primary separation process steps designed to recover organic molecules and biomass components from water.
- **Water management**
  Low capital and energy intensive solutions for dewatering and drying of biomass feedstocks, water removal and drying in pulp and paper process, and drying and removal of low levels of residual water from end products.
- **Couple Rxn/Spn/Hxt**
  Use of novel chemistries and MCPI strategies to couple heat transfer and reaction in thermal processing of biomass and/or novel applications of reactive separation technologies in biological conversion technologies such as fermentation.

Intensified Process Fundamentals

- **Scale-Out Methods**
  Scale out methodologies and models to predict performance of alternative energy input approaches for reactions and mixing and determine the suitable scale for modular manufacturing.
- **Fundamental Data Acquisition/Modeling**
  Approaches to address key issues with lack of data on fluxes, adsorption, and catalyst kinetics for wide classes of materials, enabling model development and experimental testing of novel materials as adsorbents, membranes, catalysts and their integration.
- **Predictive Models**
  Modeling capabilities to screen concepts and configurations of all types and predict optimal structures.

Modeling & Simulation

- **PI Software Tools**
  Software tools for integrated reaction and/or separation processes and/or cyclic process such as pressure swing adsorption (PSA) or temperature swing adsorption (TSA). Such tools must be widely accessible and capable of integrating MCPI solutions with existing unit operations.
- **Data Availability**
  Modeling approaches coupled with data generation and/or analysis to generate databases of physical parameters enabling design with mass separating agents.
- **PI Assessment Tools**
  Tools to assess safety, sustainability, and control in PI and MCPI applications, including tools that address unique issues of uncertainty and reduced control variable options that are present in PI and MCPI applications.

Module Manufacturing

- **Intensified Components**
  Intensified components that drive down the cost of module pre assembly, transportation, and installation, while driving significant energy savings in chemical processes.
- **Standard Designs**
  Design approaches that limit the amount of non-recurring engineering — during systems integration and installation — needed to support customized modules. This could include standard modules that enable economies of mass production and/or designs that enable incremental capacity additions.
- **Distributed Processing**
  Module design and manufacturing approaches to enable distributed chemical processing. These will provide new paths to capital cost reduction and innovative techniques for maintenance and remote access and monitoring.

www.aiche.org/rapid

A Public-Private Partnership Between U.S.DOE EERE & AIChE®

RAPID
Transforming Process Industries

RAPID Education & Workforce Development Roadmap

OBJECTIVE: Establish a technical education and workforce development program that will leverage existing resources to train and educate the workforce, who can then research, develop, design and operate processes that incorporate new process intensification (PI) technologies and modular process designs widely within U.S. industry.

Undergraduate Students

GOAL
Create PI & MCPI awareness in future generations of engineers by augmenting the standard (ABET accredited) engineering curriculum with modules and educational content introducing PI & MCPI topics relevant to industry.

WHAT SUCCESS LOOKS LIKE
500 students* trained in PI & MCPI per year by 2019.

METHOD
eLearning course modules, face-to-face courses in conjunction with major meeting, intern program.

Graduate Students

GOAL
Generate PI & MCPI expertise so that graduate students can connect the relevance of PI & MCPI research to industrial applications and create new opportunities and applications for MCPI.

WHAT SUCCESS LOOKS LIKE
500 students* trained in PI & MCPI per year by 2019.

METHOD
eLearning course modules and face-to-face courses in conjunction with major meeting.

Faculty

GOAL
Update and educate university faculty on PI & MCPI concepts and introduce eLearning modules created by RAPID for integration into their curriculum.

WHAT SUCCESS LOOKS LIKE
50 education/training professionals per year in MCPI by 2019.

METHOD
Face-to-face course in conjunction with major meeting on the Fundamentals of PI & MCPI.

Technicians/Operators

GOAL
Introduce technicians and operators to PI & MCPI, provide awareness of new processes that include PI & MCPI as well as emphasizing process safety.

WHAT SUCCESS LOOKS LIKE
At least 50 professionals** per year in MCPI by 2019.

METHOD
eLearning course modules.

Professional Engineers R&D | Design | Production

GOAL:
Train the current generation of engineers on how to utilize and deploy PI & MCPI technology in order to improve process efficiency.

WHAT SUCCESS LOOKS LIKE:
At least 50 professionals** per year in MCPI by 2019.

METHOD
eLearning course modules, face-to-face courses in conjunction with major meeting, and promotion through Manufacturing Extension Partnerships (MEPs).

Management

GOAL
Create high level awareness of the benefits of PI & MCPI from a techno-economic analysis perspective.

METHOD
Presentation material (includes relevant case studies, sustainability, and safety); short videos aimed at executives.

PI & MCPI Body Knowledge Drives Content Development Topics and Priority

Crosscutting areas identify content topics, levels of understanding for each audience member as well as high-priority topic areas. It includes four sub-sections:

- PI & MCPI Overview
- Modeling & Simulation
- Intensified Process Fundamentals
- Module Manufacturing

Chemical & Commodity Processing · Natural Gas Upgrading · Renewable Bioproducts

Application areas drive relevancy for content and provide industrial applications, examples and case studies within content and include the three sub-sections above.

* Students defined as undergraduate and graduate students.
** Professionals defined as technician, operator or engineer.

www.aiche.org/rapid

RAPID
Transforming Process Industries

2017 Call for Proposals

- Overall Results
  - 73 submissions
  - 42 "lead" institutions
  - ~$30M (fed) request for 2018
- Proposals review process completed during Sept/Oct
- Announcement of selected projects November 2nd, 2017

# of Submissions	Submissions*
Chemical Commodity Processes	13
Intensified Process Fundamentals	13
Modeling and Simulation	6
Module Manufacturing	7
Natural Gas Upgrading	18
Renewable Bioproducts	16
	73

Summary Data - RAPID 2017 Proposal Call

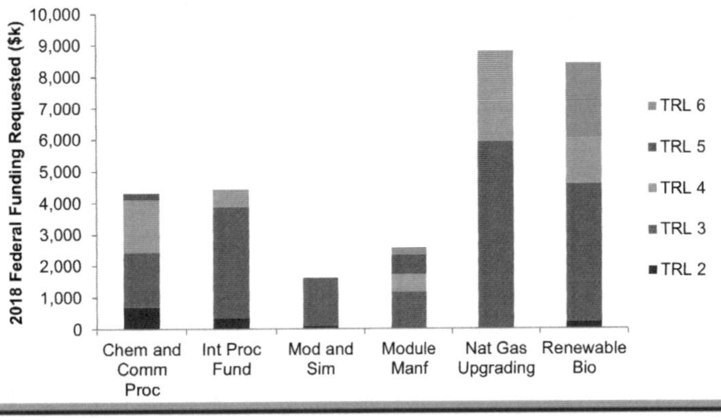

RAPID
Transforming Process Industries

Dynamic Intensification of the Operation of Dividing Wall Columns

DOW

TEXAS
The University of Texas at Austin

Innovation:
- Demonstrate that a dynamically controlled process (non-steady state) creates opportunity for improved efficiency

Project Outcomes
- Develop theoretical basis for periodic operation of intensified processes
- Experimental proof of concept using dividing wall column (DWC) pilot plant available at UT

Institute Target Addressed
- Develop tools and technologies to reduce the cost of deploying modular chemical process intensification in existing processes

Fall 2017 Update
- Theoretical framework for general value/screening of periodic operation tested
- Initial screening of DWC periodic operation assessed

RAPID
Transforming Process Industries

Adsorptive Nitrogen Rejection from Natural Gas

PRAXAIR

Georgia Institute of Technology

Innovation
- Utilize breakthrough rate dependent adsorbent to modularize conventional PSA process

Project Outcomes
- Field demonstration of portable, modular PSA system

Institute Target Addressed
- 1,000 hours of pilot scale operating time for a modular and intensified process with reduced capacity , improved energy efficiency, and lower emissions/environmental waste

Fall 2017 Update
- System installation, shakedown, and testing to demonstrate adsorbent performance in a PSA pilot system underway
- Manufacturability studies progressing - potential adsorbent manufacturer identified.
- Early 2018 tasks - additional work on TEA and H2S tolerance testing at Georgia Tech

15

RAPID
Transforming Process Industries

Autothermal pyrolysis of lignocellulose wastes to sugars & other biobased products

easy energy SYSTEMS

IOWA STATE

Innovation
- Couple intensified reaction + heat transfer design and novel process chemistry to modularize sugar production from biomass

Project Outcomes
- Establish MCPI benefits of autothermal pyrolysis, leading to design of a commercial scale modular system

Institute Target Addressed
- MCPI RD&D resulting in a doubling of energy productivity ($ Returned/Energy Consumed) by a combination OPEX and CAPEX improvements.

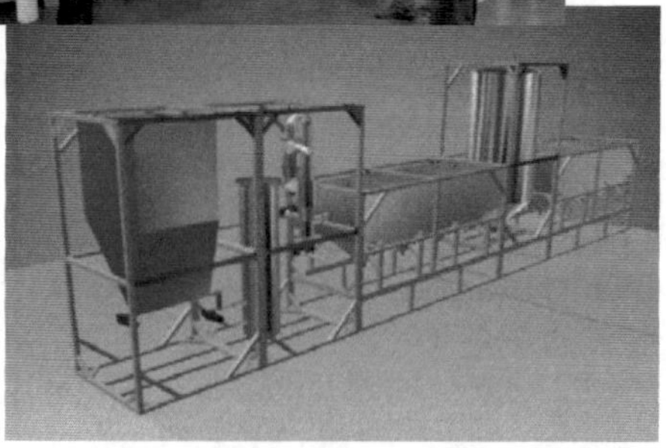

Fall 2017 Update
- Demonstrate sugar yields in excess of BP1 targets on 1 kg/h pyrolysis reactor under auto thermal conditions
- Completed continuous pilot scale pretreatment of multiple feedstocks
- Continuing to work on longer time pilot scale testing

16

Mfg Supply Chain Development for the Modular Solar-Thermochemical Conversion Platform

RAPID Transforming Process Industries

Oregon State University OSU — Pacific Northwest NATIONAL LABORATORY

Innovation

- Coupling material processing approaches and manufacturing techniques that are uniquely well suited to small scales to generate low cost modular components

Project Outcomes

- Develop novel additive manufacturing techniques and materials that enable modular designs

Institute Target Addressed

- Scale-out manufacturing MCPI, with a unit cost reduction of >20% with each doubling in cumulative module manufacturing production

Fall 2017 Update

- OUS has developed unvalidated costs models for key manufacturing steps
- PNNL has established cost targets for key components
- OSU is working in-house capability to test new designs based on selective alloying
- Targeting 1Q2018 for new tool completion in support of novel manufacturing techniques

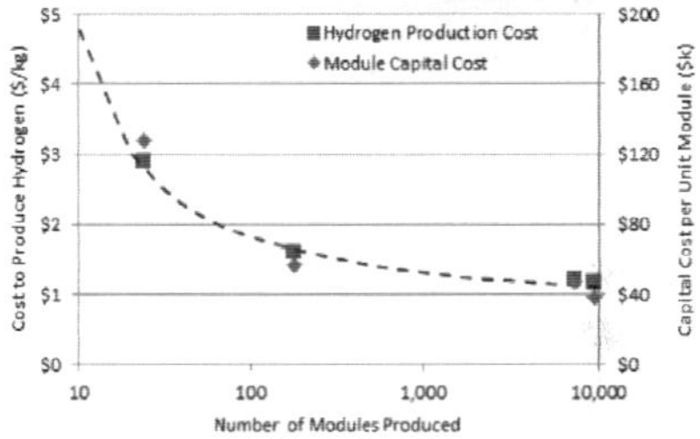

Estimated capital and product costs for individual STARS modules as a function of module mass production.

17

Plans for 2018

RAPID Transforming Process Industries

- Build the RAPID community
  - Member profiles, webinars, newsletters, website, networking opportunities
  - Conferences / face-to-face meetings (AIChE Spring / Annual)
- $30+M invested in R&D projects (fed funds + cost share)
- DOE (external) peer review
- Updated roadmaps; Gaps and new areas explored
- Project calls
- Training courses and educational webinars
- Increase membership and "reach"
  - New addition to our team: State MEPs*

* MEP = Manufacturing Extension Partnership

18

RAPID
Transforming Process Industries

To learn more about RAPID:

www.AICHE.org/RAPID

KareF@aiche.org

JameB@aiche.org

19

493251 Modeling and Simulation â€“ a Key Component in Enabling Process Intensification

Abstract:

The RAPID Manufacturing Institute aims to enable the development, de-risking, and implementation of process intensification across the chemical process industry. We will discuss critical needs in modeling and simulation in this area by highlighting the successes and challenges associated with using modeling and simulation to drive RAPIDâ€™s aims.

493349 Modeling and Simulation Challenges for Process Intensification

Abstract:

Modeling and Simulation Challenges for Process Intensification

Efstratios N. Pistikopoulos[1,2], David Sholl[3], M. M. Faruque Hasan[1,2], Salih E. Demirel[1,2] and Yuhe Tian[1,2], (1)Artie McFerrin Department of Chemical Engineering, Texas A&M University, College Station, TX, (2)Texas A&M Energy Institute, Texas A&M University, College Station, TX, (3)School of Chemical & Biomolecular Engineering, Georgia Institute of Technology, Atlanta, GA

The recently established Rapid Advancement in Process Intensification Deployment (RAPID) Manufacturing Institute aims to transform the U.S. industry by developing the next generation of modular, intensified, and innovative manufacturing technologies and processes. The RAPID Modeling and Simulation Focus Area will in particular address the development of strategies and tools for the multi-scale representation, modeling, optimization, and control of intensified systems. In this presentation, we provide an overview of the challenges towards a systematic synthesis, modeling, and simulation approach for process intensification, as well as an exclusive review of process intensification synthesis approach methods in the open literature, such as the ones included in [1-5]. We also introduce an integrated framework for process intensification incorporating operability, control and safety (illustrated in Figure 1).

Figure 1: An Integrated Framework for Process Intensification Incorporating Operability, Control and Safety

[1] Demirel, S. E., Li, J., & Hasan, M. F. (2017). Systematic process intensification using building blocks. *Computers & Chemical Engineering*. http://doi.org/10.1016/j.compchemeng.2017.01.044.

[2] Baldea, M. (2015). From process integration to process intensification. *Computers & Chemical Engineering*, *81*, 104-114.

[3] Tula, A. K., Babi, D. K., Bottlaender, J., Eden, M. R., & Gani, R. (2017). A computer-aided software-tool for sustainable process synthesis-intensification. *Computers & Chemical Engineering*. http://doi.org/10.1016/j.compchemeng.2017.01.001

[4] Papalexandri, K. P., & Pistikopoulos, E. N. (1996). Generalized modular representation framework for process synthesis. *AIChE Journal*, *42*(4), 1010-1032.

[5] Ismail, S. R., Proios, P., & Pistikopoulos, E. N. (2001). Modular synthesis framework for combined separation/reaction systems. *AIChE Journal*, *47*(3), 629-649.

494520 The Sustainable Synthesis-Design-Intensification of Chemical and Biochemical Processes

Abstract:

With the demands for efficient utilization of energy and water, better protection-control of the environment and correct-consistent use of raw materials, it is no longer sufficient to convert a selected raw material to a desired product, making profit and ensuring operability. The synthesized-designed process must now also be sustainable, incorporating not only economic feasibility but also environmental impact constraints, energy-water usage and/or waste constraints, resources availability constraints, product demand-price constraints and, many more. That is, the demands for new, innovative and sustainable processes need to be addressed. Process intensification has been identified as an important means to achieve this demand through integration with process synthesis-design in the early stage development of any chemical and biochemical process.

The presentation will highlight the issues related to defining the measures of sustainability through which new alternatives can be evaluated. Here, emphasis is given to the concept of sustainability, which is relative and therefore, a reference case first needs to be established. That is, compared to the base case, how sustainable are the new and innovative chemical and biochemical process designs?

A systematic multi-stage and multi-dimension computer aided framework which provides the necessary work-flow, the corresponding data-flow, the associated methods and the needed computational tools will be presented.

Through case studies involving chemical and biochemical processes, the application of the framework and the associated methods-tools to determine more sustainable solutions will be illustrated. It will be shown that significant improvements and non-tradeoff sustainable can be obtained through this integrated approach of combining process intensification as an essential element of process synthesis-design.

2017 AIChE Annual Meeting

October 29 - November 3, 2017
Minneapolis Convention Center, Minneapolis, MN

Democratizing Energy Technology

Dane A. Boysen, PhD
Modular Chemical, Inc.

October 30, 2017

Overview

- How did RAPID get established?
- Why do we go big?
- Why is scale-up a problem?
- How do we break the hegemony of scale-up?

2011-2014

arpa·e

Advanced Research Projects Agency • ENERGY

2015-2016

gti®

GAS TECHNOLOGY INSTITUTE

Innovation Challenge

FINANCING ECONOMIES OF SCALE

GTL PLANT

(boe/d)	($/bpd)	plant cost
1	47,000k	$47M
10	10,000k	$100M
100	2,200k	$220M
1k	470k	$470M
10k	100k	$1,000M

AMMONIA PLANT

(ton/d)	($/tpd)	plant cost
1	10,000k	$10M
10	4,500	$25M
100	2,000k	$200M
1k	900k	$900M
10k	400k	$4,000M

calculated from "2/3 scaling law"

Innovation Challenge

FINANCING ECONOMIES OF SCALE

GTL PLANT

(boe/d)	($/bpd)	plant cost
1	47,000k	$47M
10	10,000k	$100M
100	2,200k	$220M
1k	470k	$470M
10k	100k	$1,000M

AMMONIA PLANT

(ton/d)	($/tpd)	plant cost	
1	10,000k	$10M	R&D demo
10	4,500	$25M	
100	2,000k	$200M	pilot demo
1k	900k	$900M	commercial
10k	400k	$4,000M	

calculated from "2/3 scaling law"

R&D Spending

OFFICE OF FOSSIL ENERGY BUDGET

U.S. DEPARTMENT OF ENERGY

FY 2015	$561M
FY 2014	$570M
FY 2013*	$495M
FY 2012*	$337M
FY 2011	$434M
FY 2010	$660M
FY 2009	$876M
FY 2008	$465M

Pilot chemical and power plants require **more than $200M**

How can we innovate with vastly inadequate funding?

*continuing resolution

FY2016 U.S. Senate Energy and Water Appropriations Bill

"The Committee recognizes that stranded, flared, and vented natural gas is the result of low natural gas prices that make transporting it uneconomic. As topics for additional Clean Energy Manufacturing Institutes are evaluated, the Secretary is encouraged to consider modular chemical processing as a way to address the issue of natural gas flaring and enable advanced manufacturing applications in the oil and gas industry."

https://www.congress.gov/congressional-report/114th-congress/senate-report/54/1

RAPID

RAPID ADVANCEMENT IN PROCESS INTENSIFICATION DEPLOYMENT INSTITUTE

U.S. DEPARTMENT OF ENERGY | Energy Efficiency & Renewable Energy

ADVANCED MANUFACTURING OFFICE

December 2016, the US Department of Energy announced the establishment of the 10th Manufacturing USA Institute, representing a critical step in the federal government's effort to double U.S. energy productivity by 2030. The Rapid Advancement in Process Intensification Deployment (RAPID) Institute is focused on addressing the barriers listed above to enable the development of breakthrough technologies to boost energy productivity and energy efficiency through manufacturing processes in industries such oil and gas, pulp and paper and various domestic chemical manufacturers. RAPID will leverage approaches to modular chemical process intensification (MCPI)—such as combining multiple process steps such as mixing, reaction, and separation into single more complex and intensified processes—with the **goal of improving productivity and efficiency, cutting operating costs, and reducing waste.**

https://www.aiche.org/rapid/about

RAPID Successes

1. Quick to get up and operational
2. Quality R&D projects in the que
3. Good industry engagement

Challenges for RAPID

1. No clear problem statement (goal ≠ problem)
2. No plan for establishing standards (modular = standards)
3. Not clear how progress will be tracked (need baseline)

RAPID Challenges

1. No clear problem statement (goal ≠ problem)
2. No plan for establishing standards (modular = standards)
3. Not clear how progress will be tracked (need baseline)

Why do we go big?

Energy Technology Today

PRICE TAG
$1-10 BN

COAL POWER PLANT

PRICE TAG
$0.5-5 BN

AMMONIA PLANT

PRICE TAG
$5-50 BN

GTL PLANT

Economies of Unit Scale

GAS-TO-LIQUIDS PLANTS

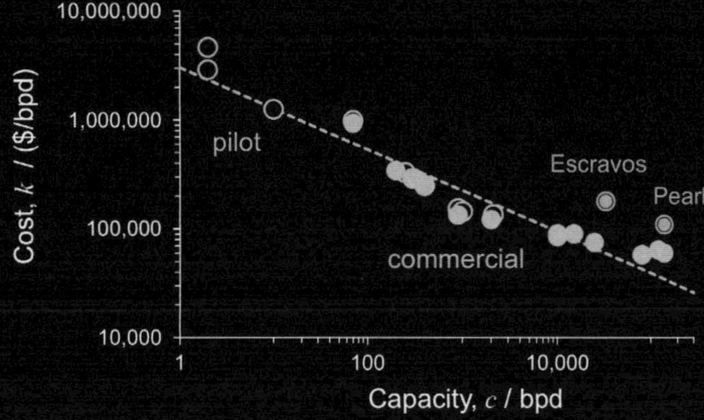

"2/3 scaling law"

$$k(c) = k(c_0)\left(\frac{c}{c_0}\right)^{\alpha}$$

Sources: (1) PJA Tijm. Gas to liquids. Fischer-Tropsch, advanced energy technology, future's pathway. Feb 2010; (2) C. Kopp. The US Air Force Synthetic Fuels Program. Technical Report APA-TR-2008-0102. (2008)

Why do we scale-up?

ARGUMENT

capital cost $\propto$ area [L^2]
capacity $\propto$ volume [L^3]
capital cost / capacity $\propto$ [L^2]/[L^3]

FLAW

pipe thickness ignored
hoop stress (constant) $\propto t / r$
capital cost $\propto l \cdot r^2$
capital cost / capacity $\propto$ constant

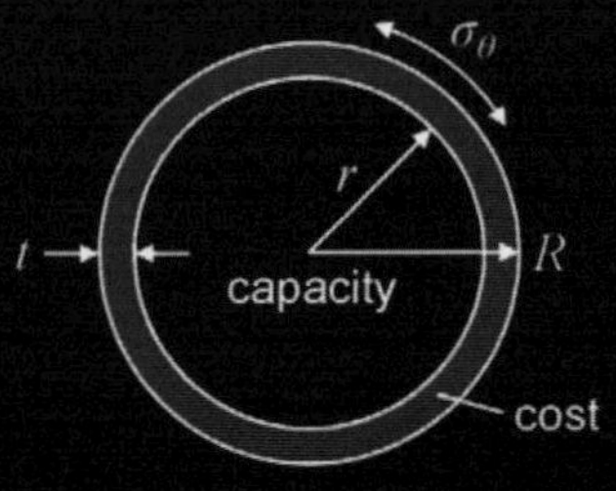

pipe cross-section

Why is scale-up a problem?

#1

BARRIER TO INNOVATION

#2

HUGE CAPITAL NEEDS

MEGA-PROJECT ECONOMICS

RAND Study:

- 52 mega-projects
- $0.5B and $10B (1984 dollars)
- average over budget = 90%

Source: E.W. Merrow. Understanding the outcomes of megaprojects: a quantitative analysis of very large civilian projects, The RAND Corporation, Santa Monica, CA, 1988.

Capital Resources

U.S. COMPANIES BY REVENUE, 2007

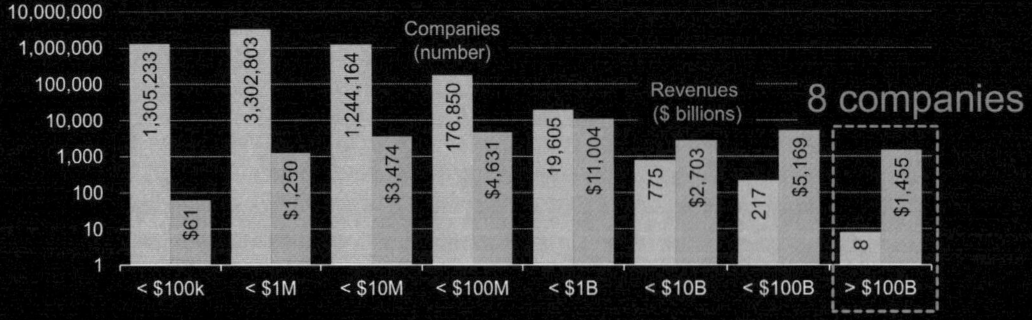

How many U.S. companies can finance a $1 billion plant?

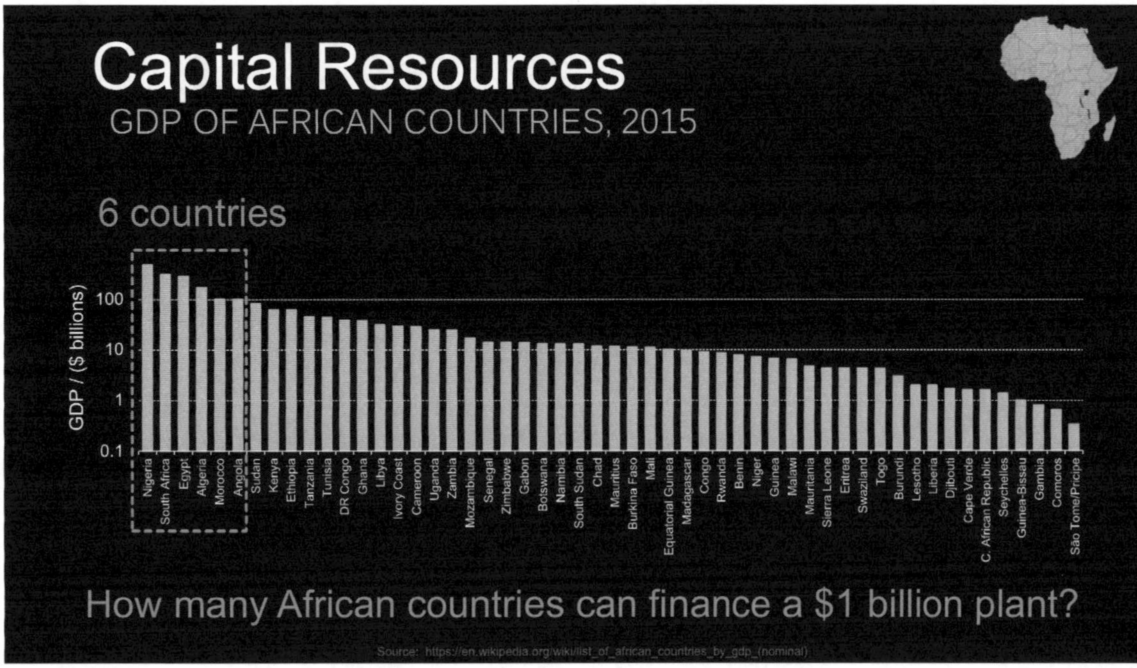
Capital Resources
GDP OF AFRICAN COUNTRIES, 2015
6 countries
GDP / ($ billions)
100
10
1
0.1
Nigeria
South Africa
Egypt
Algeria
Morocco
Angola
Sudan
Kenya
Ethiopia
Tanzania
Tunisia
DR Congo
Ghana
Libya
Ivory Coast
Cameroon
Uganda
Zambia
Mozambique
Senegal
Zimbabwe
Gabon
Botswana
Namibia
South Sudan
Chad
Mauritius
Burkina Faso
Mali
Equatorial Guinea
Madagascar
Congo
Rwanda
Benin
Niger
Guinea
Malawi
Mauritania
Sierra Leone
Eritrea
Swaziland
Togo
Burundi
Lesotho
Liberia
Djibouti
Cape Verde
C. African Republic
Seychelles
Guinea-Bissau
Gambia
Comoros
São Tome/Pricipe
How many African countries can finance a $1 billion plant?
Source: https://en.wikipedia.org/wiki/list_of_african_countries_by_gdp_(nominal)

#3
DISTRIBUTED CHALLENGES

Developing World
DISTRIBUTED MARKETS

AFRICA
17% WORLD POPULATION
1% WORLD FERTILIZER USE

In 1999, Uganda farmers bought urea for $600/ton, global market price was $100/ton, why?

- Market size (< 1% global market)
- Transport cost (>$50/ton, 30% total)
- Finance cost ($300k, 1 kton)

Source: World Bank, 2015

Isaac Mkalia, 20, checks his mobile phone in Kojiado district, near the Tanzanian border (Photograph: Sven Torfinn/Oxfam)

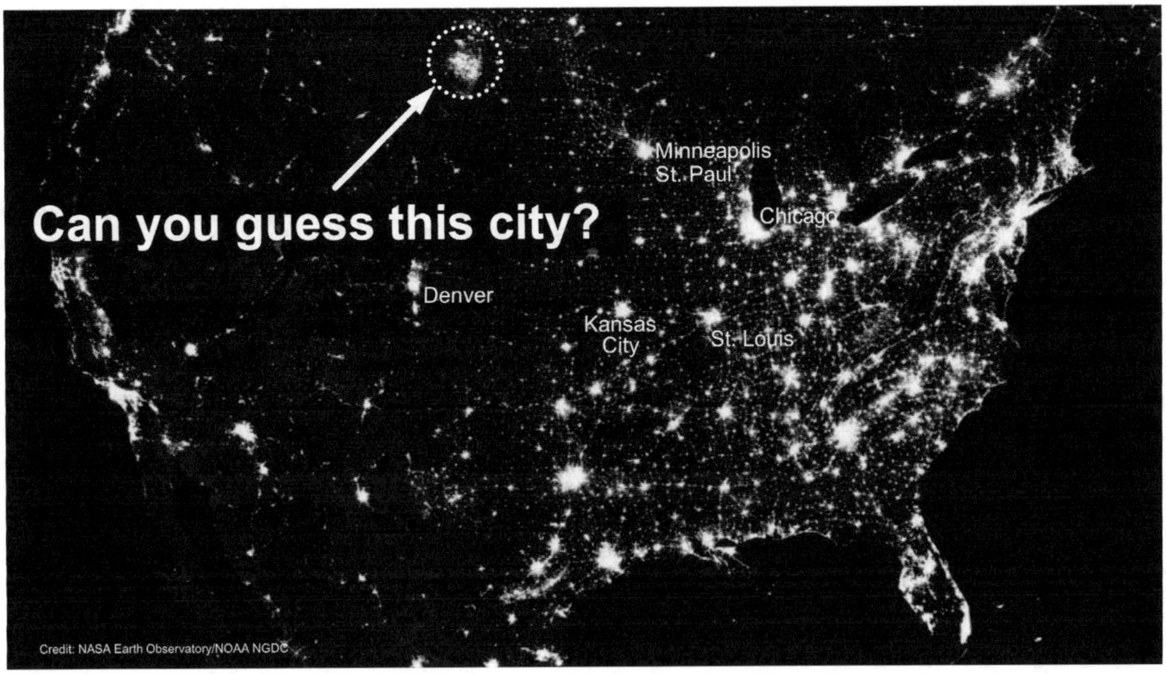
Can you guess this city?
Minneapolis St. Paul
Chicago
Denver
Kansas City
St. Louis
Credit: NASA Earth Observatory/NOAA NGDC

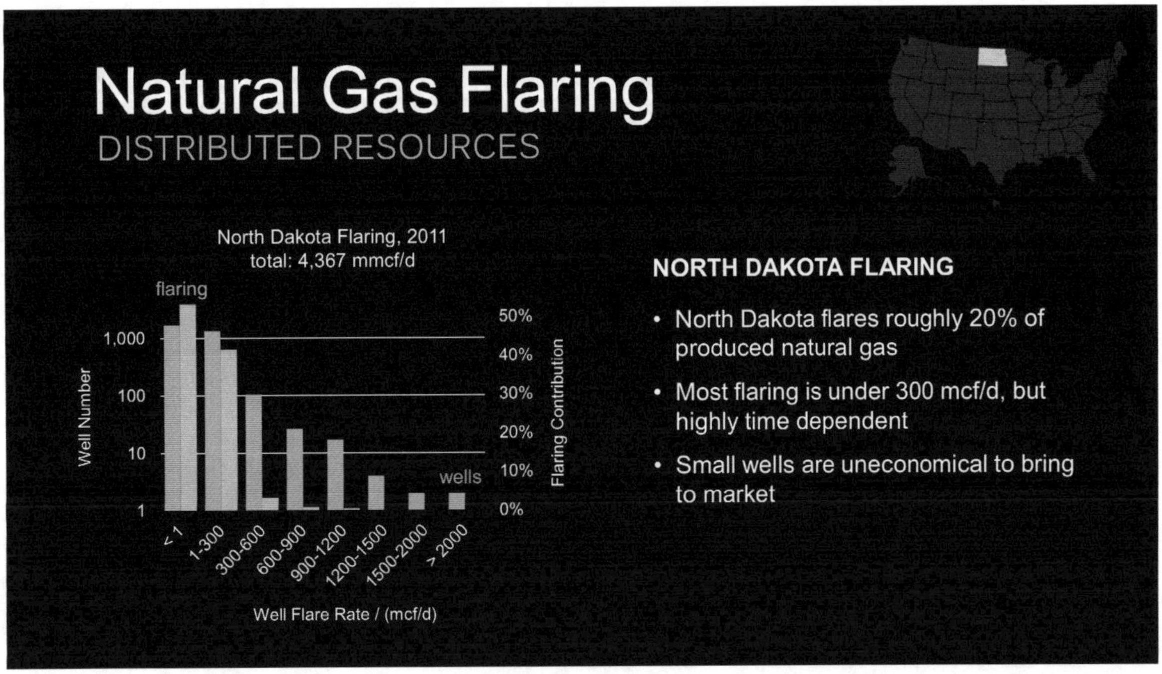
Natural Gas Flaring
DISTRIBUTED RESOURCES
North Dakota Flaring, 2011
total: 4,367 mmcf/d
flaring
wells
Well Number
1,000
100
10
1
50%
40%
30%
20%
10%
0%
Flaring Contribution
< 1
1-300
300-600
600-900
900-1200
1200-1500
1500-2000
> 2000
Well Flare Rate / (mcf/d)
NORTH DAKOTA FLARING
• North Dakota flares roughly 20% of produced natural gas
• Most flaring is under 300 mcf/d, but highly time dependent
• Small wells are uneconomical to bring to market

How do we break the hegemony of scale-up?

Democratizing Technology

HISTORIC EXAMPLES

1450	1908	1950	1973	1977
Gütenburg Press	Ford Model T	ISO Shipping Container	Motorola DynaTAC 8000X	Commodore PET
INFORMATION	TRANSPORTATION	SHIPPING	COMMUNICATION	COMPUTATION

Democratizing Technology

WHAT ARE THE DEFINING FEATURES?

1. Empowers the little guy
2. Never developed by the incumbent
3. Displaces entrenched incumbent
4. Levels the playing field
5. Leverages capital of the many
6. Modular, mass produced, standardized
7. Fast innovation cycles

…and there are many examples

guns	*…displaced*	*swords*
Gutenberg press	*…displaced*	*illuminated manuscripts*
iso-containers	*…displaced*	*bulk shipping*
cellular phones	*…displaced*	*land lines*
personal computers	*…displaced*	*central computing*
automobiles	*…displaced*	*horses*
photovoltaics	*…displaced*	*solar thermal*
electric-arc furnaces	*…displaced*	*blast furnaces*
e-commerce	*…displaced*	*store fronts*
televisions	*…displaced*	*theatres*
record players	*…displaced*	*live musicians*

Economies of Unit Number

FORD MODEL T, 1909-1916

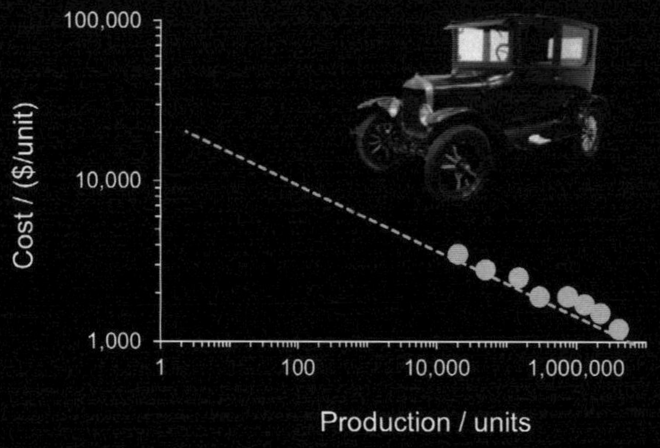

"experience learning"

$$k_n = k_1 n^{\log_2 \varepsilon}$$

Sources: (1) PJA Tijm. Gas to liquids. Fischer-Tropsch, advanced energy technology, future's pathway. Feb 2010; (2) C. Kopp. The US Air Force Synthetic Fuels Program. Technical Report APA-TR-2008-0102. (2008)

Modular Design

MODULAR VS INTEGRAL

MODULAR

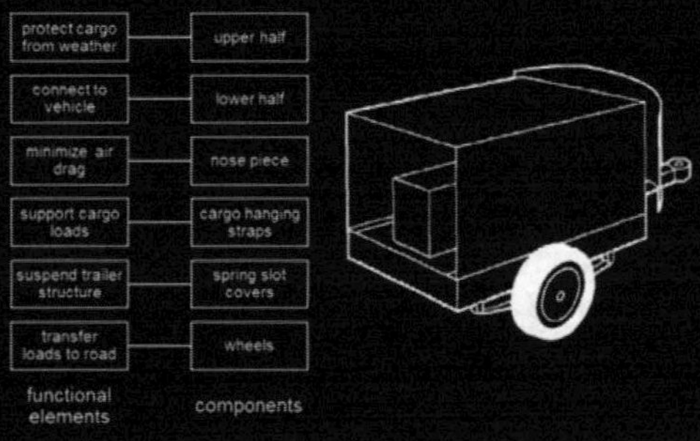

INTEGRAL

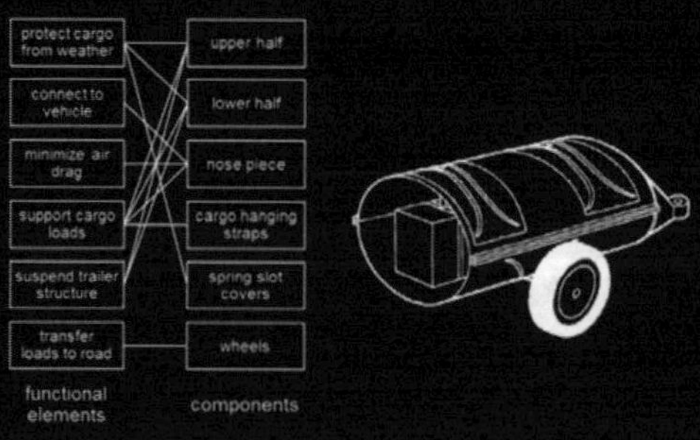

Modular Design

MODULAR VS INTEGRAL

MODULAR

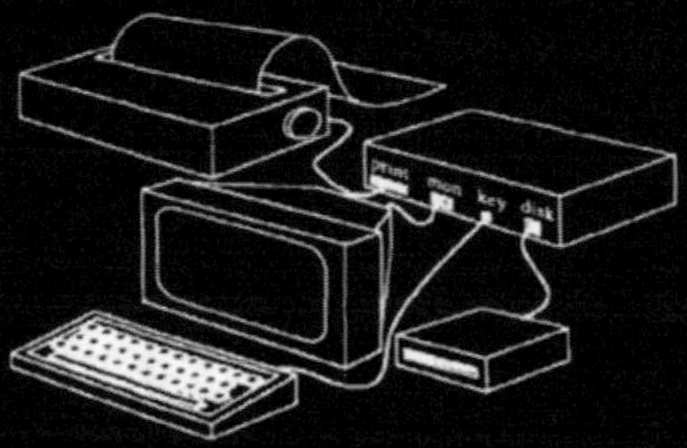

INTEGRAL

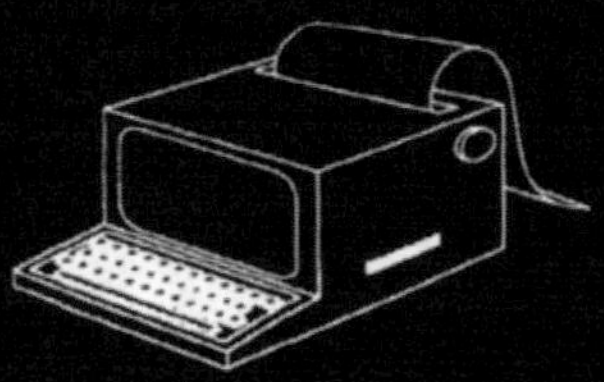

Small Modular

OIL REFINERY (WORLD)

- 7 TW petroleum refining
- 700 plants
- $500/kW capex

AUTO ENGINES (U.S.)

- 1.3 TW motive power
- 250 million engines
- $50/kW capex

Small Modular

GAS TURBINES (U.S.)

- 0.2 TW electricity
- 5000 gas turbine generators
- $1000/kW capex

AUTO ENGINES (U.S.)

- 1.3 TW motive power
- 250 million engines
- $50/kW capex

What about the democratization of chemical processing?

not if, but
when, who, how

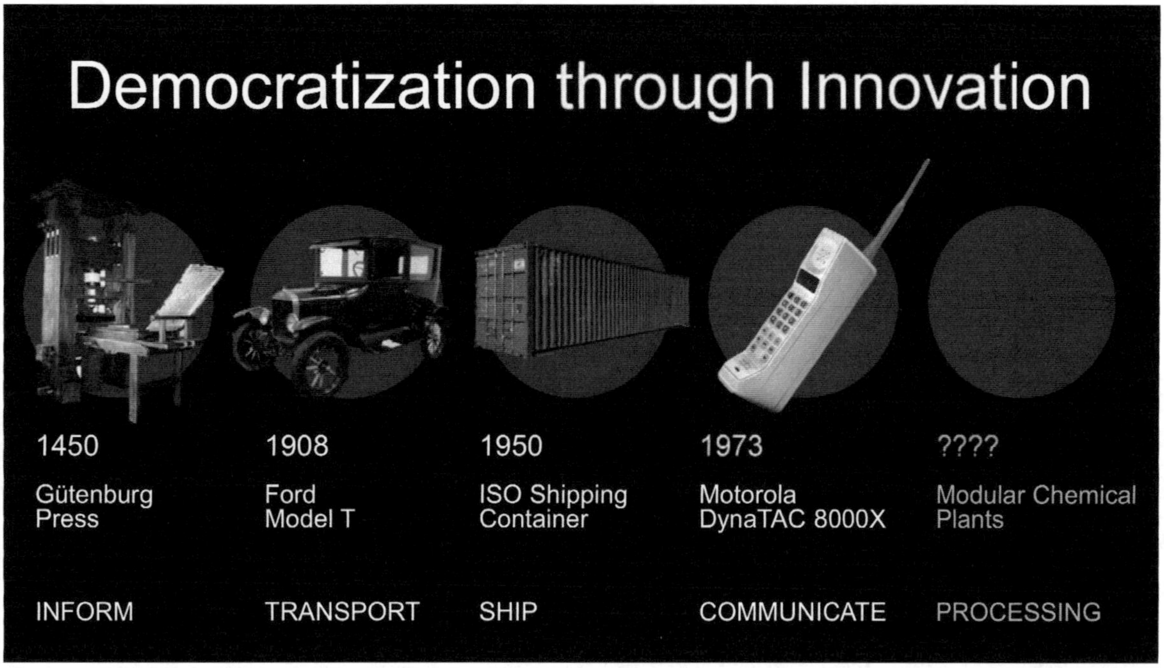

Isaac Mkalia, 20, checks his mobile phone in Kojiado district, near the Tanzanian border (Photograph: Sven Torfinn/Oxfam)

Thank You

Dane A. Boysen
dane.boysen@gmail.com

506291 Flex-to-Stretch Electronics

Abstract:

Stretchable electronic devices maintain electrical connections when subjected to stress or strain and are therefore useful for enabling electronics with new mechanical properties. However, these devices often require power or electrical signals from non-stretchable components. Connecting flexible and stretchable substrates is a possible solution that maintains the electrical connection within a medium that can withstand deformation and prolonged use. The ability to create a robust electrical connection between these mechanically disparate components may enable new types of hybrid devices. In this study we present a basic method to fabricate such new hybrid devices, utilizing liquid metals that form fantastic stretchable interconnects because they are both conductive and fluidic. We characterized the mechanical and electrical properties of these hybrid devices. While we focus on silicon elastomers and liquid metals, there are numerous other methods of fabricating stretchable devices that share many properties with the devices in this work.

506292 Systematic Analysis of Cloud Point and Crystallization in Fatty Acid Ethyl Ester Biodiesel Mixtures

Abstract:

Saturated biodiesel components, fatty acid methyl esters (FAMEs) or fatty acid ethyl esters (FAEEs), crystallize at high temperatures, which limits their commercial application. FAEEs are particularly advantageous biodiesel components because they can be renewably sourced and have lower cloud points (CP) than FAMEs with the same carbon-chain length. This research systematically evaluates FAEE cloud point and crystallization data to aid in the development of FAEE biodiesel blends with practical cold flow properties. The effects of cold-flow improving additives dimethyl azelate (DMA) and triacetin (TA) on FAMEs and FAEEs are also analyzed. Neat FAEE CP increases with carbon-chain length and decreases with greater degree of unsaturation, and saturated/saturated FAEE mixtures show partial cocrystallization. Additives lowered the CP, but did so via different mechanisms. Thermodynamic ideal solution modeling accurately predicts binary FAEE CP with and without additives within 2% of experimental values.

Feasibility Study of Ionic Liquid Desalination Design

Material Physics and Applications Division, Los Alamos National Laboratory
PI: Zachary Cosenza RLM: George Scott Goff

LA-UR 17-26682

Introduction

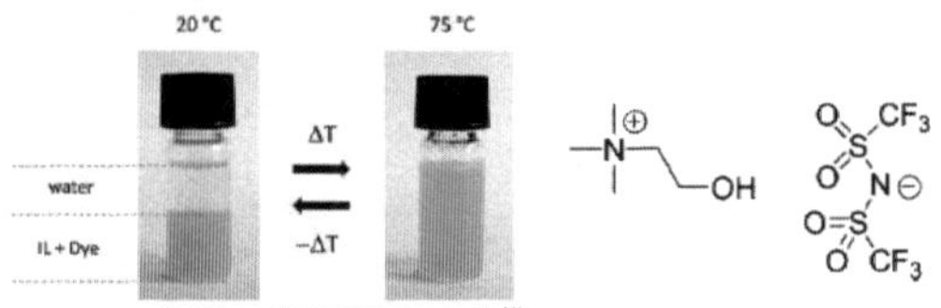

Courtesy of Nockemann et al. [1]

- Ionic liquids (ILs) are a class of room temperature salts in liquid form.
- Water has been shown to be soluble in ILs with T dependence, while the IL remains insoluble.
- Temperature may be used to drive water from a saline into the IL and back out at modest T.
- Current desalination methods (reverse osmosis and multiple-effect evaporation) run at high T or P
- Proposal is to **develop liquid-liquid extraction system using ILs** and saline.

Baseline System

Continuous Process uses Octanoic Acid as solvent

Batch Process (N = 1) uses [Choline][TF2N] as solvent

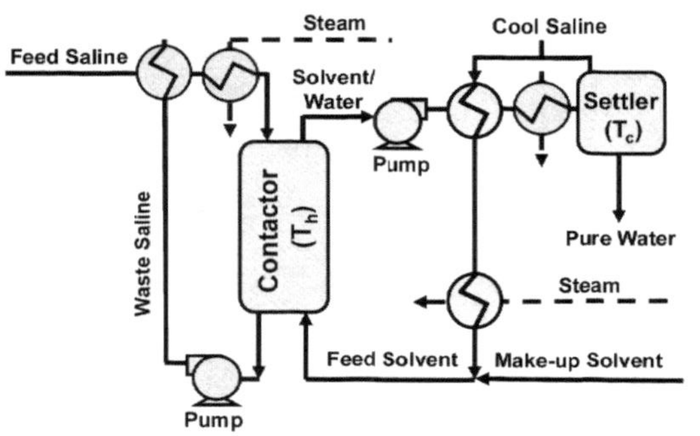

Major Process Parameters

Temperature – Separation efficiency is driven by the difference in contactor (T_h) and settler (T_c) temperatures.

Solvent – The process efficiency is limited by equilibrium, and in turn by the solvent used.

Saline Input – It has been shown [3] that equilibrium is altered by saline concentrations, with higher salt concentrations reducing the driving force.

Methods

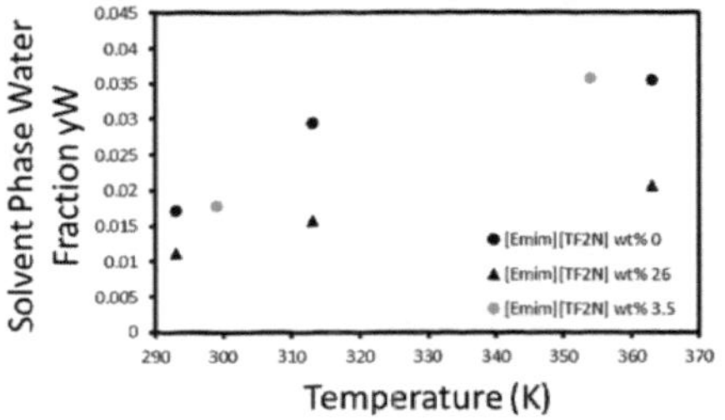

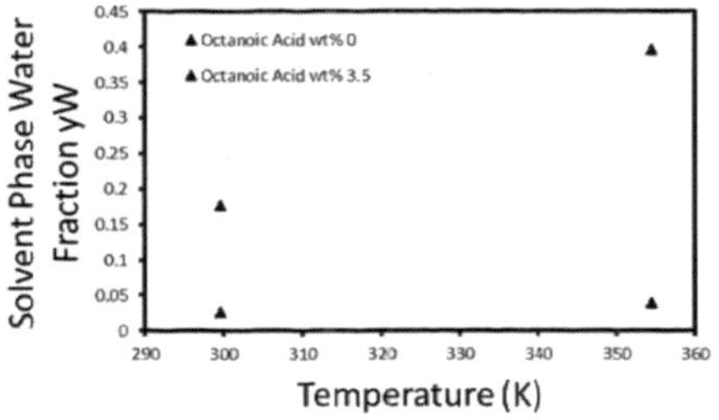

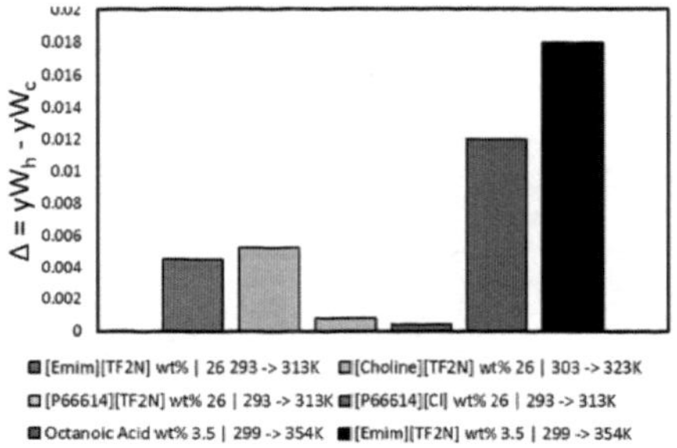

Analysis

Operating Line (OL):

$$\frac{V_m}{F} = \mathcal{N} * \iota_m + \frac{V_1}{F} - \mathcal{N} * \iota_1$$

Boundary Conditions:

(i) Set $\mathcal{N}$ (slope)
(ii) Set inlet water conc.
(iii) Pinch Point solution
(iv) Set inlet solvent conc.

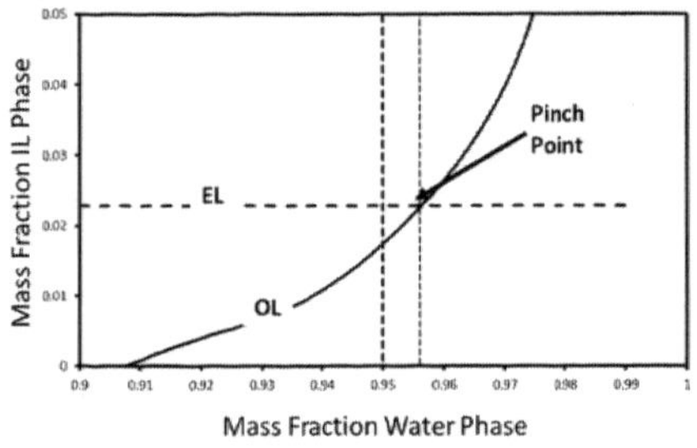

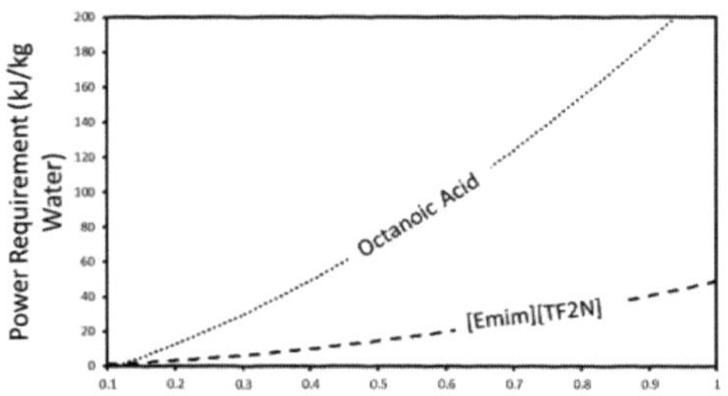

Energy Requirements OA and Emim, 353K with L = 3.0 kg/s 3.5 wt% water baseline

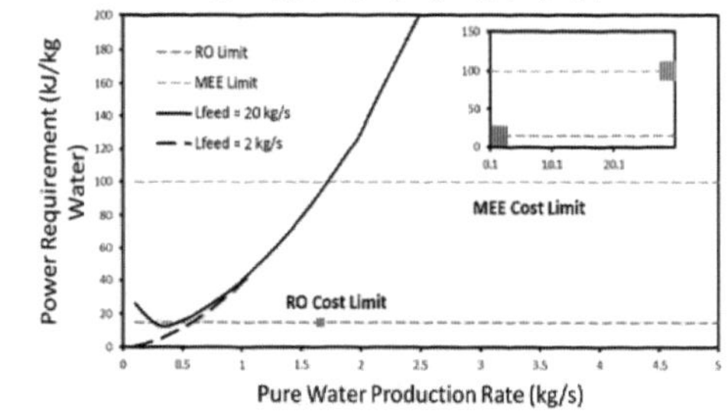

Cost Curve w/ Parameter Variation, RO ■ [14, 23] and MEE ■ [14, 24] based on pure water production (353K) and 3.5 wt% water

Conclusions

- Like the conclusion reached by Sanap et al [2], the continuous process is highly pinched at N = 1 at all upper process temperatures.
- Compared to the current literature work (Octanoic Acid as a solvent), [Emim][TF2N] has higher extractive and cost efficiency.
- If coupled with waste heat the marginal energy cost is close to that of RO, but far from MEE.
- Data on droplet size, mixing times, mass and heat transfer coefficients, and fixed costs would help scale-up process beyond basic thermodynamics.

References

[1] Nockmann, *J. Phys. Chem.* Ed 113. 1429-1437 (2009)
[2] D.B. Sanap et al, *Desalination* Ed. 357 150-162 (2015)
[3] Bajpayee, Anurag Directional Solvent Extraction Desalination, MIT 2012, *MIT Libraries*
[4] Zhong Yujiang, *Environ. Sci. Tech.* Ed. 50 1039-1045 (2016)

Acknowledgements

Los Alamos NATIONAL LABORATORY

506320 Preparation and Characterization of Shape Memory Assisted Self-Healing Coatings

Abstract:

Shape memory assisted self-healing (SMASH) coatings have the ability heal nearly completely after damage with the application of heat, making them useful for corrosion resistance in steel structures (1). This summerâ€™s project sought to expand upon existing research to develop an efficiently healing and transparent SMASH coating that could be used in technological applications, such as on cell phone or computer screens (2) . The materials tested were comprised of electrospun poly(Îµ-caprolactone) (PCL) fiber mats coated in epoxy with differing glass transition temperatures (Tg). The healing efficiency of the samples were tested by damaging each one with a razorblade and healing them over varying times and heats in an oven. The transparency of each sample was observed visually before and after heating, a technique which will be refined later in the research process. Once the most ideal combination is found, further experimentation will be conducted to determine whether the ideal epoxy/PCL combination is also one that can withstand multiple damage-healing cycles to the same spot without becoming opaque. References (1) Nejad, H. B. Thermally/Mechanically Responsive Polymeric Composites with Shape Memory/Self-Healing Properties. Syracuse University 2015, 79-95. (2) Luo, X. Thermally Responsive Polymer Systems for Self-Healing, Reversible Adhesion and Shape Memory Applications. Syracuse University, ProQuest Dissertations Publishing 2010. 3437600.

506372 Defluoridation of Ground Water Using Impregnated Aluminum

Abstract:

Groundwater is an important source of drinking water and it is estimated that more than 200 million people worldwide are drinking groundwater with fluoride concentrations greater than the WHO guideline value of 1.5 mg/L. The majority of these cases occur in the developing world. fluoride has long been found to have a beneficial effect on dental health, although when present in drinking water at concentrations much above the WHO guideline value (and national drinking water limit for most countries) of 1.5 mg/L, long term use can result in development of dental fluorosis or at its worst, crippling skeletal fluorosis. The effects on health depend on a number of factors including dose from other sources of fluoride (toothpastes, food), age, gender and general health status.

506536 Ordering pH-Responsive Polymer-Grafted Nanoparticles in Flow Coating Process

Abstract:

Poly(acrylic acid) (PAA)-grafted silica nanoparticles with different grafting densities and brush lengths were synthesized via surface-initiated ATRP, and their deposition patterns on silicon substrates were investigated by systematically varying the speed of moving substrates, pH values, and free polymer poly(vinylpyrrolidone) (PVP) concentrations in aqueous solution. Brush-brush interactions are tuned by varying the PAA chain conformations with pH and also with the addition of free PVP. Different deposition patterns were observed for nanoparticles at low grafting density and long brush length at a velocity of 0.4 mm/min and a PVP concentration of 0.24 mg/ml. At pH 11, nanoparticles assembled into well-regulated lines parallel to the moving direction of the blade. However, at pH 9, nanoparticles showed stronger repulsion, which led to Rayleigh-instability, and a dot pattern formation. At pH 3, particle attractions get strong due the hydrogen bonding between PAA and PVP, hence Bénard cell patterns were observed. The mechanisms of hydrogen bonding between two different polymers and also physical bridging within particles are demonstrated to organize the grafted nanoparticles in a flow-coating process. Our findings will enable the easy fabrication of patterned particle domains for functional polymer films.

NCEES

advancing licensure for engineers and surveyors

EFFECTIVE AND EFFICIENT USE OF THE FUNDAMENTALS OF ENGINEERING EXAM FOR OUTCOMES ASSESSMENT

AICHE 2017 ANNUAL MEETING

David Whitman, PhD, PE
NCEES Past-President
University of Wyoming, Retired

NCEES
advancing licensure for engineers and surveyors

Overview

- Applicability of the FE exam for assessment
- ABET student outcomes that can be measured
- Computer-based FE exam format
- Various methods to use FE exam results for outcomes assessment
- Self-study examples and closing the loop
- Questions and answers

NCEES

Applicability of the FE Exam for Assessment

- It is a direct method of objective assessment with comparisons of institutional results against national results.
- Assessment does not use pass rates but how students perform on individual exam areas.
- Because there are over 55,000 FE examinees per year, it provides high reliability.
- The FE can be used as an assessment tool with a pool of all graduates or with a self-selecting pool.

NCEES

ABET Outcomes Assessment Possible with FE Exam

a) An ability to apply **knowledge of mathematics, science, and engineering**

b) An ability to design and conduct experiments, as well as to **analyze and interpret data**

c) An ability to **design a system, component, or process** to meet desired needs

NCEES

ABET Outcomes Assessment Possible with FE Exam (cont.)

(e) An ability to **identify, formulate, and solve engineering problems**

(f) An understanding of **professional and ethical responsibility**

(k) An ability to use the techniques, skills, and modern **engineering tools** necessary for engineering practice

NCEES

Computer-Based Testing (CBT) Format

- The FE became a CBT exam in January 2014
  - Shorter in time and fewer questions
  - Presented in a different format

NCEES

CBT Format (cont.)

- Length
  - The appointment time at test centers is 6 hours.
    - Tutorial–8 minutes
    - Nondisclosure agreement–2 minutes
    - Exam time–5 hours, 20 minutes with a 25-minute scheduled break after approximately 55 questions
    - Brief post-exam survey
  - Total of 110 questions

NCEES

CBT Format (cont.)

- Testing opportunities
  - Testing windows
    - January–March
    - April–June
    - July–September
    - October–December
- Test center locations
  - Pearson VUE test centers

NCEES

CBT Format (cont.)

- *FE Reference Handbook*
  - Provided electronically with the exam as a searchable PDF
  - Available for free download and for purchase as a hard copy at http://ncees.org/engineering/fe/
- *NCEES Examinee Guide*
- Computer-based practice exams

NCEES

CBT Format (cont.)

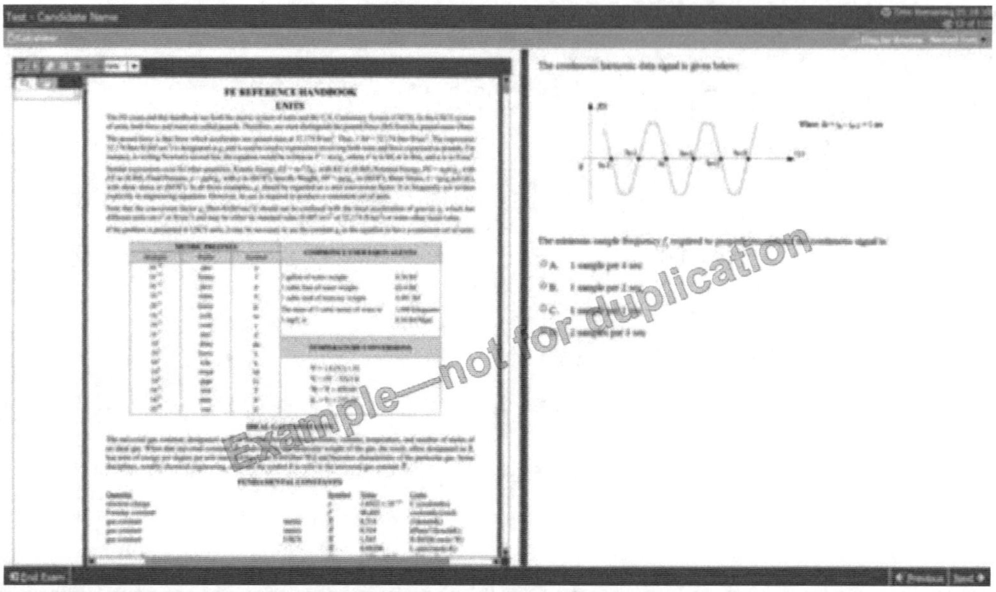

NCEES

CBT Format (cont.)

- Content of the exam
  - 7 free-standing discipline-specific exams
    - Chemical, Civil, Electrical and Computer, Environmental, Industrial and Systems, Mechanical, Other Disciplines
  - FE no longer has a common morning portion (breadth module)
    - Selected exam topics that were previously on the breadth module are now tested within each discipline-specific exam.

NCEES

FE Chemical Engineering Exam Specification

- Mathematics
- Probability and Statistics
- Engineering Sciences
- Computational Tools
- Materials Science
- Chemistry
- Fluid Mechanics/Dynamics
- Thermodynamics
- Material/Energy Balances
- Heat Transfer
- Mass Transfer and Separation
- Chemical Reaction Engineering
- Process Design and Economics
- Process Control
- Safety, Health, and Environment
- Ethics and Professional Practice

Additional details can be found at http://ncees.org/engineering/fe/

NCEES

CHE Example of Engineering Design

13. Process Design and Economics

A. Process flow diagrams and piping and instrumentation diagrams

B. Equipment selection (e.g., sizing and scale-up)

C. Cost estimation

D. Comparison of economic alternatives (e.g., net present value, discounted cash flow, rate of return, expected value and risk)

E. Process design and optimization (e.g., sustainability, efficiency, green engineering, inherently safer design, evaluation of specifications)

NCEES

So, what actual data are available, and what can you do with the data?

NCEES

Subject Matter Report

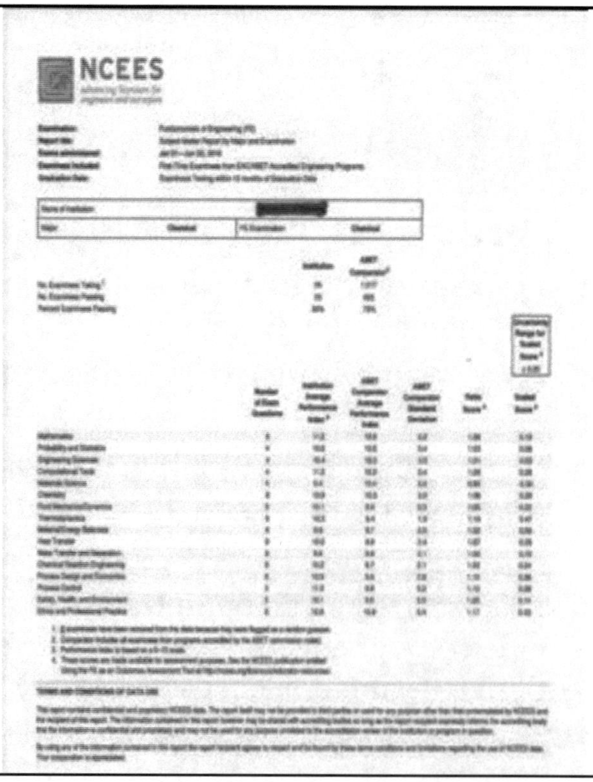
NCEES

Subject Matter Report (cont.)

- Reports are generated twice a year.
  - In July for the January–June testing period (Spring)
  - In January for the July–December testing period (Fall)
- The report is specific to
  - An institution,
  - Students within an engineering degree program at that institution, and
  - The discipline-specific exam that those students completed.

NCEES

Subject Matter Report (cont.)

- Data are provided for all examinees testing within 12 months of graduation (either before or after graduating).
- Only first-time takers are included.
- Random guessers are removed from the report.
- National performance data, with standard deviation information, are also provided for the same degree program and same discipline-specific exam.

NCEES

Subject Matter Report (cont.)

- For each topic, the students' performance is given as a Performance Index on a scale of 0–15.
- The Performance Index is <u>indirectly</u> related to the average number of questions answered correctly.
- This is necessary because each examinee receives a different set of questions within each topic area.
  - Any one examinee's set of questions may be harder or easier than another examinee's set.

NCEES

Subject Matter Report (cont.)

- Getting the data
  - NCEES sends links to reports directly to an institution via email.
  - If you don't know, NCEES can tell you who receives your institution's reports.
  - Reports also include information on the specific institution's examinees who took the FE or PE exam more than 12 months after graduation.

NCEES

The specifics of using the FE exam for outcomes assessment

NCEES

Getting Started

- **Faculty** should be involved.
- Identify areas of strength.
- Acknowledge areas that are not emphasized.
- Set program-specific goals for each area.

NCEES

Suggested Analysis Techniques

- Choose your longitudinal time basis.
  - Performance from multiple examination windows
  - Academic year performance
- Choose your presentation method.
  - Ratio method
  - Scaled score method

NCEES

Ratio Method

- The ratio score is simply the ratio between the program's performance index (P.I.) in any topic area and the P.I. of the comparator performance.
  - Ratio score = Program P.I./Comparator P.I.

NCEES

Comparison of Ratios by Subject Area

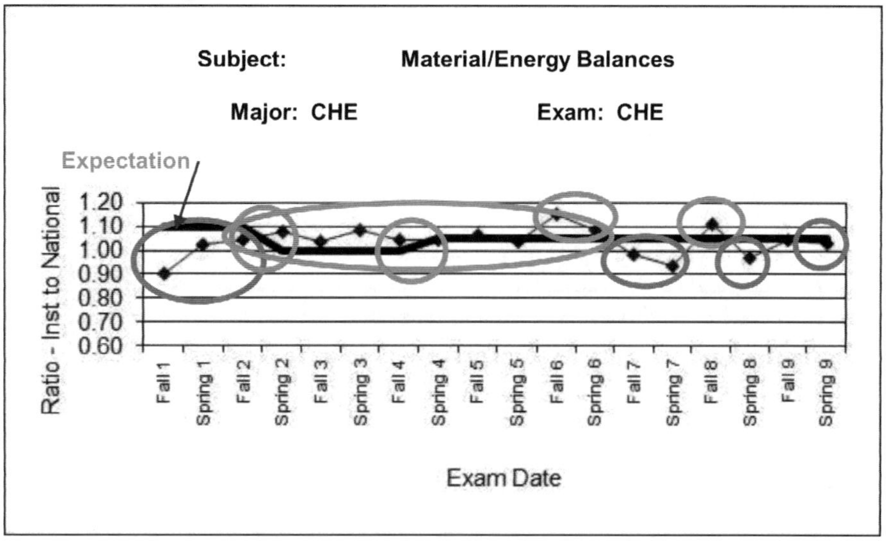

NCEES

Comparison of Ratios by Subject Area

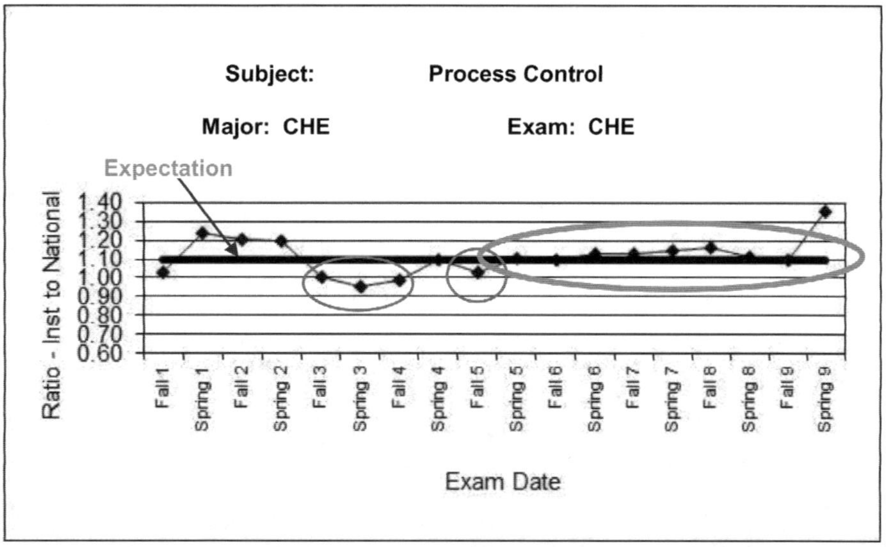

NCEES

Comparison of Ratios by Subject Area

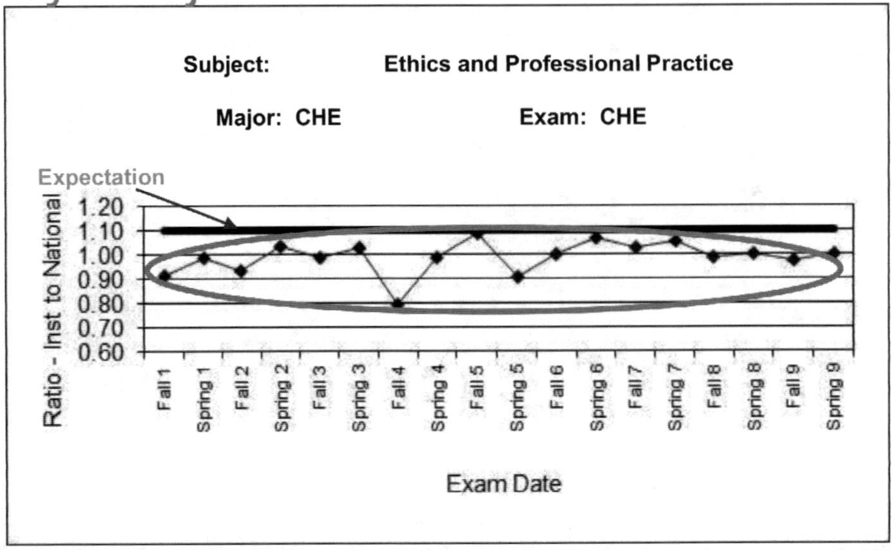

NCEES

Effect of Academic Year Averaging

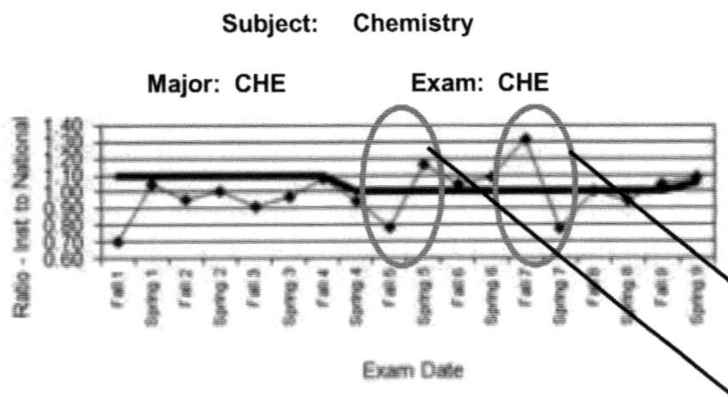

Causes a smoothing of the data—note the Fall/Spring 5 and Fall/Spring 7 results compared to the AY 5 and AY 7 academic year results.

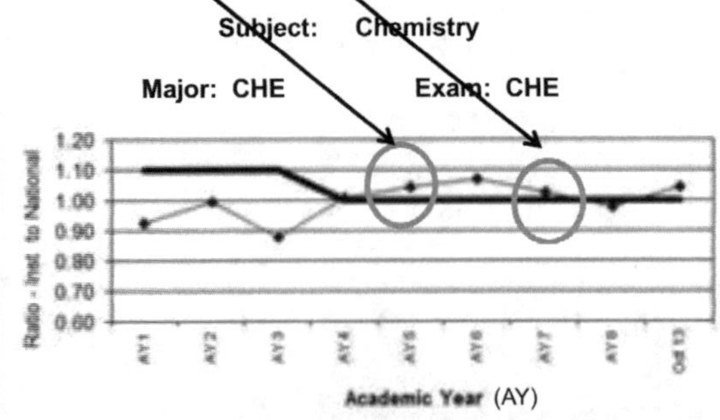

NCEES

Scaled Score Method

Define Scaled Score (S.S.)

$$S.S. = \frac{\text{PI for Univ X} - \text{PI comparator}}{\text{PI comparator standard deviation}} \pm \frac{1}{\sqrt{\text{\# of takers at Univ X}}}$$

NCEES

Comparison Using Scaled Score by Subject Area

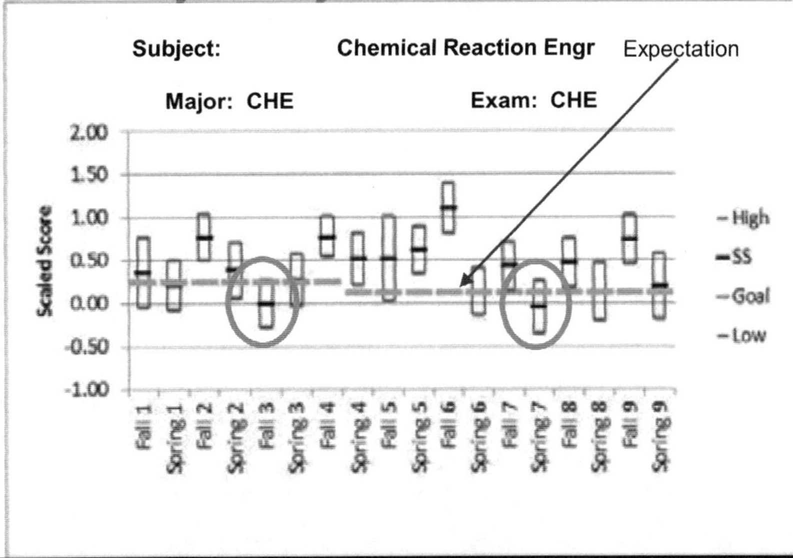

NCEES

Comparison Using Scaled Score by Subject Area

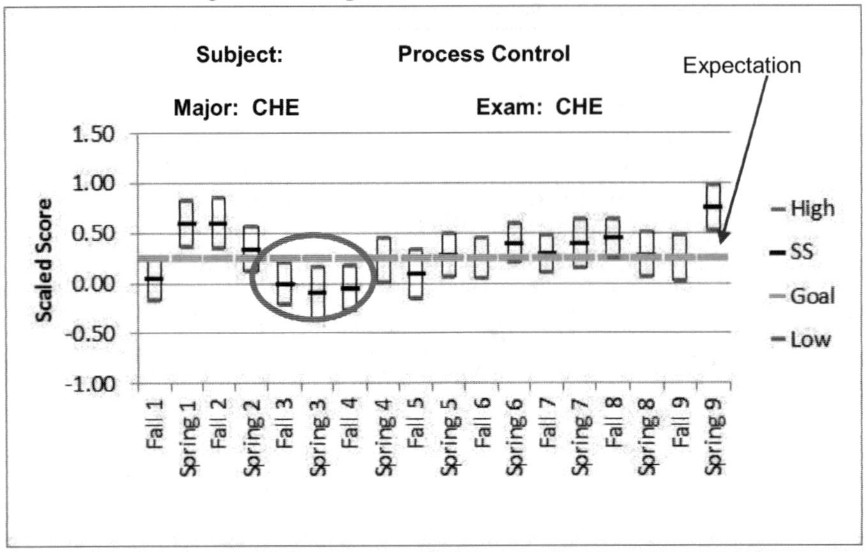

NCEES

Academic Year Averaging

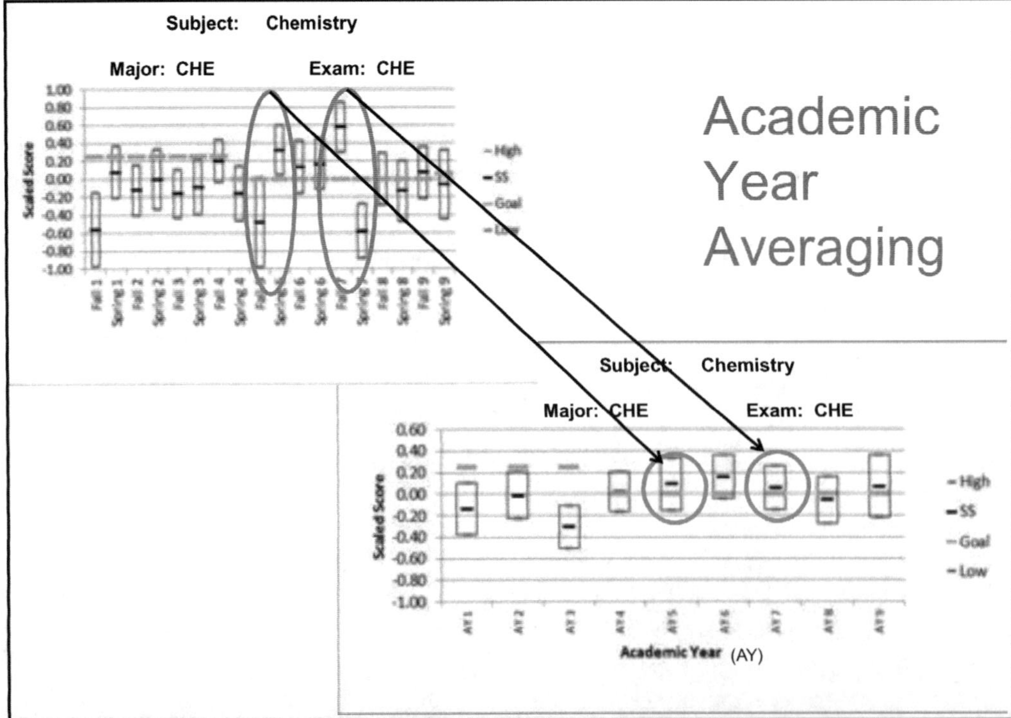

NCEES

Examples of Assessing Some of the ABET (a)–(k) Outcomes

NCEES

Ability to Measure ABET (a) Math and Science

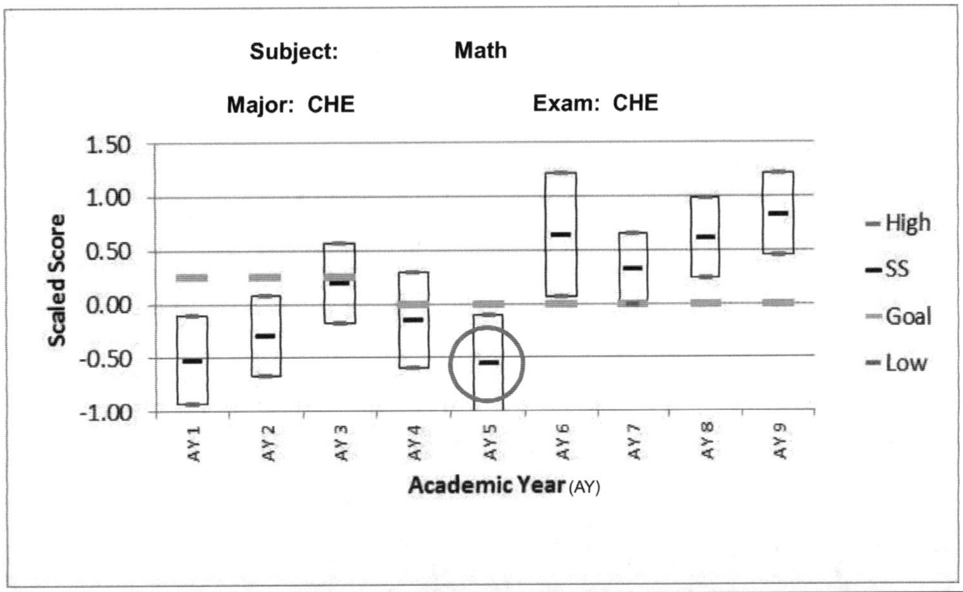

NCEES

Ability to Measure ABET (c) Design

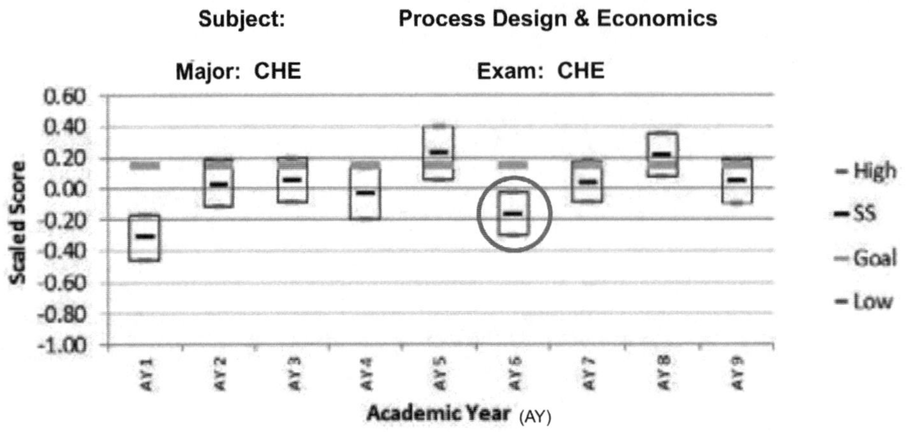

NCEES

Ability to Measure ABET (e) Solve Engineering Problems

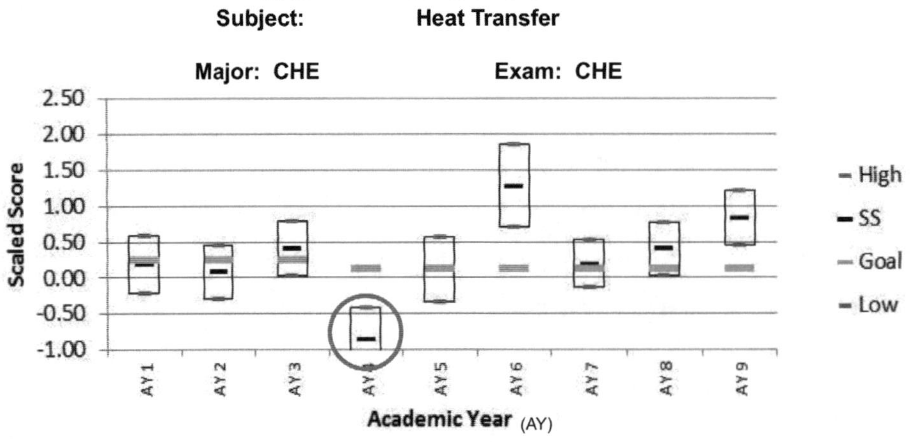

NCEES

Ability to Measure ABET (k) Engineering Tools

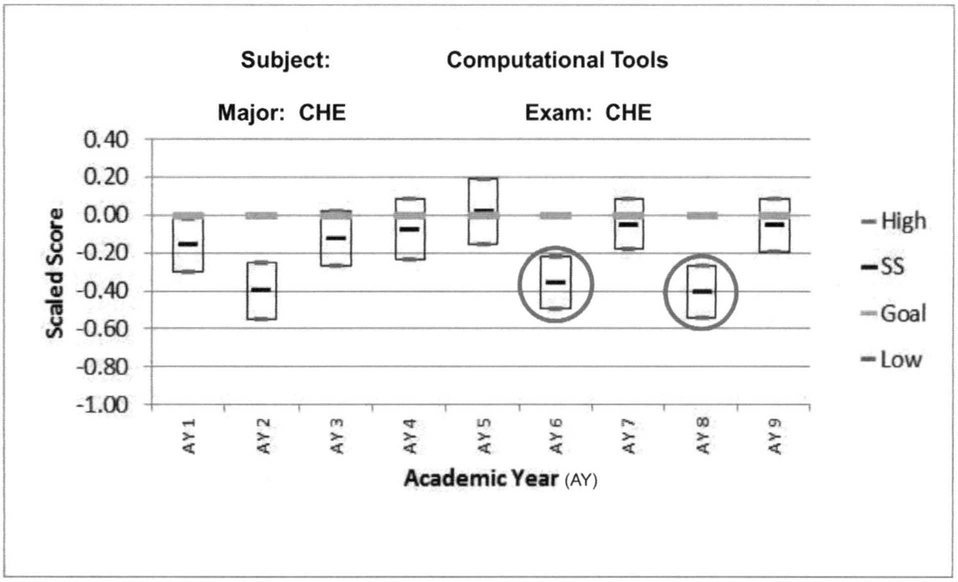

NCEES

Conclusions

- Discipline-specific assessment information can be gleaned from this exam.
- FE exam provides a direct, quantitative assessment technique.
- The Subject Matter Report provides comparative data.
- NCEES sends a link to the Subject Matter Report directly to your institution via email.

NCEES

Conclusions (cont.)

- The FE exam is one effective assessment tool to be used as part of your institution's full assessment package.

NCEES

Additional Resources

For more information on reports, email fereports@ncees.org

NCEES

Additional Resources

For further information, contact:

Cheryl Warren, Ph.D., P.E.
NCEES Exam Development Engineer
800-250-3196, ext. 5472
email: cwarren@ncees.org

NCEES

Questions?

NCEES

Dr. John Wagner P.E.

McKetta Department of Chemical & Bioprocess Engineering

Trine University

My Background

- Passed the FE as a senior in ChE at University of Colorado – 1987
- Worked at Shell Oil– Deer Park TX
- Rice University for Ph. D. – 1994
- Accepted a Faculty Position at Trine University
- Passed PE in 1997
- Have been on the NCEES FE Committee since 1998

How we use the FE Exam fits into to the Assessment in ChE at Trine

Assessment Tool	Student Outcomes											
	a	b	c	d	e	f	g	h	i	j	k	l
FE Exam	X		X		X						X	X
CAE (Internal Comprehensive Exam)	X	X	X		X							X
Exit Interview /Survey	X	X	X	X	X	X	X	X	X	X	X	X
PAB - Professional Advisory Board						X	X			X		
Math Exam - Internal Comprehensive	X											
SAChE Student Program												X
Key Assessment Courses												
Process Measurements Lab												X
Numerical Methods											X	
Unit Operations Lab I		X										X
Unit Operations Lab II		X		X								
Kinetics	X											
Design I	X				X						X	X
Design II			X			X	X	X	X	X		

Color Code

Program Educational Objective 1	-	Technical Competency
Program Educational Objective 2	-	Problem Solving, Design, Communication and Team Skills
Program Educational Objective 3	-	Professionally Obligated

FE Exam is our only External, Objective Quantitative Assessment Tool

	Objective	Subjective
Internal	CAE Math Exam	Exit Interview Key Assessment Courses
External	FE Exam SAChE Certificate	PAB

Assessment Type	Student Outcomes											
	a	b	c	d	e	f	g	h	i	j	k	l
Objective, External	X		X		X						X	X
Objective, Internal	X	X	X		X							X
Subjective, External						X	X			X		
Subjective, Internal	X	X	X	X	X	X	X	X	X	X	X	X

FE Exam Results, 2010-2013

		2010		n=	9	2011		n=	12	2012		n=	18	2013		n=	18	4 yr Average	
AM Subject	#Q	Avg	Nat. Avg	Dev	Scaled	Avg	Nat. Avg	Dev	Scaled	Avg	Nat. Avg	Dev	Scaled	Avg	Nat. Avg	Dev	Scaled	Scaled	AM Subject
Mathematics	19	58	66	-8	-0.46	63	74	-11	-0.70	65	79	-14	-0.99	67	76	-9	-0.59	-0.717	Mathematics
Engineering Prob & Stats	8	61	65	-4	-0.20	69	69	0	0.00	47	54	-7	-0.33	62	61	1	0.05	-0.12	Engineering Prob & Stats
Chemistry	11	85	89	-4	-0.34	89	85	4	0.31	80	85	-5	-0.39	77	79	-2	-0.15	-0.158	Chemistry
Computers	8	71	76	-5	-0.29	69	76	-7	-0.40	68	77	-9	-0.60	81	82	-1	-0.06	-0.338	Computers
Ethics & Business Prac	8	77	76	1	0.05	80	73	7	0.43	80	85	-5	-0.33	82	80	2	0.13	0.036	Ethics & Business Prac
Engineering Econ	10	82	66	16	0.76	66	65	1	0.06	81	71	10	0.53	78	65	13	0.59	0.485	Engineering Econ
Eng Mechanics (Statics)	8	64	58	6	0.27	40	41	-1	-0.04	60	58	2	0.10	51	57	-6	-0.27	-0.02	Eng Mechanics (Statics)
Eng Mechanics (Dynamics)	5	78	68	10	0.42	48	62	-14	-0.54	36	42	-6	-0.23	40	42	-2	-0.08	-0.147	Eng Mechanics (Dynamics)
Strength of Materials	8	36	39	-3	-0.17	39	40	-1	-0.05	31	38	-7	-0.35	44	40	4	0.21	-0.081	Strength of Materials
Material Properties	8	67	65	2	0.11	61	60	1	0.05	51	56	-5	-0.25	69	62	7	0.35	0.061	Material Properties
Fluid Mechanics	8	69	71	-2	-0.09	69	62	7	0.28	63	66	-3	-0.16	69	72	-3	-0.15	-0.054	Fluid Mechanics
Electricity and Magnetism	11	56	55	1	0.05	50	55	-5	-0.23	45	61	-16	-0.77	44	50	-6	-0.31	-0.381	Electricity and Magnetism
Thermodynamics	8	51	54	-3	-0.13	58	64	-6	-0.30	76	76	0	0.00	82	73	9	0.45	0.058	Thermodynamics
Score	120	65	66	-0.3	-0.049	63	65	-2.3	-0.12	62	68	-5.8	-0.36	66	66	-0.3	-0.04	-0.159	Score
PM Subject																			PM Subject
Chemistry	6	57	58	-1	-0.05	85	77	8	0.40	56	70	-14	-0.65	78	74	4	0.24	-0.052	Chemistry
Mat/Energy Balances	9	75	72	3	0.16	72	73	-1	-0.05	70	68	2	0.09	82	73	9	0.45	0.186	Mat/Energy Balances
Chem Thermodynamics	6	63	62	1	0.05	64	58	6	0.30	65	65	0	0.00	64	59	5	0.21	0.139	Chem Thermodynamics
Fluid Dynamics	6	83	79	4	0.22	43	48	-5	-0.27	53	61	-8	-0.32	65	57	8	0.32	-0.023	Fluid Dynamics
Heat Transfer	6	74	74	0	0.00	71	73	-2	-0.10	68	80	-12	-0.60	73	71	2	0.09	-0.181	Heat Transfer
Mass Transfer	6	46	47	-1	-0.05	69	60	9	0.42	61	69	-8	-0.37	50	50	0	0.00	-0.036	Mass Transfer
Chem Reaction Engr	6	65	68	-3	-0.13	81	77	4	0.17	62	69	-7	-0.30	68	71	-3	-0.14	-0.123	Chem Reaction Engr
Proc Dsgn Econ Opt	6	72	70	2	0.10	85	75	10	0.50	73	72	1	0.05	89	76	13	0.60	0.326	Proc Dsgn Econ Opt
Computer Usage and Chem	3	78	67	11	0.41	61	67	-6	-0.23	69	75	-6	-0.23	89	68	21	0.70	0.168	Computer Usage and Chem Eng
Process Control	3	41	35	6	0.23	92	68	24	0.90	80	68	12	0.40	89	75	14	0.53	0.517	Process Control
Safety Health and Environ	3	85	81	4	0.17	50	53	-3	-0.13	54	61	-7	-0.21	76	75	1	0.04	-0.053	Safety Health and Environ
		67	66	1.7	0.079	71	67	3.6	0.161	64	69	-4.6	-0.206	74	67.65	6.1	0.264	0.064	
Score	60																		

Trine was not a testing site starting 2014. We stopped requiring the FE and our participation dropped to only 3 or 4 students/year

Students graduating in 2019 will once again be required to take the FE

NC STATE UNIVERSITY

Dr. Lisa G. Bullard, PE

Chemical and Biomolecular Engineering
North Carolina State University

NC STATE UNIVERSITY

My Background

- Passed the FE as a senior at NC State
- PhD → Eastman Chemical Company (Engineering & Construction Division)
- Took/passed the PE after two years of industry experience (plus 2 from PhD)
  - "Unwritten rules" about promotion
  - Changes in state regulations
  - Company support
- Faculty member at NC State

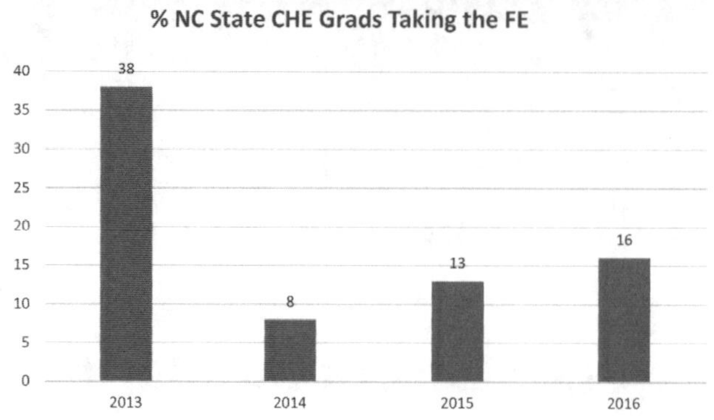

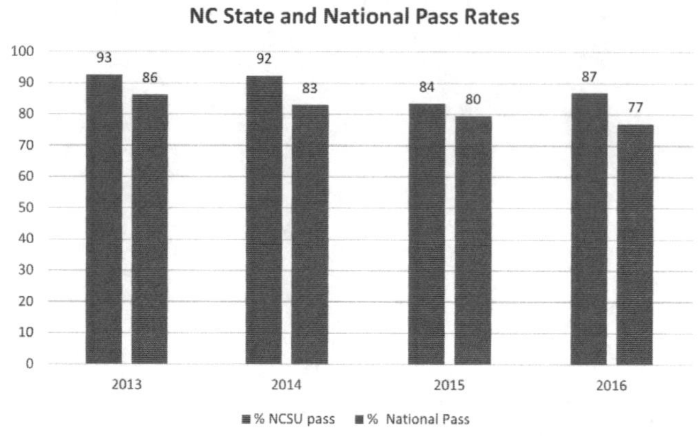

NC STATE UNIVERSITY

Issues/Questions

- Do companies hiring our students expect/promote the PE?
  - Industry
  - Functional role
- When is the best time for students to take the exam – in school or after graduation?
- Increased cost to students
- Resources for test preparation

CFD Role in Understanding Mixing Processes

IACChE's James Y. Oldshue Lecture

Prof. José Roberto Nunhez

FEQ- UNICAMP
Tel (019) 35213967
E-mail: nunhez@feq.unicamp.br

17AIChE
Annual Meeting, Minneapolis, MN

UNICAMP

PRESENTATION TOPICS

- INTRODUCTION
- MIXING AND CFD
- SOME EXAMPLES
- CFD AND OPTIMIZATION
- NUSSELT EQUATIONS AND CFD
- CONCLUSION

17AIChE
Annual Meeting, Minneapolis, MN

INTRODUCTION

UNICAMP

17AIChE
Annual Meeting, Minneapolis, MN

INTRODUCTION

UNICAMP

Laboratory of Computational Fluid Dynamics **(LCFD)**. State University of Campinas, Campinas, Brazil

MAIN RESEARCH

- CFD and Experimental studies in stirred tanks
- CFD applied to FCC process;
- CFD applied to HDT process;
- CFD applied to effluents dispersion in rivers;

17AIChE Annual Meeting, Minneapolis, MN

INTRODUCTION

UNICAMP

MAIN ACTUAL PROJECTS

HYDRATE FORMATION WITH THE NETmix STATIC MIXER/MICROMIXER

- CFD and Experimental studies (NETmix)

MIXING IN LARGE STORAGE PETROCHEMICAL PETROLEUM TANKS

- CFD and Experimental studies (PIV)

COKE FORMATION INSIDE THE TUBES OF PETROCHEMICAL HEATERS

- CFD

INTRODUCTION | MIXING AND CFD | EXAMPLES | CFD AND OPTIMIZATION | NUSSELT EQTS | CONCLUSION | 5

17AIChE Annual Meeting, Minneapolis, MN

INTRODUCTION

UNICAMP

We will first see a brief introduction on how Mixing developed as we know Mixing today and how CFD started to be used for the understanding of Mixing processes

INTRODUCTION | MIXING AND CFD | EXAMPLES | CFD AND OPTIMIZATION | NUSSELT EQTS | CONCLUSION | 6

MIXING AND CFD

Mixing is not recent

Mixing bowl (600 BC)

Source: François Bertrand Plenary Lecture at WCCE8

17AIChE Annual Meeting, Minneapolis, MN

MIXING AND CFD

UNICAMP

GEORGIUS AGRICOLA

DE RE METALLICA

TRANSLATED FROM THE FIRST LATIN EDITION OF 1556

A—Trench. B—Vat into which the salt water flows. C—Ladle. D—Small bucket with pole fastened into it.

A—Large vat. B—Plug. C—Small tub. D—Deep ladle. E—Small vat. F—Caldron.

MIXING AND CFD

MIXING AND CFD

- It was only after the second world war that research in mixing received a new impetus;
- At this time Mixing is considered to be an unit operation and rules for scaling up as well as Design principles are developed;
- Researches such as Rushton, Costich, Oldshue and others appear, laying the basis for much of what we know about mixing today.

17AIChE
Annual Meeting, Minneapolis, MN

MIXING AND CFD

UNICAMP

- Rushton, J.H., Costich, E. W. and Everett, H. J., Power characteristics of mixing impellers 1,*Chem. Eng. Progr*. **46**, 395-404, (1950). ***(383)***
- Rushton, J.H., Costich, E. W. and Everett, H. J., Power characteristics of mixing impellers 2,*Chem. Eng. Progr*. **46**, 467-79, (1950). ***(312)***

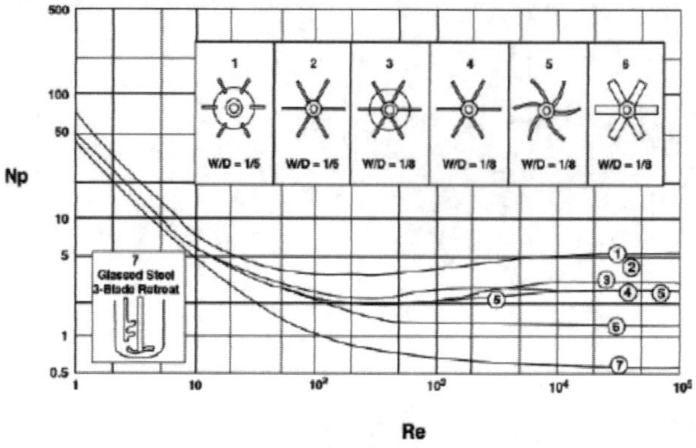

INTRODUCTION | MIXING AND CFD | EXAMPLES | CFD AND OPTIMIZATION | NUSSELT EQTS | CONCLUSION | 11

17AIChE
Annual Meeting, Minneapolis, MN

MIXING AND CFD

UNICAMP

- François Bertrand presented in 2009 some very interesting information about the first works using CFD in Mixing at the **IX WCCE** in Montreal in his plenary lecture entitled “A mix of computational Techniques for a better mix.”;
- The first CFD work in Mixing uses the Finite Difference Method and was presented by Prof. Hiraoka.

INTRODUCTION | MIXING AND CFD | EXAMPLES | CFD AND OPTIMIZATION | NUSSELT EQTS | CONCLUSION | 12

17AIChE Annual Meeting, Minneapolis, MN

MIXING AND CFD

UNICAMP

- NAMF Mixing VII (Henniker, New Hampshire, 1979)

Thursday, August 16, 1979

9:00 a.m. - 12:30 p.m. SESSION VII: NUMERICAL METHODS FOR SIMULATING MIXING

Session Chairman: G. K. Patterson
University of Missouri-Rolla

"Numerical Simulation of Flow Behavior and Heat Transfer in an Agitated Vessel"

A procedure for numericla simulation of flow and heat transfer in an agitated vessel is developed based on a two dimensional tangential flow model. The results explain the dependence of the velocity distribution on the Reynolds number in the turbulent flow regime, and the resultant turbulent diffusivity is in good agreement with experimental data.

S. Hiraoka and I. Yamada
Nagoya Institute of Technology, Japan

L. T. Fan
Kansas State University

Source: Plenary Lecture "A mix of computational Techniques for a better mix", by François Bertrand

17AIChE Annual Meeting, Minneapolis, MN

MIXING AND CFD

UNICAMP

The first investigations using CFD were two-dimensional and used the FDM

- Hiraoka, S., Yamada, I and Mizoguchi, K., 2D Model Analysis of Flow Behaviour of Highly Viscous Non-Newtonian Fluid in Agitated Vessel with Paddle Impeller, J. Chem. Engng Japan, 12, 56-62 (1978);
- Kuriyama, M., Arai, K. and Saito, S., Numerical Solution for the Flow of Highly Viscous Fluid in Agitated Vessel with Anchor Impeller, AIChE J., 28, 385-391 (1982);
- Bertrand, J. and Courdec, J. P., Agitatiopn de Fluides Pseudoplastiques par an Agitateur Bipale, Can, J. Chem. Engng, 60, 738-747 (1982)

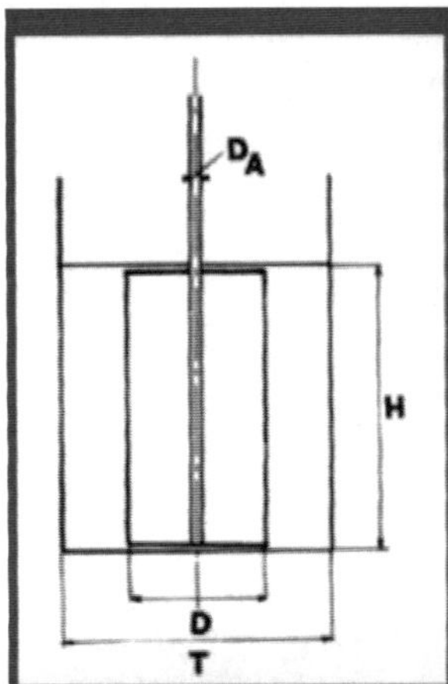

Source: Plenary Lecture "A mix of computational Techniques for a better mix", by François Bertrand

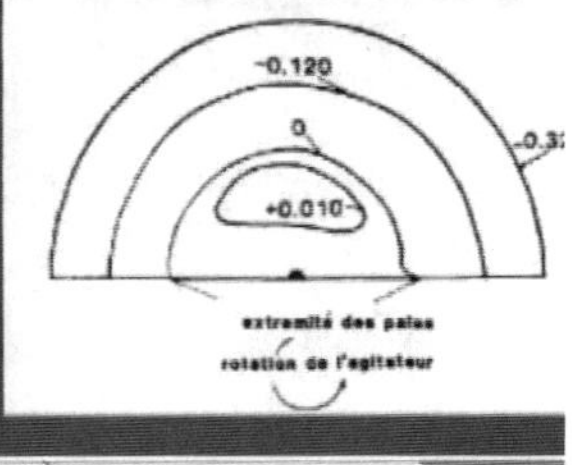

MIXING AND CFD

The first 3D CFD Model was presented at NAMF Mixing XII (Potosi, Missouri, 1989)

Source: Plenary Lecture "A mix of computational Techniques for a better mix", by François Bertrand

Wednesday August 9, 1989

Chairman: G. Tatterson

"Circulation Time Distribution in a Turbine - Stirred Tank"
A. M. Gerlak and R. V. Calabrese, Univ. of Maryland

"Gas-Liquid Reactors with Axially Pumping Impellers"
A. Bakker, H. E. A. van der Akker, Laboratorium von Fysische Technologie

"Study on Mixing Time and Turbulence Parameters in Fermenters for Culturing Mammalian Cells"
Z. Zhibing and D. Gance, East China Univ. of Chem. Tech.

"3-Dimensional Modeling of Mixing with Helical Ribbon-Screw Impeller"
R. Lacroix, E. Brito de la Fuente, P. A. Tanguy and L. Choplin, Laval Univ.

"Analytical Solutions Using Local Turbulance Models for Predicting Circulation and Mixing in Bubble Columns"
R. G. Rice and K. G. Anderson, Louisiana State Univ.

MIXING AND CFD

The first Mixing session with only CFD works was presented at NAMF Mixing XII (Santa Barbara, California, 1993)

- The sliding mesh technique is presented at this conference;
- Works using the Finite Element Method are presented;
- Turbulence is also investigated.

Source: Plenary Lecture "A mix of computational Techniques for a better mix", by François Bertrand

Thursday, June 24, 1993

7:30 a.m. BREAKFAST

9:00 a.m. - 12:00 noon 8.0 CFD
Chair: Dr. Kathleen Barton, Dow Chemical
Chair: Prof. Philip Wood, McMaster University

8.1 **"3-Dimensional Modeling of Unbaffled Mixing Tanks Using Rotating Coordinates"**
S. Strand *, T. Plumley, S. Wood and D. Leng, The Dow Chemical Company

8.2 **"A Sliding-Mesh Technique for Simulation of Flow in Mixing Tanks"**
Chin-Yuan Perng and Jayathi Y. Murthy*, Fluent Inc.

8.3 **"Modeling of a Mixing Tank With a High Recirculation Flow"**
Midey Chang-Mateu * and Sudhir Joshi, Rohm and Haas Co.

8.4 **"Progress Towards Reallistic Turbulence Predictions With CFD"**
Joe Hannon, BHR Group

8.5 **"Finite Element Flow Simulations in a SMRX Static Mixer"**
P. Tanguy *, F. Bertrand and R. Lacroix, Laval University and Rheotek, L. van Dijk, T. Meyer and A. Renken, École Polytechnique Fédérale de Lausanne

8.6 **"Numerical Simulation of Buoyant Turbulent Flow Using Low Reynolds Number Turbulence Models"**
G. Numberg, P. Wood * and M. Shoukri, McMaster University

8.7 **"Numerical Modelling of the Flow In Static Mixers"**
André Bakker * and Julian B. Fasano, Chemineer, Inc. and

17AIChE Annual Meeting, Minneapolis, MN

MIXING AND CFD

UNICAMP

The first 3D CFD Model with heat transfer is presented in 1993 at the AIChE Annual Meeting in St Louis, MI

[1] Wednesday, November 10, 8:30-11:30
Pavilion Salon D
Marriott Pavilion

Symposium
Industrial Mixing and Scale-Up
Sponsored by Group 3—Contacting & Particulates, Operations & Processes
Area 3a—Mixing
Co-Sponsored by Industrial Advisory Board

Mark Melli, *Chair*
Procter & Gamble
4550 Este Ave
Cincinnati, OH 45232

Keith J. Carpenter, *Co-Chair*
Zeneca FCMO
P.O. Box 42, Hexagon House
Blackley
Manchester, M9 3DA United Kingdom

The Influence of Geometrical Factors on the Optimum Design of Stirred Tank Reactors • *8:55 AM* • *Paper No. 1b* • J.R. Nunhez [SPEAKER] and C. McGreavy, University of Leeds, Leeds, England

INTRODUCTION | MIXING AND CFD | EXAMPLES | CFD AND OPTIMIZATION | NUSSELT EQTS | CONCLUSION | 17

17AIChE Annual Meeting, Minneapolis, MN

MIXING AND CFD

UNICAMP

SOME RESULTS OF 1993 (LAMINAR FLOW)

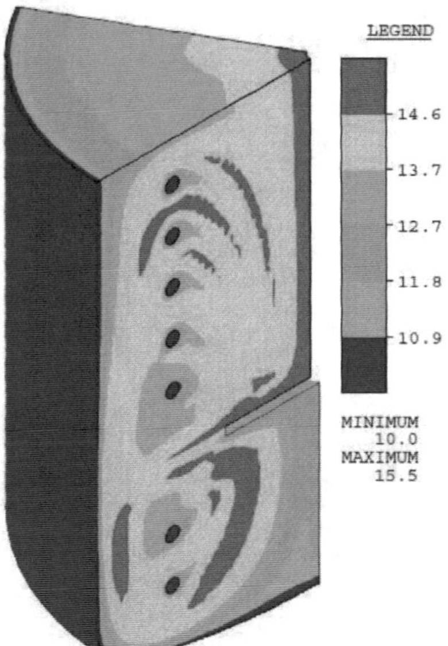

INTRODUCTION | MIXING AND CFD | EXAMPLES | CFD AND OPTIMIZATION | NUSSELT EQTS | CONCLUSION | 18

17AIChE
Annual Meeting, Minneapolis, MN

MIXING AND CFD

UNICAMP

There are several ways in which CFD helps us to better understand mixing processes:

1. By providing detailed information of the flow, it is possible to investigate:
   a. regions in the mixing equipment that is not working satisfactorily (dead zones);
   b. differences in temperature, concentration, pressure drop, temperature and other factors that influence mixing.

17AIChE
Annual Meeting, Minneapolis, MN

MIXING AND CFD

UNICAMP

2. CFD can test conditions that many times cannot be tested experimentally, such as highly exothermal reactions at run-off conditions;

3. Geometries and internals can be tested without the need of an experimental apparatus (CFD model needs to be tested and “validated” first).

INTRODUCTION | MIXING AND CFD | EXAMPLES | CFD AND OPTIMIZATION | NUSSELT EQTS | CONCLUSION | 20

MIXING AND CFD

- Some examples will be given showing how CFD help to better understand mixing processes.
- One of the examples was obtained in the internet. Some others are models developed at the Computational Fluid Dynamics Laboratory
- After we will consider two special ways in which CFD proves to be a valuable tool to understand mixing processes.

CFD Role in understanding Mixing Processes

- Product with a mixing problem in its formulation

EXAMPLES

Non-uniformity in a solid suspension

Process information:

Product:	solid suspension
Liquid specific gravity:	1225 kg/m^3
Solid density:	1250 kg/m^3
Final viscosity of the product:	350 cP
Final solids concentration:	40 mg/ml
Tank volume	10 m^3

17AIChE
Annual Meeting, Minneapolis, MN

EXAMPLES

UNICAMP

Initial configuration

- Two low shear impellers - dimeter: 0.7 m
- One PBT 45 - diameter: 0.3 m
- Centered impeller, no baffles

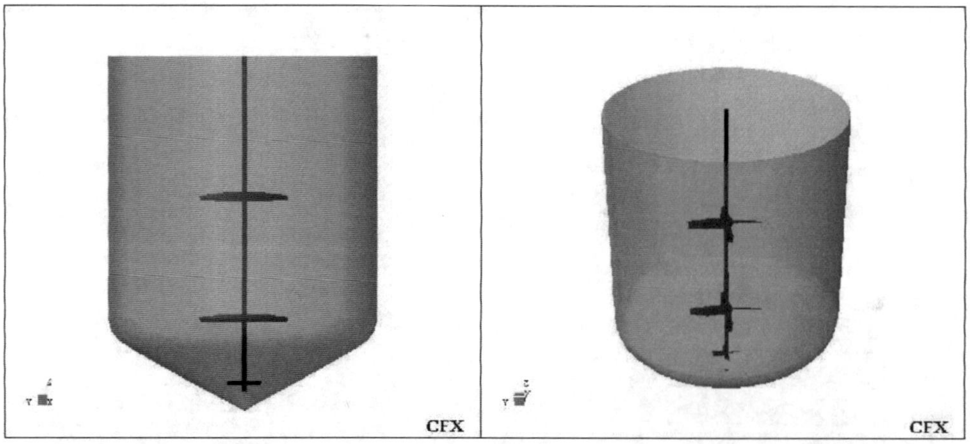

Problem: solid concentration out of specification limits

17AIChE
Annual Meeting, Minneapolis, MN

EXAMPLES

UNICAMP

Initial configuration

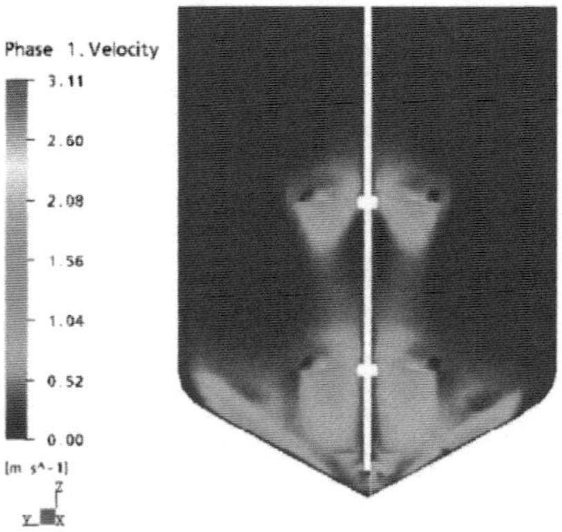

CFX

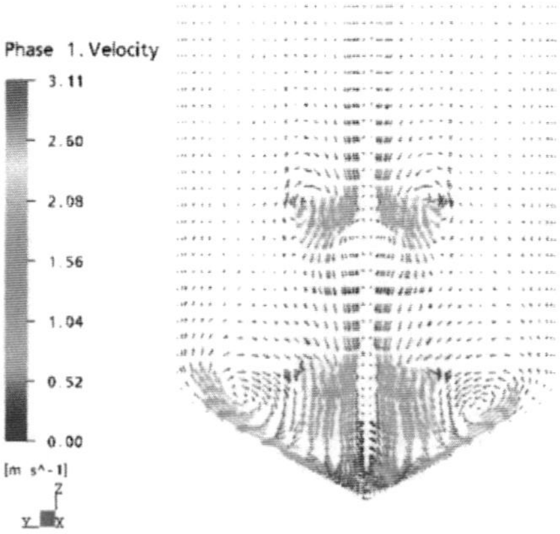

CFX

Concentration:
maximun: 66.65 mg/ml
minimun: 37.50 mg/ml

Impellers unable to promote an homogeneous velocity distribution inside the tank. Tank bottom is innapropriate and impellers are small.

17AIChE
Annual Meeting, Minneapolis, MN

EXAMPLES

UNICAMP

Final configuration

- Three impellers with diameter of 1.15 m
- One impeller close to the bottom head with diameter of 0.7 m
- Three baffles to minimize tangential flow.

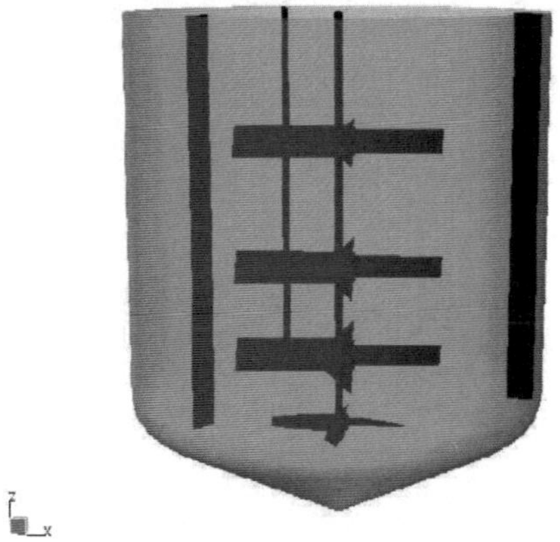

CFX

17AIChE
Annual Meeting, Minneapolis, MN

EXAMPLES

UNICAMP

Initial configuration   Final configuration

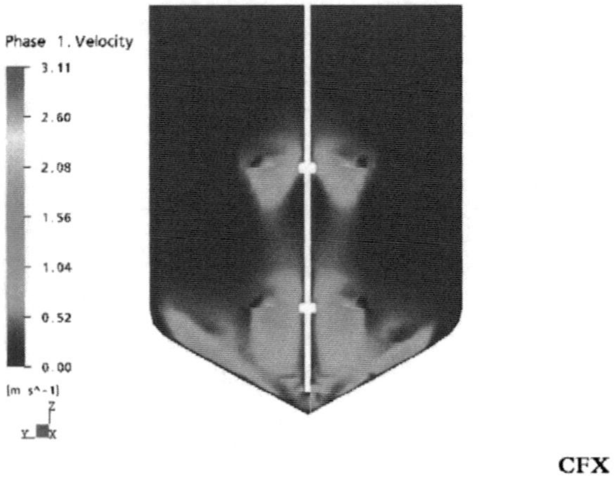

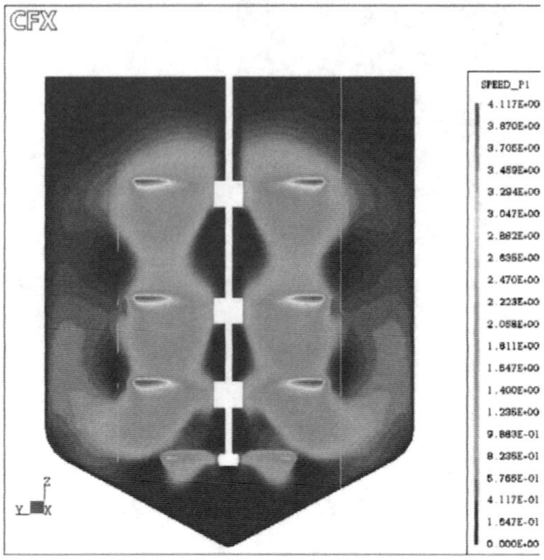

17AIChE
Annual Meeting, Minneapolis, MN

EXAMPLES

UNICAMP

Comparison:

	Initial configuration	Final configuration
maximun:	66,65 mg/ml	42.56 mg/ml
minimun:	37,50 mg/ml	38.45 mg/ml
average:	40,00 mg/ml	40,00 mg/ml

INTRODUCTION | MIXING AND CFD | EXAMPLES | CFD AND OPTIMIZATION | NUSSELT EQTS | CONCLUSION | 28

Conclusion

- This example is good because it shows a problem we find in industries. The L-CFD team pointed out that tank bottom should be replaced and impellers resized. However, they only allowed impellers to be changed;
- Many tank designs in industry are far from being ideal and CFD helps showing their inadequacy;
- The changes in this example were implemented and the system operated as predicted by the CFD model;
- CFD specialists would never suggested a geometry as the final configuration. It is however the best our CFD team could suggest with the limitations imposed.

CFD Role in understanding Mixing Processes

- Fan blade design improved with the use of CFD

Dr Nicolas Spogis
Professor at PUC – Campinas
Department of Chemical Engineering
E-mail: nicolas.spogis@gmail.com

Prof. José Roberto Nunhez.

UNICAMP

EXAMPLES

- Determine the operating curve and flow characteristics of an industrial fan with the use of CFD;
- Study the flow characteristics at several prescribed flow conditions
- Study the sound characteristics
  - Level of sound pressure;
  - Level of sound power;
  - Harmonics.

EXAMPLES

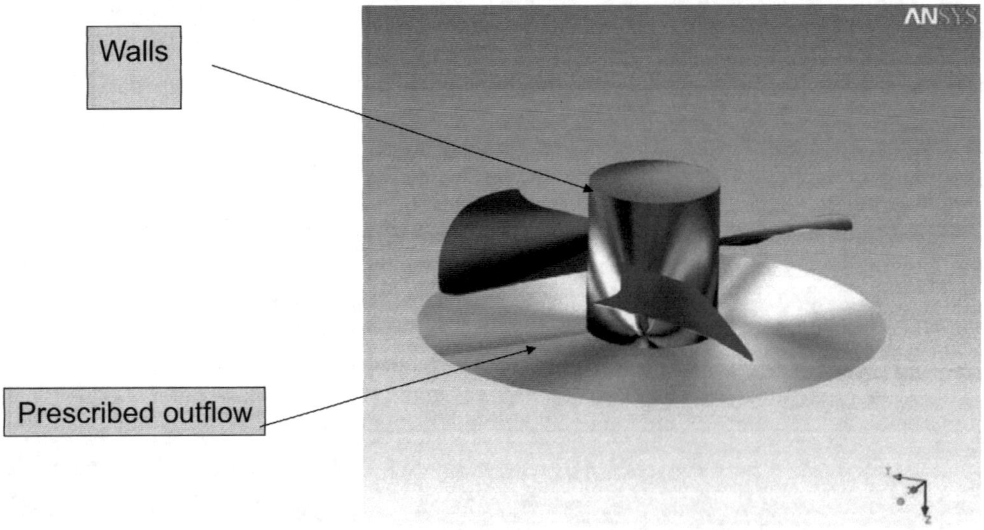

EXAMPLES

UNICAMP

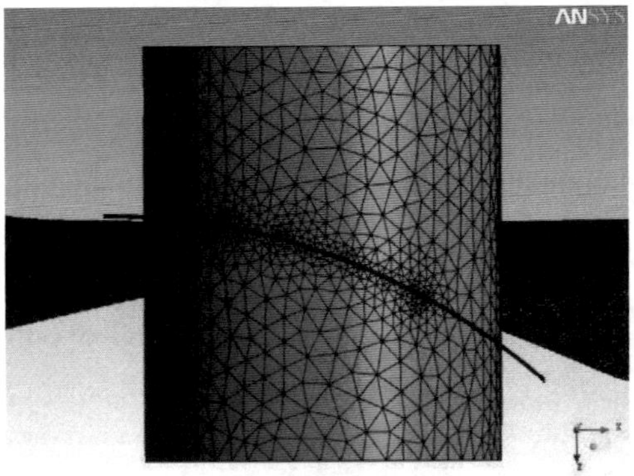

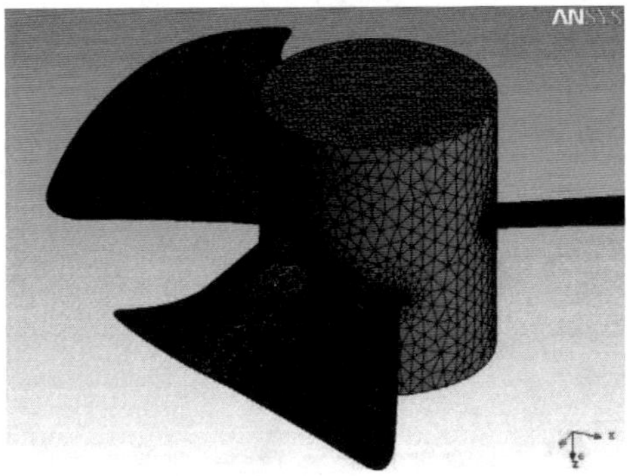

EXAMPLES

UNICAMP

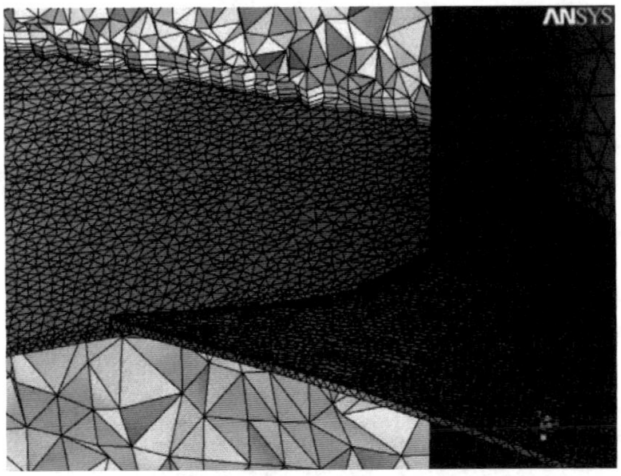

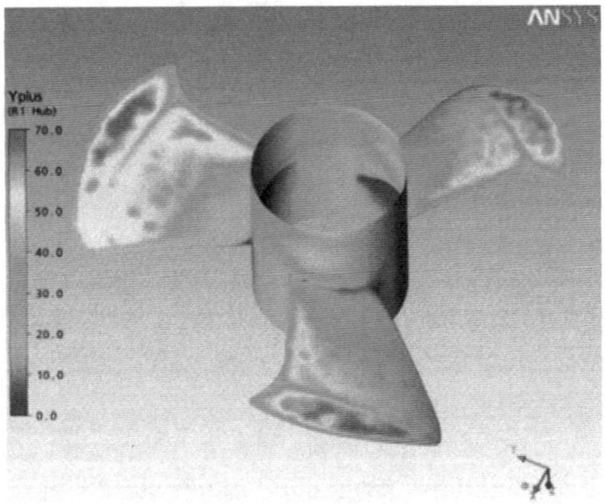

17AIChE
Annual Meeting, Minneapolis, MN

EXAMPLES

UNICAMP

Pressure at blades

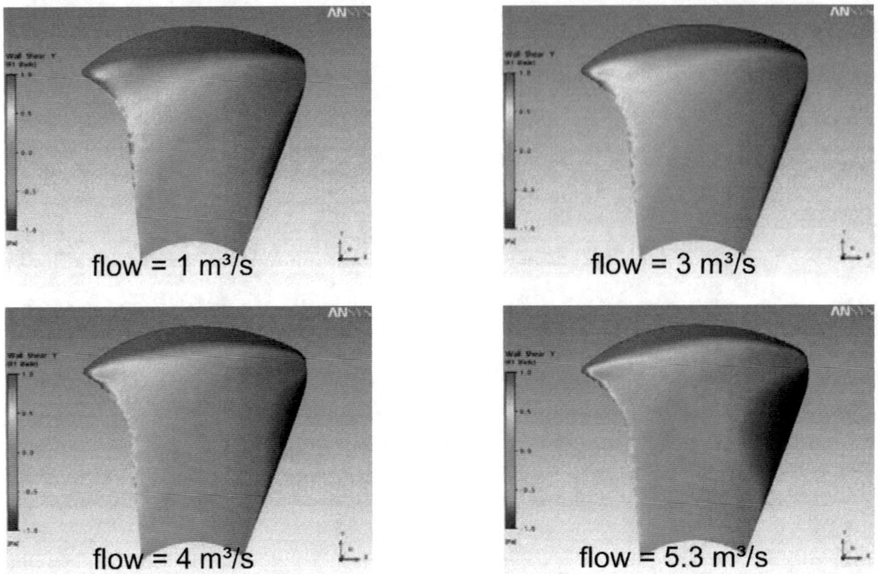

17AIChE
Annual Meeting, Minneapolis, MN

EXAMPLES

UNICAMP

Vortex

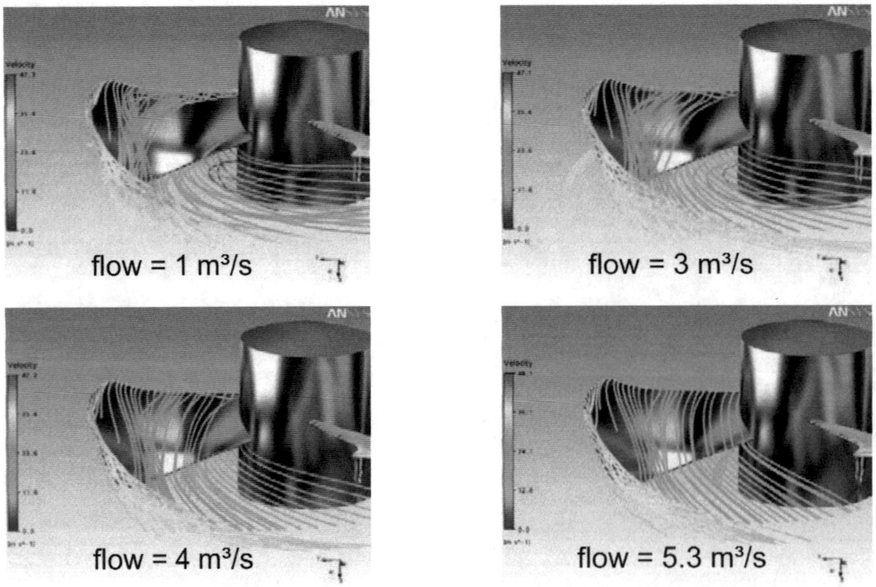

17AIChE
Annual Meeting, Minneapolis, MN

EXAMPLES

UNICAMP

Conclusion

- CFD study showed a need to twist the tip of the blade;
- Noise was considerably reduced as the frequency study carried out with CFD demonstrated;
- The CFD changes were fully implemented;
- This is a good CFD study example because it was done with the engineer who initially developed the fan blade. CFD showed how air flow could be improved.

CFD Role in understanding Mixing Processes

- New impeller developed based on CFD (BT 6- Chemineer)

(Source: http://www.bakker.org/cfmbook/bt6-book.pdf)

17AIChE Annual Meeting, Minneapolis, MN

EXAMPLES

UNICAMP

- For several decades the 6-blade Rushton impeller was the standard impeller for gas dispersion;
- Radial flow impeller
- Six flat blades on a disk
- Suitable for gas dispersion and liquid-liquid dispersion
- Turbulent power number range: 4.5-6.2
- Trailing vortices in single-phase flow
- Cavitation in gas-liquid flow

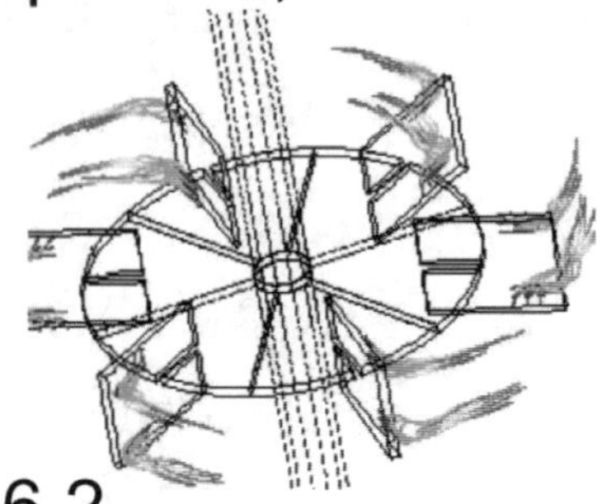

17AIChE Annual Meeting, Minneapolis, MN

EXAMPLES

UNICAMP

- In the 1980´s John Smith and his research group used the concept of curved blades and developed the Smith turbine. The curvature of the blades are symmetrical in relation to the impeller disk;
- Lightnin R130, Philadelphia's "Smith Turbine."
- Turbulent power number range: 2.8-3.2.

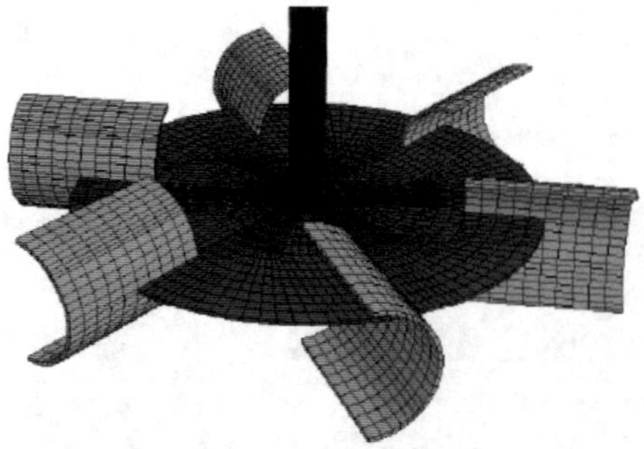

EXAMPLES

- The proprietary Scaba and ICI impellers had deeper concave blade and a sharp back edge of the blades (1988-1993);

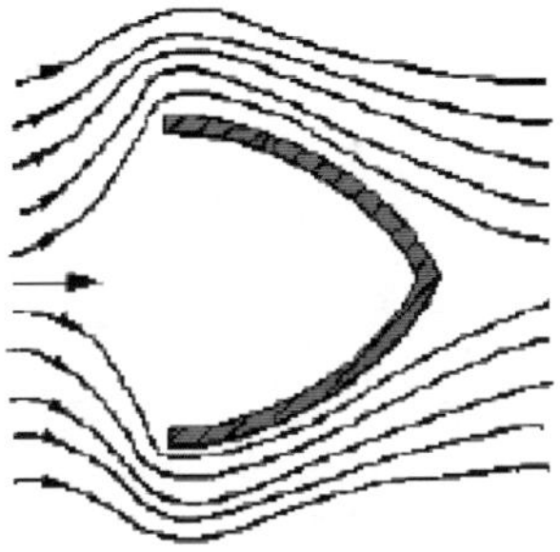

US Patent 4,779,990 (1988)

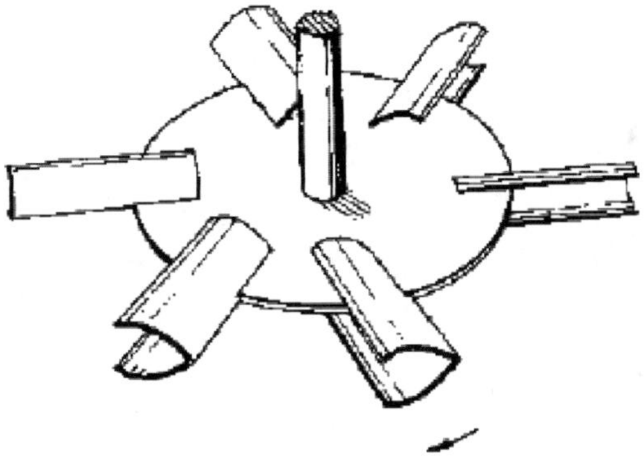

US Patent 5,198,156 (1993)

EXAMPLES

- If gas flow is asymmetric, as pointed out by Bakker, why would blade shape be symmetric?
- Bakker and his group worked on several kinds of blade configuration (more than 20) and the final design shows a blade curvature different on top and bottom, being longer on top.
- Rising gas is captured by longer upper portion of blade and dispersed from inside the deep blades

EXAMPLES

- Bakker and his group reasoned that if gas enters from below the blade and rises, why not leaving the upper blade portion greater than the lower one keeping the gas for a longer time?;

17AIChE
Annual Meeting, Minneapolis, MN

EXAMPLES

UNICAMP

Bakker ***BT-6 Performance Characteristics***

- Low turbulent power number (Po = 2.3);
- Power number is constant for impeller Reynolds numbers greater than 1,000;
- Flat power draw curve under gassed conditions;
- Superior gas dispersion capabilities.

EXAMPLES

Rushton | CD-6 | BT-6

N_P [4,5 – 6,2] | N_P [2,8 – 3,2] | $N_P = 2,3$

EXAMPLES

Conclusion

- CFD study and a series of experiments led the group to provide a new impeller design with improved gas dispersion characteristics;
- Power draw was considerably reduced.

17AIChE Annual Meeting, Minneapolis, MN

EXAMPLES

UNICAMP

- ✓ Two examples will be shown in which CFD has been used successfully in different ways to obtain important Mixing data.
- ✓ In the first example, optimization is used as a tool to help select an impeller for solid suspension. This is the first and only work in which optimization is coupled with CFD to obtain a mixing impeller;
- ✓ The second example shows how CFD can be used to obtain the heat transfer coefficient in stirred tanks. Experimental data of Oldshue and Greton (1950) for helical coils in a tank stirred by a Rushton turbine are used to validate the procedure.

INTRODUCTION | MIXING AND CFD | EXAMPLES | CFD AND OPTIMIZATION | NUSSELT EQTS | CONCLUSION | 47

17AIChE Annual Meeting, Minneapolis, MN

EXAMPLES

UNICAMP

- ✓ The arrangement suggested by Street and McGreacy and Nunhez and McGreavy is tested;
- ✓ The model is then applied to a tank with helical coils but stirred by two Rushton impellers. There is no Nusselt equation in the literature published for this configuration;
- ✓ The methodology is also applied to a jacketed tank reactor and CFD results are compared with available experimental data.

INTRODUCTION | MIXING AND CFD | EXAMPLES | CFD AND OPTIMIZATION | NUSSELT EQTS | CONCLUSION | 48

CFD Role in understanding Mixing Processes
Design of a high efficient hydrofoil through the use of computational fluid dynamics and multi-objective optimization
Nicolas Spogis, Ph.D.
Professor at PUC – Campinas
Department of Chemical Engineering
E-mail: nicolas.spogis@gmail.com
Prof. José Roberto Nunhez.
Velocity
4.0
3.6
3.1
2.7
2.2
1.8
1.3
0.9
0.4
0.0
[m s^-1]
UNICAMP
ESSS
17AIChE
Annual Meeting, Minneapolis, MN
CFD AND OPTIMIZATION
UNICAMP
AIChE
PROCESS SYSTEMS ENGINEERING
Design of a High-Efficiency Hydrofoil Through the Use of Computational Fluid Dynamics and Multiobjective Optimization
N. Spogis
ESSS, Engineering Simulation and Scientific Software, Rua do Rocio, 423 10 andar conj.1001/1002, Vila Olímpia, CEP: 04552-000, São Paulo, SP, Brazil
J. R. Nunhez
School of Chemical Engineering, State University of Campinas, UNICAMP, P.O. Box 6066, 13083-970, Campinas, SP, Brazil
DOI 10.1002/aic.11804
Published online May 14, 2009 in Wiley InterScience (www.interscience.wiley.com).
INTRODUCTION
MIXING AND CFD
EXAMPLES
CFD AND OPTIMIZATION
NUSSELT EQTS
CONCLUSION
50

CFD AND OPTIMIZATION

ELSEVIER

CrossMark

Available online at www.sciencedirect.com

ScienceDirect

Procedia Engineering 138 (2016) 403 – 412

Procedia Engineering

www.elsevier.com/locate/procedia

"SYMPHOS 2015", 3rd International Symposium on Innovation and Technology in the Phosphate Industry

Optimizing a Hydrofoil for Industrial Use

Spogis, N.[a] and Nunhez, J. R.[b]*

[a] *ESSS, Rua do Rocio, 423 10 andar conj.1001/1002, Vila Olímpia – CEP: 04552-000 – São Paulo– SP – Brazil.*

[b] *UNICAMP - Faculdade de Engenharia Química, Departamento de Engenharia de Processos, Universidade Estadual de Campinas, Cidade Universitária Zeferino Vaz, Av. Albert Einstein, 500, CEP: 13083-970 – Campinas – SP – Brazil.*

17AIChE
Annual Meeting, Minneapolis, MN

CFD AND OPTIMIZATION

UNICAMP

An optimization procedure was applied to the design of an impeller to illustrate how this approach can lead to more efficient mixing processes in general. The suspension of solid particles in a stirred tank was used to illustrate the methodology.

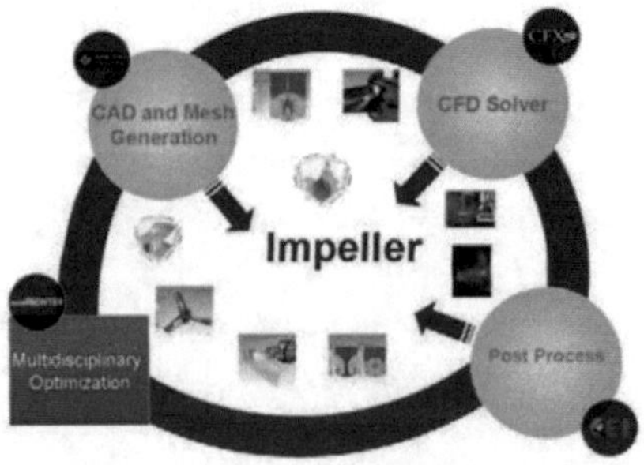

17AIChE Annual Meeting, Minneapolis, MN

CFD AND OPTIMIZATION

UNICAMP

→ 44 years →

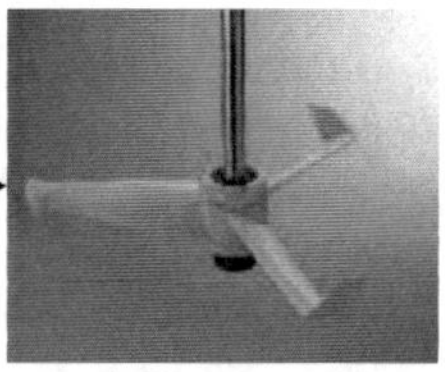

1950

LIGHTNIN Mixers - 1994

INTRODUCTION > MIXING AND CFD > EXAMPLES > CFD AND OPTIMIZATION > NUSSELT EQTS > CONCLUSION > 53

17AIChE Annual Meeting, Minneapolis, MN

CFD AND OPTIMIZATION

UNICAMP

- Design of a high efficiently hydrofoil:
  - Pure Axial Flow
  - Lack of Radial Flow
  - Torque Reduction
  - Saving Costs

- Through the use of:
  - Computational Fluid Dynamics
  - Multi-Objective Design Optimization

CFX

modeFRONTIER

INTRODUCTION > MIXING AND CFD > EXAMPLES > CFD AND OPTIMIZATION > NUSSELT EQTS > CONCLUSION > 54

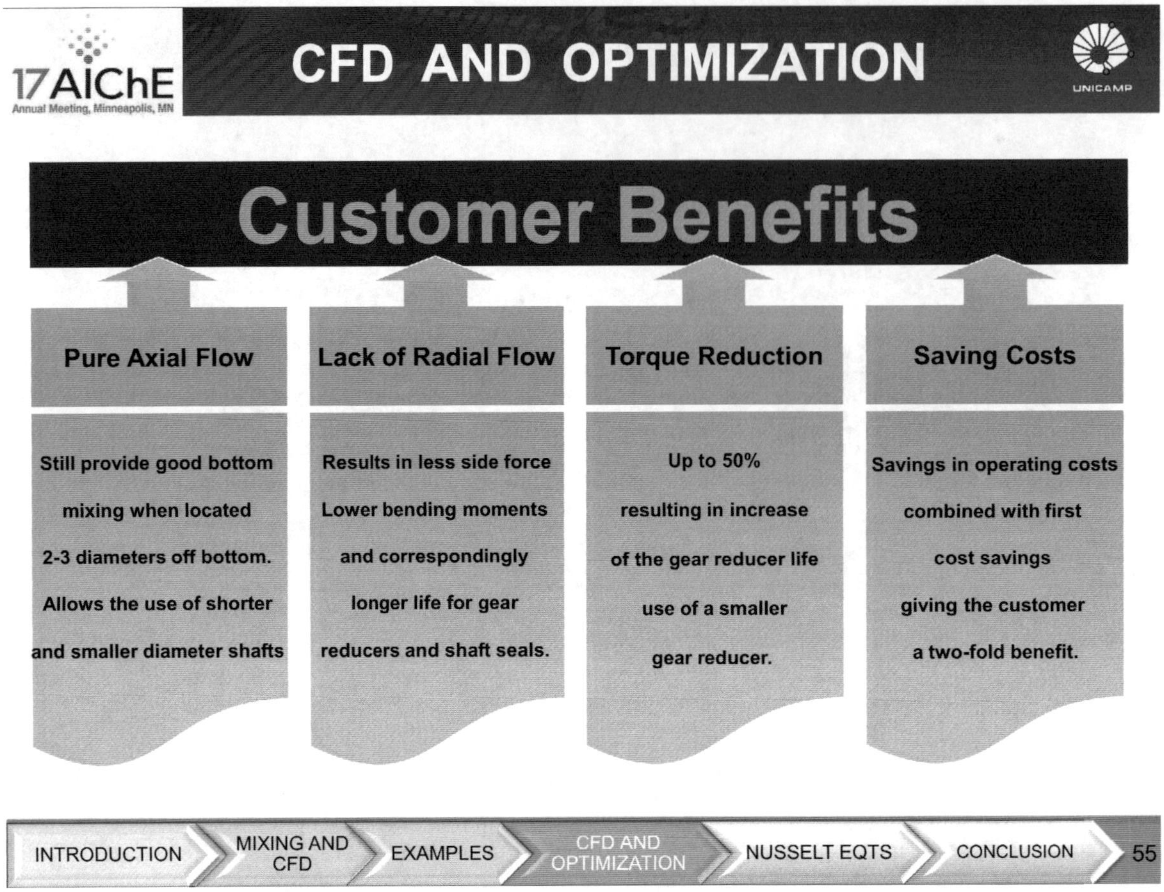
17AIChE
Annual Meeting, Minneapolis, MN
CFD AND OPTIMIZATION
UNICAMP
Customer Benefits
Pure Axial Flow
Still provide good bottom mixing when located 2-3 diameters off bottom. Allows the use of shorter and smaller diameter shafts
Lack of Radial Flow
Results in less side force Lower bending moments and correspondingly longer life for gear reducers and shaft seals.
Torque Reduction
Up to 50% resulting in increase of the gear reducer life use of a smaller gear reducer.
Saving Costs
Savings in operating costs combined with first cost savings giving the customer a two-fold benefit.
INTRODUCTION
MIXING AND CFD
EXAMPLES
CFD AND OPTIMIZATION
NUSSELT EQTS
CONCLUSION
55

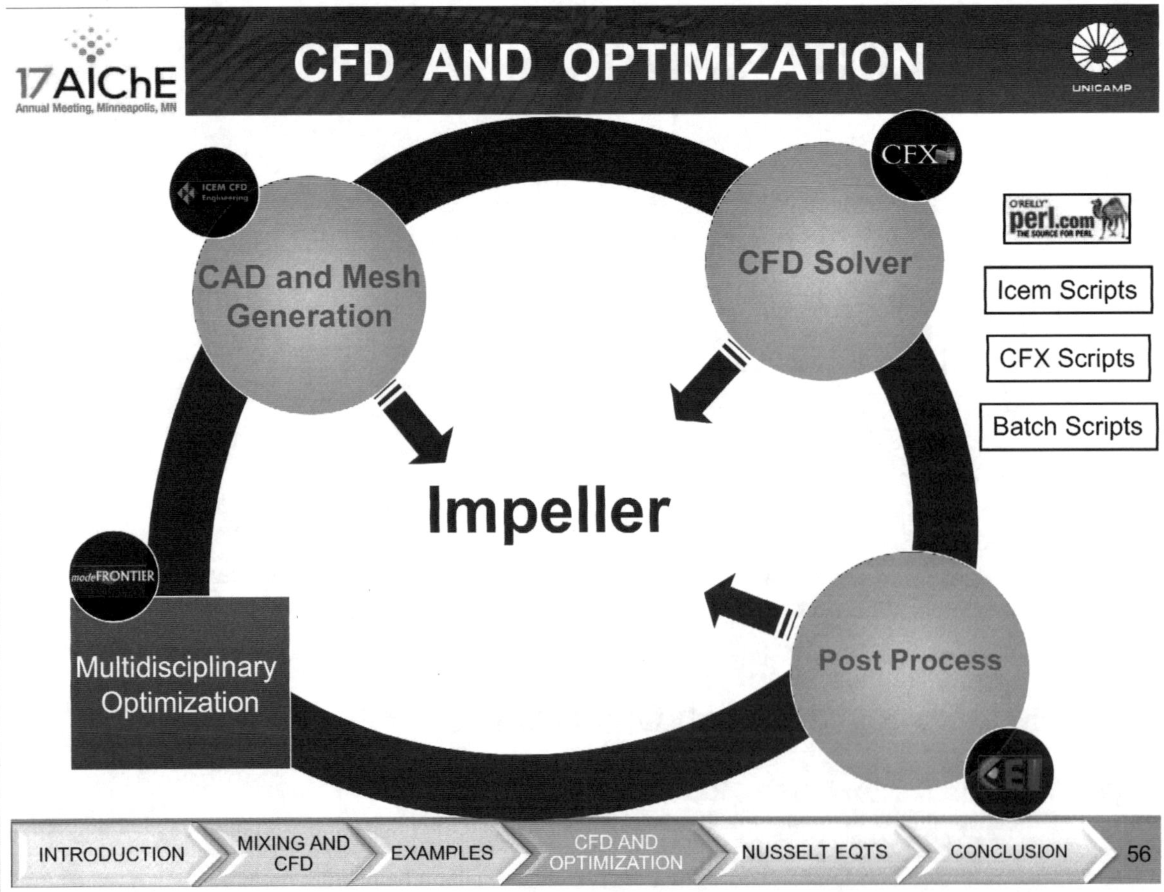
17AIChE
Annual Meeting, Minneapolis, MN
CFD AND OPTIMIZATION
UNICAMP
ICEM CFD
CAD and Mesh Generation
CFX
CFD Solver
perl.com
Icem Scripts
CFX Scripts
Batch Scripts
Impeller
modeFRONTIER
Multidisciplinary Optimization
Post Process
INTRODUCTION
MIXING AND CFD
EXAMPLES
CFD AND OPTIMIZATION
NUSSELT EQTS
CONCLUSION
56

17 AIChE Annual Meeting, Minneapolis, MN

CFD AND OPTIMIZATION

UNICAMP

Energy Level	Process Type	P/V Relation (kW/m³)
Low level	Weak solid suspension Low viscosity fluids	0.2 kW/m³
Moderately	Heat transfer Gas dispersion	0.6 kW/m³
High Level	Heavy solid suspension High gas dispersion Emulsification	2 kW/m³
Higher Level	High viscosity fluids Pasta	4 kW/m³

17 AIChE Annual Meeting, Minneapolis, MN

CFD AND OPTIMIZATION

UNICAMP

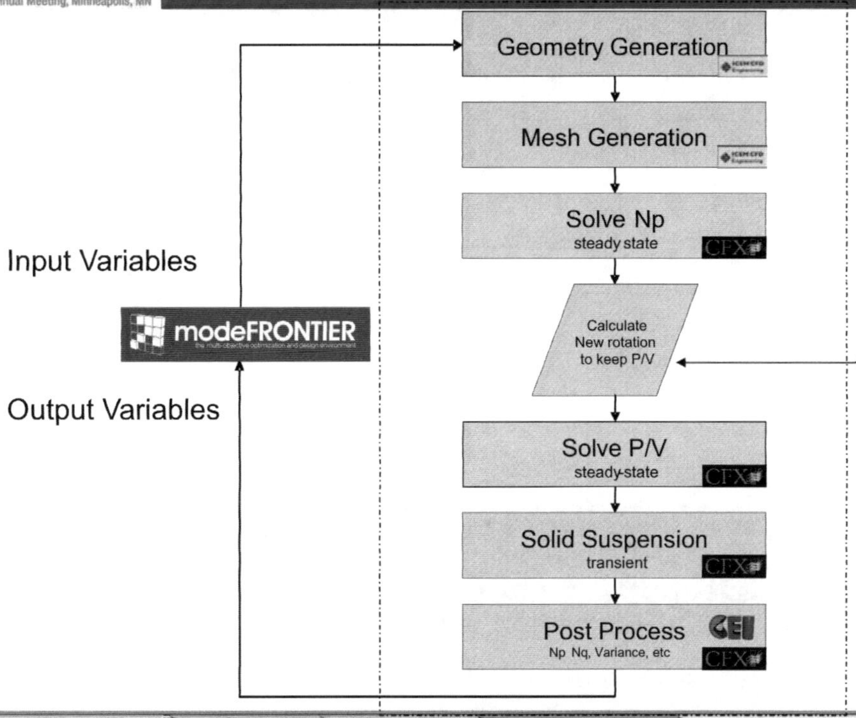

$$Np = \frac{P}{\rho . N^3 . D^5}$$

$$N = \sqrt[3]{\frac{P}{\rho . D^5 . Np}}$$

CFD AND OPTIMIZATION

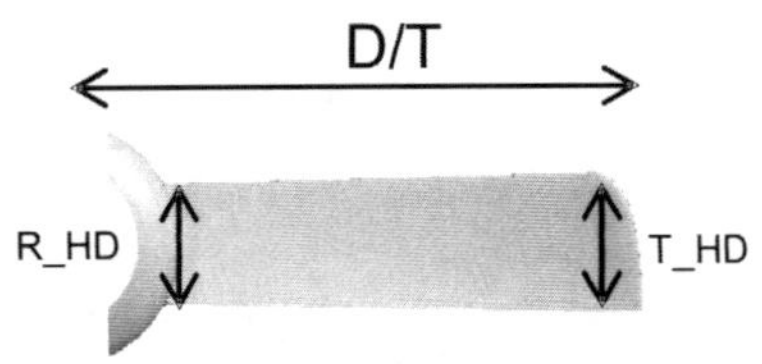

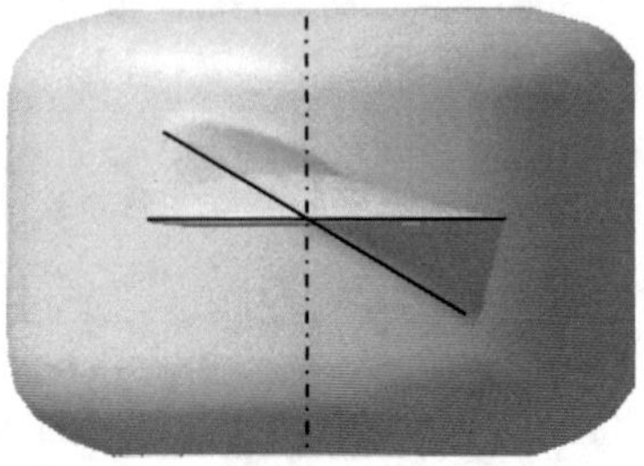

CFD AND OPTIMIZATION

Variable	Minimum Value	Maximum Value	Discrete / Continuum
Impeller diameter	0.4	0.5	Continuum
Root chord	0.1	0.2	Continuum
Tip chord	0.1	0.2	Continuum
Root chord angle	20 degrees (related to rotation axis)	70 degrees (related to rotation axis)	Continuum
Tip chord angle	30 degrees (related to rotation axis)	95 degrees (related to rotation axis)	Continuum
Root profile	DAE11, S1223, E387, FX 63-137 17 different profiles		Discrete
Tip profile	DAE11, S1223, E387, FX 63-137 17 different profiles		Discrete

17AIChE
Annual Meeting, Minneapolis, MN

CFD AND OPTIMIZATION

UNICAMP

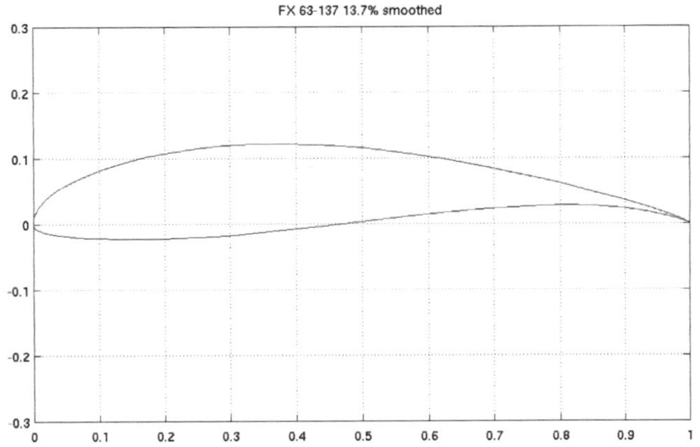

INTRODUCTION | MIXING AND CFD | EXAMPLES | CFD AND OPTIMIZATION | NUSSELT EQTS | CONCLUSION | 61

17AIChE
Annual Meeting, Minneapolis, MN

CFD AND OPTIMIZATION

UNICAMP

- **Vessel solid concentration variance**
  - Heavy solid suspension
  - Pure Axial Flow
  - Lack of Radial Flow

$$s^2 = \frac{1}{n-1}\sum_{i=1}^{n}\left(C_i - \overline{C}\right)^2 \Downarrow$$

- **Pumping effectiveness**
  - Torque Reduction
  - Saving Costs

$$P_{efect} = \frac{N_q}{N_p} \Uparrow$$

INTRODUCTION | MIXING AND CFD | EXAMPLES | CFD AND OPTIMIZATION | NUSSELT EQTS | CONCLUSION | 62

CFD AND OPTIMIZATION

A Sobol's algorithm was chosen to generate an initial population of 3000 individuals (impellers), than 50 individuals (impellers) were filtered by a D-Optimal algorithm in order to reduce the preliminary investigation computational time.

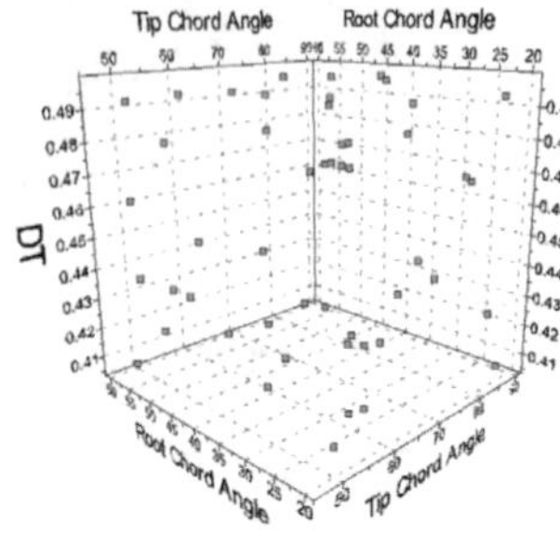

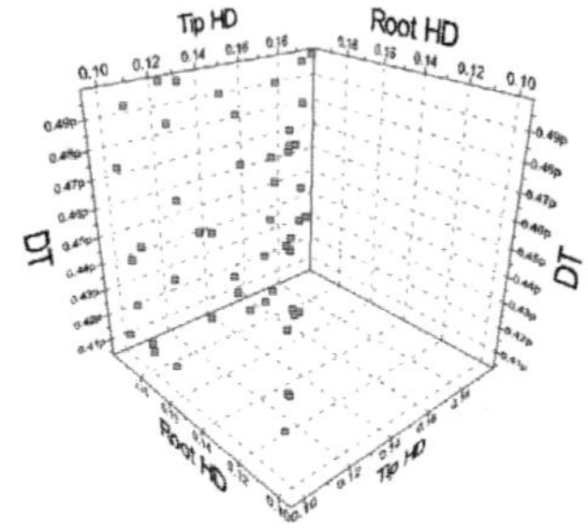

17AIChE
Annual Meeting, Minneapolis, MN

CFD AND OPTIMIZATION

UNICAMP

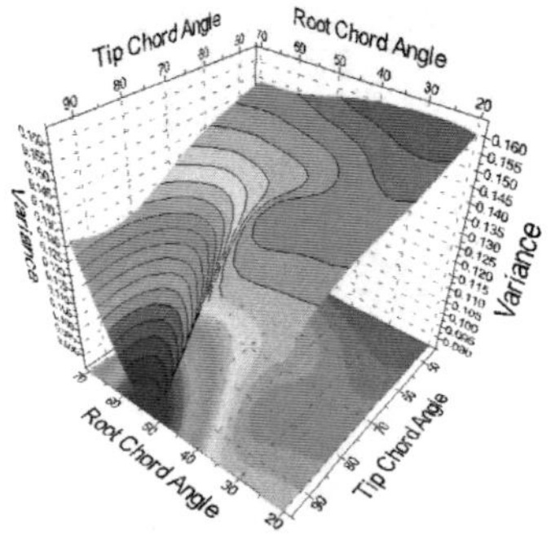

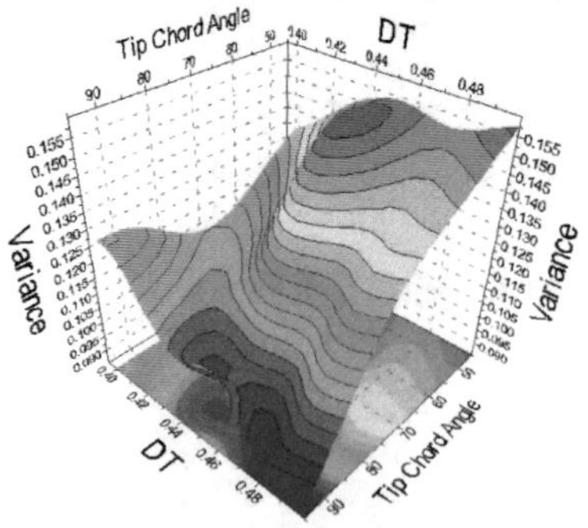

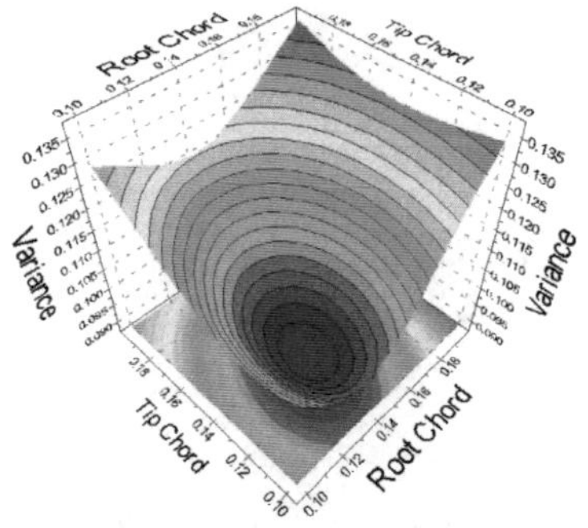

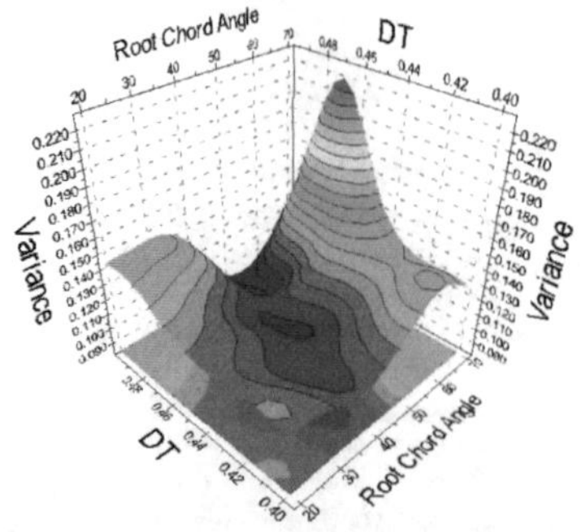

17AIChE
Annual Meeting, Minneapolis, MN

CFD AND OPTIMIZATION

UNICAMP

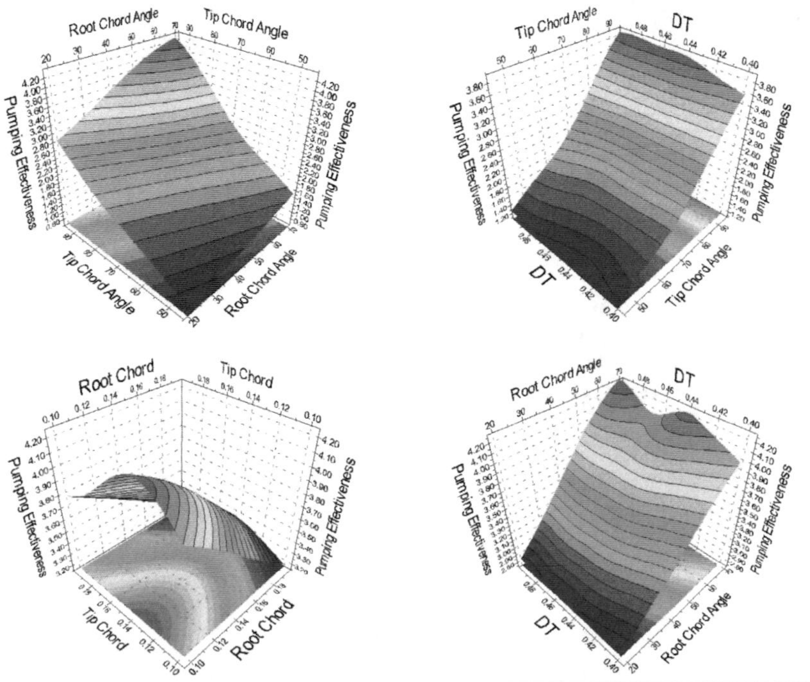

17AIChE
Annual Meeting, Minneapolis, MN

CFD AND OPTIMIZATION

UNICAMP

Optimized Impeller

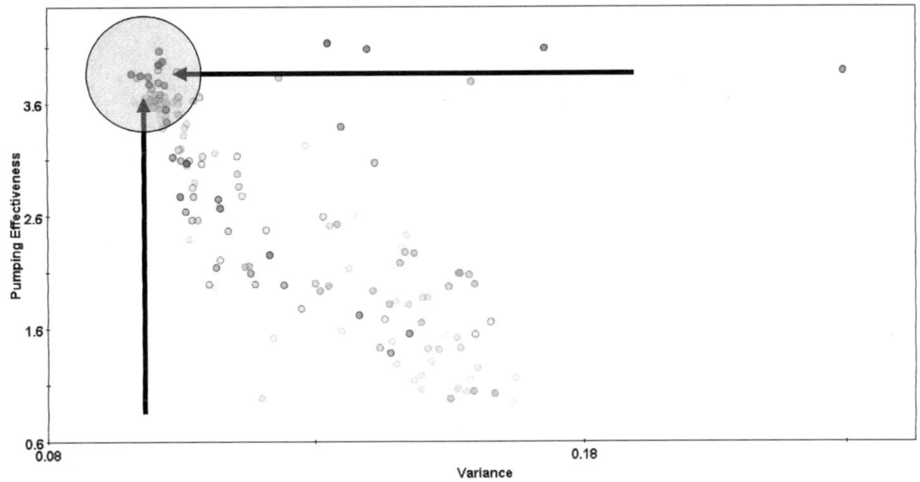

17AICHE Annual Meeting, Minneapolis, MN

CFD AND OPTIMIZATION

UNICAMP

- Multi-Objective Genetic Algorithm (MOGA-II)
- MOGA's search method also has very interesting aspects:
  - it allows global solutions to be found
  - it especially guarantees an actual multi-objective optimization, where the Pareto frontier is defined in the end.
- Reproduction operators:
  - Classical crossover;
  - Directional crossover;
  - Mutation;
  - Selection.

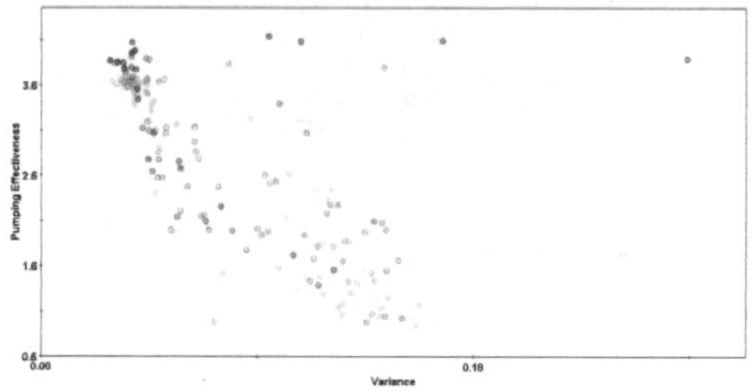

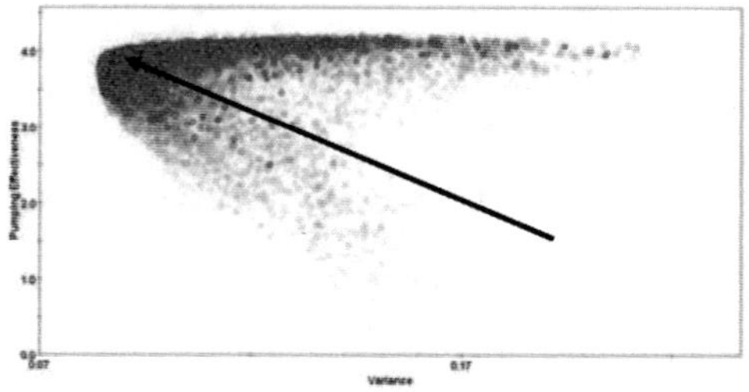

17AICHE Annual Meeting, Minneapolis, MN

CFD AND OPTIMIZATION

UNICAMP

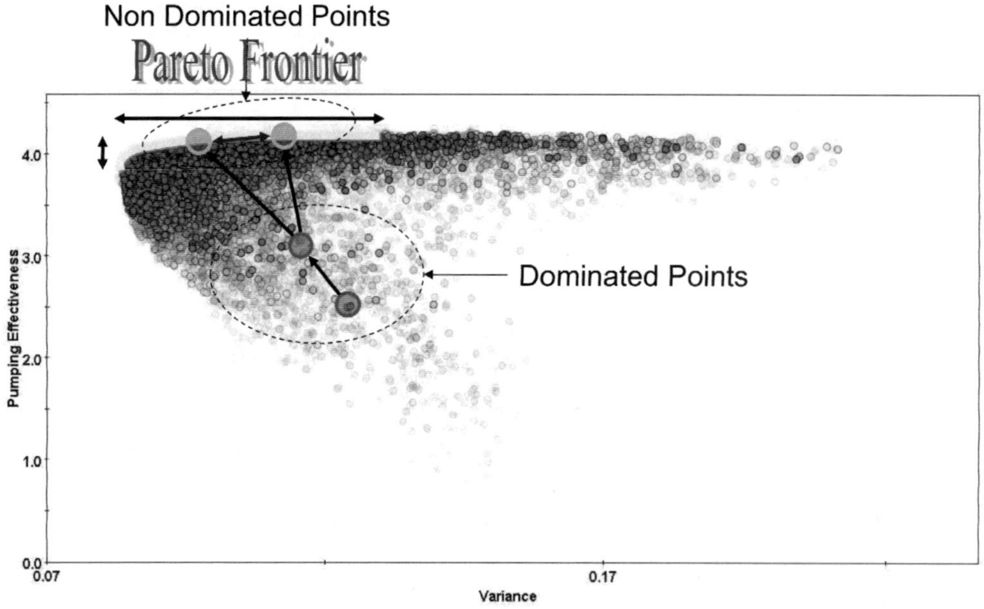

INTRODUCTION | MIXING AND CFD | EXAMPLES | CFD AND OPTIMIZATION | NUSSELT EQTS | CONCLUSION | 68

17AIChE
Annual Meeting, Minneapolis, MN

CFD AND OPTIMIZATION

UNICAMP

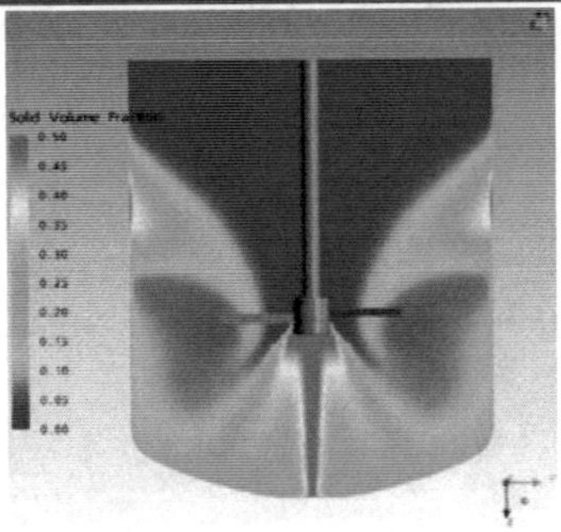

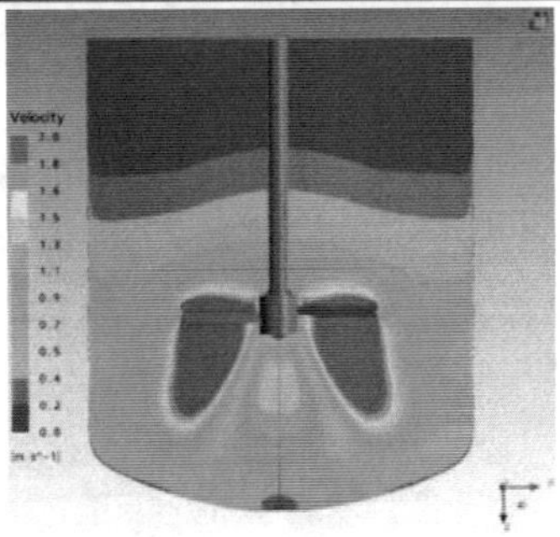

Problems:

- Average axial Flow
- Average radial flow
- High Np – Torque
- High power consumption
- Low solid suspension

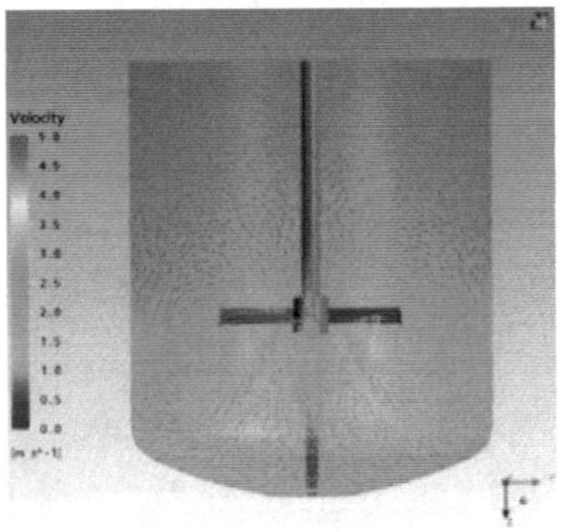

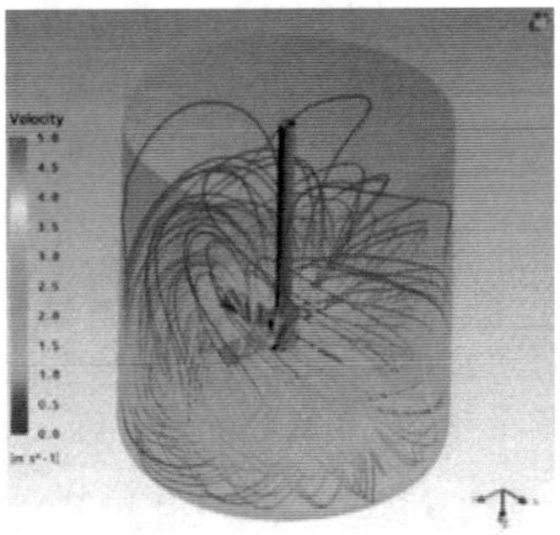

17AIChE
Annual Meeting, Minneapolis, MN

CFD AND OPTIMIZATION

UNICAMP

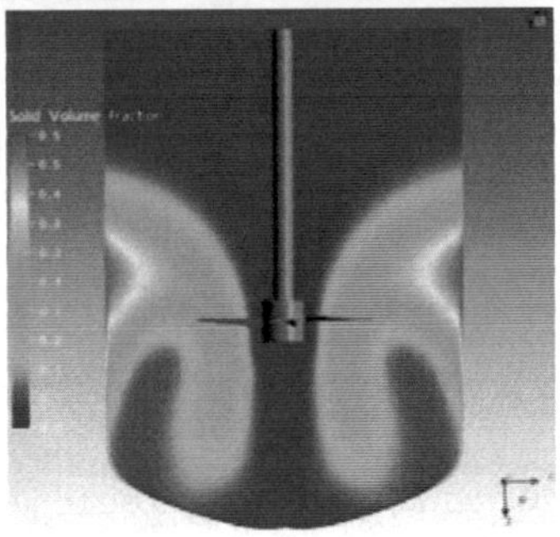

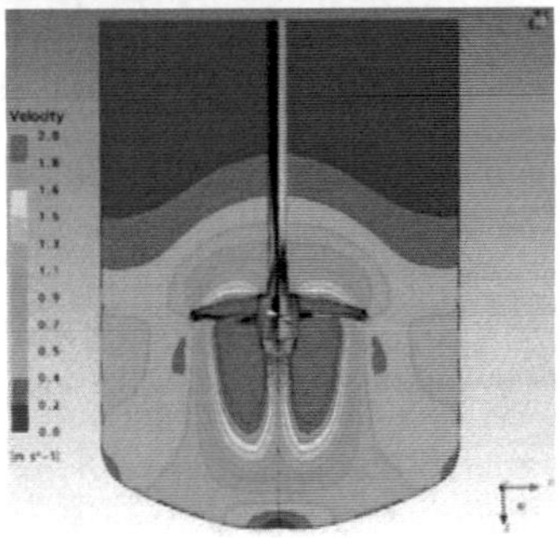

Solutions:

- Pure axial Flow
- Lack of radial flow
- Low Np – Torque
- Low power consumption
- High solid suspension

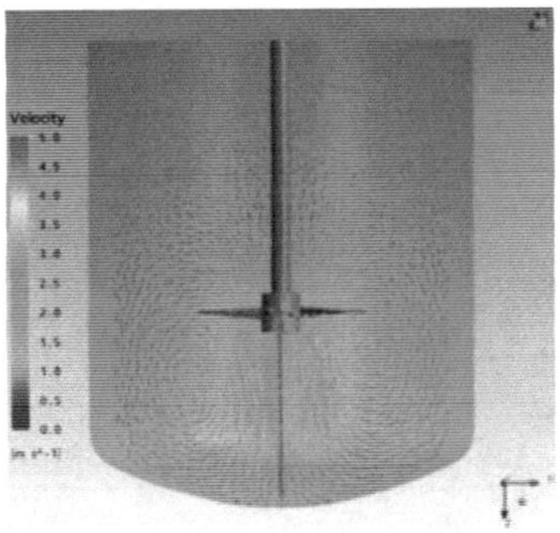

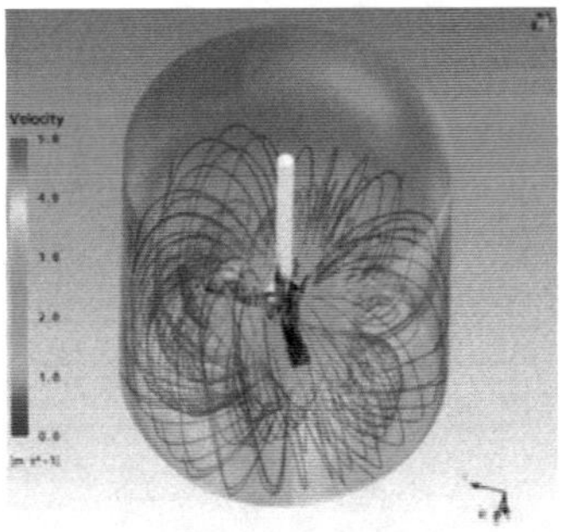

CFD AND OPTIMIZATION

UNICAMP

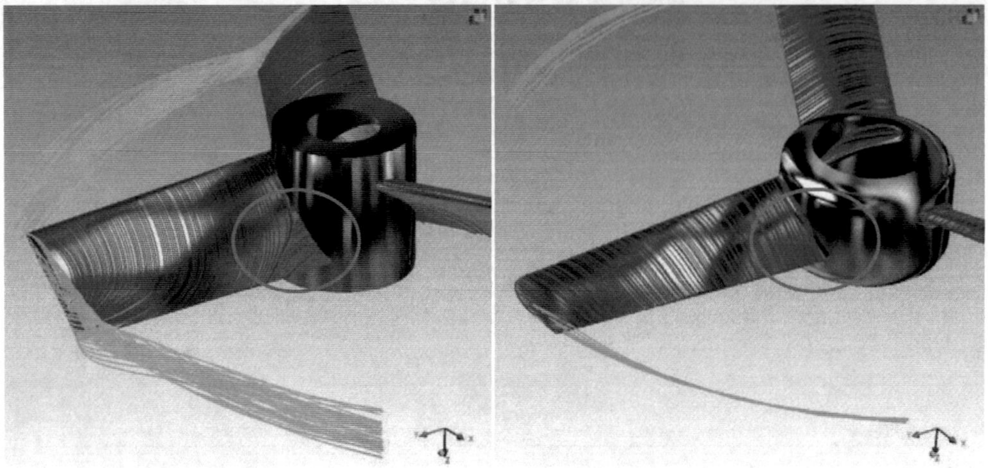

Problems:

- High Hub Separation
- High Tip Vortex
- High Shear

Solutions:

- Low Hub Separation
- Low Tip Vortex
- Low Shear

CFD AND OPTIMIZATION

UNICAMP

- The experimental tests were carried out in an agitation system that consists of:
  - Vertical Vessel with torospherical 10% bottom;
  - Structure in stainless steel 316L;
  - Baffles;
  - SEW Electric Motor - 1,5 kW;
  - Control panel;
  - MAGTROL torquimeter;
  - Fluid used on tests: water at 25°C.

17AIChE
Annual Meeting, Minneapolis, MN

CFD AND OPTIMIZATION

UNICAMP

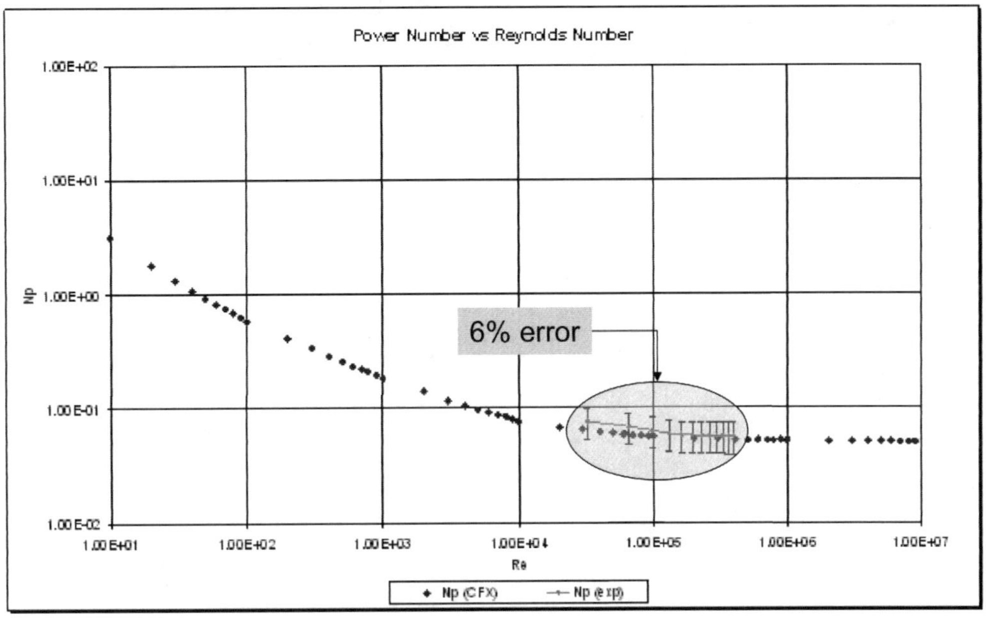

17AIChE
Annual Meeting, Minneapolis, MN

CFD AND OPTIMIZATION

UNICAMP

Optimized Impeller

N=290 RPM ->P = 6.09 W

N=390 RPM ->P = 10.54 W

PBT45 Impeller

N=172 RPM ->P = 6.06 W

N=232 RPM ->P = 11.23 W

CFD AND OPTIMIZATION

CFD AND OPTIMIZATION

- Solid concentration variance was reduced by **48.5%**
- Power consumption reduced by **84.4%**
- Pumping effectiveness increased by **410.2%**
- When compared to the performance of a pitched blade impeller (45 degrees constant Tip Chord Angle – PBT45).

CFD AND OPTIMIZATION

17 AIChE Annual Meeting, Minneapolis, MN — UNICAMP

- When this impeller was first presented in 2007 at NAMF XXI, two experienced consultants pointed out that impellers for industrial applications should have a power number greater than 0.2. The power number of the impeller developed is 0.085
- Even though very economical, the flow number was also very low. It was only 0.218. Ideally, it should be over 0.5
- The problem was run again with a further restriction that $N_p > 0.2$

CFD AND OPTIMIZATION

17 AIChE Annual Meeting, Minneapolis, MN — UNICAMP

- **Vessel solid concentration variance**
  - Heavy solid suspension
  - Pure Axial Flow
  - Lack of Radial Flow

$$s^2 = \frac{1}{n-1}\sum_{i=1}^{n}\left(C_i - \overline{C}\right)^2 \quad \Downarrow$$

- **Pumping effectiveness**
  - Torque Reduction
  - Saving Costs

$$P_{efect} = \frac{N_q}{N_p} \quad \Uparrow$$

- **Power Number restriction**
  - Not "too high" impeller speeds
  - Greater stability

$$N_p > 0.2$$

CFD AND OPTIMIZATION

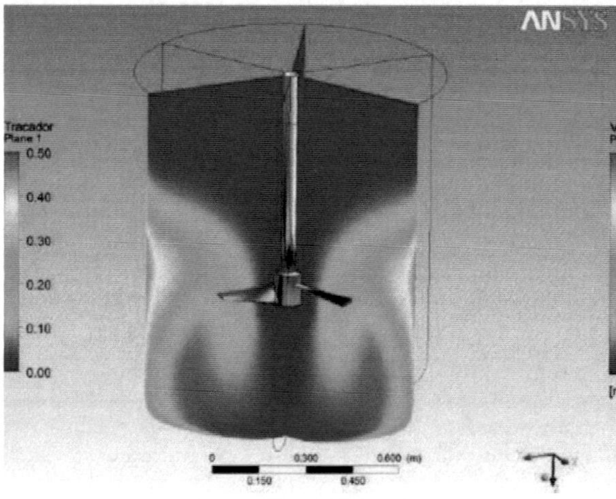

Solid distribution (b)

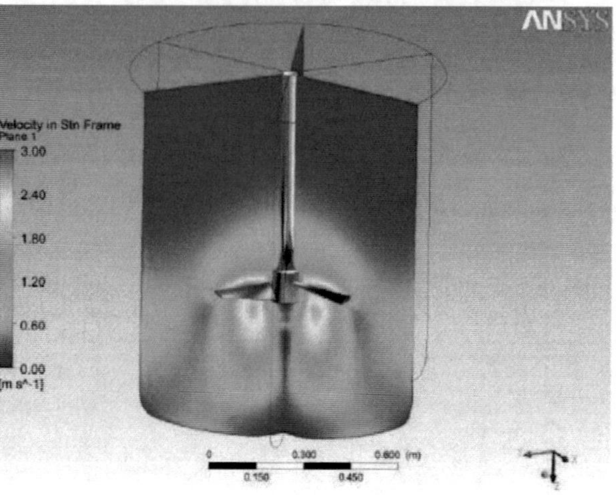

Velocity contour plot (d)

CFD AND OPTIMIZATION

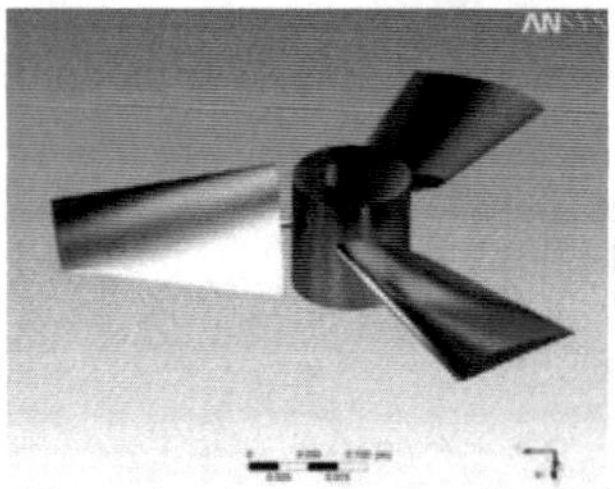

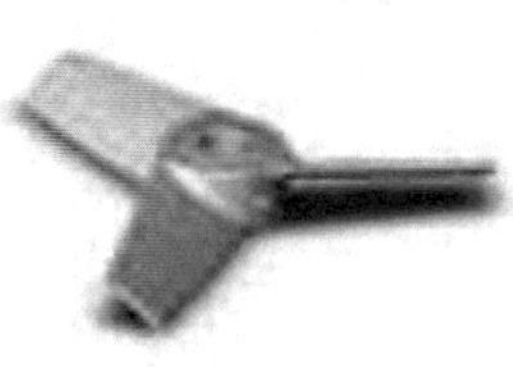

Prototype L-CFD	HE-3 Chemineer	Lightnin A310	PBT 45°
$N_P = 0,272$	$N_P = 0,22$	$N_P = 0,3$	$N_P = 1,27$
$N_Q = 0,675$	$N_Q = 0,47$	$N_Q = 0,56$	$N_Q = 0,79$
$\xi = 2,48$	$\xi = 2,14$	$\xi = 1,87$	$\xi = 0,62$

$$\xi = N_q/N_p$$

17AIChE
Annual Meeting, Minneapolis, MN

CFD AND OPTIMIZATION

UNICAMP

- The methodology provided arrived at very efficient impeller configurations;
- The impeller obtained with the further restriction of $N_p > 0.2$ seems to be a promising impeller. We however learnt that it would be ideal to have someone with industrial experience to develop an impeller that is stable and appropriate for industrial applications;
- If there is any impeller manufacturer willing to work along with us to obtain a new impeller, please contact me after this session.

CFD Role in understanding Mixing Processes

Nusselt Equation using a CFD model

Ronald Jaimes Prada, Ph.D.

Maira Puridade Marques da Silva

Silvia Araújo Daza

José Roberto Nunhez

UNICAMP

17AIChE Annual Meeting, Minneapolis, MN

NUSSELT EQUATION USING CFD

UNICAMP

A CFD model based on a detailed estimation of the temperature profile at the boundary layer has been developed to obtain an estimation of the heat transfer coefficient and the Nusselt Equation.

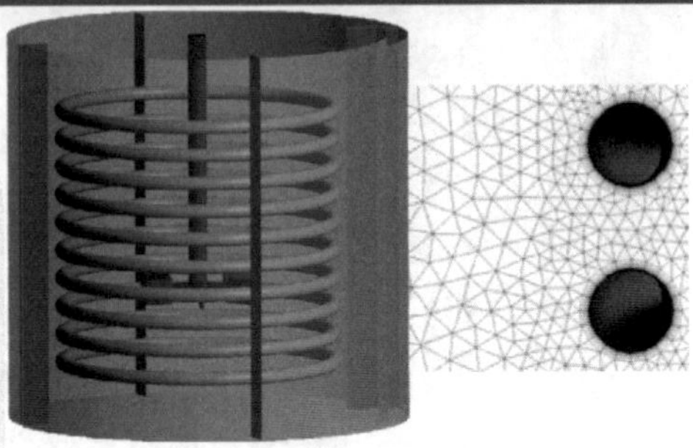

Experimental Correlation by Oldshue and Gretton

$$Nu = 0.17(Re)^{\frac{2}{3}}(Pr)^{\frac{1}{3}}\left(\frac{\mu}{\mu_W}\right)^{0.14}\left(\frac{D}{T}\right)^{0.1}\left(\frac{d}{T}\right)^{0.5}$$

CFD Correlation

$$Nu = ?$$

INTRODUCTION | MIXING AND CFD | EXAMPLES | CFD AND OPTIMIZATION | NUSSELT EQTS | CONCLUSION | 83

17AIChE Annual Meeting, Minneapolis, MN

NUSSELT EQUATION USING CFD

UNICAMP

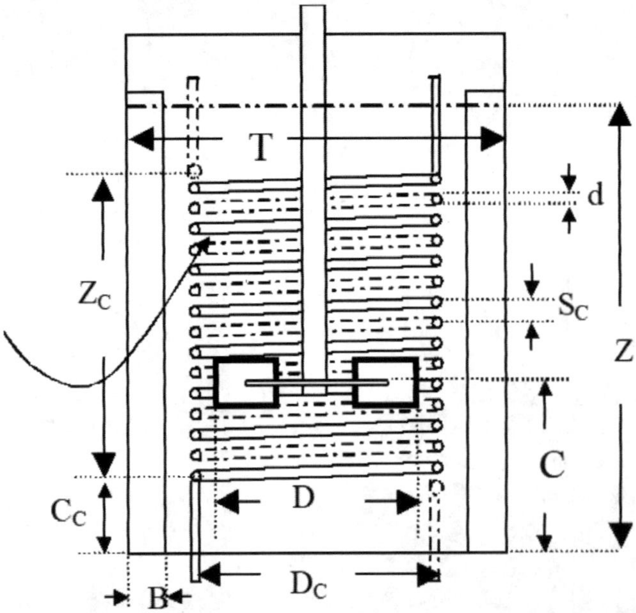

Helical coil arrangement suggested by Oldshue and Gretton

INTRODUCTION | MIXING AND CFD | EXAMPLES | CFD AND OPTIMIZATION | NUSSELT EQTS | CONCLUSION | 84

NUSSELT EQUATION USING CFD

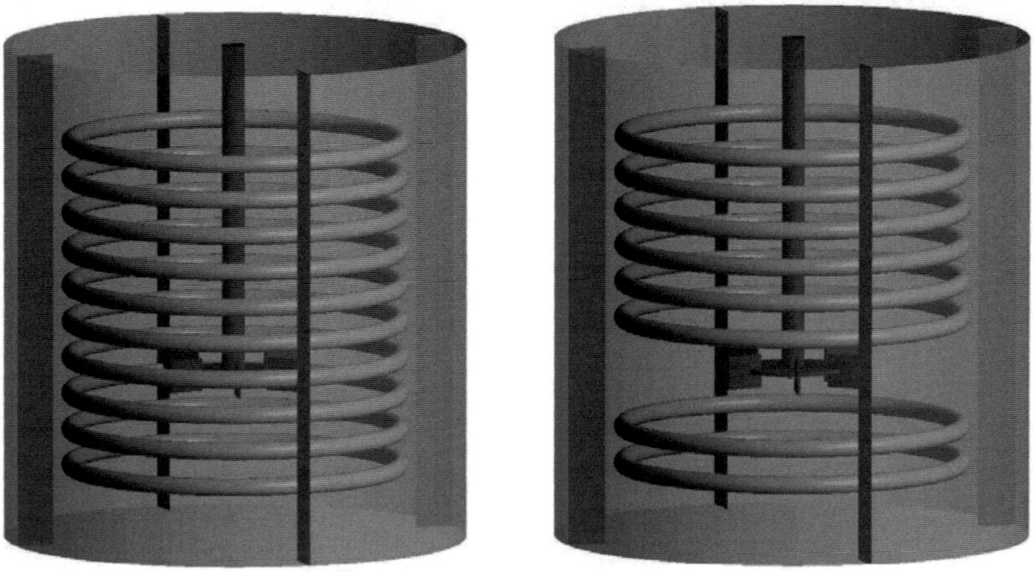

Geometry by Oldshue and Gretton and arrangement suggested by Street and McGreavy and Nunhez and McGreavy

NUSSELT EQUATION USING CFD

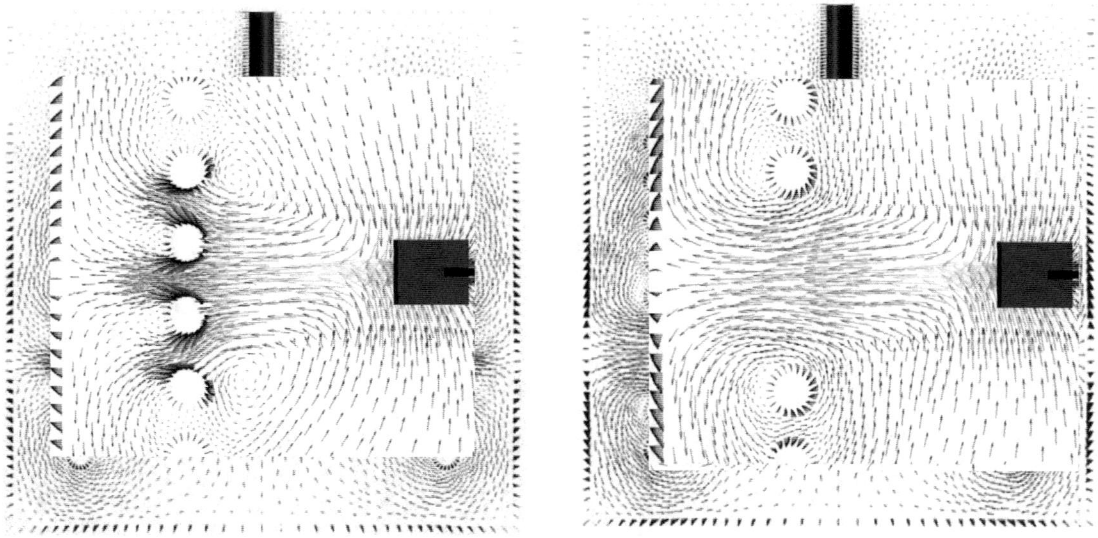

NUSSELT EQUATION USING CFD

Newton´s law of cooling

$$h_o = \frac{q}{(T_w - T_f)}$$

T_f is average bulk temperature

At the very laminar portion of the boundary layer

$$-\boldsymbol{k}_f \frac{\partial T}{\partial y}\bigg|_{y=0} = h_o(T_w - T_f)$$

$$h_o = \frac{-k_f \frac{\partial T}{\partial y}\Big|_{y=0}}{(T_w - T_f)}$$

17AIChE Annual Meeting, Minneapolis, MN

NUSSELT EQUATION USING CFD

UNICAMP

Is CFD able to rightly capture the heat and momentum changes inside the boundary layer to correctly predict the heat transfer coefficient?

17AiChE Annual Meeting, Minneapolis, MN

NUSSELT EQUATION USING CFD

UNICAMP

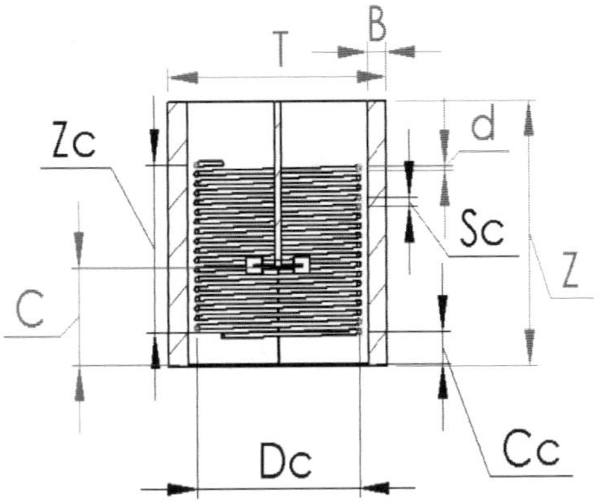

Figure. Geometry Oldshue e Gretton (1954)

Experimental equipment

- *Tank:* 48-in all sizes
- *4 Baffles*
- *Turbine type Rushton*
  12-in; 16-in; 20-in; 24-in e 28-in
- *Flow regime*
  R_e: 400-1,5 X10^6
- *Coils*

Table. Coil dimensions

Tube	d (in)	D_c(in)	D_o(in)	Z_c(in)	C_c(in)	S_c(in)
Cooper	7/8	35.125	36	31.5	7	7/4
Stainless steel	7/4	34.25	36	31.5	7	7/2;7

17AiChE Annual Meeting, Minneapolis, MN

NUSSELT EQUATION USING CFD

UNICAMP

Tetrahedral Mesh (Meshing, Ansys CFX 14.0)

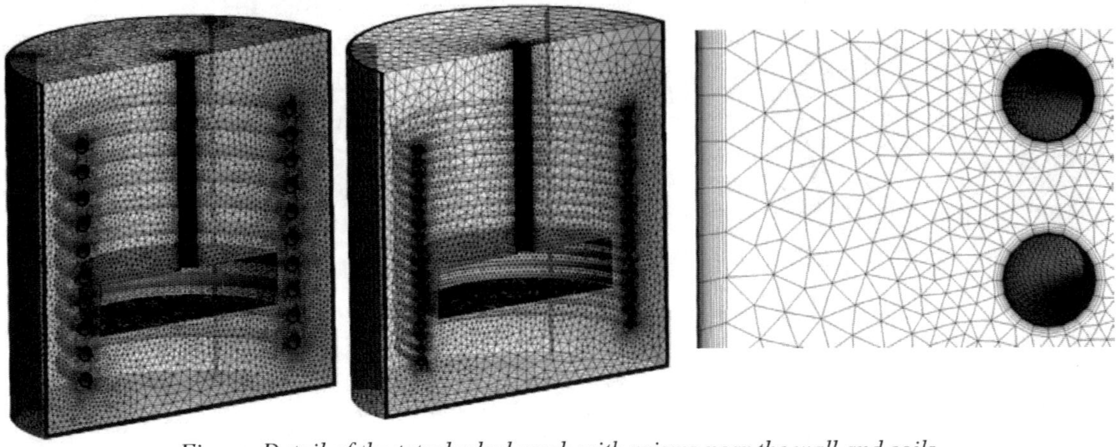

Figure. Detail of the tetrahedral mesh with prisms near the wall and coils

Number of elements and control volumes in the stationary domain 5,0 X10^6 and 9,0 X10^6 approx.
0,1 < y^+ <1 and 20 prisms in the boundary layer

17AIChE Annual Meeting, Minneapolis, MN

NUSSELT EQUATION USING CFD

UNICAMP

Underwater pipeline (riser) Transverse force coefficient
(70 elements in the boundary layer)

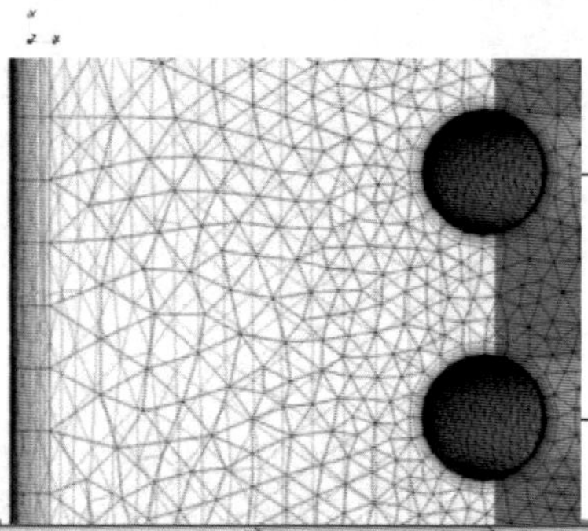

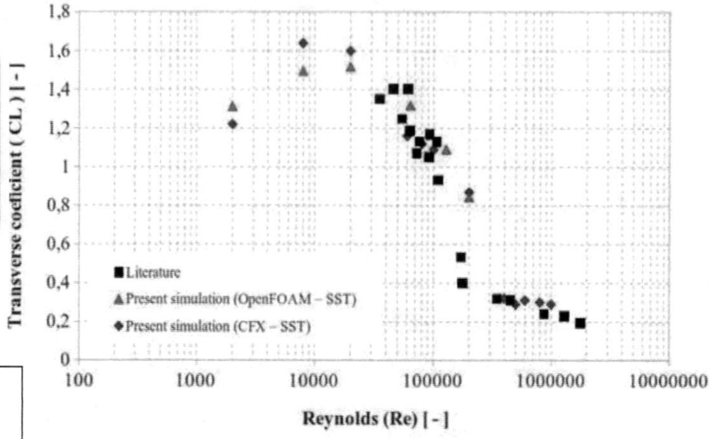

$y^+ < 0.1$ and 20 prisms in the boundary layer of the coil

17AIChE Annual Meeting, Minneapolis, MN

NUSSELT EQUATION USING CFD

UNICAMP

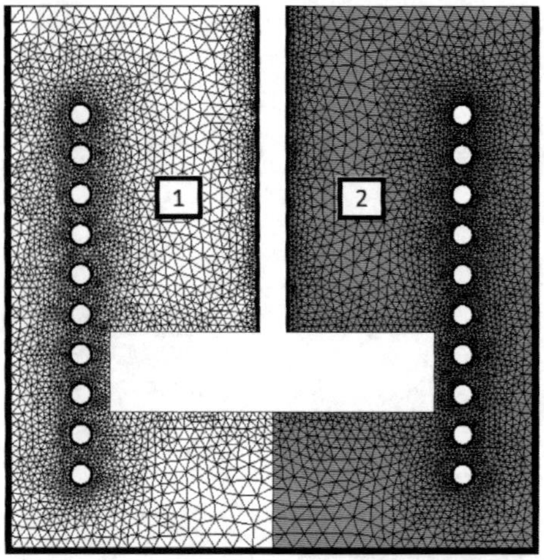

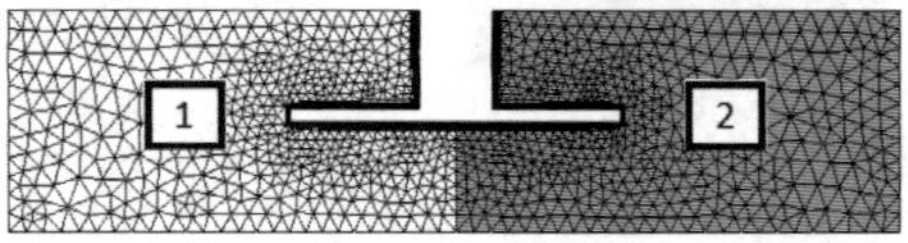

Boundary Conditions	
Wall and Bottom	No Slip Wall, adiabatic
Baffles	No Slip Wall, adiabatic
Coils	No Slip Wall Alternate coils were used for heating (constant flux = 160,000 W/m2) and cooling (constant temperature = 5 °C)
Free surface	Free Slip Wall
Impeller and shaft	MFR (No Slip Wall)
Interface	Interface fluid-fluid General Grid Interface – GGI Rotational Periodicity

Physical properties of the fluid

Fluid flow	Phase	Temperature	Pressure
Single Phase	Water and other fluids	25 °C	1 atm

17AIChE Annual Meeting, Minneapolis, MN

NUSSELT EQUATION USING CFD

UNICAMP

Levels of tested mesh

Parameters	Mesh levels				
	1	2	3	4	5
Average value y^+	25	6	0,6	0,1	0,06
Elements of the tank	1,1 x 10^6	2,9 x 10^6	4,8 x 10^6	4,8 x 10^6	5,0 x 10^6
h_o [W/m^{2}K]	2975	3910	4673	4760	4747
% Variation of h_o	-	31,26	19,36	1,72	0,15
Q [W/m2]	2,2 x 10^5	2,9 x 10^5	3,5 x 10^5	3,6 x 10^5	3,5 x 10^5
% Variation of Q	-	31,44	19,52	1,86	0,27

Mesh Independencia criteria **OK**

Operating conditions for the mesh independence test

Fluid	Water
N (rpm)	200
D (in)	16
Re	6.2 x 10^5
d (in)	1 3/4

INTRODUCTION | MIXING AND CFD | EXAMPLES | CFD AND OPTIMIZATION | NUSSELT EQTS | CONCLUSION | 93

17AIChE Annual Meeting, Minneapolis, MN

NUSSELT EQUATION USING CFD

UNICAMP

Effect of y^+ on the heat transfer coefficient (h_o)

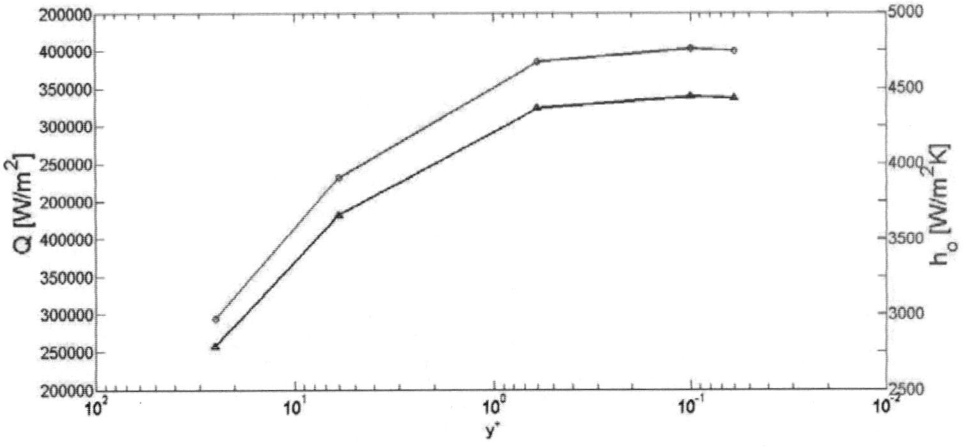

Figure. Variation of the estimate of h_0 and Q as a function of y^+

INTRODUCTION | MIXING AND CFD | EXAMPLES | CFD AND OPTIMIZATION | NUSSELT EQTS | CONCLUSION | 94

17AIChE Annual Meeting, Minneapolis, MN

NUSSELT EQUATION USING CFD

UNICAMP

Experimental work by Oldshue and Gretton

Number of runs 106

This CFD simulation work

Number of runs 24

17AIChE Annual Meeting, Minneapolis, MN

NUSSELT EQUATION USING CFD

UNICAMP

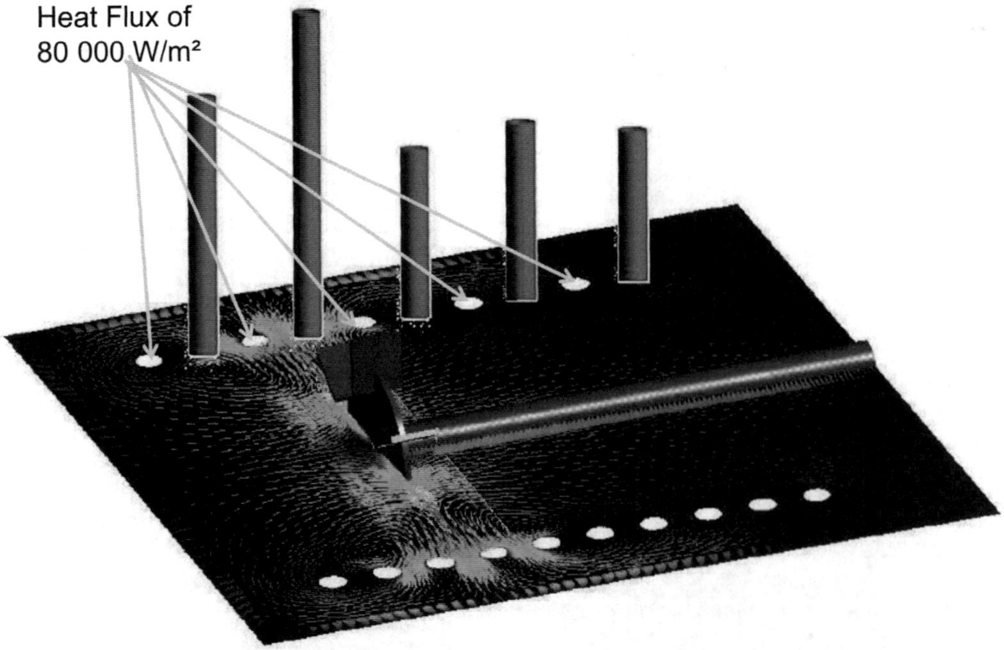

INTRODUCTION | MIXING AND CFD | EXAMPLES | CFD AND OPTIMIZATION | NUSSELT EQTS | CONCLUSION | 96

17AIChE Annual Meeting, Minneapolis, MN

NUSSELT EQUATION USING CFD

UNICAMP

Table. Results obtained by the CFD model

Simulated Case Nº	Fluid	Impeller diameter D (in)	Coil diameter d (in)	Tank Diameter T (in)	Impeller Velocity N (rpm)	Re	Pr	(D/T)	(d/T)	Nu
1	Water	16	1,75	48	100	66684	26,5	0,333	0,036	190
2	Water	16	1,75	48	200	133368	26,5	0,333	0,036	318
3	Water	16	1,75	48	300	200052	26,5	0,333	0,036	433
4	Glycerine	16	1,75	48	100	1731	1671,6	0,333	0,036	83
5	Glycerine	16	1,75	48	200	3462	1671,6	0,333	0,036	128
6	Glycerine	16	1,75	48	300	5194	1671,6	0,333	0,036	159
7	Mineral Oill	16	1,75	48	100	5065	588,2	0,333	0,036	129
8	Mineral Oil	16	1,75	48	200	10130	588,2	0,333	0,036	176
9	Mineral Oil	16	1,75	48	300	15195	588,2	0,333	0,036	205
10	Water	12	1,75	48	200	75020	26,5	0,250	0,036	186
11	Water	20	1,75	48	200	208388	26,5	0,417	0,036	480
12	Water	24	1,75	48	200	300078	26,5	0,500	0,036	669
13	Water	28	1,75	48	200	408440	26,5	0,583	0,036	869
14	Glycerine	12	1,75	48	200	1948	1671,6	0,250	0,036	107
15	Glycerine	20	1,75	48	200	5410	1671,6	0,417	0,036	182
16	Glycerine	24	1,75	48	200	7791	1671,6	0,500	0,036	222
17	Glycerine	28	1,75	48	200	10604	1671,6	0,583	0,036	263
18	Mineral Oil	12	1,75	48	200	5698	588,2	0,250	0,036	151
19	Mineral Oil	20	1,75	48	200	15828	588,2	0,417	0,036	282
20	Mineral Oil	24	1,75	48	200	22792	588,2	0,500	0,036	334
21	Mineral Oil	28	1,75	48	200	31023	588,2	0,583	0,036	388
22	Water	16	0,875	48	200	133368	26,5	0,333	0,018	197
23	Glycerine	16	0,875	48	200	3462	1671,6	0,333	0,018	86
24	Mineral Oil	16	0,875	48	200	10130	588,2	0,333	0,018	152

17AIChE Annual Meeting, Minneapolis, MN

NUSSELT EQUATION USING CFD

UNICAMP

Case Nº	Fluid	D (in)	d (in)	T (in)	N (rpm)	Re	Pr	(D/T)	(d/T)	Nu
1	Water	16	1,75	48	100	66684	26,5	0,333	0,036	190
2	Water	16	1,75	48	200	133368	26,5	0,333	0,036	318
3	Water	16	1,75	48	300	200052	26,5	0,333	0,036	433
4	Glycerine	16	1,75	48	100	1731	1671,6	0,333	0,036	83
5	Glycerine	16	1,75	48	200	3462	1671,6	0,333	0,036	128
6	Glycerine	16	1,75	48	300	5194	1671,6	0,333	0,036	159
7	M. Oil	16	1,75	48	100	5065	588,2	0,333	0,036	129
8	M. Oil	16	1,75	48	200	10130	588,2	0,333	0,036	176
9	M. Oil	16	1,75	48	300	15195	588,2	0,333	0,036	205

Case Nº	Fluid	D (in)	d (in)	T (in)	N (rpm)	Re	Pr	(D/T)	(d/T)	Nu
22	Water	16	0,875	48	200	133368	26,5	0,333	0,018	197
23	Glycerine	16	0,875	48	200	3462	1671,6	0,333	0,018	86
24	M. Oil	16	0,875	48	200	10130	588,2	0,333	0,018	152

17AIChE Annual Meeting, Minneapolis, MN

NUSSELT EQUATION USING CFD

UNICAMP

Case Nº	Fluid	D (in)	d (in)	T (in)	N (rpm)	Re	Pr	(D/T)	(d/T)	Nu
10	Water	12	1,75	48	200	75020	26,5	0,250	0,036	186
11	Water	20	1,75	48	200	208388	26,5	0,417	0,036	480
12	Water	24	1,75	48	200	300078	26,5	0,500	0,036	669
13	Water	28	1,75	48	200	408440	26,5	0,583	0,036	869
14	Glycerine	12	1,75	48	200	1948	1671,6	0,250	0,036	107
15	Glycerine	20	1,75	48	200	5410	1671,6	0,417	0,036	182
16	Glycerine	24	1,75	48	200	7791	1671,6	0,500	0,036	222
17	Glycerine	28	1,75	48	200	10604	1671,6	0,583	0,036	263
18	M. Oil	12	1,75	48	200	5698	588,2	0,250	0,036	151
19	M. Oil	20	1,75	48	200	15828	588,2	0,417	0,036	282
20	M. Oil	24	1,75	48	200	22792	588,2	0,500	0,036	334
21	M. Oil	28	1,75	48	200	31023	588,2	0,583	0,036	388

17AIChE Annual Meeting, Minneapolis, MN

NUSSELT EQUATION USING CFD

UNICAMP

Parameter settings: Generalized Reduced Gradient Method (GRG Nonlinear)

Average deviation
10,7

$$Nu = 0{,}32\left(\frac{ND^2\rho}{\mu}\right)^{0{,}67}\left(\frac{C_p\mu}{k}\right)^{0{,}37}\left(\frac{D}{T}\right)^{0{,}32}\left(\frac{d}{T}\right)^{0{,}59}$$

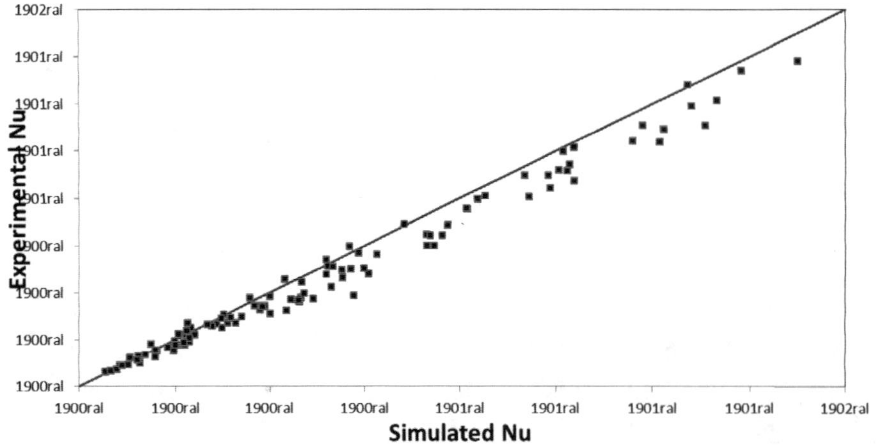

Figure. Dispersion of results in the calculation of the Nu by the CFD correlation

17AIChE Annual Meeting, Minneapolis, MN

NUSSELT EQUATION USING CFD

UNICAMP

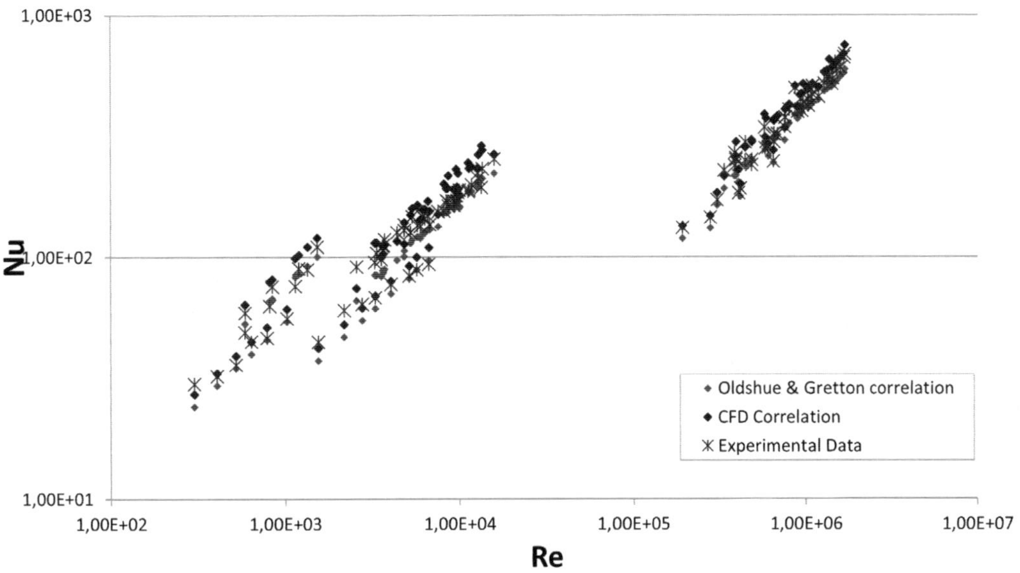

Figure. Nu vs. Re for the predicted data by the experimental correlations and CFD

17AIChE Annual Meeting, Minneapolis, MN

NUSSELT EQUATION USING CFD

UNICAMP

CONCLUSIONS OF THIS WORK

- The accuracy of the computational results for heat transfer depends strongly on the mesh refinement near the heat transfer surface (helical coil);
- The main contribution of this work is the development of a model able to predict the heat transfer in a tank jacket with a view to obtain a Nusselt equation;
- The Nusselt number obtained in the model was compared with the experimental work of Oldshue and Gretton showing an average deviation of 10%.

17AIChE Annual Meeting, Minneapolis, MN

NUSSELT EQUATION USING CFD

UNICAMP

FURTHER DEVELOPMENTS OF THIS WORK

- The geometry proposed by Street and McGreacy and Nunhez and McGreavy were compared in terms of heat transfer;
- A Nusselt equation for a system with internal helical coils and two Rushton impellers (used in industries) was proposed;
- A Nusselt equation for a jacketed tank reactor was obtained and results were compared with the experimental data of Strek (1963);
- A similar, but modified methodology is being developed for half-pipe coil jackets.

17AIChE Annual Meeting, Minneapolis, MN

NUSSELT EQUATION USING CFD

UNICAMP

$$Nu = 0.17(Re)^{\frac{2}{3}}(Pr)^{\frac{1}{3}}\left(\frac{D}{T}\right)^{0.1}\left(\frac{d}{T}\right)^{0.5}$$

$$Nu = ?$$

Experimental geometry used by Oldshue & Gretton

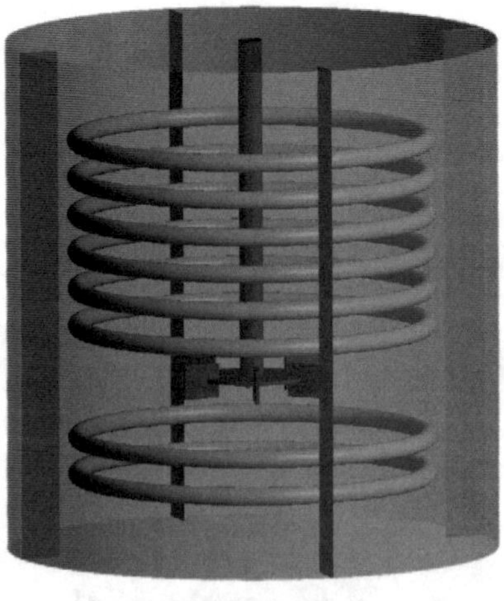

Proposed geometry

17AIChE Annual Meeting, Minneapolis, MN

NUSSELT EQUATION USING CFD

UNICAMP

Heat Flux of
80 000 W/m²

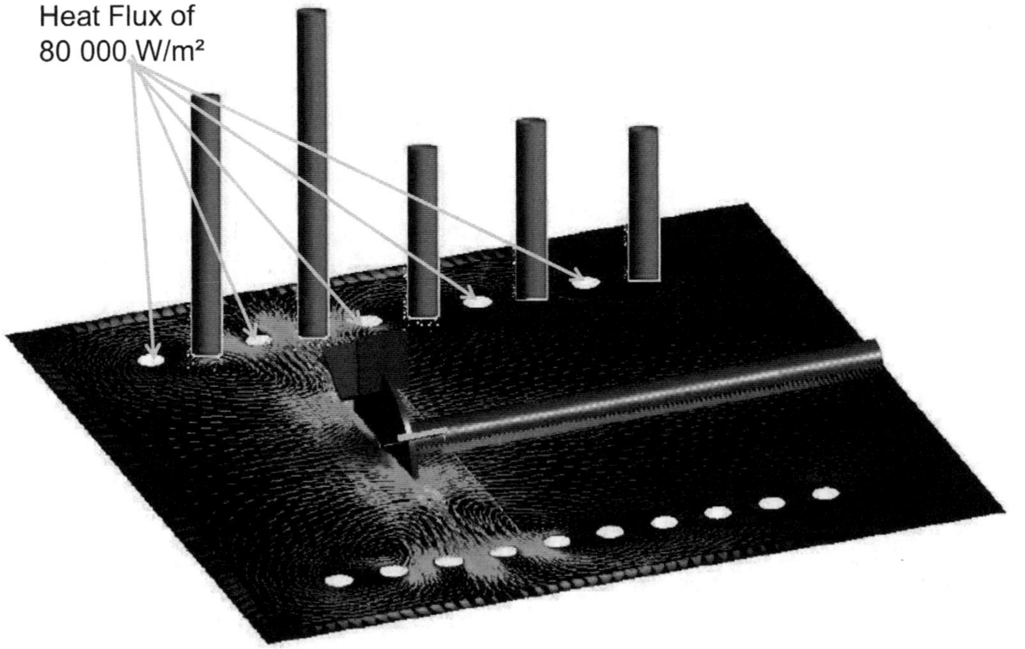

INTRODUCTION > MIXING AND CFD > EXAMPLES > CFD AND OPTIMIZATION > NUSSELT EQTS > CONCLUSION > 105

17AIChE Annual Meeting, Minneapolis, MN

NUSSELT EQUATION USING CFD

UNICAMP

Heat Flux of
80 000 W/m²

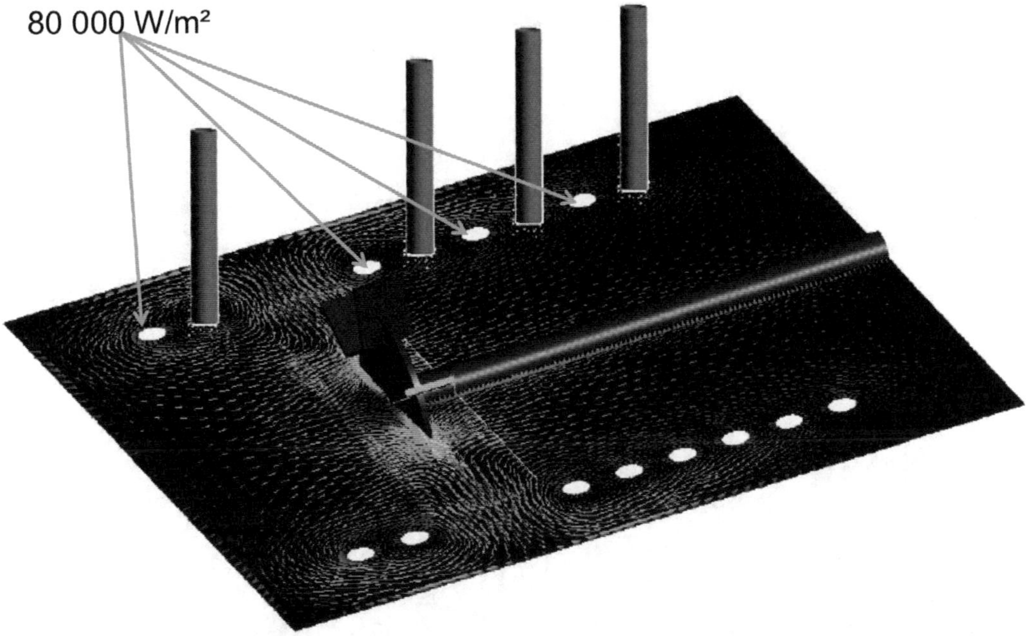

INTRODUCTION > MIXING AND CFD > EXAMPLES > CFD AND OPTIMIZATION > NUSSELT EQTS > CONCLUSION > 106

17AIChE Annual Meeting, Minneapolis, MN

NUSSELT EQUATION USING CFD

UNICAMP

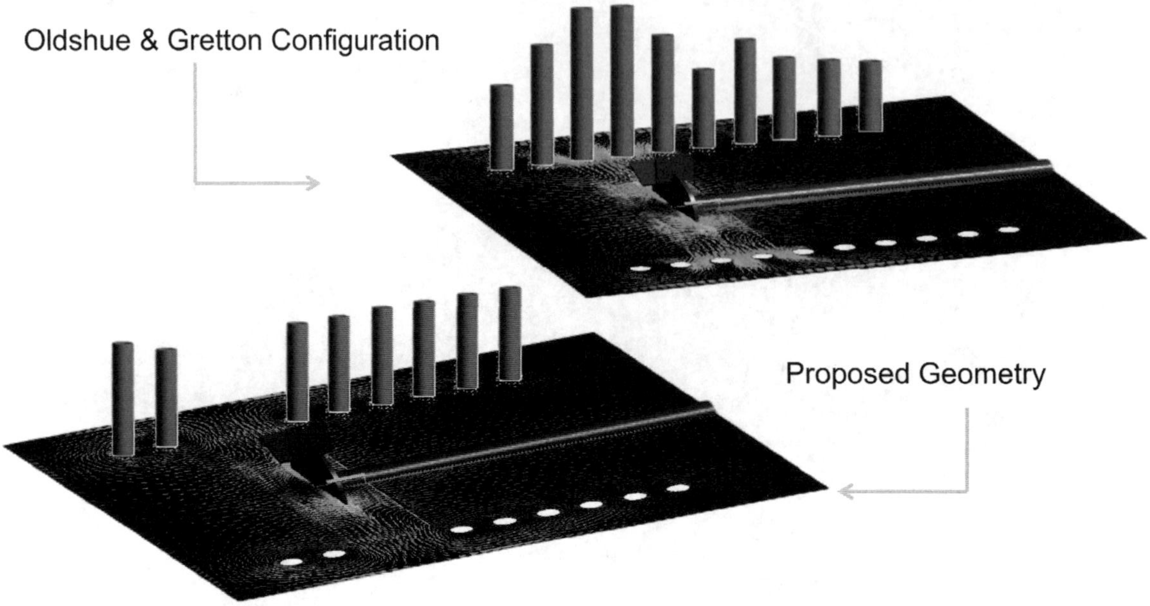

Figure. Heat flow Distribution

17AIChE Annual Meeting, Minneapolis, MN

NUSSELT EQUATION USING CFD

UNICAMP

Oldshue & Gretton geometry (Experimental)

$$Nu = 0.17(Re)^{\frac{2}{3}}(Pr)^{\frac{1}{3}}\left(\frac{D}{T}\right)^{0.1}\left(\frac{d}{T}\right)^{0.5}$$

Oldshue & Gretton geometry (Simulated)

$$Nu = 0.36(Re)^{\frac{2}{3}}(Pr)^{\frac{1}{3}}\left(\frac{D}{T}\right)^{0.3}\left(\frac{d}{T}\right)^{0.6}$$

Proposed geometry (Simulated)

$$Nu = 0.54(Re)^{\frac{2}{3}}(Pr)^{\frac{1}{3}}\left(\frac{D}{T}\right)^{0.21}\left(\frac{d}{T}\right)^{0.78}$$

17AIChE Annual Meeting, Minneapolis, MN

NUSSELT EQUATION USING CFD

UNICAMP

- The proposed geometry provides better fluid circulation in the regions away from the impeller and also in the region between coils and wall. Moreover, the heat flow distribution in the proposed geometry is more uniform. However, the amount of heat removed from the coil in front of the impeller is very high in the original design by Oldshue and Gretton and this geometry is superior in terms of heat removal;

- (Jim was right);

- This work shows that new geometries can be tested by numerical prototypes in terms of heat transfer before being tested experimentally (if needed).

17AIChE Annual Meeting, Minneapolis, MN

NUSSELT EQUATION USING CFD

UNICAMP

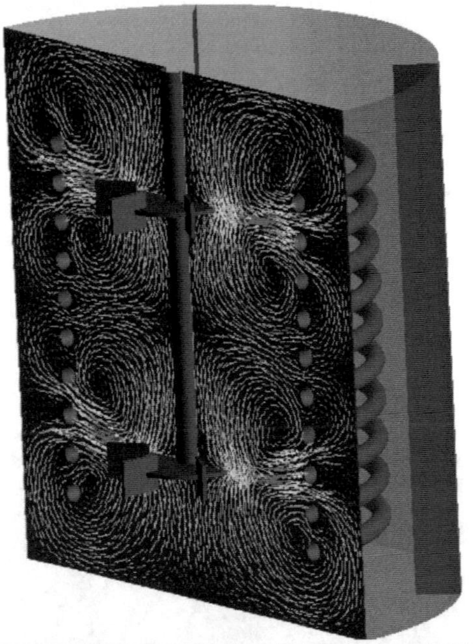

NUSSELT EQUATION USING CFD

- A much more developed velocity distribution is obtained when two Rushton impellers are used

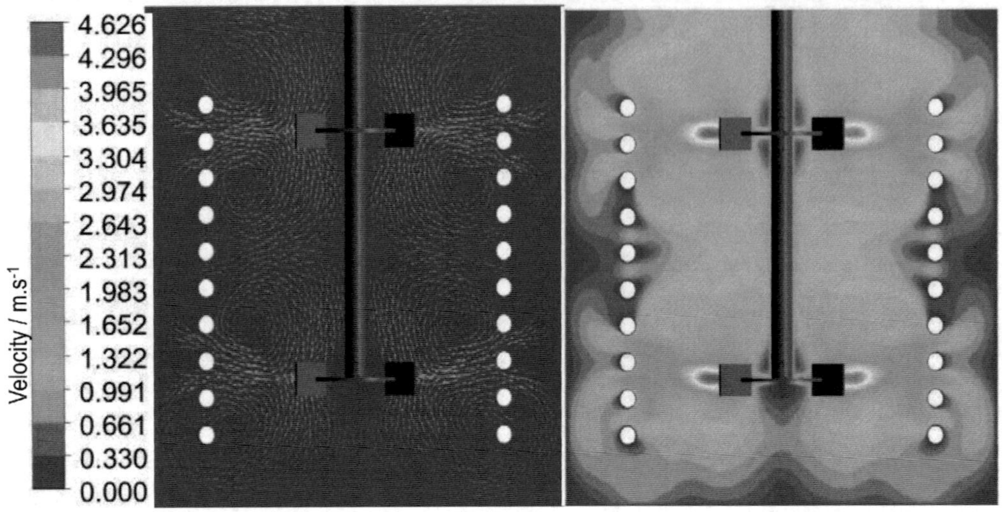

Fluid – glycerine and impeller speed of 300 rpm

NUSSELT EQUATION USING CFD

- Influence of impeller speed on heat flux

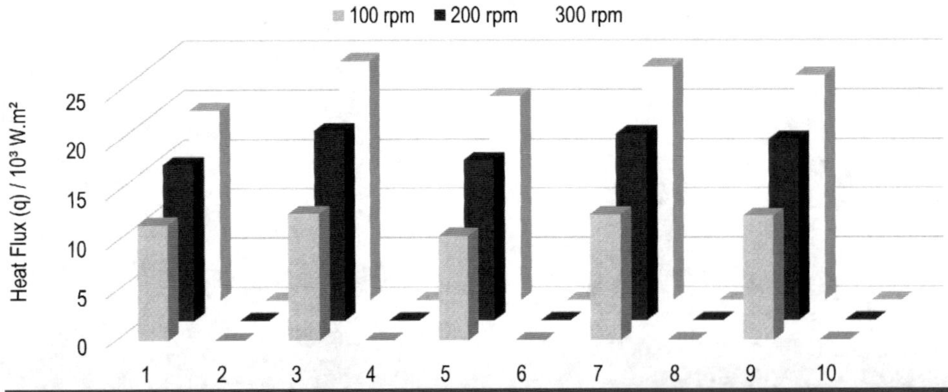

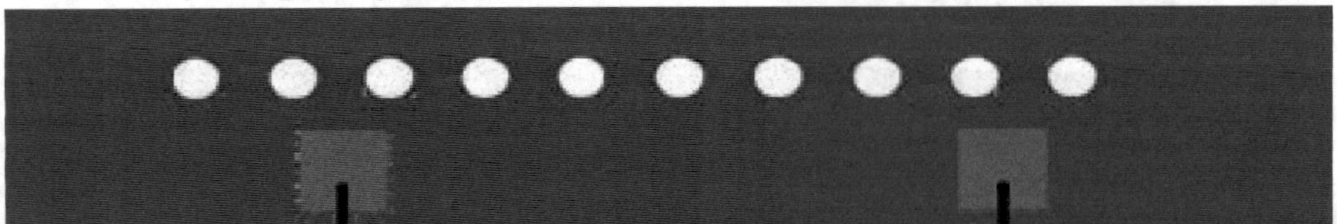

17AIChE Annual Meeting, Minneapolis, MN

NUSSELT EQUATION USING CFD

UNICAMP

Nusselt number correlation

$$Nu = 0{,}56(Re)^{0{,}575}(Pr)^{0{,}354}\left(\frac{D}{T}\right)^{0{,}65}\left(\frac{d}{T}\right)^{0{,}277}$$

INTRODUCTION | MIXING AND CFD | EXAMPLES | CFD AND OPTIMIZATION | NUSSELT EQTS | CONCLUSION | 113

17AIChE Annual Meeting, Minneapolis, MN

NUSSELT EQUATION USING CFD

UNICAMP

- The authors intend to publish this work soon;
- No experimental results were found in the literature for helical coils with two Rushton impellers.
- We would be more than happy to share our results with impeller manufacturers that probably have experimental data on this tank arrangement. We could even publish results together

INTRODUCTION | MIXING AND CFD | EXAMPLES | CFD AND OPTIMIZATION | NUSSELT EQTS | CONCLUSION | 114

17AIChE
Annual Meeting, Minneapolis, MN

NUSSELT EQUATION USING CFD

UNICAMP

Nusselt number correlation from simulated data for a jacketed vessel

$$\mathrm{Nu} = \frac{h_o T}{\kappa} = K\left(\frac{D^2 N}{\mu}\right)^a \left(\frac{C_p \mu}{\kappa}\right)^b \left(\frac{D}{T}\right)^c \left(\frac{C}{T}\right)^d$$

INTRODUCTION > MIXING AND CFD > EXAMPLES > CFD AND OPTIMIZATION > NUSSELT EQTS > CONCLUSION > 115

17AIChE
Annual Meeting, Minneapolis, MN

NUSSELT EQUATION USING CFD

UNICAMP

Experimental work by Strek

Number of runs 85

This CFD simulation work

Number of runs 15

INTRODUCTION > MIXING AND CFD > EXAMPLES > CFD AND OPTIMIZATION > NUSSELT EQTS > CONCLUSION > 116

17AIChE
Annual Meeting, Minneapolis, MN

NUSSELT EQUATION USING CFD

UNICAMP

Nusselt number correlation from simulated data - ***This work***

$$Nu = \frac{h_o T}{\kappa} = 1.22\left(\frac{D^2 N}{\mu}\right)^{0.651}\left(\frac{C_p \mu}{\kappa}\right)^{0.333}\left(\frac{D}{T}\right)^{0.116}\left(\frac{C}{T}\right)^{0.157}$$

Nusselt number correlation from experimental data - ***Strek 1963***

$$Nu = \frac{h_o T}{\kappa} = 1.02\left(\frac{D^2 N}{\mu}\right)^{0.66}\left(\frac{C_p \mu}{\kappa}\right)^{0.33}\left(\frac{D}{T}\right)^{0.13}\left(\frac{D}{T}\right)^{0.12}$$

INTRODUCTION | MIXING AND CFD | EXAMPLES | CFD AND OPTIMIZATION | NUSSELT EQTS | CONCLUSION | 117

17AIChE
Annual Meeting, Minneapolis, MN

NUSSELT EQUATION USING CFD

UNICAMP

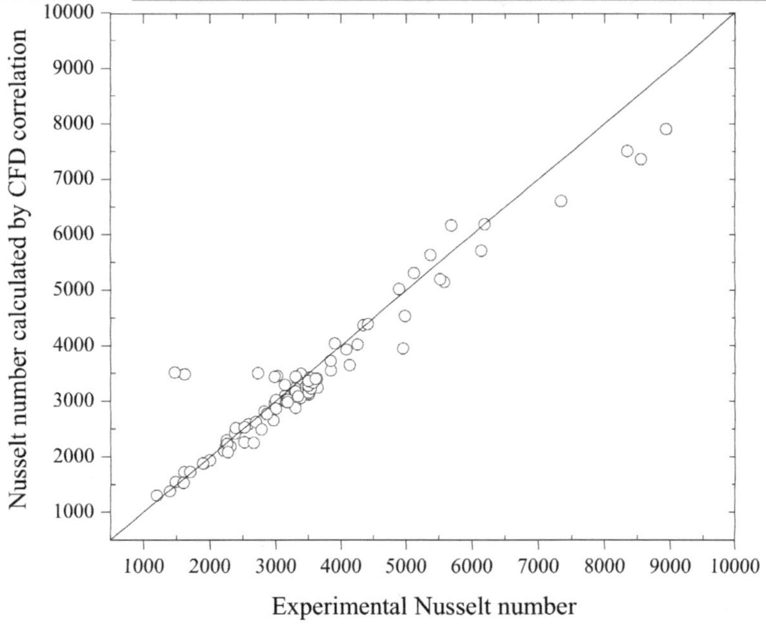

Average deviation: 7.5 %

Standard deviation: 11.8%

Comparison between experimental data and CFD Nusselt number correlation

INTRODUCTION | MIXING AND CFD | EXAMPLES | CFD AND OPTIMIZATION | NUSSELT EQTS 118 | CONCLUSION | 118

17AIChE
Annual Meeting, Minneapolis, MN

CONCLUSIONS

UNICAMP

- CFD has proved to be an invaluable tool for a better understanding of mixing;
- Some examples were given showing how CFD has helped to improve existing mixing processes;
- A different approach in which a real optimzation method is coupled to a CFD model was successfully implemented and should be explored to develop new impellers and processes.

17AIChE
Annual Meeting, Minneapolis, MN

CONCLUSIONS

UNICAMP

- CFD has proved to be an invaluable tool for a better understanding of mixing;
- Some examples were given showing how CFD has helped to improve existing mixing processes;
- A different approach in which a real optimzation method is coupled to a CFD model was successfully implemented and should be more explored to develop new impellers and processes

17AIChE Annual Meeting, Minneapolis, MN

CONCLUSIONS

UNICAMP

- Another different approach was developed to predict the heat transfer coefficient (and hence a Nusselt Equation);
- Nusselt equations were predicted without experiments.

17AIChE Annual Meeting, Minneapolis, MN

ACKNOWLEDGEMENTS

UNICAMP

❖ I thank God for everything I am and have;

❖ I thank the help of my wife Silmara, who always supported me in all my academic endeavors;

❖ I thank all the students and researches involved in all my research works. Without them, I would not be here;

❖ I thank ABEQ, IACChE and AIChE for the indication of my name for this distinction;

❖ I thank all sponsoring agencies and companies who helped supporting financially my research.

L-CFD – Laboratório de Fluidodinâmica Computacional

José Roberto Nunhez
Professor of Chemical Engineering
FEQ- UNICAMP
Tel (019) 35213967
E-mail: nunhez@feq.unicamp.br

UNICAMP

Using a Structured Approach to Efficiently Use the Brains of Other to Make Problem Solving More Productive

Jack Hipple, Principal

TRIZ and Engineering Training Services LLC

Tampa, FL

813-994-9999

www.innovation-triz.com

In our attempts to problem solve and produce novel and breakthrough ideas, we frequently rely on psychological stimulation techniques and processes. These are effective for simple problems but only make use of the knowledge base of the participants. They may be "fun" techniques, but they do not take advantage of the global knowledge base available to us.

Around 1990, a powerful "left brained" problem solving approach, long practiced in the former Soviet Union, came to the US and is now used by many Fortune 500 companies to deal with their most difficult engineering and design challenges. This process and technology is known as "TRIZ", a Russian acronym for "Theory of Inventive Problem Solving". It uses the data base of breakthrough patents, and the methods and principles used in them, to provide a "left brained" algorithm useful for problem solving. It, in effect, brings the other inventors of the world into your problem solving session and avoids the barriers to problem solving such as ego, acronyms, and fancy language that reinforces our belief that the problem we have is special and unique.

The TRIZ problem solving algorithm, sometimes known as ARIZ (Algorithm for Inventive Problem Solving), contains several key aspects. First is the very basic concept of parallel universes of invention and using the most basic of terminology to search for potential solutions in parallel universes not previously seen, due to the use of company or industry specific language. For example a search for information about fraud would produce far fewer ideas than one based on the term, "substitution of one thing or another". Second, we envision and describe, in several different ways, the Ideal Final Result—without worrying about how we will get there. This is quite different than the normal engineering approach of looking for a 10-20% improvement. Then we look for available resources that our system and its sub-system and super-system might have available to achieve this "IFR". The TRIZ list of resources is far broader than the traditional chemical engineering look at material resources. We look at time (before, during, after), fields, field conversion opportunities, and the negative aspects of our system. This latter aspect turns out to be a surprisingly powerful tool for problem solving.

These basic steps are often sufficient to solve a problem, but often there are contradictions that need resolved. The resolution of contradictions is the key to most of the top 5% of unique patents, and analysis of these patents shows a limited number of inventive principles that are repeatedly used in these patents. TRIZ organizes these inventive principles in a useful and retrievable way, allowing their use in other problems with major contradictions that need resolved.

This workshop will overview this process and show by many examples where it has been applied.

487650 AIChE Engage: Your Next Stop for Brainstorming in the Process of Problem Solving or Innovating

Abstract:

Being an AIChE member provides you the opportunity to easily connect and build relationships with other chemical engineers in the world even if you donâ€™t have the time or ability to meet them in person. The newest benefit of membership, AIChE Engage, makes doing this even easier and more interactive than before. AIChE Engage will be your next stop for brainstorming in the process of problem solving or innovating.Â You can get two benefits through AIChE Engage:

Â 1.There is an extensive search engine which enables you to find other members by their names, companies, geographic locations, job functions and industries. There are also focused AIChE member directories of Fellows, Professionals, Young Professionals, Graduate Students, Undergraduate Students and members of the divisions, forums and committees. After youâ€™ve found the members you are looking for, you can send them an invitation to communicate via Engage or through another channel.

2.Perhaps the new feature of Engage that you will find most exciting is Engage Discussion Central. Professional members now have the opportunity to post and participate in discussions with colleagues all over the world on topics related to chemical engineering. Seeing the discussions is easy as they are sent directly to your e-mail address in a daily digest.

Therefore, AIChE Engage has become and will be a mobile-friendly platform for the global collaboration with your fellow members to get academic and professional accomplishments, career achievements and other distinguishing industrial breakthrough. This presentation will provide you the introduction on how to apply the methodology of TRIZ problem solving and strategic planning process with AIChE Engage.

508515 Re-situating the Professional Formation of Engineers at Oregon State University

Abstract:

Re-situating the Professional Formation of Engineers at Oregon State University

Jim Sweeney, Milo Koretsky, Susan Nolen, Michelle Bothwell, Susannah Davis, Devlin Montfort

While industry has long called for changes to how engineers and computer scientists are prepared, much of the core coursework remains virtually the same as it was 50 years ago. To address this mismatch between industry and education, the National Science Foundation has funded a set of projects that aim to *REvolutionize Engineering and Computer ScienceDepartments* (RED), serving as leaders in institutional change. We describe early results and ongoing work in a RED project in the School of Chemical, Biological, and Environmental Engineering (CBEE) at Oregon State University. We seek to re-situate the faculty/staff/student community and curricular/co-curricular student learning experience. Our work is centered on the creation of a holistic, inclusive, and professionally-based learning environment for our students. We are working to promote change through establishing a culture of inclusion and shifting student learning environments from highly sequestered activities to more realistic and consequential work that more closely resembles actual engineering workplace. These efforts require an important and concurrent shift in unit culture (e.g. values, norms, policies and procedures).

504781 ExxonMobil's Diversity and Inclusion Efforts – Creating a Premier Global Workforce that Works to Help Power the World's Progress

Abstract:

At ExxonMobil, the diversity and inclusion of thought, skill, knowledge and culture make ExxonMobil more competitive, more resilient and better able to navigate the complex and constantly changing global energy business. Our diversity and inclusion efforts are comprehensive. They range from attracting and developing a premier global workforce within a work environment that values unique perspectives, encourages individual growth and treats all employees with respect, to educational partnerships and community development initiatives designed to improve the quality of life where we live and work.

This presentation is a view of ExxonMobil's technology focus areas and the Diversity and Inclusion efforts being used to create a global workforce ready to address them.

504171 Engineering Amine-Modified Silicates for CO_2 Separations and Catalysis

Abstract:

Worldwide energy demand is projected to grow strongly in the coming decades, with most of the growth in developing countries. Even with unprecedented growth rates in the development of renewable energy technologies such as solar, wind and bioenergy, the world will continue to rely on fossil fuels as a predominant energy source for at least the next several decades. The Intergovernmental Panel on Climate Change (IPCC) has stated that anthropogenic CO2 has contributed measurably to climate change over the course of the last century. To this end, there is growing interest in new technologies that might allow continued use of fossil fuels without drastically increasing atmospheric CO2concentrations beyond currently projected levels. In this lecture, I will describe the design and synthesis, characterization and application of new aminosilica materials that we have developed as cornerstones of new technologies for the removal of CO2 from dilute gas streams. These chemisorbents efficiently remove CO2 from simulated flue gas streams, and the CO2 capacities are actually enhanced by the presence of water, unlike in the case of physisorbents such as zeolites. Interestingly, the heat of adsorption for these sorbents is sufficiently high that the sorbents are also capable of capturing CO2 from extremely dilute gas streams, such as the ambient air. Indeed, our oxide-supported amine adsorbents are quite efficient at the direct "air capture" of CO2 and we will describe our investigations into development of new "air capture" technologies as well. Finally, the amine-modified silica materials are known to be efficient catalysts in coupling reactions important in organic synthesis, such as aldol and nitroaldol condensations. Inspired by biological catalysts that make and break bonds using cooperative organocatalytic sites, chemocatalysts designed to promote cooperativity between amines and weakly acidic sites are shown to be highly effective catalysts. Catalyst design elements that strongly impact kinetic cooperativity will be presented.

Introduction to the Fundamentals of Project Management

Abstract ID: 481742
Paper # 330a
Dr. Eldon R. Larsen
Marshall University

In this paper we discuss several important concepts which will be helpful throughout the rest of the tutorial. First, we discuss the very important concept of *principles*. Then, we will seek to understand what we really mean when we say we have a project—in other words, "What is a project?" Finally, we will introduce the most important keys to successful project management, which will really be the foundation for much of the rest of the tutorial.

Principles: The Foundation

To begin, we need to understand the concept of what a *principle* is and how we can use that concept to strengthen our ability to manage projects. In our discussion of principles, the following definitions help us to understand some important ideas applicable to project management.

Tools. Tools are instruments, concepts, procedures, or techniques used to help us accomplish work, either physical or mental. A Gantt chart, to be discussed in a later paper, would be an example of a project management tool. Tools are not our masters; they are intended to be our helpers or servants. Tools can be applied in many ways and they can be misapplied in many ways. Tools need to be maintained properly in order for them to continue to help us effectively. We use tools in applications of rules, policies, and principles.

Rules. Rules are prescribed guides, procedures, or obligations for conduct or action and are sometimes binding by regulation or legal precept. An example project management rule might be that we don't have employees work overtime for more than two weeks straight without a two-day rest break before the next overtime period. Good rules are based upon some kind of principle(s), while bad rules probably have little or no foundation in principles. Rules occasionally need to be changed, especially as times and circumstances change. In our example overtime rule, a true emergency might require an exception.

Policies. Policies are very much like rules. In fact, they are often interchangeable in definition. They often consist of management procedures based on material interests. They might involve a definite course of action selected from among alternatives and in light of given conditions to guide and determine present and future decisions. An example policy might be that we will give an employee one day off if he or she needs to attend the funeral of an immediate family member. Like rules, policies do change and often need to change with time and with new circumstances. There might need to be exceptions to the example funeral policy, if the funeral was more than a given distance from the workplace. Like rules, good policies are based upon some kind of correct principle(s), while bad ones are not.

Theories. Theories tend to be derived from various analyses of facts and their relation to one another. They usually are plausible or scientifically acceptable general concepts offered to explain phenomena. They usually are based on observations and are: hypotheses assumed for the sake of argument or investigation; suppositions or conjectures; or abstract thoughts or speculation. Because theories are not facts, but rather are attempts to explain facts or observations, theories often change or need to change, especially when new facts are obtained or when new observations occur. It is not uncommon to read in the newspaper of some theory that has changed due to new information which someone has obtained. In project management, we occasionally encounter various management theories. Some are well-based on extensive evidence, while others are less-well based.

Values. Values are concepts or perceptions which we highly rate or highly regard—something we prize. They may or may not be correct perceptions, but to each of us they are important in our decisions and actions. Religions, philosophies, personal ideals, organizational ideals, and some morals might fall under the category of values. For example, it might be that a person highly values self-reliance to the point that he or she feels it is important to become wealthy before truly being self-reliant. His or her value is that he or she needs to become wealthy. However, it is not unusual for people and organizations to change their values over time, even if only a little, as a result of experience, knowledge, and choices they make in life. In our example value of the need to become wealthy, the same person may encounter life in such a way to cause him or her to re-evaluate the desire to become wealthy and replace it with a desire to sacrifice of higher causes. Thus, one's values do occasionally change.

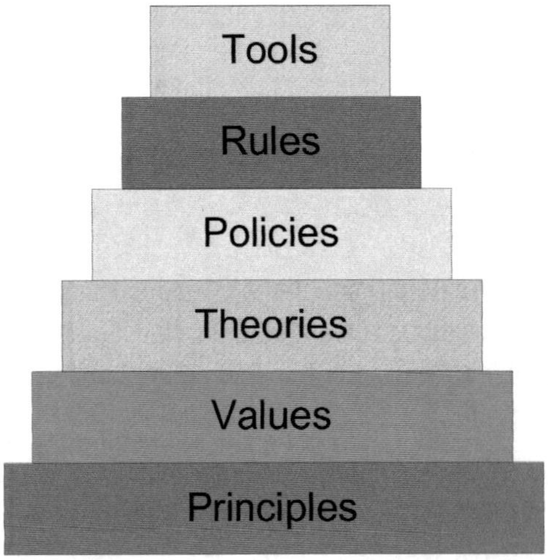

FIGURE 1Good Project Management Should Be Founded Upon Principles

Principles. Principles are fundamental or foundational unchanging truths or natural laws that cannot be broken. Principles are different from all the other concepts and definitions above in that *they do not change* with time or circumstance and they are true for everyone; regardless of background, religion, or philosophy; regardless of whether we choose to ignore them. They are comprehensive and they are self-evident. A few examples of principles are given in Table 1.

Principles are not the same as the *application* of principles. Applications of principles might change a little or a great deal as time or circumstances change.

For example, it is important that we love people—this is a principle. In some cases this might mean giving something to someone in need, like giving food to a child who cannot provide for himself. In other cases, loving an adult might be applied by *not* giving him something he needs, particularly if it is important for that adult to learn how to obtain that thing through his own skills and initiative, so that he can grow and become more self-reliant—note that the importance of growth is also a principle. In the examples of the child and of the adult, love is evident, but the practice or applications of the principle of love are very different.

Table 1: Examples of Principles

The need to love and care about other people
The necessity of integrity
The importance of honesty
The importance of service to other people
The importance of human bonding
The need for patience
The need for good communication
The need to be loved
The need for individual growth
The need for harmony with one's conscience

We must note that while we can try to ignore principles, they never go away, they are always there. We can try to say that they are not true, but that does not change that they are true; and when we try to break principles, we must suffer the natural consequences, like we do when we try to break natural laws.

Because principles are so very fundamental, unchanging, and universally applicable, it is vital that we learn them, understand them, and learn to apply them correctly. This is true in project management and in any other aspect of life.

Throughout the use of this text, you are encouraged to focus special attention on learning the *principles* of project management. Although we will be considering a number of procedures, tools, and policies or rules, all these can all change over time—they will change as a function of circumstances. The principles however *will not* change.

If the reader deeply understands the principles of project management, which will never change, then he or she will have a firm foundation upon which to build various different applications of those principles, according to the needs of the situation. Then, we can also understand how to make better adjustments or exceptions to rules and policies as needed.

What is a Project?

Before we go any farther, we really ought to understand what a project is. This might seem trivial, but it is not. In fact many individuals and teams have started out to accomplish a "project" only to find later they really did not have any clear idea of what was to be accomplished and when.

There are many people and organizations who have attempted to develop a definition for a *project*. Because it covers the basics of all projects, we will use the following definition: A project has two main characteristics: (1) it is temporary, or time limited; and when completed (2) a project provides a unique product, service, or result, or in other words the project has one or more specific objectives. See Table 2.

PMI® (The Project Management Institute®) says it this way: "A project is a temporary endeavor undertaken to create a unique product, service, or result." (PMBOK3, p. 17)

Table 2: The Essential Characteristics of a Project

1. Temporary; or in other words the project is time limited
2. Provides a unique product, service, or result; or in other words the project has one or more specific objectives.

Temporary. *Temporary* means that the project is time limited—it has a beginning and a end. It is the project that is temporary, not the product, service, or result—these normally last much longer than the project. The project cannot go on forever, even if some people treat it that way. Some projects might end in a matter of hours or days, while others may take many months or years, but each project must end; and it is important that the project stakeholders have a relatively clear expectation for when the project is to end.

We say "relatively clear expectation" because, while for some projects we can give a nearly exact end date or time, there are projects for which we can only give a very approximate range for the end time. This applies particularly to new technology projects that are highly uncertain and have high risks involved, like many R&D (Research and Development) projects.

For example, we might say that project A, which is designed to find a way to prevent wide-spread bird-flu infections, must finish within the next three to four years. In this case, although the end-date is not exact, we do have a clear expectation that we cannot go over four years, and that ideally we should try to finish within three years.

The bottom line is this: whatever the situation, there should be a clear expectation as to the time frame for the project to start and end, along with any tolerances in those start and end times.

Unique. By the time a project is completed, a that project should provide one or more unique products, services, or results—it should have accomplished one or more specific objectives. *Unique* does not mean that similar projects have not been done before, but rather that every project is unique in some way.

The project must have specific objectives to be accomplished. Without specific objectives, the project team's work will be inefficient and ineffective. The less specific the objectives are, the greater the likelihood that there will be much wasted effort, increased rework, and missed opportunities. In addition, if the objectives are not clear, the project stakeholders will have a difficult time agreeing on when the project has been completed and whether or not the project has provided the product, service, or result each stakeholder anticipated.

Non-project Work. Work that falls outside of project work is referred to as being *operational work* or *maintenance work*. These types of work tend to be ongoing and repetitive. At home for example, the work of washing the laundry, cooking the meals, washing the dishes, and cleaning the house is very repetitive and ongoing—it seems it never ends! Although people might say that cleaning the house is a project, for the purposes of project management we normally would not consider that type of work as *project work*, but rather we consider it as *operational work* or alternatively *maintenance work*, whichever you prefer—it is not unique and really is not temporary.

What is a Project Manager?

The obvious answer to the question, "What is a project manager?" is that a project manager is someone who manages projects. Perhaps we should re-phrase the question into the following: "What characteristics make up a good project manager?" Table 3 gives a list of characteristics that the author has compiled over the years which cover most of the aspects of what is needed to be a good project manager. The list is large and yet probably lacking. Good project managers need to have many talents and many excellent character traits if they are to be successful project managers and project leaders. Many of the characteristics in Table 3 are covered in detail later in this text.

Table 3: Characteristics of a Good Project Manager

Maintains high integrity
Excellent communicator and listener
Trustworthy
Honest and ethical
Friendly and courteous
Respectful *and* respectable
Organized
Teacher and builder of people
Patient
Accountable and asks for accountability
Cares about people
Flexible when needed; not flexible when shouldn't be
Motivator
Professional in interactions with others
Good decision maker
Planner, with other stakeholders
Good manager of time
Good record keeper
Has good technical and managerial credibility
Familiar with the organization's culture, policies, and procedures
Good diplomatic skills
Receptive to suggestions
Skilled in developing team solutions and facilitates group participation
Doesn't ask others to do what he or she is not willing to do
Leads rather than pushes
Values the differences in others and their perspectives
Seeks synergy and win/win solutions
Helps others set goals in support of the project and organizational objectives

Keys to Successful Project Management: The Vital Three

Let's suppose that you are nearing the end of this project management text. You meet a friend in the elevator who says, "I understand you attended a tutorial about project management. What have you learned?" You have until you exit the elevator, about 20 seconds, to tell your friend what you have learned. What would you say? Well you don't know right now because you haven't completed this presentation. So, I am going to tell you what I hope you will say. You see, I am going to tell you right now what I consider to be the three most important concepts in this tutorial. In the elevator, you would likely only have time to tell your friend the three most important concepts anyway. By me telling you now what I want you to say, I am hoping that you will pay special attention to these items throughout the course of your study. The three most important keys to successful project management are given in Table 4.

Table 4: The Vital Three Keys to Successful Project Management

First, *the importance of* ***people*** *in project management!*

Second, *the importance of excellent definition of* ***project objectives****!*

Third, *the importance of excellent* ***planning*** *and* ***re-planning****!*

There are many concepts, processes, and principles important in project management and they are all important, but I consider these three to be the most important keys. If anyone of these is missing or weak, the entire project will suffer tremendously. Sometimes I refer to these as the *vital three* because they are so preeminently important. You can often succeed by neglecting some other aspects of project management, but you cannot ignore any one of the vital three—they are essential.

References

Covey, Steven R., *The Seven Habits of Highly Effective People*.

PMI. *A Guide to the Project Management Body of Knowledge (PMBOK® Guide)*, fourth edition, The Project Management Institute, Newtown Square, Pennsylvania, 2008.

The Importance of People in Project Management

Abstract ID: 481743
Paper # 330b
Dr. Eldon R. Larsen
Marshall University

Of the many important aspects of project management, the most important key to successful project management is people. When we speak of project management, we often think of Gantt charts, time lines, budgets, resource constraints, quality challenges, performance measures, and so on. While these are important, they are neither meaningful nor effective without people. This paper presents and discusses several fundamental concepts regarding the importance of people in project management.

Introduction

In 1993 when I first became deeply involved with formally studying and applying the principles and tools of project management, I made an appointment to visit a man who was well respected at for his project management expertise. He was Charles Brod, the Chair of the UCC (Union Carbide Corporation) Project Management Council. This council was made up of the most experienced project managers in the corporation.

I had been asked by UCC to lead a team to develop and implement project management principles, techniques, and tools specifically for research and development, an area often not accustomed to the rigors of formal project management. I had been on many research and development project teams and product or process commercialization teams, once serving on twelve different teams at the same time. However, I had little formal project management training at the time. I knew I could use Charles's insights for this effort I had been assigned.

I did some preparation and wrote questions to ask Charles, so as to make the most of our visit. As I entered his office at the appointed time, and before we sat down, I said, "Charles, I know that people are important in project management..." Before I could finish my sentence he pointed his finger at me and said, "You stop right there! People aren't just important in project management, people are project management!" (Brod, 1993) I just stood there stunned for a few seconds as my brain rapidly processed what Charles had said: "People are project management!"

As my mind raced through my many experiences in the many projects I had worked on, I knew almost instantly that what Charles had just said was not only true, but was the verbalization of one of the most important principles of project management. In comparison to that statement, little else mattered in that interview.

Principle:

People are project management!

People Are Project Management

Jack Gido and James Clements (Gido and Clements, 1999) say it this way: "It is the people—not the procedures and techniques—that are critical to accomplishing the project objective. Procedures and techniques are merely tools to help people do their jobs."

As people progress in their careers, many become project managers. These people are often highly trained technically and tend to be very analytical—this helps them to be good technical problem solvers. However, technical professionals often have had little formal training which will enable them to effectively lead and manage people. Thus, they could be weak people-problem solvers. Yet, this is exactly what they are being expected to do as they move into project management or other management positions.

People in new project management positions might well ask themselves the following questions: "What do seasoned project managers wish they had known on their first project management assignment?" "What are the responsibilities?" "What are the people interactions that take place?" Following are some of the things new project managers would have liked to know, based on research by J. Pinto and O. Kharbanda (Pinto and Kharbanda, 1995).

- Realizing that conflicts occur as a natural part of team development,
- Knowing who are the project stakeholders and what are their agendas,
- Realizing that organizations are political,
- Understanding that project management is leader-intensive and you must be flexible,
- Realizing that customer satisfaction is essential for project success,
- Understanding that you must be a motivator, coach, peacemaker, and conflict manager,
- Understanding that your team develops attitudes based on your emotions,
- Knowing that you always need to ask "What if…"; don't be comfortable with the status quo, and
- Learning that above all you must plan, plan, plan!

All of the above learnings involve *people interactions*!

Project managers have responsibilities among other things to (1) define and clarify objectives, (2) lead and manage, (3) plan and re-plan, (4) organize, and (5) control and influence. Again, all of these responsibilities involve people interactions.

The following sections of this paper discuss and make suggestions for some of the multiple interactions among people involved with projects. The interactions discussed are

- Stakeholders and stakeholder management
- Project manager interacting with project team members,
- Project manager interacting with senior managers,
- Project manager interacting with line mangers,
- Project manager interacting with the other employees within his or her company,
- Project manager interacting with the general public outside of his or her company,
- Project manager and team interacting with the customer, and
- Project manager and team interacting with contractors and other internal or external service providers.

Stakeholders and Stakeholder Management

All projects have project stakeholders. Examples of potential stakeholders are the customer, the project sponsor, senior managers, the project manager, the project team, organizational line managers, contractors and subcontractors, service and trades personnel, landlords and/or leaseholders, suppliers, recipients of products and technology, government entities and regulators, and possibly the surrounding populace.

A wise project manager will do what is needed to identify all significant stakeholders. The project manager will recognize that some stakeholders that might not seem significant at the present, but they might become significant and so will need to be monitored. One major mistake many project managers make is not to do a careful analysis of all stakeholders.

Stakeholders can be either positive stakeholders or negative stakeholders. Generally, positive stakeholders will benefit from the project being a success and will want it to succeed. Negative stakeholders will likely not benefit from the project's success, and so will want it to fail. Some stakeholders that currently are positive might change and become negative, and vice versa.

You might need the full support and cooperation of some stakeholders in order for the project to succeed. Therefore, you should always ask yourself the following questions: "Who will benefit from this project? Who might be hurt by this project? Who might otherwise be impacted by this project? Who will be key in helping this project to succeed? Who else might I have forgotten about?"

Because of the many concerns, needs, and expectations of stakeholders, it is essential that the project manager conduct an effort to identify the stakeholders, their expectations, their potential impact on the project, how the project impacts them, what is needed from them, and so forth. To do this, the project manager develops a *Stakeholder Register*.

Table 2.1 shows an example of a possible stakeholder register. Here we identify the stakeholders, their positions in the internal or external organizations, their power and authority, their location, how to contact them, whether they are positive or negative stakeholders, their relationship to the project, what influences they might have upon and for the project, their needs—must have, their wants—like to have, and what concerns or fears they might have about the project. We also very clearly identify their requirements for the project and their expectations beyond their requirements. We clearly identify our strategies for managing their requirements and expectations so that changes are clearly analyzed, controlled, approved or not, and communicated. Finally, we clearly define communication needs for each stakeholder.

The rest of this paper discusses some specific needs and helps in dealing with specific types of stakeholders.

Project Manager Interacting with Project Team Members

Managers are not necessarily the same as leaders. A manager has assigned people that he or she is to manage. Leaders are different. There are no leaders without followers, and people choose to follow those they consider to be their leaders. Leadership and management are all about working with people.

Much like a music conductor (Gido and Clements, 1999) leads an orchestra to play a concert, the project manager leads the project team to accomplish the project objectives. The project team, like an orchestra, must be unified and in tune (aligned), as they follow their leader. For there to be harmony, the team must want to follow the leader—this is a vital people interaction principle. The leader and team must care enough about each other to make this happen.

Project managers and the project team members interact in many ways. They do this as they define and clarify project objectives; plan and re-plan; make and accept task assignments; delegate and expect accountability; manage and resolve conflict; reward and discipline; generally communicate; give performance feedback; motivate and inspire; and develop personnel. Following are some suggestions for project managers to help them to achieve success in these interactions with team members.

- When practical, involve the project team in refining the project objectives with the customer;
- Assure that the project objectives, constraints, and deliverables are clear and that they are understood by all team members;
- Always involve the project team in project planning, then...
- Seek commitment from team members;
- When challenges arise, genuinely ask for input from team members;

Table 2-1 Example Stakeholder Register

STAKEHOLDER REGISTER

Stakeholder Characteristics

Stakeholder Name	Position	Power & Authority	Location	Contact Information	Category: Pos./Neg.

Stakeholder Relations and Needs, Wants, and Fears

Stakeholder Name	Relation to the project	Potential Influences	Needs	Wants	Fears

Stakeholder Requirements, Expectations, Strategies, and Communications

Stakeholder Name	Stakeholder requirements statements based on needs, wants, fears, and other information	Stakeholder additional expectations with respect to and for the project	Strategies and issues or concerns for managing the requirements and expectations	How, when, and what to communicate with this stakeholder

- Delegate as much as you can to team members, according to each individual's readiness to accept the added responsibility;
- Build and develop your followers; give them growth opportunities, but don't overwhelm them;
- Whenever feasible, minimize unnecessary bureaucracy;
- Learn what motivates and inspires each team member;
- Ask for accountability, while helping team members to get the authority and other resources they need in order to be able to be accountable;
- Work with line management to assure that team members have up-to-date tools;
- Reward or discipline team members sooner, rather than later;
- Make project meetings and communications valuable—don't waste people's time; (Larsen, 2001)
- When conflict arises, manage it quickly and seek win/win resolutions;
- Make a project communications plan and use it; and
- Keep accurate and up-to-date project records—this helps to minimize miscommunications and unnecessary rework.

Project Manager Interacting with Senior Managers

There are many reasons that a project manager needs to interact with senior managers and to nurture these relationships. Senior managers

- Assure business goals are being achieved,
- Often act as the interface between the customer and key project personnel,
- Provide essential support and guidance for projects, and
- Often act as project sponsors or champions.

It is absolutely essential that effective interactions occur between the project manager and senior managers. If this interaction with senior management is not effective, support for the project could wane, misunderstandings will almost surely occur, project objectives and priorities will either be unclear or insufficiently acted upon, and the project team could easily feel itself drifting without a good sense of direction or support. Some of the interactions that often occur between the project manager and senior managers have the following purposes.

- Define project objectives and deliverables,
- Clarify and align business and technical objectives,
- Clarify company policies and procedures,
- Clarify project scope and project risks,
- Clarify critical success factors,

- Clarify key project priorities or drivers and tolerances in the triple constraints of time, resources, and performance, (Larsen, 2002) and
- Finalize and approve the project charter.

In order for the above interactions between the project manager and senior managers to be effective, consider the following suggestions.

- Be honest, maintain integrity, be respectful;
- Don't be afraid to ask for or give clarification or more information as often as is needed;
- Give and ask for the accurate, whole picture, not just the good news;
- Be concise in communications, but be clear;
- Do all you can to prevent surprises;
- Be sure to thoroughly discuss and understand all important project risks, tolerances, and priorities;
- When discussing concerns, focus on issues rather than people;
- Wisely use senior management to help with conflict and risk management when their help is truly essential; and
- Never backbite your senior managers—it won't help and will likely make things worse.

Project Manager Interacting with Line Managers

Line managers frequently play a vital role in contributing to project success or failure. In many organizations it is the line managers who select whom in their groups to assign to a given project. In many organizations it is also the line manager who actually pays the employees' salaries on a team. This can cause a definite allegiance preference for the line manager over the project manager when it comes to deciding which boss to follow when getting mixed direction signals. In organizations where line managers appoint team members, these managers

- Usually provide the people to work on the project;
- Assure that their people have proper training;
- Monitor the workload of their people;
- Often take care of the administrative needs of their people (payroll, safety certification, annual reviews); and
- Are usually much more technically expert than is the project manager.

From the above list it should be obvious that line managers can be vital to your success! Some suggestions for working with line managers are

- In the very early stages of the project, begin communicating with the line managers of people who might be selected to be on the project team;
- Assure that the project objectives, deliverables, priorities, drivers, and tolerances are clear to each line manger who will be providing resources or services;
- Develop and nurture a close working relationship—you need line managers' support, advice, wisdom, and buy-in;
- Always keep line managers informed of the project progress, concerns, challenges, and needs;
- Keep line managers aware of personnel achievements, performance issues, or needs pertinent to their areas of responsibilities;
- Listen sincerely to the concerns and advice of line managers—if you don't, you will pay for it; and above all
- Communicate well and regularly. (Larsen, 2003)

Project Manager Interacting with the Company and General Public

Depending on the project, a project manager's interaction with the company as a whole or with the general public can range from non-existent to frequent and crucial. Corporate or public perception of you and your project can have the potential to make or break your project—if this applies to you, nurture appropriate interpersonal relations with all the care you can give. Following are suggestions.

- Learn what major concerns the company or the public is likely to have and, as appropriate, be fully prepared to address them honestly and openly. If you can't be open because of secrecy agreements or other legal constraints, make that clear to those who are concerned.
- Always be prepared! Know exactly what is happening on the project. Then plan and practice how to communicate with those concerned. Make sure senior management, the customer, and the project team know what you will present and that they have given you their suggestions and support.
- Try to put yourself in the shoes of the people with whom you need to communicate. Seek to understand them (Covey, 1989).
- Always be ethical in your behavior and your words.
- Do not lose your cool! It takes many good interactions with people to make up for one bad one.

Project Manager Interacting with the Customer

The interaction the project manager has with the customer is perhaps the most important interaction of all, and could have the greatest impact on customer satisfaction—the real measure of project success. The customer, better than all others, usually knows most what he or she wants at the end of the project. Customer perceptions play a strong project-success role. Effective people-interaction is necessary to transfer customer expectations to the project manager and team.

Following are suggestions to help in this interaction.

- Unless the nature of the project prevents this, the project manager and team need close interaction with the customer as early in the project as possible—this is when exists the greatest potential to affect the final project outcome. Make this happen if possible.
- Maintain open channels of communication throughout the entire project.
- Early interactions between the project manager, team, and customer should include developing a very clear understanding of the
  - ✓ Project objectives,
  - ✓ Project priorities, triple constraint drivers, and tolerances,
  - ✓ Project's critical success factors,
  - ✓ Project's scope and deliverables,
  - ✓ Project background and justification, and
  - ✓ Any key concerns and assumptions.

Project Manager Interacting with Contractors and Other Internal or External Service Providers

Outsourcing of project work has increased greatly over the past decade. Flannes and Levin (2001) point out that those temporary workers sometimes "contribute a greater depth and breadth of experience." It might even be that the project manager is outsourced. The presence of contractors and other service providers, either part of the project team or not, presents important people-interaction challenges. Some suggestions follow.

- Remember, just as with the other key people interactions we have discussed, constant and nurtured communication is absolutely essential.
- Work early to address any cultural, technological, or other barriers that might impede this needed communication—make a plan to do this.
- Don't treat people not on the project team as second-class citizens. Value them. Care about them. Seek to understand their perspective. Your efforts will pay big dividends! Many of them could be project stakeholders.
- When dealing with contractors or people not directly on the project team, record-keeping and communication become even more important to avoid misunderstandings, re-work, and unwanted problems later on. Make sure important communications are in writing, and then follow up personally to discuss any questions or make clarifications. If changes are needed, put them in writing and get necessary approvals.

- Read all contracts thoroughly and make sure you understand them.
- If the other party is writing the contract, make a quick draft yourself of the contract before you see theirs. This will help you to have thought through what to look for and what questions to ask when you review their contract. After it is too late, many people wish they had more fully realized what they were signing!
- If you are writing the contract, ask the other party what they would like to see in it. Get input from other people who will be working on the project and people who have experience with contracts. Then, have a legal expert help you do the final contract to avoid problems you might miss. Make sure senior management signs-off on whatever the contract says. If you all work together, you are more likely to achieve win/win results.

People and Project Success

Notice how vital are the importance of people interactions and the perceptions of people, in addition to technical performance, as expressed in the following definition of project success from a study of approximately 650 projects (Baker, Murphy, and Fisher; 1974).

> "If the project meets the technical performance specifications and/or mission to be performed, and if there is a high level of satisfaction concerning the project outcome among key people in the parent organization, key people in the client organization, key people on the project team, and key users or clientele of the project effort, the project is considered an overall success."

This definition of project success places a large emphasis on perception—a people feeling. Notice also that schedule and budget are not explicitly stated as part of project success, although they undoubtedly somewhat affect perception and feeling. Nonetheless, such a definition certainly exemplifies the great importance of people in measuring and accomplishing successful projects.

Summary and conclusions

Good people interactions are vital to successful project management and must be carefully nurtured. These interactions permeate essentially every aspect of the project and are also the most time-consuming responsibility of the project manager. The above discussion should clearly demonstrate to the reader that people are indeed the most important key to successful project management. Because of the importance of people, a project manager needs to learn to sincerely care about each person associated with the project, and nurture these people interactions to achieve success. Charles Brod summarized it well, "People are project management!"

References

Baker, Bruce N., David C. Murphy, and Dalmar Fisher, "Factors Affecting Project Success," Ch. 35 of David I. Cleland and William R. King (eds.), Project Management Handbook, 2nd Ed., Van Nostrand Reinhold, New York, 1988.

Brod, Charles, Personal communication with the author, 1993.

Covey, Stephen R., The Seven Habits of Highly Effective People—Restoring the Character Ethic, Simon & Schuster, New York, 1989.

Flannes, Steven W. and Ginger Levin, People Skills for Project Managers, Management Concepts, Vienna, Virginia, 2001.

Gido, Jack and James P. Clements, Successful Project Management, South-Western College Publishing, Cincinnati, Ohio, 1999.

Larsen, Eldon R., "Not Another Meeting!—Planning and Conducting Effective Meetings," Proceedings of the ASEM National Meeting, October 13, 2001.

Larsen, Eldon R., "Start Projects Right—How to Develop Effective Project Objectives," Proceedings of the ASEM National Meeting, October 4, 2002.

Larsen, Eldon R., "Communicating With the Boss and Other Public Things," Management Topical: The Engineer in Transition to Management—Tutorial, Published in the Proceedings of the topical at the AIChE National Meeting, March 31-April 3, 2003.

Pinto, J. and O. Kharbanda, "Lessons for an Accidental Profession," Business Horizons, March-April 1995 (quoted by Gido and Clements, p. 83; 1999, cited in this paper).

Communication—A Better Understanding

Abstract ID: 481744
Paper # 330c
Dr. Eldon R. Larsen
Marshall University

Today's business organizations often seem less caring, more informal, and having less long-term mutual-commitment. Our interactions are ever-increasingly more digital. Hence, the need for excellent and clear communications between employees, managers, peers, customers, and the outside world will continually become more vital to long-term successful enterprises. This paper presents principles, concepts, and tools to help the reader gain a new vision of communication and how to do it better in today's challenging environments.

Additionally, this paper presents ideas on what is truly needed for good communication—what foundation concepts are necessary if we really hope to successfully communicate. The paper discusses how communication takes place and reviews the very important concept of *empathic listening* as defined by best-selling author Stephen Covey (2004). Finally, it is important to make a communications plan when we need to manage communication.

Introduction

Communication is one of the most important concepts and tools in human life. It is also one of the most used and abused words in human interactions. Its real meaning is rarely fully understood. Significantly, when one really understands the meaning of communication, a paradigm shift often occurs. With this paradigm shift, we start to look at our interactions with other people in a whole new, deeply meaningful way. We open the door to truly effective human interaction.

We use communication in essentially every interaction we have with other people, be it verbal, written, silent symbolism, body language, pictures, sounds, or other ways. Yet, it is also communication, miscommunication, poor communication, or lack of communication which frequently gets us into challenging situations and can cause frustrating emotions, at least when we communicate in the way that people commonly define communication.

In interviews with hundreds of project managers which my graduate students have conducted over more than a decade, many of these project managers state that some of their greatest challenges come from working and communicating with people. Why do we have so much trouble with communication? Why is it such a challenge? What are we missing?

Someone once said, "I know that you believe you understand what you think I said, but I am not sure you realize that what you heard is not what I meant." (Dinsmore, 1990) Does this sound familiar? If so, you are not alone. It happens all the time.

This paper presents important ideas and concepts that can help. One of the most important concepts discussed is what the word *communication* really means.

A better understanding of the root meaning of *communication* can cause a very important paradigm shift, to the point that the word takes on a whole new meaning. It can give us new vision. This help opens the door to better communication for us all.

Do We Really Need Good Communication?

Project managers spend the vast amount of their time communicating in some way or another. They are constantly providing direction, asking for information, clarifying objectives, seeking for understanding, and giving feedback.

Whether you are an employee or a new or seasoned manager, it is important for you to realize that (1) bosses want to be kept informed, (2) bosses don't like surprises, and (3) bosses depend on communication to help them to plan and to organize, motivate, and control or influence people—actually the same could be said for most of us, not just bosses. Good communication is like the lifeblood of a successful project manager or a successful leader.

Project managers, team members, stakeholders, and others all struggle when we use poor communication techniques. Communication in its many-varied facets consistently comes up as a major managerial issue.

It is commonly stated that the three most important factors to consider in real estate are (1) location, (2) location, and (3) location. In management, especially in project management, the three most important factors to consider are (1) communication, (2) communication, and (3) communication.

What Is Communication?

What does the word *communication* really mean? Why ask this question about a word so commonly used? Everyone knows what we mean by communication—don't they? *Do we*? Let's look a little deeper.

High school English teachers often tell their students to look at the "roots" of a word in order to really understand it. These teachers often split words into various prefixes, suffixes, and main-word fragments, and then look at the Latin or other linguistic meanings of these word fragments to determine the true meaning of words.

Let's do this root-understanding-exercise for the word *communication*. We will look at the roots in French, rather than Latin, because the roots form common French words. *Communication* is in fact a French word derived from Latin, and is spelled the same in both French and English.

Comm/un/ication

First, we break communication into three word fragments: *comm, un,* and *ication*. *Comm* comes from *comme*, which is a separate French word that means "Like" or "As." *Un* in French means "One." Finally, *ication* essentially means "The process of becoming." When we put this all together the fundamental meaning of the word comes to light:

Communication = The process of becoming "like one" or "as one."

Let's say it again: Communication means the process of becoming like one or as one! It is the process of becoming unified!

Do we realize what this is saying? Think about the implications of the root-word definition! Communication is not just sharing or transmitting information. Communication is not just sending out a message. Communication is not just giving instructions. It is not just telling someone what you think or know. In fact, in and of themselves none of these are true communication according to the root-word definition. They can't be; they are one-way.

By the root-word definition above, communication almost always must be at least two-way. How else can we become like one? To be effective, this two-way interaction often needs to be very interactive and iterative. The only exception is the rare case when the receiver of the information just happens to fully agree with the sender of the information. Even then how do we know that communication has been achieved, unless we both interact and *communicate*?

Note that "The process of becoming like one" does NOT mean becoming the "same." We can be unified without being the same. In fact, we will be stronger and can synergize more effectively if we value differences, and yet are unified. It is a bit like having two pieces of a two-piece puzzle. The two pieces are not the same, but when they fit together and become *like one*, then they make a clear and complete, unified whole picture. Separately they are greatly lacking. We need to seek to become unified for project success.

In summary, true communication is a process of coming-together, in common understanding, vision, feeling and purpose, so that we truly understand each other to the point that we are becoming like one.

Principle:

Communication = The process of becoming like one

For most people, this concept, this original root-word definition, or could we even call it a *principle*, is a paradigm shift. It is a whole new way to look at what we are really trying to do when we say we want to communicate. It changes the whole picture or vision of what we are trying to accomplish. It changes our whole approach. The objective is not to just shove our opinions down someone else's throat! *We are trying to become like one.*

As a side note: If you physically try to shove something down someone's throat, you will likely get a response like gagging or vomiting. Think about that next time you try shoving your opinion down someone else's throat—it simply won't be pretty! It won't work, no matter how well intentioned you might be. This is not the way to communicate.

With this new vision—that we are seeking to become unified—we are now ready to realize that our very motivation in communication must change. Communication is not a one sided paradigm; it is two-sided at least, and it is serious and amazing. It involves us in a whole new perspective, not just our own perspective, and for a whole new purpose. How will we start communication anew with this new paradigm? We now are ready to investigate this more.

What Is Needed to Truly Communicate?

On what foundation are you to build if you want to truly communicate, if you want to become like one or as one with someone else? There are two main foundation building blocks I will discuss here.

1. Honestly and Mutually Care About Each Other

First, to truly communicate, you, and those with whom you are attempting to communicate, need to honestly and mutually care about each other. You must mutually respect each other and be considerate of each other's needs and perspectives. Without this mutual caring, communication efforts will quickly deteriorate to our unfortunately more modern one-way transfer of information we incorrectly call communication.

This presents a new challenge: Since you can't control the other person with whom you are trying to communicate, can this idealistic *communication* really happen? Isn't this a lost cause?

The answer is that you must start with yourself—only you can decide that you will really care about the other people around you, not just for your own self-interest, but for theirs too.

When you start with yourself, when you change your own motivations, almost miraculous interactions start to happen. For some strange reason, others seem to start to care about you too! They want more to work with you and to desire the oneness in purpose and vision and understanding that you emanate and that you too desire.

Thus, communication becomes a very personal issue. It is not an issue of better technology. "Communication among people depends not on technology, but rather on forces in people and their surroundings. It is a process that occurs within people." (Gibson, et al., 1997)

2. Mutually Align

Second, to truly communicate you also need to mutually align. Alignment means that you and those with whom you work all get pointed in the same direction, so to speak. You work toward the same goals, see the same vision and have a desire to work together toward achieving that vision. See Figure 1. One big question is "How do we accomplish this alignment of people?"

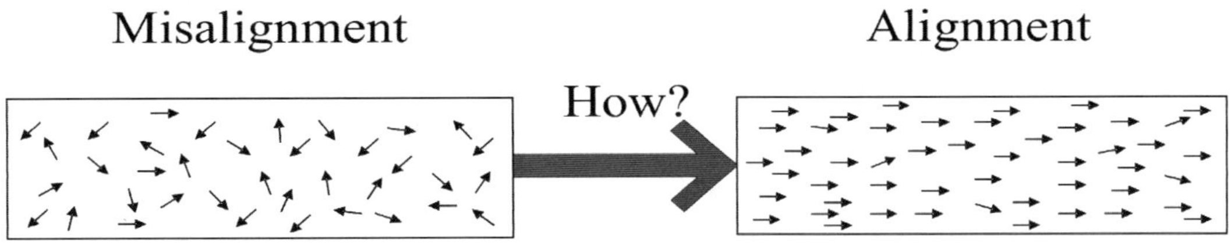

FIGURE 1 Transforming People from Misalignment to High Alignment

Let's look at an analogy to better understand personal alignment needed for better communication. When we take an automobile in for an alignment, Figure 2, what we mean is that we want to get all the wheels pointing in the same direction within certain tolerances. If the wheels are in poor alignment, the consequences are increased tire wear, steering difficulty, lower fuel efficiency, and extra wear on other parts of the car. Good wheel alignment is very important!

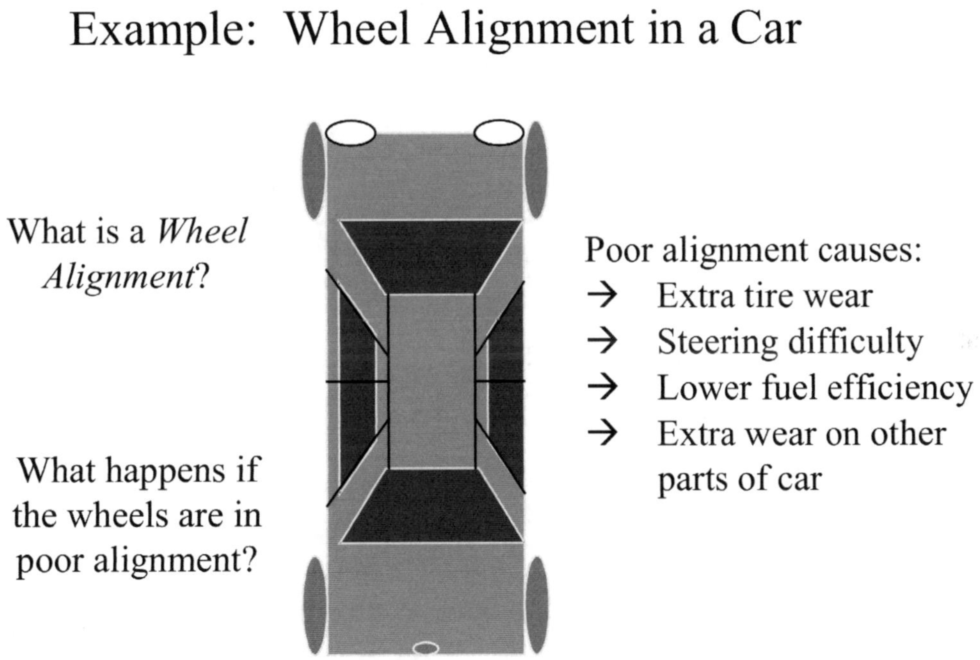

FIGURE 2 Good Alignment is Important for Cars—It is for People, Too!

Alignment of team members toward project objectives is equally important. Like the wheel-alignment analogy, poor team member alignment can lead to extra wear on the team members, difficulty to steer or manage the team, poor work efficiency, and even eventual wear on the larger organization. In contrast, a well-aligned team runs well, is much easier to manage, effectively and efficiently carries out the team's work, and improves the well-being of the organization as a whole.

Another analogy gives some insight into various ways we might try to achieve alignment. In this case, Figure 3, we use an iron bar that is not magnetized. In this bar the iron atoms are all spinning with dipole moments pointing randomly in an infinite number of directions. If we were to align enough of these iron atoms' dipoles, by use of some form of work, we could magnetize the iron bar.

This magnetization could be done by electrical induction, where we coil an insulated electric wire around the bar and pass a direct current through the wire. It could also be done by placing a permanent magnet next to the iron bar, with the work to align the previously random iron atoms coming from the permanent magnet. A third way to magnetize the iron bar would be to position the bar in a north-south orientation and pound on it with a hammer.

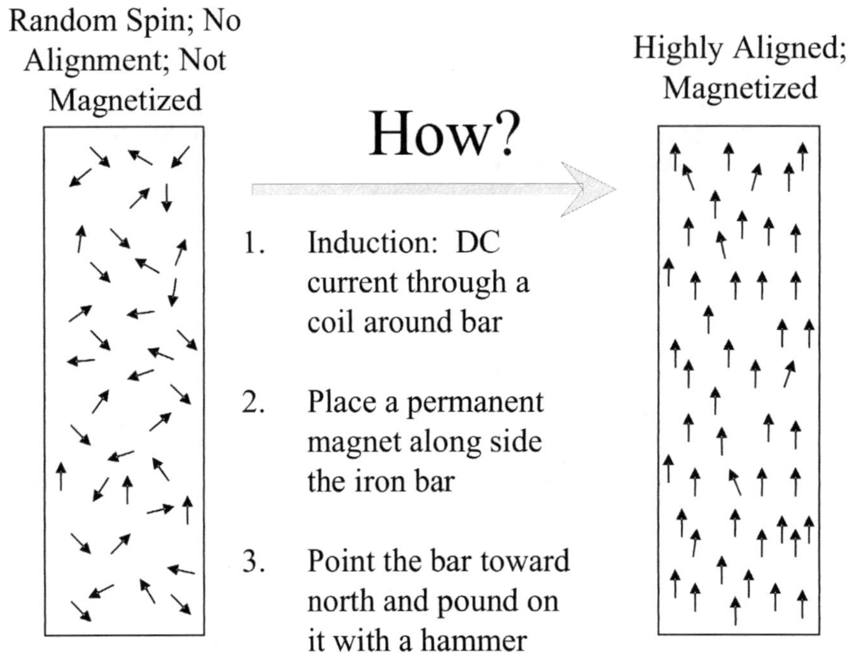

FIGURE 3 Alignment of Atoms in an Iron Bar

A group or team of people might be analogously aligned like the atoms in the iron bar. The team analogy to the iron bar is shown in Figure 4. Again, there are several ways that we might try to enhance team member alignment.

Similar to induction, we could put the team under the influence of some good interpersonal training and team-building exercises. Under this influence, the team is *induced* to start making better efforts to work together and to better understand each other.

Alternatively, we might expose the team to the example of people who have learned to permanently align themselves in order to achieve group objectives. These example-emanating people could show our team how to align by their good examples, or they might even become members of our team or work alongside us as mentors or partners. By their example and leadership, we learn how better to communicate and unify.

A less desirable approach would be to have the boss come in, point us in the right direction, and then "hammer" us into alignment by, for example, threatening us with dire consequences if we don't align.

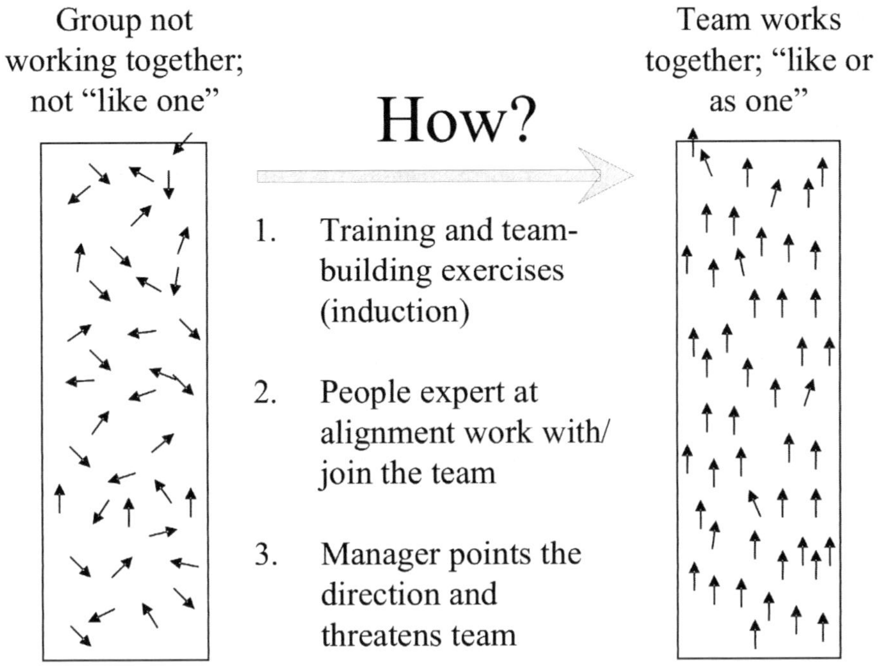

FIGURE 4 A Group of People Aligns into a *Team*

Must a group or team be perfectly aligned? No. In fact, a small degree of misalignment can have a few important benefits: (1) different perspectives, and the raising of concerns and alternatives, often aid our understanding, (2) creativity can be enhanced as we learn to accept and value differing perspectives, (3) strength can come from learning to manage minor personal-interaction challenges, and (4) people are more likely to feel they are in an environment of acceptance and freedom to communicate openly when their differences of opinion are valued.

There is a fifth benefit to having some lack of alignment: you might find, as you try to communicate with those who are not aligned, that there are some very important reasons those people are not aligned. For example, they might know something you don't, but they have been hesitant to share their knowledge. Through your sincere, caring communication efforts, you might very well find that the majority of the team is not pointed in the correct direction and that the few misaligned people actually were headed where the whole team really needs to be going!

On the other hand,, too much misalignment can cause (1) increased dissention, (2) decreased team efficiency, (3) decreased team effectiveness, (4) more arguments and backbiting, (5) lack of team spirit, and (6) increased chance of destructive behavior, and perhaps even attempts to sabotage the project. All of these problems are signs we are not really communicating—we are not becoming "like one." Therefore, high levels of misalignment in teams or individuals within a team must be dealt with, hopefully in a way that achieves better communication. Good use of the principles and concepts in this paper will help. If, after sincere application of these concepts, the matter cannot be rectified, *then* we might need to look at more drastic actions of removing a person from the team; but this should be a last resort, as this can have many negative consequences.

How Does Communication Take Place?

There are various modes and mediums through which we communicate. Examples are written words, speaking, body language, face-to-face meetings, telephone, voicemail, electronic mail, letters, memos, video-conferencing, and groupware. Of the above, *speaking* is one of the most common and important. Speaking is also a mode of communication where we often have difficulty. For example (Dinsmore, 1990), do you understand what former presidential aid Ron Ziegler meant in this statement:

> *You should not interpret by my use of "least likely" that ultimately, or when the final decision is made, that that may not be the decision, but what I am saying is that it is only one of the matters under consideration and the decision has not been made.* (Rees, 1979)

The above quotation made sense to the speaker, but is a little difficult for most readers or listeners to understand. How can we do better? Is there some model that will help us to understand better the process by which communication takes place? If so, it might give us an idea of how to improve communication.

The Effective Communication Model

One communication model that is found in many textbooks is called the Communication Process. (Gibson, et al., 1997) In this model there is a Sender who says some message in a certain way (medium) to a Receiver. The Receiver in turn gives feedback, indicating what effect the message had on the Receiver. The sending and feedback paths form a loop. This is shown in Figure 5.

Between various steps there are events involving encoding and decoding of the messages, which can muddle the message somewhat. These encoding and decoding events can be filtered or selectively influenced by personal perspective, bias, emotion, lack of understanding, errors, etc. Along the path there is also background noise caused by such things as disturbances, interruptions, side conversations, static, etc.

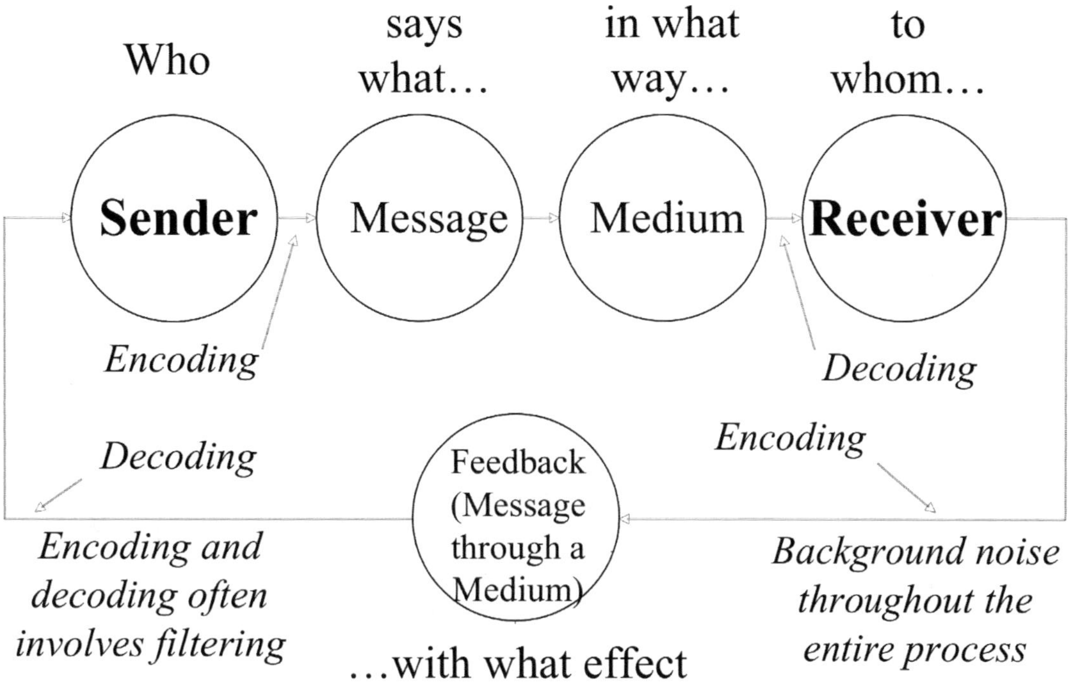

FIGURE 5 A Very Common Communication Process Model

At first glance the Communication Process model seems adequate, but in reality it tells only half the story. Communication is not just one way with some feedback—this is not how we usually become "like one" or "as one."

Instead, a more correct communication model shows that both of the people in the communication loop have information to share, both need feedback, and they must interact until they come to a common understanding—until they become like one. Figure 6 shows this improvement in the model, which I call the *Effective Communication Process Model*, which includes two-directional interaction. It is founded on principles of *mutual caring and alignment*.

To summarize, communication takes place through two-way interaction, with all participants having the opportunity to give their views as Senders and to give feedback as Receivers. In addition, effective communication must be founded upon mutual caring and alignment.

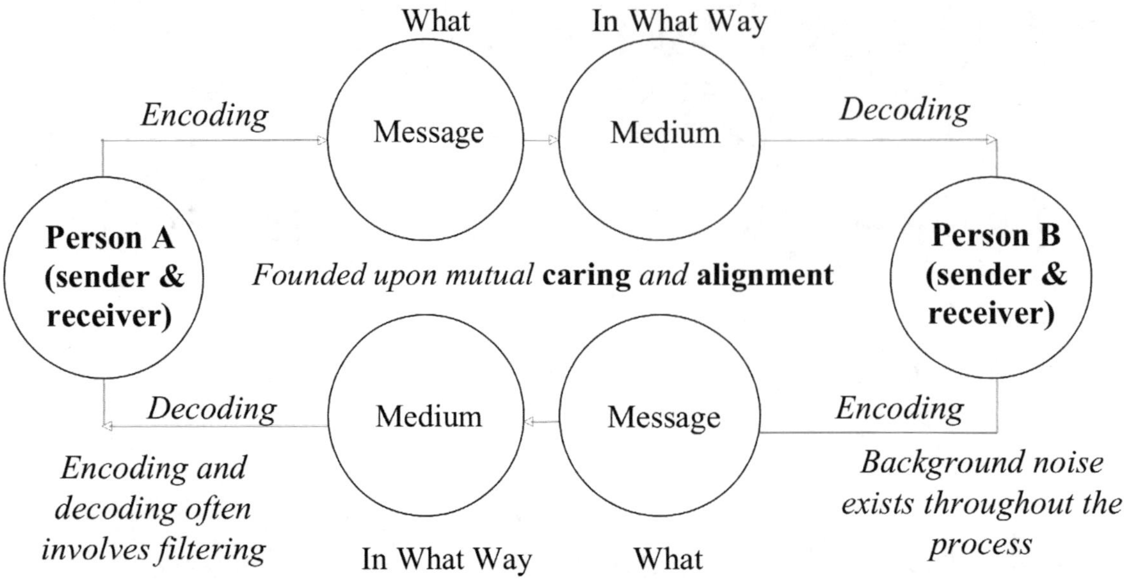

FIGURE 6 Two-Directional Effective Communication Process Model

Empathic Listening

Jack Gido and James P. Clements truly say, "The heart of communication is not words, but understanding. Not only to be understood, but also to understand. Half of making communication effective is listening." (Gido and Clements, 1999)

Much has been written about how to listen better, particularly on the topic of so-called Active Listening. But one of the most important listening concepts, when it comes to true communication as defined above, is the concept of *Empathic Listening*. (Covey, 2004)

What is Empathic Listening? Empathic Listening is not Active Listening, which is skill based. Instead, Empathic Listening is character and principle based. It is Active Listening along with sincerely caring about the other people with whom we are trying to communicate.

Empathic Listening = Active Listening + Sincere Caring

In Stephen Covey's bestselling book *The Seven Habits of Highly Effective People*, the fifth of the seven habits is

Seek first to understand, then to be understood." (Covey, 2004)

This habit is key to successful communication! We need to care so much about the other people with whom we are trying to communicate that we are willing to try to understand them, even before we try to have them understand us. We do this to increase *our* understanding, to get a different perspective, to learn new information, and to show the other person we truly care about him or her.

In addition, by understanding the other person, we greatly increase the possibility that we can synergize with them—to take the best of both of us and achieve things neither of us could have done by ourselves.

For many people this is a major behavioral change, but it's worth it! This change brings rewards far beyond most people's expectations—almost miraculous changes and rewards in some cases. Empathic Listening is probably the most important part of Covey's fifth habit—"Seek first to understand, then to be understood." Again, this form of listening must be built upon sincere caring for other people.

Covey has said, "If you use techniques or skills alone, I may sense duplicity, manipulation, or insincerity—I might wonder about your motives... I might feel over time that you really don't care so much about me as what you can get out of me."

Empathic Listening is based on truly caring about others. Someone once said, "If you really want to understand me, don't just listen to what I say, listen to what I feel."—This adds real meaning. Understanding other people, and then seeking to have them understand us, is vital to achieving the goal of becoming like one or as one.

To summarize, we need to Empathically Listen and seek first to understand, then to be understood.

Poor Listeners versus Good Listeners, and Miscellaneous Communication Barriers

Paul Dinsmore lists some typical characteristics of poor listeners and good listeners as follows: (Dinsmore, 1990). These are summarized in Table 3.1. Notice that you, the listener, are in control of most of the above good or bad listening habits!

Watch out, there are even more barriers to effective communication. For example, when trying to communicate, consider the following barriers in addition to those already listed above (some of the following are taking from Gibson, et al., 1997):

- Value judgments
- Semantics problems
- Jargon—technical, business, etc.
- Low credibility—low trust & confidence levels
- Proxemic behavior—physical distance between individuals (0-1.5 ft = intimate; 1.5-4 ft = personal; 4-12 = social; >12 feet = public)
- Assumptions
- Time pressures
- Differences in perception
- Emotions, either preceding or during communication
- Selective listening—daydreaming or preoccupations
- Thinking of counter-arguments while someone else is speaking
- The communication *medium*—telephone, email, etc.

- English as a second language
- Memory
- Distractions
- Too much formality
- Too big of group to properly interact to achieve unity

In spite of the fact that most of the above barriers are somewhat familiar to most of us, we still struggle with communications. Klaus Krippendorf says, "Despite the tremendous advances in communication and information technology, communication among people in organizations leaves much to be desired." (Gibson, et al., 1997)

Table 3.1: Poor versus Good Listeners (Dinsmore, 1999)

The Poor Listener:

- Always interrupts
- Jumps to conclusions
- Finishes my sentences
- Is inattentive; has wandering eyes and poor posture
- Changes the subject
- Writes everything down
- Doesn't give any response
- Is impatient
- Loses temper
- Fidgets with pen, pencil, or paper clip nervously

The Good Listener

- Looks at me while I'm speaking
- Questions me to clarify what I'm saying
- Shows concern by asking questions about my feelings
- Repeats some of the things I say
- Doesn't rush me
- Is poised and emotionally controlled
- Reacts responsibly with a nod of the head, a smile, or a frown
- Pays close attention
- Doesn't interrupt me
- Keeps on the subject until I've finished my thoughts

Plan to Communicate

In order to put effective communication into action, one additional tool can be very helpful: a communication plan. Someone once said, “There are three kinds of people in the world: those that make things happen, those that watch things happen, and those that wonder what happened.” A plan helps you to “make things happen.”

When making a communication plan, include (1) what to communicate, (2) with whom to communicate, (3) frequency and when to communicate, (4) format (memos or reports, for example), (5) how to communicate—the medium: formal or informal meeting, email, newsletter, voicemail, etc., and (6) who will do the particular communication effort. When making the plan, seek input from the key people who will likely be involved in the needed communication process. Their input can prove invaluable by helping you to understand their needs. A simple format for developing a communications plan is shown in Figure 7.

COMMUNICATIONS PLAN FOR PROJECT:______________________					
What to Communicate?	To Whom?	Frequency?	Format?	How? (Medium)	Who Will Do It?

FIGURE 7 Example Format for a Project Communications Plan

Summary

In summary, remember: (1) good communication is critical to success, (2) communication is the process of becoming “like one” or “as one,” (3) to communicate well, you have to truly care about other people enough to sincerely want to understand them, (4) “Seek first to understand, then to be understood”—empathically listen, and (5) make a communications plan and use it!

References

Covey, Stephen R., *The Seven Habits of Highly Effective People—Restoring the Character Ethic*, Simon & Schuster, New York, 2004.

Babcock, Daniel L. and Lucy C. Morse, *Managing Engineering and Technology—An Introduction to Management for Engineers*, 3rd ed., Prentice Hall, New Jersey, 2002.

Dinsmore, Paul C., *Human Factors in Project Management*, revised edition, AMACOM, American Management Association, New York, 1990.

Gibson, James L., John M. Ivancevich, and James H. Donnelly, Jr., *Organizations—Behavior, Structure, Processes*, Irwin/McGraw-Hill, Boston, Massachusetts, 1997.

Gido, Jack, and James P. Clements, *Successful Project Management*, South-Western College Publishing, Cincinnati, Ohio, 1999.

Mintzberg, Henry, *The Nature of Managerial Work*, Harper & Row, Publishers, Inc., New York, 1973, p. 39.

Rees, Nigel, ed., *Quote…Unquote*, St. Martin's Press, New York, 1979, p. 79.

Shainis, Murray J., Anton K. Dekom, and Charles R. McVinney, *Engineering Management—People and Projects*, Battelle Press, Columbus, Ohio, 1995, p. 33.

Planning and Conducting Effective Meetings

Abstract ID: 481745
Paper # 330d
Dr. Eldon R. Larsen
Marshall University

Meetings can be viewed as necessary evils, valuable experiences, inspirational events, or intolerably boring curses. The very word *meeting* can cause various emotions to arise within us. This paper presents various rules and suggestions that lead to more effective and more enjoyable meetings. The approaches are founded on common sense and solid principles of management and human relations that are frequently overlooked. The rules and suggestions focus on reaching meeting objectives, effectively using agendas, nurturing good human relations, and ending meetings with an effective plan of action.

Introduction

Many people abhor meetings. Perhaps you have had similar feelings; many people do. Why do people dislike meetings? Common complaints are that meetings are too long, not well organized, pointless, and a waste of time. Some of the major complaints (Dressler, 1996) about meetings are as follows:

- Their purpose is unclear
- Participants are unprepared
- Key people are absent or missing
- The conversation veers off track
- Participants don't discuss issues, but instead dominate, argue, or take no part at all
- Decisions made during the meeting are not followed up

A survey of 78 project management professionals indicated that meetings were the major time consumer in their daily schedule, those meetings taking up an average of slightly less than 2 hours per day. "The survey showed that the participants attended over six meetings per week. In those meetings an average of 25 percent of the time was spent engaging in off-the-track banter, generalized complaining, lobbying for pet causes, and other unproductive activities. When respondents were asked, 'What do you think contributed to the eventual non-productiveness of the last meeting you attended?' the three most prevalent responses were (Dinsmore, 1990, 139-140)

- Improperly planned meeting (no written objective or detailed agenda)
- Inept leadership
- Undisciplined participants"

When experience with meetings is not good, people can actually get a negative conditioned response to the very word *meeting*. Yet, meetings can and do play an important part toward facilitating progress and promoting understanding among people—a key to successful management and leadership. How can this be accomplished more effectively?

Rules and Suggestions

To plan for and to conduct effective meetings, project managers and meeting participants generally need to follow ten basic rules. When used *appropriately*, these rules lead to success. They are founded upon solid principles of management and good human relations.

Rule One. *Clearly identify the meeting objective. And, if the meeting is formal, write the objective and distribute it to all concerned in advance.* This is a vital, and perhaps the most important, rule for conducting effective meetings.

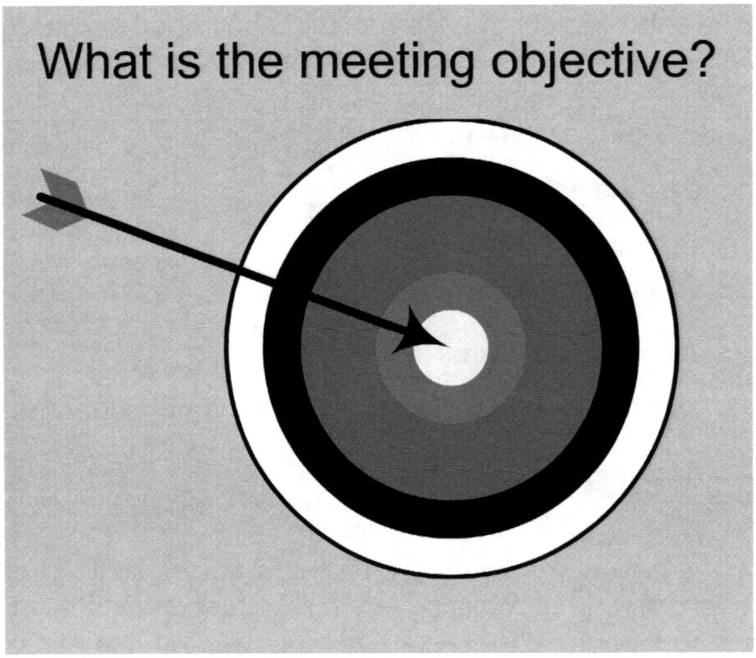

FIGURE 1. Clearly Identify the Objective of Every Meeting

Each semester in my graduate project management courses, I ask my students if they are given the objectives of the meetings they are asked to attend. For every one hundred students, I normally have less than five who are given the meeting objectives. This is disturbing when one considers that the majority of my students are full-time employees in industry, where productivity is of great concern.

When deciding to hold a meeting, a leader must determine among other things why the meeting is important and what is to be accomplished. The leader should ask, "Is this meeting necessary?" and "What is the objective or purpose of the meeting?" Answering these questions will essentially answer whether it is also necessary to meet. Though the meeting objectives might seem obvious to the leader, they frequently are not obvious to the participants. If the meeting objective is obvious, then it should not be difficult to write that objective; yet this could indeed be a challenge to many meeting participants. If you can't write down the objective, you don't truly understand it.

Principle:

If a meeting doesn't have a clearly understood purpose, don't have it! If it is going to waste people's time, either change it or cancel it!

Some generic areas around which meeting objectives might be formulated include, for example,

- Solving problems
- Brainstorming new ideas or solutions, etc.
- Making decisions
- Sharing information (this can be a weak reason to hold a full-group meeting, since much information sharing can be done through e-mail, personal conversations, or sub-group meetings)
- Giving feedback (again, this too can be a weak reason to hold a full-group meeting, as much feedback if likely to more appropriate on a one-on-one basis)
- Planning
- Selling
- Teaching
- Conducting design reviews

Before we leave the first rule for effective meetings, there is an important point to understand. Objectives are much more effective if all key people involved agree generally on those objectives. Taking time in advance of the meeting to clarify and reach agreement on meeting objectives with all key players can go a long way toward avoiding difficulties in the meeting itself. A few advantages of doing this include

- Everyone is focused on what is to be accomplished

- Participants are less likely to come with hidden agendas
- The meeting agenda will more likely be clearly oriented to accomplishing the meeting objectives
- Agenda items not contributing to the meeting objective can and should be dropped
- The agenda items can be more easily prioritized
- The participants have a greater sense of purpose and understanding—they will behave more like a team

Rule Two. *Clearly define the meeting agenda, focused solely on accomplishing the meeting objective.* For formal meetings this agenda normally should be agreed upon in advance by the meeting participants, along with the meeting objective.

The meeting agenda should not include items that do not contribute to the meeting objective. Usually items that do not contribute to the meeting objectives can easily be communicated by some other means, such as one-on-one interaction, e-mail, or telephone conversations. Many such items also do not involve all the meeting participants anyway. Invite only those who will benefit directly from, or contribute directly to, the meeting.

Effective agendas include not only the topics to be discussed, but also who is expected to lead each topic's discussion or presentation and how much time is allotted for that agenda item. Allocate and prioritize time based on importance first, and then urgency—if an agenda item is not important, drop it. Urgency does not equal importance. As Covey (2004, 150-151) explains, "*Urgent* means it requires immediate attention…Urgent things act on us…We react to urgent matters…*Importance*…has to do with results. If something is important, it contributes to your mission, your values, [and] your high priority goals." For an excellent discussion on the difference between *urgent* versus *important*, see Covey (2004, 149-168).

Use wording to describe clearly the agenda items so that participant understanding is well focused. To further clarify an agenda topic, the leader might want to consider writing down the desired expectations next to each agenda topic. For example, if an agenda item is "Forecasts of business income," the leader might add the expectation note that "Please be prepared to decide as a result of these forecasts whether to increase production of XYZ." Making expectations clear will greatly enhance the effectiveness and efficiency of the meeting.

The items on the agenda should be prioritized with the most important items being covered as early as possible. Then, if the meeting appears to be running behind, the less important agenda items can be postponed to a later meeting without serious consequences.

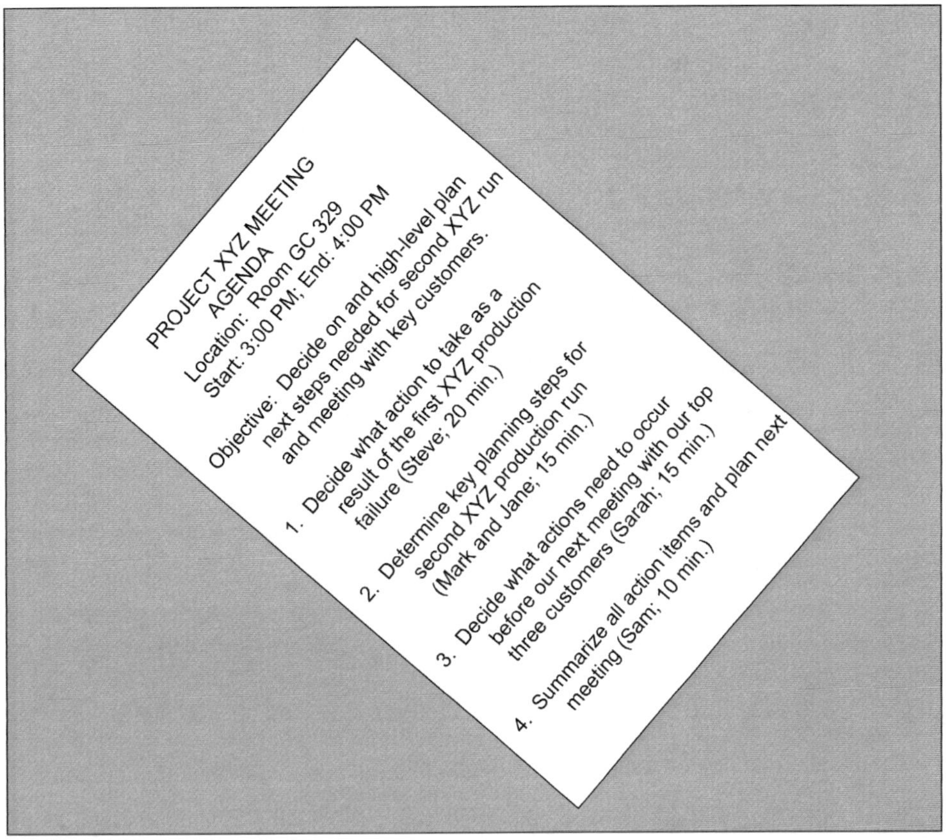

FIGURE 2. Clearly Define the Meeting Agenda to Accomplish the Objective

Be sure to schedule enough rest breaks as part of the agenda. These breaks should occur at least every one and a half hours or less, as longer periods between breaks will tend to decrease meeting effectiveness. Ten-minute breaks are usually sufficient, and should be no more than fifteen minutes except in unusual circumstances. Start promptly at the end of the break!

When distributing the agenda, include any needed background or supplementary reading material and data that will help participants prepare effectively for the meeting. Do not include material that does not directly contribute to the meeting objective.

Rule Three. *Have clear start and end times for the meeting, and treat these as a contract.* Most meetings have a clear start time. It should be firm—start on time! Doing otherwise will lead to poor meeting behavior and punish those who try to be on time.

Unfortunately, many leaders do not plan meetings to have a clear end time. This is a mistake made by meeting leaders who are not sensitive to the courtesies that ought to be shown to other participants in the meeting—their time is important and they, too, need to plan their day. To do otherwise will lead to frustration for those involved.

Once the meeting agenda has been determined and it includes time allotments for each agenda item, there should be little difficulty in deciding on when the meeting should end. Wise meeting planners include a little "contingency time" toward the end of the meeting; but this is *not* for sloppiness—it is for unforeseen events which honestly could not have been planned for. Of course other factors must also be considered such as travel schedules, meal times, and so forth. But, once the schedule is set, stick to the overall schedule. This means don't overrun the end time—you have a contract, treat it that way. On the other hand, few people, if any, will mind if the meeting ends early.

FIGURE 3. Have Clear Start and End Times, and Treat as a Contract

This does not mean that individual agenda items cannot be adjusted mid-stream. The agenda is simply a plan to reach the meeting objectives. The plan is your servant, not your master—it is there to help you! Some adaptation of the agenda might be needed to reach the objectives. Caution is in order, however—the agenda should not be so flexible that control and discipline are lost and the meeting objectives are not reached.

A final note on the time contract: It is common for the meeting to approach the end time with agenda items still to be covered. As pointed out already, the consequence of dropping off the last few agenda items will be small if the agenda has been prioritized. Still, there is frequently a temptation, especially among meeting leaders, to feel that they must get through the agenda and, therefore, they don't want to quit. The decision to run over time is *rarely* a good one. Some consequences of running overtime include

- Loosing participant attention
- Hastiness leading to poor decisions
- Loss of trust for future time contracts
- Ill will toward the leader

Sometimes we feel that most of the participants will not mind staying beyond the stop time, and we can let the one or two people who need to leave do so. While this may indeed be the case, remember that there is a reason for all participants to have been invited, including those one or two who can't stay. Otherwise they shouldn't be there. When you tell someone that it is all right for them to leave and the rest of the group will continue, you basically say, "We don't need you," or else, "We no longer need your input." Even if you state that you don't want them to get these messages, they frequently will.

This can cause feelings of loss of respect and can cause resentment, even though these results are not intended.

People who are team players will want to be involved with the discussions and decisions—don't leave them out. Conversely, the remaining team members might well feel resentment at the pressure to stay in the meeting when others are allowed to leave. It is nearly always better to keep the time contract and to schedule another follow-up meeting if it is truly needed. There may be rare exceptions, but the key word here is *rare*.

Rule Four. *Be prepared for the meeting, both in content and in process.* It is the responsibility of the meeting leader to assure that the meeting will be valuable for the participants. Proper pre-meeting planning and preparation for both content and meeting process are essential steps leading to success in making the meeting valuable.

First, *content preparation*. The agenda should be issued enough in advance so that each participant will have adequate time to be content-prepared as needed for his part. Participants should be expected to come prepared with any useful visual aids, handouts, demonstrations, reports, data analysis, and so forth, as appropriate. Use of visual aids should be encouraged, since they will help participants to better communicate and to stay focused on their assigned topic without rambling. In addition, visual aids usually greatly help in understanding.

Meeting participants should expect each other's assignments to be completed on time, and they should be prepared themselves. The expectation of disciplined preparation should be taught and used well. However, the leader should remember that this is a two-way responsibility and that he must do his part in working with the other participants in advance in a win/win approach so that these expectations can be achieved—help them prepare in any way that is needed. The leader especially must be prepared or the others will quickly see hypocrisy and often choose to follow the leader's bad example.

Additional step to take before the meeting, which will improve content preparation, include

- Consider the need for some participants to make *pre-presentations* to key decision makers before the meeting; some presentations might need to be made before the meeting to key people to prevent embarrassing surprises, to enhance time for understanding and communication before the meeting begins, and to get early feedback and buy-in. Sometimes this is called *pre-positioning.* It can be quite powerful in making meetings more effective. Meeting energy will be more focused, decision-making will be more efficient, and deadlocks will be fewer or overcome.

- As needed, determine and clarify the roles and responsibilities of all participants. This could be done one-on-one between the leader and each participant before the meeting. Some example roles and responsibilities are given in Table 4.1.

- Engender expectations among meeting participants that the meeting success will be measured mainly by how well the meeting objectives are achieved in a matter satisfying to the participants.

- Anticipate potential interpersonal communication challenges and work toward alleviating these *before* the meeting.

- Assure that all handout material is prepared in an organized way for all participants.

Second, *process preparation.* Process preparation involves assuring adequate facilities and supplies are provided to make a comfortable meeting, as well as being prepared to run the meeting in a pleasing way. There are many considerations with respect to being process-prepared for the meeting.

You should plan the meeting in a way that will maximize participation toward reaching the meeting objective; when possible, consider trying to decrease presentation time and increase time where all can participate together.

The following process items, as appropriate, should be checked well enough in advance of the meeting so that changes can still be made as needed.

- Make meeting room arrangements, including, but not limited to:
  - ✓ Reserve the right-sized meeting room and additional rooms as needed
  - ✓ Assure room temperature is properly adjusted and will not be distracting; find out *in advance* how to get it adjusted if needed
  - ✓ Assure seating arrangement fosters participation; can squeaky seats be oiled?
  - ✓ Test visual-aid equipment in advance to assure it works properly; focus projectors; check visibility from all participant locations
  - ✓ Assure miscellaneous supplies are available, such as tape, markers for white boards and flip charts, sticky notes, pencils, notepads, extension cords, flip charts, video-tape player, lap-top computer, push-pins, name-tags, microphones, projectors and screens, erasers, and pointers
  - ✓ Arrange for refreshments as appropriate, well in advance and then confirm
  - ✓ Assure adequate trash containers are available
  - ✓ Check for adequate lighting and learn to adjust it as needed
- In advance of the meeting, not during, arrange for someone to take minutes and clarify expectations for those minutes
- Are there other environmental distractions that could cause participants difficulty in paying attention or make the meeting seem too formal and not relaxed? Noise from adjoining facilities or from in-room heating and air-conditioning equipment, for example, can be unexpectedly detrimental.
- Understand the decision-making authority of the participants; find out who has authority relative the meeting objectives and find out how the group can tie into that authority quickly and effectively if need be.
- Understand what financial arrangements need to be made for the meeting and how these will be executed. Then, make sure the arrangements are made.

Table 1: Some Example Roles and Responsibilities

Example *Group-Member* Roles:

- Be prepared; fulfill all assignments on time
- Empathically listen; Communicate, communicate, communicate
- Help plan group work
- Set objectives for and plan work in alignment with the group objectives and plans
- Help establish and then obey group conduct rules
- Actively contribute toward good teamwork
- Be clear, concise, and don't dominate; but be a valuable contributor

Example Leader Roles (in addition to Group Member):

- Lead, but don't dominate
- Be responsible for meeting objectives and agenda
- Complete all pre-meeting arrangements
- Build teamwork environment; praise contributions
- Work to empower group members
- Set the example in true communication
- Lead the team in reaching the meeting objectives
- Be responsible for meeting follow-up, help people account for responsibilities

Example Minute-Recorder Roles (in addition to Group Member):

- Summarize the essence of important discussions, decisions, and reasoning
- Avoid direct quotations with names
- Clearly summarize discussion perspectives, to avoid re-hashing of these points
- Issue minutes promptly to all invitees, following approval by the leader

Example Facilitator Roles, sometimes the Leader, or another group member:

- Help plan the meeting to maximize participation and reaching meeting objectives
- Seek to decrease presentation time and increase team participation time
- Arrive early; keep the discussions on track; eliminate side conversations
- Empathically listen and help others to do the same
- Value the differences in people; then protect people from being attacked
- Help establish and enforce meeting rules agreed-upon by the group
- Facilitate processes for taking needed actions
- Assure clarity of decisions and assignments, who will do them, when they are to be completed, and to whom to report their completion
- Help leader maintain proper meeting environment and assure all participate
- Be the timekeeper
- Help assure that the minute-recorder is recording the needed information
- Make an "Issues Parking Lot" flip chart with issues to discuss later
- Lead a brief meeting-process evaluation: what went well; what to improve

- Finally, clearly understand how you will conduct the meeting and deal with possible challenges. To help prepare for this *process* part of meeting planning, consider:

  ✓ If this will be the first time the group has met, draft proposed rules of group conduct for the group to revise and achieve consensus on during that first meeting; also, orientation kick-off materials should be prepared for each participant. If the rules have already been established at an earlier meeting, print a list of the meeting rules for all participants.

  ✓ Decide how you will deal with unexpected issues brought up during the meeting, but which are not on the agenda—for example, will these issues be placed on an "Issues Parking Lot" flip chart to be later discussed off-line, at the end of the meeting, or at a later meeting?

  ✓ Will you allow some potential alterations to the agenda mid-meeting, and, if so, how will this be done? For example, the group might reprioritize the agenda under your direction.

  ✓ Consider win/win approaches to group discipline, if the need arises. If discipline is negative, you will lose the participant and you will likely lose the respect of everyone else present.

  ✓ How will you work to reach consensus on issues during the meeting; also, what will you do if the group cannot reach consensus?

- **Rule Five.** *Near the beginning of the first meeting, the group should establish its own rules of conduct; these rules should be revisable by the group at future meetings upon request and by consensus.* Note that the word *consensus* implies general or harmonious agreement of the group members—they might not each agree one hundred percent, but they do agree willingly to go along with the group's rules toward achieving the group objectives. Note also that some of the group's meeting rules might need to have clearly defined consequences if they are broken. Some possible examples of group meeting rules of conduct are shown in Table 4.2.

- Although it is possible for groups to function without agreed-upon rules for meeting conduct, "Common, or at least harmonious, goals or purposes are…not criteria of groups, but of *effective* groups." (Hersey et al., 2001, 318)

- While considering meeting rules, don't forget that one of the primary objectives a leader must have is to build the group and to positively develop the individual group members. All meetings, and especially the first one, should strongly engender member participation and teamwork. Group and individual morale is critical to achieving group objectives. Overly oppressive rules can backfire. The group members need to feel welcome, needed, and honored to be present.

Rule Six. *Understand the fundamental meaning of the word* communication *and apply it well and continuously.* The leader and meeting members should use this understanding before, during, and after the meeting. The word *communication* means literally "the process of becoming like one or as one." By this definition communication cannot be one-way. The process of becoming like one involves much give-and-take, much effort to try to put oneself in the other person's shoes, or as Covey (2004, 237) says, "Seek first to understand, then to be understood." It is truly a coming together in understanding and purpose.

Before, during, and after group meetings this concept of communication, if applied diligently, will reap tremendous rewards toward having meetings be truly effective. This application of communication will aid a group to build true team spirit and unity in working toward and achieving a common goal. It is also a builder of human dignity and effective human interpersonal relations.

TABLE 2: Example Meeting Rules of Conduct (select less than 7)
• We arrive early; start, and end on time
• We will work to achieve the meeting objectives
• We will come prepared and stay on track
• A person missing more than two of four consecutive meetings may be dropped from the group, subject to a majority vote
• To meet, 75% of the group must be present
• We strive to accommodate all members when scheduling group meetings
• We will follow good behavior etiquette and courtesy toward each other and strive to achieve win/win results as much as possible
• Meeting minutes issue within three working days
• We practice Empathic Listening; Seek first to understand, then to be understood
• We will not dominate discussions, but we will actively participate
• We will not do or say things that discourage other members from participating
• We build on each other's ideas
• One person talks at a time; no side conversations
• No gossip
• We don't make remarks that are sexist, racist, prejudiced, sarcastic, profane, or negative to any religious, social, or other organization or person
• We evaluate things, issues, and processes, not people, unless the meeting purpose is specifically to evaluate personnel for feedback and building
• Ideas belong to the group, not the individual
• The group speaks with one voice after decisions are made; leave united
• Turn off cell phones, beepers, and alarms
• Keep your cool. If you lose it, you are wrong

Rule Seven. *Show by example and teach group members how to empathically listen to each other.* Again, as in Rule 6 above, this applies before, during, and after the meeting. Empathic listening is essential to communication as described above. It must be based on sincerely caring about the other people involved; otherwise it will be perceived to be fake or manipulative. One must seek to really care about other people. "The…listener's full attention is on the content of the message *and the intention* of the speaker." (Hersey et al., 2001, 305; italics added) Listening for the speaker's feelings is just as important as listening for the facts the speaker presents. See Figure 4.

Following are some suggestions you and your group members can use to improve empathic listening.

- *Listen for the concepts and details of the message*. Make an effort to really listen precisely what the speaker is saying.
- *Listen for the feelings of the speaker.* Try to perceive the speaker's feelings, desires, and emotions an about what is being said through the way that the message is delivered—try to really put yourself in the other person's shoes and try to see things they way they see them.
- *Respond to the feelings of the speaker.* Sincerely reflect to the speaker what you recognize and understand the feelings expressed to be.
- *Watch for the speaker's verbal and nonverbal messages.* Attempt to identify the real messages the speaker may be expressing.
- *Reflect back to the speaker the concepts and feelings you are hearing*. In your own words try to share with the speaker what you think the speaker said and felt. Listen to how the speaker responds to your reflection and interact iteratively until you both understand each others messages and feelings. (Based in part on Hersey et al., 2001, 306, quoting Rogers, 1951, and on Covey, 1989)

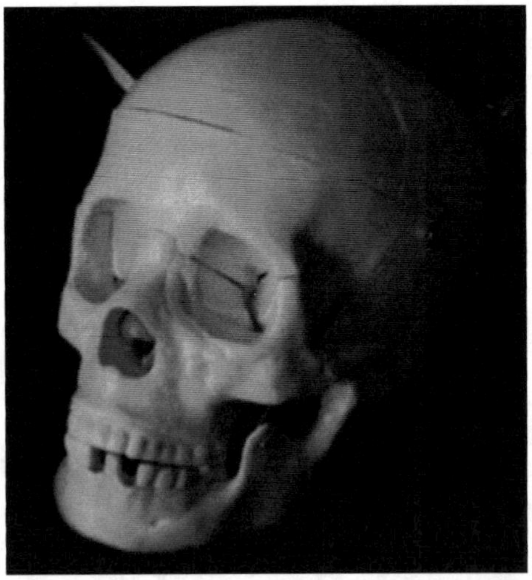

FIGURE 4. I've Been Waiting a Long Time for You to Listen to Me

The advantages of empathic listening are many. Here are a few.

- You will better understand the perspective of the other person, which will place you in a better position to cooperate,
- People will know you care about them,
- You obtain feedback,
- You gain better perspective and understanding,

- Mutual acceptance and respect are engendered,
- You will more easily achieve win/win solutions,
- You will more easily achieve synergy among the group members (Covey, 1989),
- Your group will be less likely to make costly mistakes due to misunderstandings,
- Your group will be more empowered due to better understanding and knowledge, and
- Your group will have better morale.

Many years ago I went to my manager's supervisor with a concern about how we could increase the availability of current technical literature for the engineers in our company. As I entered the supervisor's office, he shut the door behind me, sat down at a desk with me, and made me feel at ease. We had a little meeting. As I explained my concern, the supervisor took out a notebook and carefully noted my comments. He asked clarifying questions to make sure he understood. He asked for my suggestions and feelings. He summarized what he thought he heard me say and made certain that he understood correctly. Finally, he seemed genuinely appreciative that I would give him these comments and suggestions. As I left this man's office, I felt both that he had fully understood and that he cared. I also remember thinking that it really didn't matter to me anymore whether the supervisor followed my recommendations or not. Because I felt so much that he really understood me, I felt I could trust his judgment to do what was best. He had won the battle by empathic listening, whether he took any action or not.

Rule Eight. *Create a supportive and task-oriented environment during the meeting.* This rule will be accomplished as the Rules 1-7 above are followed, coupled with a dedicated effort by all meeting members to accomplish the meeting objectives. The leader and, if present, the facilitator play the crucial roles in accomplishing the benefits of this rule. They need to stay focused on the objective and care about the people! A leader who does not follow this rule will be the major cause for a miserable meeting; conversely, a leader who applies it properly will almost always have an effective meeting and one people can positively remember.

A task-oriented environment will help the meeting stay on track toward the meeting objective. Sometimes it is useful to have the team participate directly in the effort to stay on track. At the beginning of one meeting I attended, the leader handed to each member a sheet of paper with the words "We Are Off Track" in large bold letters. Anytime any group member felt the discussion was going off track, that member could hold up his sign. The result was both fun and effective.

In a supportive and task-oriented environment, meeting participants help each other, rather than hurt each other. For example, it would be inappropriate speak degradingly to or to yell at someone in front of other people. While there could be very *rare* exceptions, it is almost always not good practice to discipline someone in front of other people.

It is not appropriate to "launch missiles and throw grenades" during a meeting. Launching a verbal attack on someone in front of other people alienates them and makes a fool of you. If you do this you will lose the respect of the other meeting participants as well. If you lose your cool, you are wrong. People have been asked to participate in the meeting for a reason—or at least they have if you are following good meeting rules. Do not alienate them or they will not be able to fulfill their purpose in the meeting—they might be there in body, but they will not be there in spirit and will likely stop contributing. Thus, everyone loses, especially you!

In a supportive and task-oriented environment, meeting participants don't hold "Two meetings at the same time"—meaning that they don't carry on side conversations during a meeting. See "Don't hold two meetings at the same time." (STORY)

FIGURE 5. Create a Supportive and Task Oriented Environment

Don't come with hidden agendas to a meeting. You are out of place to do this. If you do not agree with the agenda, try to resolve this is one-on-one with the meeting leader, *well before* the meeting. Occasionally, if several of the potential meeting participants do not agree with the agenda, they could all meet in advance with the meeting leader to try to work out a compromise. In any case, when the meeting begins the time has passed to come with your own agenda.

In a supportive and task-oriented environment meeting participants sincerely phrase critiques by beginning with words such as "I wish I understood better…" or "Tell me more about how…" By doing this you might find that one of the following happens: (1) As the person explains more about their idea, they may actually convince themselves that it isn't what they initially thought, (2) As the person explains more, you might realize that it isn't what you initially thought, (3) As the person explains more, he feels so glad you are actually listening that he doesn't feel a need to pursue that idea forcefully—if you don't listen he is likely to pursue it forcefully, (4) you might find that you actually agree, or (5) you might get an new idea inspired by the other person's explanation, which idea you might not have thought of otherwise. The bottom line is to be supportive in your interactions with other people and their ideas.

Finally, in a supportive and task-oriented environment, all meeting participants should work to achieve the meeting objective(s) and not waste people's valuable time.

Rule Nine. *Close the meeting effectively.* The following hints can be very effective in closing the meeting.

- Assure that assignments and other action items are clearly defined. Use action verbs in the description of all assignments—action verbs tell what action to take and they are powerful
- Clearly decide who will do each assignment
- Clearly decide when each assignment is to be completed. If at all possible, the person who is to do an assignment should be the one to decide when that assignment will be completed
- Decide to whom the person doing an assignment will account for its status and completion—this is almost always forgotten, but can be a very helpful clarification leading to high effectiveness
- If this group is to meet in the future, have the group do the following three actions
    - ✓ Draft an objective and agenda for the next one, two, or possibly even three meetings—these are only drafts but will keep the group focused on the long-term vision of where it is going and give it a fuller sense of purpose and the need for timely performance. Note: This will be too time-consuming if built from scratch in the meeting. In advance, the meeting leader should have prepared rough drafts of these future meeting objectives and agendas for the team to revise as appropriate; otherwise this step will become disliked by the team.
    - ✓ Set the time and place of the next meeting. Do this after the objective and agenda are drafted for the next meeting(s) so that the group can identify how soon they should meet.
    - ✓ Critique the current meeting. What went well? What could be done to make it better? —This critique should not last longer than five to ten minutes or the group will not want to do it again. Keep it positive! I usually ask the group to complete the following two phrases: "I liked…" and "I wish…"

Rule Ten. *Follow up.* The meeting leader needs to build trust, accountability, and respect by following-up on assignments and action items in a caring but timely way as appropriate for each meeting member and each task to be completed. There are two objectives here:

(1) build teamwork toward accomplishing the assigned tasks and (2) build the meeting members so that they progress based on their individual situations.

Recommendations

How does one go about implementing the above rules? At first, it could seem overwhelming. It takes time and experience to implement these rules correctly. However, if meeting participants will keep the end in mind—that they want to reach their objectives in a way in which everyone wins—then they will have made the first and most important step. For many people good meeting management will be an easy and exciting step forward; for others it will be a culture shock. In all cases it is needed.

The above rules, lists, and suggestions are really tools and principles that need to be applied according to the needs of the situation and the people. Small steps at first are appropriate so that people can learn them and buy-in to the better ways. It will not take long for meeting participants to see the benefits and the need to change. To run too fast into full implementation of these rules could cause rejection due to misunderstanding and the natural tendency by many of us to hold to the *status quo*. Care deeply about your people, but persistently move forward toward helping them to have more effective meetings. Most people will welcome the needed changes.

Conclusion

Conducting effective meetings takes a concentrated effort, but the return on the effort invested will pay big dividends both during the meeting and for the long term. Objectives will be clearer and will be achieved. Morale will be higher, as will productivity. There is really little or nothing to lose by doing it right, but a great deal to gain. Meeting members will want to come back again, rather than dreading the words, "Another meeting."

References

Covey, Stephen R., *The 7 Habits of Highly Effective People—Restoring the Character Ethic,* Simon & Schuster Inc. (1989).

Dinsmore, Paul C., *Human Factors in Project Management,* AMACOM (1990).

Dressler, C. 1996. Cited by J. P. Lewis, *Team-Based Project Management,* AMACOM (1998). Cited also by Karl A. Smith, *Project Management and Teamwork,* McGraw-Hill (2000).

Gido, Jack, and James P. Clements, *Successful Project Management,* South-Western College Publishing (1999).

Hersey, Paul, Kenneth H. Blanchard, and Dewey E. Johnson, *Management of Organizational Behavior—Leading Human Resources,* 8th ed., Prentice Hall (2001).

Meredith, Jack R., and Samuel J. Mantel, Jr., *Project Management—A Managerial Approach,* John Wiley & Sons, Inc. (2000).

Rogers, Carl, *Client-Centered Therapy: Its Current Practice, Implications, and Theory,* Houghton Mifflin (1951). Cited by Hersey, Paul, Kenneth H. Blanchard, and Dewey E. Johnson, *Management of Organizational Behavior—Leading Human Resources,* 8th ed., Prentice Hall (2001).

The Importance of Excellent Definition of Project Objectives

Abstract ID: 481746
Paper # 330e
Dr. Eldon R. Larsen
Marshall University

Effective project objectives are vital for successful project management. They tell us what we want the end service, result, or product of a project to be. Surprisingly, most project management texts go little beyond acknowledging the importance of good objectives. The *"How to"* is either missing or given only very brief discussion. This paper presents some basic principles and tools helpful in setting effective objectives. The tools help address questions around project scope, decision-making trade-offs, and tolerance between what is ideal and what is failure. These same tools aid in communication between the project team and other key players so that they all understand the intended project outcome.

Introduction

Clear, specific, effective objectives give us the best picture of a project's desired destination. "Objectives are the aims or end points toward which activity is directed." (Babcock, 1991) Someone once said, "I always wanted to be successful, but I guess I should have been more specific. I should have defined success."

When working on projects, what do the project objectives really mean? Do we really understand them? Do other people understand them the way we do? How will we know when the objectives have been achieved? Is there any room for variation if we have trouble reaching the ideal? Have key people agreed upon the objectives? What constraints are we facing? Which constraints are most important and which least important?

These and other questions are the topic of this paper as we investigate one of the most important parts of starting projects right—how to develop effective project objectives.

Tool 1: Make Objectives SMART

One very helpful tool for developing effective project objectives is the acronym SMART, described in **Figure 1**. Project objectives need to be specific, measurable, agreed-upon, realistic, and time-oriented. A weakness in any one of these areas can be either like a hidden time bomb ticking away and ready to blow up at any time, or like a cancerous problem that slowly grows until it either kills the project or requires it to have major surgery and expensive life-saving resuscitation. Of the many variations on what the letters in the acronym SMART stand for, the author believes the one presented in **Figure 1** to be most useful. It is based on Smith (1991), but the author has expanded the definitions for clarification.

S	**S**pecific:	Targeted, focused, clear, concrete, understandable, documented, not abstract, not general, not overly complex
M	**M**easurable:	You can tell when you've reached it, targets are clear, tangible, verifiable, and quantified if at all possible
A	**A**greed-Upon:	Everyone involved is aligned; owner, senior management, project leader, project team, and other key people have approved and committed, preferably in writing
R	**R**ealistic:	Challenging, but attainable; not impossible to achieve nor viewed that way; consistent with resource and time constraints; not too rigid to accommodate changing priorities
T	**T**ime-Oriented:	Clear time frame for project completion

Figure 1. Objectives Should Be SMART

Specific. A project objective that is *specific* will be targeted, focused, clear, concrete, understandable, documented, not abstract, not general, and not overly complex. What we are trying to do with the objective statement is to be very good artists and to paint a very clear picture for everyone involved with the project. That artistic picture should not be like the impressionistic artistic style such that the words appear pleasing from a distance, but when you get close up there is not enough detail to understand what is intended. Instead the style needs to be more along the realistic artistic style, where both far-away and close-up examinations provide good understanding.

Part of being specific is to put the objective in writing. Sometimes for small projects we feel this is not worth the trouble. However, the act of putting an objective in writing forces us to clarify it. If you can't write it down, you don't understand it.

Measurable. The project objectives also need to be *measurable*. How will we know when we have achieved the desired outcome? We need to be able to measure this in some specific way. The objectives need to be clear, tangible, verifiable, and quantified if at all possible. "You can't prove a project's success if there aren't any measures for it." (Frye and Mason, 2001)

I once worked on a research project for approximately four years with a very fine team of people. We met regularly and were very familiar with what was happening on the project. Yet, after four years I started hearing comments in the team meetings that in effect said, "Are we finished now?" Some members thought the answer was "Yes." Others thought "No." Upon examination of the objectives, we could not tell. The objectives were neither clear enough nor measurable enough to allow us to make that decision—after four years!

In making objectives measurable, one key is to decide both *what* you want to measure and *how* you will measure it. The *what* and *how* both need to be specific enough so that everyone knows when they are met. Sometimes the measurable characteristics of the project objectives are called metrics. Regardless of the terminology, we need to answer the *what* and *how* measurement questions.

For example, one of the objectives of a project might be to build a sailboat that is thirty-six feet long. One part of the *what* might be the dimension thirty-six feet. *How* we can measure the objective might be with a tape measure. This example is fairly easy.

Another example might be that we have an objective for the people at our family reunion to enjoy the food provided. The *what* is a little harder to quantify here, to enjoy the food. The more difficult question is *how* to measure that enjoyment. We could have an end-of-reunion critique meeting in which the food is discussed; we could give everyone a written survey; or, we might come up with another more creative method of measurement. The point is that if we do not define how we will measure the objective, there is a good chance it will not be measured and we will not be able to determine if the objective has been achieved.

Let's look at one last example of a measurable objective. A hotel manager might have as an objective, "To increase the hotel owner's satisfaction with my management performance." Although this is a very nice desire, the objective is neither very specific nor very measurable. First, the hotel manager needs to define better what would affect the owner's satisfaction. Second, the manager needs to define better what the word *increase* means, as this could be anything from a very small change to a very large one. What does the owner want? What can the manager provide? As an example, perhaps a more specific and measurable objective might be, "To increase the average hotel room occupancy from the past value of eighty-five percent to a new value of at least ninety-five percent during the months of June through August of this year." This revised objective is much more measurable and specific.

Three cautions are important when making objectives measurable. First, watch out for common words that sound desirable, but are not specific and not measurable. A few examples are the words *good, excellent, pretty, high quality, long lasting, preferred,* and *successful*. To correct the problem with these words, simply express in quantifiable terms what you really mean. For example, the phrase "We will provide excellent help-desk service," might be changed to something like, "We will answer the telephone for ninety-five percent of all help-desk calls within 20 seconds and we will provide solutions acceptable, according to the customer, to at least ninety percent of the problems within one working day."

The second caution is to be careful not to skew your objectives just so you can easily measure them. This can lead to disaster just as badly as can the original immeasurable objectives. The challenge is to re-express your real objectives in a way that they are both clear and measurable as much as is realistically possible.

The third caution is to be very careful that what you choose to measure, and how you measure it, do not cause unwanted performance side effects. Sometimes while driving desired behaviors, we inadvertently drive new, unwanted behaviors. There is an old saying that you get what you measure.

Agreed-upon. The project objectives need to be *agreed-upon*. This means that everyone involved with the project is aligned and that, preferably in writing, the project owner, senior management, project leader, and other key people have approved and committed themselves to the project objectives.

Some project owners give the project team a set of objectives and simply expect them to go to work and achieve them. While this can work in some cases, experience shows that participant buy-in is usually very important for effective organizations. This is true as well for project organizations and teams.

By allowing a project manager, and where practical the project team, to have a chance to work with the project owner on the definition of the objectives, a far better set of project objectives is probable, and participant buy-in is more likely and much higher. In the long run, this extra effort pays big dividends and is more likely to produce win/win situations for the owner and the project participants.

When an owner gives a team project objectives with which they might not fully agree, and without any way to iteratively improve those objectives through two-way communication, that owner unintentionally is also often asking for ineffective results, inefficiencies due to poor morale or poor understanding, and possibly even project failure.

If used properly, the tools and concepts presented in this paper can actually be of great help in improving communication between the key people involved with a project and go a long way to facilitating agreement upon objectives.

Realistic. Project objectives should be *realistic*, meaning that they are challenging but attainable. They should not be impossible to achieve nor be perceived that way. They should be consistent with resource and time constraints. And, order to accommodate changing priorities when essential, they should not be too rigid. "Generally, the more difficult the goal, the higher the level of performance. But...if and when a goal is perceived as so difficult that it's virtually impossible to attain, the result is often frustration rather than achievement." (Gibson, et al., 1997)

One caution is worth mentioning with respect to making objectives challenging: Occasionally a project owner will allow conflict to exist within the project objectives, perhaps even intentionally, thinking that higher achievement will result. This is normally not a positive motivation for teams. Gibson, et al. (1997) report that a recent study "examined conflict as a goal-setting variable. Both a laboratory experiment and a field study of college professors suggested that conflicting goals can lead to decreases in performance...This research points out the need to pay attention to the total situation."

In summary, project team members will be better motivated both as individuals and as team members when objectives challenge them in a positive way, but do not overwhelm them due to either real or perceived lack of attainability.

Time-oriented. Finally, project objectives need to be *time-oriented*, meaning that there is a clear time frame for project completion. Sometimes, particularly in research and development, teams are given project objectives without any idea as to what the time-limit expectations are. Although there are situations where time limits are difficult to define, even in those cases an approximate time *range* in which a project needs to be completed should be describable. For example, a research project outcome and endpoint might be uncertain, but funding cannot continue indefinitely. In such a case, we might have to say something like "This project is to be completed within the next two to three years." Although some people might feel uncomfortable with this flexible time range, it is much better than saying nothing! And when there is high uncertainty, an ending time *range* in not inappropriate. That range can be narrowed down as time progresses and the project develops. We will talk more about *time* as part of our discussion of the triple constraints, below.

Tool 2: Make Full Use of the Triple-Constraints

Another important concept and tool is that of the triple constraints, illustrated in **Figure 2**. A constraint is something that imposes a limit or boundary. In project management, we normally speak of three major constraint criteria: the triple constraints. Although there are some variations of this concept, the author has found the one shown to be generally very useful, the three general constraints on the project being *time*, *resources*, and *performance* criteria. Let's examine each one.

Time Constraint. The *time* constraint, already introduced briefly above, is relatively easy to understand; at least that could be our impression. However, perhaps it is this impression of easiness that causes people occasionally to overlook adequately describing it when they write project objectives. The time constraint is very important. It basically involves identifying when the project is to be completed, at least roughly. Associated with the time constraint is how long it will take to do the project—the project duration. It is not the purpose of this paper to go into the many tools useful for planning a project and estimating project durations. The point here is that it must be done.

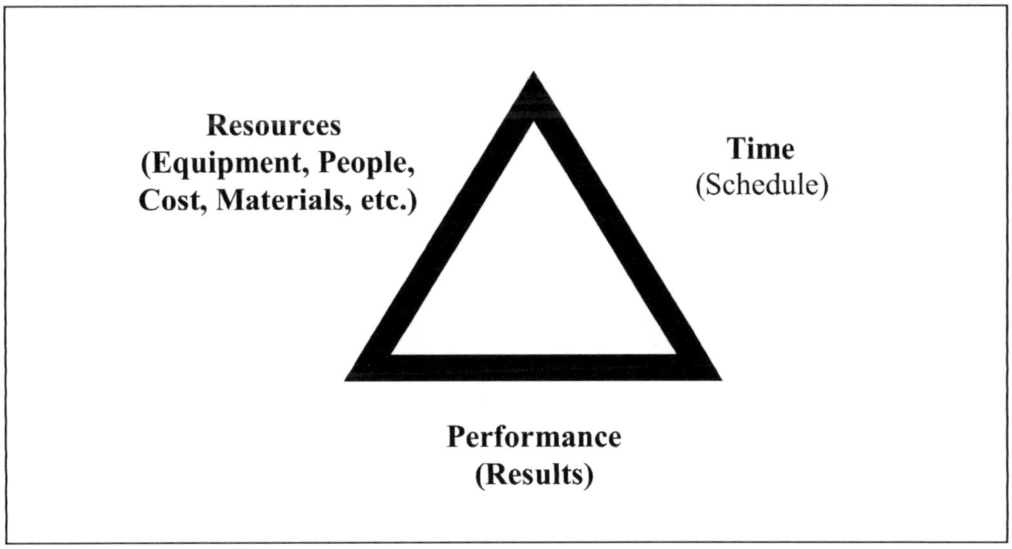

Figure 2. The Triple Constraints

It is not uncommon for the owner of a project to give a deadline for a project without due consideration as to whether the deadline is reasonable. Normally, to determine an initial, realistic time limit estimate, at least some high level planning of the entire project is needed. This initial time estimate can be refined later, and should be, but setting time limits without at least high-level overall project planning is not only unwise and likely to cause much frustration, it can also be disastrous to the project.

There are cases where a deadline is fixed due to reasons beyond the control of either the owner or the project team. For example, legal or regulatory implications, market competitions, or business and financial constraints might dictate a deadline. When this is the case, it is still very important to do at least a high-level plan to see if the deadline has a realistic chance of being achieved. Not doing this can cause a large amount of wasted money, effort, and time.

Resources Constraint. The *resources* constraint is a little more complex. Sometimes it is simply stated as cost or budget, but in reality it includes also items like people, equipment, materials, facilities, etc. When a project owner and/or project manager focuses only on budget as the resource constraint, the other resources of people, equipment, materials, etc., can be neglected until problems have had a chance to develop to the point that project execution is hampered. The message here is that all the project resources and their interactions need to be considered as a constraint on the project.

Performance Constraint. The third general constraint is *performance*, or results. It includes many possible specifications and is sometimes, but not consistently, referred to as project *scope*. Sometimes performance is defined as the project technical objectives, but in reality it can include much more, or, depending on the project, it could have nothing to do with anything technical at all. Examples of some of the items that might be considered for inclusion in the performance are given in **Figure 3**. Care must be taken not to neglect these possible aspects of performance when writing the project objectives. Some of the items in **Figure 3** were taken from Bitz (1991).

- Deliverables
- Quality
- Production rate
- Product specifications
- Customer satisfaction
- Stakeholder satisfaction
- Labor relations
- Safety
- Health
- Environment
- Work processes
- Quantity
- Reliability
- Operability
- Modularity
- Flexibility
- Maintainability
- Community relations
- Corporate image
- Product cost

Figure 3. Some Example *Performance* Constraints

Interrelation of the Triple Constraints. Note that the triple constraints are not independent of each other. A needed change in the time constraint can affect resources. A change in the resources can affect our ability to meet a time constraint, or cause a need to change that time constraint. Changes in performance criteria due to new knowledge or new situations can cause a need to adjust both the resource constraint and the time constraint. Therefore, use caution when modifying constraints to assure that the effects on the other constraints are well understood and accounted for.

Triple Constraints Are Project Drivers. There is a general understanding of the triple constraints by many project managers, but one of the most important uses of the triple constraints is often either unrealized or neglected—the concept of the triple constraints being the project drivers. By *drivers* we mean: Which one of the triple constraints (time, resources, or performance) is most important, i.e., the *primary driver;* which one is least important, i.e., the *weak driver*; and which one is in the middle, i.e., the *middle driver*? (Dobson, 1993) Do we really need to care about this? Aren't time, resources, and performance all important? Do we really have any choice? The answer to all of these questions is, "Yes."

The truly vital question to understand when discussing the importance of the drivers is, "Why do we care?" We care because at some point in the project we are going to have to make vitally important decisions with respect to the project, and the correct decisions are going to depend on the driver priority or importance relationship.

All three of the triple constraints are important for all projects, but they are *not* all three *equally* important in any project! The trick is to determine which constraint is most important, which least important, and which is in the middle for the particular project concerned. How do we do this?

Well, we could ask the project owner which driver is most important, time, resources, or performance. In response the owner is likely to say something to the effect, "What do you mean by asking me this? They are all essential and I expect you to give me them all. Now go back to work." This approach will usually fail. The problem is that even the project owner usually does not think in these terms. You will have to help him.

One way to do this is by asking the project owner a variety of trade-off questions. It is best first to patiently explain to the owner what you mean by the triple constraints. Tell the owner that you know that each of the constraints is very important. Then, tell the owner that because you really want to manage the project so that he or she is satisfied, there will come times during the project that you will have to make important decisions, and you want to do your best to make the right choices. Then, give the owner some example general trade-off questions that you might be faced with. Don't get too detailed; stay general—you probably don't really know at this point what problems you will run into and what decisions you might face. Right now we are talking hypothetical. What you are trying to do is to weigh each possible pair in the triple constraints—time vs. resources, time vs. performance, and resources vs. performance.

For example, you might say something to the effect, "Although we don't want this to happen, suppose as the project progresses we realize that in order to finish by the project due date, we must hire some people to work overtime, thus increasing the cost a little, and causing us to go over budget a little. Would you rather have us do that or would you rather we stay on budget and finish the project late by a week or two?" The project owner normally will be able to give you an answer to this question that gives you a feeling for whether resources or time is more important—he might even explain the reason, which would be very helpful insight that you did not have earlier.

Then, you could ask a trade-off question between the performance and the time; then, the performance and the resources, and so forth. You should ask enough different questions so that you get a clear feeling for the proper order of the triple constraints: which is the primary driver, the middle driver, and the weak driver.

Finally, state clearly to the owner what you have perceived to be the constraint priorities and get verification that the owner is in agreement. Putting it in writing would be best.

Driver Cautions. There are several cautions to be aware of when determining the triple constraint drivers.

First, there are things that can develop during the course of a project that cause the order of the drivers to change. Watch out for this possibility. Frequent communication with the owner and the other key project personnel is vital to help you to constantly be aware of which constraint is most important, etc. A change in driver priorities should only occur when absolutely necessary, and then only with agreement among the key decision makers on the project, such as the owner, the project manager, and other key leaders. Such a change should always be rapidly communicated to all people involved with the project.

Second, you should not purposely plan the project with the intent that the order of the drivers will change. For example, at the beginning of the project you don't say that for the first half of the project we will have time be the primary driver and then for the second half we will have performance be the primary driver. If you and the owner have agreed that performance will be the primary driver at the end of the project, then you plan the entire project from the beginning as if performance is always the primary driver. Then, if the primary driver must change, it changes, and you re-plan accordingly.

Third, you might find that two of the three drivers are both extremely important, with the third being less important. Does this mean that you should declare a tie and have two primary drivers and one weak driver, with no middle driver? The answer is "No!" Ask more trade-off questions between the two drivers that you think are tied. Even though one of them is to be designated a middle driver, this does not mean that it cannot be a very, very close second to the primary driver. It can. But when problems arise in the project and decisions have to be made, you must know beforehand which driver is on top.

Finally, as already discussed above, the triple constraints, and thus the drivers, are not independent of each other. Always be aware of how a change in one affects the others.

Tool 3: Project Tolerance Diagrams

Let's suppose the owner of your project specifies that the project is to be completed by July 4, 2003. Our impulse might be to treat this as the deadline and do everything in our ability to reach it, even if it means hiring people on overtime or cutting a few corners in performance to do it. On the other hand, we might tell ourselves that the owner just pulled July 4 "Out of the air," so to speak, as the date that he would like to have, but it is not essential and that sometime soon after July 4 should be all right—it is ideal but not really that essential. Which assumption is correct? Well, it could be either one. Do we really need to know this? Absolutely! The answer could mean the difference between project failure or not.

A tool to help us with this problem is called the *Tolerance Diagram*, shown in **Figure 4**.

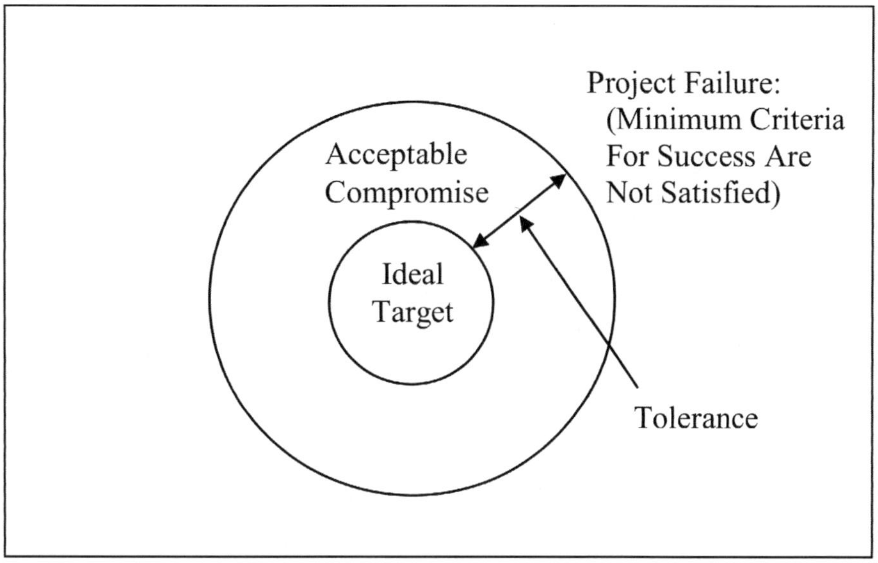

Figure 4. A Simple Tolerance Diagram

Three zones exist on the Tolerance Diagram in **Figure 4**: The ideal target zone, the acceptable compromise zone, and the failure zone. The ideal target zone describes what we would really like to have, if at all possible within the constraints on the project. The acceptable compromise zone is what we must have, even if we can't have the ideal. If we can't have at least the acceptable compromise, then we move into the failure zone and the project must terminate, or another alternative must be determined, including the possibility of a different or a significantly re-defined project. We call **Figure 4** a Tolerance Diagram because it illustrates how much tolerance or flexibility there is between our ideal desires and what might be considered failure. The primary driver will tend to have the least flexibility or tolerance, while the weak driver will tend to have the greatest flexibility or tolerance.

Let's look at an example. Suppose our only family car has broken down and this time it really is just not worth it to get the car fixed—we need to purchase a replacement. We have certain limitations or constraints, one of which is the money to purchase a car. It would of course be nice to have a new red Ferrari, but the cost is out of the question—it would cause failure in our credit rating within a couple of months. What we need is car that will reliably start and will get us to work each day—this we must have. After some financial figuring and some emotional stretching, we decide that we would like to spend no more than \$5,000 on the car—this is our ideal target, a cost less than \$5,000. We also decide that if we really can't find a car for a cost less than $5,000 that meets some minimum reliability standard (reliability is a performance constraint), we would be willing to go as high as $7,000, but absolutely no higher. Our acceptable compromise for the money constraint is therefore a cost that is between $5,000 and $7,000. Failure is to have a cost greater than $7,000.

People familiar with various problem-solving techniques will note some similarity between the one-sided Tolerance Diagram described above and the suggestion by Kepner and Tregoe (1981) that there are *wants* and *musts* that need to be identified when trying to evaluate alternative solutions to problems. Although Kepner and Tregoe's wants and musts are not exactly the same respectively as the ideal target and the acceptable compromise zones of the Tolerance Diagram, the detailed analytical approach they use to weigh various alternatives could conceptually be applied to the Tolerance Diagram, and interested readers are referred to their work.

Figure 4 is a *one-sided* Tolerance Diagram, meaning that going from best to worst is one-directional. Rather than the simple one-sided Tolerance Diagram, some project specifications will require a *two-sided* Tolerance Diagram, as illustrated in **Figure 5**. In this case the ideal target is a range in the middle of a project specification or constraint. On either side of this ideal target range are ranges deviating to one side or the other, which are still acceptable, but not ideal, giving therefore two acceptable compromises. Going outside of these two acceptable compromise ranges causes the project to move into unacceptable ranges, which again we call failure.

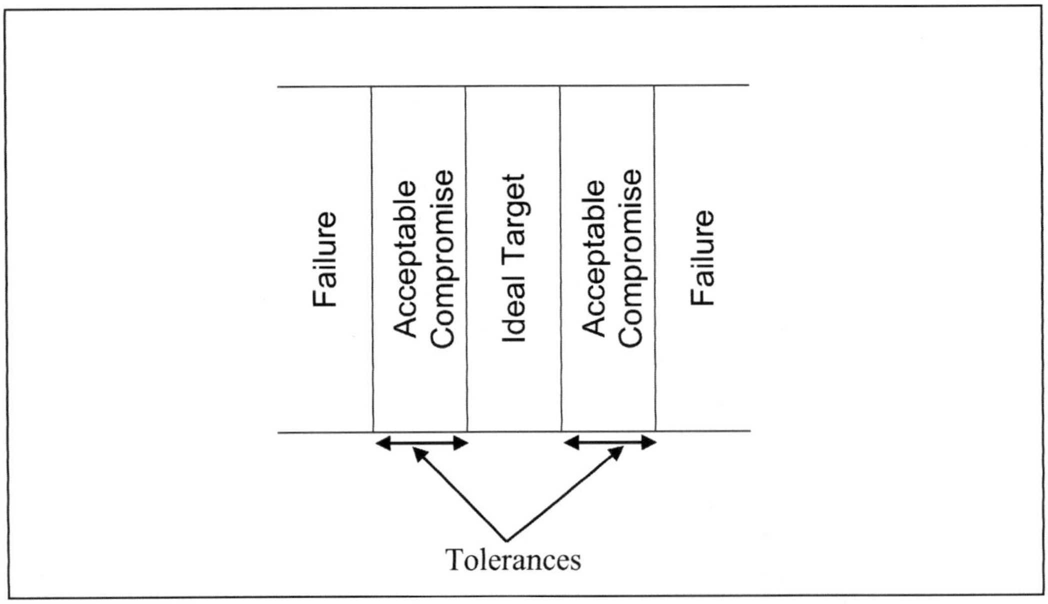

Figure 5. Two-Sided Tolerance Diagram

Let's look at an example of a two-sided Tolerance Diagram. Suppose that we are trying to make a new polyethylene polymer product that meets a certain density specification. We might want to have the polymer density be 0.950 g/cm^3. It will not be possible consistently to make that polymer with a density of exactly 0.950 g/cm^3, so instead we might specify that our ideal target range is 0.948 g/cm^3 to 0.952 g/cm^3. Although this is the ideal target, if we really must we could accept as a compromise polymer as low in density as 0.946 g/cm^3 or as high as 0.954 g/cm^3. Therefore, one acceptable compromise range would be 0.946 g/cm^3 to 0.948 g/cm^3 and the other would be 0.952 g/cm^3 to 0.954 g/cm^3. Any density less than 0.946 g/cm^3 or greater than 0.954 g/cm^3 would be failure.

Care should be taken when developing project objective specifications and identifying constraints to understand whether we need a one-sided or a two-sided Tolerance Diagram. Many projects will use both, some specifications or constraints using one-sided Tolerance Diagrams and the other specifications or constraints using two-sided Tolerance Diagrams.

Generally there should be a minimum of three Tolerance Diagrams for every project: one for each of the triple constraints of time, resources, and performance. However, usually you will need more than three. This is because the resources constraint and the performance constraint each have several aspects to them.

Looking at performance for example, we might have one Tolerance Diagram for some quality specification, another Tolerance Diagram for meeting some safety objectives, and yet another Tolerance Diagram for a new product production rate.

The author has seen an approach similar to Tolerance Diagrams receive very favorable reviews by the project team on a very large successful industrial capital product. The team reported in an end-of-project review that it had never before seen better-defined project objectives.

One of the arguments sometimes given against using the tolerance diagram concept is that perhaps people will not work as hard if you tell them the acceptable compromise in addition to the ideal target. This argument is weak for the following reasons.

First, if people have worked for you before, they have already figured out if you have habitually told them solely the ideal target, almost always to compromise it later on. In this case it is better to tell them the acceptable compromise up front rather than for them to make one up in their minds and be continually second guessing you.

Second, if you have a proper reward structure that rewards project teams for attaining the ideal objective as often as possible, they will respond to the rewards.

Third, people learn to trust you better when they know you are fully communicating with them and telling them the whole story—telling them both the ideal target and what is an acceptable compromise, if such a compromise is needed. This way the team and the owner can establish a more truly honest working relationship.

And fourth, it is much better for a team to know up front that an objective is either the ideal target or else the least you can accept, rather then for them to second guess this and have a 50% chance of being wrong, causing everyone a lot of frustration, wasted time and effort.

Some Additional Considerations

When developing project objectives there are additional considerations that can be important. **Figure 6** gives some of these considerations.

There can be times when project objectives have been given to the project leader and team, without due consideration to the material presented. In these situations the objectives will frequently be lacking in quality and need to be refined or re-done in order to be effective. There are times, in other words, when the project objectives really need to be re-negotiated.

- Verify that you are solving the right problem
- Talk to experts and those who might do the work
- Consider more than the first impression of what would be ideal for each of the triple constraints
- Are there better technologies or more cost-effective ones?
- What is changing in the market or new technology environment?
- How will key personnel affect this project?
- Document all the assumptions you are making
- Explore possible alternative solutions or objectives
- Prepare a draft set of objective statements, then test this draft with key people
- Refine the draft objectives and again review them with the key people
- Make sure that you have alignment of business and technical objectives
- Make sure that success metrics have been defined
- Assure that project implementation expectations are clear with all parties involved
- Once the objectives are finalized, constantly remind everyone involved of those objectives
- The longer the project duration, the more likely the project objectives might change or evolve—be alert for this possibility and manage the changes carefully through full communication

Figure 6. Some Additional Considerations for Developing Project Objectives

As Knutson and Bitz (1991) have stated well, this negotiation step "is a result of one of two sets of circumstances: (1) You approach the client and senior management for the first time in order to negotiate a schedule deadline and a budget because these were not established when the project was initiated, or, more commonly, (2) you realize that there is an incompatibility between the preliminary plan and the project objectives that cannot be eliminated by negotiating estimate revisions…In the latter case, the negotiations are more complex…There are four alternatives for management to consider:

1. Alter the schedule and cost objectives
2. Develop a reduced set of technical objectives (referred to as "Performance" objectives in this paper)
3. Implement a combination of the first two alternatives
4. Cancel the project

Regardless of the alternative you recommend and senior management decides on, there are only two acceptable outcomes to this process: (1) a plan acceptable to you, the project team, the client, and senior management or (2) cancellation of the project. There is one other possibility, however—one that we do not encourage: taking on the project on a best-efforts basis, without changing the project objectives."

If the procedures and tools suggested in this paper are followed and used, the above difficulties caused by poor project objective definition will be minimized or nearly eliminated altogether.

Once we have well defined, written goals or objectives, they need to be published for all people involved with the project. As Rees (1991) has said, "Publish the goals. Make them visible for all to see. Do not expect people to remember them if they are not discussed and referred to often. Do not bury them in occasional memos and documents. Instead, post them in meeting rooms on flip charts or posters; use stickers or other visual reminders to keep the goals in front of everyone. Refer to the goals often in memos, presentations, and meetings. Whenever possible, use the goals to guide a decision."

Finally, even after all the above work to do objective development right, to varying degrees the project objectives need to remain dynamic. How much the objectives are dynamic will depend on many things, and each project is unique. But the project objectives "should be reevaluated as the environment and opportunities change." (Rue and Byars, 2000) **Figure 7** gives some example factors that have the potential to cause objectives to change significantly.

- Market competition
- Significant changes in the national or corporate economy
- New senior management
- New discoveries
- New regulations
- Unknown technical difficulties in first-time scale-up to commercial production

Figure 7: Example Factors Potentially Changing Project Objectives Significantly

One important caution is appropriate here: When project objectives change very significantly, it is often much better to declare the old project dead and to officially start a new project with the new objectives. This is because the old baggage of plans and expectations from the original project can hinder the revised project to the point of making it inefficient and ineffective. It is very hard to change the direction of a speeding train locomotive.

Summary

In summary, defining good project objectives as early as possible is one of the most important project development steps to occur. Presented in this paper are a number of important tools and concepts to help project owners and project managers with this extremely important responsibility. Particularly we have discussed in detail SMART objectives, the triple constraints, the concept of prioritizing the project drivers, and the clarification that comes through use of Tolerance Diagrams.

All of these tools and concepts both help to facilitate and require much better multi-way communication between the project owner, project manager, project team members, and other key project personnel. The resulting improvement in project objectives definition will significantly raise the potential for project success.

References

Babcock, Daniel L., *Managing Engineering and Technology—An Introduction to Management for Engineers*, Prentice Hall (1991).

Bitz, Ira, *Senior Project Management*, Course Syllabus prepared for The American Management Association, revision number 3 (1991).

Dobson, Michael, from discussion at a Fred Pryor Seminar entitled *Project Management* (1993).

Frye, Chris and Holly Mason, "Project Management Simplified," Available from Internet: <http://www.webtechniques.com/archives/2001/09/frye/> (cited March 15, 2002)

Gibson, James L., John M. Ivancevich, and James H. Donnelly, Jr., *Organizations—Behavior, Structure, Processes*, Irwin/McGraw-Hill (1997).

Kepner, Charles H. and Benjamin B. Tregoe, *The New Rational Manager*, Princeton Research Press (1981).

Knutson, Joan and Ira Bitz, *Project Management—How to Plan and Manage Successful Projects*, AMACOM (1991).

Rees, Fran, *How to Lead Work Teams: Facilitation Skills*, Pfeiffer & Company (1991).

Rue, Leslie W. and Lloyd L. Byars, *Management—Skills and Application*, 9th ed., Irwin/McGraw-Hill (2000).

Smith, Larry A., *Basic Project Management*, from course materials reprinted with permission for use in American Management Association seminars (1991).

Overview of Project Planning

Abstract ID: 481747
Paper # 330f
Dr. Eldon R. Larsen
Marshall University

Effective, excellent project planning is one of the vital three keys to successful project management, the other two vital keys being the importance of people and the importance of excellence in defining the project objectives.

When we say *planning* we are including *re-planning* as well. Re-planning is just a word we use to indicate the fact that the nature of planning should be ongoing continually throughout the life of the project. In fact, one very fine project manager I once met said that re-planning is really synonymous with controlling the project. Although people might say such a statement is over-simplified, in many ways this is a good way to think about the need to continually plan and re-plan.

To further illustrate an important point about planning, I use the following story. During the Persian Gulf War in the early 1990s the United States used some new, high-tech weaponry. Yet that alone was not enough to win the war, and it had only part to do with how the war was executed. In fact, many people recognized that the execution of the war was being successful in large measure due to some excellent leadership and decision being made by those in charge.

One day during the Persian Gulf War, at a news conference, a reporter asked the following question, recognizing how difficult it must be to plan and conduct a war when the enemy was continually moving around, hiding in difficult terrain, and moving scud missiles around on the back of trucks. The question was something to the effect: "General Schwarzkopf, how do you go about executing this war when the enemy is continually doing things that are not in your plan?" General Norman Schwarzkopf, quoting another famous general, Dwight D. Eisenhower, from World War Two, said in response that "Plans are worthless." Then, following a pause, continuing to quote, he said, "Planning is essential."

What did Generals Schwarzkopf and Eisenhower mean when they said that plans are worthless, planning is essential? In essence what they were saying is that plans are not sacred and they are not fixed! Many events and changes occur, new knowledge is learned, situations change, changes in leadership occur, competitors make unexpected decisions or announcements, people issues arise, and so on and so on and so on, all combining to make it completely unrealistic to think that somehow a plan can remain fixed! And yet there are a very large number of managers today who do not seem to understand these realities.

It is not *the plan* that adds value in project management! Rather, it is the planning *process* that adds value! It is through the *process of planning* that we have to think about all the various aspects of the project; we have to think about the various options and trade-offs; we have to think about what competitors might be doing; we have to think about alternative courses of action and the associated consequences; and we have to keep doing this throughout the entire project!

For a manager or anyone else to expect a project team to make a plan and then to religiously stick to it exactly with no adjustments is project tyranny. A manager who does this will cause his people to hate planning. Plans will become like strait-jackets, or like a noose around people's necks. This does not mean that the project objectives change or are flexible, although in some cases they might have to be, but rather it means that the way we achieve those objectives will continually have to be adjusted, modified, updated, and revised. Planning and re-planning are a way of life for project managers if they want to be successful. In fact, without re-planning, it is very likely that the project objectives will not be achieved! The bottom line to remember is this: In order to achieve project objectives, it is the continual *process* of planning that is essential and adds value, not the plan itself!

Throughout this presentation, we will be discussing many very important project management planning tools to help us in the process of project planning. These tools are very important and useful. Still it is not the tools themselves that are sacred, but the value of the tools comes through the process of using these tools. Use them appropriately.

The project planning tools in this presentation follow one another in a logical pattern. This pattern might be thought of as a cycle of project planning. The cycle is iterative.

When a project is first being planned, it might be necessary to iterate through the project planning cycle several times in order to get the initial plans and project objectives to be internally consistent with each other. During this initial iterative process through the planning cycle, an organization might find that it must make some adjustments to the original project objectives. This might happen when the planning cycle reveals better estimates for the project duration, project costs and other resource requirements, and the project performance expectations, and those estimates are not compatible with the project objectives. As adjustments are made in the objectives and the consequences of those adjustments are analyzed in the project planning cycle, an internally consistent plan and objectives can be achieved, or the project might need to be cancelled or significantly re-thought.

The project planning cycle generally consists of the following major steps:

→ Project charter development and elaboration
    - Project objectives definition
    - Identification of critical success factors
    - Definition of project standards
    - Identification of project leader and team
    - Elaboration of project deliverables
    - Other charter concerns

→ WBS development and elaboration

→ Network diagram development and elaboration

→ Estimates of resources, including costs, and durations for project tasks

  - Note: Once this information is developed, a check for best-logic-consistency should be made in the network diagram

→ Development of resource loading information (resource histograms)

  - Note: Once this information is developed, a check for best-logic-consistency should be made in the network diagram and the succeeding estimates step

→ Development of project cash flow, cumulative cash flow, and overall budget

→ Development and elaboration of plans for managing procurement, quality, risks, communications, scope, integration, human resources, time, and costs

→ Development and elaboration of contingency plans

  - During and following development and elaboration of contingency plans, it is best to check for best-logic-consistency starting back from the WBS development through this step involving contingency plans, and all the steps in between (WBS, network, estimates, resource loading, cash flows and budget, and the other management process plans just before this contingency planning step)

→ Cycle back to check for internal consistency with the project charter

The above steps in the project planning cycle are shown graphically in Figure 1. After the initial iterations through this cycle which take place during the latter part of project initiation, the project team, its leader, and all key project stakeholders should have a good idea of the realistic expectations for the project. This does not mean that all will go as initially planned—it will not—but the team will now be prepared the best it can to move forward with project execution, with a much better chance of having a happy and successful project closeout for all parties involved.

Yet, the use of the project planning cycle has really only begun when project execution starts. The project manager must regularly check and monitor all calculation, assumptions, estimations, and so forth in the planning cycle. This is the heart of re-planning and the ability to control the final outcome of the project.

Summary

The project planning cycle contains a series of steps that build one upon another. Each step in the cycle is important and essential, and should not be skipped except in the most unusual circumstances, such as in a very small project that possibly does not merit a full use of these tools and processes. A wise project manager will give sincere consideration to the concepts discussed in this presentation. Doing so will help him to achieve true excellence in this, the third vital key to successful project management: Excellence in project planning and re-planning.

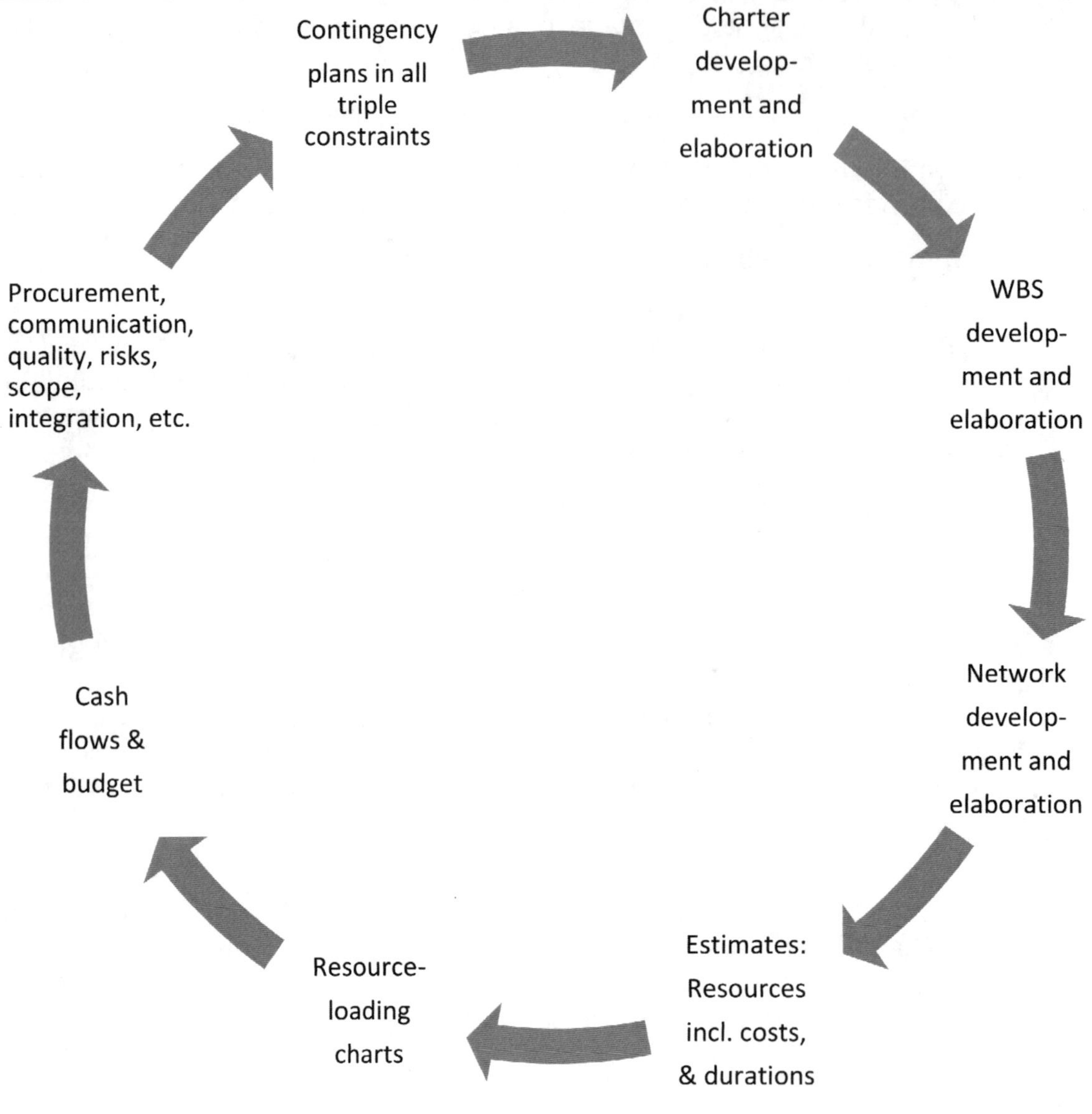

Figure 1. The Project Planning Cycle

495042 Application of Basic Project Management Skills to Small Scale Gas to Liquid Projects

Abstract:

Management of small gas to fuel projects in impoverished communities is an area of interest in Africa. Several projects do not come to completion due to not so proper project planning and execution. With the goal of producing Project managers who are able to complete such complex projects on time and on budget without sacrificing project quality and professionals preparing to transition into a new career in project management we given special hands on training in small gas to liquid projects. Also a group of professionals without formal training who are managing large, complex projects or smaller, more tactical projects were also given the same aid. The training equipped participants with the principles and practices of disciplined project management while addressing the unique challenges of gas to liquid projects. Participants apply the techniques in a small group setting using a realistic case study. Almost 100% gained a comprehensive overview of the skills, knowledge and tools needed to effectively manage projects.

493227 The Women Event: Engaging High School Girls and Their Parents in STEM

Abstract:

The Women's Outreach in Materials, Energy, and Nanobiotechnology (WOMEN) Event has been organized annually by Cornell University's Chemical and Biomolecular Engineering Graduate Women's Group (CBE Women) since 2010. The mission of the WOMEN Event is: (1) to provide important information about STEM fields, careers paths in engineering, and the college application process to female high school students, and (2) to develop leadership skills of female CBE graduate students who design and direct the WOMEN Event. Key features of the WOMEN Event that set it apart from other similar outreach programs aimed at women include targeting the event to high school students from rural areas and devoting a significant portion of the program to STEM careers and the college experience.

Every year, 20 to 40 10th grade girls and their parents are invited to Cornell for the WOMEN Event. Over the past 8 years, more than 200 girls have participated. These girls are from rural upstate New York, where 66% of women do not attend or complete college (statusofwomendata.org). In these areas, girls encounter fewer female role models in STEM and their parents may be unable to offer guidance on STEM career options or how to prepare for them. The WOMEN event program includes hands-on lab modules, college admission information sessions, talks highlighting STEM career opportunities, and a panel discussion with current Cornell students, all free of charge. Engaging lab modules created specifically for the WOMEN event illustrate topics in chemical engineering, including materials design, bioengineering, and process development. One unique aspect of our programming is engagement with parents. While students participate in lab activities, the parents participate in an information session on college financial aid and admissions. The parents also listen to talks from women faculty to gain insight into the many paths STEM careers can take. In the final laboratory module, parents and their daughters conduct experiments together. By reaching out to both girls and their parents, we hope to not only interest the girls in STEM but also enable their parents to support them in growing that interest into a career. After the 2016 event, 90% of the girls indicated they intend to enter a STEM field and the number of attendees interested in an engineering career doubled. While 4 past participants of the event have gone on to study engineering at Cornell, this is not intended to be a recruitment event. We instead aim to provide an experience that encourages girls to pursue their STEM passions wherever that leads them.

In addition to inspiring the girls who attend, the WOMEN Event provides a valuable learning experience for the graduate student volunteers. Five graduate students have given conference presentations about the WOMEN event. In 2017, half of the volunteers and leaders of the WOMEN event were women; a notable achievement in a department that has 30% women MS and PhD students. Volunteers have received recognition through the departmental Hooey Award, Cornell's Cook Award, and the College of Engineering Alumni Association Award for Student Group Leadership.

500867 Comparison of Web-Based and Lecture-Based Training Approaches to Educate High-School Students with Simulink Modeling Skills

Abstract:

The math performance of US high-school students was ranked 38th out of out of 71 countries in the most Programme for International Student Assessment (PISA) in 2015 [1]. In addition, a report from President's Council of Advisors on Science and Technology shows that US would have one million deficits in technical workers in STEM fields if the STEM education is not improved in the next decade [2]. Therefore, it is urgent for US educators to create new approaches to attract more high-school students in the STEM fields, especially in math, which deals with challenging equation and symbolic operations. One way to address this issue is providing interesting modeling projects along with a powerful software that students can reply on for math operation. We did an "experiment" in Summer 2014 by introducing MATLAB Simulink to a local high-school junior student and providing him a project in which he developed Simulink Model for a microbial fuel cell (MFC). That project was quite successful, as the student presented his work in ASEE 2015 Mid-Atlantic Spring Conference [3]. The success of that project is based upon the following two factors: 1) MATLAB Simulink provides friendly user-interface so that students can build ordinary differential equation (ODE) models in Simulink like playing Lego; 2) the microbial fuel cell project deals with converting organic compounds in waste water to bioenergy.

On the basis of the feedback from the student, we further extended our previous projects to web-based training modules to attract and train more high-school students. The web-based training idea comes from another ASEE project of ours in which we prepared biology review and virtual experiment/training videos to enhance learning in biochemical engineering courses [4]. In the web-based training modules, two videos were developed to educate students step by step on how to build two ODE models, one for water tank and the other for the continuous flow stirred-tank reactor (CSTR) [5], in MATLAB Simulink. A video was developed to introduce students the background information and application examples of MFC. Students then constructed the ODE model of a microbial fuel cell, which consists of four differential equations, 9 constrained equations and 25 parameters. The instructor offered Skype meetings to help students in their projects. The developed training modules were also delivered to another group of students using the traditional lecture-based teaching approaches. Anonymous surveys were given to both web-based and lecture-based groups of students. Survey results showed that the web-based approach was effective in educating high-school students with Simulink modeling skills. On the other hand, the lecture-based group of students had slightly better performance than the web-based group.

References:

[1] webpage: http://www.pewresearch.org/fact-tank/2017/02/15/u-s-students-internationally-math-science/

[2] President's Council of Advisors on Science and Technology, Engage to excel: producing one million additional college graduates with degrees in science, technology, engineering, and mathematics (Executive Office of the President of the United States, 2012).

[3] C. Ekaputra, Z. Huang. "Educating High School Students in Process Simulation and Control with a Simulink-Based Controller Design for Microbial Fuel Cells. ASEE Mid-Atlantic Spring 2015, Villanova, PA, 2015.

[4] J.J. Elmer, N.K. Comolli, W.J. Kelly, Z. Huang. "Preparation of Biology Review and Virtual Experiment/Training Videos to Enhance Learning in Biochemical Engineering Courses," 2015 ASEE Annual Conference and Exposition, Seattle, Washington, June 2015.

[5] Dale E. Seborg, Thomas F. Edgar, Duncan A. Mellichamp, Francis J. Doyle, III. Process Dynamics and Control, Third Edition, Wiley, 2010.

501376 Nanotechnology & Engineering Grand Challenges

Abstract:

This talk highlights outreach activities on potential nanotechnology solutions to the NAEâ€™s Engineering Grand Challenges developed as part of an NSF Nanotechnology Undergraduate Education grant and a Math Science Partnership. Most of the activities were designed for school classrooms or summer camps, but they have also been adapted for teacher training and public outreach events such as NanoDays and university open houses. This talk will provide information about these activities and evaluations of their impact on participants' science and engineering knowledge and attitudes. The value of creating adaptable activities which are easily implemented by undergraduate and graduate engineering students to reach a range of outreach needs and maximize impact while minimizing stress will be highlighted. The incorporation of consumer grade technology tools into activities will also be highlighted. Specific examples that will be described include activities on reverse engineering the brain and renewing and improving urban infrastructure.

501605 Going Beyond Demonstrations to "Chose Your Own Adventure" Engineering Experiences for Service-Learning K-12 Outreach Opportunities for 3rd Year Engineering Students and Enhanced Student Engagement for 1st Year Engineering Students

Abstract:

The enthusiasm of STEM teachers and early exposure to STEM fields has been shown to strongly influence a student's selection of engineering as a major/career. Due to the somewhat abstract and often hazardous nature of chemical engineering compared to other engineering fields, it is often difficult to create engaging chemical engineering experiences for K-12 students. To help address this issue, my students and I have created over 70 engineering adventure experiences which have been presented to over 6,000 K-12 students. We have focused on hands-on interactive demonstrations that use as many senses as possible; however, feedback from our experiences has inspired us to modify our approach to involve a higher level of active learning. Here we present our efforts to create open-ended, hypothesis-driven, "chose your own adventure" chemical engineering experiences that are safe, fun, inexpensive, and easily tailored to elementary, middle, or high school students. Survey response data from 3rd year engineering students who developed and presented these experiences is presented demonstrating their desire to be involved in service-learning opportunities and the impact it had on their understanding of basic engineering concepts.

Inspired by the above experience, we have also created a "Chose Your Own Adventure" Engineering Experience for 1st Year Engineering Students. The experience is constrained to address an Engineering Grand Challenge as Described by the National Academy of Engineering and students must complete the Engineering Design Cycle. However, beyond this and approval from risk management, the experience is completely open ended. To maximize intrinsic motivation and engagement, student teams are assigned based on interest in a given Engineering Grand Challenge. Here we also report the impact of such an open-ended experience during the 1st Year on student engagement and excitement for continuing in an engineering major.

494840 Building Block Air Quality Sensors

Abstract:

Communicating chemical engineering research to the general public can be difficult, but hands-on activities can help to both engage the layperson and bridge the gaps in understanding. We have developed an effective teaching module to introduce students to the science of air quality and the ways in which we, as engineers, may quantify it.

This module has students design build and test light scattering air quality sensors, using an arduino board and several low-cost electronic components. The sensor housings are assembled out of Legos or Lego-like building blocks, allowing students to attempt many design types. Once assembled the sensors are tested with mist or smoke, and assessed for their ability to detect changes in air quality and their susceptibility to noise. They may also be tested for other design considerations, such as material cost. The module needs approximately a one-hour class period to execute and can be used as a hands-on project for teams of up to 4 students. The difficulty of the module can be easily scaled from the middle school to the undergraduate level, and this activity could be appropriate for an introductory undergraduate lab.

This work is part of a NSF project (#1648756) to place air quality sensors in K-12 schools and create a real-time air quality map. K-12 students in this project are trained to become citizen scientists, and this module has been used to introduce them to the relevant sensor concepts; the commodity sensors integrated into a custom printed circuit board, which these classrooms will be hosting use the same sensing concepts as the student-built Lego sensors. We have performed this module for over 100 students and have received very encouraging feedback and significant attention from local and national press, leading to increased interest in participation in this research.

Skin Layer Formation during Drying of Latex Films

Hao Huang(hah213@lehigh.edu)[1,3], **H. Daniel Ou-Yang**[2,3], **Willie Lau**[4] and **Mohamed S. El-Aasser**[1,3]
[1]Department of Chemical and Biomolecular Engineering, [2]Department of Physics, [3]Emulsion Polymers Institute, *Lehigh University, Bethlehem, PA 18015*
[4]Beijing Oriental Yuhong Waterproof Technology Co., Ltd, *Beijing 100123, China*

Abstract

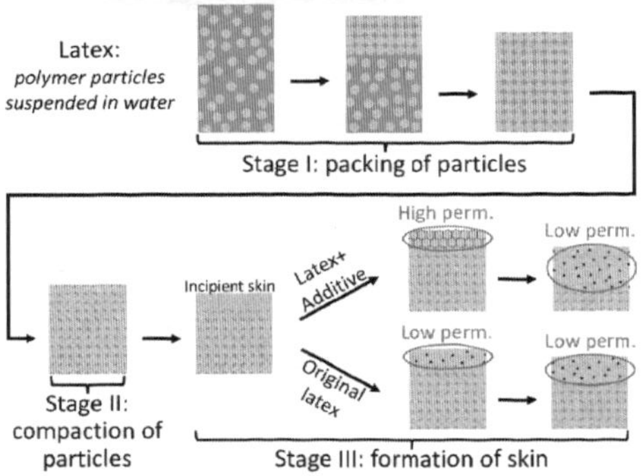

Waterborne latex coatings for protection and waterproofing purposes require film thickness in the millimeter range. During drying of such thick latex films, the top surface of film can get dried early and form a transparent polymer skin while beneath it is still wet and opaque. Skin layer impedes water evaporation and extends the film drying time considerably. To study the mechanism of skin layer formation, optical coherence tomography (OCT), analytical balance and video camera were integrated ("OCT-Gravimetry-Video" method) to simultaneously monitor the film's thickness and internal structure, the weight loss and the visual appearance as a function of drying time. Three drying stages for skin layer formation were observed: I) packing of particles, II) compaction of particles and III) formation of skin. In Stage I, particles accumulate and become packed from top to bottom of film, and the packing layer thickness increases linearly with time; the drying rate during this stage is equal to that of pure water. In Stage II, packed particles become further consolidated while the drying rate is still close to that of pure water. In Stage III, particles on the top of film become deformed and coalesced into a transparent skin, whose thickness increases with about square root of time until the film is totally dried; the drying rate during this stage is dramatically reduced and determined by the permeability and thickness of skin.

In coating applications, it is anticipated to lose as much water as possible before skin impedes drying rate. Some water-soluble additives (such as surfactants) were found to lower the water content trapped in film beneath skin. OCT also showed that surfactant additive can accelerate growth of skin thickness and the whole film becomes transparent earlier. We propose that: an additive, which can retard coalescence between deformed particles, makes skin permeable in the initial short time of skin formation. This allows more water evaporation before particles become fully coalesced. This study is expected to provide a guideline in additive selections to solve the drying problem of thick films.

Introduction

- Latex, a colloidal system of polymer particles suspended in water
- Transformation from colloid into a continuous polymer film after being dried
- Waterborne paints with low volatile organic compound, >10 million tons consumed per year
- Latex film formation process during drying: 1) concentrating and packing of particles with water evaporation, 2) deformation of particles by interfacial tensions and Van der Waals force, and 3) coalescence of particles by inter-diffusion of polymer chains.

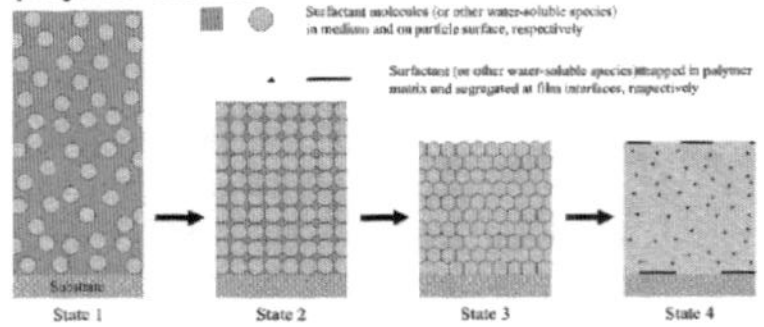

Skin layer formation

During drying, particles accumulated on the top surface of film deformed and coalesced early to form a polymer skin, while beneath skin is still wet.

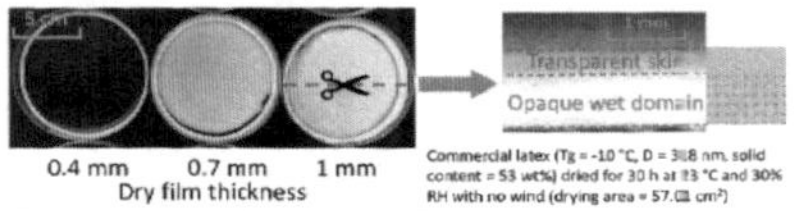

Conditions for skin formation

- Vertical inhomogeneity - Péclet number:

$$Pe = H\dot{E}/D_o = 6\pi\mu R_p H\dot{E}/(kT)$$

where D_o is particle diffusion coefficient, H is wet film thickness, $\dot{E}$ is drying rate, μ is liquid viscosity and R_p is particle radius.

- Ratio of deformation time to drying time:

$$\bar{\lambda} = \eta R_p \dot{E}/(\gamma H)$$

where η is polymer viscosity (a negative correlation with T_{room}-Tg) and γ is air/liquid interfacial tension. A skin forms when $Pe > 1$ and $\bar{\lambda} < 1$, or when film is thick and particles are soft (Tg<T_{room}).

Materials and Methods

Latexes

- Commercial LMA(lauryl methacrylate)/BA(butyl acrylate)/MMA(methyl methacrylate) latex with minor amount of MAA (Methacrylic acid), stabilized by an ionic ethoxylated surfactant, with Tg=-10 °C and particle diameter = 318 nm (PDI=0.044)
- Synthesized EHA(2-ethylhexyl acrylate)/MMA latex with minor amount of MAA, stabilized by sodium lauryl sulfate (SLS), with Tg=-19 °C and particle diameter = 173 nm (PDI=0.052)

Water-soluble additives

- Surfactants: SLS, Triton X-100
- Polymeric oligomers: polyethylene glycol (PEG, Mw = 1 kg/mol), hydroxyethylcellulose (Natrosol™ 250), ammonium polyacrylate (Mw = 2-3 kg/mol)

Drying curves under standard condition

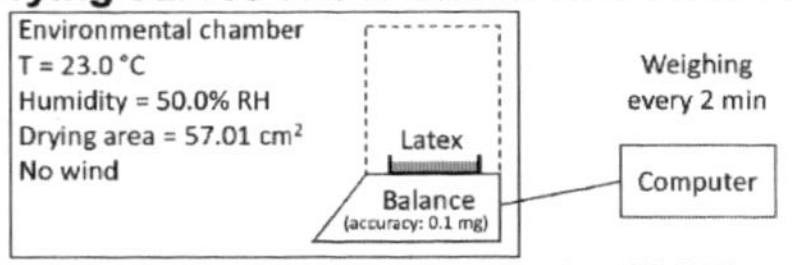

Optical coherence tomography (OCT)

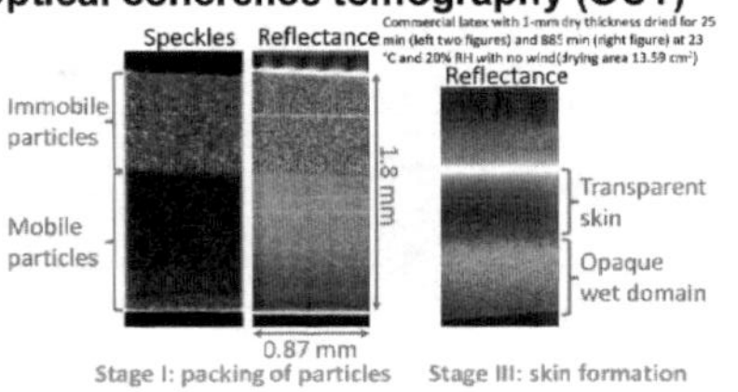

OCT-Gravimetry-Video method

Results and Discussion

Skin formation process

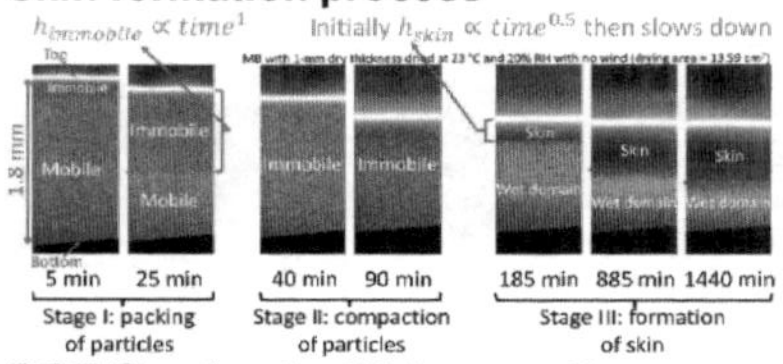

OCT-Gravimetry-Video results

- Commercial latex

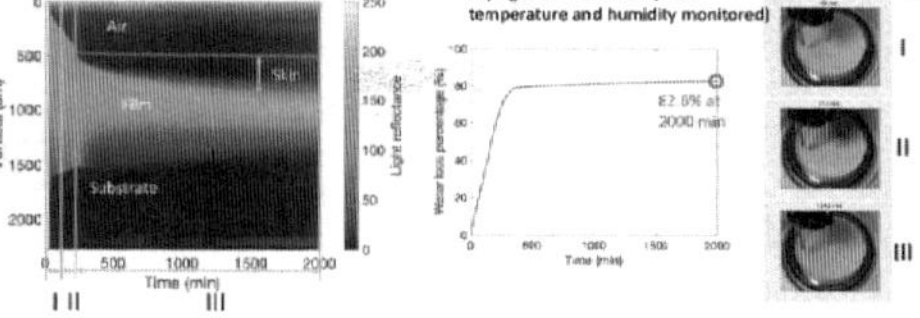

- EHA/MMA latex

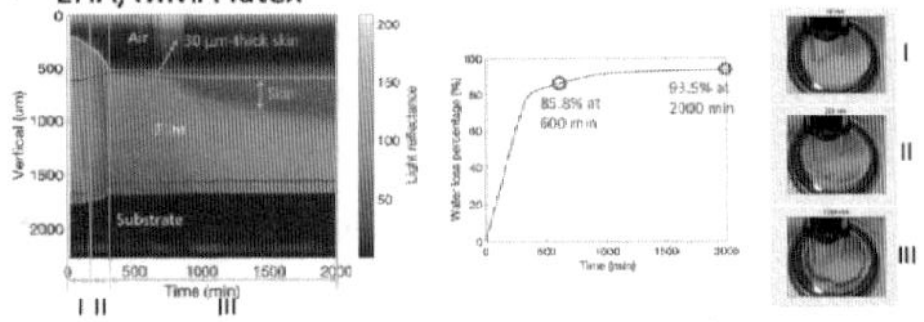

Mathematical expression of skin thickness

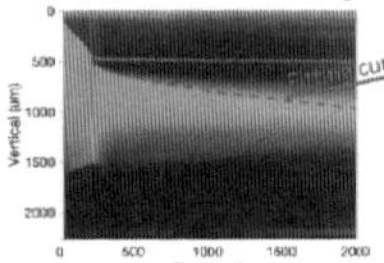

$$h_{skin}(t) = \sqrt{\frac{2WVP \cdot p_0(1 - RH_a)}{\rho_s k_w^*/(1 - k_w^*)}} \cdot t^{1/2}$$

WVP: water vapor permeability of skin
RH_a: environmental relative humidity
k_w^*: water content under skin

Effect of surfactant (SLS)

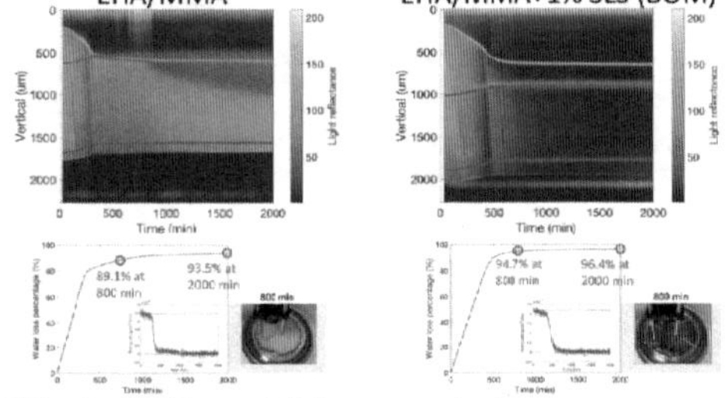

Effect of other additives on drying curves

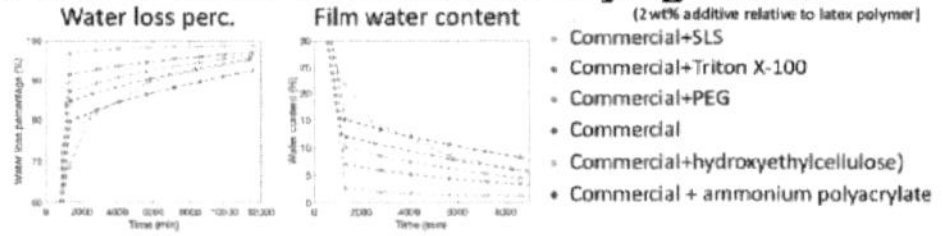

Conclusions

- OCT-Gravimetry-Video method is developed to study the drying process of waterborne latex coatings. Internal structure of film, water evaporation rate and visual appearance of film can be monitored simultaneously changing with time.
- Skin layer formation comprise three stages: particles packing, particles consolidation and skin formation.
- Surfactant additives show positive effects on promoting drying through skin by impeding coalescence of particles.

References

[1] A. F. Routh et al., Ind. Eng. Chem. Res. 40, 4302 (2001).
[2] F. T. Carter et al., Langmuir. 30, 9672 (2014).
[3] M. Wojtkowski, Appl. Opt. 49, D30 (2010).

Acknowledgement

We thank Oriental Yuhong Co., Ltd for financial support. We thank Mr. Yongyang Huang and Prof. Chao Zhou for OCT imaging. We thank Prof. James Gilchrist, Dr. Eric Daniels and Mr. Joseph Rokowski for discussions.

506549 Improving Automated Model Reconstruction Across Phylogenetically Diverse Genome-Scale Metabolic Models

Abstract:

The Department of Energy Systems Biology Knowledgebase (KBase) is a platform designed to solve the grand challenges of Systems Biology. KBase has implemented bioinformatics tools that allow for multiple workflows including genome annotation, comparative genomics, and metabolic modeling. KBase now also includes a comprehensive database of over 80K reference genomes (approximately 5K complete genomes) from NCBIâ€™s RefSeq. We have selected a phylogenetically diverse set of approximately 1000 genomes and built draft genome-scale metabolic models constructed using the ModelSEED pipeline implemented in KBase. In constructing these models, we were interested in improving the gene functional role â€" reaction templates and accuracy of biomass compositions used by ModelSEED when producing a draft model. We have curated the existing ModelSEED templates by both fixing and introducing new gene â€" reaction mappings. Previous ModelSEED biomass compositions often excluded metabolites that are essential for the viability of many organisms, while simultaneously including some non-essential metabolites. These errors in the biomass compositions lead to errors in the growth conditions and gene knockout phenotypes predicted by the models. To correct these problems, we conducted a sensitivity analysis on our full set of models, using a biomass composition that includes all possible essential biomass precursors. We also improved the ModelSEED gap filling database with new rules to restrict the addition of non-biological significant reactions. We then used this data to improve all of our models, validating our improved models with a diverse set of growth and knockout phenotype data. Our improved models are now available for download, viewing, and comparison in KBase.

Investigation of CO_2 desorption performance in tri-solvent blends (MEA-AMP-PZ) with and without catalyst

Xiaowen Zhang [a], *Helei Liu* [b]*, *and Zhiwu Liang* [a,b]*

[a] *Joint International Center for CO_2 Capture and Storage (iCCS), Provincial Hunan Key Laboratory for Cost-effective Utilization of Fossil Fuel Aimed at Reducing CO_2 Emissions, College of Chemistry and Chemical Engineering, Hunan University, Changsha 410082, PR China*

[b] *Clean Energy Technologies Research Institute (CETRI), Faculty of Engineering and Applied Science, University of Regina, Regina, Saskatchewan, S4S 0A2, Canada*

Abstract

In this work, the regeneration characteristics of 6 M MEA-AMP-PZ (monoethanolamine, 2-amino-2-methyl-1-propanol and piperazine) tri-solvent blends with different AMP/PZ molar ratios with four different solid acid catalysts (H-ZSM-5, γ-Al_2O_3, SAPO-34 and SO_4^{2-}/TiO_2) and without catalyst at 369 K were studied in terms of CO_2 desorption rate and heat duty and compared with conventional 5 M MEA. For the blank runs, the results showed that the tri-solvent blends of 3 M MEA-2.5 M AMP-0.5 M PZ performs a best CO_2 desorption performance among all the tested amines, which reduced the heat duty by 51.9% as compared with 5 M MEA without catalyst (100%). With regard to catalyst, the best blend with H-ZSM-5 provided the combination with the greatest performance in CO_2 desorption, which decreased the heat duty by 61.6% in comparison with 5 M MEA without catalyst. Based on the results of this work, the combination of solid acid catalyst with tri-solvent amine blend is proven to be an alternative method for the improvement of the energy efficiency of the CO_2 capture process.

Keywords: novel amines tri-solvent blends; catalytic CO_2 desorption; heat duty reduction; CO_2 capture.

* Author for correspondence:
Dr. lhl0925@hotmail.com (Hl Liu) & Email: zwliang@hnu.edu.cn (Zw Liang)

1. Introduction

Carbon dioxide (CO_2) is widely considered as a predominant contributor of greenhouse gases. Thus, it is an urgent to mitigate CO_2 from atmosphere. Chemical absorption technology using amine based solvent for capturing CO_2 has attracted the most attention because of its cost-effectiveness, maturity and absorption large volumes of flue gas. However, a large quantity of energy is demanded during CO_2 stripping which accounts for about 70-80% of the total cost of the CO_2 capture system. The high energy penalty accompanying the high running cost is the biggest barrier for post-combustion CO_2 capture by using chemical solvent[1].

Considerable research efforts have been devoted to reducing the regeneration energy requirement. Recently, some researchers have reported that the development of amines tri-solvent blends for CO_2 capture is an effective method. Because it could not only dramatically utilize the potentials of the individual amine, but also address their individual problems. Nwaoha et al.[2] studied the tri-solvent blend with two absorption rate promoters (i.e. piperazine (PZ) and monoethanolamine (MEA)) and a hindered amine 2-amino-2-methyl-1-propanol (AMP). The results indicated that the tri-solvent blends amines had faster desorption rate and lower regeneration heat duty in comparison with the commercial amine MEA. Based on their work, the tri-solvent blends amines could be considered as one of potential novel absorbents for CO_2 capture in post-combustion processes. However, the heat duty of tri-solvent blends amines is still high. Idem et al. [3] proposed that the addition of solid acid catalysts such as H-ZSM-5 into rich amine solution for CO_2 stripping is an efficient method to decrease the heat duty. However, the study of regeneration behavior of the amine tri-solvent blends with the catalysts is not available. Hence, the investigation on the tri-solvent blends amines with catalyst is interesting to present in order to further reduce of solvent regeneration energy requirement.

In the present study, the test of MEA-AMP-PZ tri-solvent blends with different AMP/PZ ratios (1, 2, 5) with four different acid catalysts (e.g. HZSM-5, γ-Al_2O_3, SAPO-34, and SO_4^{2-}/TiO_2) and without catalyst were performed in a batch reactor for understand the process of regeneration. This work also used the NMR technique to experimentally obtain the concentrations of bicarbonate for the three blend systems and then employed to explore their regeneration behavior.

2. Experimental and materials

2.1 Materials

MEA (99%), AMP (98 %) and PZ (99%) were obtained from Shanghai Aladdin Industrial Corporation, China. Three commercial catalysts: H-ZSM-5 (white cylinders solid, gamma-Al_2O_3 and SAPO-34 were acquired from Tianjin Nankai University Catalyst Co. Ltd., China. Amorphous titanium oxide (TiO_2) (99%) and sulfuric acid (H_2SO_4, 98%) were both purchased from Shanghai Macklin Chemical Reagent Co. Ltd., China. CO_2 (purity of $\geq$ 99%) was purchased from Changsha Jingxiang Gas Co. Ltd., China. Solid super-acid catalyst sulfated TiO_2 (SO_4^{2-}/TiO_2) was synthesized according to our previously work [4].

2.2 Experiment

In this work, the experiments regeneration was studied by using a batch reactor, which was displayed in Figure 1. 2 L rich amine solutions with the desired CO_2 loading were introduced into a 3 L four-necked flask, which was immersed in the oil bath (DF-101T). The rich CO_2 loading of all amines was set at their equilibrium CO_2 loading at absorption conditions of 313 K and 15.1% CO_2 partial pressure[2,3]. 25 g catalyst were put into the amine solution. The time was recorded once the amine solvent temperature reached at 338 K and the final regeneration temperature was maintained at 369 k. The CO_2 loadings of solution at time of at 0.25, 0.5, 0.75, 1, 1.5, 2, 4, 6, and 9 h were measured by using a Chittick apparatus[5]. The heat duty for the CO_2 desorption was measured by an electric energy meter.

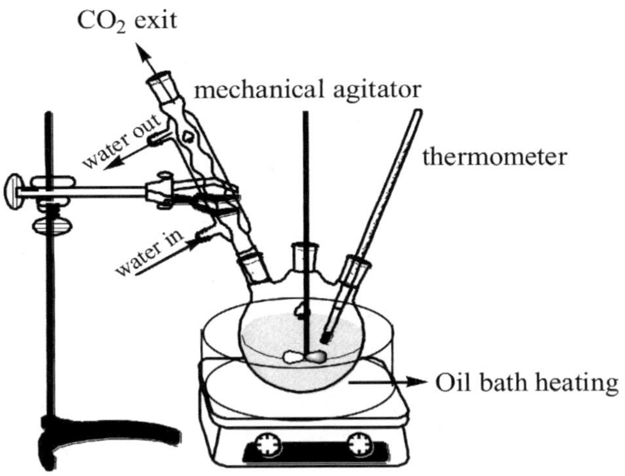

Figure 1. Schematic diagram of the batch reactor for CO_2 desorption process

2.3 Data analysis

The heat duty (H, kJ/mol) was obtained by using the following equation (1).

$$H = \frac{\text{Heat input / time}}{\text{Amount } CO_2 \text{ / time}} = \frac{\text{Electricity(kJ)}}{\text{Amount } CO_2\text{(mol)}} \quad (1)$$

The relative heat duty (RH, %) was used to present the regeneration energy of solvents in this work. The heat duty of 5M MEA without catalyst regeneration in first 2 h is considered as $H_{baseline}$ (kJ/mol). The heat duty of new systems regeneration in first 2 h was considered as H_i (kJ/mol). RH was calculated as the following equation (2).

$$RH = \frac{H_i}{H_{baseline}} \times 100\% \quad (2)$$

3. Results and discussions

3.1 CO_2 regeneration performance without catalyst

The typical CO_2 desorption profile of the four amine systems (e.g. 5M MEA, 3M MEA-2.5M AMP-0.5M PZ, 3M MEA-2M AMP-1M PZ and 3M MEA-1.5M AMP-1.5M PZ) at the regeneration temperature of 369 K is shown in Figure 2. It is clear that all the studied tri-solvent blends show a faster desorption rate in comparison with the 5M MEA. The 3M MEA-2.5M AMP-0.5M PZ system shows the fastest CO_2 desorption rate. As shown in Figure 2, the CO_2 loading decreased dramatically in the first 2 h. Therefore, the heat duty was calculated by considering the first 2 h of the regeneration process in this work.

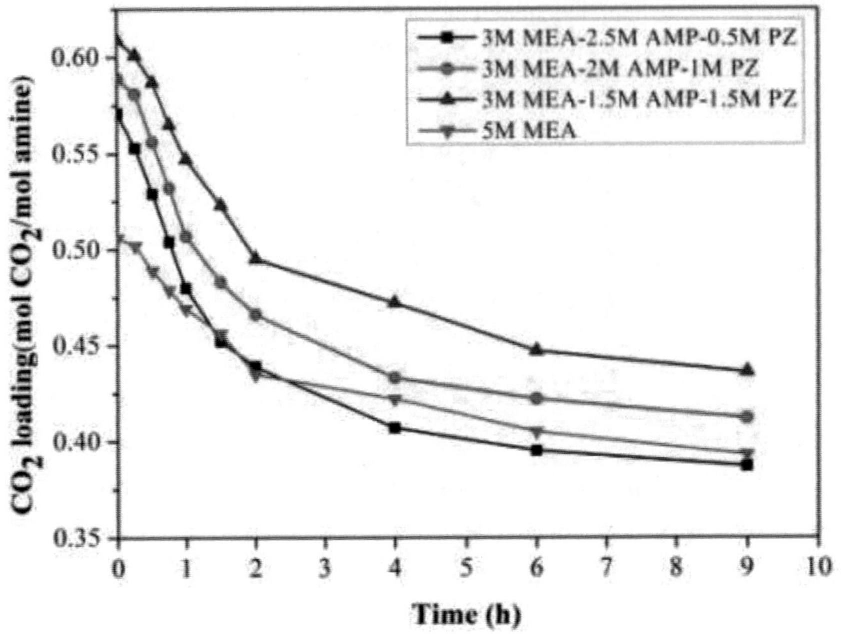

Figure 2. CO_2 desorption in amine solutions without catalyst.

Figure 3 depicts the relative energy requirement of amine solvent regeneration. It is noted that all the studied blends show much lower regeneration heat duty than 5M MEA. The 3M MEA-2.5M AMP-0.5M PZ with AMP/PZ molar ratio of 5 requires the lowest regeneration heat duty. It could be found that the heat duty decreased with the increase of concentration of the AMP. Based on this result, the system of 3M MEA-2.5M AMP-0.5M PZ could be considered as the best blend solvent in term of CO_2 desorption performance.

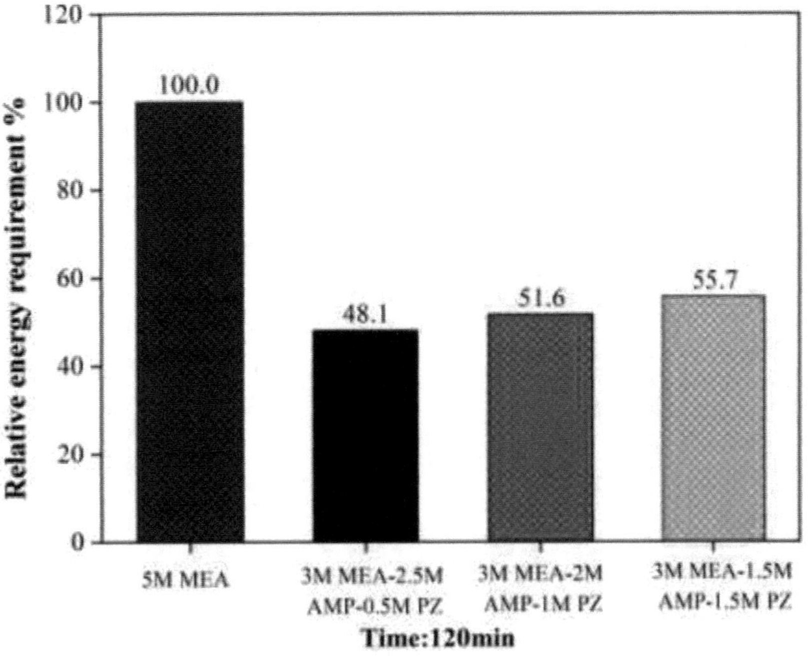

Figure 3. Relative heat duty for amine solutions without catalyst.

The comparison of the formation of HCO_3^- among these three blend systems are shown in the Figure 4. Which demonstrated clearly that the mix-1 system with AMP/PZ molar ratio of 5 produced more HCO_3^- than Mix-2 system with AMP/PZ molar ratio of 2 and Mix-3 system with AMP/PZ molar ratio of 1 at the same CO_2 loading. The presence of more HCO_3^- in the rich amine solution can enhance the deprotonation of protonated MEA ($MEAH^+$), AMP ($AMPH^+$), and PZ (PZH^+, PZH_2^+). Moreover, the HCO_3^- can act as a proton acceptor to directly release CO_2, thus, the desorption reaction become easier. And also the peotonated amines are easily to liberate their H^+ to a basic HCO_3^- instead of the neutral water (Eq. (1) - (3)) [3,6-8].

$$AmineH^+ + H_2O \longleftrightarrow Amine + H_3O^+ \quad (1)$$
$$AmineH^+ + HCO_3^- \longleftrightarrow Amine + H_2CO_3 \quad (2)$$
$$H_2CO_3 \longleftrightarrow CO_2 + H_2O \quad (3)$$
$$AmineCOO^- + H_2O_3 \longleftrightarrow Amine + HO_3^+ \quad (4)$$

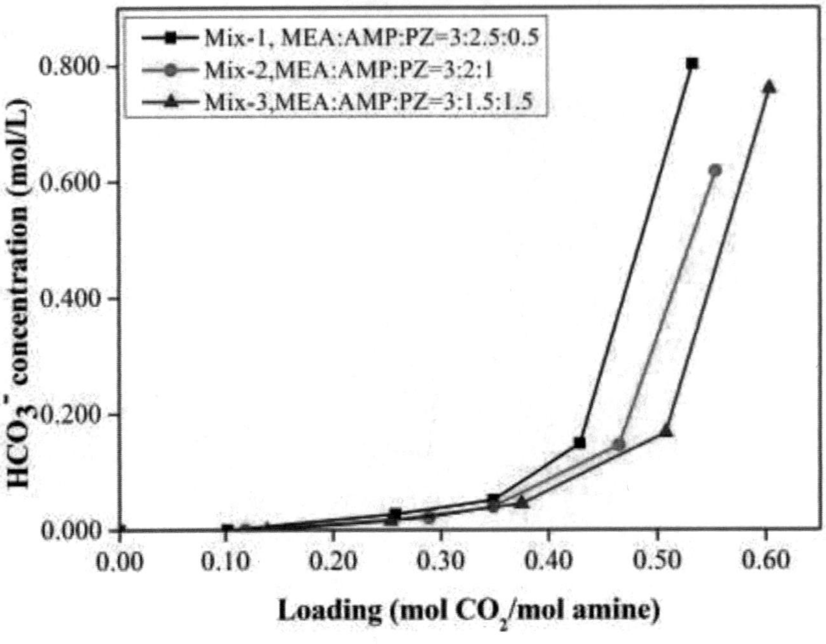

Figure 4. Concentration of bicarbonate in these three mixture systems at 293 K using NMR analysis.

With increasing the AMP/PZ molar ration from 1 to 5, the concentration of AMP increased while that of PZ decreased. This resulted in an increase in the HCO_3^- concentration, hence facilitating a faster desorption rate. Idem et al. [3] reported that the reaction (1)-(4) are strongly endothermic processes and require large energy. As previously described, the presence of sufficient $HCO3^-$ in the rich amine solution, CO_2 stripping will become easier and demand less desorption energy duo to reactions (2) and (3) will be important. For single MEA solvent, reactions (1) and (4) are predominant and require large desorption energy while for MEA-AMP-PZ tri-solvent blends reactions (1)-(4) are which contains multi pathways (involvement of reaction (2) and (3)) that demand less regeneration energy[3,8,9]. Therefore, a higher AMP/PZ molar ratio probably produces more $HCO3^-$ in the amine solution which assists offer an easier CO_2 desorption condition, hence reducing the energy penalty. This is further identification that the MEA-AMP-PZ tri-solvent blends have greatly potential to significantly decrease desorption energy in CO_2 capture applications compared with the single MEA.

3.2 CO_2 regeneration performance with solid acid catalyst

To further accelerate the CO_2 desorption rate and reduce the heat duty, the regeneration behavior of 3M MEA-2.5M AMP-0.5M PZ system studied by adding four different catalysts (e.g. H-ZSM-5, γ-Al_2O_3, SAPO-34, and SO_4^{2-}/TiO_2). The performance of CO_2 desorption curves of this system with and without catalysts are shown in Figure 4. As shown in this figure, the lean loading value of different systems declined to be the same, which means that the catalyst can only change the reaction rate but will not alter the chemical equilibrium. On the other hand, the addition of four acid catalysts play a positive effect on CO_2 desorption rate. The result proved that the catalyst can only accelerate the reaction rate without the effect on the chemical and thermodynamic equilibrium. Figure 5 displays the relative regeneration heat duty of different amine-catalyst systems. The CO_2 desorption performance in terms of the relative heat duty for different catalysts could be ranked as: H-ZSM-5 (38.4%) > γ-Al_2O_3 (40.1%) > SAPO-34 (41.0%) > SO_4^{2-}/TiO_2 (42.8%) > the blank (48.1%). The result indicated that four catalysts further accelerated the CO_2 desorption rate and reduced the heat duty in comparison with the blank test, and H-ZSM-5 showed the best catalytic performance of CO_2 desorption. The addition of acid catalyst into the rich loading amine solution results in increasing proton concentration, which could react with the carbamate (i.e. Amine-COO$^-$). Thereby, the improvement of CO_2 desorption rate and the decrease of the heat duty were observed.

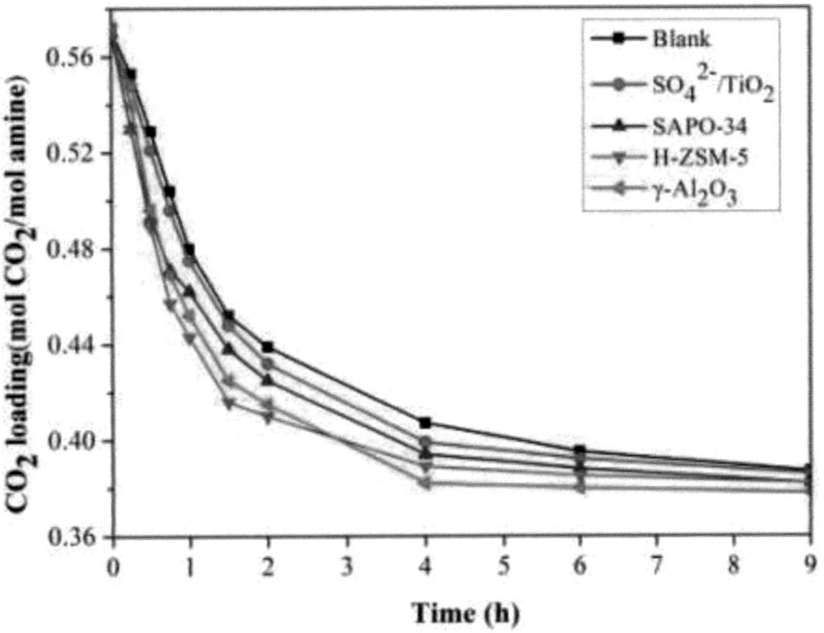

Figure 4. CO_2 desorption in 3M MEA-2.5M AMP-0.5M PZ with and without catalyst.

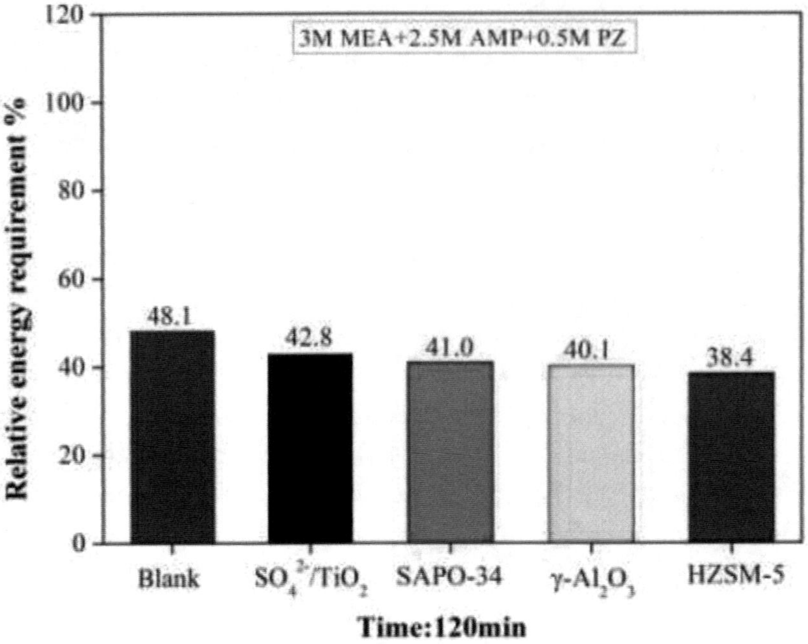

Figure 5. Relative heat duty for 3M MEA-2.5M AMP-0.5M PZ with and without catalyst.

4. Conclusion

From the results of this work, it was found that all the MEA-AMP-PZ tri-solvent blends showed better regeneration performance than the single 5 M MEA in blank runs. The 3 M MEA-2.5 M AMP-0.5 M PZ presents the greatest desorption performance, which resulted from its highest HCO_3^- concentration based on the analysis of the NMR technique. All the four acid catalysts further facilitated CO_2 desorption rate and reduced the energy consumption in the regeneration process of the blends. The best combination of catalyst and blend amine was H-ZSM-5 in combination with 3 M MEA-2.5 M AMP-0.5 M PZ. This combination generated the highest catalyst activity and amine blend influences for the fastest CO_2 desorption rate as well as the least heat duty for the rich solvent regeneration process.

Acknowledgments

The financial support from the National Natural Science Foundation of China (NSFC-Nos. U1362112, 21376067, 21476064, 21406057, 21536003 and 51521006), Innovative Research Team Development Plan (MOE-No.IRT1238), Specialized Research Fund for the Doctoral Program of Higher Education (MOE-No.20130161110025), Key Disciplines of Chemical Technology of Guangxi Universities, 2016 Open Project (Nos. 2016KLOG05, 2016KLOG11, 2016KLOG13 and 2016KLOG17), China Scholarship Council (201606135004), and China

Outstanding Engineer Training Plan for Students of Chemical Engineering & Technology in Hunan University (MOE-No.2011-40) are gratefully acknowledged.

References

1. Liang Z, Fu K, Idem R, Tontiwachwuthikul P. Review on current advances, future challenges and consideration issues for post-combustion CO_2 capture using amine-based absorbents. *Chin. J. Chem. Eng.* 2016;24(2):278-288.
2. Nwaoha C, Saiwan C, Tontiwachwuthikul P, et al. Carbon dioxide (CO_2) capture: Absorption-desorption capabilities of 2-amino-2-methyl-1-propanol (AMP), piperazine (PZ) and monoethanolamine (MEA) tri-solvent blends. *J. Nat. Gas Sci. Eng.* 2016;33:742-750.
3. Shi H, Naami A, Idem R, Tontiwachwuthikul P. Catalytic and non catalytic solvent regeneration during absorption-based CO_2 capture with single and blended reactive amine solvents. *Int. J. Greenhouse Gas Control* 2014;26:39-50.
4. Zhang X, Zhang X, Liu H, et al. Reduction of energy requirement of CO_2 desorption from a rich CO_2-loaded MEA solution by using solid acid catalysts. *Appl. Energy* 2017;202:673-684.
5. Horwitz W. Association of official analytical chemists (AOAC) methods. *George Banta Company, Menasha, WI.* 1975.
6. Liang Z, Idem R, Tontiwachwuthikul P, Yu F, Liu H, Rongwong W. Experimental study on the solvent regeneration of a CO_2-loaded MEA solution using single and hybrid solid acid catalysts. *AIChE J.* 2016;62(3):753-765.
7. Caplow M. Kinetics of carbamate formation and breakdown. *J. Am. Chem. Soc.*1968;90(24):6795-6803.
8. Zhang R, Zhang X, Yang Q, Yu H, Liang Z, Luo X. Analysis of the reduction of energy cost by using MEA-MDEA-PZ solvent for post-combustion carbon dioxide capture (PCC). *Appl. Energy* 2017;205:1002-1011.
9. Shi H, Naami A, Idem R, Tontiwachwuthikul P. 1D NMR Analysis of a Quaternary MEA–DEAB–CO_2–H_2O Amine System: Liquid Phase Speciation and Vapor–Liquid Equilibria at CO_2 Absorption and Solvent Regeneration Conditions. *Ind. Eng. Chem. Res.* 2014;53(20):8577-8591.

493490 A Highly Permeable Microporous Polyamide Membrane for Molecularsieving of Nitrogen from Volatile Organic Compounds

Abstract:

Microporous polymer membranes continue to receive tremendous attention for their energy-efficient gas separation owing to their high separation performance[1, 2]. However, it remains challenging to simultaneously achieve the solution processability of polymers along with the molecular-sieving ability of membranes for the separation of mixtures containing condensable gases, such as N_2/volatile organic compound (VOC) mixtures. Here, we report a new network microporous polyamide membrane with good molecular-sieving performance for the separation of N_2 from VOC mixture. Triple-substituted triptycene was used as the main monomer to form a fishermanâ€™s net-shaped polymer, which readily form a composite membrane by solution casting. This membrane exhibited outstanding separation performance and good stability for the molecular-sieving separation of N_2 over VOCs, such as cyclohexane. Its rejection rate reached 99.2 % with 1104 GPU N_2 permeance at 24 ^{o}C under 4 kPa. This approach opens the door for the development of microporous membranes to separate condensable gases.

[1] Jimenez-Solomon, M. F., *et al.* . *Nature Mater.* 2016 **15** 760

[2] Du, N., *et al. Nature Mater.* 2011 **10** 372

495809 Water Desalination Using Porous Organic Cage Membranes: A Simulation Exploration

Abstract:

With a comparatively low energy cost[1], reverse osmosis (RO) is estimated to account for 53% of the global water desalination capacity[2]. The cost and performance of RO processes is largely determined by the membrane used, which is primarily characterized by easy preparation, high water flux and high salt selectivity[3]. Porous organic cages (POC)[4], organic molecules with intrinsic cages, can be processed easily due to their solubility in common organic solvents. Their water stability[5] further provide the opportunity of being utilized in RO processes. In this work, we report the first proof-of-concept simulation study on the potential of a series of POCs, both crystalline and amorphous, as RO membranes for water desalination. It is revealed that the channel size and topology of POCs are two dominant factors governing desalination performance. One POC crystal, CC2, with straight channels running through the membrane, possesses a high water permeability of 2.05Ã—10^{-6} kg/(m^{2}âˆ™hr) and 100% salt rejection. These findings from the bottom-up may guide the design and preparation of POC-based membranes for high-performance water desalination.

References

[1] M. A. Shannon, P. W. Bohn, M. Elimelech, J. G. Georgiadis, B. J. Marinas, A. M. Mayes, *Nature* **2008**, *452*, 301-310.

[2] T. Mezher, H. Fath, Z. Abbas, A. Khaled, *Desalination* **2011**, *266*, 263-273.

[3] M. M. Pendergast, E. M. Hoek, *Energy & Environmental Science* **2011**, *4*, 1946-1971.

[4] aT. Tozawa, J. T. Jones, S. I. Swamy, S. Jiang, D. J. Adams, S. Shakespeare, R. Clowes, D. Bradshaw, T. Hasell, S. Y. Chong, *Nature materials* **2009**, *8*, 973-978; bJ. T. Jones, T. Hasell, X. Wu, J. Bacsa, K. E. Jelfs, M. Schmidtmann, S. Y. Chong, D. J. Adams, A. Trewin, F. Schiffman, *Nature* **2011**, *474*, 367-371.

[5] T. Hasell, M. Schmidtmann, C. A. Stone, M. W. Smith, A. I. Cooper, *Chemical Communications* **2012**, *48*, 4689-4691.

495491 A Molecular Simulation Protocol for Membrane Pervaporation

Abstract:

Pervaporation (PV) has emerged as a breakthrough technology for the separation of liquid mixtures. As a unique integration of membrane permeation and evaporation, PV offers several advantages such as low energy consumption, high separation capability and easy scaling-up. Consequently, PV has been regarded as a potential technology in chemical separation of azeotropic mixtures, thermally sensitive compounds, organic-organic mixtures and also for the removal of dilute organics from aqueous solutions.

Practically, PV is conducted with a vapor-liquid equilibria in the feed side and simultaneously a vanishingly low pressure in the permeate side. To the best of our knowledge, there is no reported molecular simulation study to mimic PV. In this study, a simulation protocol is proposed. To demonstrate, two different types of feed solutions (seawater and water-ethanol mixture) are examined through zeolitic imidazolate framework (ZIF) membranes. For each case, first, a molecular model is constructed including vapor-liquid equilibria in the feed side and an adsorbing plate on the permeate side. Then, the adsorption interaction strength in terms of É›/k_b systematically varies, and the optimal value of *É›*/k_b is identified to be 10000 K. With this optimal value, water flux remains constant despite changing the size of permeate side. Good agreement with experiment is found for the PV of seawater through ZIF-8, and high water permeability through ZIF-93 is observed for the PV of water-ethanol mixture. The proposed simulation protocol can be generally applied to other PV systems.

497755 Coordinate Immobilization of Silver Nanoparticles on Aminenated Polyethersulfone (AgNPs-APES) Composite Membrane for Prolong and Constant Silver (Ag^+) Release

Abstract:

Improving the life time of water treatment membranes can be achieved through exquisite control of active disinfectant release. The present study demonstrates a controllable and systematic novel approach to coordinately immobilize silver nanoparticles (AgNPs) on the surface of aminenated polyethersulfone (APES) composite membrane for prolonged silver (Ag^+) release to enhance the life of water treatment membrane. Aminenated-polyethersulfone was prepared by nitration of the polyethersulfone (PES) by acid treatment. The obtained nitrated polyethersulfone (NO_2-PES) was subsequently reduced using tin (II) chloride dihydrate ($SnCl_2.2H_2O$) and sodium iodide (NaI) to form aminenated-polyethersulfone APES or (NH_2-PES). AgNPs were coordinately attached on the surface of APES (AgNPs-APES) by reduction of silver nitrate ($AgNO_3$) with sodium borohydride ($NaBH_4$). The physicochemical properties of the as-synthesized membrane were examined by FT-IR, XPS, XRD, TGA, ICP-OES and SEM-EDAX analyses. It was confirmed that the AgNPs ranging from 5-40 nm were immobilized on the surface of polymer membrane. The evaluation of Ag^+ leaching over time of the obtained samples indicated that ~30% of AgNPs could be retained after 12 days from operation of 1cm × 3cm membrane strip. The extrapolation of the data curves confirmed that the membrane can be used up to 25 days. Antibacterial activities of the samples were also carried out to investigate the disinfection performance of the samples on the Gram negative bacteria (*Escherichia coli*). Generally, the as-synthesized membrane revealed 4 times longer life than AgNPs embedded polymer membranes.

497689 Use of Novel Reactor-Separator Combination (Membrane BioReactor) for Enzymatic Hydrolysis of Waste Fines and Fiber Rejects from Recycled Linerboard Paper Mills

Abstract:

Papermaking fibers, particularly waste fibers from recycling operations can be a valuable resource for biofuels (ethanol), and sustainable products such as bioplastics or biochemicals. The amount of lignin in such fibers is much lower than native wood or other plant based lignocellulosic resources eliminating the need for complex and expensive pretreatment processes. Moreover, the removal of hornified cellulosic fines from recycled pulps can accelerate water removal from the pulps on the papermachines and in the dryer sections, leading to significant productivity increases. Inter-fiber bonding in the paper sheet is improved due to a lower â€˜waste finesâ€™ content, reducing the use of strength agents like starch and improving sheet quality.

The present work develops a physico-chemical model for the kinetics of enzymatic hydrolysis of cellulosic pulp fines from recycled linerboard mills. Batch kinetics representing hydrolysis in a fixed agitated volume of the fines suspension is considered. Cellulosic fines are assumed to be spherical and porous, with the cellulase enzymes diffusing and adsorbing onto the interior pore surfaces. Enzymatic reactions result in the production of glucose which diffuses outward. The mechanistic role of inhibitory mineral (ash) fillers and the application of suitable surfactants to cover their surfaces and improve cellulose hydrolysis is considered.

Optimal usage of the enzyme catalyst is a critical part of this process. In this paper, microfiltration membranes are used to separate the enzymes and recycled so that the effective catalyst concentration is higher and overall enzyme usage is minimized. Experimental results showing the impact of flow rates, concentrations of substrate and concentrations as well as other variables such as pH and temperature are shown. Mathematical models for determining the optimal designs of both external and internal membrane bioreactors were set up for analysis.

498945 Boron-Nitride-Nanopore Membranes for Osmotic Power Harvesting

Abstract:

The use of salinity-gradient power, based on the Gibbs free energy of mixing fresh and salt water, has been under intensive investigation for clean-energy harvesting from abundant environmental resources. However, the current reverse electrodialysis (RED) systems based on polymeric ion exchange membranes (IEMs) suffer from lower power density (approximately 1-3 W/m^2), low efficiencies, and high membrane cost. A recent experimental study on a membrane containing a single boron-nitride-nanotube pore showed electric currents as high as 1.2 nA. Extrapolating this result to a macroscopic membrane, with a boron-nitride pore density of $\sim 10^{10}$ nanotubes/cm^2, would result in a power density of ~4 kW/m^2. However, no such large-surface-area vertically aligned boron-nitride-nanopore (VA-BNNP) membranes have ever been fabricated.

Here, for the first time, we demonstrate the power generation performance of macroscopic VA-BNNP membranes with high nanopore density, up to $\sim 10^8$ pores/cm^2. For the fabrication of VA-BNNP membranes, a thin hexagonal boron nitride (hBN) layer was deposited within the pores (~100 nm) of anodized alumina substrates by low-pressure chemical vapor deposition. Cross-sectional scanning-electron-microscope (SEM) images show the hBN layer (~35 nm) was uniformly deposited along the pores without excessive hBN on the top surface, ensuring that most pores remained open. In addition, the results of scanning confocal Raman spectroscopy and X-ray photoelectron spectroscopy (XPS) showed the high quality of the hBN layers in the AAO pores. We investigated the power generation of the macroscopic VA-BNNP membranes at different pH and salinity concentrations. The power generation per unit pore area increased as the salt concentration and pH increased. The highest power density of the membrane was up to ~100 W/m^2which is two orders of magnitude higher than that of RED system based on IEMs or even nanofluidic channels. These findings indicate the great potential of large-area VA-BNNP membranes as next-generation nanostructured membranes for renewable energy harvesting.

499701 Molecular Insights on the Reverse-Selectivity Potential of Room Temperature Ionic Liquid Membranes

Abstract:

Recently, room temperature ionic liquid (RTIL) membranes have attracted attention due to their excellent performance for gas separation applications. In this work, the reverse-selective potential of RTIL membranes has been investigated using molecular dynamics simulation. Since the kinetic diameter of propane (C_3H_8; 4.30 Å) is larger than that of nitrogen (N_2; 3.64 Å) and methane (CH_4; 3.80 Å), C_3H_8/N_2 and C_3H_8/CH_4 separations are good indicators of reverse-selectivity. Six imidazolium-based ionic liquids were selected for this study. The pure gas diffusivity, solubility, and permeance were determined for the RTIL membranes and a good agreement was observed for these calculated transport properties with the experimental data. In addition, based on the theoretical findings in this study, gas transport through the RTIL membranes is found to be dominated by solubility rather than diffusivity, which is a requirement for the reverse-selective membranes.

499741 Molecular Dynamics Simulation of Room Temperature Ionic Liquid Membranes for CO_2/CH_4 and CO_2/N_2 Separations

Abstract:

Room temperature ionic liquid (RTIL) membranes have shown great potential for gas separation applications due to their excellent permeability and selectivity. Molecular dynamics simulations were utilized in this work to determine the gas transport properties of CO_2/CH_4 and CO_2/N_2 mixtures in RTIL membranes. Pure gas diffusivity, solubility, and permeance, as well as perm-selectivities were calculated. Six imidazolium-based ionic liquids were selected for the study. The calculated gas transport properties were compared with the experimental data and good agreements were obtained. In addition, our findings confirm the solubility-dominated gas transport through the RTIL membranes.

501088 Ionic Liquid Based Methacrylate Polymer Membranes for Efficient Enrichment of 1,3-Propanediol from Fermentation Broths

Abstract:

The challenge of pervaporative enrichment of 1,3-propanediol (1,3-PD) from dilute aqueous mixtures was addressed by the synthesis of a novel hydrophobic imidazolium alkyl phosphate ionic liquid structure based monomer. Vinylimidazole was â€œcoupledâ€ with mercaptoethanol, converted into an alkylphosphate ionic liquid followed by esterification to the final monomer structure. Characterization and batch pervaporation was carried out on several different polymer chemistry mediated membrane structures based on this monomer, including semi interpenetrating polymeric network like structures with hydrophobic polyglycerol sebacate acrylate (PGSA). The batch pervaporation exhibited high separation factors and fair 1,3-PD fluxes compared to other hydrophobic membranes. The monomer provides recourse to high performing materials by balancing its 1,3-PD affinity with its hydrophobic nature. The excellent cost-performance trade-off makes this an attractive candidate for possible scale-up and industrial applications in multilayer modules.

492146 Water Flow inside Polyamide Reverse Osmosis Membranes: A Nonequilibrium Molecular Dynamics Study

Abstract:

Among all desalination technologies, reverse osmosis (RO) technology is the most widespread technology to obtain the freshwater capacity. To further enhance the water flux of RO membranes, it is important to understand the molecular mechanism of water transport through the membrane. We apply a steady state nonequilibrium dynamics simulation (SS-NEMD) to study the behavior of water molecules flowing inside the RO membranes, by directly applying external forces on the water molecules to simulate pressure drops, since the mass transport resistance inside the membrane is what we mostly considered.

Before measuring the relationship between pressure drops and the pure water flux, we estimated the average density along z direction of simulation boxes and pore size inside the membrane, which is in good consistency with the experimental values. Then, six pressure drops were applied on water molecules to investigate the e ect of pressure drops on the pure water flux (PWF). In each pressure drop, there is no obvious tendency of continuous increase or decrease in the PWF, we believe the simulation reached steady states. From the six pressure drops, it is obvious that the water flux increases almost linearly with pressure drops despite small fluctuations due to the random thermal motion. From that, we could calculate the water flux at the experimental condition. At last, we revealed the interactions between water molecules and functional groups. We find that, water molecules move faster around benzene rings than do around carboxyl or amino groups in the membrane.

In this work, the results confirm that RO membranes containing more benzene rings would reduce the mass transport resistance of water inside the membranes. These findings are expected to increase our understanding on the water transport behavior in RO membranes and also the rational design of advanced RO membranes with upgraded permeance.

501275 Intensification of the Enzymatic Hydrolysis of Recycled Paper Fiber Fragments Using Membrane Separations

Abstract:

Papermaking fibers, particularly waste fibers from recycling operations can be a valuable resource for biofuels (ethanol), and sustainable products such as bioplastics or biochemicals. The amount of lignin in such fibers is much lower than native wood or other plant based lignocellulosic resources eliminating the need for complex and expensive pretreatment processes. Moreover, the removal of hornified cellulosic fines from recycled pulps can accelerate water removal from the pulps on the papermachines and in the dryer sections, leading to significant productivity increases. Inter-fiber bonding in the paper sheet is improved due to a lower â€˜waste finesâ€™ content, reducing the use of strength agents like starch and improving sheet quality.

The present work develops a physico-chemical model for the kinetics of enzymatic hydrolysis of cellulosic pulp fines from recycled linerboard mills. Batch kinetics representing hydrolysis in a fixed agitated volume of the fines suspension is considered. Cellulosic fines are assumed to be spherical and porous, with the cellulase enzymes diffusing and adsorbing onto the interior pore surfaces. Enzymatic reactions result in the production of glucose which diffuses outward. The mechanistic role of inhibitory mineral (ash) fillers and the application of suitable surfactants to cover their surfaces and improve cellulose hydrolysis is considered.

Optimal usage of the enzyme catalyst is a critical part of this process. In this paper, microfiltration membranes are used to separate the enzymes and recycled so that the effective catalyst concentration is higher and overall enzyme usage is minimized. Experimental results showing the impact of flow rates, concentrations of substrate and concentrations as well as other variables such as pH and temperature are shown. Mathematical models for determining the optimal designs of both external and internal membrane bioreactors were set up for analysis.

493654 Evaluation of the Efficiency in a Set of Air Separation Units through Data Envelopment Analysis and Malmquist Productivity Index

Abstract:

Nowadays, the efficient use of energy resources is one of the main principles for industry, and it is one of the more complex challenges faced by companies. The goal treat to solve the trouble between raising living standards by continuing to expand industrial production and lessening the negative environmental impacts of industrial manufacturing activities. One of the approaches to address this challenge is to improve the energy efficiency in the core of industryâ€™s production processes, which accounts for around one-third of the worldâ€™s final energy demand. Furthermore, the use of energy by industries is often inefficient, leaving much room for improvement. For this reason, improving industrial energy efficiency will be a critical success factor for companies in order to improve cost competitiveness, brakes emissions of carbon dioxide and saves money.

Cryogenic air separation process is an old energy-intensive technology aimed to produce large amounts of technical gases with high purity standards. These pure gases can be obtained in Air Separation Units (ASUs) by cooling air until it liquefies and distilling it by means of a physical process. In the distillation, the air components (i.e., nitrogen, oxygen and argon) can be separated and recovered since they have different boiling temperatures. This process requires a very tight integration of heat exchangers and separation columns to guarantee good performance, and it uses large electricity amounts (i.e., several tens of megawatts) to cover the high compression and liquefaction needs.

In this context, we need to apply decision-making tools that allow assessing and quantifying the level of efficiency of the ASUs. In this work we propose a method to determine the efficiency in a set of real ASUs based on the non-parametric technique of Data Envelopment Analysis (DEA). The mathematical methodology of DEA has been used to assess the performance of many types of entities in many different contexts (i.e., from production and business firms, to non-profit agencies such as hospitals, universities, armies, countries, etc.) and it uses linear programming (LP) to objectively evaluate the relative efficiency of a set of units under multiple criteria conditions. In DEA, the relative efficiency of a DMU is defined as the maximum ratio of weighted sum of outputs to the weighted sum of inputs, being the efficiency less or equal than 1. The linear optimization is applied for each single unit, also called Decision Making Unit (DMU), and it provides the performance scores representing the degree of efficiency of each evaluated entity. In this work, the capabilities of DEA have been demonstrated by creating an input-oriented variable return to scale (VRS) model which allow identifying qualitatively (i.e., sources) and quantitatively (i.e., amounts) the inefficiencies in each input and output for every entity. For this, DEA creates an "efficiency frontier" formed by the efficient entities (i.e., those with efficiency scores equal than one) which serves to identify benchmarks against which such inefficient unit can be compared and, also, to define improvement targets to make efficient the inefficient units if achieved. Furthermore, the efficient units are also compared between them by solving the super-efficiency model which ranks the efficient ASUs and lets to observe differences among them.

This work is complemented by measuring the productivity change over time in each ASU by means of the Malmquist Productivity Index (MPI). This approach identifies the ratio between efficiency, as calculated by the DEA, for the same production unit in two different time periods (*i.e.*, years) to identify efficiency gains and losses in each ASU over time. The MPI can be decomposed into two components measuring the change in the technology frontier and in the technical efficiency. In this paper, both components are calculated to obtain more information from the individual Malmquist components to identify patterns of efficiency change.

The methodology of this paper is demonstrated by applying it in a set of 33 existing Air Separation Units from a gas company operating plants around the world. The MPI is illustrated to the set of ASUs during the last four years. The results provide insightful information into how the unitâ€™s performance has evolved over time, as well as how to improve the units to make the air separation technology more efficient.

496119 Plantwide Control for Maximum Throughput Operation of an Ester Purification Process

Abstract:

Plantwide control for maximum throughput operation of an ester purification process consisting of a liquidâˆ'liquid extractor (LLX), a distillation column, and material recycle is evaluated. The extractor uses water solvent and is necessary to cross a distillation boundary making pure ester product recovery feasible. Extraction capacity is the bottleneck constraint. A steady state analysis reveals the occurrence of steady state multiplicity induced infeasibility for too high fresh feed rates and for too low water feed rates. To avoid infeasibility due to the multiplicity behavior, we recommend using the extractor feed rate as the throughput manipulator (TPM) with the fresh feed rate adjusted as a makeup stream (CS1). The other control loop pairings are conventional. Rigorous closed loop dynamic results show that CS1 is significantly more robust than the conventional control structure (CS2) with the fresh feed rate as the TPM. In particular, CS1 achieves more than 25% higher maximum throughput compared to CS2 for large feed composition changes in specific directions. This is attributed to the self-regulatory nature of the recycle inventory control in CS1. In contrast, CS2 is susceptible to failure due to overfeeding-induced steady state infeasibility. The presentation will highlight the central role of control structure/strategy design for robust plant operation.

496209 Crown Ether Diols Aerosol Cross-Linked with Poly(vinyl alcohol) As Specialized Li^+ Adsorbent Nanofibers

Abstract:

Lithium-specific crown ether (CE)-based adsorbent nanofibers (NF) were successfully prepared through strategic use of CE diols, electrospinning and aerosol cross-linking. Each diol contained four ethereal oxygens for Li^+ complexation, a benzo group for a rigidified ring, and two hydroxyl groups for covalent attachments to poly(vinyl alcohol) (PVA) as the chosen matrix. The CE diol/PVA blends were electrospun into NFs to obtain structures that are favorable for Li^+ interaction. Electrospinning also eliminated CE wastage during CE/PVA NF fabrication. As a post-electrospinning treatment, aerosol glutaraldehyde (GA) acetalization not only stabilized the PVA in water (insoluble) but also effectively linked the CEs to PVA by forming CE-GA-PVA bridges. The devised cross-linking method preserved the NF structure, required short reaction time (5 h) and used low cross-linker volume (2.4 mL g^{-1} NF). From these techniques, three types of CE/PVA NFs were fabricated. The CEs slightly differ in ring size based on DFT calculations and in one-side ring sub-units (i.e. R=benzyl, neopentyl, butyl). The effectiveness of the fabrication strategy was confirmed through FTIR-ATR, ^{13}C CP-MAS NMR, FE-SEM, N_2 adsorption/desorption, and UTM. The NFs exhibited pseudo-second order rate and Langmuir-type of Li^+ adsorption. Their nearly ideal performances reveal 90-99% CE utilization through 1:1 Li^+ complexation. The NFs were highly selective to Li^+ in seawater; neopentyl bearing CE was most effective in blocking larger monovalents like Na^+ and K^+ while the dibenzo CE was best in discriminating divalents like Mg^{2+} and Ca^{2+}. Experimental selectivity trends concur with the reaction enthalpies from DFT calculations, confirming the influence of CE structures and cavity dimensions in their "size-match" Li^+ selectivity. The key advantage of the fabrication strategy is its versatility for wide application. By simply replacing the CE diols, numerous types of CE-based NF adsorbents can be developed for the recovery of precious or highly toxic metals.

This research was supported by the National Research Foundation of Korea (NRF) funded by the Ministry of Science, ICT & Future Planning (No. 2016R1A2B1009221 and 2017R1A2B2002109) and the Ministry of Education (2009-0093816).

496764 Efficient Absorption of SO_2 in Flue Gas By Environmentally Benign Functional Deep Eutectic Solvents

Abstract:

Efficient absorption of SO_2 in flue gas by environmentally benign functional deep eutectic solvents

Kai Zhang[1], Shuhang Ren[1], Yucui Hou[2], Ying Sun[1], Weize Wu[1,*]

[1]State Key Laboratory of Chemical Resource Engineering, Beijing University of Chemical Technology, Beijing 100029, China

[2]Department of Chemistry, Taiyuan Normal University, Taiyuan 030031, China

Sulfur dioxide (SO_2) mainly emitted from flue gas badly effects on environment. Deep eutectic solvents (DESs), which is a combination of a hydrogen bond acceptor (HBA) and a hydrogen bond donor (HBD) through hydrogen bond, are considered as a new kind of solvents with many advantages such as low vapor pressure, easy preparation and low price. In this work, we found DESs could be designed with a function to absorb low concentration SO_2 from simulated flue gas. Two kinds of DES, which can efficiently and reversibly absorb SO_2, based on betaine (Bet) and L-carnitine (L-car) as HBA and ethylene glycol (EG) or water as HBD were prepared with different ratios of HBA to HBD. 2% SO_2 solubility in DESs were 0.332 mol SO_2 / mol HBA for Bet + EG DES and 0.820 mol SO_2 / mol HBA for L-car + EG DES with $n_{HBA}:n_{HBD}$ = 1:3, and 0.155 mol SO_2 / mol HBA for Bet + H_2O DES and 0.606 mol SO_2 / mol HBA for L-car + H_2O DES with $m_{HBA}:m_{HBD}$ = 1:1, at 40 °C. SO_2 solubility increased with increasing SO_2 concentration and decreasing temperature. Meanwhile, the absorption capacities in DESs did not change after 5 absorption/desorption cycles. FTIR, ^{1}H NMR and ^{13}C NMR results indicated that there are strong acidâ€“base interactions between SO_2 and â’COOâ€“ on HBA. DESs based on Bet and L-car as HBA and EG or water as HBD are promising absorbents in the application of SO_2 capture.

The project is financially supported by the National Natural Science Foundation of China (No. 21176020 and 21306007), and the Research Fund for the Doctoral Program of Higher Education of China (No. 20130010120005).

Keywords: Absorption; SO_2; flue gas; Functionalization; Deep eutectic solvent

---

* Corresponding author; E-mail: wzwu@mail.buct.edu.cn; Tel./Fax: +86 10 64427603.

501584 Energy Integrated Natural Gas Liquid Recovery Process By Vapor Recompressed Internally Driven Reboiler

Abstract:

Processing of Natural gas liquid (NGL) recovery is a complex industrial process designed to clean the raw natural gas by separating the impurities and various non-methane hydrocarbons, which are value added chemicals need to separate out to fulfil the dry natural gas as pipeline quality according to the end users requirement for their process needs. Usually, NGL recovery has to be carried out at an offshore plant location, where the space and sources of energy are limited; which plays a vital role in the process economics. In this study, explored different energy integrated configurations to improve the energy efficiency and economics of the NGL recovery from natural gas. Simulated the NGL recovery from pre-treated Natura gas by considering a propane C3MR cycle as a refrigerant cycle, separated the propane, LPG and C5+ from natural gas in two column distillation unit sequence in Aspen Hysys environment and considered this case study as a base case to analyze the economics and energy efficiency of the different explored energy integrated configuration for NGL recovery. Analysed the different energy integrated schemes to enhance the energy efficiency of the NGL recovery with compared to the conventional NGL recovery. Performed the economic and energy analysis to select the better scheme for NGL recovery than conventional approach. Vapour recompressed internally driven Reboiler in petlyuk column configuration shows better performance than conventional sequence.

489984 Engineering Studies of the Effect of pH, Temperature and Protein Tertiary Structure on Î²-Lactoglobulin a and B Separation in Anion-Exchange Chromatography

Abstract:

In this presentation, we explore the effect of pH and temperature on separating a mixture of similar proteins, namely Î²-lactoglobulin A (LGA) and Î²-lactoglobulin B (LGB) in anion-exchange chromatography. The proteins are similar in size and shape, however they carry a slight difference in negative charge at basic pH. In order to separate them, a chromatography column with Sepharose Q Fast Flow was used. The proteins were separated at a range of temperatures and pH values, and the separation factor at each condition was evaluated. Along with that, to get the general trend of protein behavior on an anion exchange resin, bovine serum albumin (BSA) as model protein was run as well. The experimental results were matched to a theoretical model in order to compute the adsorption and desorption rate values of the proteins. The data demonstrated that the separation is improved at increased temperature and pH, and the data was correlated to see how the net charge of the molecule can be used to predict the separation. Furthermore, the tertiary structures of LGA and LGB are analyzed to illustrate the tertiary structure effect on the separation. We observed that LGA has a significant negative charge on the outside of the protein, which causes it to be retained more strongly. Predicting the equilibrium constant allows our theoretical model to predict the elution peak at any pH and temperature condition for this system.

492832 Nano-Cellulose Based Thin Film Nanocomposite RO Membranes with Tunable Flux Via Control of Interfacial Transport

Abstract:

Improvement of current reverse osmosis membranes poses an intricate challenge that, if solved, could affect global access to clean water. Thin film composite (TFC) membranes are used commonly in industrial desalination processes, relying on a polyamide layer for selectivity. It is desired to improve the energy efficiency of these membranes in order to lower the production cost of clean water. Recent studies in our research group involving the addition of functionalized carbon nanotubes to the selective polymer layer of existing TFC membranes have shown promising results, yielding higher water permeance (flux), improved chemical resistance, anti-biofouling properties, and mechanical property enhancement. These thin film nanocomposite (TFN) membranes are often cost-ineffective, prompting the search for an economically viable alternative. TEMPO-oxidized cellulose nanocrystals (TOCNs) have been investigated as a possible TFC additive due to their hydrophilicity, availability, and low cost. In the current work, TOCNs are deposited within the selective aromatic polyamide layer of an established TFC membrane design and subjected to reverse osmosis conditions to determine the water flux and salt rejection, and to observe the properties of the surface. Significant increases in water permeance have been observed while maintaining high salt rejection when compared to a control polyamide (PA) TFC membrane. The improved flux confirms that high aspect ratio nanoparticles can be used to improve transport in membranes via the creation of pathways for rapid transport at the interface between the particles and polymer matrix. This suggests that membrane properties can be controlled and optimized via careful control of the nanoparticle/polymer interface (i.e. through control of nanoparticle surface functionality and polymer matrix properties.

\s

494026 Synthesis of 3D Na-Embedded Carbon Nanomaterials and Their Applications in Solar Cells

Abstract:

High electrical conductivity and large accessible surface area, which are required for ideal electrode materials in energy devices, are opposed to each other in current materials. In this presentation, I will present you an invented reaction based on Na and CO to synthesize sodium-embedded carbon nanowalls. The synthesized sodium-embedded carbonâ€™s conductivity is two orders of magnitude larger than three-dimensional graphene. Furthermore, the nanowall structure has a large accessible surface area comparable to graphene. This Na-embedded carbon solved the contradiction of high electrical conductivity and large surface area in electrode materials.

496947 Electrical Energy Generation Via Reversible Chemical Doping on Carbon Nanotube Fibers

Abstract:

Chemically modified carbon nanotube fibers enable unique power sources driven entirely by a chemical potential gradient. Electrical current (11.9 µA mg^{-1}) and potential (525 mV) are reversibly produced by localized acetonitrile doping under ambient conditions. An inverse length-scaling of the maximum power as $L^{-1.03}$ that creates specific powers as large as 30.0 kW kg^{-1} highlights the potential for microscale energy generation.

497039 Observation of the Marcus Inverted Region of Electron Transfer from Asymmetric Chemical Doping of Pristine (n,m) Single-Walled Carbon Nanotubes

Abstract:

The concept of electro-organic chemistry based on voltage generated by asymmetric chemical doping (ACD) of single-walled carbon nanotube (SWNT) papers is presented. We explore 27 small, organic, electron-acceptor molecules that are shown to tune the output open circuit voltage (V_{OC}) across three types of pristine SWNT papers with varying (n, m) chirality distributions. A considerable enhancement in the observed V_{OC}, between 80 to 440 mV, is observed for SWNT/molecule acceptor pairs that have molecular volume below 0.12 nm^3 and lowest unoccupied molecular orbital (LUMO) energies centered around -0.8 eV. The electron transfer (ET) rate constants driving the V_{OC} generation are shown to vary with the chirality-associated Marcus theory, suggesting that the energy gaps between SWNT valence band and the LUMO of the acceptor molecules dictate the ET process. When the ET rate constants and the maximum *Voc* are plotted versus the LUMO energy of the acceptor organic molecule, volcano-shaped dependencies, characteristic of the Marcus inverted region, are apparent for three different sources of SWNT papers with modes in diameter distributions of 0.95 nm, 0.83 nm, and 0.75 nm. This observation, where the ET driving force exceeds reorganization energies, allows for an estimation of the outer-sphere reorganization energies with values as low as 100 meV for the (8,7) SWNT, consistent with a proposed image-charge modified Born energy model. These results expand the fundamental understanding of ET transfer processes in SWNT and allow for an accurate calculation of energy generation through asymmetric doping for device applications.

Fracturing Fluid Retention and its Effect on Fluid Flow in Microfractures of Tight Oil Reservoirs

Zhaojie Song[a,b], Liya Zhang[a,b], Jirui Hou[a,b], Qingjie Liu[c], Zhiyao Chen[a,b], Jilong Zhang[d]

[a] Enhanced Oil Recovery Institute, China University of Petroleum, Beijing 102249, China

[b] Basic Theory Laboratory of Improving Oil Recovery in Low Permeability Oilfields, Tertiary oil Recovery Key Laboratory, CNPC, Beijing 102249, China

[c] Research Institute of Petroleum Exploration and Development, PetroChina, Beijing 100083, China

[d] China International Engineering Consulting Corporation (CIECC), Beijing 100048, China

Abstract: Polyacrylamide-based friction reducer (FR) serves as one of the primary additives in fracturing fluid for tight reservoirs. Low recovery of fracturing fluid leaves much chemical residual in microfractures, which is closely related to relative permeability near fracture, flowback of fracturing fluid, production rate, etc. Some field cases also present better production performance under the circumstance of lower recovery of fracturing fluid. This work is to investigate effect of fracturing fluid residual on oil and water flux.

This study used tight sandstone core samples (φ1 in.×1 ft.) to fabricate fracture models. Each core sample was cut in half from the center, and stainless steel sheets with different thicknesses were inserted between the two halves to model fractures. The sandstone cores were used in the experiments because they could better model the adsorption-entanglement effect of polymer on fracture surfaces. Oil or brine was injected at various velocities before and after tight sandstone fracture models were fluxed with friction reducer solution, and residual resistance factors to oil and brine were specified with different fracture widths and friction reducer concentrations.

Polymer or polymer gel has the function of preferentially reducing water relative permeability in conventional reservoirs. The similar finding was reached responsible for fracturing fluid residual in microfractures of tight reservoirs. During oil/brine injection experiments in fracture models fluxed by friction reducer solution, residual resistance factors to oil ($F_{rr,oil}$) and brine ($F_{rr,brine}$) exhibited power-law characteristics through their shear rates. This phenomenon occurred because absorbed polymer layer in fractures was thinner or easier to be flushed out at higher shear rates, creating larger channels through which oil/brine can pass. The result that $F_{rr,brine}$ was larger than $F_{rr,oil}$ revealed chemical residual could selectively reduce the permeability to water more than to oil in microfractures. The reason behind it was elucidated by polymer wall effect. At the same shear rates, smaller fractures presented higher residual resistance factors. The reason was chemical residual on fracture surfaces occupied a larger portion of cross-sectional area in smaller microfractures. Besides, chemical residual grew with an increase in friction reducer concentration, resulting in a higher residual resistance factor.

Keywords: Tight oil reservoir; Hydraulic fracturing; Chemical residual; Disproportionate permeability reduction; Residual resistance factor

1 Introduction

Unconventional oil and gas resources worldwide such as tight reservoirs have been developed for many years and present good effect on oil production increment. In recent years, the exploration and development of tight reservoirs have gradually attracted much attention in China. At present, Daqing Oilfield and Tuha Oilfield have carried out tight reservoirs field tests and have provided good theoretical guidance for the development of tight reservoirs in China. Hydraulic fracturing is a necessary method for the development of tight reservoirs. Due to the lack of gel solids in fracturing fluid, longer and more complex fractures formed, and no gel residue or filter cake was left to damage the fracture conductivity. Therefore, polymer friction reducer as an important part of fracturing fluid has been widely used in fracturing construction, used to reduce well-bore friction loss and pipeline friction caused by the high pump velocities. After hydraulic fracturing of tight reservoirs, a considerable amount of fracturing fluid would remain in fractures due to the low rate of inversion. Polymer flooding field test shows that the polymer has adsorption, clogging and other characteristics, and these properties could selectively reduce the relative permeability of the water phase, with a certain amount effect of controlling water production and increasing oil production. Therefore, it is of great significance to assess the effect of polymer friction reducer on fluid flow in microfractures. Through indoor evaluation experiments, a method by which to calculate the effects of injection rate, fracture width and concentration of polymer friction reducer on residual resistance factor before and after polymer treatment were defined. The mathematical evaluation model of residual resistance factor was built to reveal the mechanisms of polymer friction reducer residual on oil and water phase permeability. This study could provide theoretical guidance for hydraulic fracturing and fracturing fluid design in tight reservoirs.

2 Experiments

Tight sandstone fracture models with different fracture widths were fabricated in this study. Oil or brine was injected at various velocities before and after tight sandstone fracture models were fluxed with friction reducer solution, and residual resistance factors to oil and brine were specified with different fracture widths and friction reducer concentrations. The variation law of the residual resistance coefficient was analyzed to study the effect of the polymer friction reducer solution on the oil and water phase permeability in the tight sandstone fracture model.

2.1 Materials

Cylindrical tight sandstone cores with a diameter of 2.5 cm and a length of 30 cm were used to fabricate tight sandstone fracture models. As Fig. 1 depicts, each core sample was cut in half from the center, and stain-less steel sheets with different thicknesses were inserted between the two halves to model fractures. In this experiment, brine was made with distilled water and the salinity was 64622mg/L. The injected simulated oil is made from paraffin oil and kerosene with a viscosity of 0.97mPa·s. Considering the range of concentration of polymer friction reducer in the fracturing fluid, the concentrations of the polymer friction reducer solution were designed as 0.05 wt%, 0.1 wt% and 0.2 wt%.

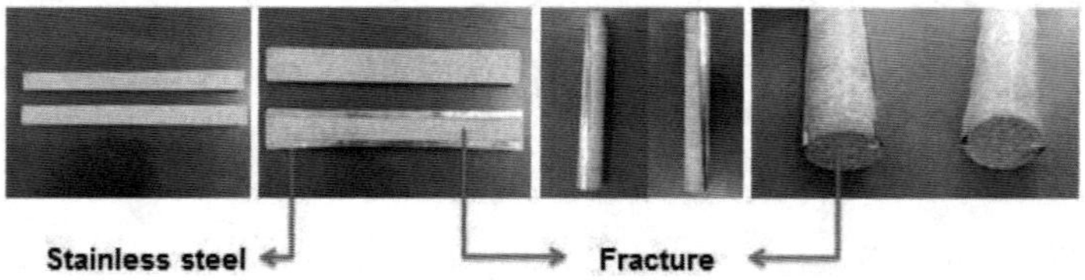

Fig. 1. Tight sandstone fracture model used in the experiments.

2.2 Experimental setup

Fig. 2 presents the flow chart of the experimental setup composed of a power driving system, a core holder, a a data acquisition system. The power driving system contained a pump (2PB00C), providing the power for injecting formation water, simulate oil and polymer friction reducer. A pressure sensor was installed in the inlet of the core holder to measure the real-time injection pressure. A data acquisition system (KL-M4000) was used to enable the fluid injection pressure to be displayed and recorded on the computer. The tight sandstone fracture model was placed horizontally in the core holder with the displacement direction from left to right. The injection pressure was monitored by a real-time monitoring system, and the fluid was collected from the core outlet. The experimental temperature was controlled by oven at 60 ℃ that is consistent with the reservoir temperature in Chang 7 layer of Changqing Oilfield.

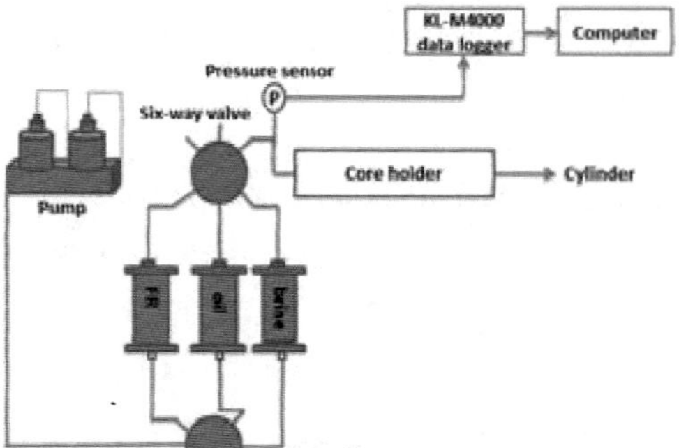

Fig. 2. Coreflooding experimental set-up.

2.3 Experimental scenario

In this experiment, the factors including injection medium, flux rate, fracture width and polymer friction reducer concentration were taken into consideration. The mechanisms responsible for disproportionate permeability reduction (DPR) were proposed. The experimental scenario was shown in Table 1.

Table 1 Experimental scenario of polymer effect on disproportionate permeability reduction to oil and water.

Scenario	Fracture width (mm)	Friction reducer concentration	Injection medium	Flux rate
1	0.05mm	0.1wt%	brine	0.15mm fracture width: 0.45, 1, 2, 3, 4, 5ml/min; Other fracture widths: 0.05, 0.1, 0.2, 0.4, 0.8, 1mL/min
2	0.05mm	0.1wt%	oil	
3	0.1mm	0.1wt%	brine	
4	0.1mm	0.1wt%	oil	
5	0.15mm	0.1wt%	brine	
6	0.15mm	0.1wt%	oil	
7	0.05mm	0.05wt%	brine	
8	0.05mm	0.05wt%	oil	
9	0.05mm	0.1wt%	brine	
10	0.05mm	0.1wt%	oil	
11	0.05mm	0.2wt%	brine	
12	0.05mm	0.2wt%	oil	

2.4 Experimental procedure

The experimental sequence during the core flooding experiments was deatiled as follows.

①The tight sandstone cores were put into the oven for 12 h, vacuumed for another 12 h and then immersed in brine for another 12 h;

②Tight sandstone fracture models were fabricated and employed in the core holder. The confining pressure was set at 2.5MPa;

③Brine or oil was injected at six incremental flux rates, Q_1, Q_2, Q_3, Q_4, Q_5 and Q_6, as depicted in Table 1. The injection pressure P_{before} was monitored and recorded when the pressure was stabilized (the change of injection pressure did not exceed 5% within three minutes);

④Polymer fraction reducer solution was injected at a constant flow rate of 0.1 cm^3/m in for 12 h;

⑤Brine or oil was injected at different flux rates in sequence to measure the stable injection pressures P_{after};

⑥Under experimental conditions, the residual resistance factors was calculated based on the change of pressure before and after the polymer friction reducer solution was injected. The formula was,

$$F_{rr} = \frac{k_{before}}{k_{after}} = \frac{P_{after}}{P_{before}}\bigg|_Q$$

where F_{rr} is the residual resistance factor for brine or oil; k_{before} is the permeability before the polymer treatment; k_{after} is the permeability after the polymer treatment; P_{after} is the injection pressure after the polymer treatment; and P_{before} is the injection pressure before the polymer treatment.

⑦Another Sandstone fracture model was fabricated to repeat step①–step⑥ under different conditions of fracture width or polymer friction reducer concentration.

3 Results and discussion

3.1 Effect of flux rate on $F_{rr,brine}$ and $F_{rr,oil}$

In this study, the effects of flux rate on brine/oil residual resistance factor were analyzed. During the brine/oil-injection experiments, six flux rates from 0.05 ml/min to 1 ml/min or from 0.45ml/min to 5ml/min (Table

1) were designed for each sandstone fracture model. The F_{rr} can be obtained at different flux rates and corresponding shear rates. Under the condition of the same fracture width and polymer concentration, Fig. 3 illustrated the effects of flux rate on brine/oil residual resistance factor after polymer treatment.

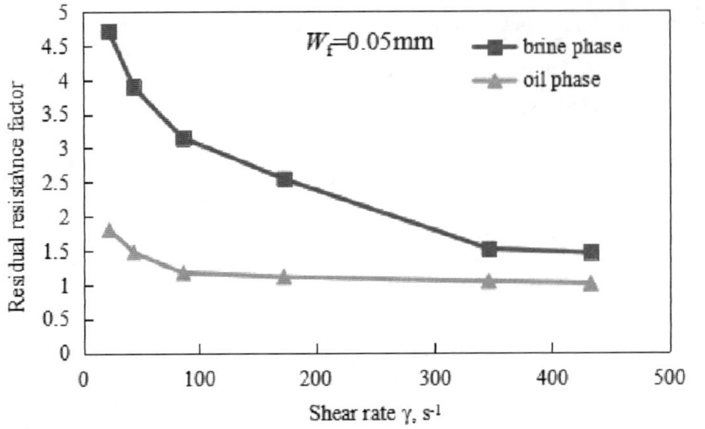

(a) W_f=0.05mm

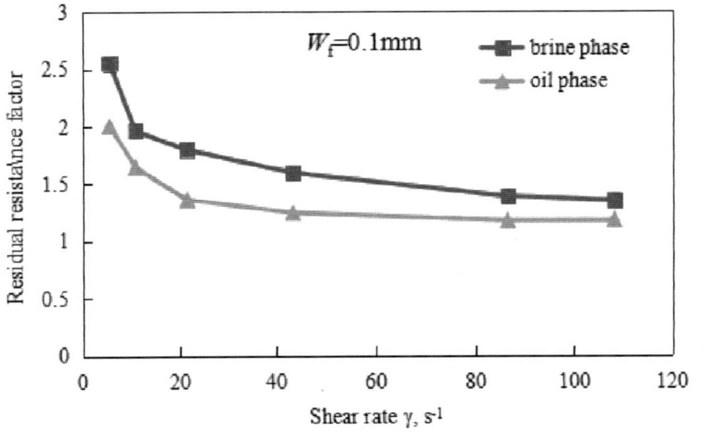

(b) W_f=0.1mm

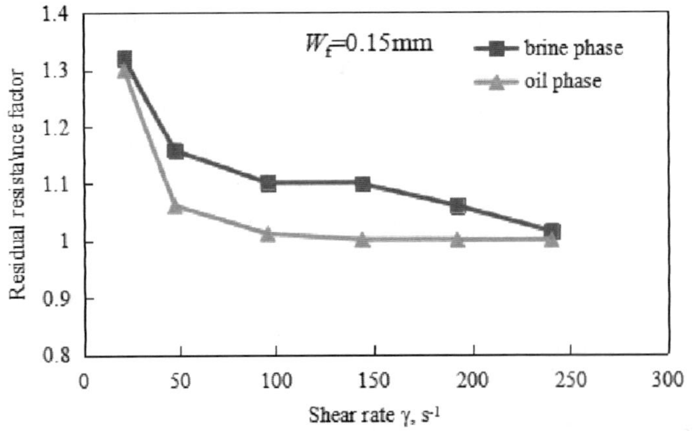

(c) W_f=0.15mm

Fig. 3. Residual resistance factor to brine and oil decreased as shear rate increased at low shear rates, and tended to stabilized at high shear rates.

The figures showed similar trends for the three sandstone fracture models with different fracture widths. At lower water flow rates, the residual resistance factor to brine and oil decreased as the shear rate (flux rate) increased. However, F_{rr} tended to stabilize at high shear rates. Oil/brine gradually flushed more and more polymer out of fracture models as flux rate increased; on the other hand due to the elasticity and plasticity of polymer, the change of fluid flow rate will cause the shear migration of the polymer chains. The boundary layer was thicker at low shear rates and thinner at high shear rates. In addition, higher oil/brine shear rates squeezed polymer coating on fracture surfaces more, creating larger channels through which oil/brine can pass.

For different fracture widths, the residual resistance factor to brine was always higher than the residual resistance factor to oil, indicating polymer coating in the fracture could preferentially reduce water phase permeability. The reason behind that was explained as ①The polymer layer remained in the fracture greatly changed the physical properties of the core and reduced the fracture conductivity as a whole; ②The polymer was shown to have a lower blocking efficiency in oil-wet micromodels than water-wet micromodels; ③The presence of the polymer adsorbed on pore walls may induce a lubrication effect that favors the flow of oil through the center of the pore channels and attenuate pore-wall roughness; ④A majority of experiments have been confirmed that the selective action of the polymer.

3.2 Effect of fracture width on $F_{rr,brine}$ and $F_{rr,oil}$

In order to investigate the effect of fracture width on the residual resistance factor, tight sandstone fracture models with different fracture widths of 0.05mm, 0.1 mm and 0.15mm were fabricated. On the basis of the same polymer concentration, polymer friction reducer solution was injected at a flow rate of 1ml/min. Fig. 4 illustrated the effects of fracture width on brine/oil residual resistance factor after polymer treatment. With the increase of fracture width, the resistance of polymer adsorption layer to fluid flow decreases and the residual resistance factor decreases gradually. Correspondingly, the polymer solution remaining in the microfracture is more difficult to be washed out by the subsequent water or oil flux, resulting in a relatively thick polymer adsorption layer in the microfracture. Superficial FR velocity is higher in a smaller fracture at the same FR solution injection rate. Therefore, polymer bridging effect enhanced. In addition, chemical residual on fracture surfaces occupied a larger portion of cross-sectional area in smaller microfractures, resulting in higher resistance to fluid flow.

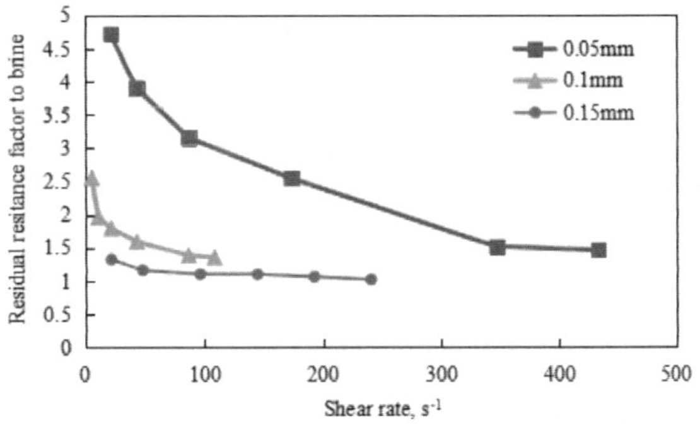

(a) Residual resistance factor to brine

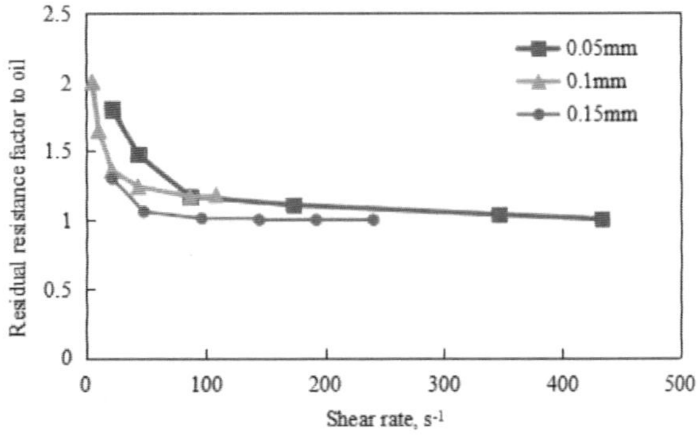

(b) Residual resistance factor to oil

Fig. 4. A smaller fracture exhibited a larger residual resistance factor (a) to brine and (b) to oil at the same shear rate.

In order to better analyze the relationship between the residual resistance factor and the shear rate, the log-log plot of residual resistance factor was draw in Fig. 5.

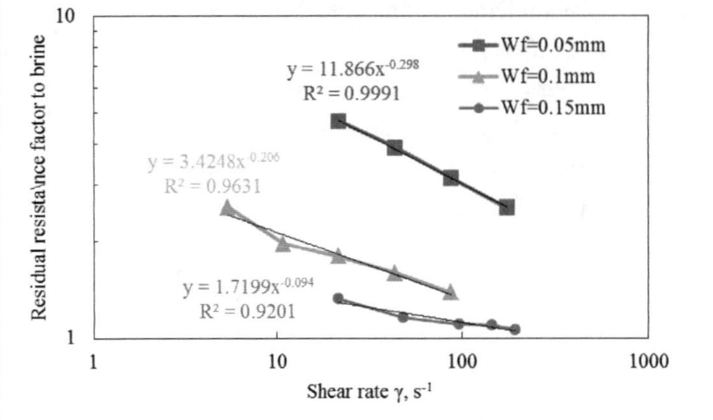

(a) Residual resistance factor to brine

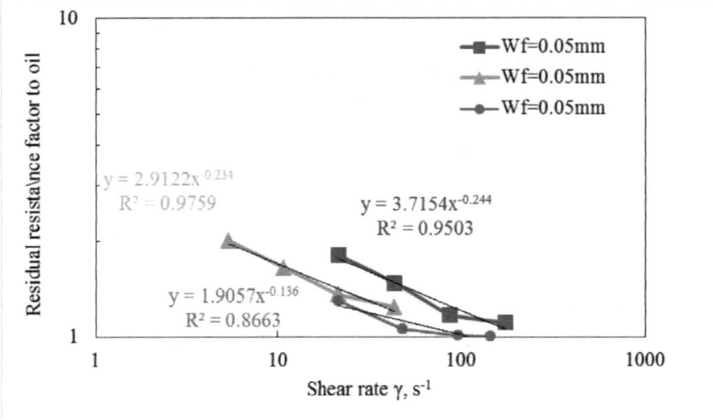

(b) Residual resistance factor to oil

Fig. 5. At low flux rates, (a) residual resistance factors to brine and (b) to oil exhibited power-law characteristics through their shear rates.

The relationship between residual resistance factor and shear rate could be well fitted using a power-law equation,

$$F_{rr} = C \cdot \gamma^{-b}$$

Where γ is shear rate, and C and b is experimental-related coefficients. The fitting results were shown in Table 2.

Table 2 The mathematical fitting results between residual resistance factor and shear rate.

	Fracture width (mm)	C	b	R^2
Residual resistance factor to brine	0.05	11.866	0.298	0.999
	0.1	3.425	0.206	0.963
	0.15	1.720	0.094	0.920
Residual resistance factor to oil	0.05	3.715	0.244	0.950
	0.1	2.912	0.234	0.976
	0.15	1.906	0.136	0.866

The coefficient b represents the sensitivity of residual resistance factor with the change of shear rate. As illustrated in Table 2, a smaller fracture presented a larger sensitivity index b. This was due to the fact that the polymer bridging effect was enhanced in a narrow fracture, and the thickness of the polymer adsorbed layer was relatively thicker compared to that in wider fracture. Therefore, the smaller the fracture width, the greater degree of the polymer damage.

3.3 Effect of friction reducer concentration on $F_{rr,brine}$ and $F_{rr,oil}$

In order to investigate the effect of friction reducer concentration on the residual resistance factor, a tight sandstone fracture model with a fracture width of 0.05mm was used, and the friction reducer solutions with three concentrations of 0.05%, 0.1%, 0.2% were prepared. Fig. 6 illustrated the effects of friction reducer concentration on brine/oil residual resistance factor after polymer treatment. With the friction reducer concentration increased, residual resistance factor showed an increasing trend. Higher concentration solutions display a larger injection pressure gradient, and the pressure gradient decreased as fluid flow rate decreased. Chemical residual grew with an increase in friction reducer concentration and occupied a larger portion of cross-sectional area of microfractures, resulting in a higher residual resistance factor.

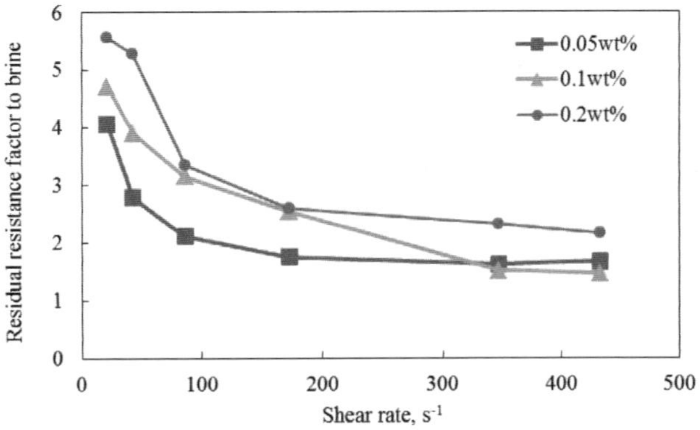

(a) Residual resistance factor to brine

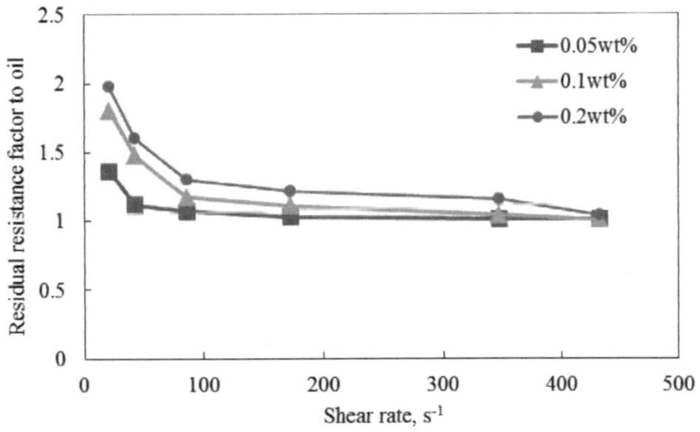

(b) Residual resistance factor to oil

Fig. 6. Higher friction reducer concentration presented a higher resistance to (a) brine flux and (b) oil flux.

4 Conclusions

(1) Low recovery of fracturing fluid leaves much chemical residual in microfractures of tight oil reservoirs, resulting in a resistance to brine and oil flux;

(2) Residual resistance factor to brine and oil decreased as flux rate and fracture width increased and FR concentration decreased;

(3) The fracturing fluid system with low interfacial tension and low flowback velocity are recommended after the fractures were generated in tight oil reservoirs.

Acknowledgements

The authors are highly appreciative of the financial support for our work from PetroChina Innovation Foundation (2015D-5006-0209) and Science Foundation of China University of Petroleum, Beijing (2462014YJRC053, 2462015YQ1105).

References

[1] Zaitoun A, Kohler N. Two-phase flow through porous media: Effect of an adsorbed polymer layer. In: SPE Annual Technical Conference and Exhibition, Society of Petroleum Engineers, Houston, Texas; 1988.

[2] Barreau P, Lasseux D, Bertin H, Glenat Ph, Zaitoun A. Polymer adsorption effect on relative permeability and capillary pressure: Investigation of a pore scale scenario. In: International Symposium on Oilfield Chemistry, Society of Petroleum Engineers, Houston, Texas; 1997.

[3] Liang J, Sun H, Seright RS. Why do gels reduce water permeability more than oil permeability? SPE Reservoir Engineering, 1995, 10(4): 282-286.

[4] Sparlin DD, Hagen RW. Controlling water in producing operation-part 5. World Oil, 1984, 137-142.

[5] Liang J, Sun H, Seright RS. Reduction of oil and water permeabilities using gels. In: SPE/DOE Enhanced Oil Recovery Symposium, Society of Petroleum Engineers, Tulsa, Oklahoma; 1992.

[6] White JL, Goddard JE, Phiilips HM. Use of polymers to control water production in oil wells. Journal of Petroleum Technology, 1973, 25(2): 143-150.

[7] Barreau P, Bertin H, Lasseux D, Glenat Ph, Zaitoun A. Water control in producing wells: Influence of an adsorbed-polymer layer on relative permeabilities and capillary pressure. SPE Reservoir Engineering, 1997, 12(4): 234-239.

[8] Zaitoun A, Bertin H, Lasseux D. Two-phase flow property modifications by polymer adsorption. In: SPE/DOE Improved Oil Recovery Symposium, Society of Petroleum Engineers, Tulsa, Oklahoma; 1998.

[9] Zitha PLJ, Botermans CW. Bridging-adsorption of flexible polymers in low permeability porous media. SPE Production & Facilities, 1998, 13(1): 15-20.

[10] Elmkies Ph, Bertin H, Lasseux D, Murray M, Zaitoun A. Further investigations on two-phase flow property modification by polymers: Wettability effects. In: SPE International Symposium on Oilfield Chemistry, Society of Petroleum Engineers, Houston, Texas; 2001.

[11] Elmkies Ph, Lasseux D, Bertin H, Pichery T, Zaitoun A. Polymer effect on gas/water flow in porous media. In: SPE/DOE Improved Oil Recovery Symposium, Society of Petroleum Engineers, Tulsa, Oklahoma; 2002.

[12] Grattoni CA, Luckham PF, Jing XD, Norman L, Zimmerman RW. Polymers as relative permeability modifiers: Adsorption and the dynamic formation of thick polyacrylamide layers. Journal of Petroleum Science and Engineering, 2004, 45(3-4): 233-245.

[13] Blanchard V, Lasseux D, Bertin H, Pichery T, Chauveteau G, Tabary R, Zaitoun A. Gas/water flow in porous media in the presence of adsorbed polymer: Experimental study on non-Darcy effects. SPE Reservoir Evaluation & Engineering 2007; 10(4): 423-431.

[14] Denys K, Fichen C, Zaitoun A. Bridging adsorption of cationic polyacrylamides in porous media. In: SPE International Symposium on Oilfield Chemistry, Society of Petroleum Engineers, Houston, Texas; 2001.

[15] Zitha PLJ, Darwish MMI. Effect of bridging adsorption on the placement of gels for water control. In: SPE Asia Pacific Improved Oil Recovery Conference, Society of Petroleum Engineers, Kuala Lumpur, Malaysia; 1999.

[16] Zhang H, Bai B. Preformed-particle-gel transport through open fractures and its effect on water flow. SPE Journal, 2011, 16(2): 388-400.

[17] Bai B. Preformed particle gel for conformance control. Final Report, Contact No. 07123-2, Missouri University of Science and Technology/US DOE, Rolla, Missouri, March 2012.

[18] Song Z, Liu L, Wei M, Bai B, Hou J, Li Z, Hu Y. Effect of polymer on disproportionate permeability reduction to gas and water for fractured shales. Fuel, 2015, 143: 28-37.

[19] Song Z, Liu L, Hou J, Bai B, Su W. Effect of polymer on gas flow behavior in microfractures of unconventional gas reservoirs. Journal of Natural Gas Science and Engineering, 2015, 23: 26-32.

[20] Sun Y, Wu Q, Wei M, Bai B, Ma Y. Experimental study of friction reducer flows in microfracture. Fuel, 2014, 131: 28-35.

497070 Synthesis of Lithium Carbonate Nanoparticles Using an Upscaled Microfluidic Reactor

Abstract:

Lithium carbonate nanoparticles were synthesized using a CHEMTOR fiber reactor. The reactor operates as a massively arrayed microfluidic environment and can be industrially scaled for high throughput. The manufacturing platform is tunable and uses traditional manufacturing equipment. We have successfully fractionally precipitated a mono-disperse lithium carbonate nanoparticle solution in water with particles as small as 5.9 nm in diameter and in continuous flow. The solvent was a CO_2-saturated water solution and the anti-solvent was isopropanol. The effects of process parameters such as temperature ,surfactant concentration and process flowrate on the particle size and dispersity are reported. The product was characterized using dynamic light scattering (DLS). These nanoparticles could potentially be used as a sacrificial template in fuel cell manufacture in order to improve the efficiency of catalyst usage. Future work will include mass transfer models under continuous flow to determine where the nanoparticles may be forming along the length of the plug reactor.

497131 Bijel Derived Nanocomposite Membranes for Advanced Separations

Abstract:

New approaches to fabricate separation membranes are needed to replace energy intensive industrial separation processes. Here, we introduce bicontinuous interfacially jammed emulsions (bijels) as templates for micro-, ultra- and nanofiltration membranes. These polymer/nanoparticle composite membranes have exceptionally high nanoparticle loadings (here up to 50 wt-% silica nanoparticles) with virtually the entire membrane surface covered by nanoparticles. Maximum organic solvent resistance of the membranes is achieved by employing highly cross-linked polyacrylates. The structurally well-defined, asymmetric membranes facilitate distinguished control over membrane flux and selectivity, enable the formation of stimuli responsive membranes, and can be easily surface modified to introduce antifouling features. They achieve 100 % separation of 15 nm gold particles from an aqueous dispersion. Moreover, we reveal for the first-time intermediate stages of the membrane structure-formation-dynamics by scanning electron microscopy. Our findings provide mechanistic insights also to formation routes of conventional membranes obtained by nonsolvent induced phase separation. The present work lies the foundation for advanced nanocomposite membranes with potential applications in water treatment, industrial separations and catalytic membrane reactors.

497570 Preparation of Nanoporous Silica with Agnps at the Core and Curst to Control the Ag+ Ion Release and Enhance the Antibacterial Properties

Abstract:

The present study demonstrates a novel, facile and one-pot approach to develop silica nanoparticle with silver at its core and crust. A modified stöbers method was used to make silica-AgNPs@ (Core&Crust). A significant reduction in the size of silica nanoparticle was seen, clearly 2-5 nm AgNPs was uniformly distributed on the surface while10-20 nm AgNPs were seen in the center. A typical mesoporous silica from stöbers method was transformed to nanoporous silica by this modified stöbers method. Nanoporous silica was advantageous for slow and consistent silver release which was confirmed by Ag+ ion release test. Antibacterial activities of the samples were carried out to investigate the disinfection performance of the samples on the Gram negative bacteria (*Escherichia coli*) using disk diffusion and LB-agar method. Silica-AgNPs@ (Core&Crust) showed prolonged silver release for more than 20 days and improved antibacterial properties even after 5 days of incubation.

497834 Development of Yttrium Nanoparticle/PVA Modified Psf Membrane and Application in Decontamination of Arsenate from Waters

Abstract:

In this study, a novel yttrium nanoparticle/PVA modified PSF membrane was developed for arsenate removal from contaminated waters. The porous PSF membrane as a supporting media was first fabricated by a phase inversion method with hydrophilic PAA polymer as a pore-forming additive. The porous PSF membrane was then coated by a PVA thin layer on the surface, and finally loaded by yttrium nanoparticle (YNP) through an in-situ precipitation method using ammonia vapor. The coating of PVA and YNP on the membrane surface can effectively affect the membrane porosity and water flux. Spherical YNP was uniformly distributed on the surface and matrix of membrane, leading to an excellent adsorption ability and extremely rapid adsorption kinetics to the membrane. The adsorption isothermal data were better done by the Langmuir than the Freundlich equation. The maximum adsorption capacity of 35.04 mg-As/g was achieved at pH 7.0. The adsorption performance of the membrane had no obvious difference in the pH range of 4.0-8.0 and reached the optimal value at pH 5.0. About 74% of ultimate adsorption capacity was achieved in the first 30 min. The adsorption kinetics followed the pseudo-second-order and intraparticle diffusion models. The filtration studies at the initial arsenate concentration of 94.8 Î¼g/L showed that the treatment capacities of the composite membrane at pH 7.0 and 5.0 were 278 and 637 L/m^2, respectively. The XPS analysis indicated that ligand exchange between hydroxyl groups bonded to yttrium atoms and arsenate could be proposed as the main adsorption mechanism.

498157 Combined Molecular Confinement and Metal-Support Interface Effects for Control of Hydrodeoxygenation Selectivity on Porous Pd@TiO_2

Abstract:

Combined molecular confinement and metal-support interface effects for control of hydrodeoxygenation selectivity on porous Pd@TiO_2

Bingwen Wang[a], Jing Zhang[b], Will Medlin[b] and Eranda Nikolla[a]

[a]Chemical Engineering Department, Wayne State University

5050 Anthony Wayne Drive, Detroit, MI, 48202

[b]Chemical Engineering Department, University of Colorado

3415 Colorado Ave., UCB 596, Boulder, CO, 80303

Alcohols with aromatic substituents are key intermediates formed from deconstruction of biomass[1]. Selective deoxygenation of these alcohols is often desirable to produce fuels or fuel additives. Two probe molecules that have received attention are benzyl alcohol and furfuryl alcohol. Under hydrogenation conditions, various products can be formed from these reactants. The hydrodeoxygenation (HDO) products, toluene and 2-methyl furan, are the desired products for many fuels applications. Efforts to design catalysts capable of selective HDO of such oxygenated aromatic compounds have generally focused on two approaches: combination of two materials with complementary activity for hydrogenation and deoxygenation[2], and establishing control over aromatic-surface interactions[3].

In this study, we combine these approaches by successful encapsulating Pd nanoparticles (NPs) within a porous TiO_2 film of controllable pore sizes to achieve high HDO selectivity, while maintaining high catalytic activity (Figure 1)[4]. Catalyst selectivity was found to be a strong function of both the presence of Pd-TiO_2 interfacial sites and the pore size of the TiO_2 shell. Moreover, experiments measuring hydrogenation rates for olefins of different molecular size strongly suggested that small pores hindered adsorption of aromatics in a "flat-lying" configuration that has been associated with non-selective C-C scission.

SHAPE * MERGEFORMAT

Figure 1. (a) Scanning transmission electron micrograph (b) bright field transmission electron micrograph and (c) pore size distribution of synthesized Pd@TiO_2. (d) Selectivity and (e) turnover frequency for HDO of furfuryl alcohol over

different Pd catalysts.

REFERENCES

[1] A. M. Robinson, J. E. Hensley, and J. W. Medlin, ACS Catal. 6 (2000) 5026.

[2] P. M. De Souza, R. C. Rabelo-Neto, L. E. P. Borges, G. Jacobs, B. H. Davis, T. Sooknoi, D. E. Resasco, and F. B. Noronha, ACS Catal. 5 (2015) 1318.

[3] C. H. Lien, and J. W. Medlin, J. Phys. Chem. C 118 (2014) 23783.

[4] J. Zhang, B. Wang, E. Nikolla, and J. W. Medlin, Angew. Chem. Int. Ed. (2017), in press.

499241 Examining Effects on Bending Elasticity and Structure of Phospholipid Bilayer Membrane in Presence of Embedded Surface Functionalized Inorganic Nanoparticles

Abstract:

Nanoparticles (NPs) have unique optical, electronic and thermal properties and have found applications in biomedicine, gene therapy, targeted drug delivery, bioimaging, biosensors, catalysis and a plethora of other important applications. They are also an attractive option due to the straightforward synthesis, stability and precise control of their size and morphology during synthesis. The greatly enhanced use of these engineered nanoparticles has raised a question, whether there are potential health risks and side effects involved due to the unintentional interactions of biological entities with nanoparticles. The interactions between engineered nanoparticles and biological entities like biomembranes can also be intentional, as in the case of designing novel stimuli responsive drug delivery vehicles, imaging and diagnostics agents. Previous researches conducted in this regard have shown the toxic effects of NPs are dependent on the concentration, size, morphology, nature of surface functionalization and charges on the nanoparticles. Among the many toxic effects observed due to nanoparticle membrane interactions, formation of nanopores and unzipping of membrane bilayer are a few. It has also been shown that the membrane disruption due to nanoparticles depends on the phase state of the lipid bilayer. Toxicology studies have mostly been conducted on a symptomatic approach and there has been no clear consensus on the mechanism related to these effects. There is a dire need to study the interaction of these particles with cell membranes systematically and elucidate the interactions under various conditions of bilayer phase states and nanoparticle concentration, shape and surface functionalization.

The cell membrane comprises of a lipid bilayer and is the first biological entity encountered by the nanoparticle. The lipid bilayer comprising of phospholipids are ~5 nm thick. In this treatise, we have studied the interactions of gold nanoparticles (AuNPs) and phospholipid bilayers. We have not changed the surface functionality or morphology of the AuNPs, but only its mean diameter and concentration. We have focused on hydrophobic ligand stabilized gold nanoparticle having diameter less than the bilayer thickness (3 nm) and diameter more than the bilayer thickness (6 nm). A mixture of phospholipids (zwitterionic and anionic) was used to synthesize unilamellar vesicles in the 100 nm diameter size range. These vesicles are convenient models for studying cell membrane properties as most eukaryotic cells comprises of similar lipids. During synthesis of these vesicles in the presence of the hydrophobic nanoparticles, the 3 nm and 6 nm diameter AuNPs preferentially embed into the hydrophobic acyl region of the lipid bilayer. We hypothesize that the 3 nm AuNPs embed easily in the bilayer without compromising its integrity significantly, but the 6 nm AuNPs causes significant disruption of the bilayer due to a mismatch.

To answer our hypothesis, we have examined the bilayer thickness and membrane bending elasticity by utilizing Small Angle Neutron Scattering (SANS) and Neutron Spin Echo (NSE) spectrometry. Bending elasticity is a mechanical property that controls the thermal fluctuations of bilayer. We have attempted to explain the influence of AuNPs of different sizes and concentration on the bending elasticity of the model bilayer at a range of temperatures from the fluid to the gel phase. Most research focuses on measuring structure of bilayers by static methods only like SANS and x-ray scattering. NSE spectroscopy is the most suitable method for our studies because it is a dynamic method and is ideal for measuring thermal fluctuations in lipid bilayers because of its correlation times and length scales overlap with cell membrane fluctuations.

We have supported these findings with various other techniques. We have studied the changes in phase behavior of the bilayer in presence of AuNPs by calorimetric methods like differential scanning calorimetry. We have examined the transport across the bilayer of carboxyfluorescein with temperature to elucidate the effect on diffusion across the bilayer due to the presence of AuNPs in them. We have employed other fluorometric methods to characterize the microenvironment of the bilayer in presence of AuNPs. Diphenylhexatriene is a fluorescent probe which changes the polarization of fluorescence emission with change in lipid ordering in the bilayer. This can be used to measure the transition temperature of the bilayer in presence of AuNPs by correlating membrane fluidity with the phase state. Additionally, Laurdan fluorescent probe has been used to determine the diffusion of solvent inside the bilayer due to increased fluidity beyond its transition temperature. This measurement can also be employed to measure the fluidity induced if any in the presence of the AuNPs.

500545 *In situ* Isolation of Bacteria Using Microfluidic Devices

Abstract:

The microbial world is mostly unexplored, much less understood. Identifying and studying new bacterial species can yield new information about diseases, result in the discovery of new drugs, and reveal the missing pieces of the microbial world. New species are difficult to study however because most organisms do not grow in standard laboratory growth media. To address the current limitations to isolating and cultivating bacteria, a nanofluidic device was developed to isolate bacterial species based on their size in complex samples.

A polydimethylsiloxane (PDMS) polymer microfluidic device containing seven sub-micrometer constrictions with new non-parallel wall geometries was developed. The constrictions were smaller than the diameter of a bacterial cell in one or two dimensions. Two devices were placed in river water samples collected from the Charles River in the summer. When bacterial cells entered the main entrance of the device, fresh food in the isolation chambers created a gradient, which chemotactically attracted microorganisms toward the constrictions. The microorganism was trapped in the non-linear constrictions as it tried to reach the chamber. The cells divided, and each progeny advanced further through the constriction. After several divisions cells from a single species should enter the isolation chamber through the constriction.

The two devices isolated 11 different bacterial species using a total of only 14 constrictions and chambers: *Brevundimonas nasdae*, *Brevundimonas staleyi*, *Devosia psychrophila*, *Microbacterium maritypicum*, *Pseudomonas extremaustralis*, *Pseudomonas lurida*, *Pseudomonas marginalis*, *Pseudomonas rhodesiae*, *Pseudomonas trivialis*, *Pseudomonas veronii* and *Sphingobium xenophagum*. In contrast, 22 different bacterial species were obtained using conventional plate culture techniques, but required multiple dilutions to isolate.

Of the 11 species isolated using the microfluidic device, only one species overlapped with the bacteria collected with the conventional technique. The microorganisms collected with the devices are generally considered not to be abundant species in river water. With increasing numbers of isolation chambers tested, the species richness is expected to increase as well.

This study validates the use of a microfluidic device to capture and isolate bacterial cells not obtained through plate culture techniques, giving us further bacterial diversity to study. Future work includes looking at different sample sources, such as bacteria isolated from soil and swamps.

501511 Tailoring Pore Topology to Polymorphism By Engineering Metal Oxide Interfaces during Templating of Nanostructure Materials

Abstract:

The rational engineering of porous metal oxide catalysts requires identification of versatile strategies for simultaneously tailoring properties ranging from porosity to polymorphism across a diverse materials palette. The low surface energies characteristic of most oxides enable increased atomic surface mobility that makes oxides susceptible to low-temperature structural (chemical, topological), and, thereby, functional reconfiguration. In this talk, we will introduce a phenomenon [1] whereby the interface between metal oxides can be exploited as a means for preserving both the structure and the traditionally metastable metal oxide polymorphs at temperatures well in excess of those where complete transformation to the coarsened, thermodynamically stable polymorph occurs in bulk powders. Specifically, we will employ titania as a surrogate metal oxide and will demonstrate the preservation of the traditionally metastable anatase crystal structure along a two-dimensional oxide interface at temperatures well in excess of those that normally trigger a full polymorphic transformation to rutile in higher dimensionality crystalline powders.

We will then extend this polymorphic stabilization phenomenon to cases of three-dimensionally distributed interfaces. Specifically, we will show how convectively assembled colloidal crystal templates, comprised of size-tunable (*ca.* 15-50 nm) silica nanoparticles, enable versatile sacrificial templating of three-dimensionally ordered mesoporous (3DOm) metal oxides (MO_x) at both mesoscopic and microscopic size-scales [2]. Beyond scaffolding porous materials with high surface areas and tunable pore sizes, we will identify a synergistic, template-mediated interfacial and volumetric mechanism for stabilizing metastable 3DOm metal oxide polymorphs upon high-temperature processing and for tuning fractional polymorphism [3]. Mechanistic investigations suggest that this polymorph stabilization is derived from the combined effects of the template-replica (MO_x/SiO_2) interface and simultaneous interstitial confinement that limits the degree of coarsening during high temperature calcination of the template-replica composite. We will employ ZrO_2 (i.e. tetragonal ZrO_2) as a surrogate metal oxide for establishing the generalizability of 3DOm templating for tailoring both the mesostructure (i.e. pore size and surface area) and the crystallography (i.e. polymorphism) of porous oxides. This work ultimately identifies a facile yet versatile templating strategy for realizing nanostructured oxide catalysts and supports with (i) surface areas that are more than an order of magnitude larger than untemplated control samples, (ii) pore diameters and volumes that can be tuned across a continuum of size-scales, and (iii) selectable polymorphism.

References

[1] D. G. Gregory, L. Lu, C. J. Kiely, M. A. Snyder, *J. Phys. Chem. C*, **121**(8) (2017) 4434-4442

[2] M. A. Snyder, *MRS Bulletin*, **41**(9) (2016) 683-688.

[3] D. G. Gregory, Q. Guo, L. Lu, C.J. Kiely, M. A. Snyder, *submitted* (2017).

502365 Supported, Homogeneously Alloyed Bimetallic Nanoparticles By Electrostatic Adsorption

Abstract:

Supported, Homogeneously Alloyed Bimetallic Nanoparticles by Electrostatic Adsorption

Andrew P. Wong, Qiuli Liu, and John R. Regalbuto

Department of Chemical Engineering, University of South Carolina

ABSTRACT

Bimetallic catalysts are often important for chemical reactions due to their enhanced performance over monometallic catalysts [1]. Incorporating a secondary metal into a system can benefit the catalytic system through ensemble, bifunctional, and electronic effects. The most prevalent method for synthesizing supported catalysts is dry impregnation (DI), also known as incipient wetness, in which a solution containing a specific amount of metal precursor is added to just fill the pore volume of the support [2]. While the procedure is simple and the metal loading is precisely known, limitations of DI preparations often include larger particle sizes and less intimate metal-support interactions. Moreover, bimetal catalysts produced through DI often exhibit a poor degree of alloying between the metals.

Our goal was to demonstrate a simple, effective method to synthesize highly dispersed, well alloyed bimetallic nanoparticles using strong electrostatic adsorption (SEA). SEA is achieved by charging the support by controlling the pH of the solution. A metal precursor that is oppositely charged can then be adsorbed onto the surface of the charged support. Using a simultaneous adsorption of either two cationic or anionic precursors allows excellent bimetallic interactions. The combinations of multiple bimetallic systems (Cu, Co, Ni, Pd, and Pt) prepared by SEA were compared to equivalent DI catalysts. Particle sizes and the degree of alloying were determined through electron micrographs, elemental mapping of nanoparticles, and powder XRD.

Results from XRD show the bimetallic catalysts produced by SEA were highly dispersed (<2 nm), while the DI catalysts exhibited large XRD peaks (>5nm). Initial elemental maps and STEM images indicate the SEA catalysts had a higher degree of alloying compared to the DI catalyst, and particle agglomeration was avoided for the SEA catalyst. A complete comparison of the bimetal systems prepared by DI and SEA will be discussed.

References

[1] Sankar, M., et al. (2012). "Designing bimetallic catalysts for a green and sustainable future." Chemical Society Reviews **41**(24): 8099-8139.

[2] Cho, H.-R. and J. R. Regalbuto (2015). "The rational synthesis of Pt-Pd bimetallic catalysts by electrostatic adsorption." Catalysis today **246**: 143-153.

488630 One-Step Synthesis of Carbon Nanotube-Supported Fischer-Tropsch Catalysts Via Liquid Injection Chemical Vapor Deposition

Abstract:

The production of clean fuel via Fischer-Tropsch synthesis (FTS) using carbon nanotube (CNT)-supported Fe catalyst is demonstrated. Conventional synthesis method for CNT-supported catalyst requires an additional functionalization step to create sufficient anchoring sites for deposited catalyst. The functionalization process involves aggressive acid treatment that impart significant amount of defects on CNTs and change their overall morphology, as well as degrade their outstanding material properties. Here, we propose a one-step liquid injection chemical vapor deposition (CVD) for simultaneous growth of CNTs and deposition of Fe catalyst on CNT wall. Ferrocene/xylene solution is used as catalyst and carbon precursors. Temperature-programmed reduction (TPR) results reveal that reducibility of Fe is higher on CNTs than conventional oxidic supports (SiO_2/Al_2O_3). The decorated Fe catalysts on CNTs have been evaluated as catalysts in FTS and the results have been compared to convention Fe catalysts supported on CNTs, Al_2O_3, and SiO_2 prepared by incipient wetness impregnation.

490096 Characterization of Aluminum Carbide in Aluminum-Graphene Nanocomposites

Abstract:

Characterization of Aluminum Carbide in

Aluminum-Graphene Nanocomposites

Aditya Nittala[1], Frank Kraft[2], Keerti Kappagantula[1,2]

[1]Nanomaterials and Energetic Systems Lab,

[2]Center for Advanced Materials Processing,

Department of Mechanical Engineering,

Russ College of Engineering and Technology,

Ohio University, Athens Ohio 45701

Abstract

Recently, there has been an increased interest in investigating the effects of reinforcing aluminum matrix with nanoscale carbon additives such as carbon nanotubes and graphene nano-films (GNF) to make metal nanocomposites with enhanced mechanical, thermal, and electrical properties. In general, CNTs and GNFs have shown an ability to influence the macro-scale properties of the substrates they have been introduced into as reinforcements. In particular, in graphene reinforced aluminum nanocomposites, ensuring optimal material contact at the interface between the aluminum matrix and the graphene reinforcement is paramount for ensuring desired mechanical, thermal, and electrical properties of the nanocomposite. A typical interfacial phenomenon observed in aluminum/graphene nanocomposites is the formation of aluminum carbide (Al_4C_3) owing to a chemical reaction between aluminum matrix and the carbon present in the graphene film. While small quantities of Al_4C_3 have shown an improvement in the mechanical properties of the constituent nanocomposite, it is undesirable as an insulator and decreases the electrical and thermal properties of the material system. Thus, controlling the amount of Al_4C_3 formed at aluminum-graphene interface can lead to the development of nanocomposites with precisely designed mechanical, electrical, and thermal properties for specific applications.

While there is variegated information present of aluminum carbide formation, there is no literature clearly delineating the mechanism and conditions for carbide formation in aluminum-graphene nanocomposites. The present research explores the quality, quantity and topography of aluminum carbide formed in aluminum/graphene nanocomposites. Aluminum/graphene nanocomposite samples are synthesized with varying graphene content by hot-extrusion method performed at a range of temperatures from 300°C to 600°C. Aluminum carbide is characterized using electron beam microscopy and spectroscopy methods. Aluminum carbide formation in the nanocomposite is described as a function of graphene and aluminum matrix interface, graphene film quality, and the rate of energy provided to the nanocomposites during their synthesis. Additionally, the effect of aluminum carbide content on the thermal, electrical, and mechanical properties of the nanocomposite is determined. The results from the study are used to determine thermodynamic conditions conducive for aluminum carbide formation such that its quantities can be predictively anticipated and controlled to make aluminum/graphene nanocomposites with prescribed mechanical, thermal, and electrical properties.

499448 3D Vertically-Aligned CNT/Graphene Hybrids from Layer-By-Layer Transfer for Supercapacitors

Abstract:

3D Vertically-Aligned CNT/Graphene Hybrids from Layer-by-Layer Transfer for Supercapacitors

Dr. Enoch A. Nagelli[1*], Dr. Liang Huang[2], Dr. Feng Du[2], and Prof. Liming Dai[2]

[1]Department of Chemistry & Life Science

United States Military Academy, West Point, New York 10996

[2]Center of Advanced Science and Engineering for Carbon (Case4Carbon)

Department of Chemical and Biomolecular Engineering

Department of Macromolecular Science and Engineering

Case Western Reserve University, Cleveland, OH

*Corresponding Author: Dr. Enoch Nagelli, **Email:** enoch.nagelli@usma.edu

Exceptional properties such as stable crystal structure, optical transparency, and superior electronic properties from high electron mobility make graphene a nanomaterial with promising applications.[1, 2] Among the many methods employed for producing graphene films, mechanical cleavage of HOPG[3] or exfoliation of graphite crystals,[4,5] chemical vapor deposition,[6] solvent thermal reaction,[7,8] and chemical routes via carbon nanotubes (CNTs), graphite intercalation compounds (GIC) or graphite oxide (GO) are the most widely recognized and used methods to produce single- and few-layer graphene.[9-12] Moreover, the oxidation and reduction of graphite is one of the most used methods for mass production of graphene for industrial applications. However, the reduction of graphene oxide results in a decrease in hydrophilicity, which in turn leads to greater aggregation and precipitation thus, losing its unique 2D characteristics.[13] Both chemical functionalization and electrostatic stabilization have been employed to reduce the agglomeration of graphene in order to utilize its unique properties.[14]

Furthermore, there is a need to integrate graphene into 3D macroscopic architectures while maintaining its superior intrinsic properties. Graphene and carbon nanotubes have been combined to form transparent conductive films[15] and electrodes for rechargeable lithium ion secondary batteries.[16] Researchers recently designed a computational model of the novel 3D structure consisting of parallel graphene layers stabilized by vertically aligned CNTs (VA-CNTs) in between the graphene planes.[17] However, experimental fabrication of the 3D graphene/VA-CNT hybrid structure via the transfer assembly process remains challenging. In addition, on the basis of our previous work[18] on the tunable 3D pillared VA-CNT/graphene architectures through intercalated growth of VA-CNTs into thermally expanded highly ordered pyrolytic graphite (HOPG) by the pyrolysis of iron phthalocyanine, we report a novel strategy to prepare 3D graphene/VA-CNT hybrid structures by a contact film transfer process, which utilizes attractive intermolecular-surface forces e.g., van der Waals of graphene and VA-CNTs. For the first time, we demonstrate a novel yet facile technique to develop three dimensional hybrid graphene/VA-CNT nanostructure via free-standing film transfer through the hydrophobic-hydrophobic interactions. As a result, supercapacitors based on the 3D graphene/VA-CNT hybrid formed via the transfer process were developed.

References

[1] A. K. Geim and K. S. Novoselov, *Nat. Mater.*, **2007**, *6*, 183.

[2] A. K. Geim, *Science*, **2009**, *324*, 1530.

[3] K. S. Novoselov, A. K. Geim, S. V. Morozov, D. Jiang, Y. Zhang, S. V. Dubonos, I. V. Grigorieva and A. A. Firsov, *Science*, **2004**, *306*, 666.

[4] Y. Hernandez, V. Nicolosi, M. Lotya, F. M. Blighe, Z. Y. Sun, S. De, I. T. McGovern, B. Holland, M. Byrne, Y. K. Gunâ€™ko, J. J. Boland, P. Niraj, G. Duesberg, S. Krishnamurthy, R. Goodhue, J. Hutchison, V. Scardaci, A. C. Ferrari and J. N. Coleman, *Nat. Nanotechnol.*, **2008**, *3*, 563.

[5] S. Park and R. S. Ruoff, *Nat. Nanotechnol.*, **2009**, *4*, 217.

[6] J. J. Wang, M. Y. Zhu, R. A. Outlaw, X. Zhao, D. M. Manos, B. C. Holloway, *Carbon* **2004**, *42*, 2867.

[7] Q. Kuang, S. Y. Xie, Z. Y. Jiang, X. H. Zhang, Z. X. Xie, R. B. Huang, L. S. Zheng, *Carbon* **2004**, *42*, 1737.

[8] Y. Zhou, Q. L. Bao, L. A. L. Tang, Y. L. Zhong, K. P. Loh, *Chem. Mater.* **2009**, *21*, 2950.

[9] Z. P. Zhu, D. S. Su, G. Weinberg, R. Schlogl, *Nano Lett.* **2004**, *4*, 2255.

[10] Y. Wang, Z. Q. Shi, Y. Huang, Y. F. Ma, C. Y. Wang, M. M. Chen, Y. S. Chen, *J. Phys. Chem. C* **2009**, *113*, 13103.

[11] S. Stankovich, D. A. Dikin, R. D. Piner, K. A. Kohlhaas, A. Kleinhammes, Y. Jia, Y. Wu, S. T. Nguyen, R. S. Ruoff, *Carbon* **2007**, *45*, 1558.

[12] X. B. Fan, W. C. Peng, Y. Li, X. Y. Li, S. L. Wang, G. L. Zhang, F. B. Zhang, *Adv. Mater.* **2008**, *20*, 4490.

[13] X. S. Du, Z. Z. Yu, A. Dasari, J. Ma, M. S. Mo, Y. Z. Meng, Y. W. Mai, *Chem. Mater.* **2008**, *20*, 2066.

[14] X. Y. Yang, X. Dou, A. Rouhanipour, L. J. Zhi, H. J. Rader, K. Mullen, *J. Am. Chem. Soc.* **2008**, *130*, 4216.

[15] V. C. Tung, L. M. Chen, M. J. Allen, J. K. Wassei, K. Nelson, R. B. Kaner, Y. Yang, *Nano Lett.* **2009**, *9*, 1949.

[16] E. Yoo, J. Kim, E. Hosono, H. Zhou, T. Kudo, I. Honma, *Nano Lett.* **2008**, *8*, 2277.

[17] G. K. Dimitrakakis, E. Tylianakis, G. E. Froudakis, *Nano Lett.*, **2008**, *8*, 3166.

[18] F. Du, D. Yu, L. Dai, S. Ganguli, V. Varshney, A. K. Roy, *Chem. Mater.* **2011**, *23*, 4810.

499172 Oligodendrocyte Precursor Cell Maturation in a 3D Hydrogel System through the Incorporation of Drug Delivery Nanoparticles or Topographical Cues (Grad Student Award)

Abstract:

In the central nervous system, oligodendrocyte precursor cells are capable of differentiating into oligodendrocytes responsible for generating the myelin sheath. This electrically insulating layer enables neuronal axons to send electrical signals rapidly and efficiently to surrounding neurons. Following diseases or injuries this myelin sheath is broken down as oligodendrocytes are damaged and oligodendrocyte precursor cells (OPCs) are often unable to differentiate into mature myelinating oligodendrocytes in the damaged tissue, leaving patients with long-term disabilities and pain. While many biomaterial systems enable neuronal maturation and axon extension, few systems have been investigated in terms of oligodendrocyte precursor cells and the biomaterial properties that are necessary for regenerating functional, myelinating oligodendrocytes.

Here, we investigate the influence of controlled drug delivery and topographical cues on oligodendrocyte precursor cell maturation in a 3D polyethylene glycol based hydrogel. OPCs were encapsulated in poly(ethylene glycol)-dimethacrylate gels made through photoinitiation with lithium phenyl(2,4,6-trimethylbenzoyl)phosphinate. In some gels, nanoparticles releasing platelet derived growth factor-alpha (PDGFR-Î±) were encapsulated along with cells. Poly(lactic acid-co glycolic acid) nanoparticles were made using the double emulsion, W/O/W method with varying lactic acid to glycolic acid (L:G) ratios and resulting release profiles. Nanoparticles had diameters ranging from 0.959 to 3.46 micrometers, measured via scanning electron microscopy. The fastest release system was the 50:50 (L:G) 0.59IV nanoparticle which released 90% of the load in the first 24 hours. Interestingly, cells responded to faster release rates by upregulating RNA for the mature oligodendrocyte marker, myelin basic protein. This was found with a four fold increase in myelin basic protein RNA, calculated through RT-PCR, in gels containing the fastest release rate nanoparticles compared to those without nanoparticles. In addition to nanoparticles, topographical cues were incorporated into some gels through the addition of electrospun fibers. On going work is investigating how the incorporation of polystyrene electrospun fibers affect the differentiation of oligodendrocyte precursor cells as well, however it is believed that topographical cues play an important role in oligodendrocyte differentiation and maturation. In 2D cells plated on fibers are observed extending processes towards the polystyrene fibers, much like oligodendrocytes would to axons in the central nervous system. These results taken together indicate the importance of biochemical and topographical cues on the differentiation of oligodendrocyte precursor cells in 3D biomaterial systems and their potential application towards nervous system regeneration.

488727 Functionalized Graphene/Polyimide Thermal Conductivity Composites *Via* Electrospinning-Hot Press Technique

Abstract:

Highly thermally conductive polymeric composites as heat transfer materials are urgently needed in electronic fields. In our present work, the combining method of aminopropyllsobutyl polyhedral oligomeric silsesquioxane (NH_2-POSS) and hydrazine monohydrate was introduced to functionalize the surface of graphene oxide (GO), finally to obtained the chemical modified graphene (CMG), which was then performed to fabricate the corresponding thermally conductive CMG/polyimide (CMG/PI) composites *via in*-situ polymerization followed by electrospinning-hot press technique. Both the thermally conductive coefficient (*Î»*) and the glass transition temperature (T_g) of the CMG/PI composites were increased with the increasing addition of CMG. The *Î»* of the CMG/PI composite with 5 wt% CMG was significantly improved to 1.05 W/mK, 4 times higher than that of pristine PI (0.28 W/mK), and the corresponding T_g was also increased to 213°C. Meantime, the obtained CMG/PI composites also presented excellent thermal stabilities. Our work provides great potential for electronic systems, expected to realize the heat transfer by electronic components, thus to prolong service life and avoid electronic componentsâ€™ damage.

References:

1. J. Gu, C. Liang, X. Zhao, B. Gan, H. Qiu, Y. Guo, X. Yang, Q. Zhang, D. Wang, *Compos. Sci. Technol.*, 2017, **139**, 83-89.
2. J. Gu, S. Xu, Q. Zhuang, Y. Tang, J. Kong, *IEEE Trans. Dielectr. Electr. Insul.*, 2017, **24**, 784-790.
3. J. Gu, Y. Guo, X. Yang, C. Liang, W. Geng, L. Tang, N. Li, Q. Zhang, *Compos. Part A*, 2017, **95**, 267-273.
4. J. Gu, Z. Lv, Y. Wu, Y. Guo, L. Tian, H. Qiu, W. Li, Q. Zhang, *Compos. Part A*, 2017, **94**, 209-216.
5. J. Gu, Y. Guo, Z. Lv, W. Geng, Q. Zhang, *Compos. Part A*, 2015, **78**, 95-101.

490167 Multicolored Triboluminescent Composites for Wind Utilization and Lubrication Failure Warning

Abstract:

Triboluminescence (TL) refers to the phenomenon that materials could emit luminescence when stimulated by mechanical activities including friction, compress, stretch, impact, bend, etc. [1,2] Compared with the traditional photoluminescence and electroluminescence, TL could utilize the extensively existed mechanical energy in daily life, avoiding the requirements of light and electricity as the excitation sources. Therefore, TL material is considered as a new generation of energy-saving, environmentally friendly, and sustainable luminescent material, showing promising applications in lighting, displaying, imaging and intelligent sensors [3,4].
For practical applications, TL materials should be further composited with elastic matrices which could enable its non-destructive luminescence [5,6]. The basic performance of the matrices determines the applications of TL materials. In this presentation, flexible polydimethylsiloxane (PDMS) and tough epoxy was employed as the polymer matrices. The fabricating processes of both PDMS and epoxy-based composites were optimized based on the mechanical and TL performance. The luminescence color of the composites was further manipulated by adjusting the TL components and their ratios. The structure and morphology of the flexible TL composites were further regulated based on the surface texture technique. Such design of flexible TL composites allow its effective response for wind energy, which is promising for wind utilization applications, e.g., gale warning and wind displaying. For the multicolored epoxy-based hard composites, they were further assembled in the bottom of lubricating film. The morphology and spatial dispersion of TL composites were designed and optimized. The introduction of TL composites showed no negative effects on the mechanical and tribological performance of the lubricating film. When the lubricating film wear to the TL layer in the bottom, it could emit luminescence driven by the friction behavior from mechanical equipment. The TL signals could be easily captured by naked eyes or simple detection equipment, presenting the self-warning activity to avoid major accidents and economic loss caused by lubrication failure.

Acknowledgements

The authors would like to thank the support from CAS Pioneer Hundred Talents Program.

References

[1] A. J. Walton, Advances in Physics, 1977, 26, 887-948.
[2] D. Peng, B. Chen and F. Wang, ChemPlusChem, 2015, 80, 1209-1215.
[3] J.-C. Zhang, C.-N. Xu, S. Kamimura, Y. Terasawa, H. Yamada and X. Wang, Optics Express, 2013, 21, 12976-12986.
[4] X. Wang, H. Zhang, R. Yu, L. Dong, D. Peng, A. Zhang, Y. Zhang, H. Liu, C. Pan and Z. L. Wang, Advanced Materials, 2015, 27, 2324-2331.
[5] C.-N. Xu, X.-G. Zheng, M. Akiyama, K. Nonaka and T. Watanabe, Applied Physics Letters, 2000, 76, 179-181.
[6] S. M. Jeong, S. Song, S. K. Lee and N. Y. Ha, Advanced Materials, 2013, 25, 6194-6200.

498980 Mimicking Nature: Mechanical Properties of Ultrastretchable, Silica-Based Synthetic Spider Webs Fabricated Via 3D Printing

Abstract:

Spider-webs are one of the most fascinating and precisely engineered architectures seen in nature. The structure of a web consists of a spiral supported with spokes, all weaved using elastic silk. Ability of this thin net to capture high velocity objects is not only a way to obtain food for spiders, but also has been adapted into our lives, where similar manmade nets mitigate accidents in fields like construction, sports, and aviation by acting as a barrier. As 3D printing becomes a common and facile approach to pattern traditional as well as novel materials into intricate geometries, designing and mimicking spider webs is starting to attract scientific attention. The wide range of existing inks allows us to print complex 3D structures: from common small household tools to commercial assemblies, such as jet engines or buildings with multiple rooms; however, there is still a need for inks with unique nature-imitating functional properties for structural applications. We have used 3D printing to fabricate synthetic spider webs, with an in-house synthesized polyurethane composite as an ink to mimic spider silk properties.

By reacting polyurethane with i) ethylene glycol, ii) hydroquinone, and iii) fumed silica particles, we have synthesized three types of polymers with varying tensile properties. We have quantified the glass transition temperature and rheological properties of the polymers. Both films and fibers prepared with polyurethane-silica composites were highly stretchable, with an average strain above ~4000 % at break of the fibers. We propose that these tensile properties originate from the covalent and non-covalent 3D crosslink system between the elastic polymer and rigid silica domains. The polyurethane-silica composite was cast into films, wet spun into fibers and used as an ink for 3D printing at room temperature using a customized 3-dimensional pneumatic dispensing robot. We 3D printed grids and spider-webs with increasing number of interconnected junctions, relating mechanical properties to the architecture. This in depth understanding of the structural-functional-property relationship of elastic nets will enable future design of complex structures, requiring advanced functional materials for high tech applications.

501716 Zwitterionic Conjugated Polymers and Their Application in Biosensing

Abstract:

As a group of emerging biomaterials, conjugated polymers have attracted significant interests for diagnosis, imaging and therapy. Conjugated polymers have been used as core components to improve communications between bioelectronic devices and biological systems by delivering/detecting lower electrical signals. However, conjugated polymers are originally designed for non-biological and non-aqueous systems and have several intrinsic drawbacks. Nonspecific adsorption of biomacromolecules on bioelectronic devices reduces their sensitivity and performance, and triggers foreign body response that eventually leads to the failure of embedded devices. In this presentation, I will show you how we designed, synthesized and characterized novel zwitterionic conjugated materials to address the challenges associated with the conventional conjugated polymers. Through these studies, we developed a better understanding of the structure-function-property relationships of zwitterionic materials, which will help us to design advanced and high-performance materials for a variety of biomedical applications in the future.

Novel Environmentally Benign Hydrogel: Nano-silica Hybrid Hydrolyzed Polyacrylamide/ Polyethyleneimine Gel System for Conformance Improvement in High Temperature High Salinity Reservoir

Yifu Long, Changqian Zhu

Department of petroleum engineering, Missouri University of Science&Technology

Abstract

In order to address the conformance problem in high temperature and high salinity/hardness reservoir, an environmentally benign in-situ gel was developed based on hydrolyzed polyacrylamide (HPAM) and polyethyleneimine (PEI). Moreover, silica nanoparticle (SNP) was embedded to improve the performance of HPAM/PEI gel. To achieve the optimum formulation of HPAM/PEI-SNP gel, a comprehensive investigation was first conducted with the respect of gelation time, gel strength, and thermal stability. In addition, the particle size of SNP was determined by Dynamic Light Scatter (DLS) and the microstructure of HPAM/PEI-SNP gel was characterized via Scanning Electron Microscopy (SEM).

The optimum formulation, containing the polymer/crosslinker ratio of 10,000-2,000 using high molecular weight HPAM (10-15 million Dalton) with a low hydrolysis degree of 3-7%, was obtained. Given the compatibility of comparability of SNP and formation water, LS nano silica with an SNP concentration of 30 wt% was selected as hybrid resource. The results indicated that the gelation time of novel gel system was prolonged to 132 hours attributed to the introducing of nanomaterial. Taking advantage of nano-composite, the strength of HPAM/PEI-SNP gel maintained as Grade G for 312 h. Furthermore, the HPAM/PEI-SNP gel exhibited an excellent thermal stability at 85 °C for 660 hours without syneresis. Subsequently, SEM images confirmed the successful incorporation of SNP in the 3-dimentional network of hydrogel, supporting an effective surface modification.

This work demonstrated that the novel environmentally benign HPAM/PEI-SNP gel system can be used as a potential plugging agent for the conformance improvement in high temperature high salinity reservoirs.

Introduction

The existence of fractures, high-permeability channels or thief zones in reservoirs (Figure 1) is unfavorable and detrimental to oil recovery. To alleviate these conformance problems, gel treatment was implemented for a conformance improvement. Upon decades of using, it has been proven an economical, effective, and applicable technology. In term of gel treatment, liquid-based or colloid-based polymeric agent was injected into the target zone to seal or plug the thief-zone, fractures or high-permeability channels.

Polyacrylamide (PAM) or hydrolyzed polyacrylamide (HPAM) which could be cross-linked by inorganic or organic cross-linkers are the major polymers deployed for the plugging agents. The inorganic cross-linker such as Cr(III) acetate could bond with carboxylate groups ($-COO^-$) on HPAM chains through coordinative bonding (Figure 2). However, the application of inorganic cross-linker was impeded in virtue of the poor thermal stability, gelling performance and mechanical integrity of formed gel. On the other aspect, the organic cross-linker such as phenol-formaldehyde (Figure 3) could provide a stronger covalent bonding. Whereas, such a condensation cross-linking always encompassed short gelation time and environmental toxicity. Alternatively, an environmental benign gel, polyethyleneimine (PEI) hydrogel is attracting increasing attention.

Morgan (1997) reported that bulk gel formed using acrylamide/t-butyl acrylate (PAtBA) copolymer and PEI. Hardy (1998) presented that the good thermal stability of the similar gel system. Besides PAtBA system, PEI was also demonstrated a cross-linking capability with other PAM derivations such as the copolymer of acrylamide and acrylamido-2-methylpropane sulfonic-acid (Vasquez et al. 2003) and the terpolymer of acrylamido-2-methylpropane-sulfonic-acid and N, N-dimethyl acrylamide (Vasquez et al. 2005). As known, two different mechanisms are prevailing to interprate the PEI cross-linking reactions. For PAM/PEI system, the imine nitrogen from the PEI reacted with the amide groups ($-CONH_2$) via transamidation (Reddy, B, R, et al, 2003) as Figure 1 shown. For PAtBa/PEI system, it was reported that the cross-linking was induced by acid based proton exchange between the ammonium ion in PEI and the carboxylate group derived from the hydrolysis of t-butyl acrylate (Mary Hardy, et al, 1999). Due to the formation of strong covelant bonding, the thermal stability of formed gel was significantly improved. G. A. Al-Muntasheri (2006), reported that hydrogels prepared with PAM, and PEI using distill water maintained thermal-stable for 2 weeks at 120 °C. Polo, R.P.,O (2004) claimed that a series of conformance treatments based on PAtBA/PEI gel had acquired positive responses in fractured carbonates up to 129 °C and sandstones up to 82 °C.

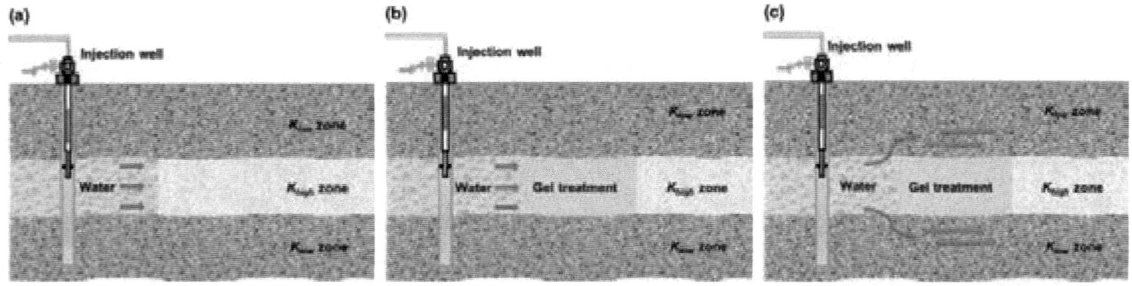

Figure 1. Schematic of gel treatment, (a) preferential flow path in heterogenous reservoir, (b) gel slurry was injected to plug high-permeability zone (c) water was diverted low-permeability zone

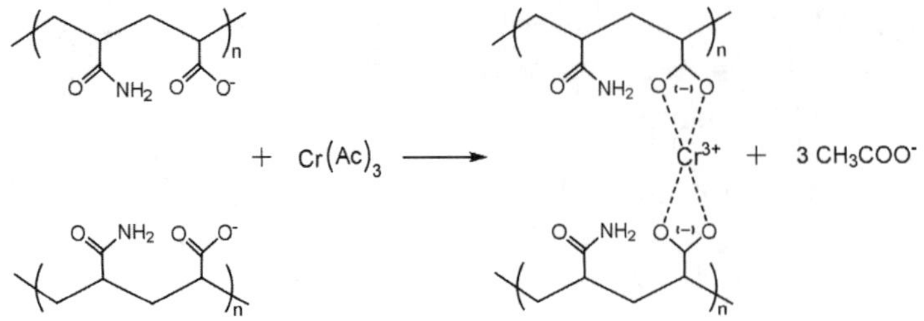

Figure 2. HPAM cross-linker by $Cr(Ac)_3$

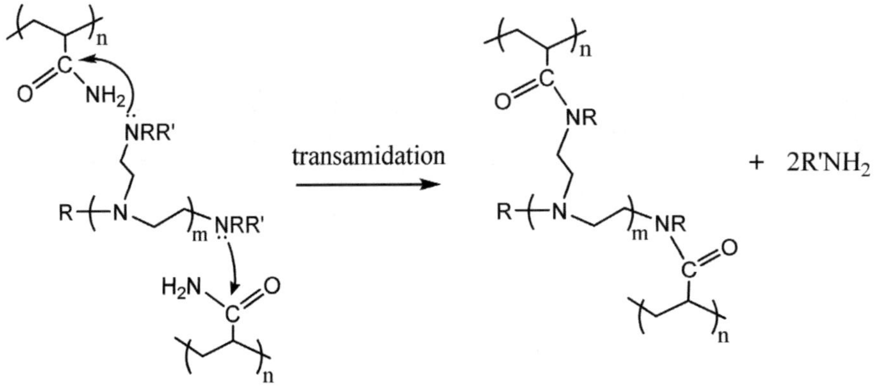

Figure 4. Mechanism of crosslinking between PAM and PEI (Reddy et al. 2003)

To alleviate the defects of pristine HPAM gel, it was an advantage to introduce silica nano-particles (SNP) for a nano-composite modification. The solution mixed by HPAM and dispersions of discrete nonporous nano-particles of silica/silicon dioxide (SiO2), was reported a prominent viscosity increase (Maghzi, A et al, 2014), due to the physical adsorption. It was also noted that a pronounced improvement of thermal stability was found in PAM/SiO2 nano-composite system (Bhardwaj et al, 2008). This phenomenon was attributed to the formation of hydrogen bonding. Other authors (Zhu et al, 2014) concluded that SNP functioned as a physical cross-linker with the presence of carbonyl-based polymer, whereupon the mechanical intergrity of polymer system was strengthened.

Albeit substantial work based on HPAM/PEI or SNP nano-composite has been conducted, as far as we know, the combination of these technologies has not been reported. In this work, these two technologies were integrated together for a development of novel environmentally benign gel system, HPAM/PEI-SNP hydrogel, for a comformance treatment in the formidable reservoir conditions.

Experimental

Material

The hydrolyzed polyacrylamide (HPAM) used in this study was provided by SNF Holding Company as a form of powder. In accordance to the disclosed content, the information of polymers has been specified in Table1. Polyethyleneimine (PEI) was purchased from Sigma-Aldrich Co., LLC as a form of viscous liquid. Three types of colloidal silica (CS), LUDOX® LS, LUDOX® TM, and LUDOX® SM were obtained from Sigma-Aldrich Co., LLC. Sodium chloride, calcium chloride, magnesium chloride, sodium sulfate, and sodium bicarbonate were all obtained as ACS grade from Fisher Chemical. Distilled water was prepared in the laboratory. All of the chemicals were used as received without further purification.

Table 1– Specification of polymers

Polymer Type	Molecular weight, dalton	Hydrolysis degree, %
AN 905	8-10*10^6	1-5
AN 907	7-10*10^6	5-10
AN 105	10-15*10^6	3-7

Table 2– Ion composition of synthetic formation water

Ions	K^+ and Na^+	Ca^{2+}	Mg^{2+}	Cl^-	SO_4^{2-}	HCO_3^-	TDS
(mg/L)	1.43×10^4	5803.08	180.46	3.28×10^4	6.59	60.41	5.32×10^4

Preparation of gelling solution

Colloidal silica was first dropwise added into the synthetic formation water. Then, the polymeric powder was slowly added along the vortex under vigorous stirring. The stirring in ambient maintained until a homoneous solution was achieved. Afterwards, PEI in designed amount was added and allowed for a stirring of 30 to 60 minutes. Finally, the glass bottles contained gelling solution was sealed well and allowed for a curing in 85°C oven.

Evaluation of gel properties

Gelation time indicated the evolution of gelant to bulk gel. In this work, gel code method (Sydansk 1988) was implemented to determine the gelation time and gel strength. At different intervals, the bottles were inverted by which the gel strength was identified with an alphabetic code from A to I. In the meantime, the gelation time was quantified as the period that gel type evolved from A to D. Beyond the gel strength and gelation time, long-term thermal stability of gel system was examined via dehydration time which was also achieved from bottle tests.

Selection and characterization of colloidal silica (CS)

Colloidal silica, as an effective and environmental-benign additive, has been widely deployed for foam stabilizer, nano-composite reagent, and conformance control. In spite of the versatility of CS, this material was notably sensitive to circumstances involved pH, monovalent/divalent cations, and temperature. Since the formation water was utilized as carrier fluid, it would be unfavorable if SNP precipitated during the mixing or pumping in actual gel treatment. To avoid this negative phenomenon, examining the compatibility of colloidal silica and formation water was imperative. In this investigation, CS was dropwise added into formation water, and thereby diluted to half and ¼ of its original concentration. After a standing of 6 h in ambient, the dispersion was visually examined. In case SNP precipitated, the corresponding CS was indicated not compatible with formation water.

To characterize the size distribution of SNP, the solid concentration of CS was first diluted to 100 ppm with distilled water. Malvern Zetasizer Nano-ZS90 (Malvern Instruments Ltd., UK) with an incident beam of 633 nm wavelength was deployed in which the detection angle and temperature was set as 90° and 25 °C respectively. The intensity size distributions were then obtained from analysis of the correlation functions by the Multiple Narrow Modes algorithm.

Scanning Electron Microscopy (SEM) characterization

Gel microstructures were characterized with Hitachi S-4700 Field Emission Scanning Electron Microscope. HPAM/PEI and HPAM/PEI-SNP gels were first frozen with liquid nitrogen which was followed by a 24 h freezing-dry process. Then, the dried hydrogel networks were placed on conductive tape that attached on the stainless-steel stub, and allowed for a spray of Au/Pd nano-particles for 2 min. SEM imaging was performed using a field emitter source via a mixed detector at 15 kV accelerating voltage, and 9.5 μA current.

Table 3– Sydansk gel code

Code	Gel type	Gel behavior in the bottle
A	No detectable gel formed	The gelant solution appears to have the same viscosity(fluidity) as the original polymer solution, and no gel structure is visually detectable.
B	Highly flowing gel	The gel appears to be only slightly more viscous than the initial polymer solution
C	Few Flowing gel	Most of the obviously detectable gel flows to the bottle cap on inversion of the bottle
D	Moderately flowing gel	A small portion(approximately 5 to 15%) of the gel does not readily flow to the bottle cap on inversion-usually characterized as a “tonguing” gel (i.e., after hanging out of the bottle, the gel can be made to flow back into the bottle by slowly turning the bottle upright).
E	Barely flowing gel	The gel slowly flows to the bottle cap and/or a significant portion (>15%) of the gel does not flow to the bottle cap on inversion.
F	Highly deformable non-flowing gel	The gel does not flow to the bottle cap on inversion (gel flows to just short of reaching the bottle cap).
G	Moderately deformable non-flowing gel	The gel flows approximately halfway to the bottle cap on inversion.
H	Slightly deformable non-flowing gel	Only the gel surface slightly deforms on inversion
I	Rigid gel	There is no gel-surface deformation on inversion.
J	Ringing rigid gel	A tuning-fork-like mechanical vibration can be felt after the bottle is mechanically tapped.

Results and discussion

The performance of AN907, AN905 and AN105 systems has been illustrated in Table 4, Table 5 and Table 6 repesctively. Note that *de* was an abbreviation for dehydration. The gelation performance of the HPAM/PEI-SNP gel has been exhibited in Table 7.

Table 4- Performace of AN907/PEI gel system

Gel composition	HPAM (wt%)	1				0.75				0.5			
	PEI (wt%)	0.1	0.2	0.3	0.4	0.1	0.2	0.3	0.4	0.1	0.2	0.3	0.4
Gel code at different time (h)	1 hr	B	B	B	B	B	B	B	B	A	A	A	B
	6 hr	A	A	A	A	A	A	A	A	A	A	A	A
	12 hr	A	A	B	B	A	A	B	B	A	A	A	A
	18 hr	A	A	B	B	A	B	B	B	A	A	A	A
	24 hr (1d)	B	B	B	B	B	B	B	B	A	A	B	B
	36 hr	B	C	D	D	B	B	B	C	A	A	B	B
	48 hr (2d)	B	D	D	E	B	C	D	D	A	B	de	de
	72 hr (3d)	C	E	F	F	B	E	de	de	B	de	de	de
	96 hr (4d)	C	F	de	de	C	F	de	de	B	de	de	de
	120 hr (5d)	D	F	de	de	D	de	de	de	de	de	de	de
	144 hr (6d)	E	de	de	de	D	de	de	de	de	de	de	de
	168 hr (7d)	E	de	de	de	E	de	de	de	de	de	de	de
	192 hr (8d)	F	de	de	de	E	de	de	de	de	de	de	de
	216 hr (9d)	F	de	de	de	F	de	de	de	de	de	de	de
	240 hr (10d)	de	de	de	de	de	de	de	de	de	de	de	de
	288 hr (12d)	de	de	de	de	de	de	de	de	de	de	de	de
	360 hr (15d)	de	de	de	de	de	de	de	de	de	de	de	de
	408 hr (17d)	de	de	de	de	de	de	de	de	de	de	de	de
	480 hr (20d)	de	de	de	de	de	de	de	de	de	de	de	de

Table 5-Performace of AN905/PEI gel system

Gel composition	HPAM (wt%)	1				0.75				0.5			
	PEI (wt%)	0.1	0.2	0.3	0.4	0.1	0.2	0.3	0.4	0.1	0.2	0.3	0.4
Gel code at different time (h)	1 hr	B	B	B	B	B	B	B	B	A	A	A	B
	6 hr	A	A	B	B	A	A	A	A	A	A	A	A
	12 hr	A	A	B	B	A	B	B	B	A	A	A	A
	18 hr	A	A	B	B	A	B	B	B	A	A	B	B
	24 hr(1d)	B	B	B	B	B	B	B	B	A	A	B	B
	36 hr	B	D	D	E	B	B	B	C	B	B	B	B
	48 hr (2d)	B	D	D	E	B	C	D	D	B	B	B	C
	72 hr (3d)	C	E	F	F	B	E	E	E	B	B	C	C
	96 hr (4d)	C	F	F	F	C	F	de	de	B	C	de	de

	120 hr (5d)	E	F	G	G	D	F	de	de	B	C	de	de
	144 hr (6d)	E	G	de	de	D	F	de	de	de	de	de	de
	168 hr (7d)	F	G	de	de	E	de	de	de	de	de	de	de
	192 hr (8d)	F	G	de	de	F	de	de	de	de	de	de	de
	216 hr (9d)	F	G	de	de	F	de	de	de	de	de	de	de
	240 hr (10d)	de	de	de	de	de	de	de	de	de	de	de	de
	288 hr (12d)	de	de	de	de	de	de	de	de	de	de	de	de
	360 hr (15d)	de	de	de	de	de	de	de	de	de	de	de	de
	408 hr (17d)	de	de	de	de	de	de	de	de	de	de	de	de
	480 hr (20d)	de	de	de	de	de	de	de	de	de	de	de	de

Table 6-Performace of AN105/PEI gel system

Gel composition	**HPAM (wt%)**	**1**				**0.75**				**0.5**			
	PEI (wt%)	**0.1**	**0.2**	**0.3**	**0.4**	**0.1**	**0.2**	**0.3**	**0.4**	**0.1**	**0.2**	**0.3**	**0.4**
Gel code at different time （h）	**1 hr**	B	B	B	B	B	B	B	B	A	A	A	B
	6 hr	B	B	B	B	A	A	B	B	A	A	A	A
	12 hr	B	B	B	B	B	A	B	B	A	A	B	B
	18 hr	B	B	B	C	B	A	B	B	A	A	B	B
	24 hr(1d)	B	B	C	C	B	B	B	B	A	A	B	B
	36 hr	C	D	D	E	B	C	C	D	B	B	B	C
	48 hr (2d)	D	E	E	E	C	D	E	E	B	C	D	D
	72 hr (3d)	E	E	F	F	E	E	E	E	C	D	E	E
	96 hr (4d)	E	G	G	G	E	E	E	F	E	E	E	E
	120 hr (5d)	F	G	G	G	E	E	F	F	E	E	E	de
	144 hr (6d)	G	G	H	H	E	F	F	F	E	E	E	de
	168 hr (7d)	G	G	H	H	F	F	F	F	E	E	de	de
	192 hr (8d)	G	G	H	H	F	G	de	de	E	E	de	de
	216 hr (9d)	G	H	H	H	G	G	de	de	E	E	de	de
	240 hr (10d)	G	H	H	de	G	H	de	de	de	de	de	de
	288 hr (12d)	G	H	H	de	G	H	de	de	de	de	de	de
	360 hr (15d)	G	H	de	de	G	H	de	de	de	de	de	de
	408 hr (17d)	G	H	de	de	de	de	de	de	de	de	de	de
	480 hr (20d)	de	de	de	de	de	de	de	de	de	de	de	de

Table 7-Performace of HPAM/PEI-SNP gel system

CS concentration	**1.25%**	**0.63%**	**0.31%**	**0.15%**	**CS concentration**	**1.25%**	**0.63%**	**0.31%**	**0.15%**
Time (h)	Gel code	Gel code	Gel code	Gel code	Time (h)	Gel code	Gel code	Gel code	Gel code
1	B	B	A	B	**324**	E	F	G	G

2	A	A	B	A	**336**	E	F	G	G
3	A	A	A	A	**348**	E	G	G	G
4	A	A	A	A	**360**	E	G	G	G
5	A	A	A	A	**372**	E	G	G	G
6	A	A	A	A	**384(16 d)**	E	G	G	G
7	A	A	A	A	**396**	E	G	G	G
8	A	A	A	A	**408 (17d)**	E	G	G	G
10	A	A	A	A	**420**	E	G	G	G
12	A	A	A	A	**432 (18d)**	E	G	G	G
18	A	B	B	B	**444**	E	G	G	G
24	B	B	B	B	**456 (19d)**	E	G	G	G
36	B	B	B	C	**468**	E	G	G	G
48(2 d)	B	C	C	C	**480 (20d)**	E	G	G	G
60	B	C	C	C	**492**	E	G	G	G
72(3 d)	B	C	C	D	**504 (21d)**	E	G	G	G
84	B	C	C	D	**516**	E	G	G	G
96(4 d)	B	C	C	D	**528 (22d)**	E	G	G	G
108	B	C	D	E	**540**	E	G	G	G
120(5 d)	C	C	D	E	**552 (23d)**	E	G	G	G
132	C	D	D	E	**564**	E	G	G	G
144(6 d)	C	D	E	E	**576 (24d)**	E	G	G	G
156	C	D	E	E	**588**	E	G	G	G
168(7d)	C	D	E	E	**600 (25d)**	E	G	de	de
180	C	E	E	E	**612**	E	G	de	de
192 (8 d)	C	E	E	E	**624 (26d)**	E	G	de	de
204	D	E	E	E	**636**	E	G	de	de
216 (9 d)	D	E	E	E	**648 (27d)**	E	G	de	de
228	D	E	E	F	**660**	E	G	de	de
240(10 d)	D	E	E	F	**672 (28d)**	E	de	de	de
252	E	E	F	F	**684**	E	de	de	de
264	E	E	F	F	**696 (29d)**	E	de	de	de
276	E	E	F	G	**708**	E	de	de	de
288	E	E	F	G	**720 (30d)**	de	de	de	de
300	E	F	G	G	**732**	de	de	de	de
312	E	F	G	G	**709 (31d)**	de	de	de	de

Effect of HPAM concentration

As shown in Figure 2, the relationship between gelation time and polymer concentration has been exhibited. Herein, HPAM concentration was selected as 1 wt%, 0.75 wt%, 0.5 wt%, in the meantime PEI concentration was at 0.4wt%. It was found that the variation of polymer concentration brought a significant effect on gelation performance. To be noted, bulk gel did not form in case AN907 and PEI concentration were 0.5 wt% and 0.4 wt%. As observed from Figure 2, the gelation time dimished with an increase of HPAM concentration. The higher the polymer concentration, the more cross-linking sites were established in polymer chains, and thus there were more opportunities for collisions between polymer molecules. In this situation, one PEI molecule might cross-link with more amide groups (-$CONH_2$) from HPAM molecules.

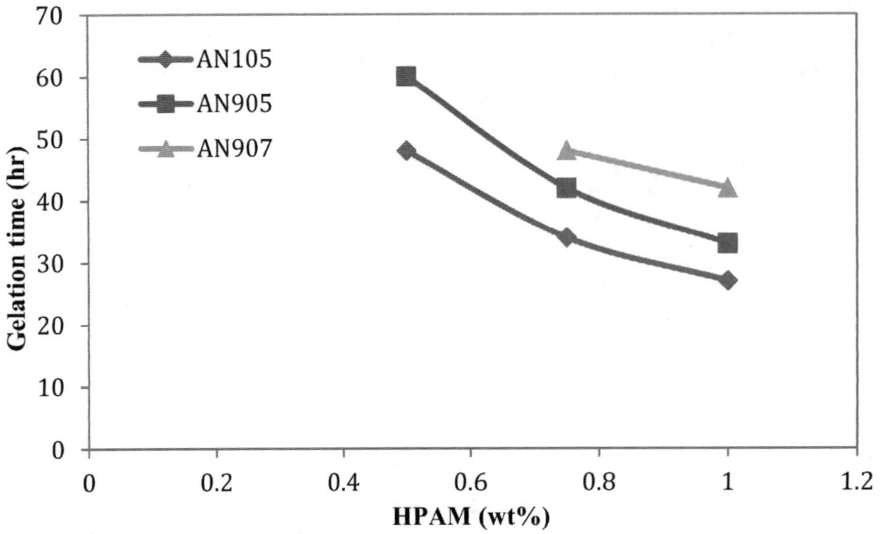

Figure 2-Effect of HPAM concentration on gelation time (PEI concentration was 0.4wt%)

According to the results, strong bulk gel was prone to forming at high polymer concentration albeit cross-linker concentration was relatively low. Taking an example of 1 wt% AN105, a non-flowing gel in code G was achieved even PEI concentration was as low as 0.1wt%. Whereas, the ultimate gel strength did not exceed gel code E once the polymer concentration diminished to 0.5 wt%. Moreover, the polymer concentration played a determinative role for gel thermal stability. In case polymer and the cross-linker concentration were 1 wt% and 0.2 wt% respectively, the gel maintained highest strength for nearly 20 days without syneresis. It was analyzed that since the polymeric network was strong and well-established, the capability for water retention was reinforced in the gel. Therefore, 1 wt% was recommended for HPAM concentration.

Effect of PEI concentration

Figure 3 showed the variation of gelation time versus PEI concentrations in which the PEI concentration differed from 0.1 wt% to 0.3 wt% while the polymer concentration was maintained at 1 wt%.

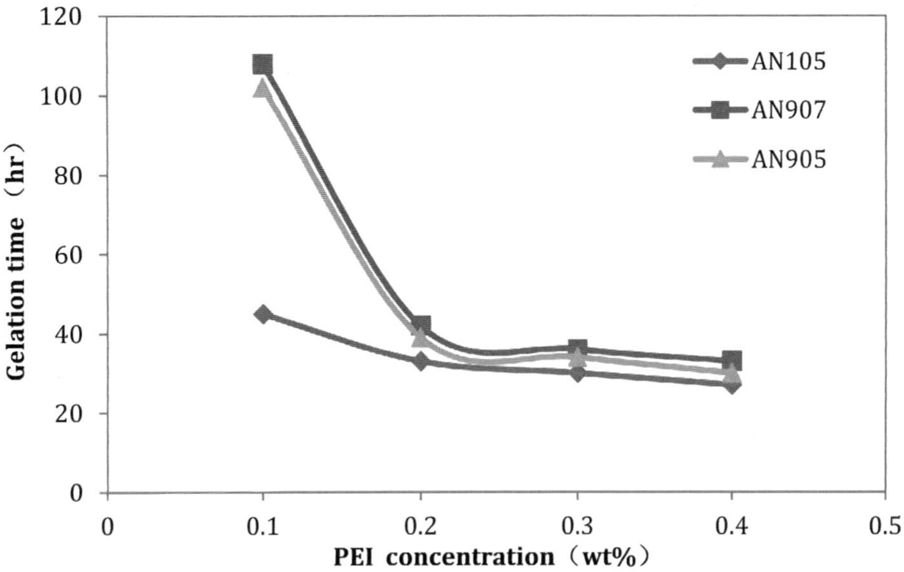

Figure 3-Effect of PEI concentration on gelation time (polymer concentration was 1wt%)

The results indicated that gelation time was in the range of 24~108 h for all cases. As shown in Figure 4, an increase of PEI concentration from 0.1 wt% to 0.2 wt% led to a repaid decrement in gelation time, especially for AN 905 and AN 907 polymer. The mechanism of above phenomenon was speculated that there were more opportunities for PEI molecule to attack the carbonyl carbon and attach to amide by increasing its concentration.

Moreover, the excessive cross-linker concentration was unfavorable for gel thermal stability. As observed from Table 6, once the concentration of PEI was above 0.2 wt%, shorter gelation time and higher gel strength have been obtained. Whereas, the gel syneresis was severe due to the excessive cross-linking and inappropriate cross-linking density. Accordingly, an appropriate concentration of PEI was recommended as 0.2 wt%. It was also noteworthy that polymer concentration was indicated an overwhelming impact on gel performance when compared with the influence of crosslinker concentration.

Effect of the HPAM molecular weight (Mw)

The results indicated that a higher molecular weight HPAM polymer resulted in a shorter gelation time as shown in Figure 4. In the meantime, higher Mw benefited gel strength thermal stability as the polymer and crosslinker concentrations were kept constant.

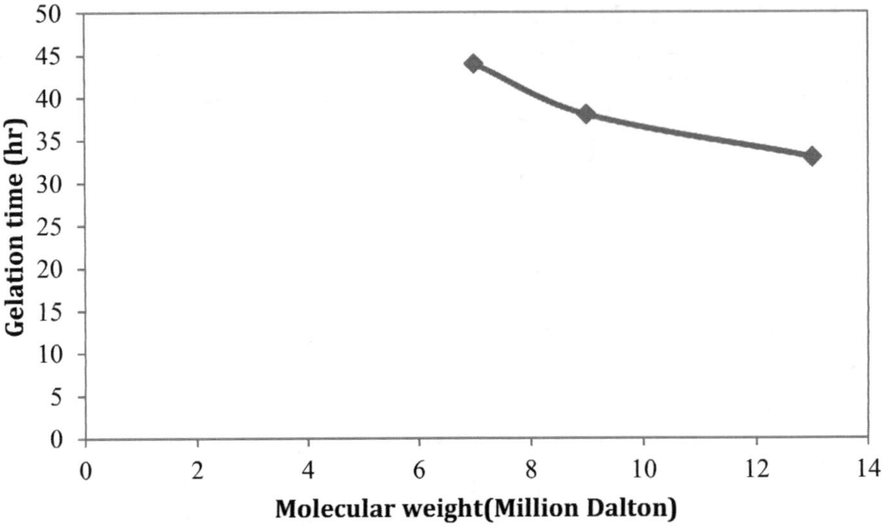

Figure 4-Effect of Molecular weight on gelation time (Polymer concentration was 1wt%, PEI concentration was 0.2wt%)

As known that abundant amide pendents distributed in the chains of HPAM, polymers in higher Mw often has more carbonyl carbon in the amide groups which could be attacked by PEI molecules. In accordance, a shorter gelation time was achieved due to a higher reacting rate of nucleophilic substitution. Furthermore, polymer with higher Mw featured with a larger coil size, leading to a formation of stiff three-dimensional structure in microscopic scale, namely a strong bulk gel in macroscopic scale. In virtually, polymer with large Mw always possessed complicated chain skeletons and large chain length, rendering a chain-chain entanglement in large magnitude. Hence, a more viscoelastic gel was obtained by increasing the Mw of polymer. According to the results of Table 4 and 5, most of the gels synthesized by lower Mw polymers, AN907 and AN905, were found a dehydration within 10 days. In comparison, AN105 based polymer gel revealed a superior property in thermal stability, denoting that a larger Mw of polymer was more favorable for gel stability. On the other aspect, for a polymer of lower Mw, its hydrodynamic volume was apt to be impaired by the ions in aqueous. As molecular chains rolls became smaller, the formed gel was rendered weak and unstable.

Selection and characterization of CS

The states of original CS(left), two-fold diluted(middle) CS and four-diluted CS(right) were shown in Figure 5. It was apparent that, compared with TM and SM, LS colloidal silica was indicated more compatible with formation water according to its homogeneous and stable state. Afterwards, the DLS characterization was conducted with LS colloidal silica, which indicated that distribution of its diameter size concentrated in 9~12 nm.

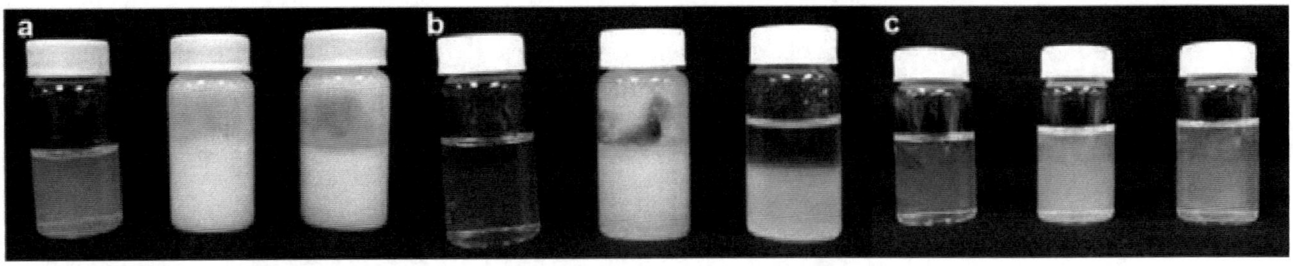

Figure 5-Compatability test of TM(a) SM(b) LS(c) CS with formation water

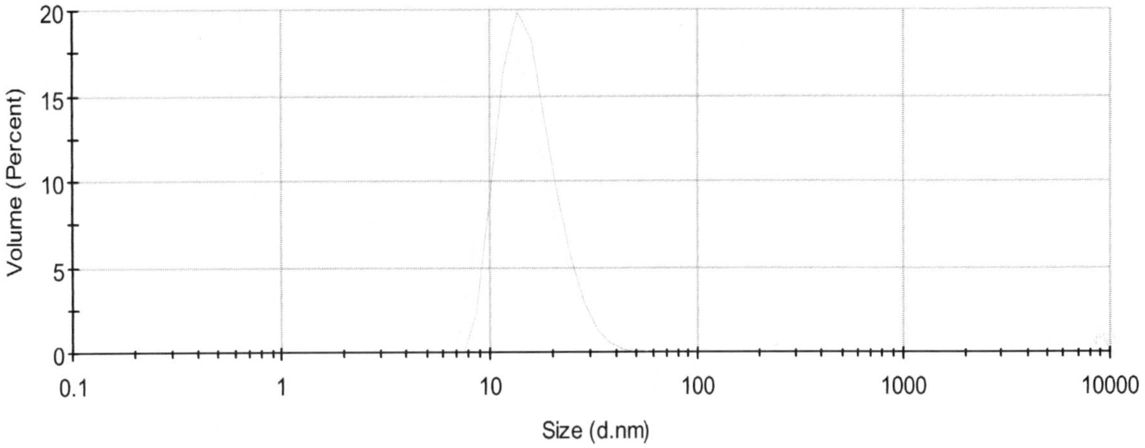

Figure 6-The particle size distribution of SNP measured by DLS

Effect of nano-particle silica concentration

Based on abovementioned results, an optimized formulation of HPAM/PEI gel that contained 1.0 wt% AN105 and 0.2 wt% PEI was achieved. One step further, SNP of various concentrations were introduced and its optimum concentration was investigated in which LS CS was added as concentrations of 0.15, 0.31, 0.63 and 1.25 wt%. It is obvious that the introducing of SNP was conducive to prolong the gelation time compared with the pristine HPAM/PEI system as the performance shown in Table 7. For instance, relying on the additions of 0.15 and 0.31 wt% SNP, the gelation time was dramatically delayed to 72 and 108 h respectively. It was also noticed that the additional of SNP enhanced the viscoelasticity of gel. The related mechanism was widely studied by other researchers (Hu et al, 2017; Zhu et al 2014). Hu (2017) proposed that the hydrogen bonding formed through the -OH groups from silica surface associated with the amide groups. More specifically, such bonding could be formed between the oxygen from silica surface and hydrogen from HPAM polymer, or between the hydrogen on silica molecules and the nitrogen or oxygen in polymer chain. Moreover, the particle-particale in the hybrid interaction existed as a form of Si-O-Si bonding, further facilitating the improvement of gel viscoelatisity.

Despite the successful modification of SNP, the excessive loading of SNP, such as 1.25 wt%, would induce a decrement of gel viscoelastisity. It was noted that there was a critical concentration for this hybrid gel system. As the concentration of SNP was lower than the critical concentration, SNPs absorbed on polymer chains through the interaction between SNP and the carbonyl groups in HPAM molecule (Haraguchi, et al, 2002).

Beyond the critical concentration, SNPs were prone to agglomerating. In this situation, the flexibility of the polymer chain was impeded and the crosslinking process was interfered. Given in the gelation time, gel strength and thermal stability, it was reasonable to conclude that the optimum HPAM/PEI-SNP gel was formulated in an SNP concentration of 0.63 wt%.

Figure 7-Apparence of HPAM/PEI-SNP gels after 20 days aging (CS concentrations were 1.25, 0.63, 0.31, 0.15 wt% from left to right)

SEM characterization

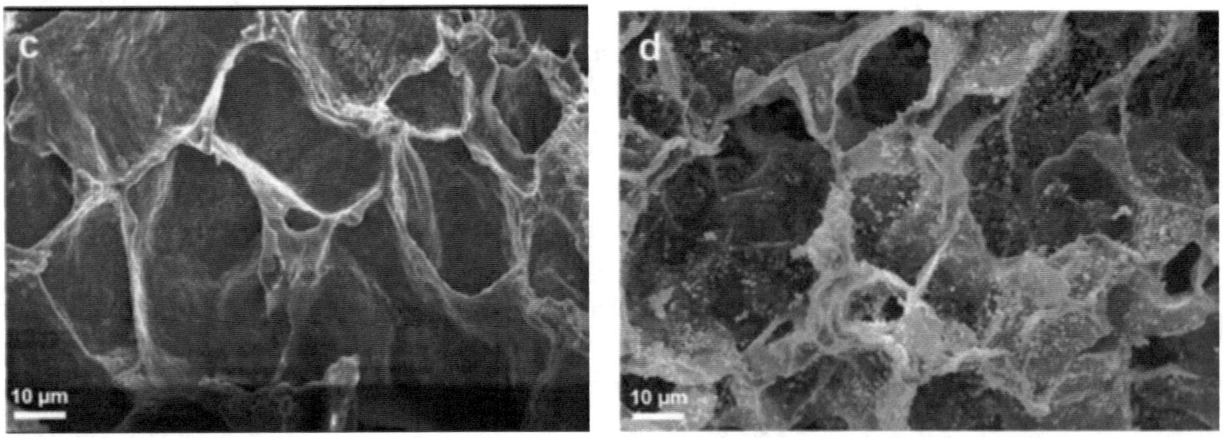

Figure 8-Microstructures of HPAM/PEI-SNP gel (a and b) and HPAM/PEI gel (c and d)

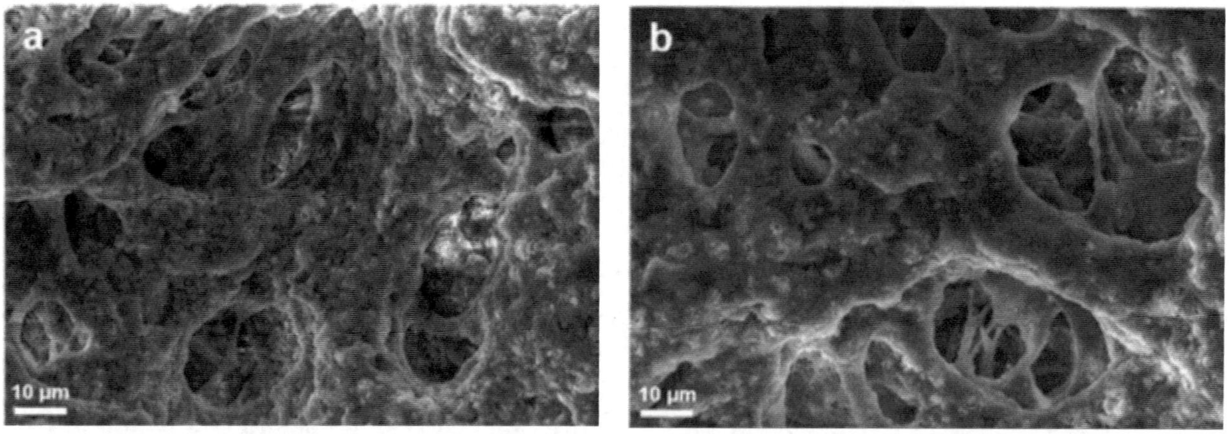

The microstructures of both HPAM/PEI and HPAM/PEI-SNP gel were illustrated in Figure 8, which showed the three-dimensional networks in hydrogel. Moreover, the Figure 8(a and b) demonstrated the successful embodiment of SNP. Comparing SEM images of HPAM/PEI and PLS, the distinction was notable that the microscale pore structure of HPAM was thin and rigid. In contrast, the microstructure of PLS gel was made up due to the attachment of SNP the polymer chain of HPAM.

Conclusion

- The gelation performance of HPAM/PEI and HPAM/PEI-SNP systems envolvING gelation time, gel strength and thermal stability was well studied.

- The optimum formuation of HPAM/PEI system was determined based on the investigations among polymer concentration, crosslinker concentration and polymer molecular weight. As a result, 1 wt% high Mw polymer, AN105, and 0.2 wt% PEI formulated in the optimum formulation of HAPM/PEI system.

- In accordance to the compatability tests, LS colloidal silica was selected as the hybrid resource for nano-composite modification. Besides, the diameter of SNP was characterized as 9~12 nm using DLS.

- Through the successful nano-composite modification that confirmed SEM characterization, the gelation time of HPAM/PEI-SNP system was prolonged to 132 h; the gel strength of G grade was maintained for 312 h; the dehydration time was delayed to 672 h. HPAM/PEI-SNP was demonstrated a potential plugging agent for conformance treatment in high temperature and high salinity/hardness reservoir.

Reference

C. Maltesh,P. Somasundaran,R.A. Kulkarni(1991). Effect of the degree of hydrolysis of polyacrylamide on its interactions with poly (ethylene oxide) and poly (vinylpyrrolidone) [J]. Macromolecules 24(21): 5775~5778.

F. Torabi, W.G. Luo, S.X. Xu. (2013). Chemical degradation of HPAM by oxidation in produced water: experimental study[C]. SPE American E&P Health Safety Security and Environmental Conference 2013.

G. Muller, J.C. Fenyo, E. Selegny. (1980). High molecular weight hydrolyzed polyacrylamides: III. Effect of temperature on chemical stability [J]. Journal of Applied Polymer Science, 25(4) 627 ~ 633.

G. Smets, A.M. Hesbain.(1959). Hydrolysis of polyacrylamide and acrylic acid acrylamide-copolymers [J]. Journal of Polymer Science, 40(136): 217~226.

J. Vasquez, L. Eoff, and B.R. Reddy and F. Civan, "Development and Evaluation of High-Temperature Conformance Polymer Systems" paper SPE 93156 presented at the 2005 SPE International Symposium on Oilfield Chemistry held in Houston, Texas, U.S.A., 2 – 4 February.

Moradi-Araghi, P.H. (1987). Hydrolysis and precipitation of polyacrylamides in hard brines at elevated temperatures [J]. SPE Reservoir Engineering, 2(2): 189~198.
Mary Hardy, Wouter Botermans, Aly Hamouda, Jarl and John Warren. 1999. The First Carbonate Field Application of a New Organically Cross-Linked Water Shutoff Polymer System. Paper presented at the 1999 SPE International Symposium on Oilfield Chemistry held in Houston, TX, February 16-19. SPE-50738-MS.

P. Albonico, M. Bartosek, T.P. Lockhart, et al. (1994). New polymer gels for reducing water production in high -temperature reservoirs[C]. European Production Operations Conference and Exhibition.

Polo, R.P.O., Monroy, R.R., Toledo, N, Eoff, L, Everett, D. (2004). Field applications of low molecular-weight polymer activated with an organic cross-linker for water conformance in South Mexico. Paper SPE 90449 presented at the SPE Annual Technical Conference and Exhibition held in Houston, TX, pp. 26–29

Pan-Sang Kang, Jong-Se Lim and Chun Huh. (2016). Artificial Neural Network Model to Estimate the Viscosity of Polymer Solutions for Enhanced Oil Recovery, Appl. Sci. 6(7)

R.D. Shupe. (1981).Chemical stability of polyacrylamide polymers [J]. Journal of Petroleum Technology, 33(8) 1513-1529.

R.D. Hutchins, H.T. Dovan, B.B. Stanford. (1996). Field applications of high temperature organic gels for water control[C]. Symposium on improved oil recovery, 419~416.

Sydansk, R. D. (1988, January 1). A New Conformance-Improvement-Treatment Chromium(III) Gel
Technology. Society of Petroleum Engineers. doi:10.2118/17329-MS

Sheng, J.J. (2011). Modern Chemical Enhanced Oil Recovery: Theory and Practice, 1st ed.; Gulf Professional Publishing: Burlington, MA, USA.

Vasquez, J., Civan, F., Shaw, T., Dalrymple, E., Eoff, L., Reddy, B.R. and Brown, D.: "Laboratory Evaluation of Conformance Polymer Systems," paper SPE 80904 presented at the 2003 SPE Production and Operations Symposium held in Oklahoma City, Oklahoma, 23-25 March.

502103 Composelector: An European H2020 Project for Integrating Multi-Scale Material Simulation and Industrial Business Decisions

Abstract:

The mission of the recently launched European H2020 project COMPOSELECTOR is to develop a Business Decision Support System (BDSS), which integrates materials modelling, business tools and databases into a single workflow to support the complex decision process involved in the selection and design of polymer-matrix composites (PMCs) by means of an open integration platform which enables interoperability and information management of materials models and data and connects a rich â€˜materials modelling layerâ€™ with industry standard business process models.

The integration of modelling and simulations techniques to support material selection process is more and more impelling in the materials science and industrial domains, due to the need of effectively designing and producing increasingly sophisticated materials, components and systems with advanced performance on a competitive time scale. In this perspective, for complex structural materials, such as PMCs, there is a particular need in industry for chemistry/physics-based materials models and modelling workflows that fulfil the following requirement: predicting relevant properties and Key Performance Indicators (KPIs) that capture the performance of materials, accounting for material internal microstructure and effects of processing.

In this scenario, the simulations techniques aims at developing, demonstrating and assessing an easy-to-use multiscale simulation suite that delivers models with feasible computational loads for the design and production of industrial products, i.e., PMCs. Specifically, the project focus on the development of high-fidelity models to predict properties and behaviors of PMCs at all scales, from atomistic to finite element calculations and the development of a multi-scale software-modeling framework which integrates these models across the scales. In this work will be presented i) the exploiting of atomistic simulation to study PMCs thermomechanical properties and their interface phenomena as well; ii) the linking atomistic and mesoscale models and mesoscale models to finite element models; and iii) the development of multiscale simulation protocols for nanostructured materials.

502708 Single Step Catalytic Conversion of Propane to Propylene Via Reactive Separation

Abstract:

There is a tectonic shift in steam cracker markets. The shale gas revolution in the United States has led to most of the ethylene crackers being shifted to use of ethane as a feedstock. This severely impedes the domestic availability of propylene leading to increased imports. In addition, new proposed ethylene crackers will use ethane as the feedstock instead of naphtha that will increase total ethylene capacity by 52% (~10 Million metric tonnes per annum) without any increase in new domestic propylene availability.

This presents opportunity for custom dedicated processes for propylene manufacturing. Direct conversion of propane to propylene is a feasible catalytic method that co-produces hydrogen. If the co-product hydrogen is removed during the process, the equilibrium shifts to right leading to highly favorable economics. The dehydrogenation of propane was examined in Aspen Plus using both the Gibbs Reactor (RGIBBS) and Equilibrium Reactor (REQUIL) models to demonstrate the beneficial effect. This paper will present experimental and modeling results of the direct propane to propylene conversion process. A catalyst membrane reactor design and development will be discussed.

505337 Crystallization and Foaming Behaviors of Modified Polypropylene by Phenyl-contained Function Group

Abstract:

Polypropylene (PP) foam has a wide application in many fields such as package, transportation, and architecture due to its excellent mechanical and thermal properties. However, PP as a semicrystalline polymer is very difficult to foaming because its very low melting strength would cause cell collapse. Hence, a meliorating foaming of PP through improving its crystallization behavior was put forward. In this work, phenyl-contained function group was grafted onto PP chain through melting grafting reaction, which showed that branched chain resulting from only benzene in sequence can increase dramatically the crystal density and decrease significantly the crystal size[1]. Length and density of the branched chain also have effect on the crystallization properties[2] because of the steric hindrance to the ordered packing of PP backbones chains, and a reduction in crystallinity was observed. A batch foaming process including melting, cooling, isothermal saturation, and foaming stages was utilized to produce PP foams. The formed small and dense crystal at isothermal saturation stage could not only act as a strong heterogeneous nucleation for PP foaming but also hinder the growth of bubbles so that microcellular PP foam could be obtained. Moreover, it could also enhance the limiting upper foaming temperature to 160 °C.

Keywords: microcellular foam, phenyl modified polypropylene, crystallization

References:

[1] Shuai Zhou, Shicheng Zhao, Zhong Xin, *Polymer Engineering & Science.* **2015**, 55, 2

[2]Fengmei Su, Youxin Ji, Lingpu Meng, Zhen Wang, Zeming Qi, Jiarui Chang, Jianzhu Ju, Liangbin Li, *Macromolecules.***2017**, 50, 5

509951 The Reaction Condition Impacts on the Performance and the Kinetic of the Reduction of Copper Oxides by Methane during Chemical Lopping Combustion

Abstract:

Earth abundant copper oxide (II) is a promising oxygen carrier for CLC and CLOU processes due to its high oxygen content and reaction rate. Nevertheless, CuO has not receive much attention from researchers due to the problem of particle agglomeration at relatively high temperatures used by CLC. In this work, we investigated the impact of several parameters related to CLC, such as CuO particle sizes, CH_4 flow rate, CH_4 concentration and competition between CLC and CLOU as a function of reaction temperature, on the performance of unsupported CuO as an oxygen carrier. The reduction of two commercial unsupported CuO samples with 50 nm and 1 μm particle sizes was investigated in a temperature range of 600 to 800 °C using dilute (5%) and pure (99.999%) CH_4 and a flow rate range from 12.5 to 250 h^{-1}. The reactivity of Cu_2O was also considered as a common intermediate of CuO reduction to Cu metal. To track the phase transformation of CuO during methane oxidation, bulk and surface analysis was performed on several samples with different reaction times at 700 °C. We have found that CuO reduction is a multistep process that is initiated by the formation of reduced phase nuclei and then proceeds by a shrinking core mechanism. Increasing the reaction temperature and CH_4 concentration increase the CuO reduction rate. When the reaction temperature exceed that of CuO decomposition, specifically at 800 °C, competition between CLC and CLOU resulted in sustained 100% conversion of CH_4 to CO_2 and H_2O, allowing utilization of 50% of the CuO oxygen content with full capturing of CO_2. Collectively, these new fundamental insights into CuO reduction by methane will help to increase its feasibility for CLC or CLOU processes.

511526 Optimizing Pt loading on Three-Dimensional Carbon Foam for HER

Abstract:

The high demand of platinum is due to its high catalytic activity making it the best catalyst for many applications like hydrogen evolution reaction (HER), oxygen reduction reaction (ORR), and fuel cell applications. Pt nanoparticles were deposited by pulsing potential method on four samples of large surface area carbon foams with different amounts of charge passed. Scanning Electron Microscopy (SEM) and Histogram were used to study the particle distribution of Pt nuclei on the carbon foam surface. The results show an increase in particle density with increasing amount of Pt. The actual amount of platinum deposited was found by using inductively coupled plasma mass spectroscopy (ICP-MS). All samples are tested for hydrogen evolution activity in acid and base and found to be highly active compare to Pt wire. The overpotential values used for comparison were taken from the CV graphs, and show that samples with 50 mC cm^{-2} and 300 mC cm^{-2} have the lowest overpotentials. The stability tests were done at a current density of -100 mA cm^{-2}. The overpotential in both cases reached a stable value after around 2-3 hours.

492320 Validation of CFD Model for the Pilot Scale Mineral Carbonation Bubble Column Reactor

Abstract:

As the global warming issue attracts worldwide attention, CCU(Carbon Capture and Utilization) technology is being actively studied. Mineral carbonation process, one of the CCU technology, removes carbon dioxide in flue gas by reacting with calcium hydroxide, and use the product, calcium carbonate, as a raw material for cement. In order to study on mineral carbonation reactor, a pilot scale mineral carbonation bubble column reactor of 2.2 m in diameter and 6.0 m in height was built in Incheon, South Korea.

Diameter of the mineral carbonation bubble column reactor should be at least about 10 m to remove more than 100,000 tons of carbon dioxide in a year. However, most of the studies on bubble column reactor have been intensively studied on laboratory scale with a diameter of less than 30 cm. Few studies on large bubble column reactor only compared the experimental results with empirical equation. Even in the studies using CFD(Computational Fluid Dynamics), which can simulate bubble coalescence and breakage for calculating gas holdup and interfacial area by using PBM(Population Balance Model), the studies are limited on laboratory scale due to the enormous computational cost.

In this research, the pilot scale mineral carbonation bubble column reactor was modeled using the CFD with Eulerian-Eulerian approach. This approach has dramatically cheap computational cost versus Eulerian-Lagrangian approach coupling CFD and PBM. The CFD results for gas holdup, interfacial area and mass transfer coefficient were compared with empirical equation to obtain a reliability of the CFD model. Finally, by comparing the CFD results for the CO_2removal efficiency with actual experimental results, the CFD model was validated.

The CFD results for the three parameter and the CO_2 removal efficiency showed an error of 5, 7 %, respectively. Furthermore, correction diameter was introduced to calibrate the diameter in a large bubble column reactor. After this correction diameter was applied, an error of the CFD results for the CO_2removal efficiency was reduced to about 2 %.

This study suggests the possibility of modeling a pilot scale bubble column reactor using the CFD with computational efficiency. Based on the CFD model, four times larger bubble column reactor will be built in 2017.

495750 Sandstone Deformation By CO_2 Adsorption

Abstract:

Sandstone Deformation by CO_2 Adsorption

Sahar Bakhshian and Muhammad Sahimi

Understanding CO_2 adsorption in underground porous formations provides significant information about their potential for the geological storage of CO_2 in such porous media in order to mitigate the impact of carbon emissions on the climate. The present work aims to study adsorption of CO_2 in sandstone formations over a wide temperature and pressure range through a comprehensive model. We adopt a three-dimensional (3D) image of the Mt. Simon sandstone, the target reservoir for a pilot CO_2sequestration project conducted by the Illinois State Geological Survey, and study sorption and deformation in it.

We have developed a model based on the energetics of the system that couples gas adsorption with the host-guest interaction and structural deformation of the sandstone. The total energy of the system – the rock plus the gas – is written as the sum of three terms: The elastic energy of the material, the fluid-phase energy, and the interaction energy between the fluid (gas) and the solid matrix. By minimizing the total energy with respect to both the fluid's density and the displacement field of the solid we derive the governing equations for both. The equations are then solved by the finite-element method. Adsorption/desorption isotherms, along with the spatial distribution of the induced strain in the sandstone, are then computed. When the model was applied to sorption of CO_2 in the sandstone samples, all the reported experimental features of the phenomena were reproduced.

495909 Ccus Development in Middle China

Abstract:

China has been the second largest CO_2 emitter because the economic boom during the last 20 years. As a consequence, the controls of CO_2 emissions has become more and more important. To realize both the reduction of CO_2 emissions and enhanced oil recovery, several CCUS pilots were built in Ordos Basin which is located in middle China.

Ordos Basin is one of the largest urban agglomerations which can be regarded as the most important industrial center in middle China. Coal industry, oil refinery and steelworks are the pillar industries of Ordos Basin which can emit a large amount of CO_2. Besides, there has natural continental trap structure underground in Ordos Basin which is a geological structure that is suitable for generating and storing oil. These two reasons provided conditions for the development of CCUS in this region.

CCUS contains 3 parts: capture, transport and storage (besides utilization which is EOR in Ordos Basin). In capture part, we researched five different capture methods, including solid amine, ammonia solution, TSA (temperature swing adsorption), PSA (pressure swing adsorption) and membrane separation. Finally, considering various factors, the low-temperature methanol-wash and ammonia solution were proved to be the most suitable methods for the CO_2capture in Ordos Basin.

In transport part, we designed a transportation plan inspired by the transmission network: First, we set up several transfer stations among adjacent emission sources and connect them with branch pipelines. Pretreatment will be done in the transfer stations to unify the pressure, the temperature and the ingredient of CO_2. After that, CO_2 will be injected into the main pipeline and transported to another transfer station in supercritical. The target transfer station is close to storage locations and the supercritical CO_2 will be transformed into the status that is suitable for geological storage and EOR here. At last, CO_2 will be injected into the underground through another branch pipeline.

CO_2 underground storage suitability evaluation, the evaluation of crust stability and the suitability evaluation of reservoirs are necessary in storage part. We collected stratigraphic data and reservoir information, selected evaluation systems to finish the research. Finally, Qiaojiawa block of Jingbian oilfield, Wuqi oil-groove block of Wuqi oilfield and Tangshan block of Dingbian oilfield were proved to be the most suitable places for CO_2 underground storage and EOR. In addition, CO_2 storage potential and EOR effect will be predicted.

At last, several details of the pilots` operation status will be introduced.

495935 Study of Kinetics, Solubility, Heat of Absorption and Formation of Bicarbonate and Carbamate of Linear and Ring Diamines in CO_2 Absorption Process

Abstract:

Study of kinetics, solubility, heat of absorption and formation of bicarbonate and carbamate of linear and ring diamines in CO_2 absorption process

Rui Zhang[a], Zhiwu Liang[a]*, Qi Yang[b]*, Xiao Luo[a]

[a]Joint International Center for CO_2 Capture and Storage (iCCS), Provincial Key Laboratory for Cost-effective Utilization of Fossil Fuel Aimed at Reducing Carbon-dioxide Emissions, College of Chemistry and Chemical Engineering, Hunan University, Changsha, Hunan, 410082, P.R. China

[b]CSIRO Manufacturing, Clayton Victoria 3168, Australia

Keywords: amines; CO_2 capture; ring amines

* Author for correspondence: Dr. Zhiwu Liang Email: zwliang@hnu.edu.cn

Tel: +86-13618481627 and +86-73188573033

ABSTRACT

Climate change has caused strong concern worldwide and excessive CO_2 emission is considered to be the main contributor to global warming, therefore CO_2 capture and storage has become an important research subject. Among different CO_2 capture methods, post-combustion capture using chemical absorbents, especially amine absorbents, is regarded as a feasible method for mitigating the CO_2 emissions from many industrial sources. The most commonly studied amine absorbent is monoethanolamine (MEA). Aqueous solutions of other primary, secondary and tertiary amines and amino acid salts have also been studied to investigate their mass transfer performance, reaction kinetics and energy consumption in absorption-desorption processes. All of these absorbents, however, demonstrated some challenges for use in CO_2 capture processes. For example, primary amines such as MEA have a fast reaction rate with CO_2 during absorption but require large energy consumption during CO_2 desorption; a tertiary amine such as N-methyldiethanolamine (MDEA) requires less energy for CO_2 desorption but has a slow reaction rate with CO_2 during absorption. Other challenges for CO_2 capture processes include amine degradation, facility corrosion, and the escape of amine or other substances to the environment. These challenges, in particular high energy consumption, result in high CO_2 capture costs. To address these challenges, it is important to develop better solvents and better processes.

An ideal amine absorbent for CO_2 capture should have fast reaction kinetics during CO_2 absorption, large CO_2 cyclic capacity (the difference between rich amine CO_2 loading and lean amine CO_2 loading), and low energy consumption during desorption. The performance of amines during CO_2 absorption and stripping is strongly related to amine molecular structure. To design and develop better amine absorbents for CO_2 capture, it is necessary to understand the effect of amine molecular structure on amine performance in CO_2 capture. Yang et al. and Conway et al., have studied a number of specifically designed amine absorbents for CO_2 capture[1, 2]. These designer amines contained two or three amino groups of different types which were spaced by two or three carbons in the same molecules. Their results showed that the designer amines demonstrated a significant improvement in CO_2 cyclic capacity. Especiallly the designer amines contain a ring structure in its molecular structure, so in this work, there are 5 damines (Table 1) were selected to investigated their CO_2 capture performance. The CO_2 equilibrium solubility, the formation of carbamate/bicarbonate, the reaction kinetics and the heat of absorption also was investigated at the corresponding experimental condition. The current results we obtained now shows that the ring amine could produced more bicarbonate and less carbamate than the linear diamines as shown in Figure 1. More bicarbonate vs. less carbamate is benefits to the CO_2 desorption performance.

The other experiments about the CO_2 absorption-desorption performance will be performed in future. Through comparing the solubility, the heat of absorption and the reaction kinetics to get a good understanding on the effect of the molecular structure (ring) on the amine performance for CO_2 capture.

Table 1. The details of each amine used in this work.

Amine	Abbreviation	Mol Wt	Purity (%)	Molecular structure
Monoethanolamine	MEA	61.08	99.0	
Ethylenediamine	EDA	60.10	99.5	
N_1,N_2-Dimethylethane-1,2-diamine	DMEDA	88.15	98.0	
Piperazine	PZ	86.14	99.0	
2-Methylpiperazine	2-MPZ	100.16	95.0	
2,6-Dimethylpiperazine	2,6-DMPZ	114.19	97.0	

Figure 1. The profile of each species with the CO_2 loading changes of each amine. Solid and dash line are the fitted line for different species of each amine.

ACKNOWLEDGEMENTS

The authors are very grateful for the CSIRO PhD student scholarship to support this research and would like to acknowledge the research contribution to the CO_2 loaded amines test by the CSIRO PCC team in Australia. The authors also would like to express thanks for the financial support from the National Natural Science Foundation of China (NSFC-Nos.21536003, U1362112 and 51521006), National Key Technology R&D Program (MOST-No. 2014BAC18B04), Graduate Student Innovation Project of Hunan Province (CX2016B121). China Scholarship Council and China Outstanding Engineer Training Plan for Students of Chemical Engineering & Technology in Hunan University (MOE-No.2011-40) are also greatly appreciated.

Reference

1. Yang, Q.; Puxty, G.; James, S.; Bown, M.; Feron, P.; Conway, W., Toward Intelligent CO_2 Capture Solvent Design through Experimental Solvent Development and Amine Synthesis. *Energy & Fuels* **2016,** *30*, (9), 7503-7510.

2. Conway, W.; Yang, Q.; James, S.; Wei, C.-C.; Bown, M.; Feron, P.; Puxty, G., Designer Amines for Post Combustion CO_2 Capture Processes. *Energy Procedia* **2014,** *63*, 1827-1834.

Thermokinetic Properties and Mass Transfer of CO_2 Absorption in Aqueous Benzylamine Solvents for CO_2 Capture

Satyajit Mukherjee[1], Amar N. Samanta[1] and Syamalendu S Bandyopadhyay[2],
(1) Chemical Engineering, Indian Institute of Technology Kharagpur, Kharagpur, India,
(2) Chemical Engineering Department, Pandit Deendayal Petroleum University, Ghandhinagar, India

Extended Abstract:

In March 2017, atmospheric Carbon dioxide (CO_2) concentration level of 407 ppmv was recorded by the monitoring system of Mauna Loa Observatory, Hawaii. This is expected to rise in future largely due to massive fossil fuel firing as major source of industrial power plants application to meet the global energy demand. Global warming due to Greenhouse effect is one of the most concerned environmental effects due to accumulation of acid gases in troposphere. Reduction of heating value of natural gas, corrosion in pipes and process equipments, and catalyst poisoning in ammonia synthesis are some of the adverse effects due to the potential amount of CO_2presence in respective industrial gas streams. Mankind are at the verge of a great threat of future disaster. Scientists, environmentalist and engineers are very concerned about the possible solutions of this major environmental threat.

Amine based CO_2 absorption is the most mature technology to reduce greenhouse gas emissions by coal fired power plants. Chemical solvents remain the best solution to absorb CO_2 because of low partial pressure of CO_2 in post combustion exhaust gases from industrial operations. Aqueous alkanolamine absorbents such as monoethanolamine (MEA) and diethanolamine (DEA) are widely used for post combustion CO_2 capture. Several drawbacks are associated with conventional amines such as large energy demand to regenerate CO_2, high reaction enthalpy, low absorption capacity and high oxidative and thermal degradation. At high concentration, viscosity of MEA is found to be high which resisting the liquid phase mass transfer and high corrosion rate is also related to concentrated MEA. However, unlike secondary amines (like piperizine), carcinogenic nitrosamines are not formed directly with primary amines in the presence of $NO_x$and therefore primary amines are commercially more acceptable for PCC applications.

Recently, Benzylamine (BZA) has been identified as a potential absorbent for CO_2 removal with higher cyclic capacity similar to that of a tertiary or sterically hindered amine solution. BZA, a primary amine functional group attached to a benzyl group, is water soluble in all proportions. Aqueous BZA is readily biodegradable and its volatility is comparable to MEA. The corrosion rate of BZA is also lower than MEA. Vapour-liquid equilibrium data and kinetics parameters are the key requirements to design absorption and regeneration column. Density, viscosity and physical solubility data are needed to find kinetic parameters and thus the determination of required solvent circulation rate, and the size of the absorption column to achieve a specified CO_2 removal rate. Solubility of CO_2 in new solvents has been reported in recent years to develop new solvent system for acid gas removal. Density, viscosity and CO_2 solubility data in aqueous BZA solvent are very rare in open literature. There is no direct measurement data of kinetics of BZA available in open literature. In our recent publication, the CO_2 solubility in BZA with ENRTL modelling have been studied in detail to present BZA as potential solvent for CO_2 removal.

In this work, kinetics of CO_2 absorption in aqueous BZA has been studied in the temperature range of 303.15-333.15K. The reactive absorption rate was determined in stirred cell by pressure drop method in a reaction calorimeter system. Molar concentrations of BZA studied in this work are 2.78M, 1.87M, 0.94M and 0.47M. Thermokinetic properties have been measured which are used in kinetic measurement of CO_2 absorption in aqueous BZA solution. Pure component density and viscosity data of BZA are experimentally measured and presented by empirical model equation. The densities and viscosities of binary mixtures composed of benzylamine and water are measured over experimental temperature and molar concentration range. Experimental density and viscosity data are correlated using a Redlich-Kister type equation in terms of mole fraction and Grunberg-Nissan model with prediction error of 0.02% and 5.05% respectively. For kinetic measurement of BZA-CO_2 reaction, low CO_2 partial pressure was maintained to satisfy the criterion for a pseudo-first-order reaction condition. In pseudo-first-order reaction regime, second order rate constant for (BZA+H_2O+CO_2) system is evaluated using Zwitterions and Termolecular mechanism and correlated using Arrhenius power law expression. Hatta number and instantaneous enhancement factor are calculated and found to be satisfied the range of pseudo-first-order reaction regime condition. It can be seen that the enhanced mass transfer coefficient in liquid phase, k_LE, varies with the concentration of BZA solutions and temperature. It manifests that the CO_2 absorption rate increases with increasing BZA concentration and temperature. The experimental kinetic data is better correlated by Termolecular mechanism compared to Zwitterion mechanism. It can be obtained from the Arrhenius power law model that the reaction activation energy (E_a) of the reaction of aqueous BZA+CO_2 is 26kJ/mole, indicating lower energy barrier for the reaction of aqueous BZA solution with CO_2.

References:

Mukherjee, S., Bandyopadhyay, S. S., Samanta, A. N. (2017). Vapor-Liquid Equilibrium (VLE) of CO_2in Aqueous Solutions of Benzylamine: New Data and Modeling Using ENRTL-Equation,. Int. J. Greenh. Gas Control, 56, 12-21.

Gilles Richner, (2013). Promoting CO_2absorption in aqueous amines with benzylamine. Energy Procedia 37, 423 – 430.

Gilles Richner, Graeme Puxty, Amanda Carnal, William Conway, Marcel Maeder, Pauline Pearson, (2015). Thermo kinetic properties and performance evaluation of benzylamine-based solvents for CO_2capture. Chemical Engineering Journal 264, 230–240.

Ma'mun, S., & Svendsen, H. F., (2009). Solubility of N_2O in aqueous monoethanolamine and 2-(2-Aminoethyl-amino)ethanol solutions from 298 to 343 K. Energy Procedia, 1(1), 837–843.

Mondal, B.K., Bandyopadhyay, S.S., Samanta, A.N., (2017). Kinetics of CO_2 absorption in aqueous hexamethylenediamine. Int. J. Greenh. Gas Control 56, 116–125.

499176 Process Modeling and Experimental Studies of a Novel Micro-Encapsulated Solvent System for CO_2 Capture

Abstract:

To screen new CO_2 capture technologies and to expedite commercialization of potential technologies, process systems engineering tools can be highly useful. The U.S Department of Energyâ€™s *Carbon Capture Simulation Initiative* (CCSI) has developed a number of innovative models and tools that can be instrumental in achieving this goal[1]. Process level evaluation of a novel micro-encapsulated solvent (MECS) system is undertaken in this work by leveraging the tools developed by the CCSI.

A number of excellent solvents such as ionic liquids and phase-changing solvents have strong potential for reducing operating cost, but face significant transport issues due to their high viscosity and/or because of solids precipitation. Quite often, loadings of these solvents are kept low or their operating temperature is kept high to aid in the transport of these materials between the absorbers and desorbers. Microencapsulation of such solvents to minimize viscosity effects in transport can be instrumental in realizing the full potential of these technologies. The high-viscosity or phase-changing solvents can be encapsulated in an inert polymer shell through which mass transfer from/to the bulk still takes place. While the microencapsulation technology is more developed and prevalent in pharmaceutical and food industries, encapsulation of CO_2 capture solvents has not been investigated much. A number of studies have been recently conducted by Vericella et al.[2] and Stolaroff et al.[3] on microencapsulation of solvents for CO_2 capture. Vericella et al. (2015) have proposed polymer microcapsules with silicone as the shell material and liquid carbonate as the solvent. Stolaroff et al. (2016) reported the capability of encapsulating viscous ionic liquids for carbon capture. However, for commercial application of the MECS for CO_2 capture, selection of a suitable contactor technology and its optimal design by due consideration of mass transfer, heat transfer, and hydrodynamics is necessary.

In this work, first a detailed model of the MECS experimental setup at the Lawrence Livermore National Laboratory is developed. Experimental data under varying temperature and pressure conditions are used to obtain maximum likelihood estimate of the parameters for the mass transfer models. Heat transfer properties and hydrodynamic characteristics of the MECS system are also studied. Finally, a rigorous process model of a fluidized bed contactor is developed. An optimal design of the contactor is developed by exploiting the CCSI models and tools.

References

1. Miller, D. C., Syamlal, M., Mebane, D. S., Storlie, C., Bhattacharyya, D., Sahinidis, N. V., Sun, X. Carbon capture simulation initiative: a case study in multiscale modeling and new challenges. *Annual review of chemical and biomolecular engineering* **2014**, *5*, 301-323,.
2. Vericella, J. J. et al. Encapsulated liquid sorbents for carbon dioxide capture. Nature Communications, 6:6124, 1-7 (2015).

3. Stolaroff et al. Microencapsulation of advanced solvents for carbon capture. Faraday Discussions. DOI: 10.1039/c6fd00049e (2016).

501562 CO_2 Capture Process Dynamic Design of Experiments and Model Validation

Abstract:

With growing global CO_2 release per year to the atmosphere[1], it is critical that CO_2 capture technologies be developed and deployed rapidly. The U.S. Department of Energy's *Carbon Capture Simulation Initiative* (CCSI) recently developed a suite of advanced computational tools and models to accelerate the development and commercialization of CO_2 capture technologies. As part of CCSI, high-fidelity process models were developed to serve as a "gold-standard" reference for benchmarking solvent-based CO_2 capture processes[2]. Rigorous thermodynamic[3] and transport models[4] for MEA-H_2O-CO_2 systems have been developed. In addition, a rigorous mass transfer model was developed by simultaneously estimating the parameters for the interfacial area, reaction kinetics, mass transfer coefficients and diffusivity models by concurrently using the data from wetted wall column and packed column experiments. The resulting submodels were integrated into a steady-state "gold-standard" model in Aspen Plus and was validated with pilot plant data from U.S. DOE's National Carbon Capture Center (NCCC). Since Aspen Plus rate-based models cannot be currently exported to Aspen Plus Dynamics, a Murphree efficiencies model was developed to incorporate results from the rigorous, rate-based, steady-state model and enable satisfactory results in the Aspen Plus Dynamics platform. Hydraulics correlations previously developed by the authors were also implemented in Aspen Plus Dynamics to obtain more accurate time constants.

Dynamic test runs were also conducted at NCCC so that information about process nonlinearities could be obtained while ensuring persistence of excitation. In addition to the transient data available from the plant control system, liquid samples are analyzed every 15 min, resulting in a rich set of transient data spanning several days. A dynamic data reconciliation problem is solved to reconcile the mass and energy balances with plant data and to estimate the missing measurements. The dynamic model is found to satisfactorly predict transient behavior and is used to investigate the transient behavior of the process. The resulting model can further be used for developing an advanced control system and can offer valuable insights in the CO_2 capture behavior during plant disturbances or failures.

References

1. Global Carbon Project. Carbon budget and trends, **2016**. Available at [www.globalcarbonproject.org/carbonbudget]
2. Miller, D. C., Syamlal, M., Mebane, D. S., Storlie, C., Bhattacharyya, D., Sahinidis, N. V., Sun, X. Carbon capture simulation initiative: a case study in multiscale modeling and new challenges. *Annual review of chemical and biomolecular engineering* **2014**, *5*, 301-323,.
3. Morgan, J. C., Chinen, A. S., Omell, B., Bhattacharyya, D., Tong, C., Miller, D.C. Thermodynamic Modeling and Uncertainty Quantification of CO_2-Loaded Aqueous MEA Solutions. *Chemical Engineering Science* **2017,** (submitted).
4. Morgan, J. C., Bhattacharyya, D., Tong, C., Miller, D.C. Uncertainty Quantification of Property Models: Methodology and Its Application to CO_2-Loaded Aqueous MEA Solutions. *AIChE Journal* **2015**. *61*, 1822-1839.

Differential Permeability Reduction of CO_2 and Water by Polymer Gel in Sandstone Rocks during WAG Process

Xindi Sun, Baojun Bai

Department of Geosciences and Geological and Petroleum Engineering
Missouri University of Science and Technology, Rolla, MO 65401

Abstract

Polymer gels have been widely applied to control reservoir conformance and improve sweep efficiency for CO_2 flooding reservoirs. The plugging performance of polymer gels on water and CO_2 have been reported separately; however, few researches have been carried out about how the cycles of CO_2 and water affect the plugging efficiency of gels on CO_2 during a WAG (water alternated with CO_2) process. The objective of this study is to investigate whether a polymer gel has a different effect on water and CO_2 flow and how the cycles of water and CO_2 affect the plugging performance during the WAG process. Core flooding experiments were carried out using four sandstone cores with different permeability varied from 100 to 1200 md. After determining the permeability of a core using brine and CO_2, the Cr(III)-acetate-HPAM gelant was injected into the core until injection pressure reached stable and the cores were shut in for 24 hours to allow gelation. After that, two cycles of water and CO_2 were injected to determine the effect of gel on the plugging efficiency of water and CO_2 at different injection rates. Results showed that the polymer gel is a selective plugging material that can reduce the permeability to water much more than that to CO_2 during the first flooding cycle. The resistance to CO_2 was only 5% of that to brine. The selective plugging in low permeability cores was more prominent than that in high permeability cores. The cycles had a significant effect to polymer gel plugging performance. With increased cycles, the resistance to both CO_2 and brine significantly decreased. The resistance of the polymer gel to CO_2 became negligible only after two cycles of water and CO_2 was injected for low permeability cores. This paper will be the first time to report that a polymer gel has much less plugging efficiency to CO_2 than that to water. The WAG cycles could significantly decrease the plugging efficiency of a polymer gel to both water and CO_2. The results will help the field engineers to understand why most polymer gel treatments are not very effective or the effective term is very limited.

Introduction

After the first propose of CO_2-enhanced oil recovery (CO_2-EOR) technology in 1952 (Brownscombe et al. 1952), CO_2-EOR technique ushered in the vigorous development. Many techniques have been studied to improve CO_2 sweep efficiency, such as alternating injections of water and gas (WAG), CO_2 direct thickener (Oglesby et al., Martin et al. 1987, Raje et al. 1996, Pipes et al. 2014, Li et al. 2016), CO_2 foam (Raje et al. 1996, Lee et al. 2014, Sang et al. 2014), and chemical gels and so on. The WAG method is the most commonly used among these methods. Crogh et al. (2002) reported that WAG method successfully improves oil recovery for 3.5 MSm^3 in 5 years in Statjord field, North Sea. WAG method could improve oil recovery but it also causes a channeling problem in the long-term WAG injection fields. Channeling could cause early CO_2 breakthrough and decrease the CO_2 sweep efficiency, which significantly decreases oil recovery, therefore, limit the life of the reservoirs.

Some gel systems such as polymer gel, gelled foam (Friedmann et al. 1997), and preformed particle gel (Duan et al. 1992, Green et al. 2003, Hangx 2005, Sun et al. 2017, Wang et al. 2017), were studied in order to solve channeling problems. The polymer gel is one of the most commonly used plugging agents in the fields. Syed et al. (2014) evaluated a polymer gel composed of AO-22 polymer from SNF and Cr(III)-acetate crosslinker. The experiments were conducted using high-permeability cores and fractured cores under 50 psi at 20 to 80°C and a 99% permeability reduction to CO_2 was observed in high-permeability cores and 90% reduction in fractured cores. Seright (2006) studied the performance of Xanthan-Cr(III) polymer gel, resorcinol-formaldehyde gel, HPAM-Cr(III) gel, and colloidal silica gel on resisting water, nitrogen, and CO_2 flow in porous rocks under 900 and1500 psi. The results revealed that the all the gels had higher resistance to water flow than CO_2 flow, and they all had some residual resistance factor loss permanently during WAG process. Syed et al. (2014) also reported the disproportional permeability reduction behavior of HPAM-Cr(III) gel in other porous media, such as polyethylene, sand pack and fused silica. Martin et al. (1987) conducted an evaluation of polyacrylamide-Cr(III) and xanthan-Cr(III) gels using sandstone under 1500 psi at 40.6°C. The results indicated that both polymer gels were only effective for two WAG cycles. Woods et al. (1986) reported that an organic crosslinked polymer gel could stay stable in supercritical CO_2 under 2500 psi at 43 °C for more than 1 month. Some field applications indicated that polymer gels have mixed results on plugging CO_2 flow. Woods et al. (1986) concluded that anionic polyacrylamide gel was only effective for few WAG cycles. They also reported a successful gel treatment in Lick Creek field in Arkansas using organic crosslinked polymer to reduce excessive CO_2 production. Sun et al. (2017) state that the Cr(III)-acetate polymer gel could not only apply in vertical wells but also could be used in horizontal wells. Stevens (2000) reported a successful application of Cr(III)-acetate acrylamide polymer gel to reduce CO_2 production and improve oil recovery in Rangely Weber Sand Unit in northwestern Colorado. Seright (2006) reported a successful gel treatment using acrylamide-polymer/chromium(III)-carboxylate gel in Wertz field in Wyoming. The results implied the gel treatment was effective for nearly two years. Sohrabi et al. (2004) mentioned that gel treatments using chromium crosslinked medium and high molecular weight polyacrylamide polymer were conducted in three wells in SACROC unit in Kelly-Snyder field located in West Texas. An increase of oil recovery and a decrease of CO_2 production were observed in two well, however, another well had no pressure response after 10,000 bbls polymer gel injection.

In this paper, CO_2 flooding was conducted in sandstone cores to evaluate polymer gel plugging efficiency to water and CO_2. The experiments were designed to evaluate the impact of several factors such as WAG cycles, core permeability. This study comprehensively reports the disproportional permeability reduction polymer gel to water and CO_2 in various permeability cores.

Experiments

Materials

Polymer Gel

A Cr(III)-acetate-HPAM polymer gel as shown in **Figure 1** was selected in this study. This polymer gel is composed of SNF AN907 polymer and Cr(III)-acetate. The polymer molecular weight is 7-10 million daltons and the degree of hydrolysis is 5-10%. The polymer concentration was 5000 ppm and the crosslinker (Cr(III)-acetate) concentration was 417 ppm. The carrier fluid was 1% NaCl brine. The gelation temperature was 45 °C and the gelation time was 5 hours. The gel strength was measured using Haake Rheometer which was 9 Pa.

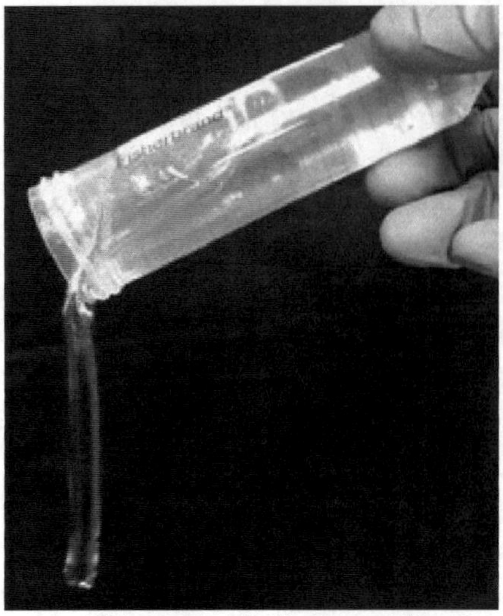

Figure 1. Polymer gel

Sandstone Cores

Four Berea sandstone cores were employed in this study. Core #1 was used to test WAG cycle effect to polymer gel plugging efficiency. Cores #1 to #4 were used to study permeability effect. All the cores were wrapped with Teflon sheets to prevent CO_2 leakage through confining sleeve rubber. Core information was listed in **Table 1**.

Table 1. Core dimensions data

Core #	**Permeability (md)**	**Length (mm)**	**Diameter (mm)**	**Porosity (%)**	**PV (mL)**
1	107	77.76	25.03	18.24	6.98
2	282	53.87	24.63	22.53	5.78
3	431	64.11	25.08	21.48	6.80
4	1225	66.11	25.06	27.55	8.98

Experimental Setup

Figure 2 was the apparatus setup, which was mainly composed of one syringe pump, 2 accumulators, and one core holder. The whole system was heated to 45 °C. One back pressure regulator was utilized to hold the system pressure over CO_2 supercritical pressure (1100 psi) during the WAG process. One of the accumulators was used for supercritical CO_2 injection and another one was used for gelant injection. Two pressure sensors monitored the differential pressure along the cores. A bypass was designed to equilibrate the pressure at the two ends of the cores during the process of pressurizing the system using CO_2. In this design, valve 3 and 4 were open before starting CO_2 flooding, therefore, the differential pressure along the core was zero, which could prevent gel extrusion caused by extremely large differential pressure. After the system pressure stabilized, valve 3 was left open and valve 4 was closed, therefore, CO_2 was injected to the cores.

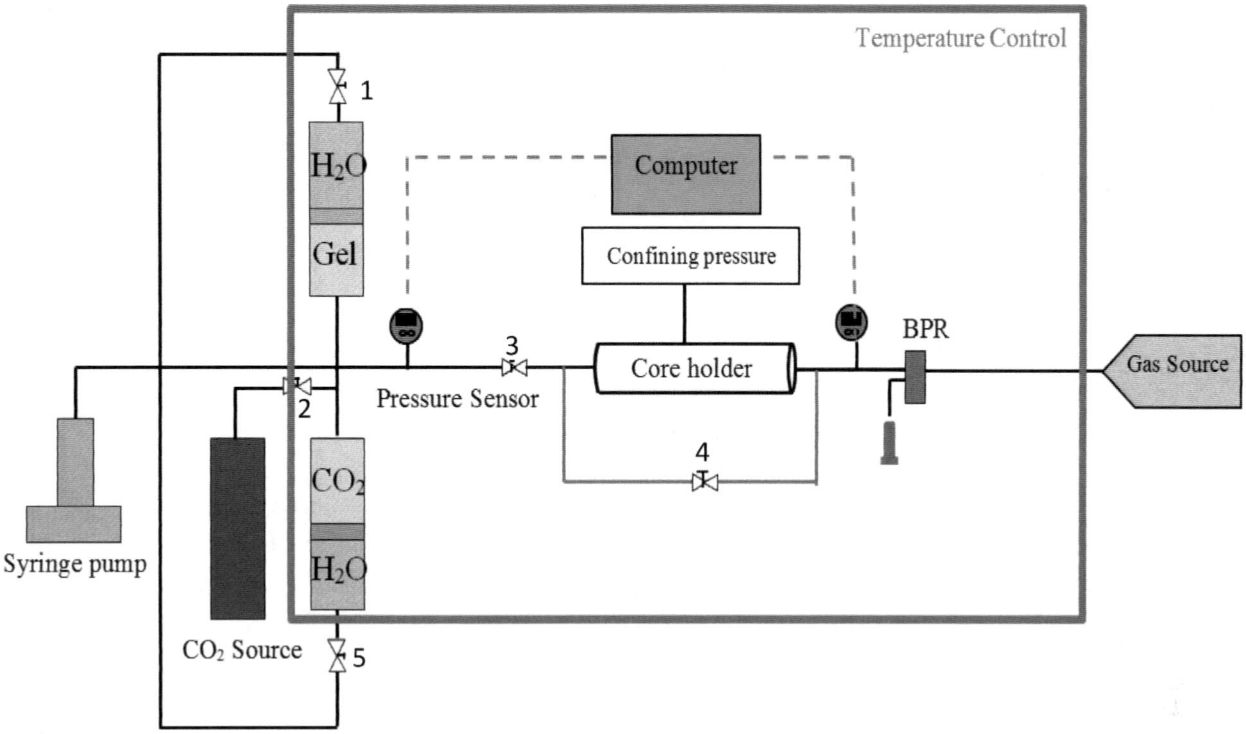

Figure 2. Apparatus setup

Experimental Procedure

1) Absolute permeability measurement

The cores were vacuumed and saturated with 1% NaCl brine. Core absolute permeability was measured using brine.

2) 1st WAG cycle

The WAG cycle started with supercritical CO_2 injection and then followed by 1% NaCl brine. Supercritical CO_2 and water were injected into the cores at constant flowrates: 0.25, 0.5, 0.75, 1 mL/min. The injection was terminated when the differential pressure along the cores stabilized.

3) Gel treatment

The gelant was injected into the cores using 1 mL/min flowrate until the injection pressure stabilized. The cores were put into a 45 °C oven for 24 hours until gelation process completed.

4) CO_2 breakthrough pressure test

CO_2 was injected at a constant pressure for 3 min, if no gas production from the outlet then increased the constant pressure by 1 psi each time. Once the gas produced from the outlet, the injection pressure was the CO_2 breakthrough pressure.

5) WAG cycles

Similar to the 1st WAG cycle, all the WAG cycles started with supercritical CO_2 flooding. The supercritical CO_2 and water were injected at constant flowrates of 0.25, 0.5, 0.75, 1 mL/min until the differential pressure became stable.

Residual resistance factor calculation

The parameter that used in this study to evaluate gel plugging efficiency was residual resistance factor (Frr). It represented the permeability reduction after gel treatment. Since it could reflect the resistance of gel to fluid flow, it was often used to characterize the gel plugging efficiency. It was the ratio of core permeability before gel treatment to the permeability after gel treatment. Under the same flowrate, Frr could be simplified as the differential pressure after gel treatment divided by the differential pressure before gel treatment. Residual resistance factor could be calculated using the following equation:

$$Frr = \frac{k_{before}}{k_{after}} = \left(\frac{P_{after}}{P_{before}}\right)_q \qquad \text{(eq. 1)}$$

Where, P_{before} is the CO_2/water differential injection pressure from 1st WAG cycle, P_{after} is the CO_2/water differential injection pressure from 2nd or 3rd injection pressure.

Results and analysis

CO_2 breakthrough pressure

CO_2 could not flow into the core if the differential pressure along the core was lower than the breakthrough pressure. However, once reaching the breakthrough pressure, the CO_2 would flow through the core and had more contact area with the polymer gel, which could cause gel dehydration and degradation, furthermore decreased the plugging efficiency. Therefore, the CO_2 breakthrough pressure was of great importance when designing gel treatment. **Figure 3** shows the CO_2 breakthrough pressure for the polymer gel in 100 md sandstone core. The breakthrough pressure was measured by injecting CO_2 at a constant pressure for 3 min and increasing the injection pressure by 2 psi each time if no CO_2 production. Once the CO_2 produced from the outlet, the corresponding constant CO_2 injection pressure was the breakthrough pressure. The CO_2 breakthrough pressure, in this case, was 8.5 psi. The breakthrough pressure gradient was calculated by using the breakthrough pressure divided by the core length, which was 33.3 psi/ft.

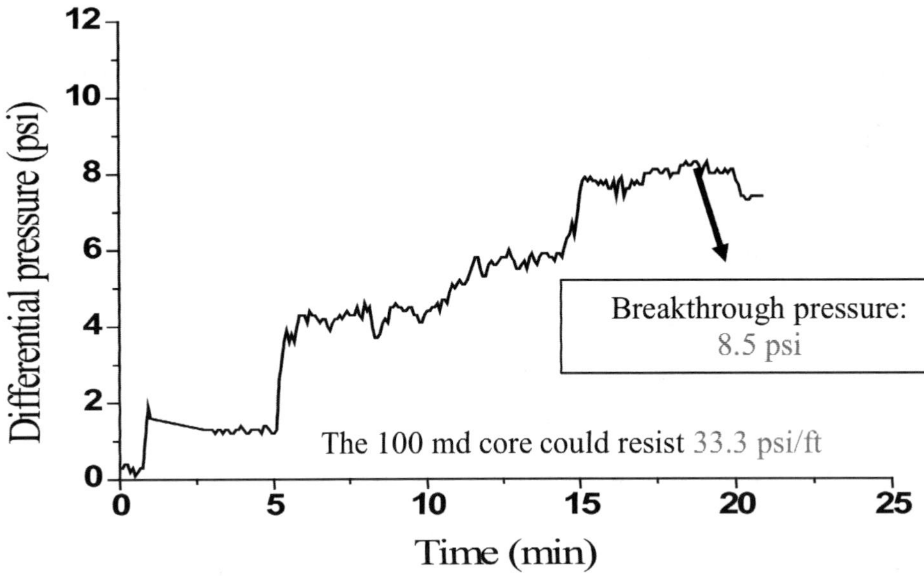

Figure 3. CO_2 breakthrough pressure

WAG cycle Effect

In order to study the WAG cycle effect on polymer gel plugging efficiency, 4 cycles of water alternated with CO_2 were injected into Core #1 (107 md core) after gel treatment. The stabilized CO_2 and water injection differential pressures for all the 5 cycles were plotted in **Figure 4** and **5**, respectively. The figures show that both CO_2 and water injection pressure increased with the increase of injection flowrate. Water injection pressure was higher than CO_2 injection pressure. After gel treatment, both CO_2

and water stabilized differential injection pressure in the 2nd WAG cycle were significantly increased, however, the CO_2 injection pressure decreased dramatically after two WAG cycles, while the water injection pressure decreased gradually with the increase of WAG cycles.

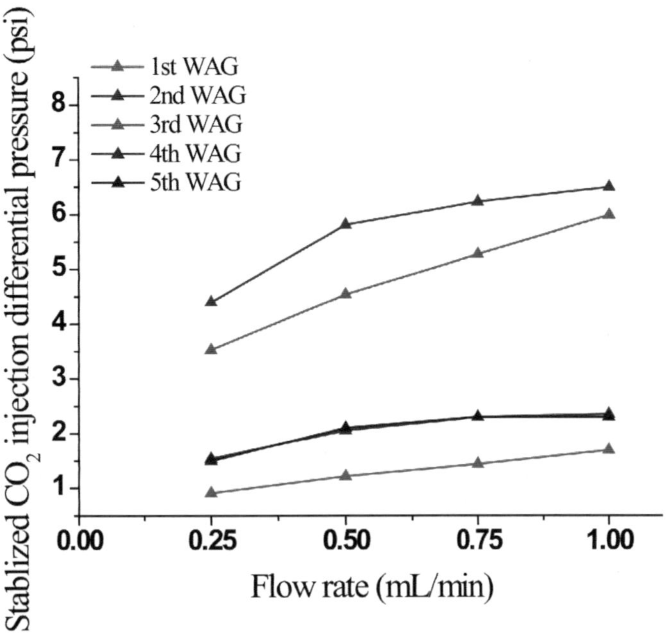

Figure 4. Stabilized CO_2 injection pressure for different cycles

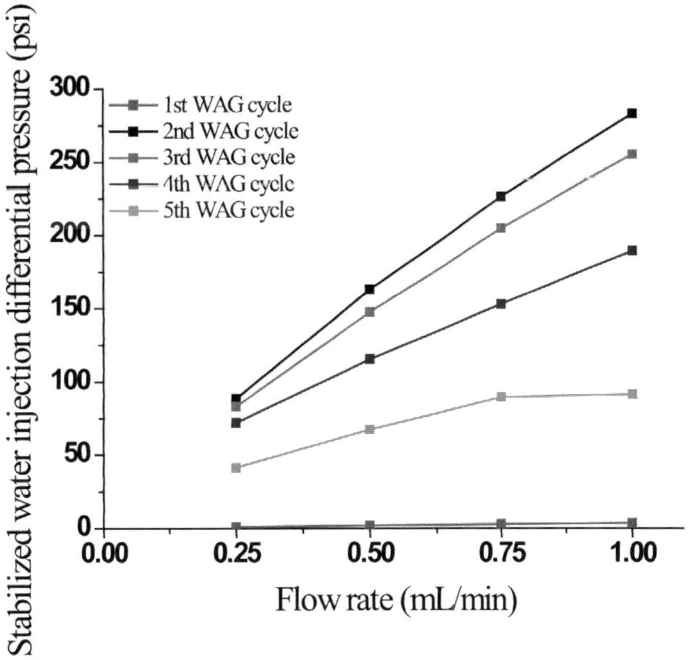

Figure 5. Stabilized water injection pressure for different cycles

The residual resistance factor of CO_2 and water were calculated based on **Equation 1** and the results were plotted in a semi-log plot as shown in **Figure 6**. In the figure, Frrw was about 10 times higher than $Frrco_2$ for all the cycles, which indicated a better performance of the polymer gel reducing water permeability than CO_2 permeability. The figure also illustrated that both Frrw and $Frrco_2$ decreased with the increase of WAG cycles; therefore, WAG cycles had a negative effect on polymer gel resistance to CO_2 and water flow. After two cycles, the $Frrco_2$ decreased to almost 1, which means the polymer gel had nearly no restriction to CO_2 flow, meanwhile, the Frrw decreased to around 50% of the original Frrw.

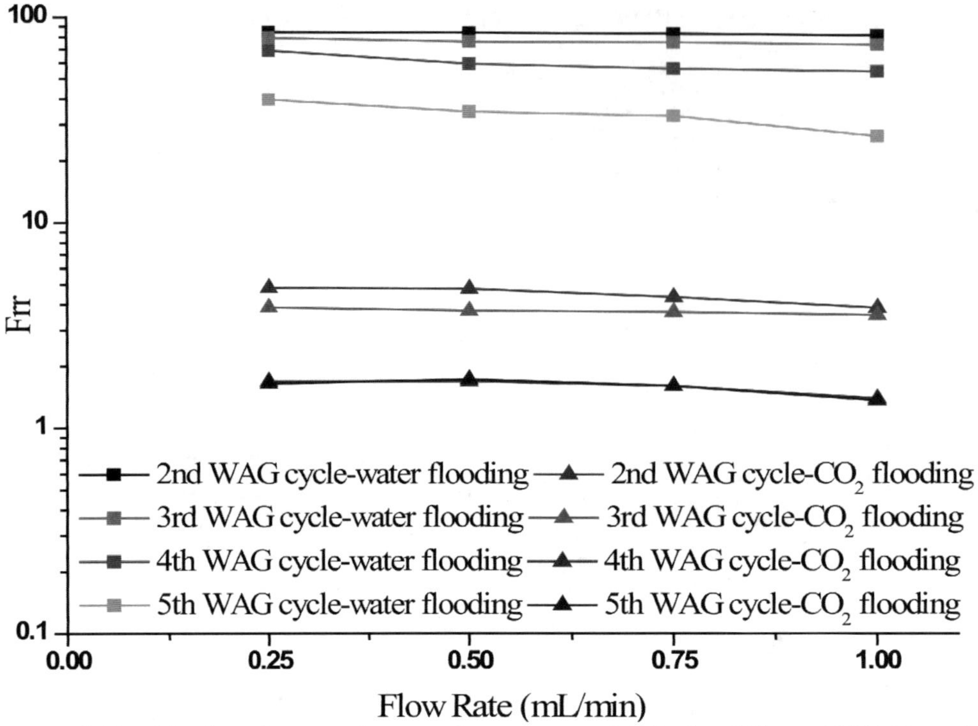

Figure 6. Residual resistance factors for water and CO_2 in different WAG cycles

Permeability Effect

Four Berea sandstone cores with different permeability (107, 282, 431, 1225 md) were used to study the permeability effect on polymer gel plugging efficiency to water and CO_2. After gel treatment, two WAG cycles were conducted. Frrw and $Frrco_2$ were calculated separately for two cycles and plotted in **Figure 7** and **8**. As observed, both Frrw and $Frrco_2$ decreased with increased permeability, which revealed that the polymer gel had a better restriction to water and CO_2 flow in low permeability rocks. As shown in **Figure 8**, the Frrw and $Frrco_2$ in the 3rd cycle are lower than those in the 2nd cycle.

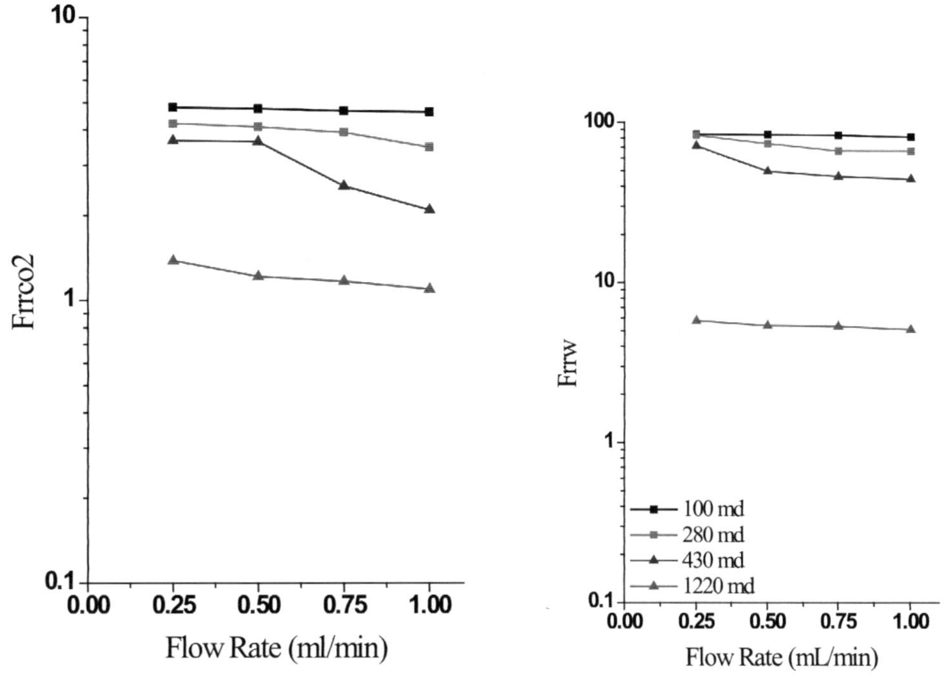

Figure 7. Frrw and $Frrco_2$ for different permeability cores in the second WAG cycle

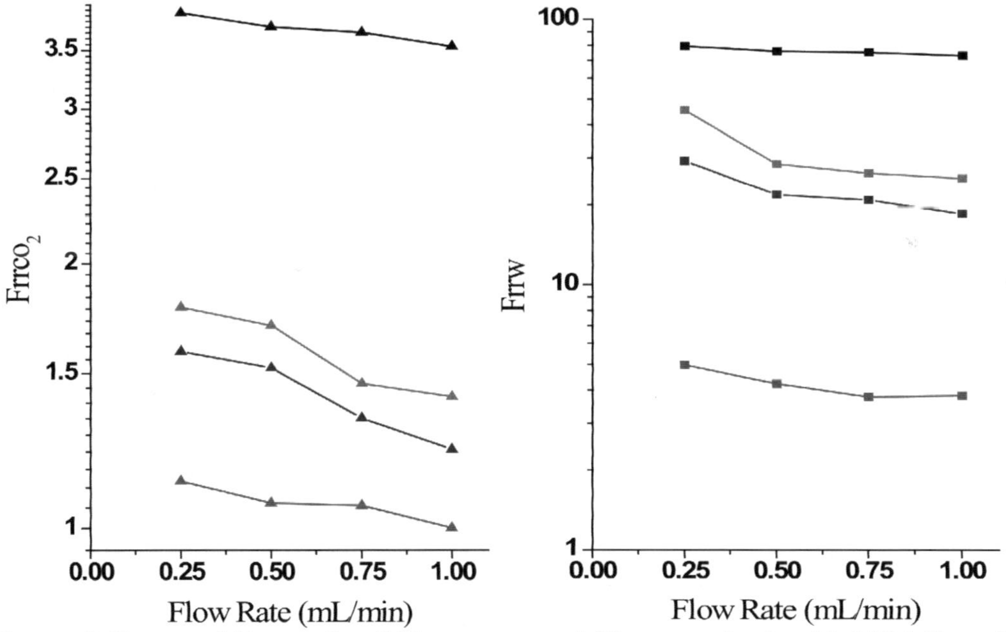

Figure 8. Frrw and $Frrco_2$ for different permeability cores in the third WAG cycle

Discussion

The results in the WAG cycle effect and permeability effect studies indicated that the polymer gel had higher resistance to water flow than to CO_2 flow, which was defined as disproportional permeability reduction (DPR) behavior. In order to evaluate the DPR behavior, the Frrw to $Frrco_2$ ratio (R_{DPR}) was calculated using **Equation 2** for different permeability cores. The Frrw and $Frrco_2$ under 1 mL/min flow rate were utilized in calculating the R_{DPR}. The higher R_{DPR} indicated a more prominent DPR behavior. As indicated in **Figure 9**, the Core #1 (107 md) had the highest R_{DPR}, therefore, the polymer gel had more DPR behavior in lower permeability cores, which manifested a better performance in resist water flow than CO_2 flow. The R_{DPR} decreased with the increase of rock permeability; therefore, the DPR behavior was more distinct in low permeability rock. Since the 1225 md core was too permeable to be plugged by polymer gel, the result of this core was not included in the analyzing. Comparing the R_{DPR} of 2nd WAG cycle and 3rd WAG cycle, the 3rd WAG cycle had lower R_{DPR} than the 2nd cycle. The difference of R_{DPR} of the 2nd cycle and 3rd cycle increased with the increase of rock permeability. Cr(III) crosslinked polymers was not stable in the acidic condition, therefore HPAM-Cr(III) polymer could not resist the acidic environment created by the supercritical CO_2, the polymer gel dehydrated and degraded during CO_2 flooding (Al-Muntasheri et al. 2006). Therefore, the CO_2 injected in the 2nd cycle damaged the polymer gel leaving a larger path in the gel for CO_2 and water flow, which resulted in an increase of R_{DPR}. In the higher permeability cores, the pore throat was larger, therefore, the CO_2 had more contact with the polymer gel, and more serious damage was caused in the higher permeability cores.

$$R_{DPR} = \frac{F_{rrw}}{F_{rrco_2}} \quad \text{(eq. 2)}$$

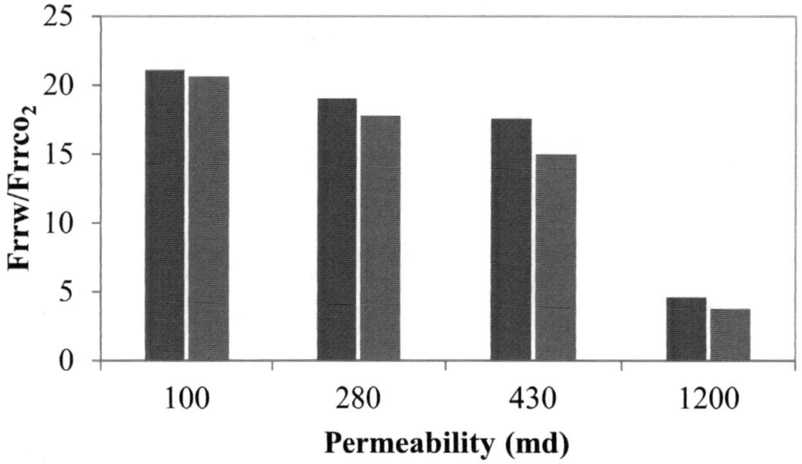

Figure 9. R_{DPR} in different permeability cores

As discussed previously, the polymer gel could not plug the CO_2 flow sufficiently as reported by Woods et al. (1986) that the polymer gel could only effect for 2 WAG cycles, nevertheless, some successful field applications were reported, the results of which can be explained by the reservoir heterogeneity and CO_2 flow diversion. The CO_2 flooding after gel treatment can be interpreted as two stages: no breakthrough and after breakthrough. If no breakthrough, CO_2 only in contact with polymer gel at the cross-section of the core, the migration of CO_2 relies on a slow rate of diffusion, the damage to the polymer gel only limited to the inlet. This is the reason that some field applications are successful. Even the breakthrough pressure is low; the breakthrough pressure gradient is 33.3 psi/ft in 100 md core. In the field, the near wellbore pressure drop might be higher than 33.3 psi/ft but the pressure gradient at in-depth is much lower. Therefore, the CO_2 only caused damage to the polymer gel near the wellbores. The polymer gel in the in-depth could still plug the CO_2 flow and divert it to other layers. However, once breakthrough, CO_2 flows through the polymer gel and the causes more damage to the gel because of the increasing of the contact area, therefore the plugging efficiency decreased dramatically after breakthrough. In order to improve the plugging efficiency, the polymer and crosslinker concentration should be designed properly in order to increase the CO_2 breakthrough pressure gradient thereby to increase the life cycle of the polymer gel.

Conclusions

A set of supercritical CO_2 flooding was conducted using four sandstone cores and polymer gels to study the WAG cycle effect, permeability effect, residual oil effect, and polymer concentration effect. The results indicated that the residual resistance factor decreased for both water and CO_2 with increasing of WAG cycles, and it also decreased with increasing of permeability. The Frrw is higher than $Frrco_2$ which revealed a disproportional permeability reduction behavior of polymer gel. The DPR behavior was evaluated by calculating the Frrw to $Frrco_2$ ratio. The ratio decreased with increased permeability and decreased with increased WAG cycles. Since the polymer gel could not withstand the acidic environment caused by supercritical CO_2, polymer gel dehydration and degradation occurred during CO_2 flooding, which decreased the polymer gel plugging efficiency. If the differential pressure along the core is lower than the breakthrough pressure, CO_2 only damages the polymer gel near the inlet, however, once breakthrough, CO_2 would damage the polymer gel in the matrix, therefore the plugging efficiency would be decreased.

In the fields, the near wellbore pressure drop might be higher than the CO_2 breakthrough pressure, therefore the polymer gel would be damaged near wellbore, nevertheless, the pressure gradient decreased with increasing of the distance to the wellbore, the polymer gel in the in-depth might not be broke through, therefore the polymer gel still can divert the CO_2 to other layers to improve sweep efficiency. In order to increase the life cycle of polymer gel, the polymer and crosslinker concentration should be designed properly.

Acknowledgements

The authors would like to express their gratitude for the financial support provided by the DOE under contract DE-FE0024558.

References

1. Al-Muntasheri, G. A., H. A. Nasr-El-Din, J. Peters and P. L. J. Zitha," Investigation of a High-Temperature Organic Water Shutoff Gel: Reaction Mechanisms," *SPE Journal* 2006 11: 494-504.
2. Brownscombe, E. R., A. B. Dyes and L. P. Whorton (1952)," Method for producing oil by means of carbon dioxide," Google Patents.
3. Crogh, N. A., K. Eide and S. E. Morterud (2002)," WAG injection at the statfjord field, a success story," *European Petroleum Conference, Society of Petroleum Engineers.*
4. Duan, Z., N. Møller and J. H. Weare," An equation of state for the CH_4-CO_2-H_2O system: I. Pure systems from 0 to 1000 C and 0 to 8000 bar," *Geochimica et Cosmochimica Acta* 1992 56(7): 2605-2617.
5. Friedmann, F., T. Hughes, M. Smith, G. Hild, A. Wilson and S. Davies (1997)," Development and testing of a new foam-gel technology to improve conformance of the Rangely CO_2 flood," *SPE Annual Technical Conference and Exhibition, Society of Petroleum Engineers.*
6. Green, C., P. Creel, S. McDonald and T. Ryan (2003), "Utilization of a crystallized hydrating copolymer to modify an injectivity problem in a horizontal CO_2 WAG injector in the South Cowden Unit, Ector County, Texas—Post-treatment coil tubing acidizing stimulation—Case history," *Southwestern Petroleum Short Course, Lubbock, TX.*

7. Hangx, S. (2005)," The behaviour of the CO_2-H_2O system and preliminary mineralization model and experiments," *Ph.D. project report supported by Shell International E&P.*
8. Lee, J. J., S. Cummings, A. Dhuwe, R. M. Enick, E. J. Beckman, R. Perry, M. O'Brien and M. Doherty (2014)," Development of small molecule CO_2 thickeners for EOR and fracturing," *Tulsa, Oklahoma, USA, Society of Petroleum Engineers.*
9. Li, D. X., L. Zhang, Y. M. Liu, W. L. Kang and S. R. Ren," CO_2-triggered gelation for mobility control and channeling blocking during CO_2 flooding processes. *Petroleum Science* 2016 13(2): 247-258.
10. Martin, F. and F. Kovarik (1987)," Chemical gels for diverting CO_2: Baseline experiments," *SPE Annual Technical Conference and Exhibition*. Dallas, Texas, Society of Petroleum Engineers.
11. Martin, F. and F. Kovarik (1987)," Chemical gels for diverting CO_2: Baseline experiments," *SPE Annual Technical Conference and Exhibition, Society of Petroleum Engineers.*
12. Oglesby, K. D., D. D'Souza, C. Roller, R. Logsdon, L. D. Burns and B. J. Felber," Field Test Results of a New Silicate Gel System that is Effective in Carbon Dioxide Enhanced Recovery and Waterfloods," *Society of Petroleum Engineers.*
13. Pipes, J. W. and L. G. Schoeling (2014),"Performance Review of Gel Polymer Treatments in a Miscible CO_2 Enhanced Recovery Project, SACROC Unit Kelly-Snyder Field," *SPE Improved Oil Recovery Symposium*. Tulsa, Oklahoma, USA, Society of Petroleum Engineers.
14. Raje, M., K. Asghari, S. Vossoughi, D. Green and G. Willhite (1996),"Gel systems for controlling CO_2 mobility in carbon dioxide miscible flooding," *SPE/DOE Improved Oil Recovery Symposium*. Tulsa, Oklahoma, Society of Petroleum Engineers.
15. Sang, Q., Y. Li, L. Yu, Z. Li and M. Dong," Enhanced oil recovery by branched-preformed particle gel injection in parallel-sandpack models," *Fuel* 2014 136: 295-306.
16. Seright, R. S. (2006),"Optimizing disproportionate permeability reduction," *SPE/DOE Symposium on Improved Oil Recovery, Society of Petroleum Engineers.*
17. Sohrabi, M., D. Tehrani, A. Danesh and G. Henderson," Visualization of oil recovery by water-alternating-gas injection using high-pressure micromodels." *SPE Journal* 2004 9(03): 290-301.
18. Stevens, S. H., Kuuskraa, V. A., Taber, J. J. (2000)," Sequestration of CO_2 in depleted oil and gas fields: Barriers to overcome in imple-mentation of CO_2 capture and storage," *IEA Greenhouse Gas R&D program.*
19. Sun, X. and B. Bai," Dehydration of polyacrylamide-based super-absorbent polymer swollen in different concentrations of brine under CO_2 conditions," *Fuel* 2017 210: 32-40.
20. Sun,X. and B. Bai (2017), "Comprehensive review of water shutoff methods for horizontal wells," *Petroleum Exploration and Development*, 44(6):1-7
21. Syed, A., B. Pantin, S. Durucan, A. Korre and J.-Q. Shi," The use of polymer-gel solutions for remediation of potential CO_2 leakage from storage reservoirs," *Energy Procedia* 2014 63: 4638-4645.
22. Wang, J., H. q. Liu, H. l. Zhang and K," Sepehrnoori Simulation of Deformable Preformed Particle Gel Propagation in Porous Media," *AIChE Journal* 2017.
23. Woods, P., K. Schramko, D. Turner, D. Dalrymple and E. Vinson (1986)," In-situ polymerization controls CO_2/water channeling at Lick Creek," *SPE Enhanced Oil Recovery Symposium, Society of Petroleum Engineers.*

494517 Valuing Flexibility in CCS-Equipped Power Plants

Abstract:

Today's trilemma between avoiding carbon dioxide emissions, managing cost, and maintaining security of electricity supply requires a delicate balance in power system planning and operation. Common metrics for technologies comparison, such as the levelised cost of electricity (LCOE), assess power generation options in isolation and cannot account for integration and grid-level effects. Similarly, such metrics are unable to value specic performance characteristics such as the ability to provide ancillary services, firm capacity, or flexibility.

In this work, we present a methodology to calculate the "System Value" (SV) of a given technology to the electricity system. The SV of a particular technology is dened as the marginal change in total electricity generation cost (capital, energy, ancillary services) arising from the addition of a capacity unit of that technology. The centrepiece of the SV approach is a mixed-integer linear program (MILP) which simultaneously optimises the electricity system design as well as the hourly unit-wise dispatch schedule. We extend a general unit commitment formulation [1, 2] by including environmental and security aspects in addition to constraints capturing the detailed unit operation [3].

We show that the value of a given technology is a function of the existing capacity mix, i.e., the composition of the incumbent energy system into which it is integrated, and that this value continues to evolve as a function of the changing energy landscape. We identify the value of CCS power plants as the potential to reduce total system cost by £500-800/kW of installed CCS-equipped capacity. CCS power plants with increased flexibility (e.g., w.r.t. start-up time, minimum
stable generation, ramp rate) show the ability to accommodate higher shares of intermittent renewable power generation. This provides additional value to the electricity system by reducing start-up costs and turn-down frequency.

[1] Morales-Espana, G., Gentile, C., Ramos, A. Tight MIP formulations of the power-based unit commitment problem. OR Spectrum, pages 1-22, 2015.
[2] Staffell, I., Green, R. Is there still merit in the merit order stack? The impact of dynamic constraints on optimal plant mix. IEEE Transactions on Power Systems, 31(1):43-53, 2015.
[3] C. F. Heuberger, I. Staffell, N. Shah, and N. Mac Dowell. Levelised Value of Electricity - A Systemic Approach to Technology Valuation. In 26th European Symposium on Computer Aided Process Engineering, volume 38, pages 721-726, 2016.

Development of Potassium- and Sodium-Promoted CaO Adsorbents for CO2 Capture at High Temperatures

Ahmed, Al-Mamoori, Xin Li, Harshul Thakkar, Ali Rownaghi , and Fateme Rezaei

Department of Chemical & Biochemical Engineering, Missouri University of Science and Technology

INTRODUCTION

CO2 adsorption at high temperature has attracted a great deal of attention since it offers various opportunities in processes such as sorption enhanced reaction (SER) and integrated gasification combined cycle (IGCC). Additionally, there is no need to cool flue gas to ambient temperature and thus lose a lot of energy. Double salt adsorbents based MgO has attracted much attention by researchers due to its high capacity, rapid sorption desorption kinetic, and relative stability. However, its capacity limited with the range (300-400°C).To overcome this limitation double salt adsorbents based CaO was selected for this research.

$$CaO + K_2CO_3 + CO_2 \rightarrow K_2Ca(CO_3)_2$$

$$CaO + Na_2CO_3 + CO_2 \rightarrow Na_2Ca(CO_3)_2$$

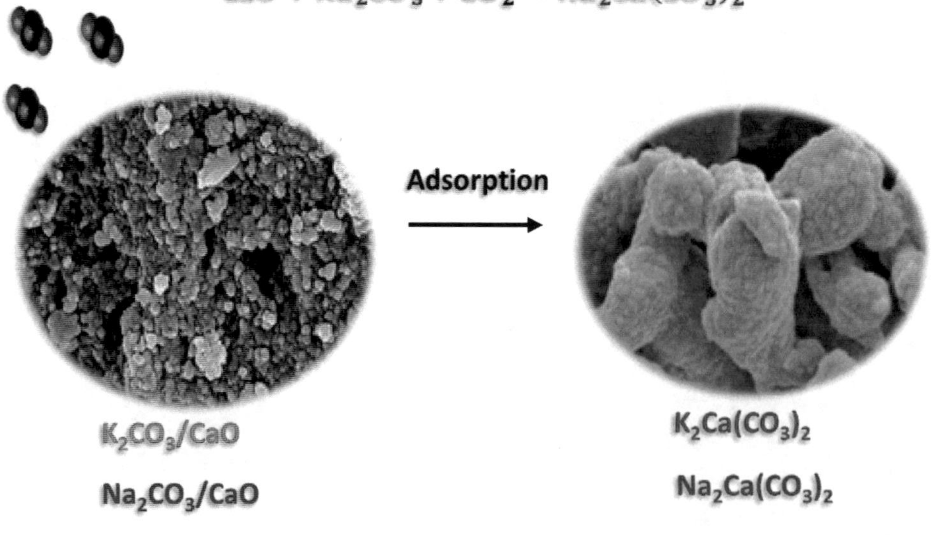

OBJECTIVE

Development of a series of double salts such as potassium promoted calcium (K−Ca) and sodium-promoted calcium (Na−Ca) adsorbents for CO2 capture that could be used in a hybrid material that can capture CO2 (from flue gas) and convert it into valuable commodity such as fuel or chemicals at high temperatures.

RESULTS

- ✓ Enhance CO_2 capacity for the bare CaO at low temperature (375 °C) from 0.4 mmol/g to 3.8 & 3.2 mmol/g when adding the optimum molar ratio of K & Na respectively.
- ✓ The molar ratio of K & Na has significant effect in terms of CO_2 capacity.
- ✓ Increasing the adsorption temperature improved not only CO_2 capacity but also the kinetics.
- ✓ The maximum capacity has been obtained 10.7 and 9.5 mmol/g at 650 °C for K-Ca and Na-Ca adsorbents respectively.
- ✓ The capacity dropped significantly at 700 °C due to the decomposition of the double salt adsorbents K-Ca &Na-Ca at this temperature.
- ✓ K-Ca showed better stability (85%) than Na-Ca (63%) after 10 cycle of CO_2 adsorption-desorption.

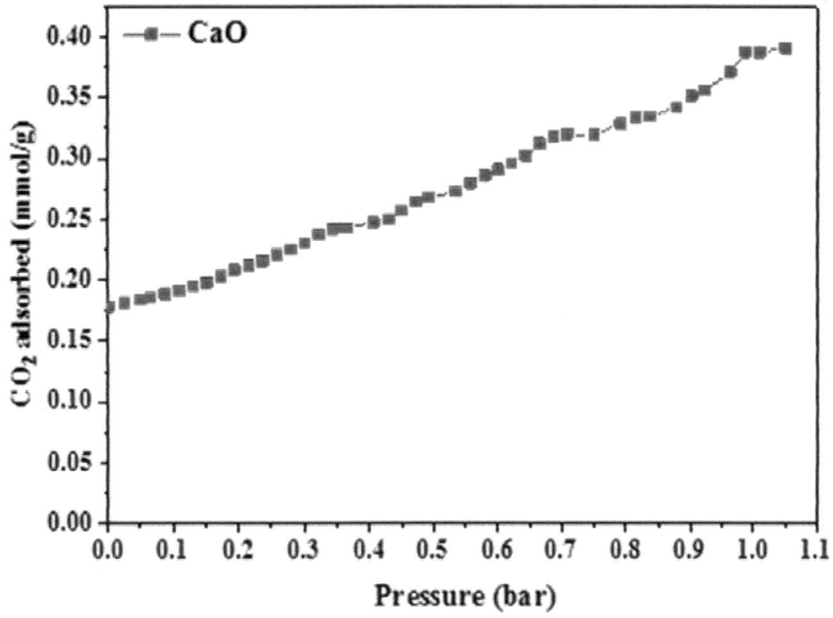

Fig.1 Adsorption isotherm of bare CaO at 375 °C.

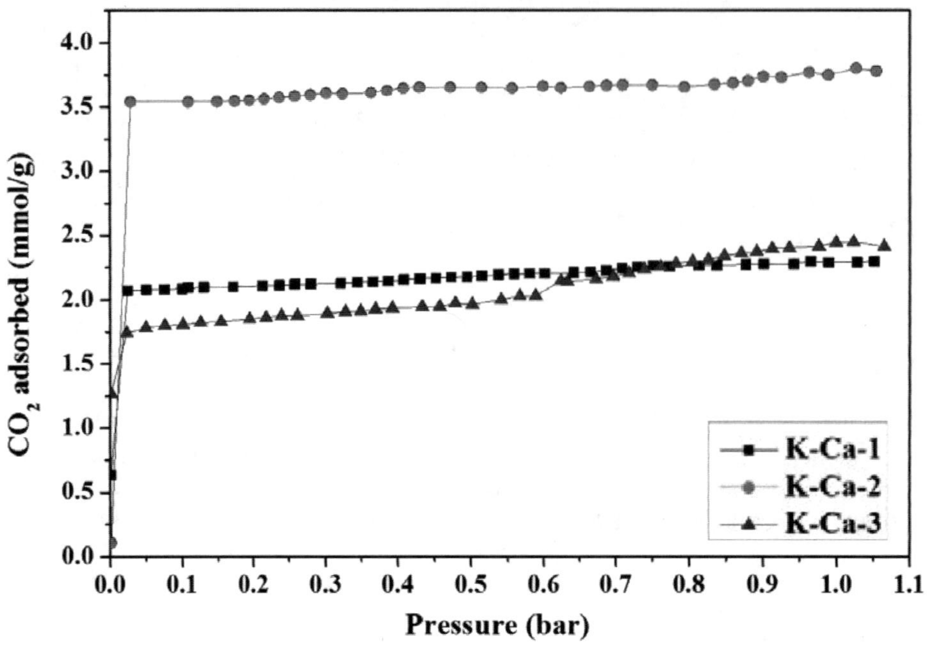

Fig.2 CO_2 adsorption isotherms for K−Ca double salts obtained at 375 °C.

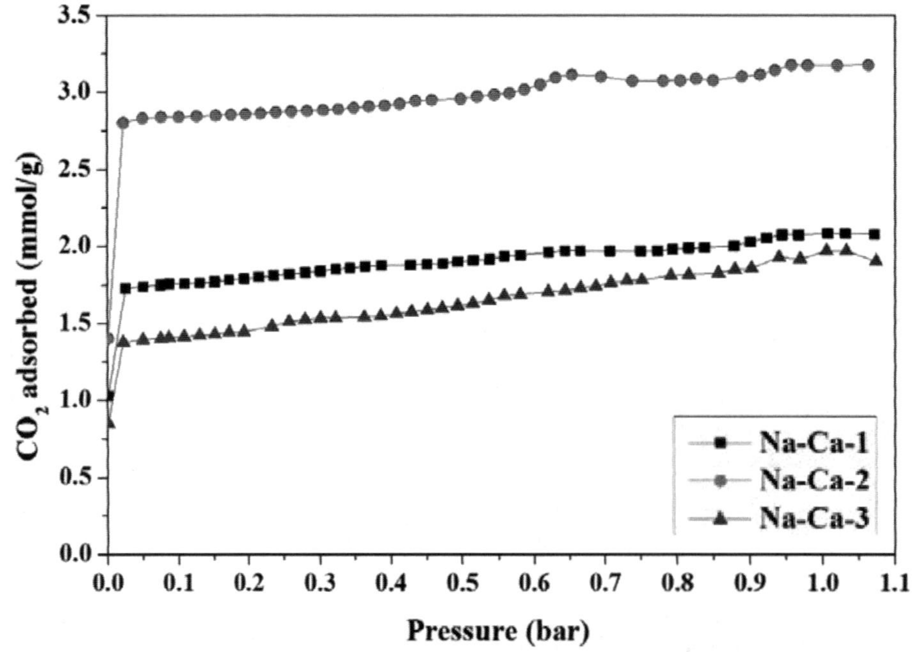

Fig.3 CO_2 adsorption isotherms for Na−Ca double salts obtained at 375 °C.

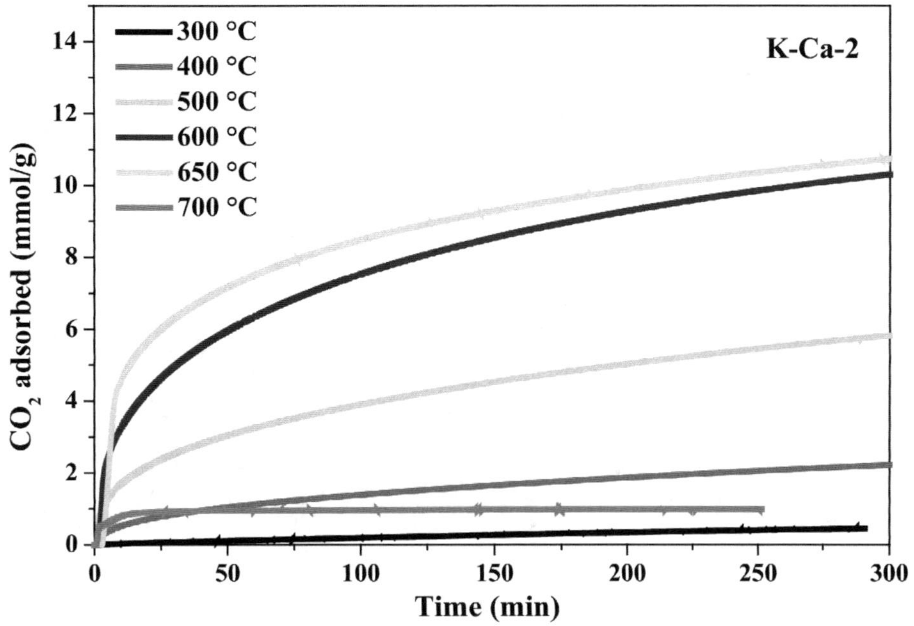

Fig.4 CO_2 uptake over K-Ca-2 double salts as a function of a temperature

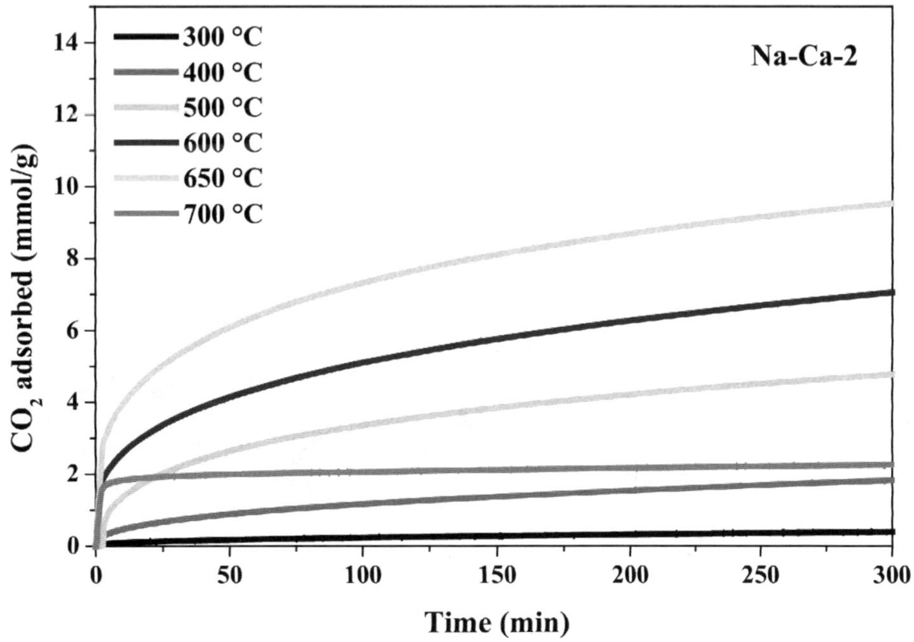

Fig.5 CO_2 uptake over Na-Ca-2 double salts as a function of a temperature

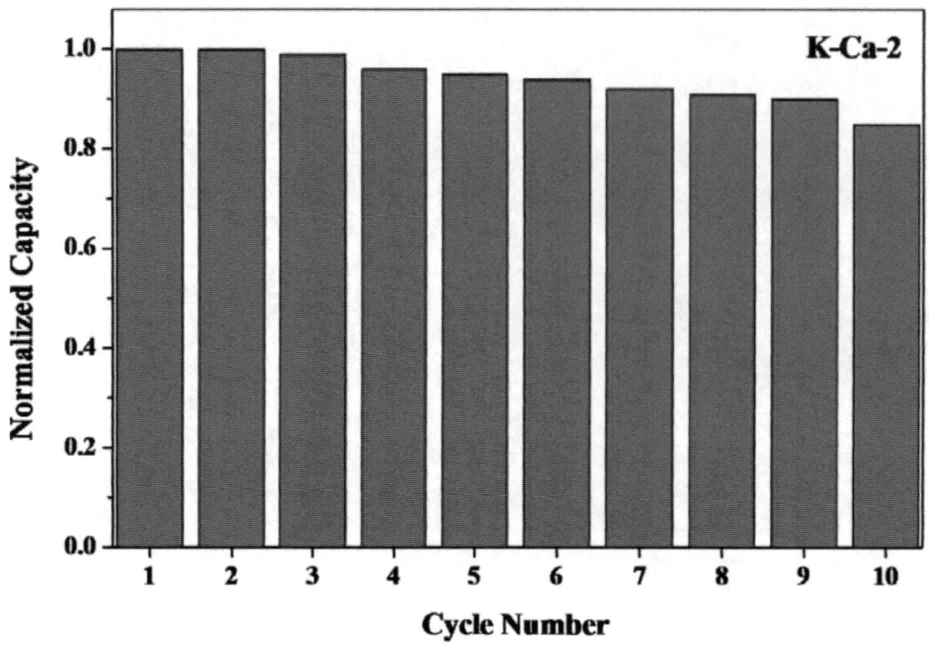

Fig.6 Normalized capacity of K-Ca-2 over 10 cycles at 375 °C

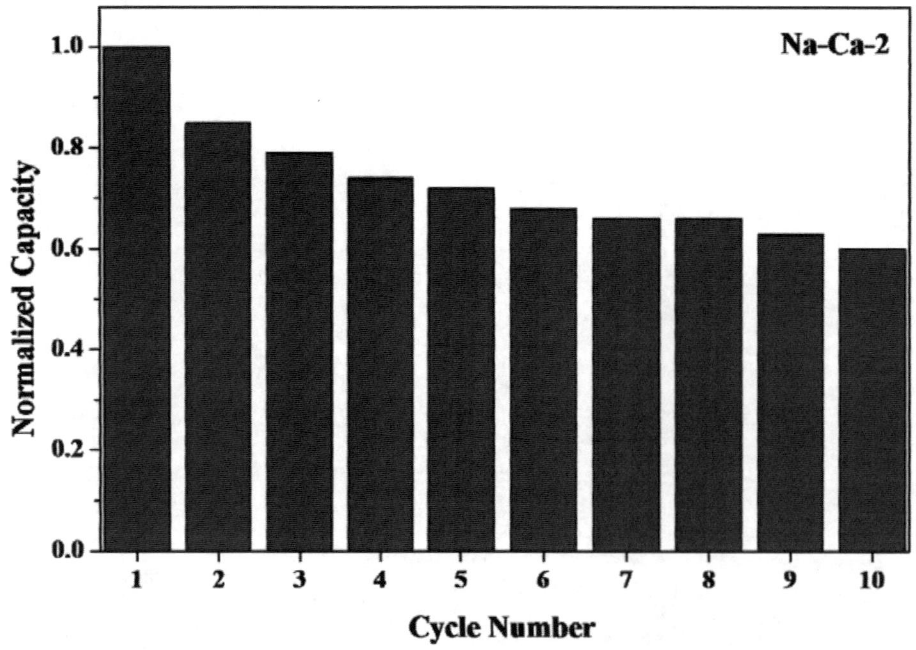

Fig.7 Normalized capacity of Na-Ca-2 over 10 cycles at 375 °C

FUTURE WORK

- Utilizing K-Ca and Na-Ca adsorbents in hybrid materials form (adsorbent-catalyst) for capture and utilization of CO_2.
- Catalytic evaluation of hybrid materials.

REFERENCES

- A. Al-Mamoori, H. Thakkar, X. Li, A.A. Rownaghi, F. Rezaei .Ind. Eng. Chem. Res., 56 (2017), pp. 8292-8300
- Lee, C. H.; Mun, S.; Lee, K. B. Characteristics of Na-Mg Double Salt for High-Temperature CO_2 Sorption. Chem. Eng. J. 2014, 258,367−373.
- Lee, C. H.; Kwon, H. J.; Lee, H. C.; Kwon, S.; Jeon, S. G.; Lee,K. B. Effect of pH-Controlled Synthesis on the Physical Properties and Intermediate-Temperature CO_2 Sorption Behaviors of K − Mg Double Salt-Based Sorbents. Chem. Eng. J. 2016, 294, 439−446.

497406 Inert-Substrate-Supported Tubular Single Cell for Direct Operation on Isooctane

Abstract:

Solid oxide fuel cells (SOFCs) are electrochemical energy conversion devices with advantages of directly utilizing hydrocarbon fuels. Direct liquid isooctane SOFCs show promising application potential due to system simplicity and low operation cost. Nickel-containing anode has been widely used as the supporting layer for SOFC single cells because of its high conductivity and good electrocatalytic property for oxidation of fuel. However, when using hydrocarbons as fuel, the nickel in the anode suffers from performance deactivation due to the carbon deposition in the anode. The interaction between carbon and nickel results in noticeable volume expansion and loss of mechanical strength. Therefore, improving the structure stability of the single cell appears as an important issue for running SOFCs on cheap hydrocarbon fuels

To enhance the mechanical and performance stability of the single cell in hydrocarbon fuels, we designed an inert-substrate-supported tubular cell with a configuration of porous yttira-stabilize zirconia (YSZ) supporter/Ni anode current collector/Ni-$Ce_{0.8}Sm_{0.2}O_{1.9}$ (SDC) anode/YSZ/SDC bi-layer electrolyte/$La_{0.6}Sr_{0.4}Co_{0.2}Fe_{0.8}O_{3-î´}$ cathode. As the YSZ displays good mechanical stability in various fuels in the fuel cell operating temperature, applying the porous YSZ as the supporting layer could enhance the structure and electrochemical performance stability of the cell in liquid isooctane fuel.

The initial electrochemical performance and performance stability of the tubular single cell has been investigated in hydrogen and isooctane as fuel. Multi-functional layers in the tubular cell were modified to reduce the ohmic and polarization resistances and improve the maximum power output. The single cell showed a good maximum power density of 298 mW/cm^2 in hydrogen fuel. When changing the fuel from hydrogen to isooctane, it maintained over 70% of the initial performance obtained in hydrogen. In addition, active reforming catalyst (Ni-Mo-Ceria-Zirconia) was introduced into the tubular channel to internally reform hydrocarbon fuels and increase the performance stability. By applying the catalyst, the single cell exhibited excellent performance stability during 12 h operation in isooctane fuel. The results suggest the feasibility of applying the inert-substrate-supported cell in liquid isooctane fuel.

500798 Molecular Dynamics Simulations of Zeolite Nanosheets for Water Desalination

Abstract:

Reverse osmosis (RO) water desalination could represent an ideal solution to the increasingly limited availability of fresh water. Such an approach to produce clean water using currently available membranes, however, requires high costs and a considerable amount of energy. Discovering new membranes possessing significantly enhanced water permeability while remaining excellent ability to reject salts is critical. Zeolite nanosheets, ultrathin-film materials with precisely controlled nanostructures as well as good chemical, mechanical, and thermal stability, can be potential RO membrane candidates. In this study, molecular dynamics simulations are used to systematically investigate 27 zeolites with a wide range of geometrical characteristics for desalination. Water permeability and salt rejection are determined from the trajectories of water molecules and salt ions through the zeolite membrane. Our calculations show that orders of magnitude enhancement in water permeability can be achieved, and zeolites with properly chosen structural topologies have a salt rejection of approximately 100%. Moreover, our study establishes the structure-performance relationship for zeolite nanosheet membranes in water desalination, leading to important guidelines for the rational design of zeolite RO membranes. Detailed analysis of free energy barriers for transport is also carried out to better understand the desalination performance of zeolite nanosheets. Overall, zeolite nanosheets have been shown computationally to provide opportunities to RO water desalination. The outcomes of this study can lead to important experimental consequences, facilitating the development of novel zeolite nanosheet membranes for water desalination.

Reference:

Jamali, S.H., Vlugt, T.J.H & Lin, L.-C.*, Atomistic Understanding of Zeolite Nanosheets for Water Desalination, *J. Phys. Chem. C*, in press, DOI: 10.1021/acs.jpcc.7b00214.

483566 Prediction of Water Uptake in Ion Exchange Membranes Using Gel Swelling Models

Abstract:

Ion sorption and diffusion in ion exchange membranes are significantly influenced by water uptake in such membranes. Therefore, fundamental understanding of the relationship between polymer structure and water uptake contributes better design of new membranes. However, there have been a smaller number of studies about water uptake prediction model describing the relationship in an ion exchange membrane field compared to the number of studies regarding ion sorption model.

In this work, predictive models for water uptake are discussed using gel swelling models which have been developed for crosslinked polymer networks with ionic functional groups. Flory-Rehner model was firstly introduced to predict swelling of the polymer networks which have ionic functional groups and polymer chains with Gaussian distribution. [1] This model has been modified by Tanaka [2] and Peppas [3] to account for the effects of diluted polymer solution and non-ideality of polymer chains for hydrophilic gels.

Strongly charged cation and anion exchange membranes were synthesized in this study. Water uptake in those membranes were measured as a function of external NaCl concentration within 0.01 M to 1.0 M, and interpreted using the gel swelling models. Water uptake predicted by Flory-Rehner model which accounted for the effect of diluted polymer solution showed good agreement with experimentally obtained water uptake. Furthermore, the model which accounted for non-Gaussian effect in addition to diluted polymer solution showed better prediction. These results suggest water uptake in ion exchange membranes can be predicted by Flory-Huggins interaction parameter between polymer networks and water, crosslinking density of polymer networks, and ion concentration in the membrane. In this presentation, water uptake prediction based on gel swelling model combined with ion sorption model [4] is also discussed.

[1] P. J. Flory, CRC press, 1953.

[2] T. Tanaka et al., Phys. Rev. Lett., 45(20), 1636-1639, 1980.

[3] N. A. Peppas and L. M. Lucht, Chem. Eng. Commun., 30, 291-310, 1984.

[4] J. Kamcev et al., Phys. Chem. Chem. Phys., 18, 6021-6031, 2016.

492020 Modelling Direct-Flow Hollow Fibre Membrane Filtration at Fixed Pump Driving Pressure

Abstract:

Key words: hollow fibre membrane, direct-flow filtration, packing density, membrane fouling, fibre diameter, pumping pressure

Direct-flow (DF) filtration operating in capped hollow fibre membrane (HFM) modules with a regular intermittent backwash has become the dominant technology for potable water treatment. Since the capped-end design ensures no outlet flow from the module, the pressure drop along the feed channel is very small and smaller diameter fibres can be used. Our previous analytical investigation [1, 2] indicated that fibres with large inner diameter and small wall thickness were to be preferred. However the reason for the clear industrial preference for fibre diameters in the tight range 0.8mm to 0.9mm inner diameter (ID) and 1.3mm to 1.4mm outer diameter (OD) was not clear. Therefore, the underlying trade-off between the fibre diameters and other key design considerations, *e.g.* the module packing density was explored. This work presents an MATLAB model for simulating a constant flux, fixed pump-driving pressure DF-HFM filtration. The model was validated against our previous mathematical model resulting from solution of the Navier-Stokes equations. The new model provides a more practical method for investigating the solution space and here we present an analysis of the filtration process, including pressure drops, velocity profiles and the module productivities. Specifically this presentation focuses on the analysis of the effects of the fibre diameter and module packing densities on the overall filtration outputs. The modelling results (Figure 1) show that at fixed pumping pressure of 10 kPa (above outlet permeate pressure), 55% to 65% packing of 1.0mm to 1.2mm OD fibres generate higher outputs than other designs. However, a slightly higher OD range would be preferred if one wants to avoid the sharp decline of filtration throughputs which occur at higher packing densities. Given that the packing density is non-uniform this can be an advantage. Additional investigations that will be part of the presentation. There will include an exploration of the effect of fibre lengths and the effect of imposed flux.

, and fibre OD upon the overall module flowrate, at fixed pumping pressure of above permeate outlet pressure. The module had an inner diameter of 165mm and a length of 1.6m.

[1] J. G. Herterich, Q. Xu, R. W. Field, D. Vella, and I. M. Griffiths, "Optimizing the operation of a direct-flow filtration device," *Journal of Engineering Mathematics,* pp. 1-17, 2016.

[2] Q. Xu, G. K. Pearce, and R. W. Field, "Pressure driven inside feed (PDI) hollow fibre filtraitons: Optimizing the geometry and operating parameters," *Journal of Membrane Science [Submitted for publication],* 2017.

504169 Biomaterials for Tissue Engineering

Abstract:

Advances in biology, materials science, chemical engineering, and other fields have allowed for the development of tissue engineering, an interdisciplinary convergence science. For the past two and a half decades, our laboratory has focused on the development and characterization of biomaterials-based strategies for the regeneration of human tissues with the goal of improving healthcare outcomes. In a collaborative effort with physicians, surgeons, and other scientists, we have produced new material compositions and three-dimensional scaffolds, and investigated combinations of biomaterials with cell populations and bioactive agents for their ability to induce tissue formation and regeneration. We have examined the effects of material characteristics, such as mechanical properties, topographical features, and functional groups, on cell behavior and tissue guidance, and leveraged biomaterials as drug delivery vehicles to release growth factors and other signals with spatial and temporal specificity. This presentation will review recent examples of biomaterials-based approaches for regenerative medicine applications and highlight future areas of growth, such as the use of tissue engineering for validation of cancer therapeutic discovery and synthetic biology.

503772 Process Systems Engineering Contributions in Pharmaceuticals

Abstract:

Healthcare is a major and rapidly growing sector of every developed economy, with 2015 world wide healthcare costs exceeding $7.7 trillion per annum. It is notable that US healthcare costs have risen to 18% of GDP. While the largest portion of the cost is for hospital, clinical, laboratory and physician services, some 10% is on expenditures for medicines. Although a modest portion of the total, the medicine component is receiving heightened attention due, first, to its immediate impact on patient pocketbooks but also to concerns about drug safety, quality and even availability. Our primary focus in this presentation will be in the drug component of healthcare.

Although the process systems engineering community has had some impact on the new drug discovery process, its most notable contributions have been to facilitate improvements in product development and manufacturing. Specifically these contributions have consisted of approaches for improving the design of processes for the production of the active ingredients, the design and development of formulated drug products and the robust operation of associated manufacturing processes. In recent years, the industry and the regulatory agencies have recognized the need for manufacturing innovation, one consequence of which is the accelerating transition from batch to continuous manufacturing. This in turn has driven changes in other components of the product life cycle, from process development to supply chain management. Moreover, this openness to change has stimulated further innovations in manufacturing methods such as 3D- and drop-on-demand printing of dosage, production of film-based products, such as roll-to-roll printing, as well as scaled-down, point of use manufacturing systems. PSE methodologies have important roles to play in all of these innovations.

Beyond these traditional product focused contributions, there also have emerged opportunities to deploy process systems methodology to facilitate the individualized administration of these products in real time or via regularly timed administration of specific dosage amounts. PSE modeling and optimization methodologies deployed in the form of automated closed loop or open-loop physician advisory systems can offer significant benefits to the patient, the clinician and the healthcare system. This bridging between the medicine and the clinical service provider points to further opportunities for PSE methodology in the healthcare services sector.

In this presentation, we will provide an overview of a selection of PSE contributions in these areas, drawing on work from the world-wide community, our collaborators and our own research group.

DEVELOPMENTS OF COAL FIRED POWER GENARATION PROCESSES IN JAPAN AFTER THE FIRST OIL CRISIS (1973)

Shigekatsu Mori, Emeritus Professor of Nagoya University, Nagoya, Japan

Introduction

Forty-four years have passed during my research contribution with coal conversion technologies. It is great honor to present this paper on the developments of various coal fired power generation processes in Japan. Additionally, recent Japanese situations of power generation system are also briefly introduced

1. Atmospheric Pressure Type Bubbling-bed Fluidized-bed Boilers

Most of Japanese coal conversion projects were started after the first oil crisis.
Until 1973, main power generation plants in Japan had been oil fired plants and therefore the replacement of oil with coal were strongly requested at this point. Then our Japanese chemical engineering group started projects to develop the fluidized-bed coal combustion process with desulfurization by lime-stone. Fundamental stages were passed through in a short term and the national project was started immediately. Both 20t/h pilot plant and also 160t/h (50 MW) demonstration plat operations were successfully achieved and then 350MW oil fired power generation plant was replaced by coal fired FBC plant in1995. To keep same power generation capacity as the original oil fired plant, stacked up structure was newly adopted as the coal fired FBC boiler plant as shown in Fig.1.

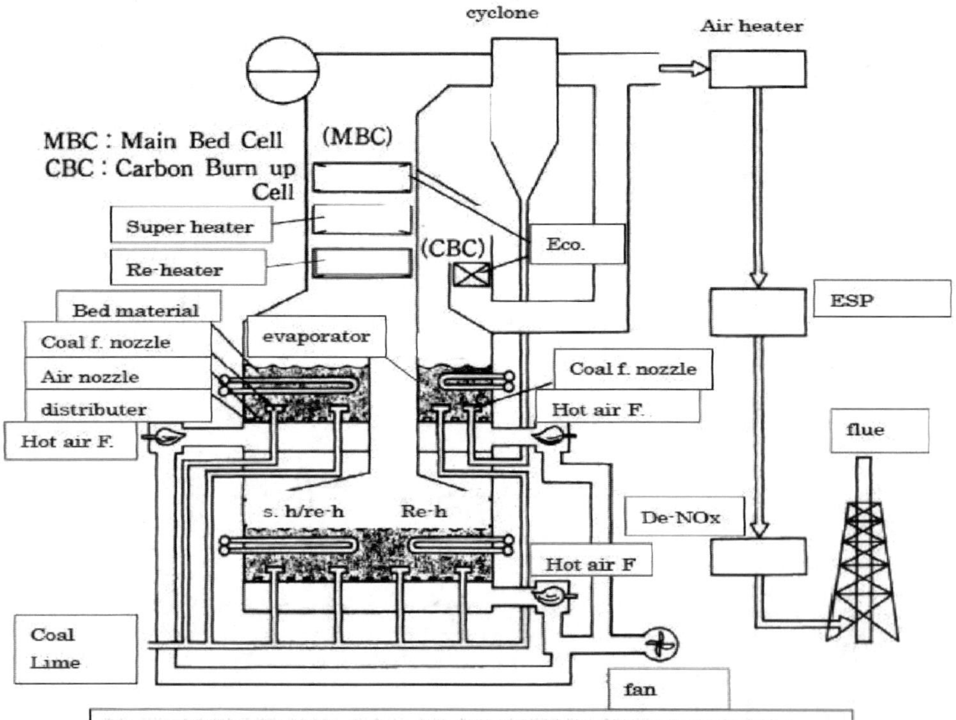

Fig.1 350MW Takehara No2 AFBC Boiler of J-Power

Excellent wasted gas treatment technologies were also established and then three single values were achieved, such as SOx and NOx concentrations in flue gas were less than 10 ppm and particulate concentration was less than 10 mg/m3.

However, around the same time, the global warming problem generated by carbon dioxide emission had become the significant worldwide social problem. Then fossil power efficiency of power generation plants had been emphasized. Since the bed temperature of A-FBC plants were limited less than 1273K, and their steam conditions were limited at the sub critical level, their power generation efficiency were limited less than 40 %. Then the development program of A-FBC coal fired power generation plant was suspended. However, bubbling fluidized bed combustion technologies have been adopted for relatively small-scale incineration plants of municipal wastes. Recently, FBC plants have been also widely expanded as power generation processes of woody biomass in Japan.

2. Development of pulverized coal-fired power generation process

In 1960's, a few pulverized coal-fired power generation plants had been operated and their maximum unit capacity were 250Mw and their steam condition were sub-critical, then their fossil power efficiency(LHV) were 37 to 39%. After the energy crisis, pulverized coal fired power plants were thoroughly developed. As shown in Fig.2, the super critical condition was achieved and their efficiency were 39 to 40% in 1983. Their maximum unit capacity was reached to 500Mw in 1981 and 700Mw in 1988. In 1996, the ultra-super critical condition was achieved and their efficiency were 41 to 43%. Their maximum unit capacity was reached 1000Mw in1990 and now it has been 1050Mw.

More important achievements of their developments were establishment the of air-pollution control processes. Such as, the electrostatic precipitator was adopted as the dust collector, the wet lime plaster method was adopted as the desulfurization process and the dry ammonia selective catalytic reduction was adopted as the denitrification process.

Fig.2 Efficiency of Pulverized Coal Fired Boiler in Japan

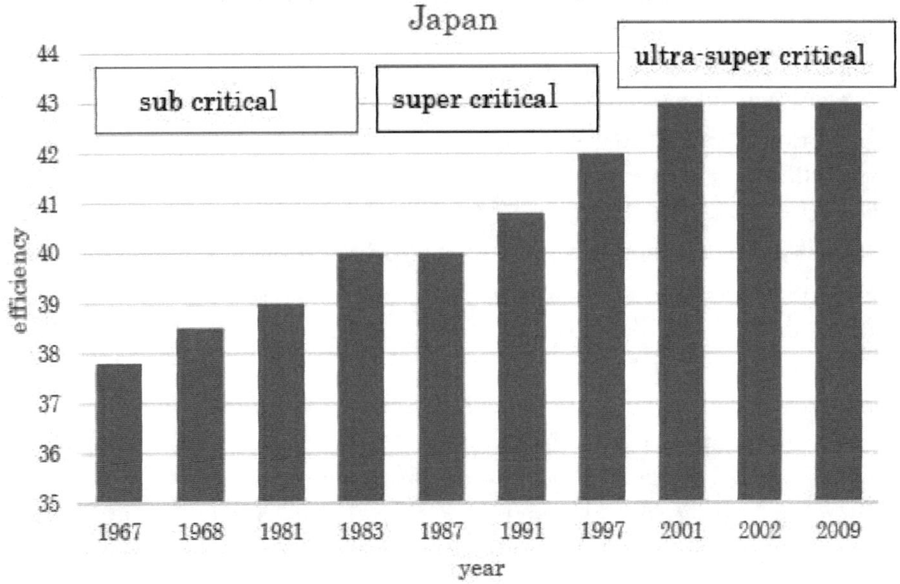

Although their attractive advantage had been its cheapest power generation cost, the most important principal problem of the coal power generation process has been its emission of carbon-dioxide gas. Developments of improved technologies of emission of carbon-dioxide have been still most important issues for any thermal power generation processes in these days.

3.Gas-turbine combined cycle power generation process of liquified natural gas

In 1984, gas-turbine combined cycle power generation process was adopted as the power generation process of natural gas in Japan. The combined cycle was constituted with gas turbine generation process and waste heat recovery steam boiler system. This engineering achievement had been extremely effective to introduce the power generation process of imported liquified natural gas(LNG) from overseas. The thermal efficiency of the gas turbine combined cycle was determined by the gas turbine inlet temperature. In 1984, the gas turbine inlet temperature was 1473 K and its thermal efficiency (LHV) was 48 %. Turbine inlet temperature was limited by heat resistant properties of turbine blades and then aggressive research works on the material development of turbine blades had been continued. Then turbine inlet temperature had been increased and then thermal efficiency had been raised up, such as; turbine inlet temperature was 1633 K and its thermal efficiency(LHV) was 49% in 1994, turbine inlet temperature was 1773 K and its thermal efficiency(LHV) was 58% in 1996, thermal efficiency(LHV) was reached 60% in 1998, turbine inlet temperature was 1873K in 2011 and its efficiency(LHV) was 62%.

4.Entrainment type coal gasification process for integrated combined cycle

Just after the successful operation of gas-turbine combined cycle power generation process of LNG, developments of integrated coal slagging gasification combined cycle(IGCC) were started. Two types of coal slagging gasification processes were developed in Japan, such as the air blown slagging coal gasification and the oxygen blown slagging coal gasification. Dry pneumatic conveying systems of coal were adopted in both plants.

4.1. Integrated air blown slagging coal gasification combined cycle[1)]

Based on the fundamental research with 2t/d process-development unit during 1983 to 1995, 200t/d pilot plant of air blown slagging gasification combined cycle was operated in 1998 to 1995. 1700t/d and 250Mw demonstration unit shown in Fig.3 was designed and constructed. Then this demonstrated operation from 2001 was successfully finished in 2012. From 2013, commercial operation had been started as 250MW Nakoso No.10 plant. Specification of this plant are mentioned as follows, such as: capacity is 250MW gross, coal consumption is approx.1700t/d, gas turbine is 1473K class(50Hz). Efficiency gross is 48%(LHV) and net is 42%(LHV). Flue gas properties are achieved as follows, such as; particulates < 4 mg/m^{3}N, SO_x < 8ppm, NO_x < 5ppm

4.2. Integrated oxygen blown slagging coal gasification combined cycle[2)]*.*

Based on the operation experience of 10t/d test plant (HYCOL plant) during 1990-1993, 150t/day of 2.5MPa pilot plant (EAGLE plant) shown in Fig.4 was constructed and operated in 2002- 2013. This entrainment slagging coal gasifier had two stage oxygen blown gasification burners in the single gasification room. Slugged coal ash was quenched in the bottom quencher by water and discharged. Since this plant was oxygen blown gasification process, both nitrogen and CO_2 concentration in the product gas was limited and then CO and H_2 were enriched.

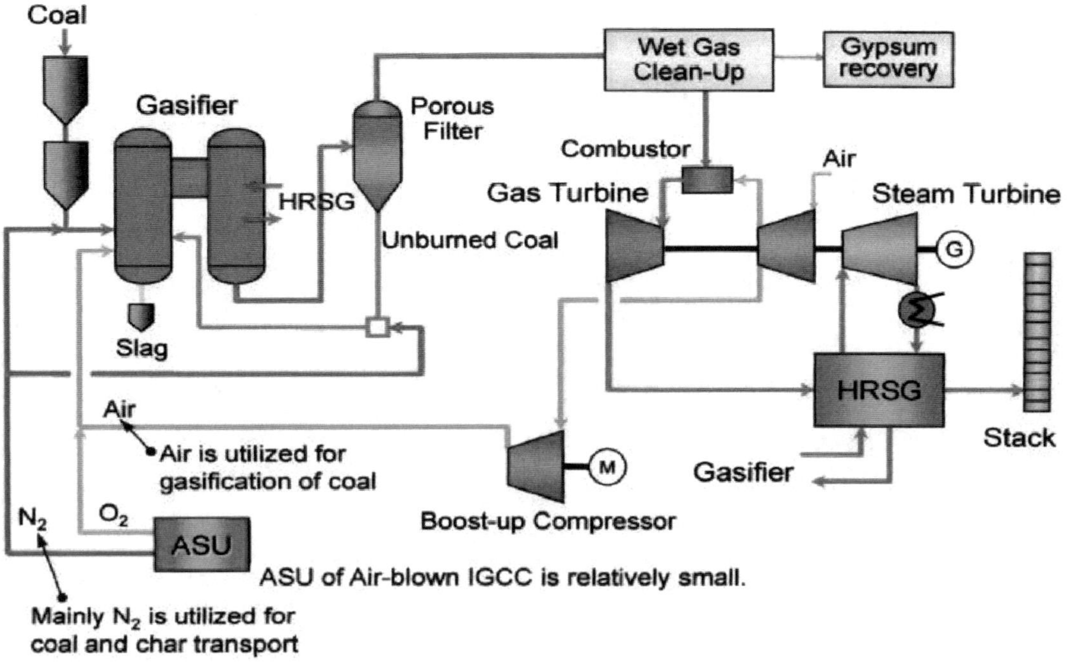

Fig.3 Schematic Diagram of NAKOSO IGCC Plant

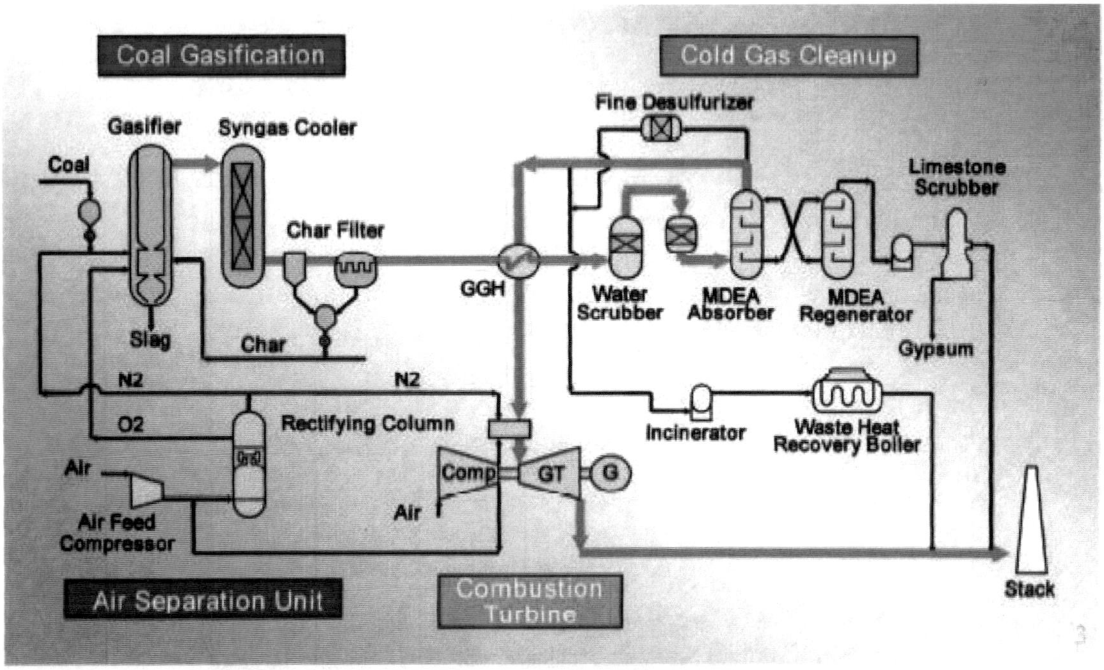

Fig. 4 EAGLE Plant of 150t/d Coal Gasification Pilot Plant

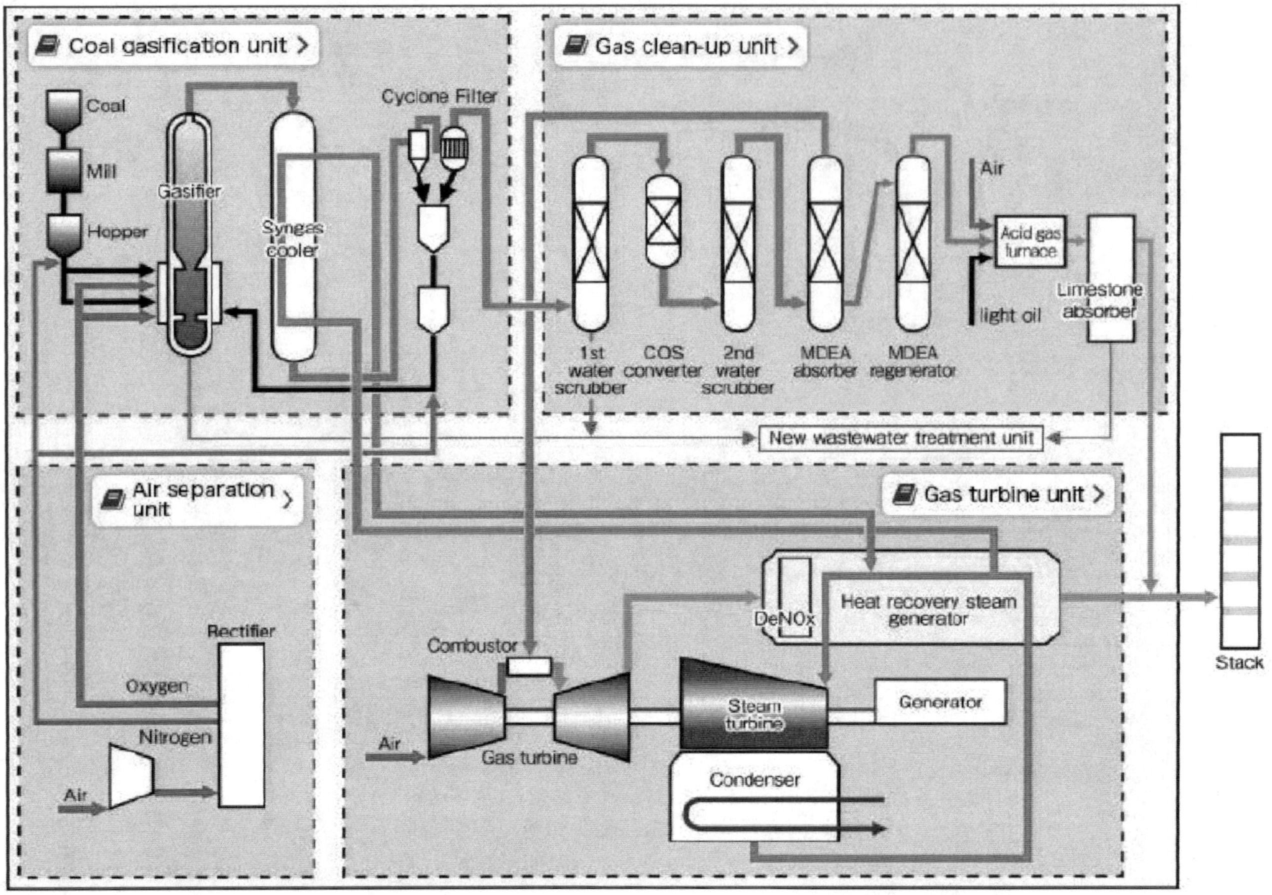

Fig.5 Osaki-Coolgen Oxygen-blown IGCC demonstration unit

Very low emission level of product gas was achieved, such as, SO_x , Halogen and Ammonia were kept less than 1ppm and dust was kept less than $1mg/m^3N$.
Based on these excellent operation experience, 1180 t/d, 166Mw demonstration plant (Oosaki-Coolgen plant)as shown in Fig.5 was constructed and gas purification test facilities were configured with CO_S conversion and removal unit and also with H_2S adsorption unit. The demonstration and semi-commercial operation was started in 2017. As the second stage, the demonstration of CO2 removal and recovery operation will be also challenged to develop the IGFC (Integrated Coal Gasification Fuel Cell Combined Cycle) and also CCS(Carbon Dioxide Capture and Storage). This Process is also applied to hydrogen production unit from coal directly.

5. Recent situations of Japanese power generation system

Here, resent situations of Japanese power system are introduced briefly. In Fig.6, trends of crude oil price(WTI) and Japanese electricity fee during these 44years after the oil crisis. Both oil price and electricity fee sharply increased during 1973~1980 and gradually decreased until 2000. Japanese electricity fee was continuously decreased further until Lehmann shock.

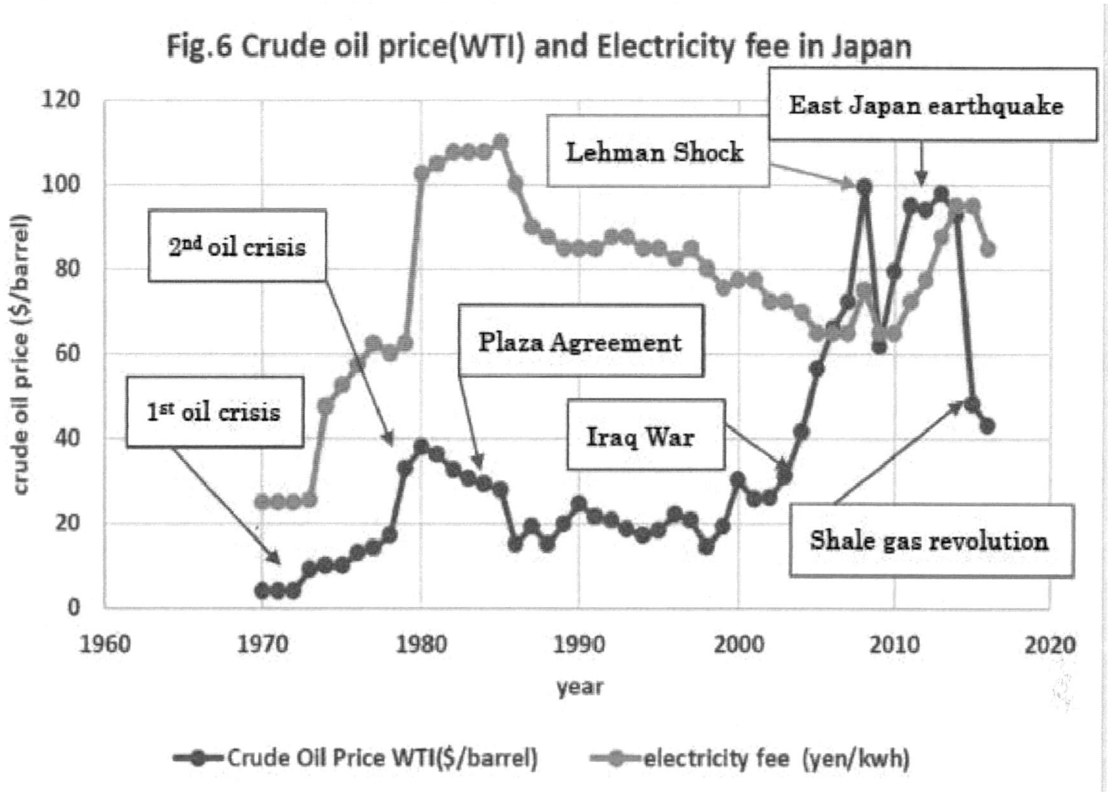

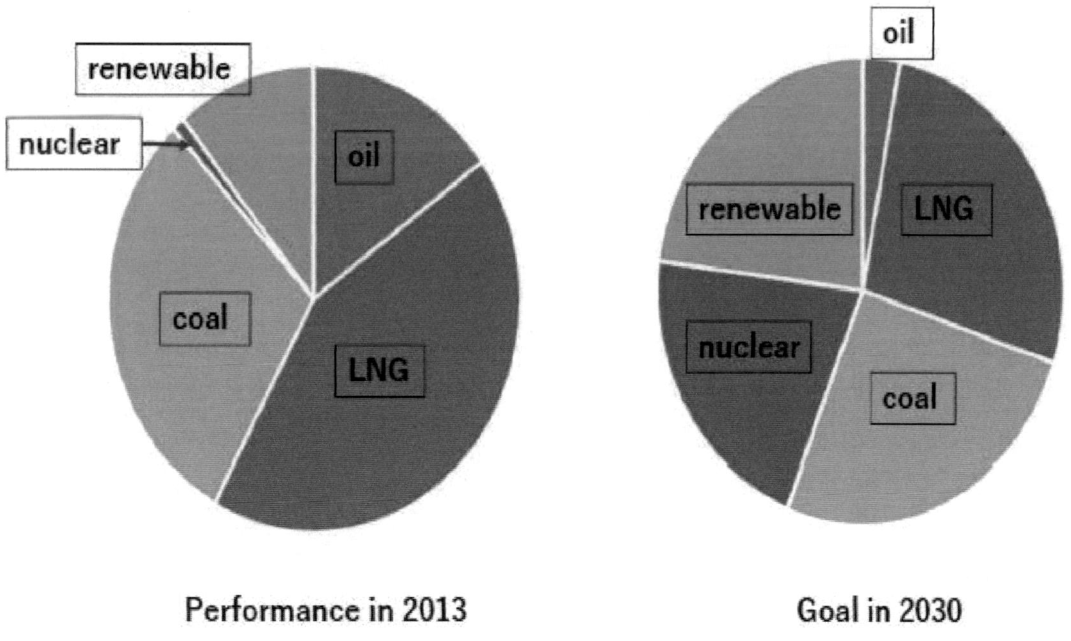

Fig.7 Power supply configuration in Japan in 2013 and its goal in 2013

The oil price was sharply increased again after Iraq War till Lehman shock. Both oil price and the electricity fee were increased again for the reason of East Japan Earthquake and the accident of Fukushima nuclear power plants. However, oil price was sharply decreased again recently by the effect of the shale gas and oil revolution.

In Japan,16 nuclear power plants were presented and total their capacity was 44,264MW, however Fukushima nuclear power plants were completely destroyed by the great tsunami caused by East Japan Earthquake disaster. All Japanese nuclear plants were shut down temporarily and then only three plants have been re-operated now. Therefore, the power supply configuration was affected directly as shown in Fig.7. Although Japanese government has promoted their restart and also determined the Japanese power supply configuration in 2030 as shown in Fig.7, major peoples have been disagreed with their re-start and full operation.

Worldwide uncertainty factors make the future energy situation extremely unclear as mentioned in this paper, then it is very difficult to clearly predict the optimum energy system in the world.

As shown in Fig.6, the world supply-demand relationship of energy resource such as oil, LNG and coal are extremely important factors in Japan since all of these resources are completely imported from foreign countries.

The importance of CO_2 reduction for the protection of the green-house problem have been well recognized and following policies are considered to be effective,

1)nuclear power generation
2)reduction of CO_2 from thermal power generation plant
3)renewable energy
4) import of hydrogen from foreign countries

However, any effective policy has not been adopted yet in japan.

Acknowledgments

Author express great gratitude to AIChE and Society of Chemical Engineering of Japan and also to Professor LS Fan and Professor Nobusuke Kobayashi for their great contribution on the organization of this session.

References

1) Nunokawa, M.: Paper presented in International Conference on Power Engineering (2013)
2) Aiso, K.: Paper presented in Subcommittee of power generation from coal, 9th China-Japan Forums on energy conservation and environment (2015)

494901 Selected Topics in Fluidization Fundamentals and Fluidized Bed Applications â€“ a Presentation Honoring Prof. Shigekatsu Mori

Abstract:

This presentation is especially prepared to celebrate Prof. Shigekatsu Moriâ€™s career long accomplishments in fluidization and fluid-particle systems. In this presentation, selected topics in fluidization fundamentals, including the minimum fluidization velocity and the bubbling bed properties as well as its modeling options in a gas-solid fluidized bed will be reviewed. The presentation will also cover several selected fluidized bed applications in the energy and environmental areas, including a fluidized bed technology which uses various sorbents to capture heavy metals at high temperatures during coal combustion or waste incineration. Finally, a potential fluidized bed technology which uses microwave energy to convert nanoparticle-embedded plastic waste into activated carbon capable of responding to magnetic field will also be introduced.

491050 Fluidized-Bed Drying Process Based on Self-Heat Recuperation Technology

Abstract:

A novel fluidized-bed evaporator for seawater desalination and biomass drying processes was proposed based on self-heat recuperation technology, in which all the heat of process streams is recirculated inside the system. Process simulation results showed that the energy consumption of the proposed process can be reduced to one-quarter than of a conventional process. However, in the case of high moisture content several problems are caused such as scale deposition on the surface of the heat exchanger, agglomeration of fluidizing particles, etc. In the present study, for the development of the drying process using the fluidized bed, evaporation experiments in a fluidized bed were conducted and the agglomeration behavior of the fluidized particles with seawater feed was investigated.

498902 The Role of Pressure Balance in Nonmechanical Device Design

Abstract:

Nonmechanical devices used for both control or automatic operation are increasingly being used in chemical processes. However, for these devices to operate successfully, conducting an accurate pressure balance is critical.

The reason for an accurate pressure balance and how the pressure balance affects nonmechanical device operation will be presented for two types of devices â€“ a nonmechanical L-valve used for control of the solids, and a loop seal for automatic control of the solids.

In the first case, how a misunderstanding of how the L-valve works led to a non-working system will be detailed. For the second case, an application of the loop seal that is essentially 180-degrees different that most loop seals are designed will be discussed. Both system problems and applications will be discussed using detailed pressure balances to illustrate the problems/solutions.

岐阜大学
GIFU UNIVERSITY

2017 AIChE Annual Meeting
October 29 - November 3, 2017
Minneapolis, USA

Particle Technology Forum
Special Session: Celebrating Prof. Mori's Career Long Accomplishments
Wednesday, November 1, 1:55pm-2:15pm

573e
Carbpon Fiber Reclamation from CFRP Waste

Gifu University, Japan
Senior Professor
Hiroshi MORITOMI
moritomi@gifu-u.ac.jp

1

岐阜大学
GIFU UNIVERSITY

Contents

1. Introduction
2. Materials and methods
3. Results and discussion
(1) Saving energy
(2) Strength of rCF and rCFRP
    1) Carbonization (pyrolysis)
    2) Two-stage thermal treatment
    3) Volume fraction of rCF
    4) Fiber length
    5) Fiber strength
    6) Ratio of rCF to vCF
    7) Resin
4. Conclusions

2

岐阜大学
GIFU UNIVERSITY

CFRP demand

CFRP : Carbon Fiber Reinforced Plastic

Carbon fiber

- Low density
- Light weight
- High strength
- Expensive

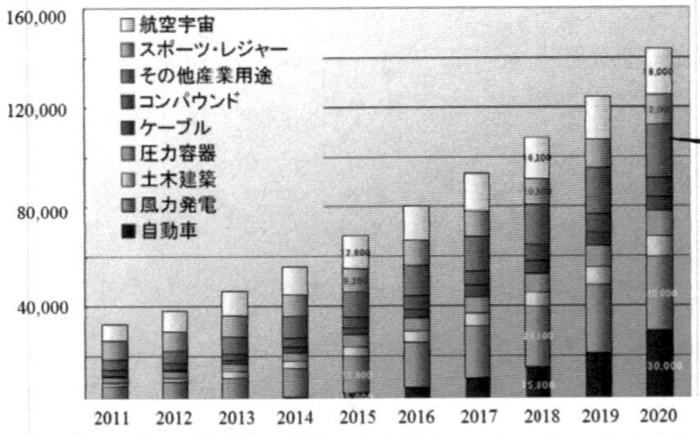

- Production in 2020 will be 3.5 times in 2012
- Increasing demand is for industrial application (wind power and vehicle, etc.)

Demand forecasting of CF production ⇨ If 10% of total virgin CF is recycled, the recycling market will be 15,000 ton/year CF.

- Higher energy consumption for virgin CF production
- Increasing of waste CFRP

Necessary to establish a recycling technology

3

岐阜大学
GIFU UNIVERSITY

Recycling technologies

Japan

Tech	Thermal	Thermal	Solvent	Super-critical	Sub-critical	Super heated	Catalytic	Electrical	Super heated
Organiz.	CFRI	CF Associ	Hitachi Kasei	Shizuoka U	Kumamoto U	JFCC	Shinshu U	Hatinohe U	Doshisha U
Type of resin	All	All	Ester	Ester	Ester	All	All	All	All
Recover	CF (Long & short) Gases	CF (Milled)	CF (Long) Chemicals	CF Resin compounds	CF Chemicals	CF	CF	CF	CF
Capacity	1,800t/y	1,000t/y	12t/y	Lab	Lab	30t/y ?	Lab	Lab	Lab

Overseas

Country	UK		Germany		USA	
Technique	Thermal (Milled)	Micro-wave Fluidized Heated fluid	Reforming?	Thermal	LP thermal LP fluid HT fluid	Micro-wave
Organization	ELG(Recycle Carbonfiber)	University of Nottingham	Fraunhofer	CarboNEXT	Adherant Technologies.	Firebird Advanced Materials
Capacity	1,200t/y	PDU?	Lab	1,000t/y	Pilot	Pilot

4

岐阜大学 GIFU UNIVERSITY

Contents

5

岐阜大学 GIFU UNIVERSITY

Present reclamation process

Recovering carbon fiber as "long fiber"

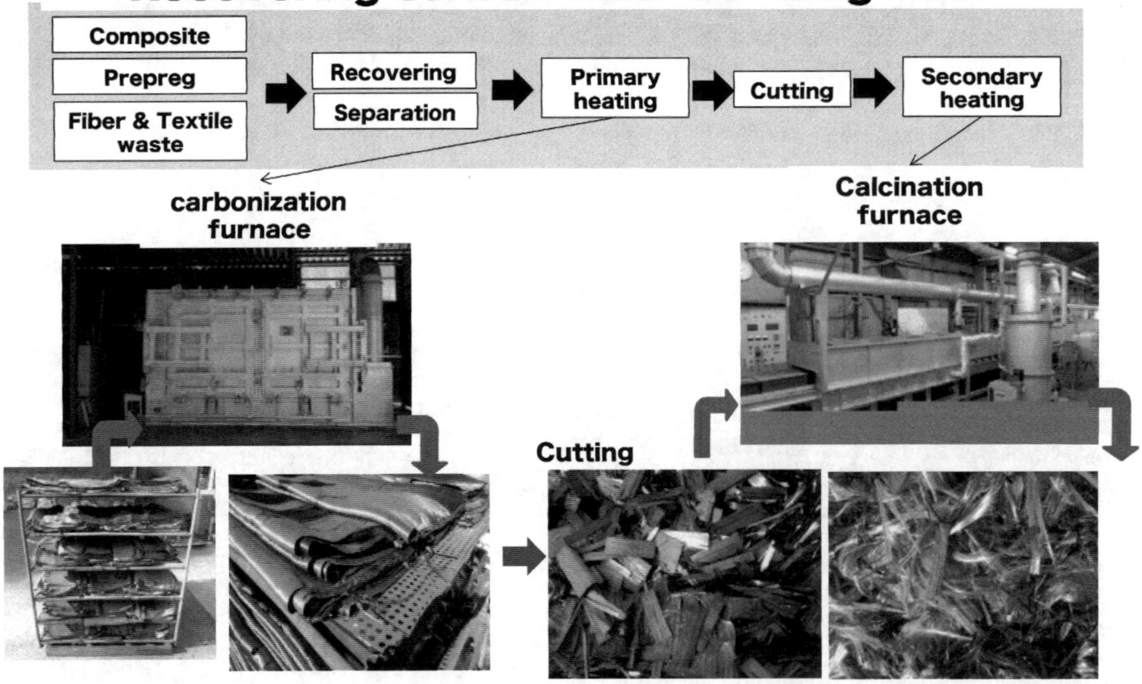

6

岐阜大学 GIFU UNIVERSITY

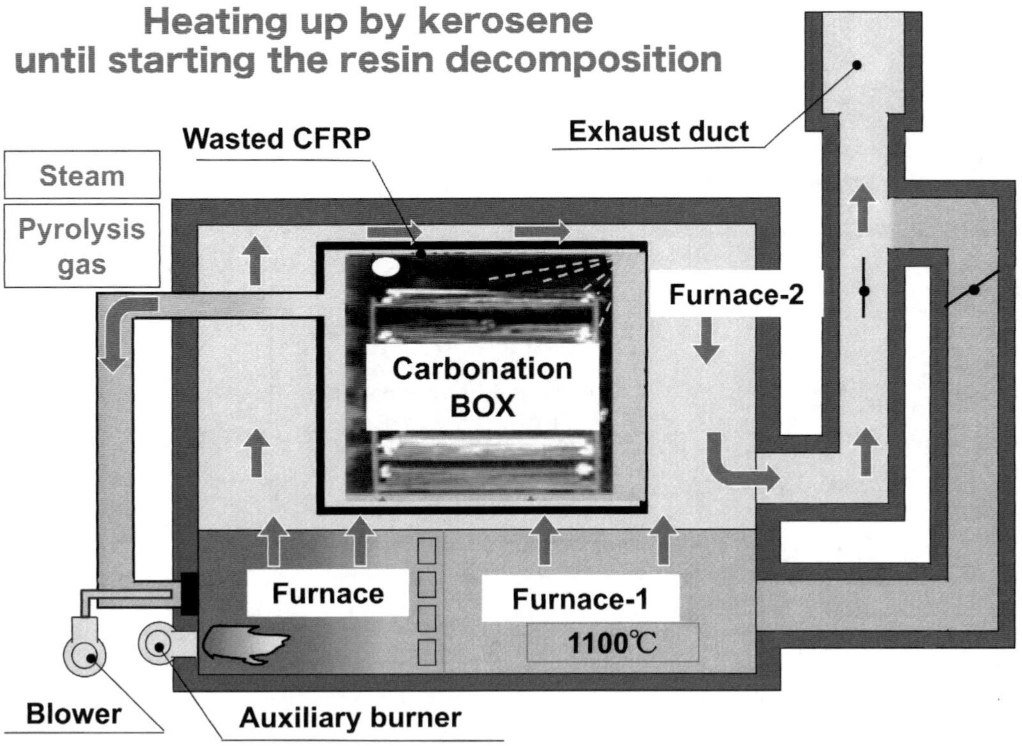

7

岐阜大学 GIFU UNIVERSITY

Secondary furnace (Lab)

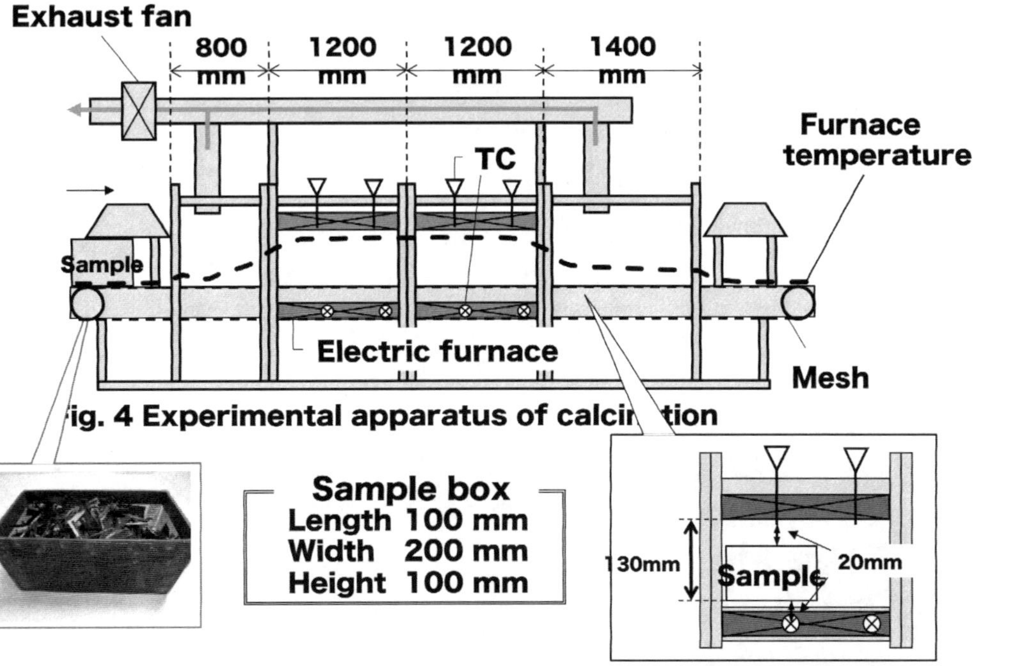

Fig. 4 Experimental apparatus of calcination

Enlarged view of experimental equipment

8

岐阜大学 GIFU UNIVERSITY

Two-staged thermal treatment Process

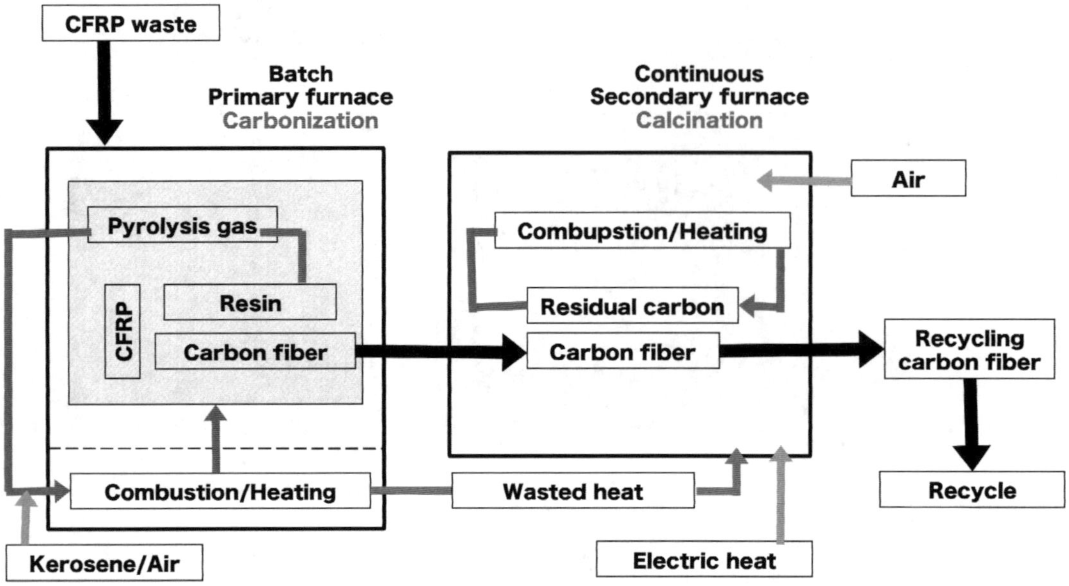

9

岐阜大学 GIFU UNIVERSITY

Contents

1. Introduction
2. Materials and methods
3. Results and discussion

(1) Saving energy

(2) Strength of rCF and rCFRP

- 1) Carbonization (pyrolysis)
- 2) Two-stage thermal treatment
- 3) Volume fraction of rCF
- 4) Fiber length
- 5) Fiber strength
- 6) Ratio of rCF to vCF
- 7) Resin

4. Conclusions

10

Operating conditions and results of carbonization process (CYT-1700)

	Test-1	Test-2	Test-3
Carbonization furnace	CYT-1700 (a unit operation)	CYT-1700 (two units consolidated)	CYT-1700 (two units consolidated)
Inside box	Usual	High temperature box Input/Output	High temperature box Input/output
Charging temperature	100°C	250°C	300°C
Superheated steam condition	200mL/min 0.3Mpa	400mL/min 0.3Mpa	400mL/min 0.3Mpa
Setting furnace temperature	650°C	650°C	650°C
Carbonization time	195min	180min	150min
Heating time to 650°C	30min	15min	10min
Energy consumption	17.3MJ/kg-CF	15.6MJ/kg-CF	12.6MJ/kg-CF

Table 2 Energy consumption of carbonization and calcination processes

	Initial		Present	
	Carbonization (CYT-1700)	Calcination (30m kiln)	Carbonization (CYT-1700)	Calcination (30m kiln)
Feeding	201kg	23.5kg/h	150 kg	35.3 kg/h
Energy consumption	18.2MJ/kg-CF	6.50 MJ/kg-CF	12.6 MJ/kg-CF	2.75 MJ/kg-CF

Ref: 24.6MJ/kg-CF at 2015

Saving energy & low cost

Recycling carbon fiber	Pyrolysis (The Japan Carbon Fiber Manufacturers Association)	Pyrolysis (Carbon Fiber Recycle Industry)	Solvent Extraction (Hitachi Chemical)
Recovering energy [MJ/kg-CF]	46.0	14.6 → 8.5	63.0 (Distillation 38.0)
Materials	Carbon fiber	Steel	Aluminum
Production energy [MJ/kg]	286	23.0	12

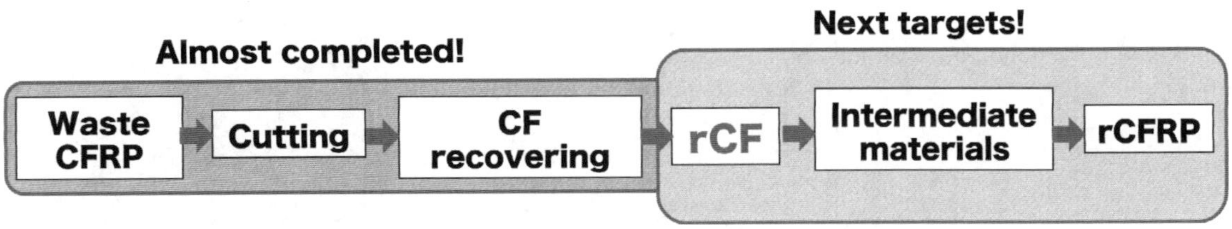

岐阜大学
GIFU UNIVERSITY

Contents

1. Introduction
2. Materials and methods
3. Results and discussion
(1) Saving energy
(2) Strength of rCF and rCFRP
    1) Carbonization (pyrolysis)
    2) Two-stage thermal treatment
    3) Volume fraction of rCF
    4) Fiber length
    5) Fiber strength
    6) Ratio of rCF to vCF
    7) Resin
4. Conclusions

岐阜大学
GIFU UNIVERSITY

Surface observation with secondary heating

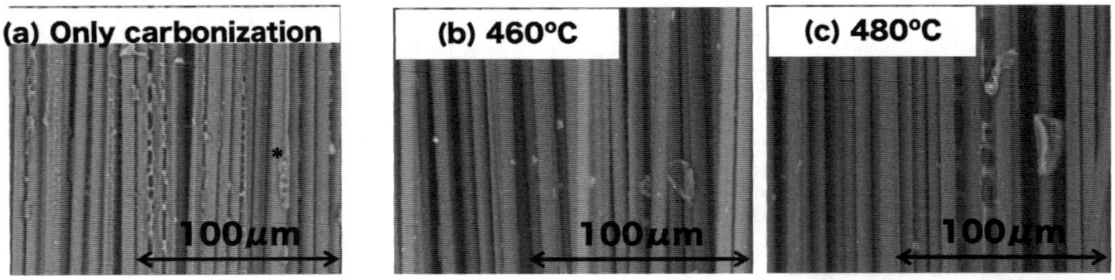

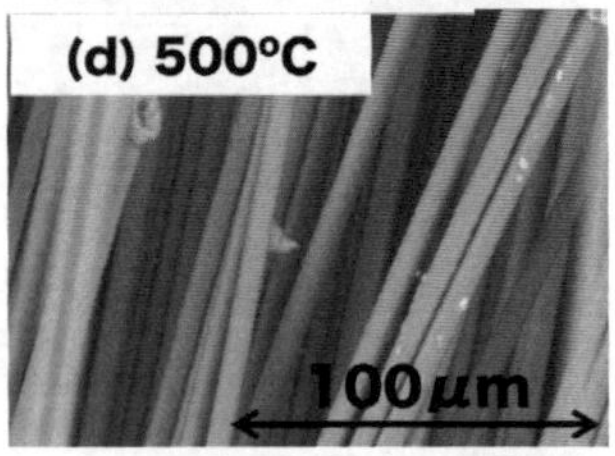

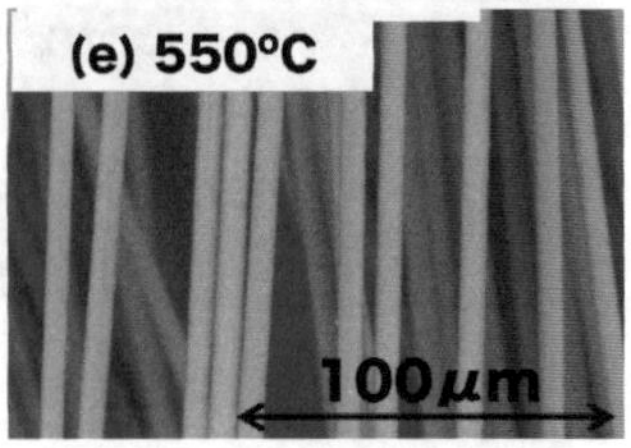

Carbonized CFRP and residual carbon

	Temperature [°C]	Carbonized CFRP [wt%]	Residual carbon [wt%]
(a)	–	100	11.6
(b)	460	93.9	5.5
(c)	480	92.5	4.1
(d)	500	9.06	2.2
(e)	550	87.4	-1.0

** Residual carbon = Carbonized CFRP - 88.*

岐阜大学 GIFU UNIVERSITY

Examination of tensile strength

Weight method

Sample	Recycled CFRP (SampleA)	Virgin CF (SampleA)
1-step Temp	25-600 °C , 210min (400 and 500 °C, 30 min hold)	Non
2-step Temp	460-500 °C 130 min	
Length	90 mm	90 mm

Single fiber

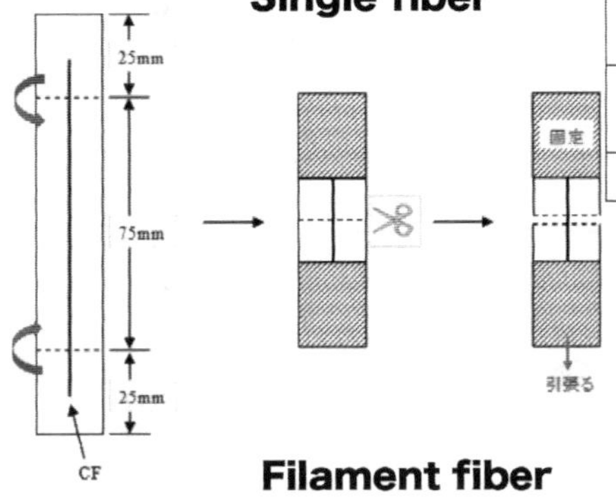

Filament fiber

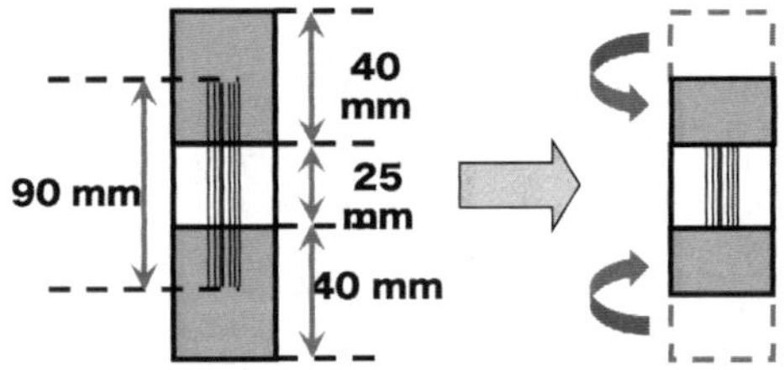

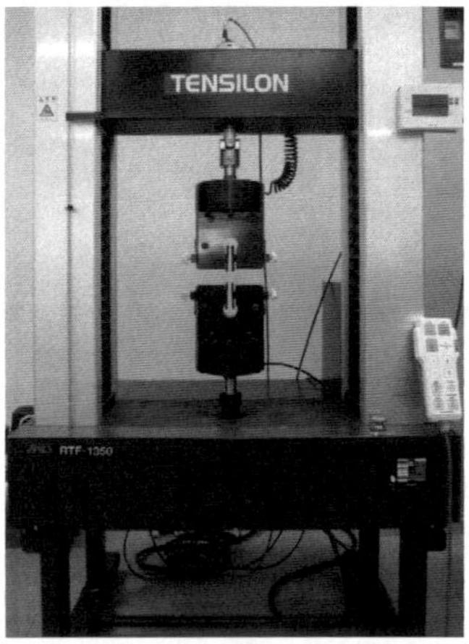

岐阜大学 GIFU UNIVERSITY

Tensile strength(1)

Effect of carbonizing temperature

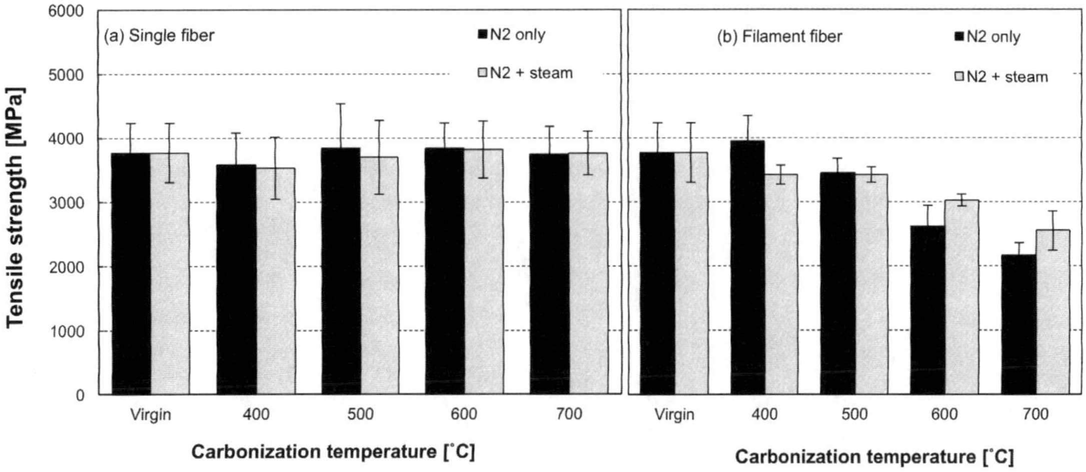

JIS R 7606

岐阜大学 GIFU UNIVERSITY

Tensile strength(2)

Effect of Calcination temperature

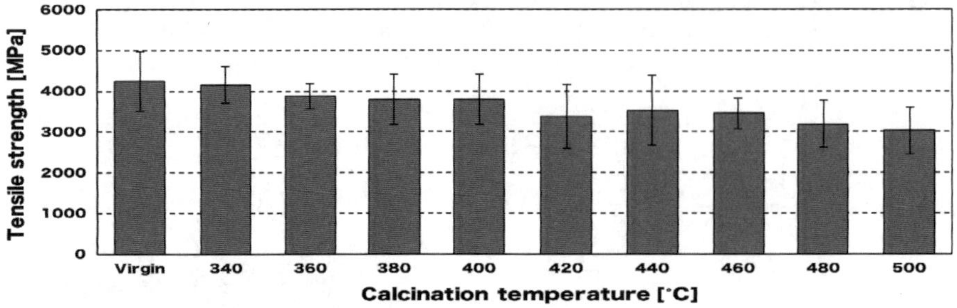

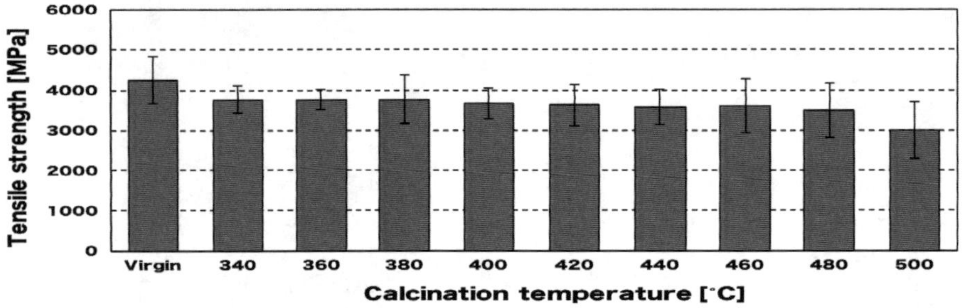

17

岐阜大学 GIFU UNIVERSITY

Contents

18

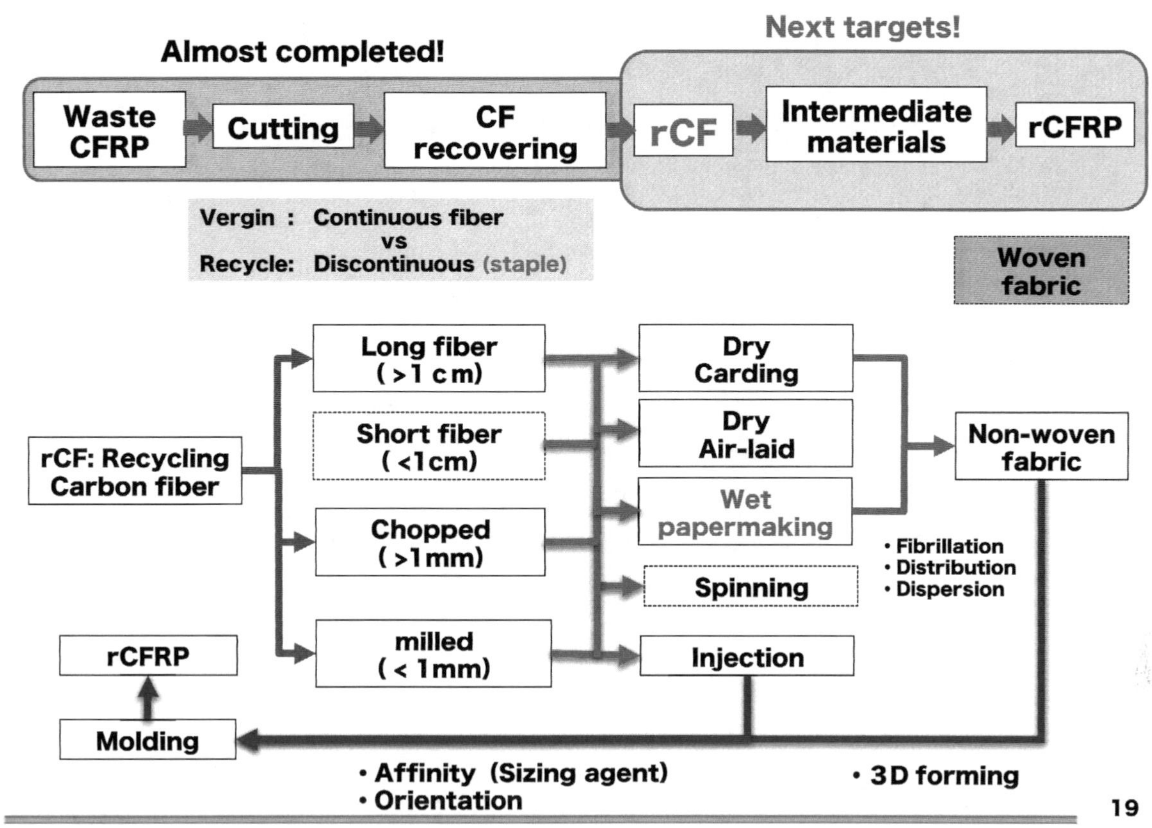

19

Tensile strength(3)

Effect of volume fraction of rCF

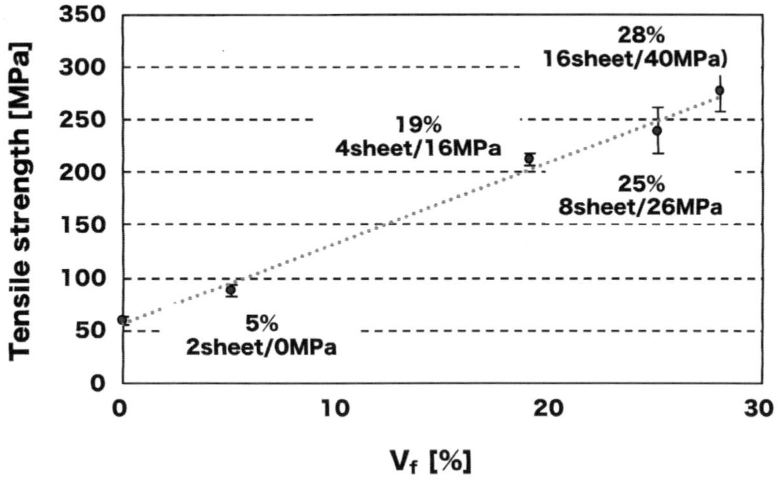

20

岐阜大学 GIFU UNIVERSITY

Tensile strength(4)

Effect of fiber length

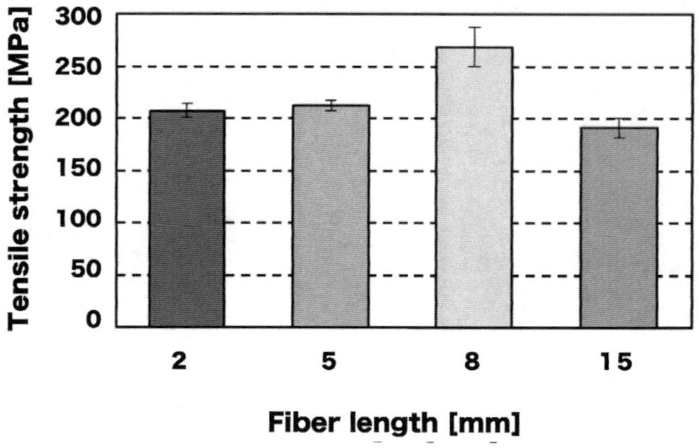

21

岐阜大学 GIFU UNIVERSITY

Tensile strength(5)

Effect of fiber strength

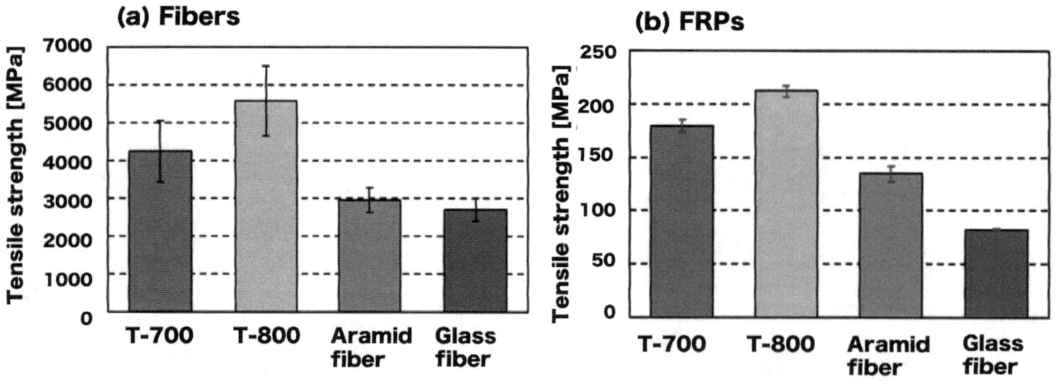

22

Tensile strength(6)

Effect of rCF ratio in paper making

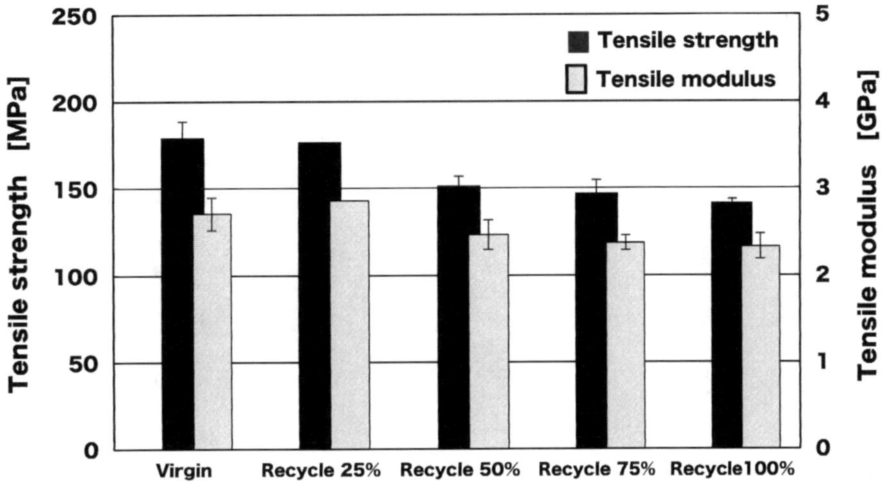

岐阜大学 GIFU UNIVERSITY

Tensile strength(7)

Effect of layer structure of vCF and rCF sheets in paper making

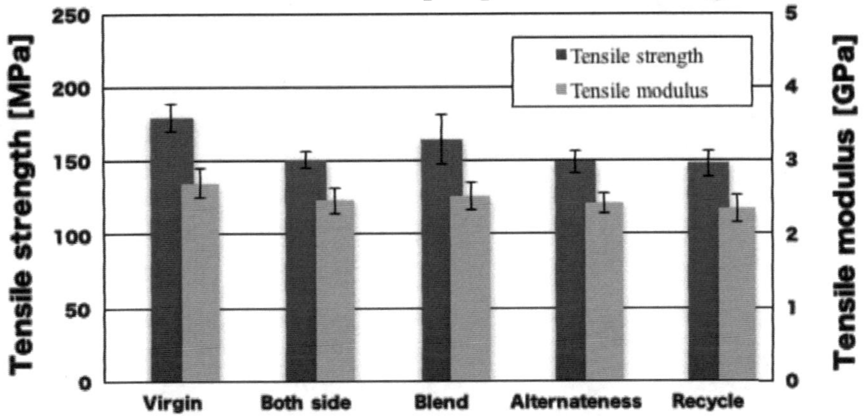

Total 8 sheet
Both side: rCF_4 sheet and vCF_2 sheet x 2
Blend: Mixed sheet with rCF and vCF_1 sheet x 8
Alternateness: Layer_rCF_sheet by layer_vCF_sheet x 4
Recycle: rCF_1 sheet x8

岐阜大学 GIFU UNIVERSITY

Tensile strength(8)

Effect of virgin cross sheet and rCF sheets in paper making

Tensile modulus

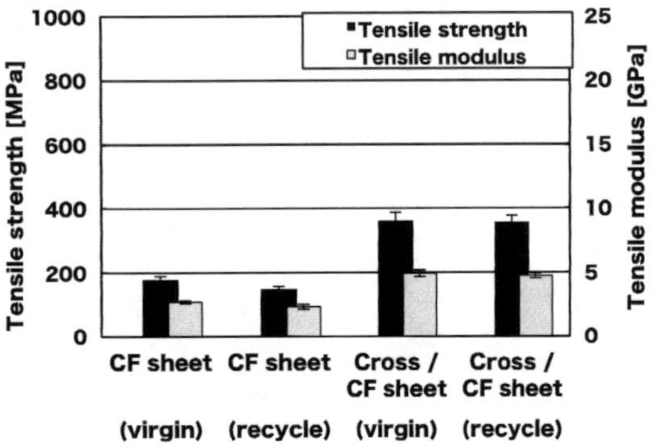

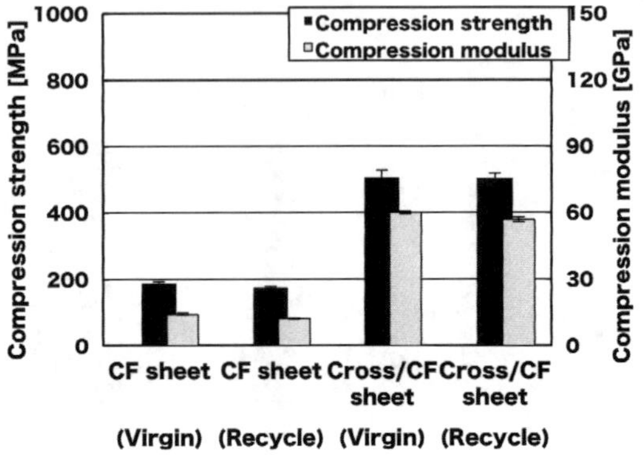

Total 8 sheet
Both side: rCF_6 sheet and cross_1 sheet x 2

25

岐阜大学 GIFU UNIVERSITY

Tensile strength(9)

Effect of reinforcement of rCF for different resins (TSP and TP)

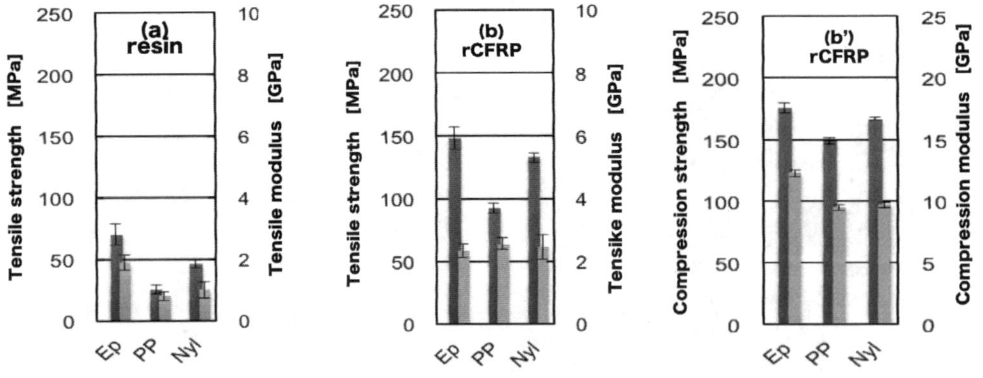

Ep: Epoxy resin(TSP)
PP: Polypropylene resin (TP)
Nyl: Nylon resin (TP)

26

Contents

Conclusions

Establishing high efficient thermal treatment method for recycling carbon fibers from CFRP

- **Achieved less than 10MJ/kg -CF of fuel consumption**
- **Achieved more than 80% strength of virgin CF**
- **Cleared the non-woven rCFRP characteristcs**

Acknowledgement

This paper is based on results obtained from a project subsidized by the New Energy and Industrial Technology Development Organization (NEDO). We would like to express the gratitude here.

岐阜大学
GIFU UNIVERSITY

Thank you
for your warm support!

Senior Professor
Hiroshi MORITOMI
moritomi@gifu-u.ac.jp

29

499412 A Nature-Inspired Approach to Aid the Understanding and Improve the Performance of Fluidized Beds

Abstract:

Gas-solid fluidized beds remain fascinatingly complex, despite many decades of research and industrial use. Chosen for their desirable mass and heat transfer properties for processes as diverse as drying, chemical production, waste-to-energy, fuel combustion and gasification, the complex fluid and particle dynamics are notoriously difficult to model, leading to a predominance of empirical approaches for scale-up and design.

Our research has focused on ways to structure fluidized beds, imposing more uniformity in mixing and bubble patterns, so that their behavior becomes more predictable, and they become easier to operate and scale-up. Biology provides us with wonderful examples of efficient, scalable hierarchical transport networks, such as the airways of the lungs and the vascular network, while regular sand ripples on dunes and beaches form despite the complexity of granular dynamics. This has inspired, respectively, the use of a fractal branching fluid injector for bubbling fluidized beds to facilitate scalability and promote more uniform mixing, and the pulsation of the gas flow to turn a chaotic bubble flow into a periodic bubble pattern.

An unexpected side-effect to our most recent work on pulsed fluidization is that it provides fundamental insights into the mesoscopic physics of fluidization, and helps to validate computational methods. Experimentation with a variety of particles and operation parameters allows us to survey the conditions for which regular bubble patterns appear. Coupling computational fluid dynamics (CFD) simulations with discrete element modeling (DEM) for the particles, helps us to understand how the patterns form, by comparing the numerical results with the experimentally observed patterns. As the patterns are sensitive to the drag law and particle friction, they also serve as an excellent fingerprint to investigate the physics of fluidization more generally.

Mitacs Accelerate Project
Final Report
Application Ref.: IT08015

Comprehensive Evaluation of NH_3 Production and Utilization Options for Clean Energy Applications

Period:
September 2016 - March 2017

Submission Date:
March 25, 2017

Principal Investigator:
Prof. Dr. Ibrahim Dincer

Intern:
Yusuf Bicer

Partner Organization:
Hydrofuel®TM Inc.

TABLE OF CONTENTS

LIST OF FIGURES

LIST OF TABLES

ACKNOWLEDGEMENT

The principal investigator and intern acknowledge the financial support provided by the Mitacs and Hydrofuel Inc.

NOMENCLATURE

e	Charge of an electron (1.60217657 × 10^{19} C)
F	Faraday constant (C/mol)
h	Enthalpy (kJ/kg)
I	Irradiance (W/m^2)
i	Current density (A/m^2)
$\dot{m}$	Mass flow rate (g/s)
M	Molarity (M)
$\dot{n}$	Mol flow rate (mol/s)
R	Resistance (ohm)
t	Time (s)
V	Volume (L)
$\dot{W}$	Work rate (W)

Greek Letters

η	Efficiency
δ	Nernst diffusion layer thickness (cm)

Acronyms

ADF	Abiotic Depletion Factor
ASTM	American Section of the International Association for Testing Materials
AOI	Angle of incidence
AR	Anti-reflective
CFC	Chlorofluorocarbon
CML	Center of Environmental Science of Leiden University
CCS	Carbon Capture Storage
CFC	Chlorofluorocarbon
CNG	Compressed Natural Gas
CPV	Concentrated photovoltaic
DB	Dichlorobenzene
DWT	Dead Weight Tonnage
DWT	Dead Weight Tonnage
DC	Direct current
EV	Electric Vehicle
EIS	Electrochemical impendence spectroscopy
GHG	Greenhouse Gas
GREET	Greenhouse Gases, Regulated Emissions, and Energy Use in Transportation
GWP	Global Warming Potential
HEV	Hybrid Electric Vehicle
IATA	The International Air Transport Association
IAM	Integrated Assessment
ICE	Internal Combustion Engine
IPCC	Intergovernmental Panel on Climate Change
IR	Infrared
ISO	International Organization for Standards
KN	Knot

LCA	Life Cycle Assessment
LDI	Lean Direct Injection
LPG	Liquefied Petroleum Gas
LNG	Liquefied Natural Gas
LHV	Lower heating value
MSWI	Municipal Waste Incineration Plant
NMI	Nautical Mile
NM	Non-Methane
NAAQS	National Ambient Air Quality Standards
NIR	Near infrared
NOCT	Nominal operating cell temperature
PAH	Polycyclic aromatic hydrocarbon
PTH	Pump to Haul
PM	Particulate Matter
PTWa	Pump to Wake
PV	Photovoltaic
PEC	Photoelectrochemical
PtB	Platinum black
PV	Photovoltaic
RSZ	Reduced Speed Zones
RER	Europe
RMS	Root mean square
SSPC	Solid-state proton conductors
SMR	Steam Methane Reforming
TKM	Tonne-kilometer
UCG	Underground Coal Gasification
UCTE	Union for the Co-ordination of Transmission of Electricity
UV	Ultraviolet
VOC	Volatile Organic Compounds
WTP	Wheel to Pump
WTH	Well to Haul

SUMMARY

Ammonia is not only a fertilizer but also an alternative fuel for transportation and power generation. It can easily be adapted to fuel both diesel generators and larger utility power plants. The ability of being produced from renewable energy sources allows on-site ammonia production where it is needed. The high hydroelectric, wind, and solar energy source potential of Canada make the energy storage attractive using ammonia. Ontario, with decreasing electricity prices, has potentials for on-site ammonia production from hydropower. Solar energy, as the most abundant source of energy in the world, can be directly utilized for ammonia synthesis via various routes such as electrochemical and thermochemical cycles.

Canada have significant potential of renewable resources such as hydropower and wind power. Ammonia can be transported via ocean tankers or pipelines. Besides being a zero emission fuel in the utilization process, about 30% greenhouse gas reduction is possible using renewable energy such as wind in the ammonia production when compared to conventional unleaded gasoline. Compared to propane, greenhouse gas production decreases about 18%. Hence in the overall life cycle, ammonia has significant environmental advantages. Even if ammonia is produced from hydrocarbons, it has similar greenhouse gas emissions with solar energy based route. It is important to emphasize that an ammonia driven passenger vehicle releases less greenhouse gas emissions than compressed natural gas (CNG), liquefied petroleum gas (LPG), diesel, and even hybrid electric vehicles. Considering vehicle and fuel production together, emissions from an ammonia-fueled passenger car is very close to electric vehicles per km traveled as even lower in a few routes. An ammonia-fueled vehicle can save 100 g of greenhouse gas per km compared to gasoline when the complete vehicle and fuel life cycle is considered. Depletion of abiotic resources is moderately lower for conventional ammonia production, which is originated from natural gas, than liquefied natural gas, diesel, petrol and propane fuels.

In this report, detailed background information about production and utilization of ammonia is presented in Chapter 1. In Chapter 2, an alternative ammonia synthesis method is experimentally realized and tested. The selected route is molten salt electrolyte based electrochemical ammonia production. In Chapter 3, the required hydrogen for ammonia synthesis is produced using photoelectrochemical route and integrated to ammonia synthesis. The experimental results are presented under different conditions. Chapter 4 explains ammonia production from hydrocarbon sources such as natural gas. The comparative results for liquefied natural gas (LNG) and ammonia are given for production and transport phases. Chapter 5 investigates the usage of ammonia in maritime applications and performs a life cycle assessment. In Chapter 6, ammonia is evaluated as a potential fuel for aviation industry and the performances of alternative fuels are comparatively presented. Chapter 7 performs a life cycle assessment for ammonia usage in road vehicles whereas in Chapter 8, ammonia is produced using different routes and decomposed on-board for hydrogen driven vehicles. On-board ammonia electrolysis is also analyzed and presented. Chapter 9 investigates the cost of photoelectrochemical hydrogen and ammonia using exergoeconomic approach and the system is scaled-up in Chapter 10 to find the total cost of hydrogen and ammonia.

CHAPTER 1:QUICK FACTS ABOUT AMMONIA

Currently about 4.8 million metric tonnes of ammonia are produced yearly in Canada. 95.4% of produced ammonia originates from natural gas via steam methane reforming technique. 2.3% is produced by coal gasification, and 2.3% is from petroleum-coke in Canada [1]. On the other hand, the U.S. yearly produces about 12 million metric tonnes of ammonia. China is the largest ammonia producer in the world with a capacity of 55 million metric tonnes [1]. Global ammonia production corresponds to about 200 million metric tonnes. Only 4% of ammonia is used directly whereas the remaining is either used as a fertilizer or as a chemical in other industrial applications [1]. 80 million metric tonnes of ammonia is converted to urea for fertilizer uses. 1 L of gasoline costs about 0.65 USD in the U.S. This corresponds to 0.84 USD per kg gasoline. In comparison, 1 kg of ammonia sells for about 0.35 USD in the U.S [2]. Hence, ammonia is a lower cost fuel in mass basis.

The United States, Mexico and Canada have now set to declare that 50% of North America's electricity will come from clean power sources by year 2025 at the Three Amigos Summit in Ottawa [3]. It is now more obvious that the importance of alternative fuels and storage mediums increases day by day. In this regard, renewable ammonia (NH_3), which is a carbon-free fuel, refrigerant and working fluid, appears to be a unique solution to overcome local and global issues. It will help achieve carbon-free economy. The most unique part of ammonia is that it can be used for multi-purpose applications, including fuel, refrigerant, working fluid, and storage media of hydrogen. The other clear advantage is that it can serve almost all sectors, ranging from transportation to residential, industrial to commercial, public to utility, and agricultural to chemical.

Ammonia as a potential fuel candidate for vehicles can also be produced using conventional hydrocarbons in a cleaner manner by implying current technologies and developments even with lower costs. Hydrogen can be produced by dissociation of hydrocarbons, which can be then converted to ammonia using nitrogen supply from air [4]. Dissociation of hydrocarbons such as methane and oil sand bitumen is a promising option especially for Canada [5,6]. Although decomposition of heavy fractions such as oil sand bitumen can bring some challenges because of various metal and sulfur contents, purification is always possible and applicable. This technology can be promising for oil sand reserves in Alberta and stranded natural gas reserves in Newfoundland and Labrador. Ammonia is an excellent fuel candidate with the ability to secure off peak power at negative or very low prices. We can convert excess electricity to ammonia, and then we can burn it in generators, internal combustion engines, fuel cells and use as fertilizer. Where and when needed, it can be transported via pipelines, trucks, and ocean tankers.

Using appropriate renewable resources and cleaner hydrocarbon utilization paths, ammonia production can be cost-effective and environmentally friendly. On-site production and utilization of ammonia brings additional significant cost and efficiency advantages.

Ammonia (NH_3):

- consists of one nitrogen atom from air separation and three hydrogen atoms from any conventional or renewable resources.
- is the second largest synthesized industrial chemical in the world.
- is a significant hydrogen carrier and transportation fuel that does not contain any carbon atoms and has a high hydrogen ratio.
- contains about 48% more hydrogen by volume than liquefied hydrogen.
- does not emit direct greenhouse gas emission during utilization

- can be used as solid and/or liquid for many purposes.
- can be stored and transported under relatively lower pressures.
- can be produced from various type of resources ranging from oil sands to renewables.
- is a suitable fuel to be transferred using steel pipelines with minor modifications which are currently used for natural gas and oil.
- can be used in all types of combustion engines, gas turbines, burners as a sustainable fuel with only small modifications and directly in fuel cells which is a very important advantage compared to other type of fuels.
- brings a non-centralized power generation via fuel cells, stationary generators, furnaces/boilers and enables smart grid applications.
- can be used as a refrigerant for cooling in the car.

1. What are the uses of ammonia?

Ammonia is considered a possible working fluid for thermodynamic cycles, working for refrigeration, heating, power or any mixture of those can be coupled with internal combustion engines, and using exhaust gasses to drive automotive absorption refrigeration system.

Ammonia has been recognized and employed as a leading refrigerant in the industrial sector due to its outstanding thermal properties, zero ozone depletion and global warming potential (GWP). Ammonia has the highest refrigerating effect per unit mass compared to all the refrigerants being used including the halocarbons. The remarkable advantages of ammonia over R-134a could be lower overall operating costs of ammonia systems, the flexibility in meeting complex and several refrigeration needs, and lower initial costs for numerous applications. Ammonia has better heat transfer properties than most of chemical refrigerants and consequently allow for the use of equipment with a smaller heat transfer area. Thereby plant construction cost will be lower. But as these properties also benefit the thermodynamic efficiency in the system, it also reduces the operating costs of the system. In many countries the cost of ammonia per mass is considerably lower than the cost of HFCs. This kind of advantage is even multiplied by the fact that ammonia has a lower density in liquid phase. Modern ammonia systems are fully contained closed-loop systems with fully integrated controls, which regulate pressures throughout the system. Ammonia is used as refrigerant highly in the refrigeration structures of food industry like dairies, ice creams plants, frozen food production plants, cold storage warehouses, processors of fish, poultry and meat and a number of other uses.

It is also stimulating to note that NH_3 is a reduction agent for the NOx typically current in combustion releases. The reaction of NOx with ammonia over catalysts produces only steam and nitrogen. An average car needs only approximately 30 mL of NH_3 per 100 km to neutralize any NOx emissions. If the vehicles run with NH_3 as fuel, this amount is unimportant with respect to the fuel tank volume.

Ammonia is used as fertilizer in the agriculture. It is also converted into urea by reacting with CO_2. The majority of growth in ammonia usage is expected to be for industrial uses and the production of fertilizer products.

It is also worth to examine the option to cool the engine with ammonia that can act as a refrigerant while it is heated to the temperature at which it is fed to the power producer (ICE or fuel cell). Optionally, the cooling outcome of ammonia, i.e., its high latent heat of evaporation, may be used to harvest some air conditioning onboard.

2. Is ammonia really a fuel?

Ammonia as a sustainable fuel can be used in all types of combustion engines, gas turbines, and burners with only small modifications and directly in fuel cells. Ammonia was initially used as a fuel for buses in Belgium in 1940s [7]. Many studies have already been performed and many applications have been implemented so far. A prototype unit for combustion which enabled liquid kerosene and gaseous ammonia to be fed, and ammonia was combusted in a gas-turbine unit. Further studies have been performed by various researchers which have proven the practicality of using ammonia as fuel [8–14]. Numerical studies of combustion characteristics of ammonia as a renewable fuel have been conducted. Ammonia can also be used a fuel blending option for current gasoline and diesel engines. Combustion and emissions characteristics of compression-ignition engine using dual ammonia-diesel fuel have been performed. Performance enhancement of ammonia-fueled engine by using dissociation catalyst has been studied. These are just a few examples to show the current progress in the ammonia utilization options in transportation applications.

3. Is ammonia a suitable fuel for transportation sector?

The storage and delivery infrastructure of ammonia is similar to liquefied petroleum gas (LPG) process. Under medium pressures (5-15 bar), both of the substances are in liquid form which brings the significant advantage because of storage benefits. Today, vehicles running with propane are mostly accepted and used by the public since their on-board storage is possible and it is a good example for ammonia fueled vehicle opportunities since the storage and risk characteristics of both substances are similar to each other. An ammonia pipeline from the Gulf of Mexico to Minnesota and with divisions to Ohio and Texas has served the ammonia industry for many years. It indicates that there is a working ammonia pipeline transportation which can be spread overall the world. The potential of ammonia usage in many applications will be dependent on the availability of ammonia in the cities. Ammonia is a suitable substance to be transferred using steel pipelines with minor modifications which are currently used for natural gas and oil. In this way, the problem of availability of ammonia can be eliminated.

4. How can ammonia be used in transportation?

Ammonia has significant potential as an alternative fuel to further the sustainable development of transportation sector. A few of the following alternatives are shown in Fig. 1 for direct ammonia usage in transportation applications.

Currently, the majority of the locomotive fleet is made up of diesel-electric locomotives, operating with either two-stroke or four-stroke prime mover diesel engines that is coupled to an electric generator. Application of ammonia fuel for internal combustion engine (ICE) with the alternative locomotive configurations direct feed, or a combination of direct feed and decomposition subcategory options will bring more sustainable solutions. Additionally, fuel cell driven vehicles and locomotives may contribute to solve the associated matters of urban air superiority and national energy security influencing the rail and transportation sector.

5. Is ammonia a clean fuel?

Compared to gasoline vehicles, ammonia-fueled vehicles do not produce direct CO_2 emission during operation. Since ammonia produces mainly water and nitrogen on combustion, replacing a part of conventional fuel with ammonia will have a large effect in reducing carbon dioxide emissions.

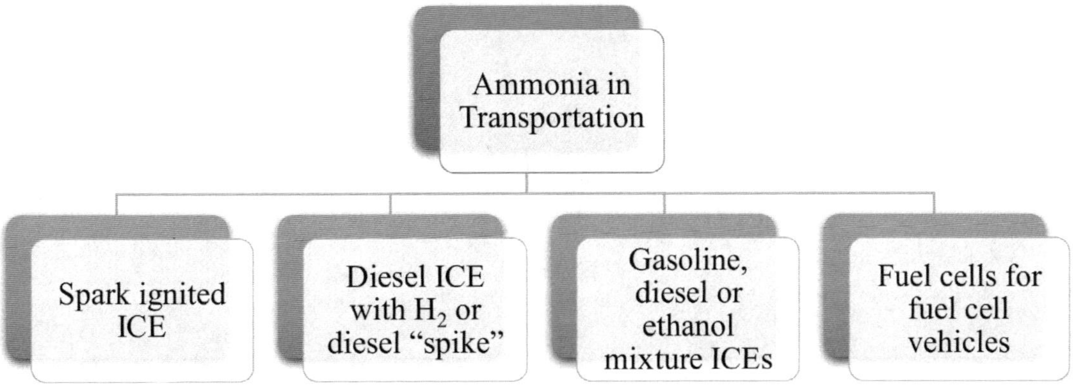

Figure 1 Ammonia utilization options in transportation sector

6. How much greenhouse gas can I save if I drive an ammonia car?

Considering a complete life cycle counting the production, transport and usage of the fuel, a diesel driven car can emit greenhouse gas emissions of about 220 g per km. Ammonia driven car can decrease this number down to about 70 g per km if it is produced from solar energy and about 150 g per km if it is produced from hydrocarbon cracking.

7. Is ammonia a cost effective fuel?

The illustrative cost comparison of various fueled vehicles is shown in Fig. 2 and 3. Considering the current market prices of the fuels, ammonia is the lowest cost fuel corresponding to about 3.1 US$ in a 100 km driving range. This shows that ammonia is a promising transportation fuel in terms of cost. There is an advantage of by-product refrigeration which reduces the costs and maintenance during vehicle operation. Some additional advantages of ammonia are commercial availability and viability, global distribution network and easy handling experience. Ammonia is a cost effective fuel per unit energy stored onboard compared to methanol, CNG, hydrogen, gasoline and LPG as shown in Fig. 2.

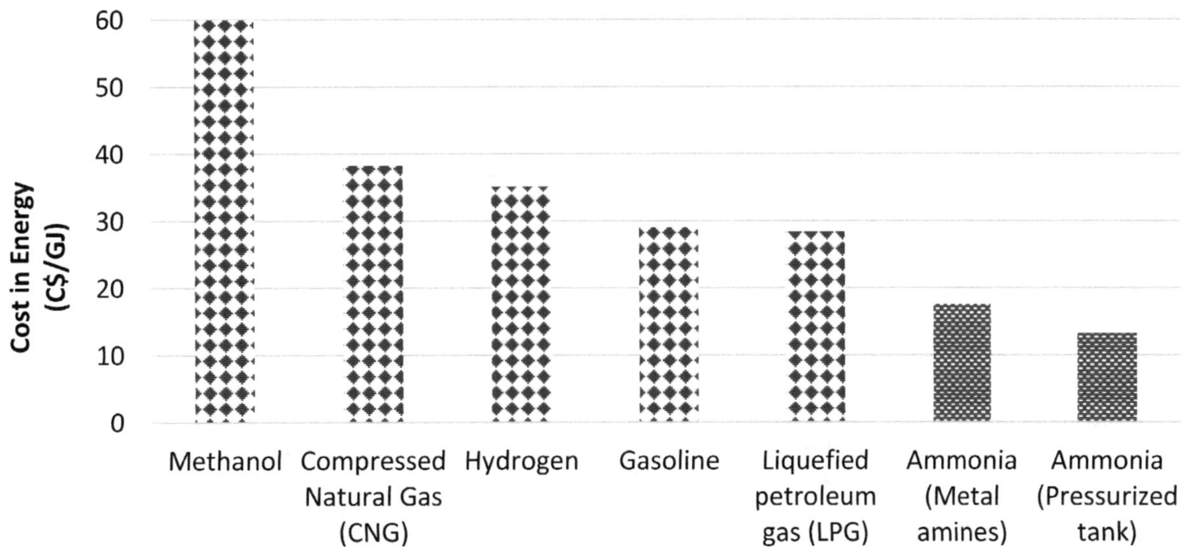

Figure 2 Comparison of various vehicle fuels in terms of energy cost per gigajoule

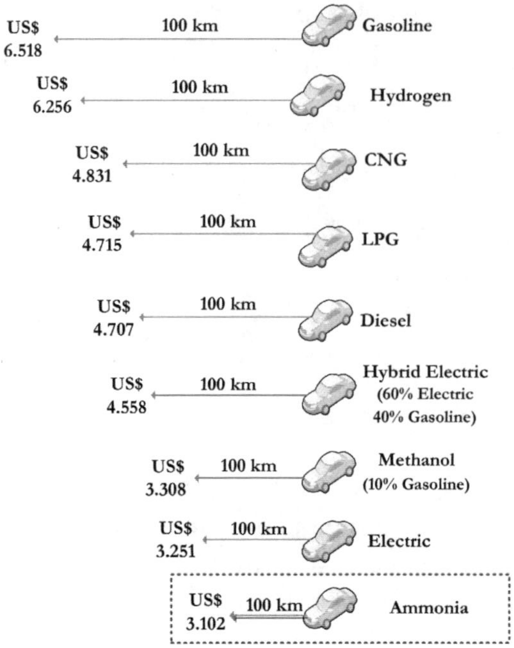

Figure 3 Comparison of driving cost for various fueled vehicles

Table 1 Comparison of ammonia with other fuels

Fuel/storage	**Pressure (bar)**	**Density (kg/m³)**	**HHV (MJ/kg)**	**HHV per Volume (GJ/m)**	**Energy per Volume (GJ/m)**	**Cost per Mass (US$/kg)**	**Cost Per Volume (US$/m)**	**Cost per Energy (US$/GJ)**
Gasoline, C_8H_{18}/liquid	1	736	46.7	34.4	34.4	1.03	754.86	21.97
CNG, CH_4/integrated storage	250	188	42.5	10.4	7.8	0.91	170.60	21.29
LPG, C_3H_8/pressurized tank	14	388	48.9	19	11.7	1.06	413.66	21.74
Methanol, CH_3OH/liquid	1	786	14.3	11.2	9.6	0.41	317.80	28.31
Hydrogen, H_2/metal hydrides	14	25	142	3.6	3	3.02	75.49	21.29
Ammonia, NH_3/pressurized tank	10	603	22.5	13.6	11.9	0.23	136.63	10.04
Ammonia, NH_3/metal amines	1	610	17.1	10.4	8.5	0.23	138.14	13.21

In Table 1, the exact fuel energy per mass is given regarding fuel's higher heating value. The volumetric energy of the fuel is found by multiplying the HHV with the density value listed in the third column. Ammonia's HHV is around half of the one of gasoline, and its density is also inferior. Therefore liquid NH_3 stores 2.5 fewer energy per unit capacity than gasoline. If the NH_3 is stored in the form of hexaamminemagnesium chloride to remove the hazard related to its toxicity, the energetic cost to pay for discharging ammonia reduces its HHV. Among alternative fuels, ammonia yields the lowest cost per energy basis. Therefore, it is important to note that low-carbon transportation projects should be eligible for compliance purposes.

8. What is the process of ammonia production?

A most common ammonia synthesis technique is recognized as Haber-Bosch process in the world. In this process, nitrogen is supplied through air separation process. Hydrogen is mainly supplied using steam methane reforming or coal gasification. However the source of hydrogen can be renewable resources. The Haber-Bosch is an exothermic process that combines hydrogen and nitrogen in 3:1 ratio to produce ammonia. The reaction is facilitated by catalyst and the optimal temperature range is 450-600°C.

Alternative production pathways are also available and under investigation including electrochemical and biological routes. These routes can easily be integrated to renewable energy sources for cleaner production. The electrochemical process can be carried out under ambient conditions or at higher temperatures depending on the type of the electrolyte material used. There are various electrochemical pathways such as molten salt, polymer membrane, liquid electrolyte etc. are intensively being researched at the moment [15–20].

The electrochemical process can be carried out under ambient conditions or at higher temperatures depending on the type of the electrolyte material used. For high temperature electrolytic routes of ammonia production, the use of waste heat from thermal or nuclear power plants or heat from renewable energy sources like solar would make the overall process more environmentally friendly.

One of the key advantages of ammonia is to be a storage medium. Renewable energy generation does not often match electrical demand which causes a requirement of storage. Green ammonia can be manufactured from surplus renewable sources, which would reduce the amount of electricity exported to neighboring jurisdictions at a negative cost.

9. What is the source of ammonia and is it cleaner than other fuels?

In terms of conventional resources, naphtha, heavy fuel oil, coal, natural gas coke oven gas and refinery gas can be used as feedstock in ammonia production. Natural gas is the primary feedstock used for producing ammonia in worldwide corresponding to about 72%. However, renewable resources can easily be integrated for ammonia production. In this way, decentralized ammonia production can be realized which further decreases the delivery cost of the fuel. Many studies have been performed to investigate the ammonia production routes and their environmental impacts. Here, some of them are briefly shown.

The production of the different fuels is compared in terms of abiotic depletion of sources as shown in Fig. 4. Ammonia fuel has the lowest abiotic depletion value compared to others although the production process may be fossil fuel based. There are multiple pathways for ammonia production. Ammonia is cleaner when produced from renewable resources. Fig. 5 shows the comparison of ozone layer depletion values for various transportation fuels. Ammonia has lowest ozone layer depletion even if it is produced from steam methane reforming and partial oxidation of heavy oil.

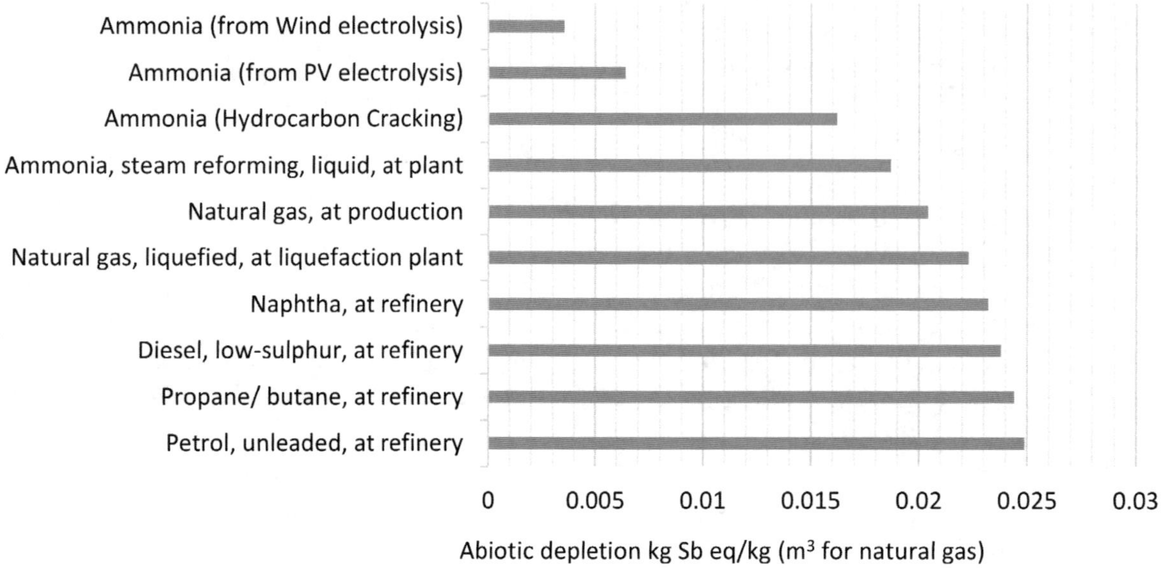

Figure 4 Abiotic depletion values during production of various fuels

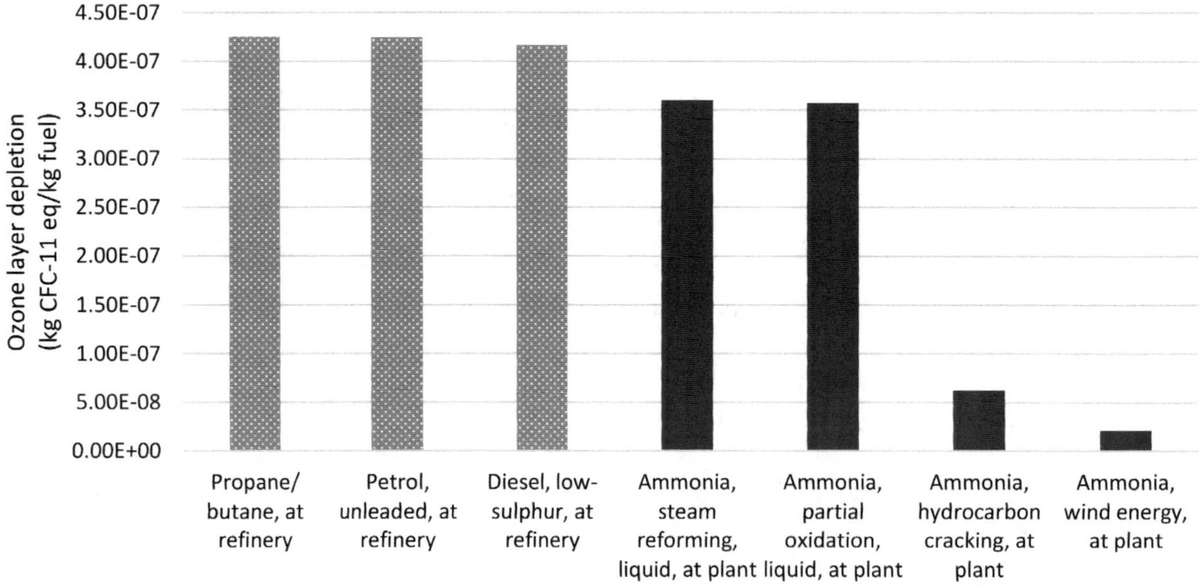

Figure 5 Ozone layer depletion during productions of various fuels

Fig. 6 compares the total greenhouse gas emissions during production of 1 MJ energy from various resources including gasoline, LPG, diesel, natural gas and ammonia. Production of 1 MJ energy from ammonia has lower emissions than gasoline, LPG, diesel, oil and natural gas.

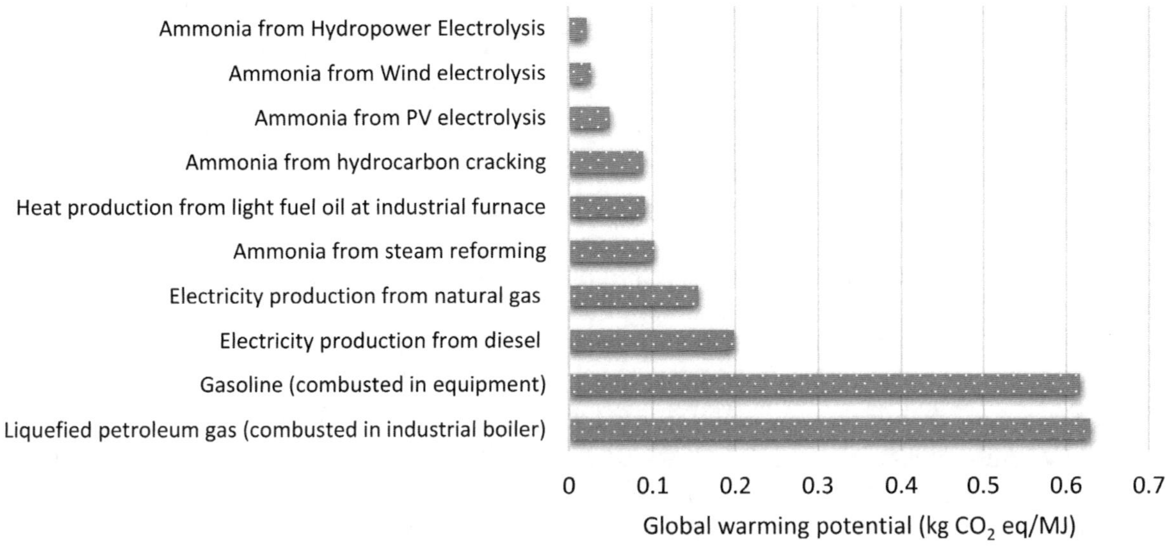

Figure 6 Comparison of global warming potential of 1 MJ energy production from various resources

Giving priority for complete conversion from fossil fuel based fuels to carbon-free fuels will bring short term and long term solutions to combat global warming. Therefore, the activities to lower the carbon intensity of conventional transportation fuels be eligible for compliance purposes

CHAPTER 2:ELECTROCHEMICAL SYNTHESIS OF AMMONIA

Natural gas is the primary feedstock used for producing ammonia in worldwide which increases the fossil fuel dependence and greenhouse gas emissions. Therefore, alternative ammonia synthesis methods are in development stage such as electrochemical ammonia synthesis which is experimentally tested in this chapter.

1. Introduction

The electrochemical process can be carried out under ambient conditions or at higher temperatures depending on the type of the electrolyte material used. For high temperature electrolytic routes of ammonia production, the use of waste heat from thermal or nuclear power plants or heat from renewable energy sources like solar would make the overall process more environmentally friendly. Ammonia production from hydrogen and nitrogen is exothermic in nature and is facilitated by high pressures and low temperatures. Thus a balance between the operating temperature, pressure and the ammonia yield needs to be proven for electrochemical system in determining ammonia production rates. Among solid state electrolyte methods, there are studies in the literature although maximum production rates are lower than liquid electrolyte based methods. A study used a proton-conducting solid electrolyte at 450°C to 700°C with Ru based catalyst. They resulted that the conversion rates are lower compared to nitrogen or steam because of the low conductivity of the working electrode [20]. In another study with a Nafion divider in aqueous 2 M KOH and a Ru on cathode allowed ammonia generation from water and nitrogen at a rate of 2.8×10^{-12} mol NH_3/s cm^2 and coulombic efficiency of 0.9% at 20°C. Also, at 90°C, a maximum rate of 2.1×10^{-11} mol /s cm^2 at 0.2% efficiency was observed [21]. Membrane based applications can be performed at quite lower temperatures than molten salt electrolyte based methods. Having Pt/C on a gas diffusion layer at both electrodes and room temperature, using Nafion as the electrolyte yielded NH_3 at a higher rate of 1.1×10^{-9} mol/s cm^2, which consumed water at the anode and air at the cathode at 0.6% coulombic efficiency [16]. In addition, there is also a molten NaOH-KOH eutectic electrolyte based ammonia fuel cell studied by researchers [22]. Licht et al. [18] obtained about 35% Faradaic efficiency where they supplied water and air to produce ammonia in molten salt electrolyte. In their further research Li and Licht [17] reported that since they used water as hydrogen source, at 200 mA/cm^2, over 90% of applied current generated H_2, rather than NH_3. In this case, hydrogen was cogenerated but required higher potentials because of water splitting voltage.

Numerous electrochemical methods have recently been developed for consideration as alternative syntheses of NH_3 [23–33]. Kugler et al. [32] studied to increase the NH_3 generation rate by applying galvanic deposition of Rh and Ru on Ti felts in which they observed higher formation rates for Rh coatings. Kim et al. [33] performed electrochemical synthesis of ammonia in molten LiCl-KCl-CsCl electrolyte by a mixture of catalysts as nano-Fe_2O_3 and $CoFe_2O_4$. Their maximum formation rate was 3×10^{-10} mol/s cm^2 where they used water and nitrogen for the reaction. Xu et al. [34] investigated generation of ammonia at atmospheric pressure and low temperature electrochemically, using the SFCN materials as the cathode, a Nafion membrane as the electrolyte, nickel-doped SDC (Ni-SDC) as the anode and silver-platinum paste as the current collector. NH_3 was produced from 25°C to 100°C temperature levels when the SFCN materials were utilized as cathode, with $SmFe_{0.7}Cu_{0.1}Ni_{0.2}O_3$ which gives the maximum rates of ammonia formation. Garagounis et al. [35] summarized the test results of studies within the last 15 years using electrolyte cells. More than 30 electrolyte materials with 15 catalysts that were used as working electrodes (cathode) were tested. The polymer Nafion yielded the highest rate of ammonia formation at a very low temperature. Nafion can also be used as a proton conductor with and a Ru/C cathode which yielded NH_3 from H_2O and N_2 at 90°C. These low operating temperatures are sought after when designing new systems because of the reduced energy input required and lower rate of decomposition of the ammonia formed. As an alternative approach, the use of oxygen ion conductors where steam and nitrogen are introduced together at the cathode should be considered. The rate of NH_3 production was however very small in a demonstration at 500°C but improved by up to two orders of magnitude at higher temperatures as reported by Skodra et al. [36]. Furthermore, NH_3 synthesis using molten salt electrolyte based systems can yield high conversion ratios similar to that of polymer based membranes. Solid-state proton conductors (SSPC) denote a class of ionic solid electrolytes which have the capability to transfer hydrogen ions (H^+) [37]. However, this method has some disadvantages such as high temperature requirement and formation of secondary phases [38–40]. A variety of factors need to be analyzed when selecting the cell material such as system operating temperature, current density, pressure, and conductivity which all affect the ammonia production rate. It is important to note that conductivity of a solid electrolyte increases exponentially with temperature and by reducing cell thickness as reported by Giddey et al. [41]. Kyriakou et al. [42] recently reported extensive literature data about low temperature, medium temperature and high temperature electrochemical NH_3 synthesis routes showing that the synthesis rates can reach up to 3.3×10^{-8} mol/s cm^2. Shipman and Symes [43] presented the recent developments in electrochemical NH_3 production and they categorized the sources of proton as water, hydrogen and sacrificial proton donors. They resulted that the techniques keeping the temperatures in the range of 100°C and 300°C may well demonstrate to be the most efficient.

H_2 can be directly utilized in the electrochemical NH_3 formation as investigated in this study. Most of the literature used water as hydrogen source which also requires water splitting process at the same time with ammonia synthesis. Specifically, for solar energy storage applications, H_2 can act as short-term storage whereas NH_3 can serve as long-term storage medium which reduces the storage losses significantly. Here, electrochemical synthesis of ammonia using H_2 and N_2 at ambient pressure in a molten hydroxide ambient with nano-Fe_3O_4 catalyst is performed. The active surface areas of the Nickel mesh electrodes are increased to allow higher formation rates. The effects of various parameters such as applied potential and current density, reaction temperature on ammonia formation rate are investigated. The nickel mesh electrodes are utilized having large surface area and the reaction temperatures are quite lower than the conventional Haber-Bosch process.

2. Experimental Investigation and Analysis

In this study, H_2 and N_2 are directly used for electrochemical synthesis of ammonia at the electrodes. N_2 receives the electrons from external power supply. Hence nitrogen gas sent via the porous nickel cathode is reduced to nitride according to the following equation:

$$N_2 + 6e^- \rightarrow 2N^{3-} \quad (1)$$

It becomes N^{3-} then after moves to the other electrode where H_2 is being supplied. Hydrogen ions combine with nitrogen ions and form NH_3 at anode electrode and shown the following equation:

$$2N^{3-} + 3H_2 \rightarrow 2NH_3 + 6e^- \quad (2)$$

The anode reaction is also achieved on porous nickel electrode.

The overall reaction is:

$$3H_2 + N_2 \rightarrow 2NH_3 \quad (3)$$

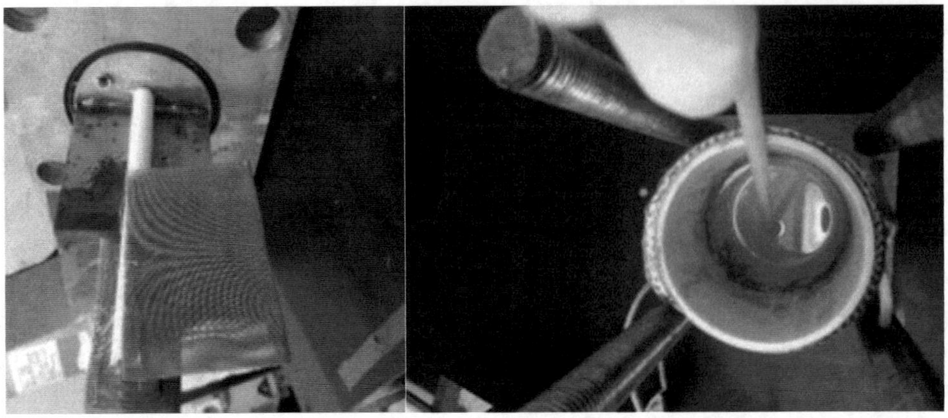

Figure 7 Nickel mesh electrodes in the reactor, molten salt in the reactor

Iron oxide (Fe_3O_4) as nano-powder (20-30 nm, 98+%) is used in the experiments as catalyst. The high surface area of the nano-Fe_3O_4 in the electrochemical synthesis is critical for the reaction to occur and to obtain higher ammonia evolution rates. The reactants, H_2 and N_2, are bubbled through the mesh over the anode and cathode, respectively. The combined gas products (H_2, N_2 and NH_3) exit through two exit tubes in chamber head space. The exiting gases are firstly measured using flowmeters and bubbled through an ammonia water trap then analyzed for ammonia, and subsequently the NH_3 scrubbed-gas is analyzed for H_2 or N_2.

As mentioned earlier, the product gases from the reactor is bubbled through an ammonia trap consisting of a dilute 500 ml 0.001 M H_2SO_4 solution, changed every 15 minutes for ammonia analysis. Ammonia concentration is determined using various techniques to confirm the results. The methods utilized are as follows: ammonia test strips, ammonia gas flowmeters, Arduino ammonia gas sensor and salicylate-based ammonia determination method as the experimental setup is shown in Fig. 7, 8 and 9. For the salicylate-based method, two different solutions are used where one of them contains sodium salicylate and the other one contains sodium hydroxide and sodium hypochlorite. In each case, redundant measurements yield similar ammonia formation values, with the observed reproducibility of methodologies.

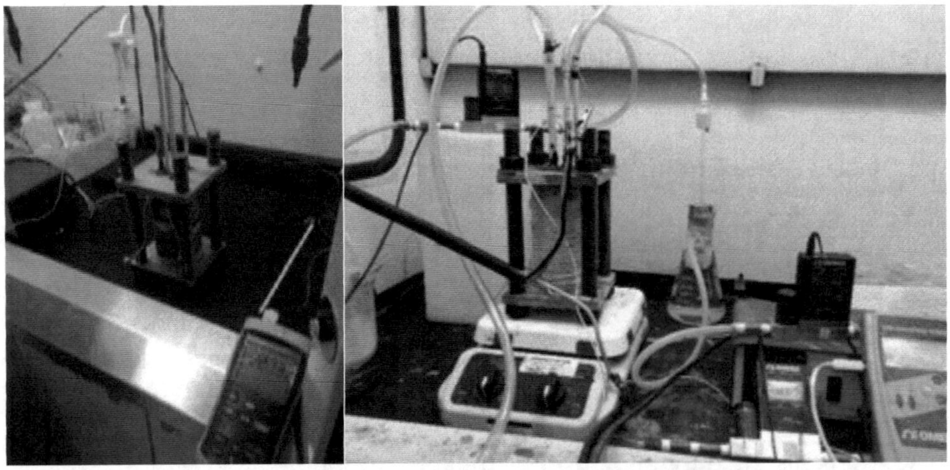

Figure 8 Heating tape used around the alumina crucible and experimental setup with flowmeters, temperature controller and tubing

In addition, the pH level of the dilute H_2SO_4 solutions are recorded before and after NH_3 trapped in the solution in order to observe the dissolved ammonia. Ammonia formation rate is calculated by converting the measured NH_3 to moles per seconds and considering the surface area of Ni electrodes as 100 cm^2. The ammonia formation rate is calculated using the following equation:

$$\dot{m}_{NH_3} = \frac{[NH_4^+] \times V}{t} \quad (4)$$

where $[NH_4^+]$ is the concentration of formed ammonia as mg/L, V is the total volume of H_2SO_4 for trapping ammonia as L and t is the time of collection.

The Faradaic efficiency is calculated based on the moles of electrons consumed compared to the $3e^-/NH_3$ equivalents produced. Thus, the Faradaic efficiency of ammonia generation process is defined as follows:

$$\eta_{\text{Faradaic}}\ (\%) = \frac{\dot{n}_{NH_3} \times F \times 3}{i} \quad (5)$$

where F is Faraday constant and i is the current density (A/cm^2).

The energy efficiency of the ammonia production process is also calculated based on lower heating values of reacted hydrogen and ammonia, nitrogen enthalpy and electrical power input as follows:

$$\eta_{\text{Energy}}\ (\%) = \frac{\dot{m}_{NH_3} \times LHV_{NH_3}}{(\dot{m}_{H_2} \times LHV_{H_2} + \dot{m}_{N_2} \times h_{N_2} + \dot{W}_{el})} \quad (6)$$

where W_{el} is the total electricity input during the experiment calculated using the total charge, applied voltage and duration.

3. Results and Discussion

The pure alkali hydroxides NaOH and KOH each melt only at temperatures above 300°C. The individual melting temperatures of NaOH and KOH are 318°C and 406°C, respectively. Among various alternatives, these two salts melt at quite lower temperatures which is a highly desired property in order to decrease external heat energy input. Based on common materials, the NaOH-KOH eutectic is of particular attention and melts at 170°C. Ammonia synthesis rates increase when the molten hydroxide (NaOH-KOH) electrolyte is mixed with high–surface area Fe_3O_4 to provide iron as a reactive surface and when nitrogen and hydrogen are present in the reactor. The molten salt medium is supplied electricity between two nickel anode and cathode electrodes. The mixture is prepared in the beginning by simply adding NaOH and KOH pellets in the reactor. After the salts melt, nano-Fe_3O_4 is added to the electrolyte and then stirred. When the mixture is ready, the lid is tightly closed and sealed. In order to yield NH_3 in the reactor, H_2, N_2 and nano-Fe_3O_4 are simultaneously needed. Table 2 shows the experimental conditions for four different runs.

Table 2 Experimental conditions for electrochemical ammonia synthesis

Experiment #	**Temperature (°C)**	**Duration (min)**	**Current density (mA/cm^2)**	**Voltage (V)**
1	210	15	2	1.4
2	255	30	3	1.5
3	215	45	2	1.3
4	220	25	2.5	1.55

Experiment 2 is performed at constant applied potential of 1.5 V whereas the others are performed at constant current in galvanostatic mode. The temperatures given in the table are average temperatures because, the temperature controller is on/off type and keeping the temperature constant brings fluctuations. For each run, different ammonia trapping H_2SO_4 solution is used. The unreacted H_2 is also measured using a hydrogen sensor embedded to Arduino board which shows the portion of H_2 which does not react.

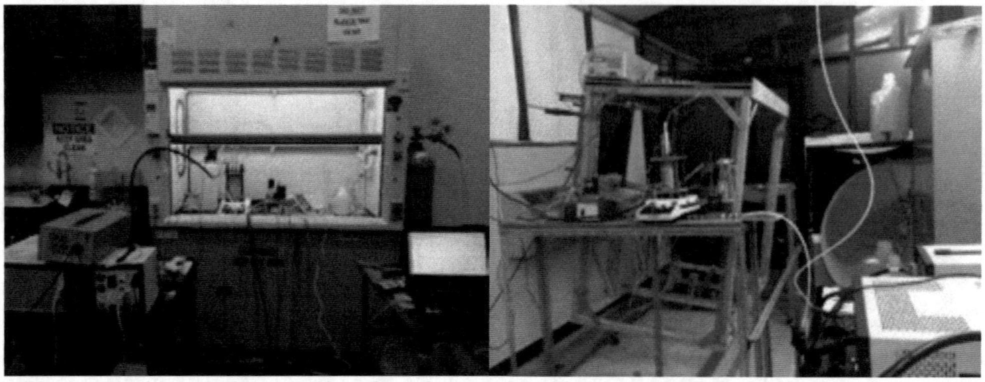

Figure 9 The experimental setup for the electrochemical ammonia synthesis in the fume hood and solar light

The required cell voltage to initiate the reaction of nitrogen and hydrogen in molten hydroxide at 210°C in the existence of nano-Fe_3O_4 is measured to be in average 1.4 V when the applied current is 200 mA between the 100 cm^2 Ni electrodes in the molten NaOH-KOH electrolyte.

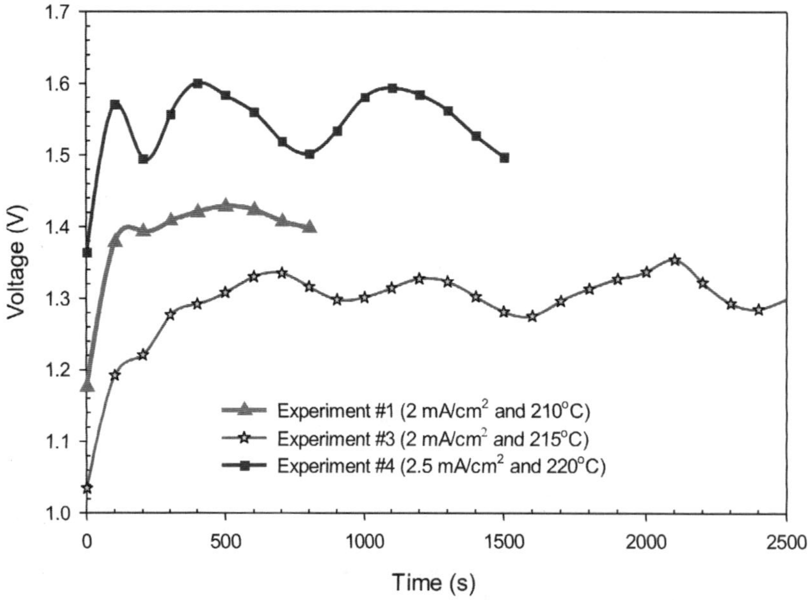

Figure 10 The relationship between voltage and time during several experimental runs at different applied currents and temperatures for electrochemical synthesis of NH_3 using N_2 and H_2 with nano-Fe_3O_4 in a molten salt hydroxide electrolyte

The potential increased to 1.54 V when the current density is increased to 2.5 mA/cm^2 at 220°C as shown in Fig. 10. At 2 mA/cm^2 and 210°C, ammonia is synthesized at a rate of 6.54×10^{-10} mol/s cm^2. At 2.5 mA/cm^2 and 220°C, the ammonia evolution rate decreased to 4.9×10^{-11} mol NH_3/s cm^2. At 215°C in a eutectic $Na_{0.5}K_{0.5}OH$ electrolyte with suspended nano-Fe_3O_4, it is observed that at 2 mA/cm^2 applied current, NH_3 is generated at a Faradaic efficiency of about 6.3%, which declines to about 0.56% at 2.5 mA cm^2 in the reactor temperature of 220°C. Constant current electrolysis at 2 mA/cm^2 and 2.5 mA/cm^2 are driven at 1.3 V and 1.54 V, respectively at different temperatures as depicted in Fig. 10. It is also observed in the experiments that even though the reactor temperature is below 200°C, ammonia is generated with a similar production rate to above 200°C. The fluctuations in the potentials are caused by the temperature on/off processes to keep the temperature constant during the experiments. When the heater is on, the required potential to drive the reaction decreases as seen in Fig. 10. The potential gradually declines from 1.6 V to 1.5 V during the experiment at constant current density of 2.5 mA/cm^2. It is observed in the experiments that lower current density and lower temperature improve the stability of the rate of NH_3 evolution.

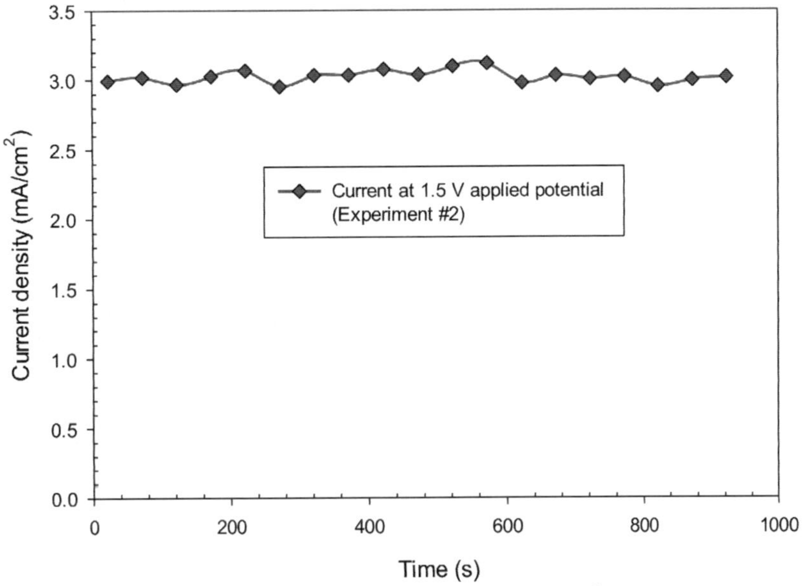

Figure 11 Current density at 1.5 V applied voltage for 100 cm^2 Ni electrodes of electrochemical NH_3 synthesis reactor

Fig. 11 shows the current density across the electrodes of the reactor when constant voltage of 1.5 V is applied. In this case, in average 3 mA/cm^2 current density is measured where the reaction temperature is considerably higher than other experiments which is about 255°C. The greater ammonia generation rate at lower voltages can be because of the lower hydrogen ion stream at the cathode which provides more time for generation of ammonia according to reaction.
Higher NH_3 synthesis rates are obtained within the first hour of the experiments. By the end of the experiment which is close to about two hours, 5.69 mL of NH_3 is formed.

The generated NH_3 is trapped and measured in a room temperature dilute H_2SO_4 trap. A non-dilute H_2SO_4 trap is also tried before the experiments reported here to understand the absorptivity of the solution. However, the ammonia readings were not successful in this case. Hence, dilute H_2SO_4 solutions are utilized for the reported experiments. The conversion efficiency is not only dependent on the hydrogen amount but also amount of catalyst available to stimulate the conversion of N_2 and H_2 into NH_3. In order to make sure that there is enough N_2 to be reacted with supplied H_2, the supplied volume of N_2 is kept quite higher than H_2.

In order to understand the current-voltage characteristics at lower temperature levels such as 200°C, the applied potentials are varied between 1.1 V and 1.5 V as shown in Fig. 12. At 200°C and 1.3 V, the average current density is 2.16 mA/cm^2 whereas it is about 2 mA/cm^2 at 215°C. The given temperatures and current densities are the average values where there are fluctuations because of the temperature controller. The obtained results for different experiments are tabulated in Table 3.

Table 3 Summary of the experimental results showing the NH_3 formation rates and efficiencies

Experiment #	NH_3 formation rate (mol/s cm^2)	NH_3 mass flow rate (g/min)	Faradaic efficiency (%)	Energy efficiency (%)
1	6.54×10^{-10}	6.67×10^{-5}	9.46	6.66
2	6.54×10^{-10}	6.67×10^{-5}	6.30	3.99
3	1.91×10^{-10}	1.94×10^{-5}	2.76	2.07
4	4.90×10^{-11}	5.00×10^{-6}	0.57	0.36

The variations of NH_3 formation rates at different current densities and temperature levels are comparatively shown in Fig. 13. The figure reveals that the temperature and current density are not the sole parameters affecting the performance of the reaction. Although the temperature is high at 2.5 mA/cm^2 current density, the NH_3 formation rate is low.

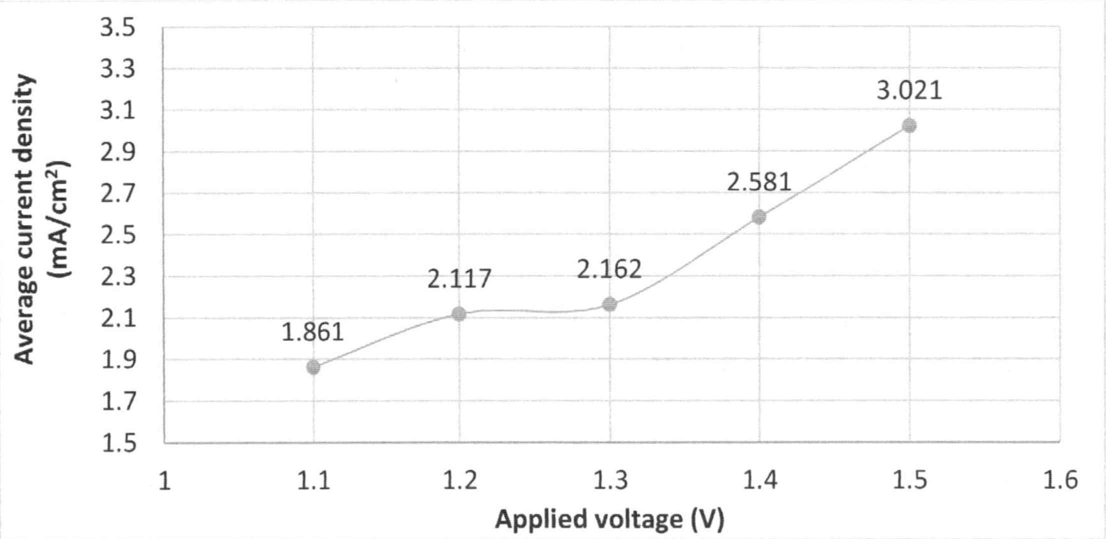

Figure 12 Applied potential-current density relations at 200°C for electrochemical NH_3 formation using N_2 and H_2 with nano-Fe_3O_4 in a molten salt hydroxide electrolyte

In contrast, at higher current densities at 3 mA/cm^2 and higher temperatures at 255°C, the NH_3 formation rate is high corresponding to about 6.6×10^{-10} mol/s cm^2. The differentiations might be caused by the catalyst saturations as well as the changes in supplied H_2 rates. The effects of catalyst quantity and type of electrodes are likely to be investigated in the future work. Furthermore, the electrochemical ammonia synthesis reactor will be integrated to photoelectrochemical hydrogen production cell to develop a clean and environmentally friendly ammonia production technique.

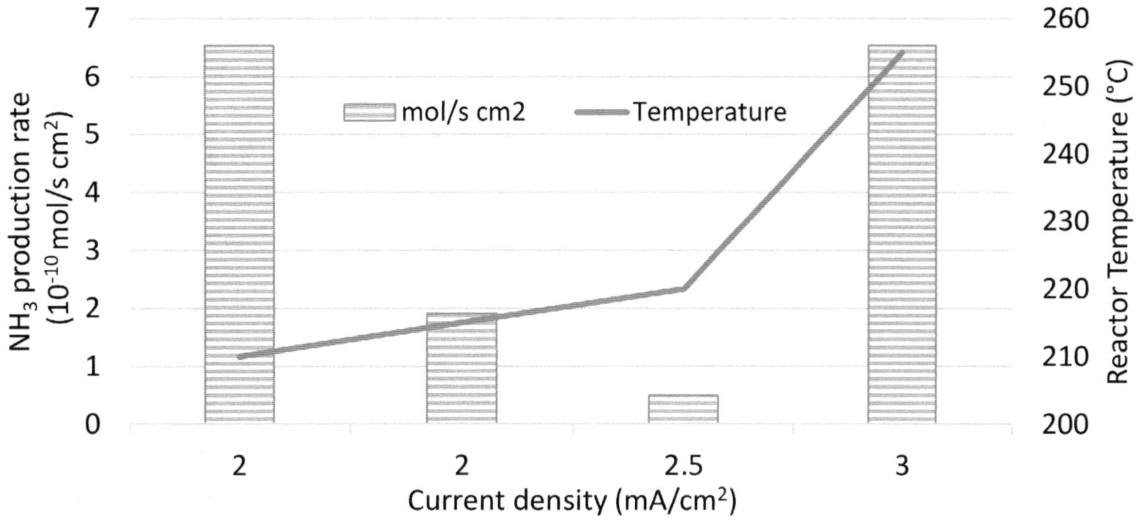

Figure 13 Change of electrochemical NH_3 formation rates depending on the applied current densities and reactor temperature using N_2 and H_2 with nano-Fe_3O_4 in a molten salt hydroxide electrolyte

4. Concluding Remarks

The electrochemical synthesis of NH_3 is a promising alternative to conventional energy intensive NH_3 production plants. Using renewable energy resources to drive the electrochemical NH_3 synthesis, the carbon footprint of current NH_3 production industry can be lowered significantly. Electrochemical NH_3 synthesis routes offer higher integrability to stand alone and distributed NH_3 production which is a carbon free fuel for various sectors. In this study, NH_3 is electrochemically generated at ambient pressure without a necessity of huge compressors using H_2 and N_2 in a molten hydroxide medium with nano-Fe_3O_4 catalyst. The reaction temperature is varied in the range of 200°C to 255°C to investigate the impact of temperature on NH_3 production rates. Having non-corrosive and high surface area nickel mesh electrodes allowed to generate more NH_3. The maximum Faradaic efficiency is calculated as 9.3% with a reaction temperature of 210°C. The NH_3 formation rate is determined to be 6.53×10^{-10} mol/s cm^2 at 2 mA/cm^2 current density.

CHAPTER 3: ELECTROCHEMICAL AMMONIA SYNTHESIS FROM PHOTOELECTROCHEMICAL HYDROGEN

The electrochemical process may be carried out under ambient conditions or at elevated temperatures depending on the nature of the electrolyte material used. For high temperature electrochemical ammonia production modes, the use of waste heat from thermal, nuclear power plants and renewable energy sources such as solar energy would make the process more environmentally friendly. In this section, we report the electrochemical synthesis of ammonia using photoelectrochemical (PEC) H_2 and N_2 at ambient pressure in a molten salt ambient with the catalyst of nano-Fe_3O_4.

1. Introduction

Assisting electrochemical process with solar energy will contribute an environmentally friendly method for ammonia production. Especially for solar energy storage applications, H_2 can act as a short term storage medium, whereas NH_3 can serve as a long term storage medium, which significantly reduces storage and transportation losses. The active surfaces of the nickel mesh electrode are 25 cm^2. The effects of various parameters such as the applied potential and current density, the reaction temperature on the rate of ammonia formation are examined. In this study, the solar light splitting and concentrator has been used for photoelectrochemical hydrogen production and electrochemical ammonia synthesis which has not been tested in the literature. The underlying motivation of this study is the potential for combining photoelectrochemical hydrogen production system with electrolytic ammonia synthesis processes to increase the solar spectrum utilization and ammonia production yield.

2. Life Cycle Assessment of Photoelectrochemical Hydrogen Based Electrochemical Ammonia Synthesis

In this section, the LCA results obtained for PEC based electrochemical ammonia production method using concentrated light are given in detail to reveal the contribution of various sub-processes. There are mainly three processes in the PEC based ammonia synthesis namely; hydrogen production from photoelectrochemical reactor, nitrogen production from air separation and electricity production from PV cells for energizing the process. The boundary of the LCA study for PEC hydrogen production is shown in Fig. 14.

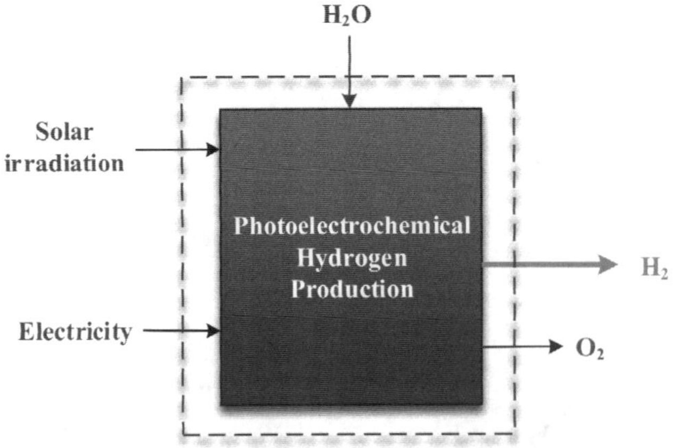

Figure 14 The boundaries of the conducted LCA for PEC hydrogen production

In the second step, electrochemical ammonia synthesis process is simulated as shown in Fig 15.

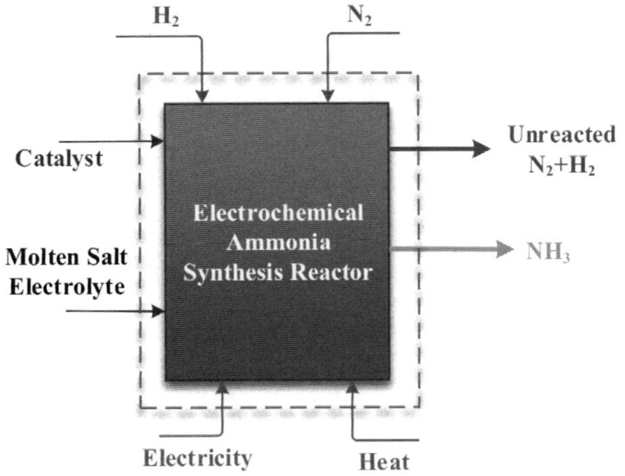

Figure 15 The boundaries of the conducted LCA for electrochemical ammonia synthesis process

The main contributor in all categories is the electricity production from PV cell as shown in Tables 4 to 6. 6.9% of total human toxicity is caused by nitrogen production process whereas 29.9% is due to hydrogen production from PEC system as listed in Table 4. Electricity production from PV is mainly responsible for remaining. There are numerous substances causing toxicity for human health such as arsenic and nickel as shown in Fig. 16.

Table 4 The shares of different sub-processes in human toxicity category for PEC (concentrated light) based electrochemical ammonia synthesis

Inflows	Flow	Unit
Total	100	%
Electricity, production photovoltaic, multi-Si	63.2	%
Hydrogen, PEC cell, PV, Concentrated Light-Integrated System	29.9	%
Nitrogen, gas, at plant	6.9	%
Potassium hydroxide	0.000102	%
Sodium hydroxide	6.96E-05	%
Iron oxide	7.06E-06	%

Arsenic and polycyclic aromatic hydrocarbons are the two fundamental toxic substances (about 64% in total) released to the environment in this method. There are mainly caused by copper and aluminum production processes for PV and support structures as shown in Fig.17. Polycyclic aromatic hydrocarbons are released due to nitrogen production from air separation plant since mix grid electricity is used.

Table 5 shows the shares of main processes contributing to abiotic depletion. Almost half of the total abiotic depletion is because of PV electricity production whereas 25% is due to hydrogen production. The molten salt electrolyte and reaction catalyst have quite small shares in total impact. Furthermore, as shown in Fig. 18, coal and natural gas are two main substances depleting in this method due to high electricity consumption in the PV cell factory and aluminum needed for support mechanism. Crude oil and brown coal have shares of 14% and 7%, respectively as shown in Fig. 19.

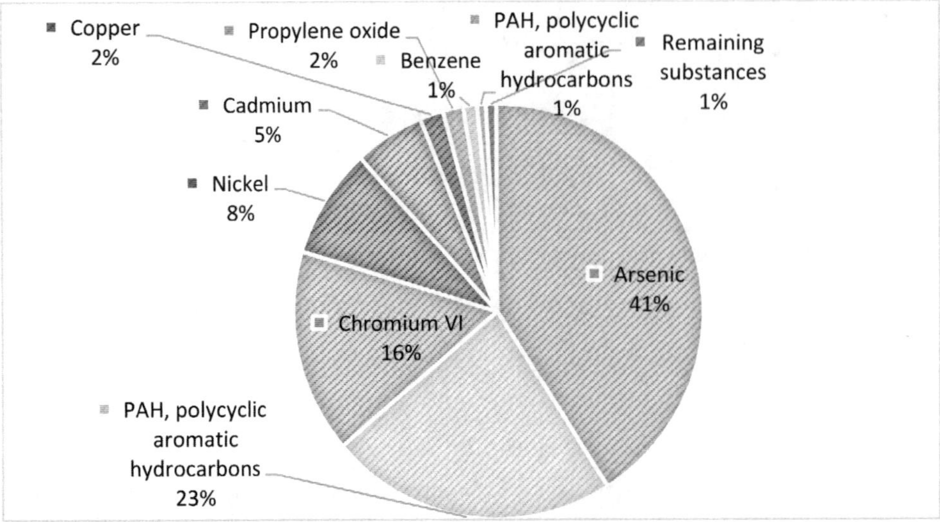

Figure 16 The share of toxic substances for PEC (concentrated light) based electrochemical ammonia synthesis

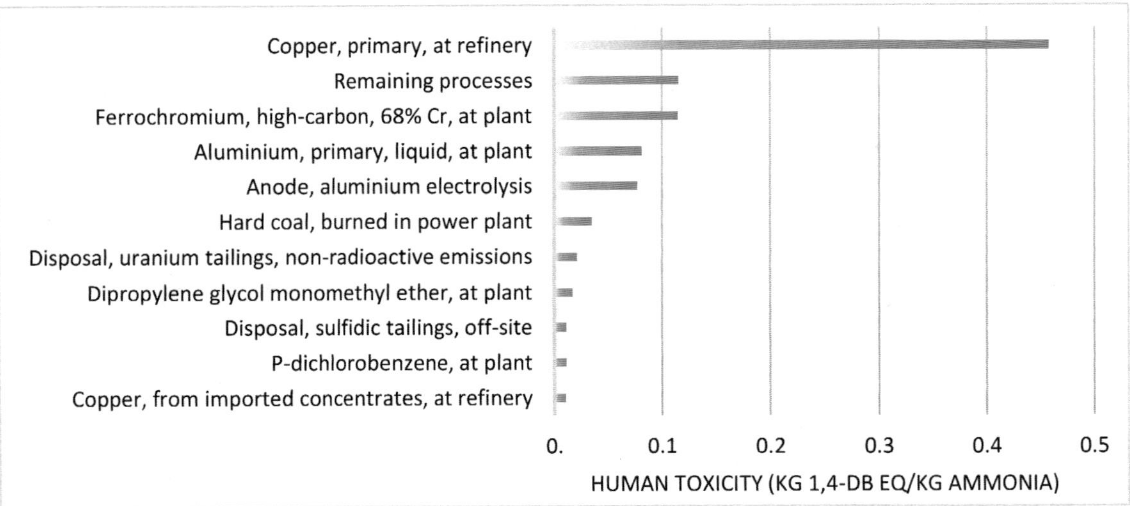

Figure 17 Contribution of various sub-processes to human toxicity potential of PEC (concentrated light) based electrochemical ammonia synthesis

Table 5 The shares of different sub-processes in abiotic depletion category for PEC (concentrated light) based electrochemical ammonia synthesis

Inflows	Flow	Unit
Total	100	%
Electricity, production photovoltaic, multi-Si	51.7	%
Hydrogen, PEC cell, PV, Concentrated Light-Integrated System	24.5	%
Nitrogen, gas, at plant	23.8	%
Potassium hydroxide	0.000256	%
Sodium hydroxide	0.000141	%
Iron oxide	1.83E-05	%

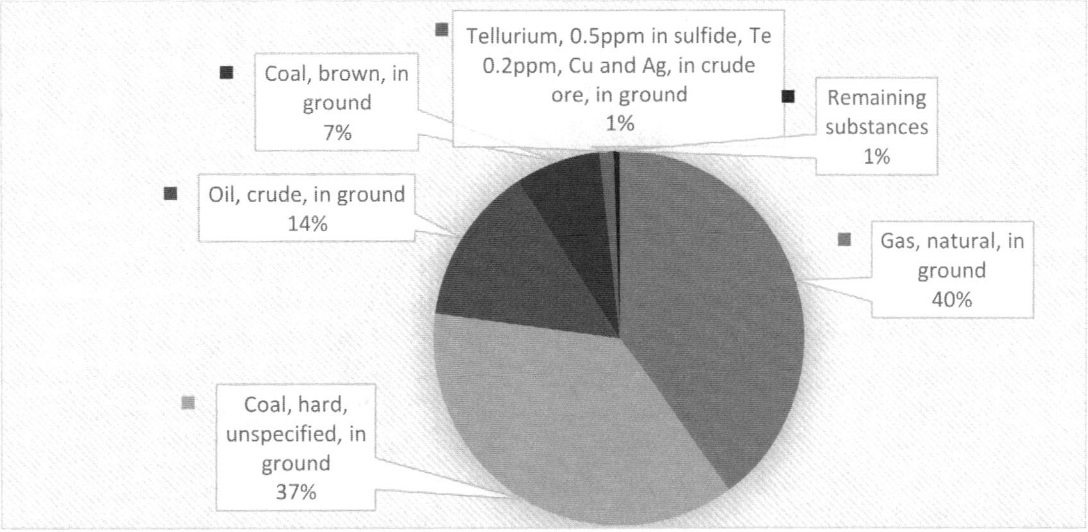

Figure 18 The share of depleting abiotic sources for PEC (concentrated light) based electrochemical ammonia synthesis

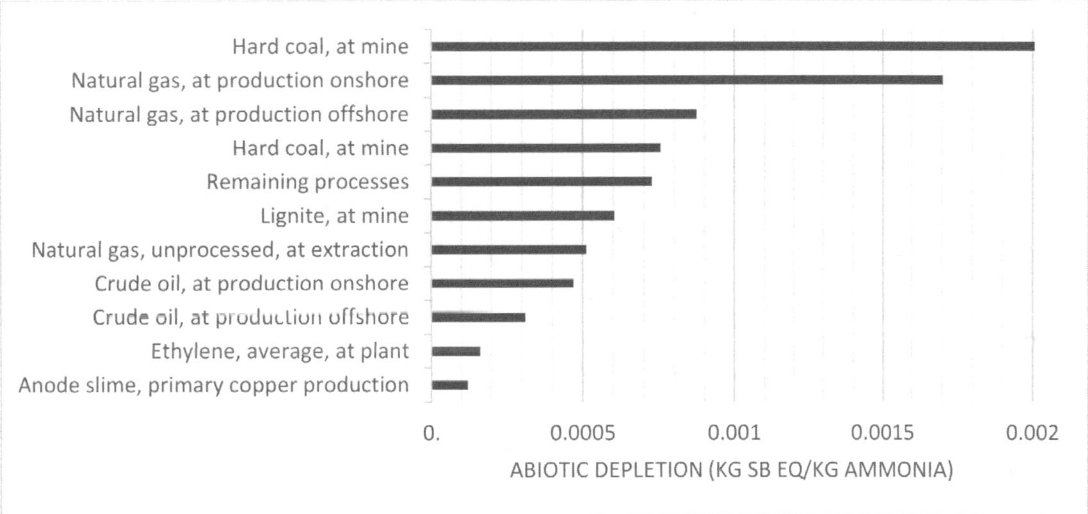

Figure 19 Contribution of various sub-processes to abiotic depletion potential of PEC (concentrated light) based electrochemical ammonia synthesis

The global warming potential of PV electricity production is responsible for almost 50% of total GHG emissions where almost 76% of PV electricity is because of PV cell production process in the factory. The shares of main processes for global warming potential are tabulated in Table 6. Global warming category includes all greenhouse gas emissions however, CO_2 is the main gas emitted to the environment corresponding to about 93% of total in the method as shown in Fig. 20. Sulfur hexafluoride (3%) and methane (2%) are the other gases contributing to total GHG emission. Sulfur hexafluoride emission is mainly due to magnesium production in the plant required for PV cell production. Shown in Fig. 21, electricity production in cogeneration plant and hard coal burned in power plant are mainly because of silicon production required for PV cells.

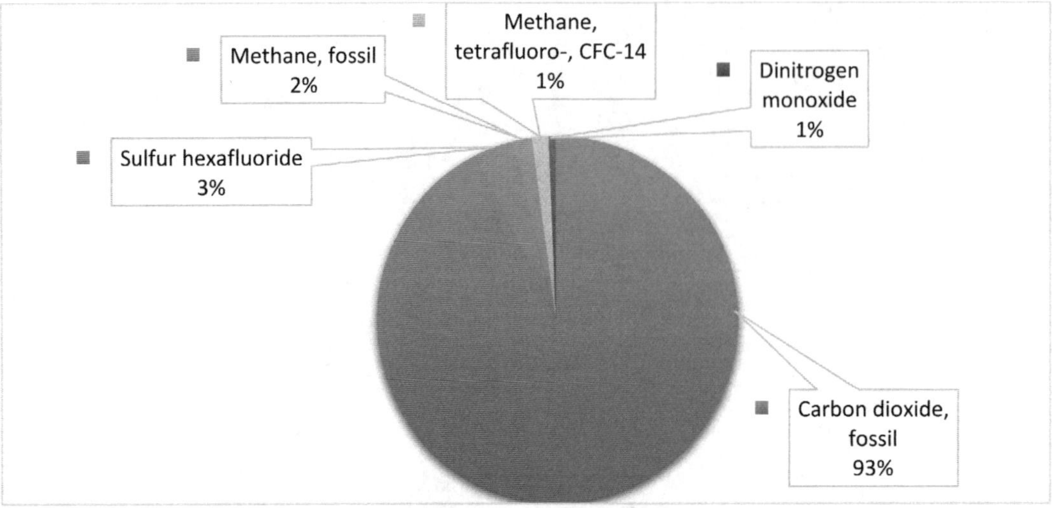

Figure 20 The share of greenhouse gas emissions for PEC (concentrated light) based electrochemical ammonia synthesis

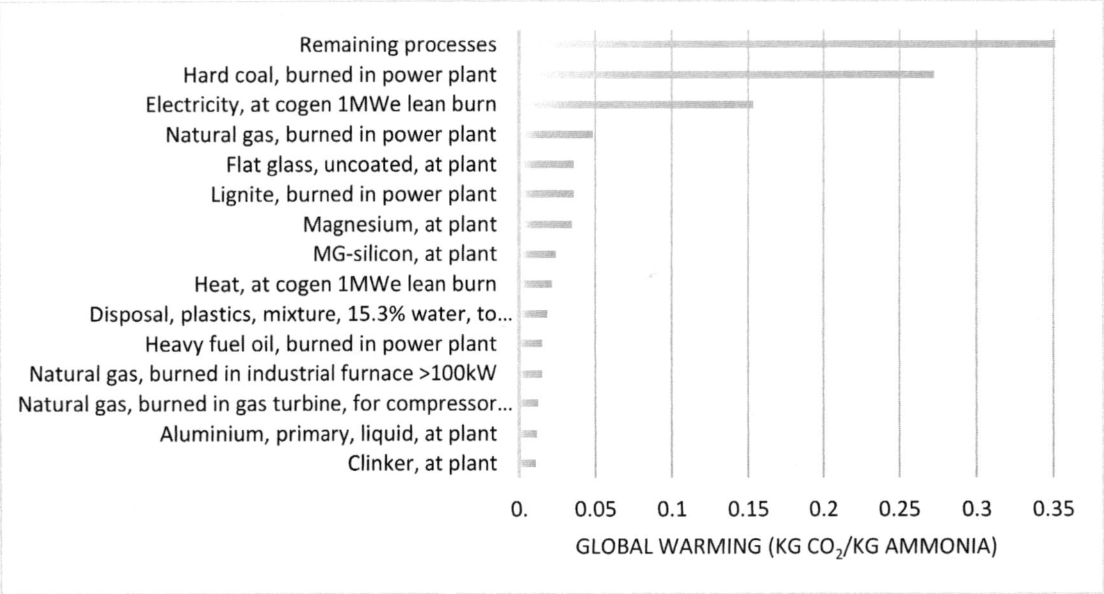

Figure 21 Contribution of various sub-processes to global warming potential of PEC (concentrated light) based electrochemical ammonia synthesis

Table 6 The shares of different sub-processes in global warming category for PEC (concentrated light) based electrochemical ammonia synthesis

Inflows	Flow	Unit
Total	100	%
Electricity, production photovoltaic, multi-Si	51.9	%
Hydrogen, PEC cell, PV, Concentrated Light-Integrated System	24.5	%
Nitrogen, gas, at plant	23.6	%
Potassium hydroxide	0.000248	%
Iron oxide	0.000143	%
Sodium hydroxide	0.000143	%

3. LCA Uncertainty Analyses Results

Defining the uncertainties within the LCA study brings more reliability of the results. The uncertainty analyses are performed in SimaPro software using Monte Carlo technique. The presented results here are only for PEC based (concentrated light) electrochemical ammonia production method using the experimental system defined in the modeling section.

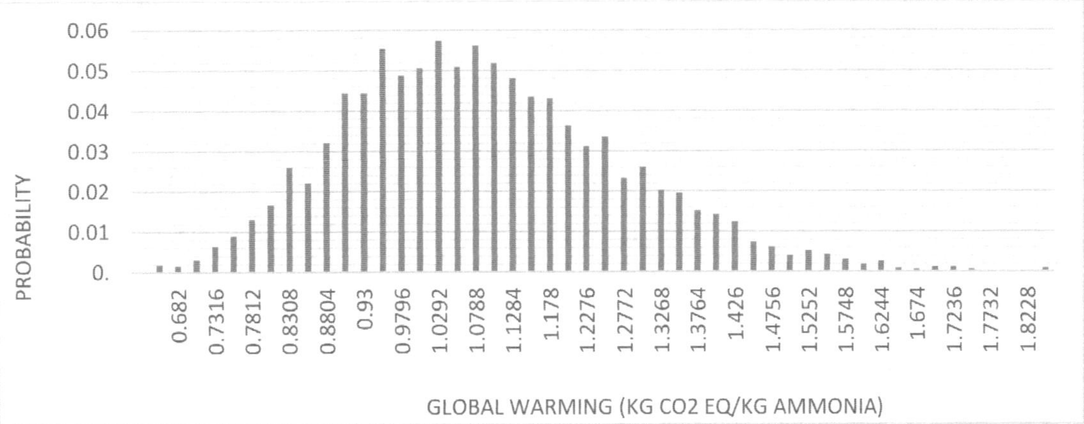

Figure 22 Probability distribution of global warming potential for PEC based (concentrated light) electrochemical ammonia production method

The confidence interval is 95% for the results. The number of runs performed for the results is 3224. The uncertainty analyses results are shown in Table 7 for the selected environmental impact categories. The mean of global warming value is 1.09 kg CO_2 eq. and standard error of mean is 0.00301 kg CO_2 eq. corresponding to 17.1% coefficient of variation which is the lowest among other categories. The highest coefficient of variance is found to be 43.9% for abiotic depletion category

Table 7 Uncertainty analyses results of PEC based (concentrated light) electrochemical ammonia production method

Impact category	Unit	Mean	Median	SD	CV (Coefficient of Variation)	97.50 %	Std.err.of mean
Abiotic depletion	kg Sb eq	0.00822	0.00746	0.00361	43.90%	0.0169	0.00773
Acidification	kg SO_2 eq	0.00637	0.00623	0.00121	19%	0.00903	0.00335
Global warming 500a	kg CO_2 eq	1.09	1.07	0.187	17.10%	1.5	0.00301
Human toxicity 500a	kg 1,4-DB eq	0.949	0.884	0.302	31.80%	1.69	0.0056
Land competition	m^2a	0.0523	0.0495	0.0167	32%	0.089	0.00564
Ozone layer depletion 40a	kg CFC-11 eq	2.75E-07	2.63E-07	7.65E-08	27.80%	4.57E-07	0.00489
Terrestrial ecotoxicity 500a	kg 1,4-DB eq	0.00104	0.000999	0.000269	25.70%	0.00168	0.00453

The probability distributions of the selected environmental impact categories are shown in Figs. 22 to 24.

Fig. 25 shows the comparison of uncertainty ranges for the different categories. This method is still in early investigation phase resulting in less reliable data for LCA inventory step. Taking into account the uncertainties of LCA results for PEC based electrochemical ammonia production method, this process can be more environmentally benign than other renewable routes.

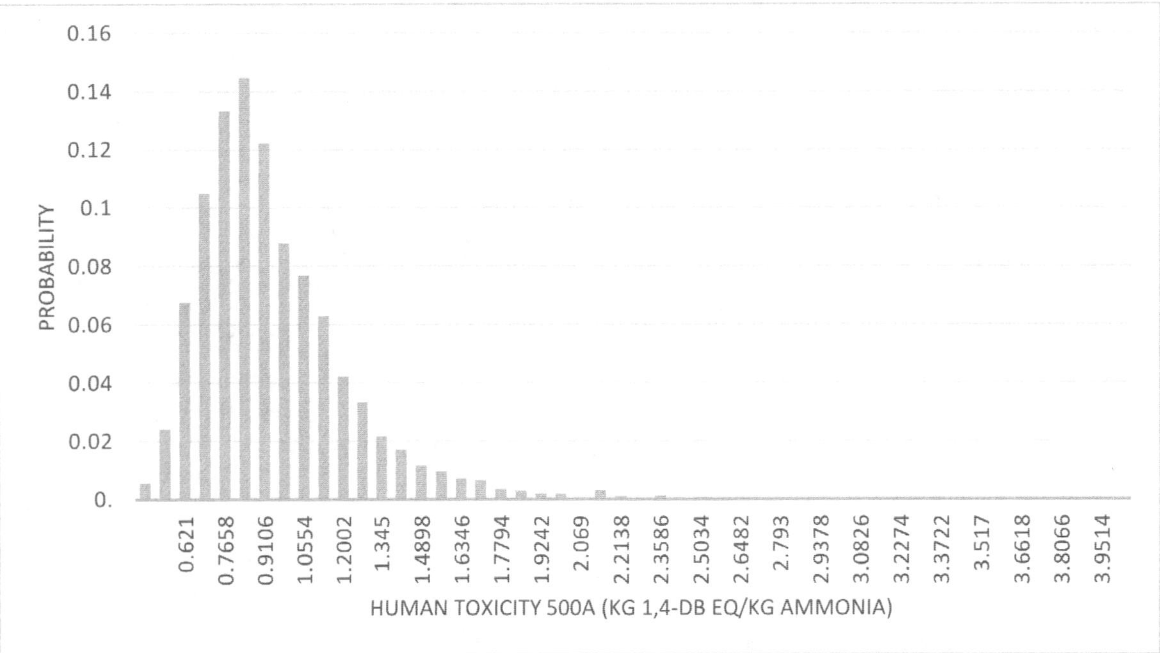

Figure 23 Probability distribution of human toxicity potential for PEC based (concentrated light) electrochemical ammonia production method

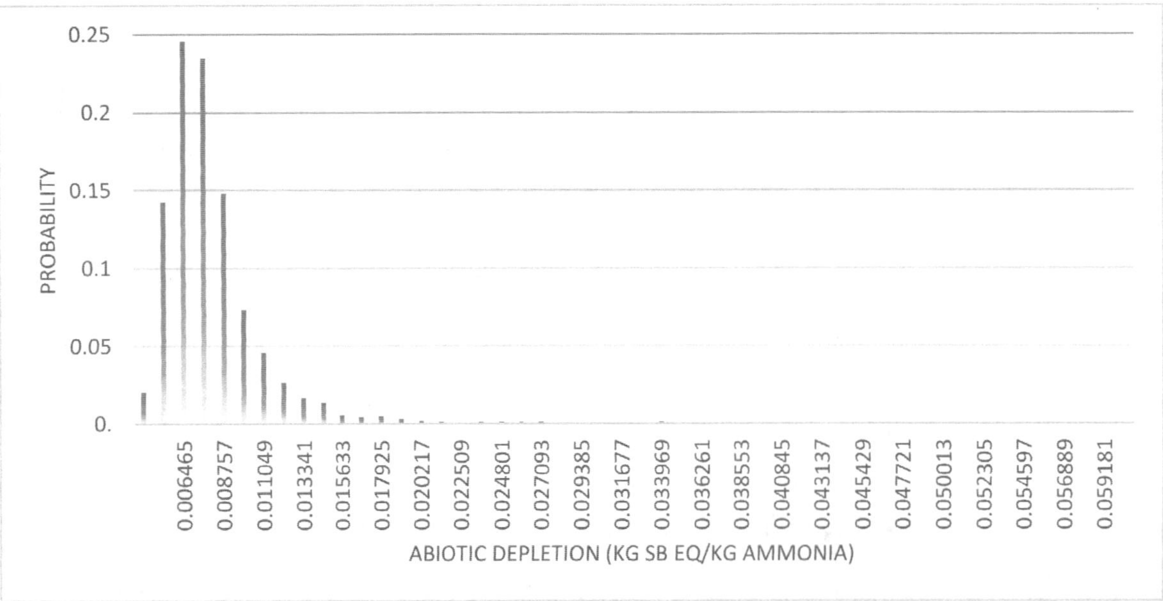

Figure 24 Probability distribution of abiotic depletion potential for PEC based (concentrated light) electrochemical ammonia production method

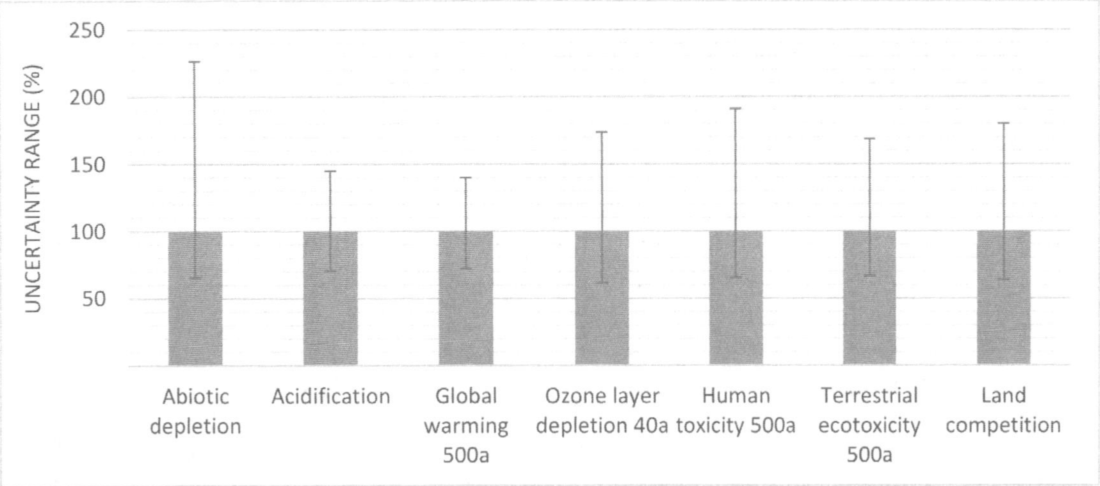

Figure 25 Uncertainty ranges of the selected impact categories for PEC based (concentrated light) electrochemical ammonia production method

4. Experimental Investigation and Analysis

PEC hydrogen generation system in this study comprises of primarily a photoelectrochemical cell with a membrane electrode assembly (Fig. 26), photovoltaic (PV) module, light source (concentrated), electricity supply and optical tools such as Fresnel lens and spectrum splitting mirrors as shown in Fig. 26. The Fresnel lens is utilized for concentrating the light. It is a periodic refractive arrangement of concentric prisms. The cathode plate is electrochemically deposited with copper oxide photosensitive material enhancing the hydrogen evolution as photocathode. In this study, the electrochemical deposition of Cu_2O onto the stainless steel cathode plate is conducted in an electrolyte solution consisting of 0.4 M $CuSO_4 \cdot 5H_2O$ and 3 M lactic acid. The solution temperature is kept constant during deposition by temperature controller where it is set to 55°C. The electrodeposition process continued about 80 minutes in total. The applied voltage for the electrodeposition process is -0.3 V vs. Ag/AgCl.

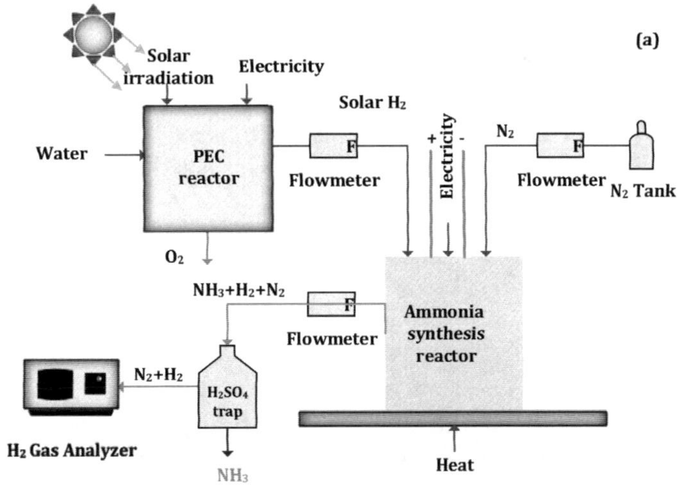

Figure 26 Experimental setup for photoelectrochemical hydrogen production integrated to ammonia synthesis

The ammonia electro-synthesis reactor comprises a nickel mesh cathode and a nickel mesh anode immersed in molten hydroxide electrolyte containing 10 g suspension of the nano-Fe_3O_4 contained in alumina crucible sealed to allow gas inlet at the cathode and gas outlet from the exit tubes as shown in Fig. 27. The reactants, H_2 and N_2, are bubbled through the mesh over the anode and cathode, respectively. The nitrogen gas flow rate is about 80 mL/min in average and hydrogen flow rate is about 10 mL/min in average during the experiments. The combined gas products (H_2, N_2 and NH_3) exit through two exit tubes in chamber head space. The exiting gases are firstly measured using flowmeters and bubbled through an ammonia water trap then analyzed for ammonia, and subsequently the NH_3 scrubbed-gas is analyzed for H_2 or N_2. The electrodes are connected externally by spot welded Ni wires.

Figure 27 The experiments under concentrated sun light for photoelectrochemical hydrogen production integrated to ammonia synthesis

5. Results and Discussion

The photoelectrochemical cell having Cu_2O coated cathode plate is tested for photoelectrochemical characterization at 1.7 V and 3 V. The obtained photocurrent densities are shown in Figs. 28 and 29 for 1.7 V and 3 V, respectively under concentrated and non-concentrated light conditions. The accumulated charges during the hydrogen production experiment at 1.7 V are calculated to be 89.9 C and 108.1 C, respectively for concentrated light and non-concentrated light. Similarly, for 3 V measurements, the accumulated charge is 555 C for concentrated light and 547 C for non-concentrated light. The maximum photocurrent densities for 1.7 V and 3 V are observed to be 0.5 mA/cm^2 and 0.25 mA/cm^2, respectively. During electrochemical ammonia synthesis, to satisfy high hydrogen production rates, the applied potential is selected to be 3 V as shown in Fig. 19. The average current and hydrogen evolution rate at 3 V are measured as 1.85 A and 14.2 mL/min and 1.82 A and 13.8 mL/min for concentrated and non-concentrated light measurements, respectively. The supplied hydrogen to the ammonia reactor is measured to be 10 mL/min in average because of possible losses in tubing. Ammonia synthesis rates increase when the molten hydroxide electrolyte is mixed with high–surface area Fe_3O_4 to provide iron as a reactive surface and when nitrogen and hydrogen are present in the reactor. The molten salt medium is supplied electricity between two nickel anode and cathode electrodes. The mixture is prepared in the beginning by simply adding NaOH and KOH pellets in the reactor. After the salts melt, nano-Fe_3O_4 is added to the electrolyte and then stirred. When the mixture is ready, the lid is tightly closed and sealed. In order to yield NH_3 in the reactor, H_2, N_2 and nano-Fe_3O_4 are simultaneously needed.

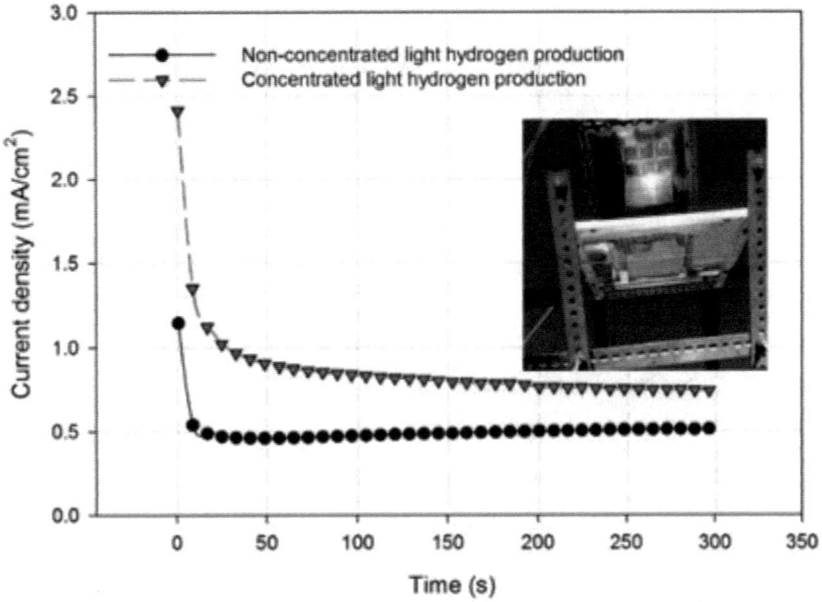

Figure 28 Photoelectrochemical hydrogen production using concentrated light and solar light splitting at 1.7 V applied potential.

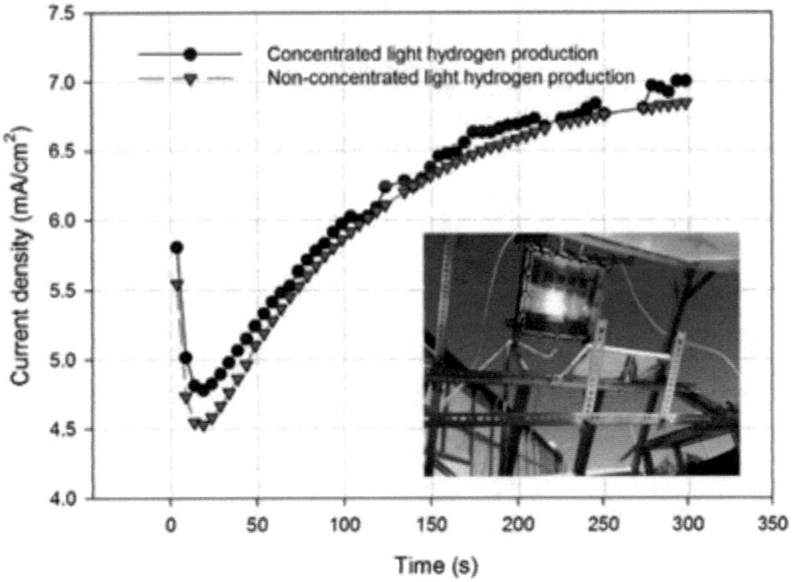

Figure 29 Photoelectrochemical hydrogen production using concentrated light and solar light splitting at 3 V applied potential during electrochemical ammonia synthesis

Table 8 tabulates the experimental conditions and yielded results for two different runs which are performed at constant current modes. Test 1 is performed at current density of 9 mA/cm^2 (0.225 A in total) and Test 2 is performed at 6.2 mA/cm^2 (0.155 A in total) current density. For Test 1, the reactor temperature is 200°C in average and for Test 2, the temperature is about 240°C. For each run, different ammonia trapping H_2SO_4 solution is used. The required cell voltage to initiate the reaction of nitrogen and hydrogen in molten hydroxide for Test 1 at 200°C in the existence of nano-Fe_3O_4 is measured to be in average 1.75 V when the applied current is 225 mA between the 25 cm^2 Ni electrodes in the molten NaOH-KOH electrolyte.

Table 8 Summary of the experimental results showing the NH_3 formation rates and efficiencies

Parameter	Test #1	Test #2
Experiment time (s)	1000	600
Mass flow rate (g/min)	0.0001125	0.00001875
Volume flow rate (mL/min)	0.1601195	0.02668659
Mol flow rate (mol/s)	1.10×10^{-7}	1.83×10^{-8}
Reactor temperature (°C)	200°C	240°C
NH_3 production rate (mol/s cm^2)	4.41×10^{-9}	7.35×10^{-10}
Current density (mA/cm^2)	9	6.2
Voltage (V)	1.75	1.2
Current (A)	0.225	0.155
Coulombic Efficiency (%)	14.17	3.43
Energy Efficiency (%)	5.50	2.45

The potential decreases to 1.2 V when the current density is to 6.2 mA/cm^2 at 240°C. In Test 1, ammonia is synthesized at a rate of 4.41×10^{-9} mol/s cm^2 whereas in Test 2, the ammonia evolution rate decreased to 7.35×10^{-10} mol NH_3/s cm^2. NH_3 is generated at a coulombic efficiency of about 14.2% at 9 mA/cm2, which declines to about 3.4% at 6.2 mA/cm^2 at 240°C. Constant current electrochemical ammonia synthesis at different applied current densities and temperature are comparatively shown in Fig. 30.

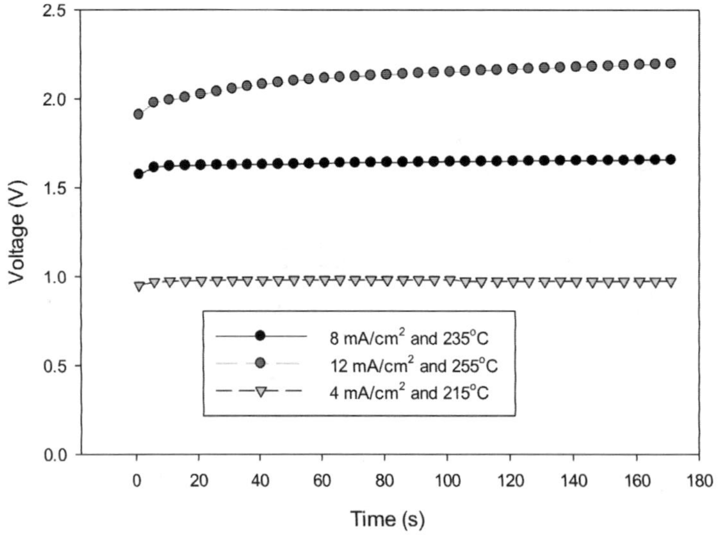

Figure 30 The relationship between voltage and time during several experimental runs at different applied currents and temperatures for electrochemical synthesis of NH_3 using N_2 and H_2 with nano-Fe_3O_4 in a molten salt hydroxide electrolyte

It is also observed in the experiments that even though the reactor temperature is below 200°C, ammonia is generated with a similar production rate to above 200°C. The potential gradually declines from 2.1 V to 1 V for the applied current densities from 12 mA/cm^2 to 4 mA/cm^2. It is observed in the experiments that lower current density and lower temperature improve the stability of the rate of NH_3 evolution.

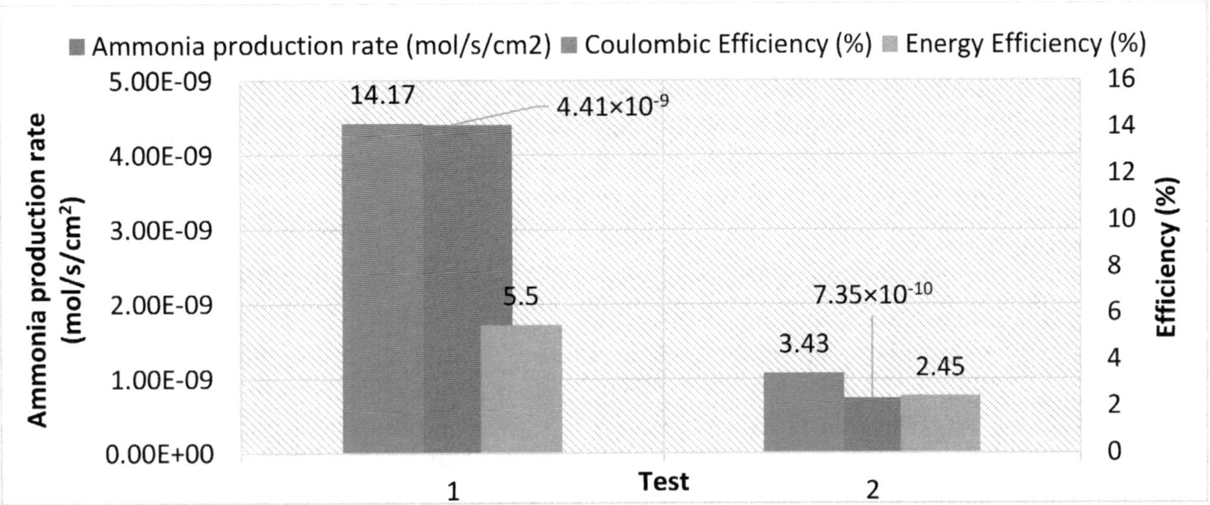

Figure 31 Coulombic and energy efficiencies of two experimental runs for electrochemical NH_3 synthesis using N_2 and H_2 with nano-Fe_3O_4 in a molten salt hydroxide electrolyte.

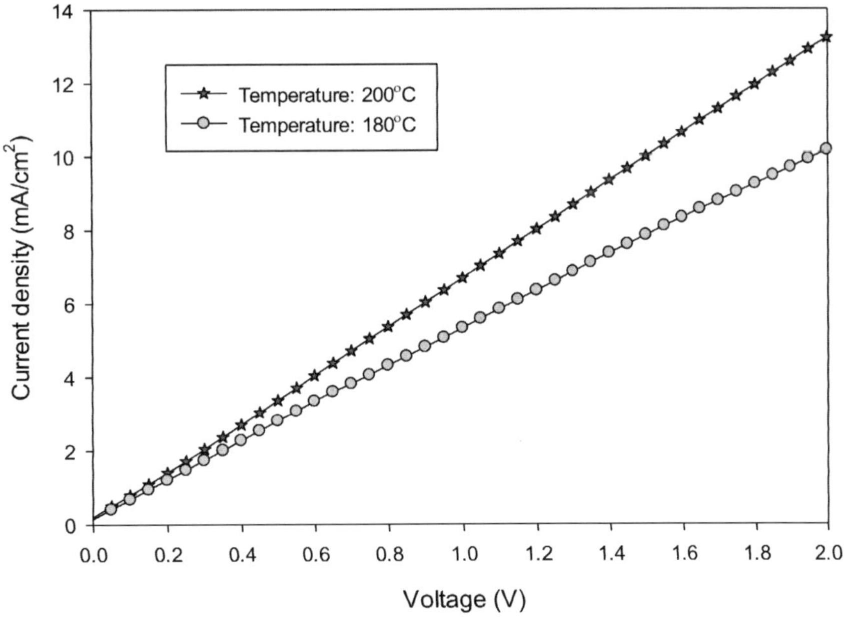

Figure 32 Applied potential-current density relations at 200°C and 180°C for electrochemical NH_3 formation using N_2 and H_2 with nano-Fe_3O_4 in a molten salt hydroxide electrolyte

The measured coulombic and energetic efficiencies of ammonia evolution in time at different temperature levels and conditions in NaOH-KOH molten electrolyte are comparatively illustrated in Fig. 31. The conversion efficiency is not only dependent on the hydrogen amount but also amount of catalyst available to stimulate the conversion of N_2 and H_2 into NH_3. The greater ammonia generation rate at lower voltages can be because of the lower hydrogen ion stream at the cathode which provides more time for generation of ammonia according to reaction. Higher NH_3 synthesis rates are obtained for Test 1 as illustrated in Fig.31. In addition, the coulombic efficiency is higher for Test 1. At higher current density of 9 mA/cm^2, the NH_3 formation rate yields about 4.41×10^{-9} mol/s cm^2. In order to understand the current-voltage characteristics at lower temperature levels such as 180°C and 200°C , linear sweep voltagram of between 0 V and 2 V as shown in Fig. 32. At 200°C and 1.2 V, the obtained current density is 8 mA/cm^2 whereas it is about 6.1 mA/cm^2 at 180°C. Hence, higher temperatures lowers the required voltage at constant supplied current.

6. Concluding Remarks

Hydrogen and ammonia are two of the most significant clean fuels in the near future. Production of these chemicals are desired to be environmentally friendly. Here, electrochemical synthesis of NH_3 is achieved using photoelectrochemical hydrogen. This method appears to be a potential alternative to conventional energy intensive NH_3 production plants especially for on-site ammonia production. The use of renewable energy resources to promote the electrochemical synthesis of NH_3, the carbon footprint of the current NH_3 production industry can be significantly reduced. The electrochemical synthesis pathways offer a high potential for NH_3 production to separate and distribute a carbon-free fuel for the various sectors. Copper oxide is electrodeposited to have photocathode. NH_3 is electrochemically generated at ambient pressure using photoelectrochemical H_2 in a molten hydroxide medium with nano-Fe_3O_4 catalyst. The reaction temperature is varied in the range of 180°C to 260°C to investigate the impact of temperature on NH_3 production rates. The maximum coulombic efficiency is calculated as 14.17 % corresponding to NH_3 formation rate of 4.41×10^{-9} mol/s cm^2. It is expected that possible problems in the electrochemical synthesis of NH_3 based on liquid electrolytes are further improved by the addition of suitable additives, optimization of the reactor configuration and more resistive materials.

CHAPTER 4: FROM HYDROCARBONS TO AMMONIA

The decarbonisation of fossil fuels, particularly, natural gas, is a promising alternative and compromises definite benefits over the use of carbon capture storage (CCS) technologies. Methane decarbonisation by pyrolysis also called as methane cracking includes the dissociation of methane (CH_4) into its molecular particles: solid carbon (C) and hydrogen (H_2). Its key benefit lies in the lack of CO/CO_2 emissions. Conversely to CCS, it substitutes the managing of CO_2 with a much lower quantity of easier-to-handle solid carbon. Hydrogen signifies a significant clean energy carrier, with an already substantial demand and capable projections for the future energy system. Moreover, carbon is hypothetically marketable as a product for both current and envisaged usages such as carbon fibres, materials and nanotechnology.

1. Natural Gas to Ammonia

Compared to natural gas, there are more environmentally friendly fuels such as ammonia. Ammonia does not emit direct greenhouse gas emissions when utilized in the vehicles. Furthermore, production process of ammonia yields lower environmental impacts compared to natural gas production. Ammonia, which is a sustainable and clean fuel, can also be produced from natural gas and hydrocarbons. Henceforth, in the ideal case, if stranded natural gas reserves can be converted into ammonia and then transported via pipelines/trucks/ocean tankers to the ports, it would have lower total environmental impact both in the production process and utilization process. Furthermore, ammonia is liquid at higher temperatures (-33°C) than natural gas (-162°C) which implies lower energy requirement in liquefaction process of natural gas.

The other option for a more environmentally friendly process can be conversion of LNG to ammonia after being produced and transported via pipelines. Natural gas can be cracked into carbon black and hydrogen using hydrocarbon disassociation technique. In this case, carbon black is also utilized as a useful output for tire, plastic etc. industry. Instead of emitting CO_2 to the environment, produced carbon black is used for various sectors, and greenhouse gas emissions are lowered. Produced hydrogen can be used for ammonia synthesis and stored in the vessels for the overseas transportation. In this manner, a cleaner alternative fuel is consumed and total greenhouse gas emissions are significantly decreased. Ammonia can be produced from any hydrogen including hydrocarbons using cracking of hydrocarbons into hydrogen and carbon. Methane is a favored option for hydrogen production from a hydrocarbon because of its high H to C ratio, availability and low cost. Furthermore, microwave disassociation of methane is a promising option for cleaner ammonia production. Methane is separated into carbon black and hydrogen. The carbon produced can be sold as a co-product into the carbon black market which could be utilized in inks, paints, tires, batteries, etc. or sequestered, stored, and used as a clean fuel for electricity production. The sequestering or storing of solid carbon requires much less development than sequestering gaseous CO_2. Ammonia can also be produced from steam reforming of methane which is a little more energy intensive method. Steam methane reforming is the conversion of methane and water vapor into hydrogen and carbon monoxide which is an endothermic reaction. The heat can be supplied from the combustion of the methane feed gas. The process temperature and pressure values are generally 700°C to 850°C and pressures of 3 to 25 bar, respectively.

There are several technological options for methane dissociation to hydrogen and carbon, which are summarized in Fig. 33. Natural gas power plant electricity can be utilized in ammonia production process from natural gas as it is illustrated in Fig. 34. Produced ammonia can be used in power plants, vehicles and also household furnaces.

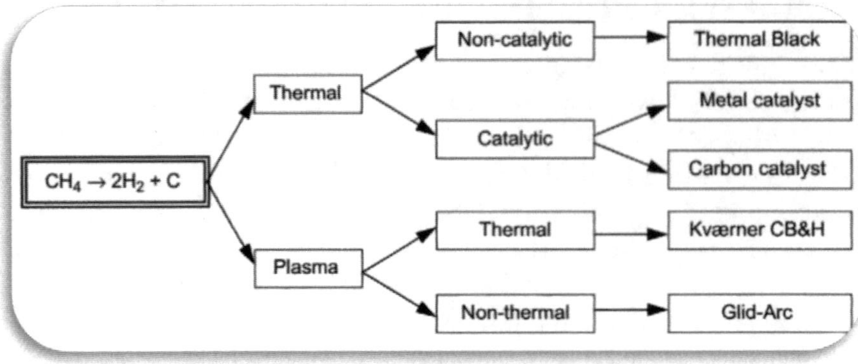

Figure 33 Main routes for decomposition of methane to hydrogen and carbon (modified from [44])

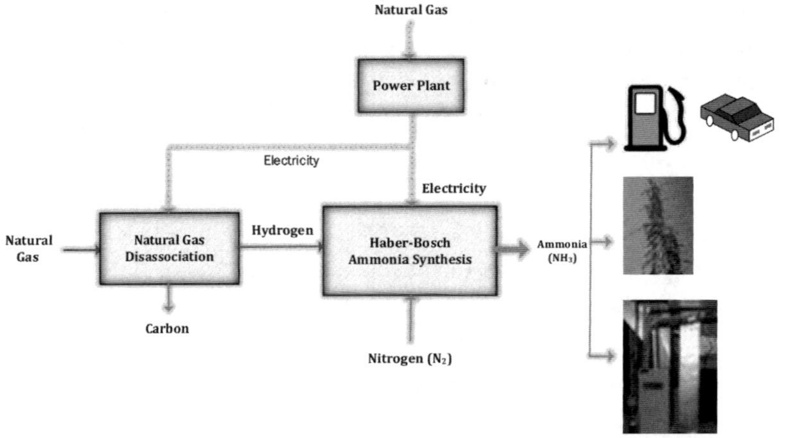

Figure 34 Schematic diagram of ammonia production from natural gas and alternative utilization options

The following Fig. 35 shows an integrated plant for ammonia and urea production which utilizes the captured CO_2 for urea production.

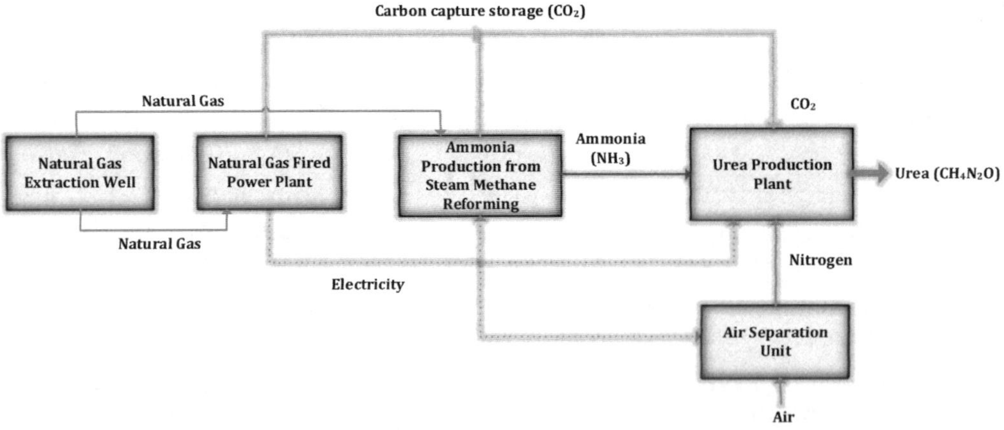

Figure 35 Illustration of ammonia and urea production from natural gas

The efficiency and CO_2 emissions from steam methane reforming, coal gasification and methane pyrolysis are comparatively shown in Table 9. Using CCS technology with steam methane reforming decreases the efficiency down to 54% which is quite similar to methane pyrolysis method. For the microwave dissociation of hydrocarbons for ammonia production, it is seen that the microwave energy may be of sufficient power and duration to cause microwave depolymerization of the high molecular weight materials such as bitumen. Microwave energy is environmentally friendly since it has no harmful effect during hydrocarbon cracking process.

Table 9 Comparison of hydrogen production technologies from fossil-fuels

Process	**Methane steam reforming**	**Coal gasification**	**Methane pyrolysis**
Reaction Heat of reaction (kJ/mol-H_2)	$CH_4+2H_2O \rightarrow CO_2+ 4H_2$ 63.25	$Coal+2H_2O \rightarrow CO_2+ 2H_2$ 89.08	$CH_4 \rightarrow C + 2H_2$ 37.43
Energy efficiency in transformation (%)	74	60	55
Energy efficiency with CCS (%)	54	43	55
CO_2 emission (mol-CO_2/mol-H_2)	0.34	0.83	0.05
Carbon production (mol-C/mol-H_2)	0	0	0.5

It is also remarkable to explain the life cycle emissions comparatively. Production of natural gas from various locations yield higher ozone layer depletion values as seen in Fig. 36. Production of fuel ammonia yields lower acidification values compared to petrol and natural gas production as shown in Fig. 37.

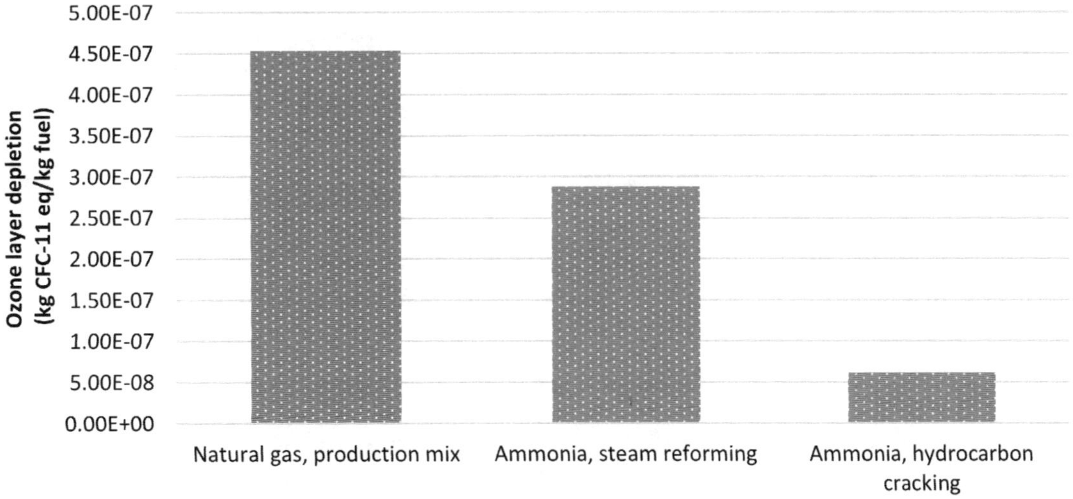

Figure 36 Ozone depletion values during production of one kg of ammonia and natural gas

Fig. 38 shows the comparative cost of ammonia production from renewable and conventional resources. Currently, steam methane reforming is the dominant method of production. However, as seen in the figure, hydrocarbon dissociation yields lower costs than low cost hydropower option and steam methane reforming method. Furthermore, hydrocarbon dissociation also produces carbon black which is a commercial commodity in the market. For example, per each kg of ammonia produced, about 0.5 kg of carbon black can be obtained from methane dissociation. If the price of carbon black is assumed to be 1 US$/kg in the market, the cost of ammonia for the hydrocarbon dissociation scenario decreases down to 0.17 US$/kg.

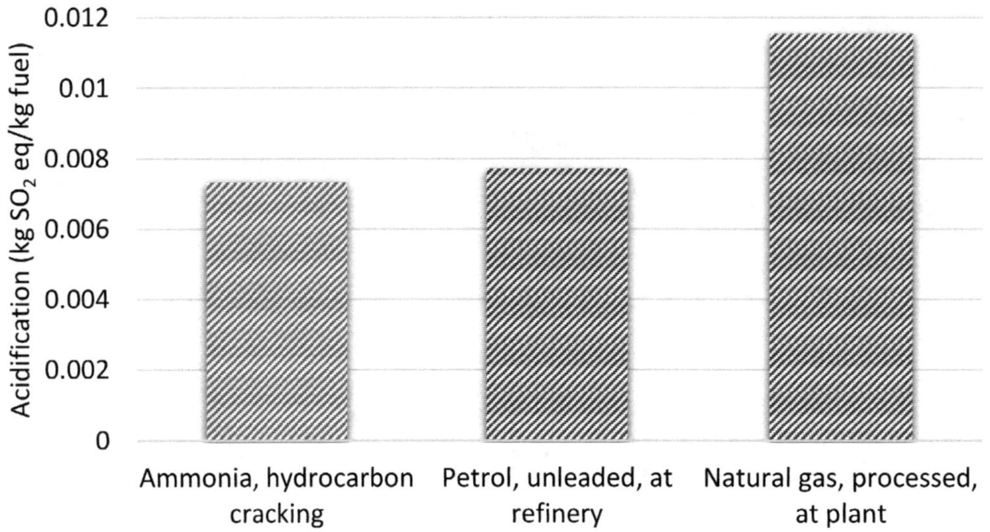

Figure 37 Acidification values of ammonia, natural gas and petrol during one kg fuel production process

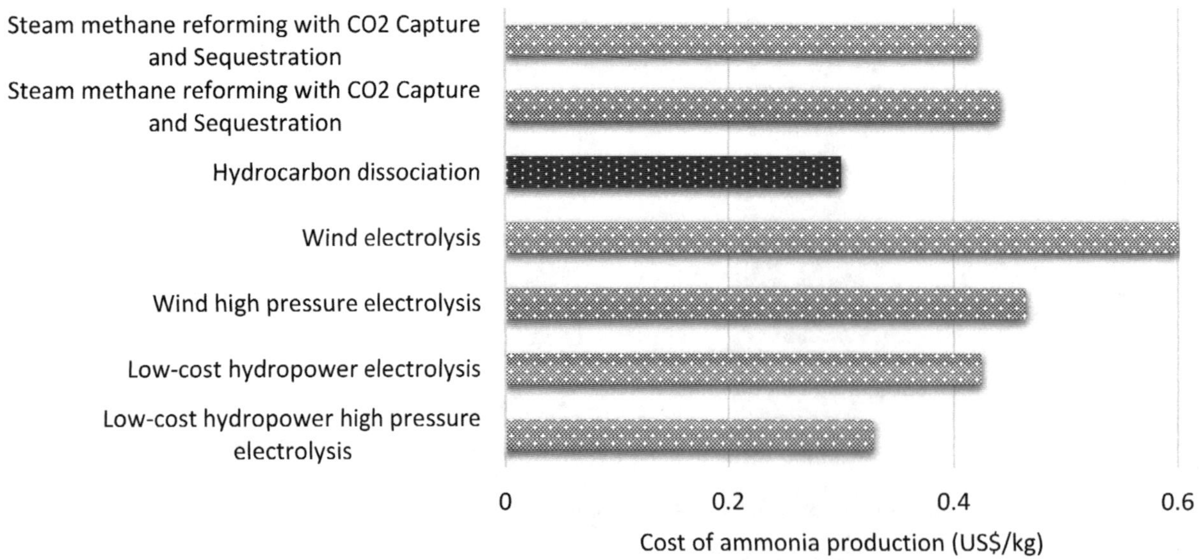

Figure 38 Comparison of cost of production for ammonia using various routes

Fig. 39 compares the total greenhouse gas emissions during production of 1 MJ energy from various resources including gasoline, LPG, diesel, natural gas and ammonia. Production of 1 MJ energy from ammonia has lower emissions than gasoline, LPG, diesel and natural gas.

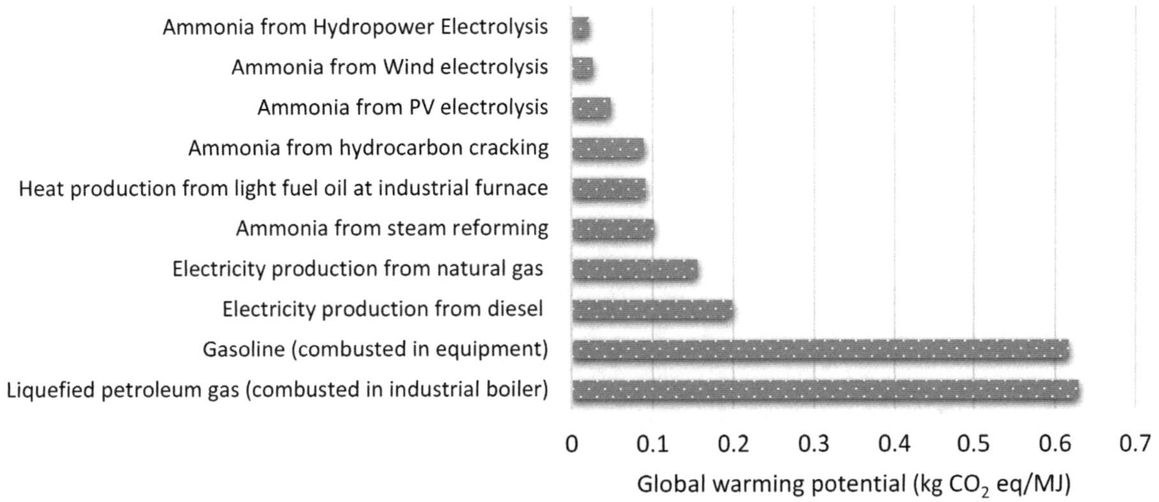

Figure 39 Comparison of global warming potential of 1 MJ energy production from various resources

2. Transport of Natural Gas or Ammonia

Much of this natural gas is considered "stranded" as it is located in regions distant from consuming markets. Liquefying natural gas and shipping it overseas provides an opportunity for these regions to economically develop their natural gas reserves. As unconventional gas production increases, the U.S. is becoming increasingly self-sufficient with respect to natural gas. Pipeline exports from Canada to the U.S. are decreasing. With plenty unconventional resources, industry is shifting its focus from importing LNG into North America to exporting LNG from North America. The export of LNG could facilitate Canadian natural gas production growth and result in significant investment, jobs and economic growth.

LNG is primarily composed of Methane (85.6% – 96.6%), Ethane (3.2% – 8.5%), and Propane (0.0% – 3.0%). The transportation of liquefied natural gas (LNG) refers to any movement or shipping of natural gas while in its liquid form. The two major methods of transporting LNG are by pipeline and vessel. As LNG requires a temperature of -160°C to remain in its liquid form, significant insulation must be incorporated into LNG pipelines in order to maintain this low temperature and ensure no re-gasification occurs. This normally includes a combination of mechanical insulation, for example glass foam and a vacuum layer. This complex insulation system makes LNG pipelines significantly more difficult and expensive to manufacture than standard natural gas pipelines [45].

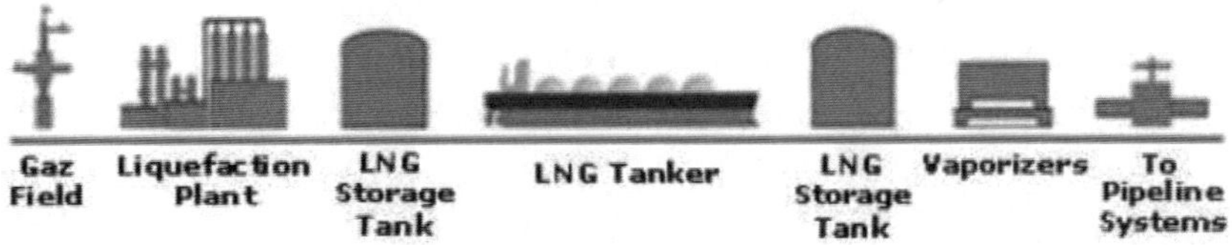

Figure 40 LNG transport processes

Interconnected elements shown in Fig. 40 include the gas field, liquefaction plant, LNG storage tank, LNG tanker, LNG storage tank, vaporizers, and pipeline systems. In LNG exporting countries, natural gas is extracted from basins and transported by pipeline to liquefaction plants. There, the natural gas is liquefied and stored. The majority of worldwide LNG exports take place at an intercontinental level, meaning that shipping LNG across the ocean is often required. This is done with the use of an LNG vessel, which transports large quantities of LNG between export and import terminals. Several types of LNG vessels exist in the industry today, with the main one being referred to as an LNG tanker. The main components of an LNG tanker are the boiler and pump rooms, a double hull for added strength, bow thrusters, and the LNG storage tanks themselves. Typically, an LNG tanker is built with 4 or 5 individual LNG tanks, as seen in Fig. 41.

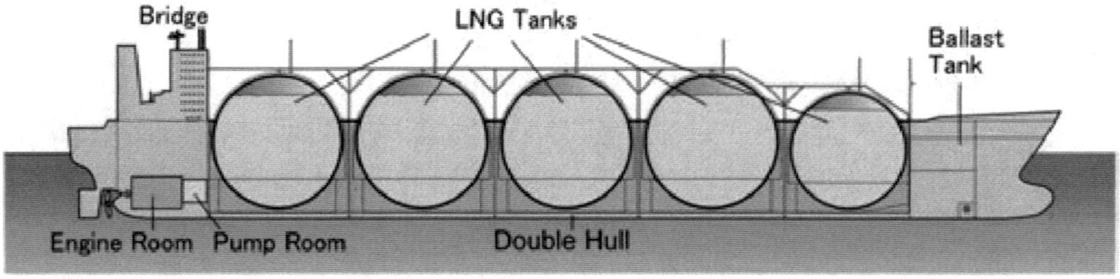

Figure 41 Illustration of an LNG tanker (adapted from [45])

Ammonia tankers are typically refrigerated ships. In addition to ammonia, they can usually transport liquefied propane gas (LPG), propylene, vinyl chloride monomer and other condensable gases. Capacities more than 84,000 m^3 are available. On the other hand, a typical LNG tanker can carry around 160,000 m^3 of natural gas on a single voyage. Ammonia is a more suitable energy carrier to be transported overseas because it is liquid at higher temperatures and require less cooling compared to natural gas.

3. Case Studies for LNG and Ammonia

Natural gas liquefaction costs include:

- liquefaction costs of about US$3.00/MMBTU,
- shipping/fuel costs of about $1.50-$3.50/MMBTU,
- the cost of the gas at about $4.00/MMBTU in a low cost region

Therefore, the total delivered cost corresponds to about $8.50-$10.50/MMBTU relative to a market price of about $12-$15/MMBTU in Europe and Asia based on $100/barrel crude oil as gas in Europe and Asia is typically priced at 11-15% of the price of crude on an energy equivalent basis [46]. The natural gas prices are also derived from another report prepared by International Gas Union [47].

Ammonia distribution costs are considered to be similar to LPG costs. The LPG is kept at approximately -48°C, which is below the boiling point of LPG (propane), which is -42°C. Ammonia distribution would require additional safety equipment, but there are likely to be cost reductions if ammonia were distributed on scales approaching those of current gasoline/LNG distribution. Hence, it seems reasonable to assume that these effects would offset each other, yielding similar costs. A study estimates LPG distribution costs to be around $0.36 to $0.55 per gasoline gallon equivalent (gge), including retail margins. Converting the $0.36/gge cost of distribution equates to approximately $0.62/kg H_2 when distributed as anhydrous ammonia [48]. Here, we employ different cases based on the production cost of ammonia and LNG. For steam methane reforming based ammonia production, the natural gas cost represents almost 58% of the total ammonia cost as shown in Fig. 42. Therefore, it is highly dependent on natural gas market price. The production cost of natural gas includes the extraction and processing.

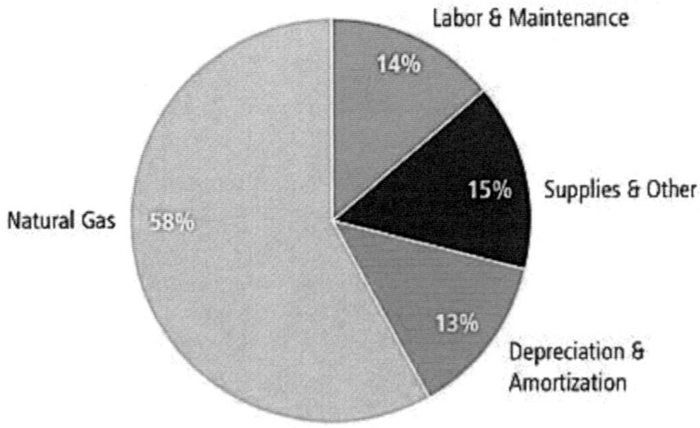

Figure 42 Cost contributions for natural gas based ammonia production plant (adapted from [49])

Case 1: Europe

In this case, ammonia is considered to be produced with a unit cost of 0.44 US$/kg [50,51]. The LNG prices are converted to kg and MJ for comparison purposes as shown in Table 10. For the conversions, 1 MMBTU corresponds to about 1055 MJ. The lower heating value of the LNG is taken as 49.4 MJ/kg [52]. The production cost is cost of gas at the liquefaction plant. After the liquefaction process, the LNG is loaded on the tankers and transported overseas. At the arrival port or plant, the LNG is regasified to be used which is an additional cost. This case represents the average natural gas and ammonia prices in Europe as the illustration of the case is shown in Fig. 43.

Table 10 Unit cost of LNG including the sub-processes for Case 1

LNG	US$/MMBTU	US$/MJ	US$/kg
Production	4.4000	0.0042	0.2060
Liquefaction	3.2000	0.0030	0.1498
Transport	4.1000	0.0039	0.1920
Regasification	0.3500	0.0003	0.0164
TOTAL	12.0500	0.0114	0.5642

The total cost of ammonia and LNG are comparatively evaluated in Table 11 and Fig. 44. The selected ammonia production route is steam methane reforming which receives natural gas as the feedstock. The production plant of ammonia and LNG are assumed to be in the same location and has a distance of 100 km to the port. The land transportation cost of ammonia is calculated based on the average costs per km where a truck's payload capacity is 16 tonne. The truck transportation cost values are taken from the report [53]. Although, there are alternative ways for the transport of ammonia such as pipelines, since there is no ammonia pipeline infrastructure at the moment in Canada, truck transportation is taken into account. It is assumed that 16 tonne liquid fuel can be carried per truck for both fuels. Here, the storage costs of these liquid fuels are not taken into account since they are considered to be transported after the production.

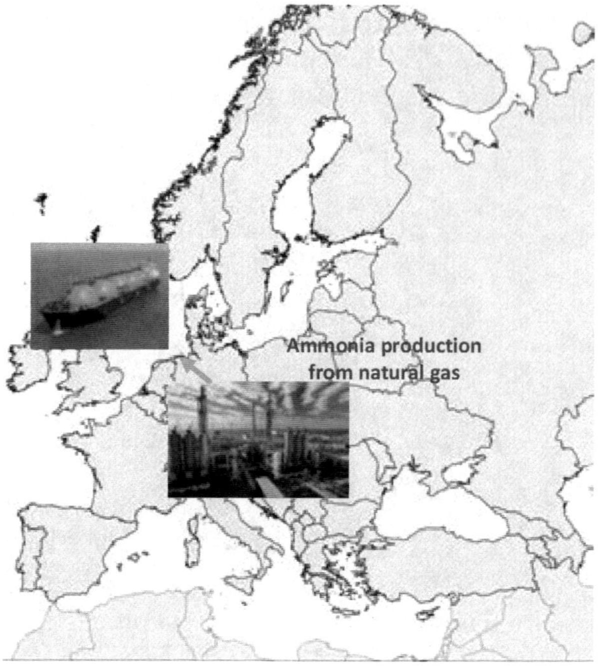

Figure 43 Illustration of the Case 1 for Europe

Table 11 Total unit cost of ammonia and LNG considering production, liquefaction and transport for Case 1

Cost contribution	LNG	AMMONIA
Production cost (US$/kg)	0.206	0.440
Liquefaction cost (US$/kg)	0.149	0
Regasification cost (US$/kg)	0.016	0
Land transportation cost (US$/kg)	0.012	0.012
Tanker overseas transportations cost (US$/kg)	0.191	0.109
Total cost (US$/kg)	0.576	0.562
Total cost (US$/MJ)	0.011	0.030

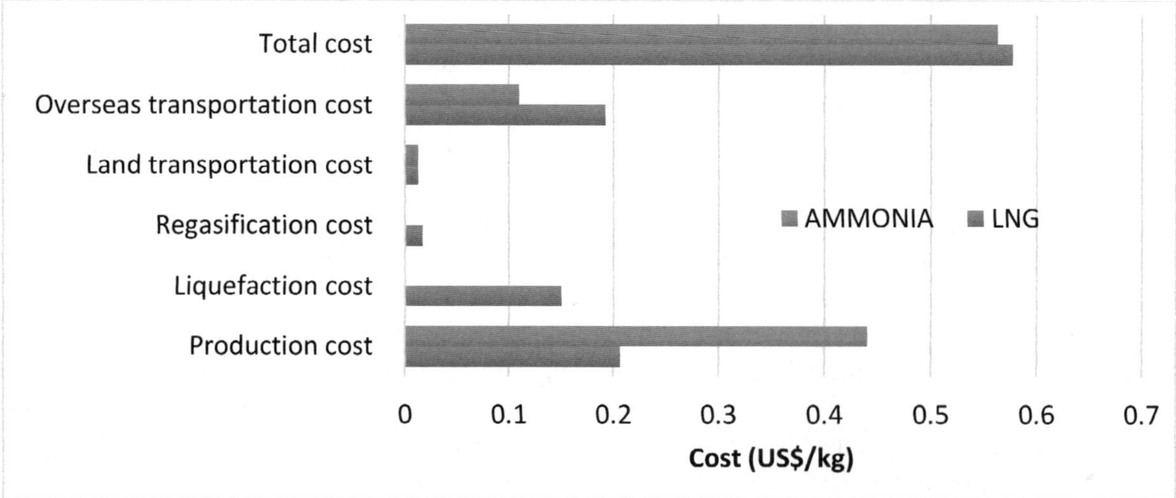

Figure 44 Contribution of sub-processes to total cost of LNG and ammonia for Case 1 in Europe

As shown in Fig. 44, the cost of ammonia at the destination port is slightly lower than LNG although the production cost of ammonia is higher than LNG. Since ammonia is liquid at the production plant, there is no need for additional liquefaction. In addition, regasification cost is neglected for ammonia.

Case 2: U.S

In this case, the production cost of ammonia is taken lower corresponding to 0.22 \$/kg [50,51]. The LNG prices are converted to kg and MJ for comparison purposes as shown in Table 12.

Table 12 Unit cost of LNG including the sub-processes for Case 2

LNG	US\$/MMBTU	US\$/MJ	US\$/kg
Production	3.8	0.0036	0.1779
Liquefaction	2	0.0019	0.0936
Transport	3	0.0028	0.1405
Regasification	0.3	0.0003	0.0140
TOTAL	9.1	0.0086	0.4261

The total cost of ammonia and LNG are comparatively evaluated in Table 13 and Fig. 45. The selected ammonia production route is steam methane reforming which receives natural gas as the feedstock. The production plant of ammonia and LNG are assumed to be in the same location and has a distance of 200 km to the port. This case represents the average natural gas and ammonia prices in the U.S. as the illustration of the case is shown in Fig. 46.

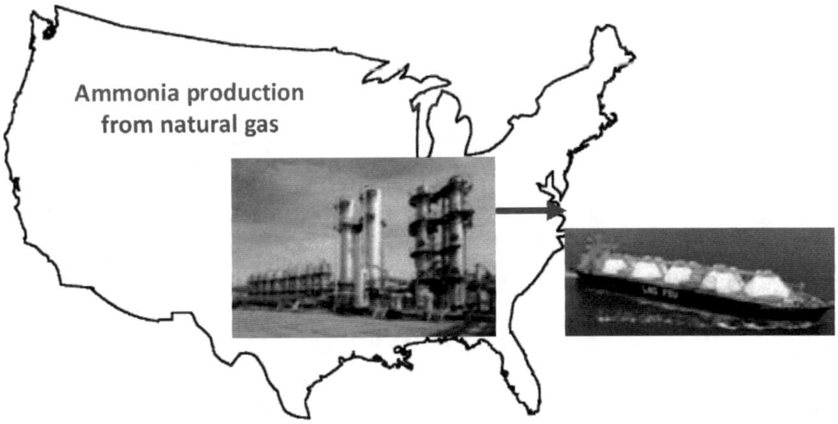

Figure 45 Illustration of the Case 2 for the U.S

As shown in Fig. 46, the cost of ammonia at the destination port is considerably lower than LNG although the production cost of ammonia is slightly higher than LNG. The difference is mainly caused by the liquefaction and regasification processes required for LNG. Because the boiling temperature of LNG is very low compared to ammonia.

Table 13 Total unit cost of ammonia and LNG considering production, liquefaction and transport for Case 2

Cost contribution	LNG	AMMONIA
Production cost (US$/kg)	0.178	0.220
Liquefaction cost (US$/kg)	0.094	0
Regasification cost (US$/kg)	0.014	0
Land transportation cost (US$/kg)	0.025	0.025
Tanker overseas transportations cost (US$/kg)	0.140	0.110
Total cost (US$/kg)	0.451	0.355
Total cost (US$/MJ)	0.009	0.019

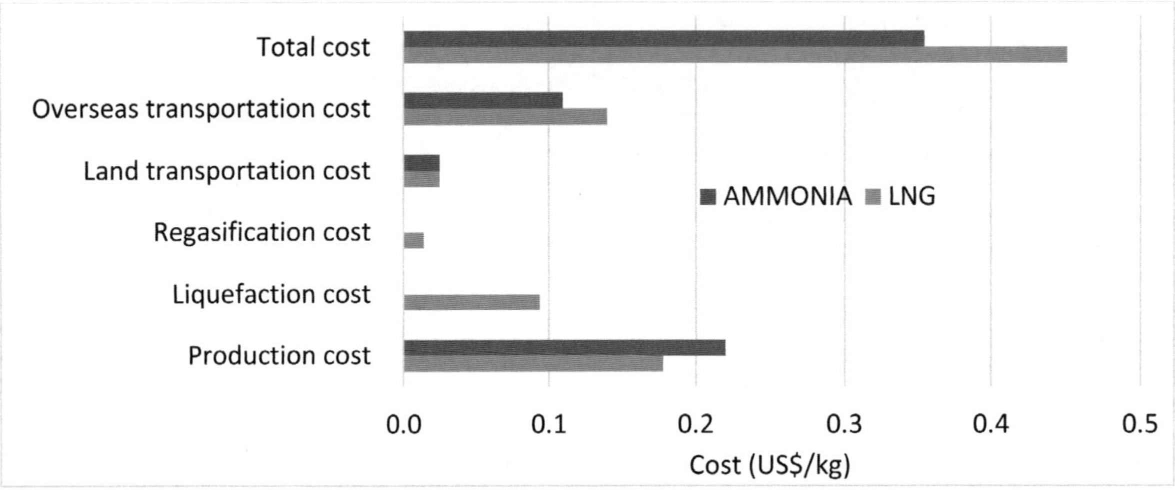

Figure 46 Contribution of sub-processes to total cost of LNG and ammonia for Case 2 in the U.S

Case 3: Middle East

In this case, the production cost of ammonia is taken lower corresponding to 0.1 $/kg [50,51]. The LNG prices are converted to kg and MJ for comparison purposes as shown in Table 14. The cost of natural gas and ammonia in Middle East is lower than other continents. This case represents the average natural gas and ammonia prices in the Middle East as the illustration of the case is shown in Fig. 47.

Table 14 Unit cost of LNG including the sub-processes for Case 3

LNG	US$/MMBTU	US$/MJ	US$/kg
Production	2	0.0019	0.0936
Liquefaction	1.5	0.0014	0.0702
Transport	1.5	0.0014	0.0702
Regasification	0.3	0.0003	0.0140
TOTAL	5.3	0.0050	0.2482

The total cost of ammonia and LNG are comparatively evaluated in Table 15 and Fig. 48. The selected ammonia production route is steam methane reforming which receives natural gas as the feedstock. The production plant of ammonia and LNG are assumed to be in the same location and has a distance of 300 km to the port.

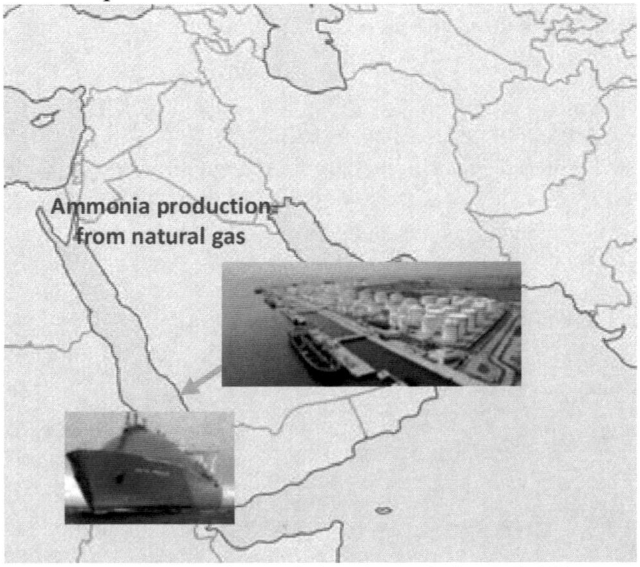

Figure 47 Illustration of the Case 3 for Middle East

Table 15 Total unit cost of ammonia and LNG considering production, liquefaction and transport for Case 3

Cost contribution	LNG	AMMONIA
Production cost (US$/kg)	0.094	0.100
Liquefaction cost (US$/kg)	0.070	0.000
Regasification cost (US$/kg)	0.014	0.000
Land transportation cost (US$/kg)	0.038	0.038
Tanker overseas transportations cost (US$/kg)	0.070	0.110
Total cost (US$/kg)	0.286	0.247
Total cost (US$/MJ)	0.006	0.013

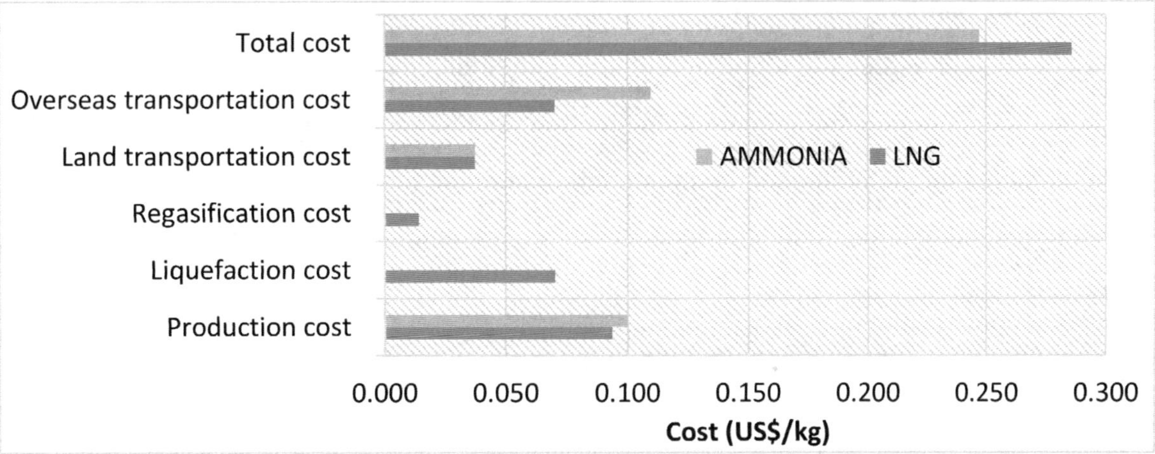

Figure 48 Contribution of sub-processes to total cost of LNG and ammonia for Case 3 in Middle East

As illustrated in Fig. 48, the cost of ammonia at the destination port is 0.04 $/kg lower than LNG. The low cost of natural gas is mainly due the massive resources available in the Middle East. This yields lower costs of liquid ammonia and LNG.

Case 4: Ontario, Canada

In this case, the production cost of ammonia is taken as 0.25 $/kg [50,51]. The LNG prices are converted to kg and MJ for comparison purposes as shown in Table 16. The cost of natural gas is quite low compared to Europe however higher than Middle East. The illustration of the case is shown in Fig. 49.

Table 16 Unit cost of LNG including the sub-processes for Case 4

LNG	US$/MMBTU	US$/MJ	US$/kg
Production	2.1	0.0020	0.0983
Liquefaction	3.0	0.0028	0.1405
Transport	2.5	0.0024	0.1171
Regasification	0.3	0.0003	0.0140
TOTAL	7.9	0.0075	0.3699

The total cost of ammonia and LNG are comparatively evaluated in Table 17 and Fig. 50. The selected ammonia production route is steam methane reforming which receives natural gas as the feedstock. The production plant of ammonia and LNG are assumed to be in the same location (nearby Toronto) and has a distance of about 800 km to the port in Quebec. This case represents the average natural gas and ammonia prices in the Canada

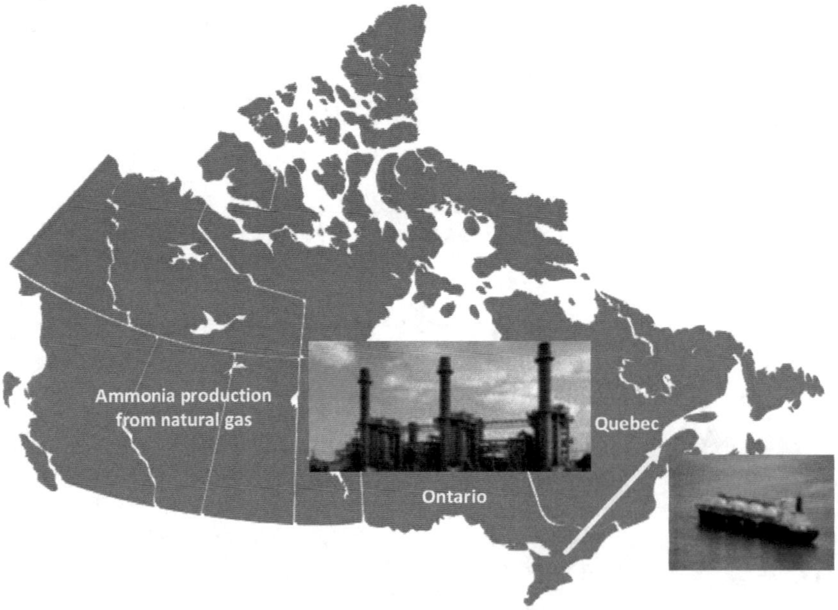

Figure 49 Illustration of the Case 4 for Ontario

Table 17 Total unit cost of ammonia and LNG considering production, liquefaction and transport for Case 4 in Canada

Cost contribution	LNG	AMMONIA
Production cost (US$/kg)	0.098	0.250
Liquefaction cost (US$/kg)	0.140	0.000
Regasification cost (US$/kg)	0.014	0.000
Land transportation cost (US$/kg)	0.100	0.100
Tanker overseas transportations cost (US$/kg)	0.117	0.110
Total cost (US$/kg)	0.470	0.460
Total cost (US$/MJ)	0.010	0.025

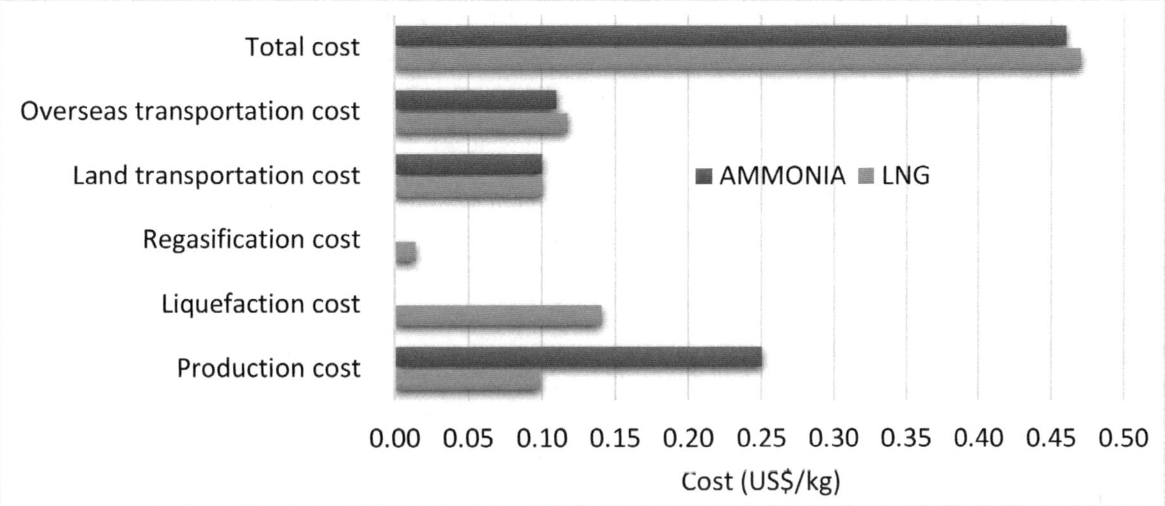

Figure 50 Contribution of sub-processes to total cost of LNG and ammonia for Case 4

As illustrated in Fig. 50, the cost of ammonia at the destination port is 0.01 $/kg lower than LNG in Ontario, Canada.

4. Oil sand to Ammonia

Microwave energy is an alternative type of energy which can be used in oil sand separation. Many of the inorganic particles in processed oil sands carry a charge, and could be influenced by electromagnetic radiation. They are excited at an altered rate than the water and bitumen when irradiated, making a temperature gradient between the different components of the oil sands. The surfactants and other forces cannot cope with this gradient and the solids are able to break free. Since all oil sands are dissimilar, a single frequency or power of microwaves are not available which functions best for all. There have been already some attempts for dissociation of bitumen using microwave energy. Pierre et al. [54] used a 915 MHz microwave to separate a 570 g sample of oil sands. This sample was exposed for 5 min at 500W and extended a last temperature of 315°C. This resulted in numerous layers comprising a bottom layer of sand with an asphaltene-like material on top. A second, noticeably bigger specimen with a volume of about 20 L was exposed for 9 min at 1500W at the same frequency. It got a temperature of 142°C where it exhibited three distinct layers. The bottom layer was typically sand, but also included other solids. The second layer contained a yellowish solution, accounting for all the water and other impurities in the oil sand. The top layer was black, viscous oil. Bosisio et al. [55] considered microwave-assisted extraction of oil sands. They conducted oil sand extraction experiments under inert atmosphere in a quartz reactor which was located in a rectangular microwave guide (WR 284 wave guide) built-in with a coupling iris and modifiable short circuit. Instance microwave power of 100W at 2450 MHz frequency was realized to the oil sands samples and the different phases of reaction were perceived. The duration of first was 10 to 15s and second-third step were 15s to 15 min. Microwave radiation of oil sands created a crude oil and also small quantities of gaseous yields. Therefore, electromagnetic heating also proved to increase yields of crude oil from 70% to 86%. Lately, Global Research Corporation technologies declared a method using over 8,700 RF microwave frequencies essential to hydrocarbon elements. About 1.2 gal of diesel fuel was extracted from a tire after microwave radiation under vacuum at different frequencies. The company requested to extract oil and gas from diverse feedstocks by microwave radiation such as oil sands, oil shale, used plastics, or rubber with little or no additional processing. Initial testing results formed great amounts of hydrogen and methane gases without CO or CO_2 contaminants [56]. Samples of Lloydminster oil sands (oil–water–solids ratio of 19:40:41) considered principally stubborn to separation were tested with exposure to adjustable frequency microwaves [57]. The tests were able to avoid the FCC regulations because of distinct apparatus to comprise the electromagnetic waves. The aim was to explore what microwave frequency is best for this type of oil sands. All samples separated into a liquid upper layer and a mostly solid lower layer. The degree of separation was calculated by evaluating the oil content of the lower layer. The lower layer of the control specimen, which was not irradiated, had 27% oil content by weight. Tests were achieved by changing frequency and exposure time. Most of the samples presented marginally lower oil content. The best result occurred when the sample was irradiated for a duration of 10 min at 6400±100 MHz, where the lower layer only had 19% oil content. It was described microwave assisted extraction to recover hydrocarbon substances from oil shale, oil sands and lignite in the U.S. Patent 4.419.214 [58][6]. The patent describes how oil sand, oil shale rock and lignite samples were irradiated in a pressure vessel with gaseous or liquefied carbon dioxide and other gaseous or vapor hydrocarbon solvents. For example, oil sand was taken into a microwave feeder pipe and irradiated at 5.8 GHz frequency. Further, BTX (benzene-toluenexylene-ethylbenzene) was pumped through the sand. The extraction solution contained green oil on the top and bitumen was obtained at the bottom. In another example, crushed oil shale rock was irradiated at 915 MHz frequency. After radiation, carbon tetrachloride was injected to extract kerogen and the projected isolated kerogen was 65% of initial organic content of the rock. It is also specified in the review article [59] that oil sands and oil shale samples irradiated to microwaves produced the required heat to dissociate bitumen components to produce a crude oil and distilled kerogen. A laboratory microwave oven was utilized to test oil sand samples from Athabasca and oil shale specimens from Green River and Sunnyside. The authors found that a 128 g sample of oil sand separated into its components when irradiated at 800W for 10 min, followed by 1500W for 15 min.

Microwave energy is environmentally friendly since it has no harmful effect during hydrocarbon cracking process. Optimized ammonia synthesis using the excess heat in Haber-Bosch (which is the most common method for ammonia production) ammonia plant for oil sand bitumen extraction which is used for hydrogen production via microwave dissociation process is possible as shown in Fig. 51.

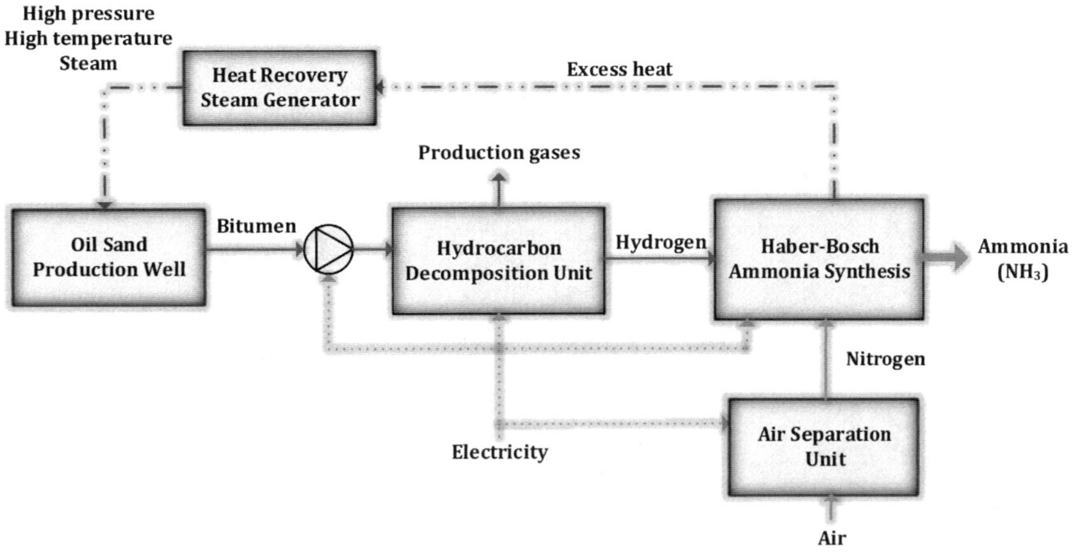

Figure 51 Schematic diagram of oil sand to ammonia plant

Fig. 52 shows the acidification potential (AP) for the selected ammonia production routes. Acidifying substances causes a wide range of impacts on soil, groundwater, surface water, organisms, ecosystems and materials. It is mainly caused by hard coal usage in the electricity grid mixture. The eutrophication category reflects the impacts of to excessive levels of macro-nutrients in the environment caused by emissions of nutrients to air, water and soil. As shown in Fig. 53, the values are close to each other corresponding to 0.0012 kg PO_4 eq/kg ammonia for hydropower route. Fig. 54 shows the ozone layer depletion (ODP) potential of the routes. It may have damaging properties upon human health, animal health, terrestrial and aquatic ecosystems, biochemical cycles and on materials. Hydrocarbon route has the lowest ODP value whereas wind has the highest since it is mainly caused by the transport of natural gas which is used in the power plants where the electricity is supplied to wind turbine production.

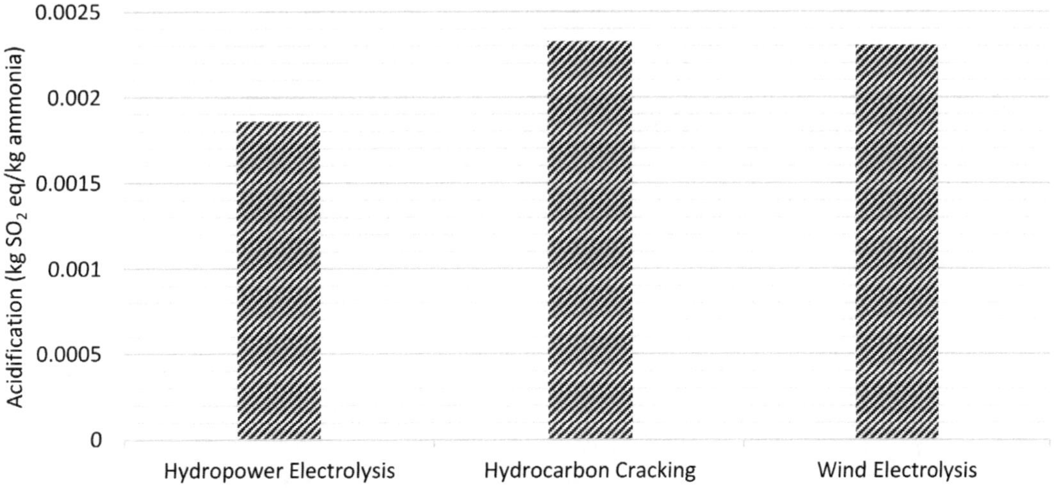

Figure 52 Acidification impact comparison of selected ammonia routes

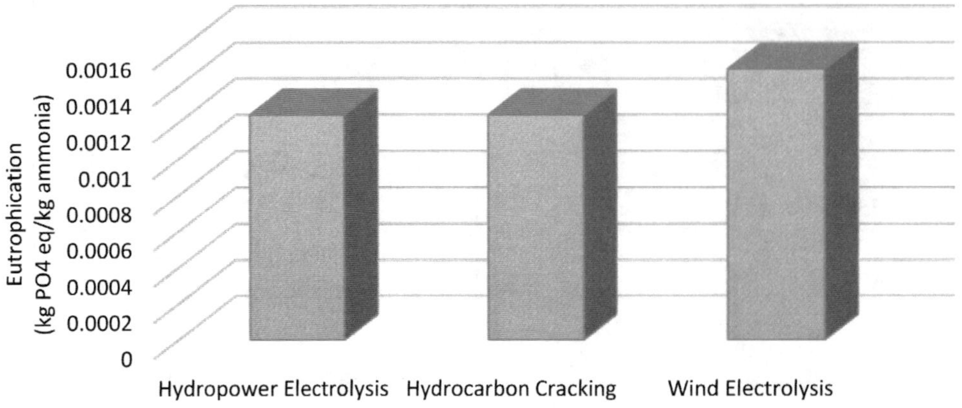

Figure 53 Eutrophication impact comparison of selected ammonia routes

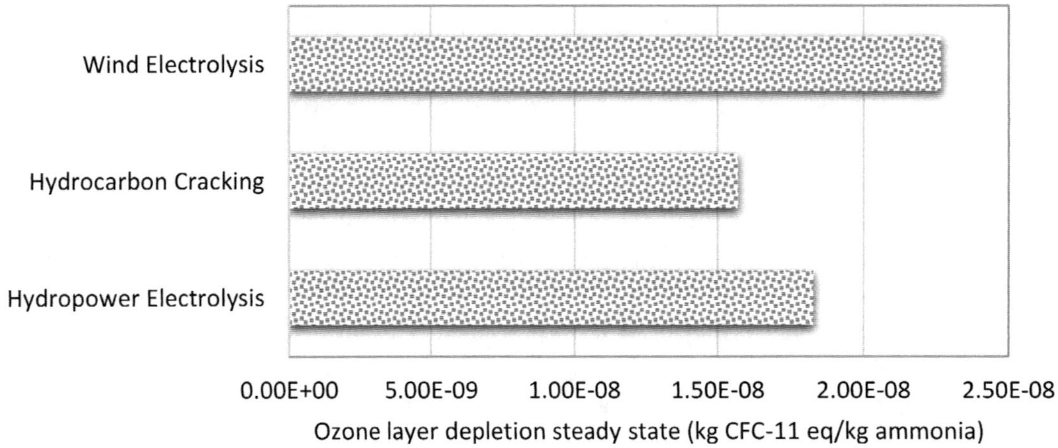

Figure 54 Ozone layer depletion impact comparison of selected ammonia routes

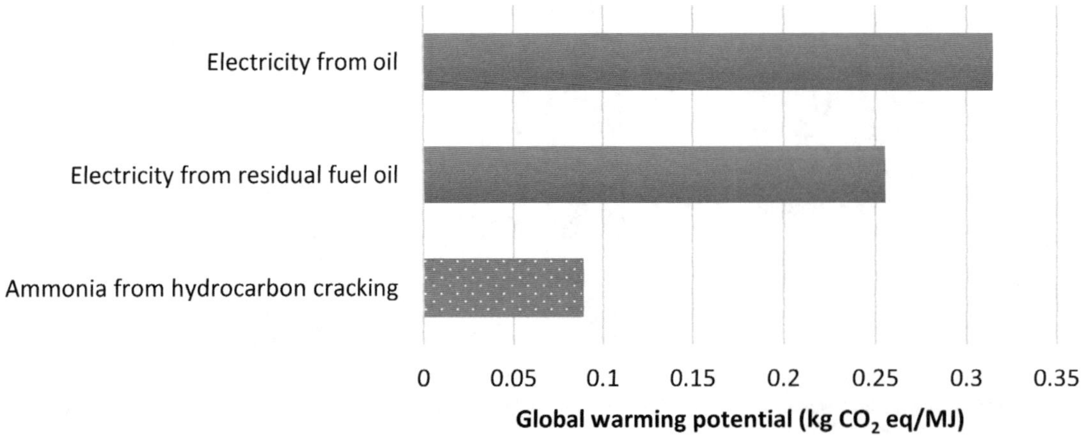

Figure 55 Comparison of 1 MJ electricity production and ammonia production

The production of 1 MJ electricity from residual oil has higher global warming potential than same amount of ammonia production from hydrocarbon cracking method as shown in Fig. 55. This result implies that by replacing ammonia as power generating fuel, total greenhouse gas emissions can be decreased significantly.

5. Concluding Remarks

Utilization of hydrocarbons in an environmentally friendly manner becomes more significant day by day. Dissociation of hydrocarbons such as methane is a promising option especially for British Columbia. Based on the extensive literature review and assessments, the following concluding remarks are noted.

- Hydrocarbons can be used as a source of hydrogen which is required for ammonia synthesis. There are various alternative pathways for hydrogen production from hydrocarbons such as thermal, non-thermal, plasma routes.
- Methane decomposition reaction is moderately endothermic process. The energy requirement per mole of hydrogen produced is considerably less than that for the steam reforming process.
- Hydrogen via thermo-catalytic dissociation of hydrocarbons represents an alternative solution. It is accompanied by the formation of carbon deposits. Methane can be thermally or thermocatalytically decomposed into carbon and hydrogen without CO or CO_2 production.
- It can be estimated that the electric energy supply needed for the cracking operation varies between 4 and 7 kWh per kg of carbon produced or between 1 and 1.9 kWh per normal cubic meter of hydrogen produced.
- Gliding arc discharge reactor is one of the highest efficient route for methane conversion which was experimentally tested by many researchers.
- H_2 production cost that can be expected from industrial methane cracking could be of the order of 1.5 $/kg and NH_3 in the range of 0.3-0.5 $/kg.
- The microwave energy can be of sufficient power and duration to cause microwave depolymerization of the high molecular weight materials such as bitumen.
- For oil sands or extremely high viscosity reservoirs, where the temperature effect on viscosity is significant, electromagnetic heating could be used as a preheating purposes. Because lower frequency waves carry less energy, heating times are considerably longer compared to the higher energy microwaves.
- Optimized ammonia synthesis using the excess heat in Haber-Bosch ammonia plant for oil sand bitumen extraction which is used for hydrogen production via microwave dissociation process is possible.
- The current ammonia retail prices continue to decrease by low natural gas prices. Current retail price is about 550 US$/ton. However, ammonia price is strictly dependent on natural gas price which can be eliminated if oil sand bitumen is utilized.
- Although hydrocarbon dissociation route is a fossil fuel based process, the technology is clean and environmentally friendly close to renewable resources in some environmental impact categories.

CHAPTER 5:AMMONIA IN MARITIME APPLICATIONS

Sea transportation constitutes a large share of global transportation. It is principally used for the transportation of goods, liquid fuels, all type of products and humans. Transoceanic tankers and freight ships need a great amount of energy for operation which is commonly provided by diesel or heavy fuel oils. In order to reduce the total greenhouse gas emissions caused by maritime transportation, ammonia is potential replacement and/or supplements for conventional fuels. Here, zero carbon fuel, ammonia, is proposed to replace heavy fuel oils in the engines of maritime transportation vehicles. Furthermore, it is also proposed to use ammonia as dual fuels to quantify the total reduction of greenhouse gas emissions. An environmental impact assessment of transoceanic tanker and transoceanic freight ship is implemented to explore the impacts of fuel substituting on the environment. In the life cycle analyses, the complete transport life cycle is taken into account from manufacture of transoceanic freight ship and tanker to production, transportation and utilization of ammonia in the maritime vehicles. Several ammonia production routes ranging from municipal waste to wind options are considered to comparatively evaluate environmentally benign methods. Besides global warming potential, environmental impact categories of marine sediment ecotoxicity and marine aquatic ecotoxicity are also selected in order to examine the diverse effects on marine environment. Being carbon-neutral fuels, ammonia, yields significantly minor global warming impacts during operation. The ecotoxicity impacts on maritime environment vary based on the production route of ammonia. The results imply that even if engines are dual-fueled with ammonia, the global warming potential is significantly lower in comparison with heavy fuel oil driven transoceanic tankers.

1. Ammonia production routes

Although steam methane reforming is currently more common technique for hydrogen and ammonia production, by developments in electrolyzer industry, water electrolysis using renewable energy resources become more attractive and practical. Hence, in this study, hydropower and wind energy sources are considered two different pathways for ammonia production. For hydrogen production, electrolysis route is employed where an electrolyzer is used requiring about 53 kWh electricity to generate 1 kg of hydrogen. The source of electricity is taken from wind and hydropower individually for both cases. Haber-Bosch process is utilized for ammonia synthesis in this study. In this process, nitrogen is supplied through air separation process. For the LCA of nitrogen manufacture, electrical work for procedure, cooling water, surplus heat and groundwork for air separation facility are taken into account. The distribution elements were obtained from the heat of vaporization and the specific heat capacity multiplied with the temperature difference from 20°C to the boiling point. SimaPro database has the values of nitrogen manufacture from cryogenic air separation method [60]. Cryogenic air separation process becomes more cost effective compared to non-cryogenic methods at the level of about 200-300 tons per day nitrogen. Since gas phase nitrogen is required in the reaction, the energy requirement is lower because liquefaction is not required. The major input to the air separation plant is electricity required to compress the air. Air is not taken as an input because of inexhaustibility. The separated CO_2 and water vapor are not evaluated as emissions in the process. The transportation is not accounted for the analyses since it is considered that the ammonia synthesis plant is located near air separation plant. The transportation of produced fuels are later considered in the analyses. Producing liquid products from air separation plant requires about two times higher energy than gaseous products. Commercial cryogenic air separation plants require electricity in the range of 0.6 to 1 kWh per kg of liquid nitrogen product. However, as mentioned earlier, gaseous product necessitates lower power input. Hence, in this study, 0.42 kWh electricity is assumed for nitrogen gas production as taken from GREET 2016 model [61]. The utilized electricity for nitrogen production is US mix grid. For hydrogen generation, hydropower electricity or wind power plant electricity is used. For the transportation of the produced ammonia and hydrogen, an average distance is assumed where 100 km is by lorry with a capacity of higher than 16 tonne and 600 km by rail transport. Dual fuel operation of vessels are also considered in the study as 50% clean fuel and 50% heavy fuel oil.

2. Methodology

The goal of this study is to investigate the environmental impacts of alternative fuels driven marine transportation tankers and ships in comparison with conventional heavy fuel oil from cradle-to-grave: global warming potential, abiotic depletion, acidification, stratospheric ozone layer depletion, marine eco-toxicity and marine sediment ecotoxicity. The motivation behind this study is to decrease the environmental impacts caused by current hydrocarbon dependent marine transportation systems. The results of this study will mainly attract marine transportation sector and academicians working in the area of clean fuel production and utilization technologies. The function of this study is assess environmental impacts per tonne-kilometer cruise travel where the functional unit is 1 tonne-kilometer.

Life cycle analysis includes all stages in a process life namely from the extraction of raw materials through the material processing, manufacture, distribution, use, and disposal or recycling. For this analysis, we account for all the stages in the life cycle of maritime transportation, including feedstock recovery and transportation, fuel production and transportation, and fuel consumption in the ocean tankers and freight ships. The exploration and recovery activities from the well to fuel production and the subsequent transportation to the pump constitute the well-to-pump (WTP) stage. The combustion of fuel during ocean vehicle operation constitutes the pump-to-hull (PTH) stage. These two stages combined comprise the well-to-haul (WTH) cycle.

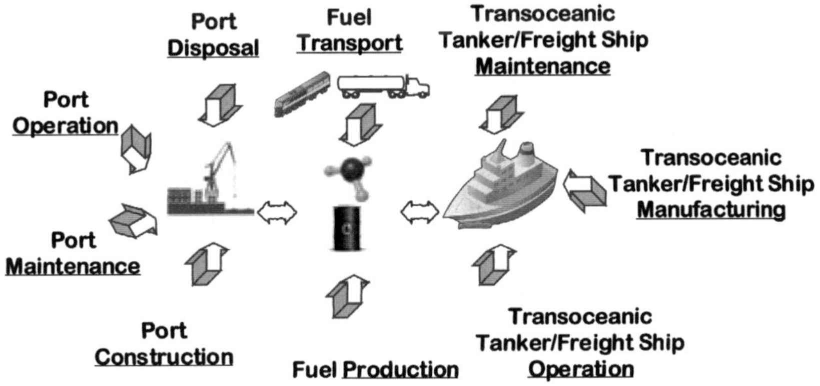

Figure 56 Well-to-haul life cycle phases of maritime transportation

The system boundaries for the LCA analyses are defined as shown in Fig. 56, namely: mining of the raw materials and extraction of the nutrients from these materials, transportation of raw materials and pre-products, generation and supply of required energy, manufacturing of the ammonia and the related field operations. Various environmental impact categories including global warming, marine sediment ecotoxicity, marine aquatic ecotoxicity, acidification, ozone layer depletion and abiotic depletion are selected in order to examine the diverse effects of switching to clean fuels in ocean transportation. The selected indicators of this method which are utilized in this study are explained as follows [60]:

Depletion of abiotic resources

This impact category is concerned with protection of human welfare, human health and ecosystem health. This impact category indicator is related to extraction of minerals and fossil fuels due to inputs in the system. The Abiotic Depletion Factor (ADF) is determined for each extraction of minerals and fossil fuels (kg antimony equivalents/kg extraction) based on concentration reserves and rate of de-accumulation.

Global warming potential

Climate change can result in adverse effects upon ecosystem health, human health and material welfare. Climate change is related to emissions of greenhouse gases to air. The characterization model as developed by the Intergovernmental Panel on Climate Change (IPCC) is selected for development of characterization factors. Factors are expressed as Global Warming Potential for time horizon 500 years in kg carbon dioxide/kg emission.

Stratospheric ozone depletion

For the reason of stratospheric ozone depletion, a superior portion of UV-B radiation spreads the earth surface because of increasing chlorofluorocarbons (CFCs). This may yield damaging impacts on human and animal health, terrestrial and aquatic ecosystems, biochemical cycles and on substances. Ozone depletion potential of several gasses are specified in kg CFC-11 equivalent per kg emission where the time span is infinity.

Acidification potential

Acidifying substances causes a wide range of impacts on soil, groundwater, surface water, organisms, ecosystems and materials. SO_2 equivalents/kg emission is used to expresses the acidification potential.

Marine aquatic eco-toxicity

Marine eco-toxicity refers to impacts of toxic substances on marine aquatic ecosystems. The characterization factor is the potential of marine aquatic toxicity of each substance emitted to the air, water or/and soil. The unit of this factor is kg of 1,4-dichlorobenzene equivalents (1,4-DB eq) per kg of emission.

Marine sediment ecotoxicity

Marine sediment eco-toxicity refers to impacts of toxic substances on marine sediment ecosystems. The unit of this indicator is kg of 1,4-dichlorobenzene equivalents (1,4-DB eq) per kg of emission.

3. Life cycle phases

Analysis of well-to-haul (WTH) of clean marine fuels is performed in this study where each step is briefly explained in this section. A tonne kilometer by shipping is defined as unit of measure of goods transport which represent the transport of one tonne by a vessel over one kilometer. Marine diesel engines are generally further categorized into two different groups as slow speed (15 knots in average) and medium speed (25-30 knots). Transoceanic tanker and freight ships are slow speed vehicles which include two-stroke cycle with crosshead engines of 4-12 cylinders. In the marine industry, these engines are used for main propulsion and constitute larger portion of installed power on the ship.

Table 18 Trip scenario for transoceanic freight ship

	Distance	Speed	Time	Load factor
Cruise	2306.041 nmi	18.006 kn	128.073 h	0.6
RSZ1	100.827 nmi	18.006 kn	5.6 h	0.6
RSZ2	25 nmi	18.006 kn	1.388 h	0.6
Hotel1	-	-	22 h	0.19
Hotel2	-	-	22 h	0.19

Each vessel type is characterized by power rating of its two engines - main and auxiliary. For each engine, fuel consumption and emission factors are defined. GREET 2016 software is utilized to find the power ratings and total energy consumptions of the selected ships [61].

In order to determine the average power consumptions, the trip is assumed to be from pacific to international ports. After determining the total trip distances as listed in Table 18 and 19, each trip was divided into segments, including transit through reduced-speed zones (RSZs). Each trip segment may have distinct fuel consumption and emission factors owing to different speeds and load factors and engine/fuel switching. At the origin and destination ports, the vessel will hotel and burn fuel dockside using mainly auxiliary engines. After a vessel leaves port, it travels in an RSZ, during which it uses a lower load factor and consumes less fuel, thereby emitting fewer pollutants, than when traveling at cruising speed. It will pass through an RSZ before hoteling at the port of destination. The GREET marine module aims to model these trip segments, representatively, for different vessel types leaving or arriving at U.S. ports in different regions [61].

Table 19 Trip scenario for transoceanic tanker

	Distance	Speed	Time	Load factor
Cruise	6177.495 nmi	15.049 kn	410.498 h	0.83
RSZ1	40.748 nmi	15.049 kn	2.708 h	0.83
RSZ2	25 nmi	15.049 kn	1.661 h	0.83
Hotel1	-	-	58 h	0.26
Hotel2	-	-	58 h	0.26

Production of transoceanic freight ship

This phase includes the processes of material production, representing the material composition of an average water vehicle used for solid goods transportation. For manufacturing, electricity and heavy oil burned in industrial furnace are included. For the transportation of materials standard distances are applied. Also, waste treatment processes for non-metal components of a water vehicle are accounted for. The exchanges of the ship manufacturing are derived from an assessment of a ship, with a load capacity of 51,500 tonne. The energy consumption in manufacturing is estimated as 50% of the cumulative energy of the used materials. The split of energies is 10% electricity and 90% heavy fuel oil [60].

Production of transoceanic tanker

Similar to transoceanic freight ship, this step includes the processes of material production, representing the material composition of an average water vehicle used for solid goods transportation. This type of vehicle has higher load capacity corresponding to 100,000 tonne. The energy consumption in manufacturing is estimated as 50% of the cumulative energy of the used materials. The split of energies is 10% electricity and 90% heavy fuel oil [60].

Maintenance freight ship and tanker

This step of the inventory includes the use of paint and emissions of the solvent of the paint as NMVOC. Consumption of lubricates are excluded. It is assumed that the ship is painted 6 times in its entire life span of 25 years. The consumption of lubricates is included in the fuel consumption for vessel operation [60].

Port facilities

Seaport facilities comprise of the construction and disposal of one of the world's biggest port in Rotterdam, Netherlands. The inventory contains the processes of material production, representing the material used in the construction phase of the port. The building activities and electricity consumption are accounted for construction and disposal phases. Emission of NMVOC are included. Also lorry transport of materials to the constructions site are taken into account. The expenditures due to construction and disposal are addressed. The data represent one seaport for both, sea and inland shipping. The material composition of the sealed industrial area is derived from the construction and disposal expenditures of a European highway. The built up is modelled as a steel building. The life time of port is assumed to be 100 years [60].

Operation and maintenance of the port

This inventory includes emissions of to water due to non-removed oil spills. In this step, the land occupation and transformation due to seaport are taken into account. The energy consumption at the port is based on assumptions for the specific electricity consumption at the port in Hamburg, Germany. The land use is further distinguished in built up area (0.6%), road area (45.6%) and water bodies (54.29%). Emission to waters include emission from production sites on the port site, which are not directly connected to the transport activities [60].

Transportation and operation of the transoceanic freight ship

In this step, the full cycle is represented by supply of the fuel, operation of the ship and transport of the goods as 1 tkm. Direct airborne emissions of gaseous substances, particulate matters, dioxins, PAHs, halogens and heavy metals are accounted for. These emissions are caused by heavy oil burning in the engines, hence they are mostly eliminated in ammonia driven ships. Also, the disposal of bilge oil and emissions of tributyltin compounds are included. Individual hydrocarbons are estimated based on the share of diesel engines of road vehicles. Heavy metals are estimated from trace elements in fuel. A distinction between distilled (28%) and residual fuel (72%) is applied. The amount of disposed bilge oil is estimated as 0.6% of the consumed fuel. The average data for steam turbine (5%) and diesel engine (95%) propulsion are considered in the study. The fuel used is for conventional ships is heavy fuel oil and is representative for slow speed engine types about 15 knots. The data represents solid bulk transport of about 40,000 dwt (deadweight tonnage) where the ship is driven by steam turbine and diesel engines [60]. The power ratings of the main engine and auxiliary engines are about 37.5 MW and 8.3 MW, respectively. The average energy consumption for the freight ship is calculated to be 0.214 MJ per mile ton [61].

This step refers to the entire transport life cycle namely; the operation of vessel; production of vessel; construction and land use of port; operation, maintenance and disposal of port;

production and transportation of fuel to the port. Port infrastructure expenditures and environmental interventions are allocated based the yearly throughput (0.37). The vessel manufacturing is allocated based on the total kilometric performance corresponding to about 2,000,000 km and its transport performance. Since transport activity requires loading and unloading, for each transport activity two ports are required [60].

Transportation and operation of the transoceanic tanker

In this step, transoceanic tanker is considered instead of freight ship. Different than the transoceanic freight ship, the average data for steam turbine (62%) and diesel engine (38%) propulsion are considered here. The fuel consumption and emissions are representative for slow speed engine types and a tanker size between 50,000-300,000 dwt. Similarly, the inventory refers to the entire transport life cycle where the functional unit is one tkm. The port infrastructure expenditures and environmental interventions are allocated based the yearly throughput (0.27). The tanker manufacturing is allocated based on the total kilometric performance of 3,920,000 km and its transport performance [60]. The power ratings of the main engine and auxiliary engines are about 15 MW and 2.85 MW, respectively. The average energy consumption for the tanker is calculated to be 0.056 MJ per mile ton [61].

4. Results and Discussion

The methodological analyses are conducted in SimaPro and GREET 2016 software for the evaluation of LCA. The LCA is implemented by employing the impact assessment method of CML 2001.

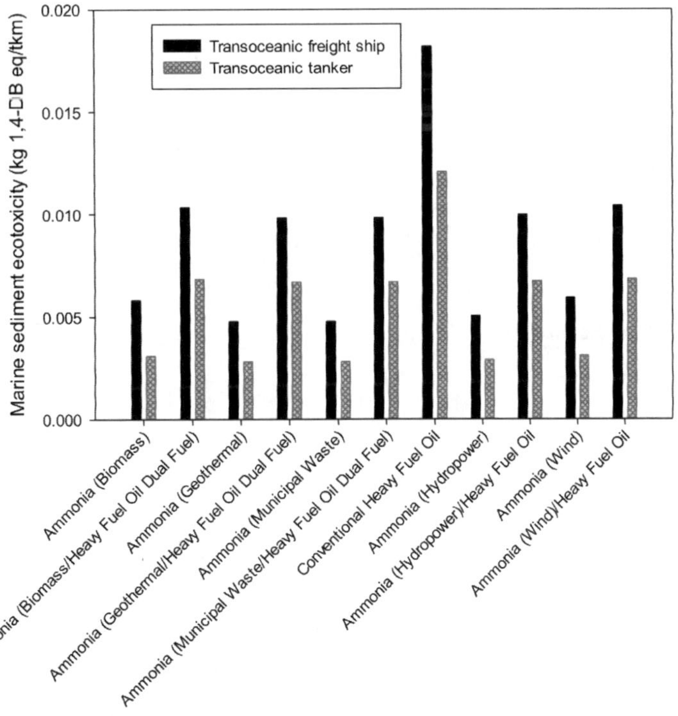

Figure 57 Marine sediment ecotoxicity values of transoceanic tanker and transoceanic freight ship per tonne kilometer for ammonia and conventional heavy fuel oil

The toxic substances on the marine sediment and aquatic environment are the main concerns of marine sediment ecotoxicity and marine aquatic ecotoxicity categories. 1,4-dichlorobenzene equivalents/tkm is used to express each toxic substance. Among the selected fuels, the conventional heavy fuel oil for both tanker and freight ship have the greatest damage on marine sediment and aquatic environment as shown in Figs. 57-61.

In general transoceanic freight ship have higher impact values than tanker because of mainly higher energy consumption rate per tonne-kilometer. Using ammonia (wind energy) in transoceanic tanker as dual fuel with heavy fuel oil lowers the ecotoxicity level about 47%.

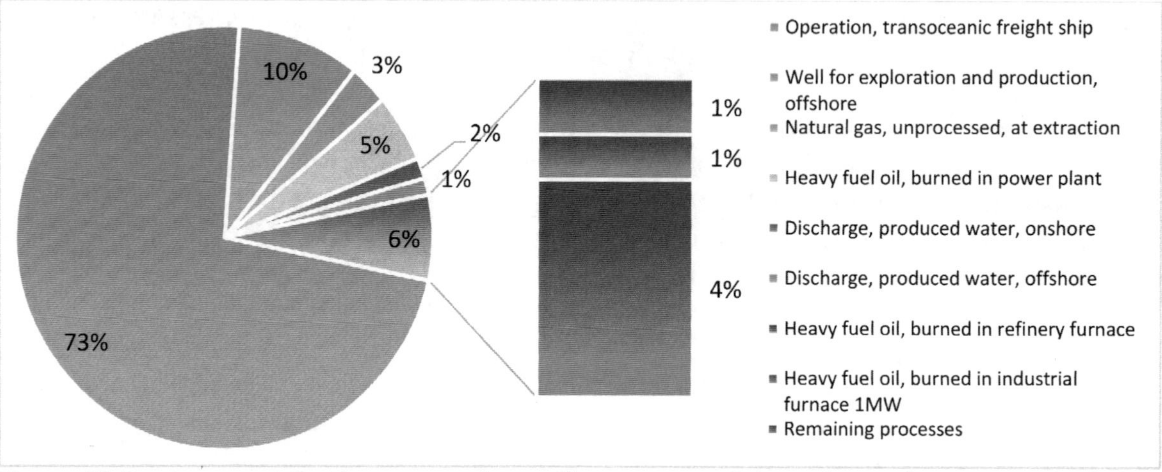

Figure 58 Process contributions to marine sediment ecotoxicity values of transoceanic freight ship driven by dual fuel (50% ammonia from hydropower and 50% heavy fuel oil)

The heavy fuel oil combustion significantly contributes to many of the environmental impacts. Energy production from fossil fuels yield high GHG emission. The contribution of different processes to ecotoxicity of marine sediment and marine aqua is illustrated in Figs. 58 and 59.

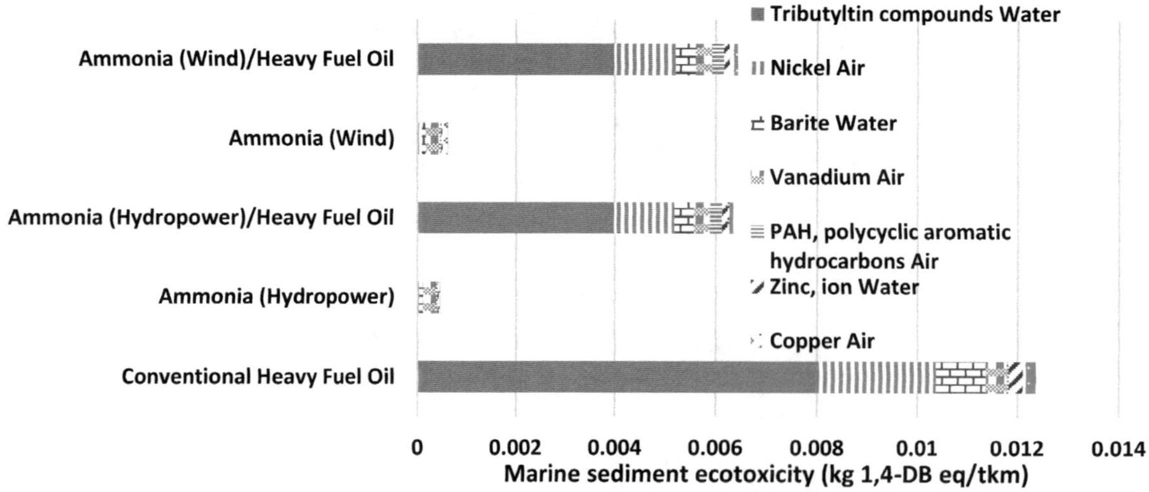

Figure 59 Emissions causing marine sediment ecotoxicity for transoceanic tanker driven by various fuels

Operation of the freight ship is responsible about 73% of marine sediment ecotoxicity for ammonia/heavy fuel oil driven ship whereas exploration and offshore production of heavy oil represents the 10%. This is due to crude oil production which is then used in heavy fuel oil refinery and transported to heavy fuel oil regional storage to be combusted in tanker.
Fig. 60 comparatively shows the materials and emissions causing ecotoxicity. Most of them are related to tributyltin compounds emitted to water because of bottom paintings of the ships. The heavy metals such as nickel and vanadium are due to combustion of heavy fuel oil which can been as the second highest contributor in dual fuel and sole heavy fuel oil options.

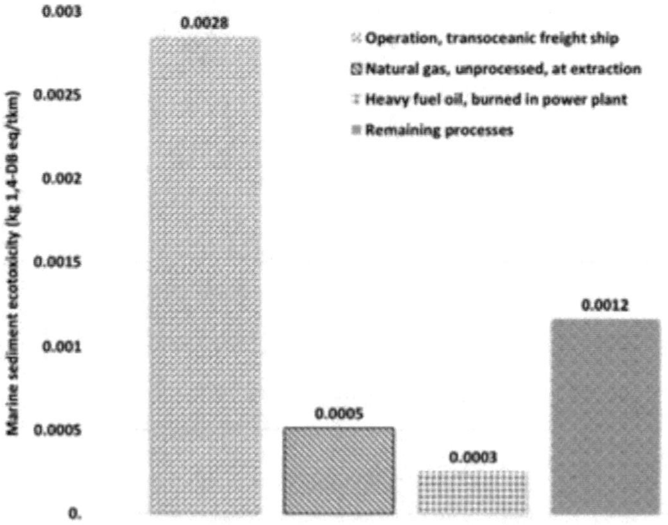

Figure 60 Process contributions to marine sediment ecotoxicity values of transoceanic freight ship fueled by sole ammonia from municipal waste

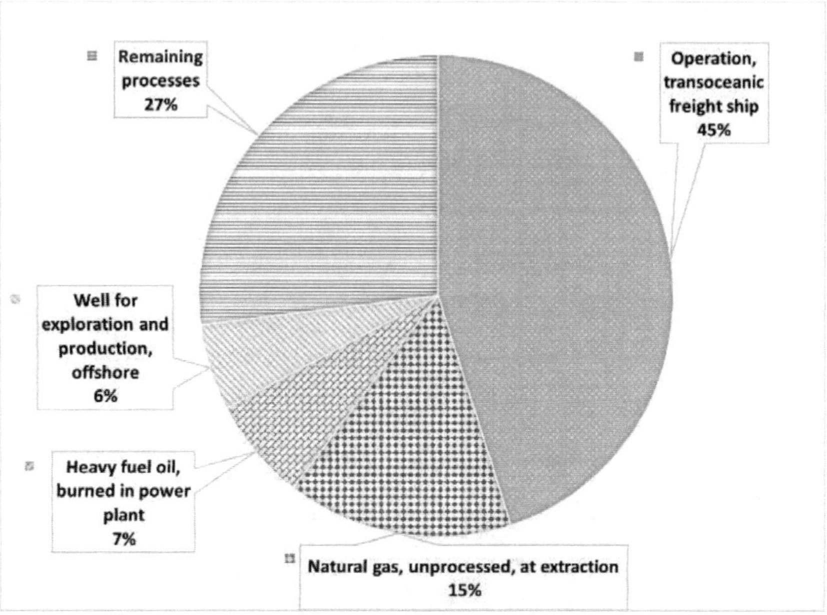

Figure 61 Process contributions to marine aquatic ecotoxicity values of transoceanic freight ship fueled by sole ammonia from geothermal energy

The contributions of different processes to ecotoxicity of marine sediment and marine aqua are shown in Figs. 61, 62 and 63. Operation of the freight ship is responsible about 45% of marine aquatic ecotoxicity as for sole ammonia fueled ship whereas exploration and offshore production of heavy oil represents the 6% and natural gas extraction represents 15% of total. This is due to natural gas and oil fired power plants which is then used for nitrogen production plant. For biomass based ammonia driven freight ship, barium, tributyltin compounds and vanadium are top three substances causing marine aquatic ecotoxicity as shown in Fig. 62 where barium has an impact corresponding to 0.0017 kg 1,4-DB eq per tonne-kilometer. Some of them are related to tributyltin compounds emitted to water because of bottom paintings of the ships

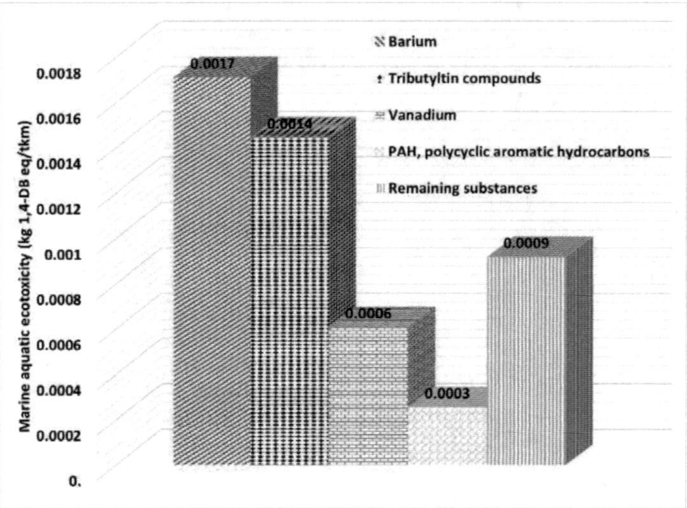

Figure 62 Process contributions to marine aquatic ecotoxicity values of transoceanic freight ship fueled by sole ammonia from biomass energy

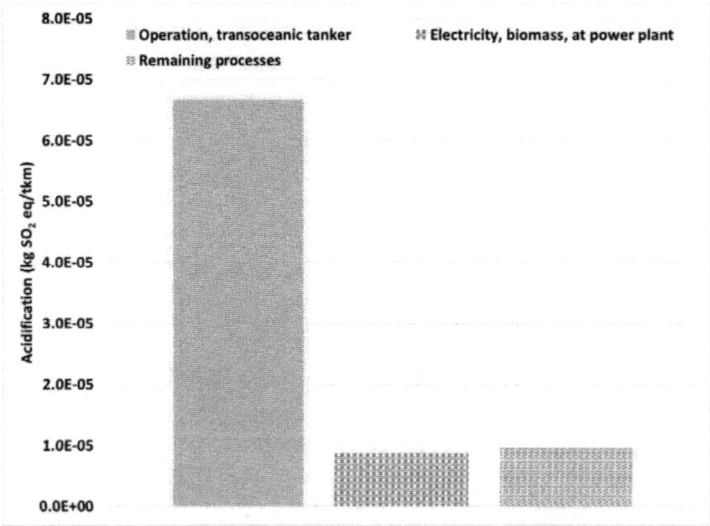

Figure 63 Process contributions to acidification values of transoceanic tanker fueled by ammonia from biomass energy and heavy fuel oil

The source of SO_2 emission is predominantly the operation of tanker and freight ship (96.8%) for conventional heavy fuel oil tankers and ships. It is also high for dual fuel processes as illustrated in Fig. 63.

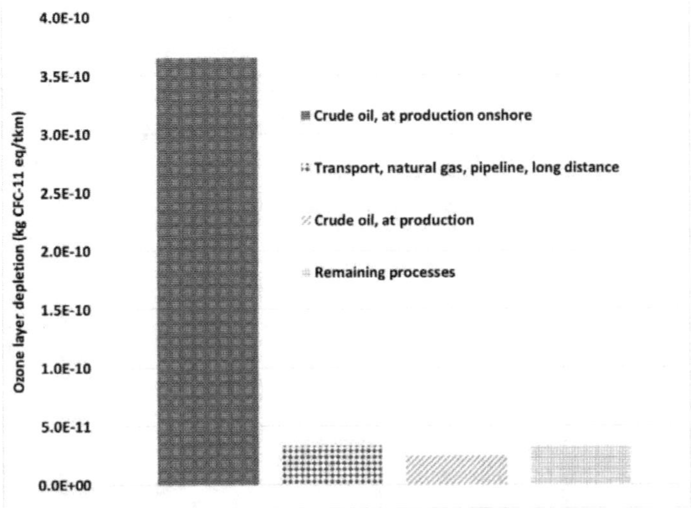

Figure 64 Process contributions to stratospheric ozone layer depletion of transoceanic freight ship by dual fuel (50% ammonia from municipal waste and 50% heavy fuel oil)

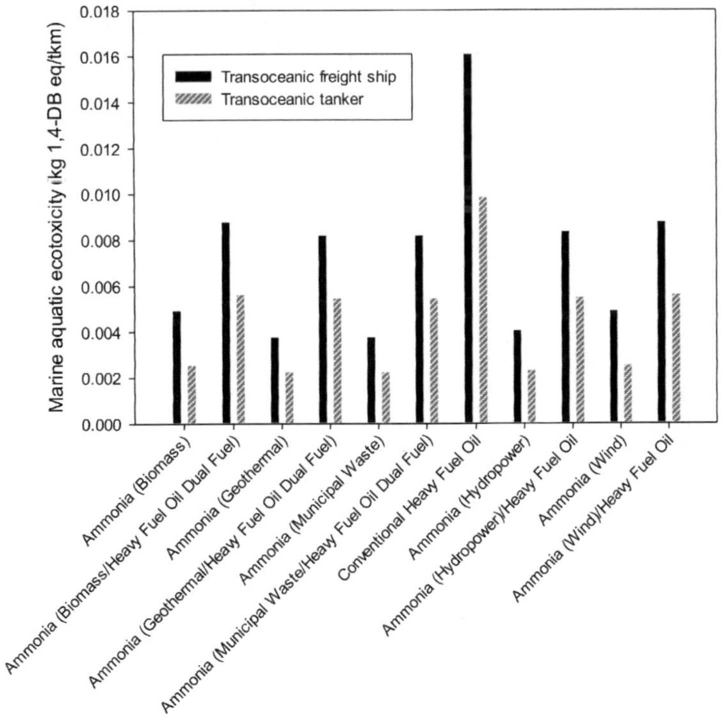

Figure 65 Marine aquatic ecotoxicity values of transoceanic tanker and transoceanic freight ship per tonne kilometer for ammonia and conventional heavy fuel oil

For ammonia and heavy fuel oil driven transoceanic freight ship, crude oil production has the highest share corresponding to about 81.2% as shown in Fig. 64. Operation and maintenance of the port is responsible for about 16.6% of overall ozone layer depletion whereas the manufacturing of the ship constitutes only 2.2%. Two main substances causing ozone layer depletion are bromotrifluoromethane (Halon 1301) and bromochlorodifluoromethane (Halon 1211) corresponding to about 1.98E-10 and 3.31E-11 kg CFC-11 eq/tkm, respectively. The marine aquatic toxicity results are shown in Fig. 65.

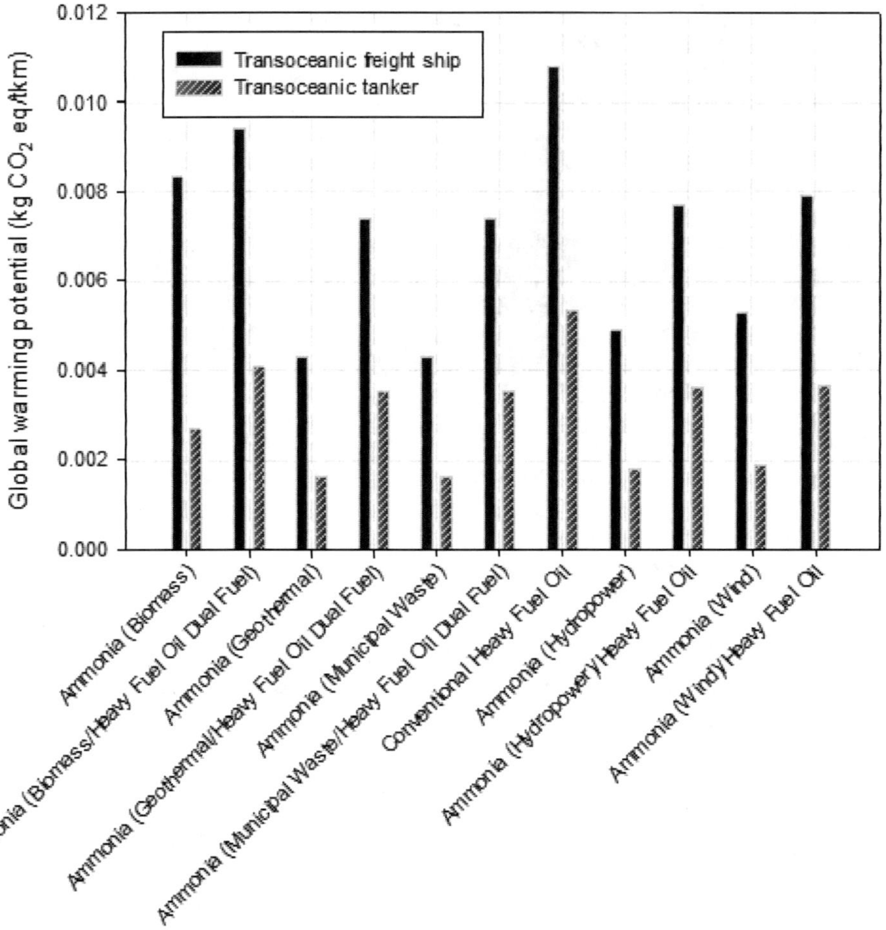

Figure 66 Global warming potential of transoceanic tanker and transoceanic freight ship per tonne kilometer for ammonia and conventional heavy fuel oil

In terms of global warming potential, ammonia (municipal waste) driven transoceanic tanker and freight ship yield the lowest greenhouse gas emissions in the entire life cycle corresponding to 0.0041 and 0.0016 kg CO_2 eq. per tonne-kilometer for freight ship and tanker, respectively as shown in Fig. 66. However, it is very high (about 10 times) for the conventional heavy fuel oil as shown in Fig. 66. The highest values after sole heavy fuel oil are 0.0079 kg CO_2 eq/tkm and 0.0036 kg CO_2 eq/tkm for ammonia (wind)/heavy fuel oil combination, respectively for freight ship and tanker. Using ammonia as dual fuel in the marine engines can decrease total greenhouse gas emissions up to 34.5% per tkm. Similarly, sole ammonia driven transoceanic tanker releases about 0.0018 kg CO_2 eq/tkm greenhouse gas compared to 0.0055 kg CO_2 eq/tkm for sole heavy fuel oil tanker. Process contributions to global warming potentials transoceanic freight ship driven by conventional heavy fuel oil are shown in Fig. 67.

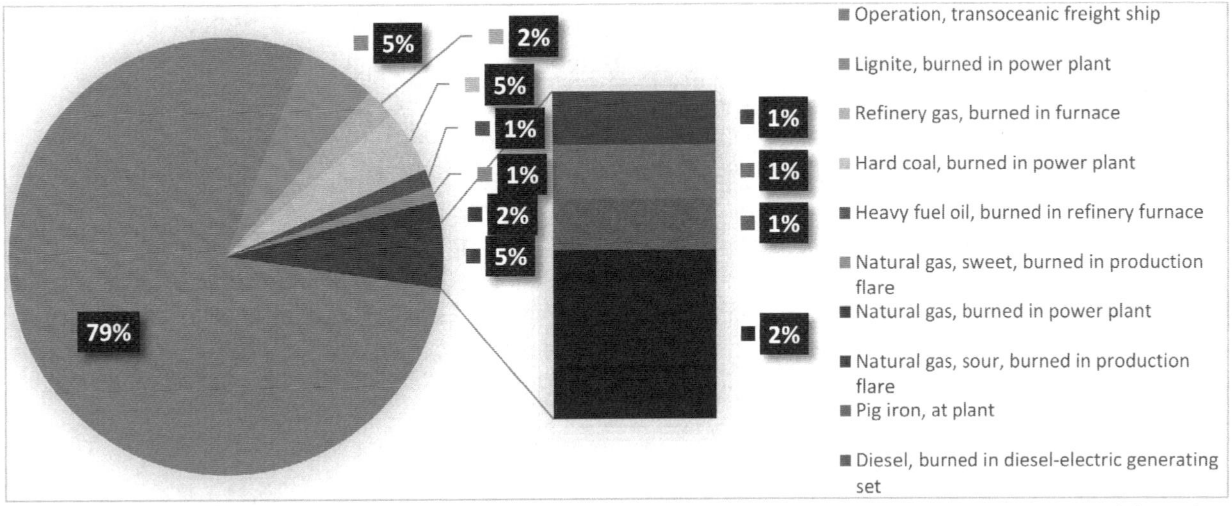

Figure 67 Process contributions to global warming potentials transoceanic freight ship driven by conventional heavy fuel oil

For ammonia (hydropower)/heavy fuel oil driven tanker, total GHG emissions are caused by mainly operation of tanker (64%), maintenance and operation of port (31%) and manufacture of tanker (5%). For the operation of tanker, 49.3% of is caused by heavy fuel oil and only 11.4% is by ammonia utilization.

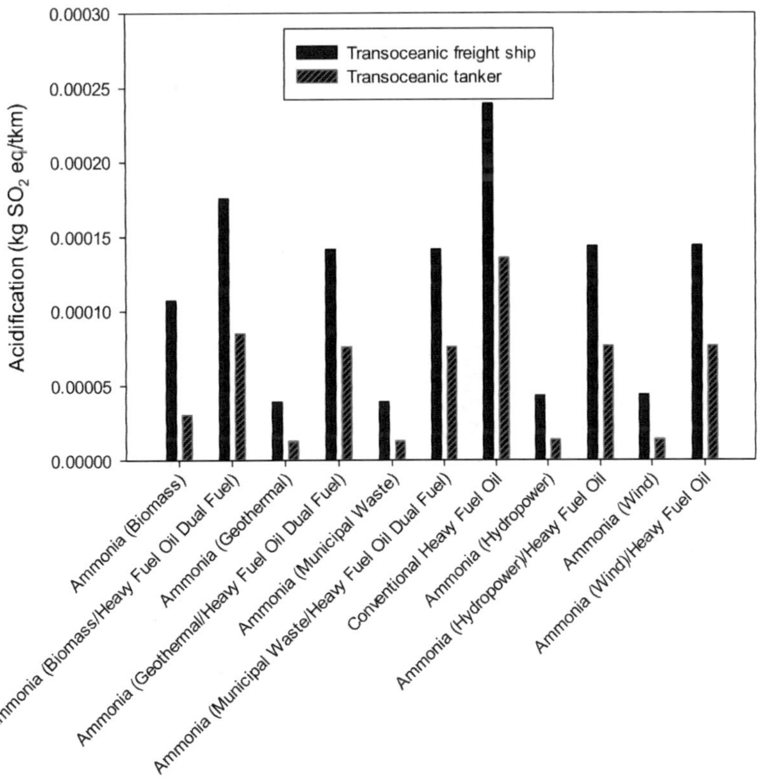

Figure 68 Acidification values of transoceanic tanker and transoceanic freight ship per tonne kilometer for ammonia and conventional heavy fuel oil

The acidification values of heavy fuel oil driven transoceanic tanker and freight ship are mainly caused by SO_2 and NO_x emissions which corresponds to more than 90% of overall acidification value. The source of SO_2 emission is predominantly the operation of tanker and freight ship (96.8%). This is caused by the sulfur content of the heavy fuel oil hence it is mostly eliminated in ammonia as seen in Fig. 68. Acidification values are lowest for hydrogen fueled tankers. The combustion of diesel and heavy fuel oil have the high impact hence, particularly transportation processes with railway and trucks create high emissions leading to higher acidification values. The variation in this impact category for ammonia is because of considerable differences in transportation distances of the hydrogen and ammonia to the port. Because, hydrogen has higher energy content pre mass, the number of transports from production plant to the seaport is lower yielding lower diesel combustion in railway and lorry.

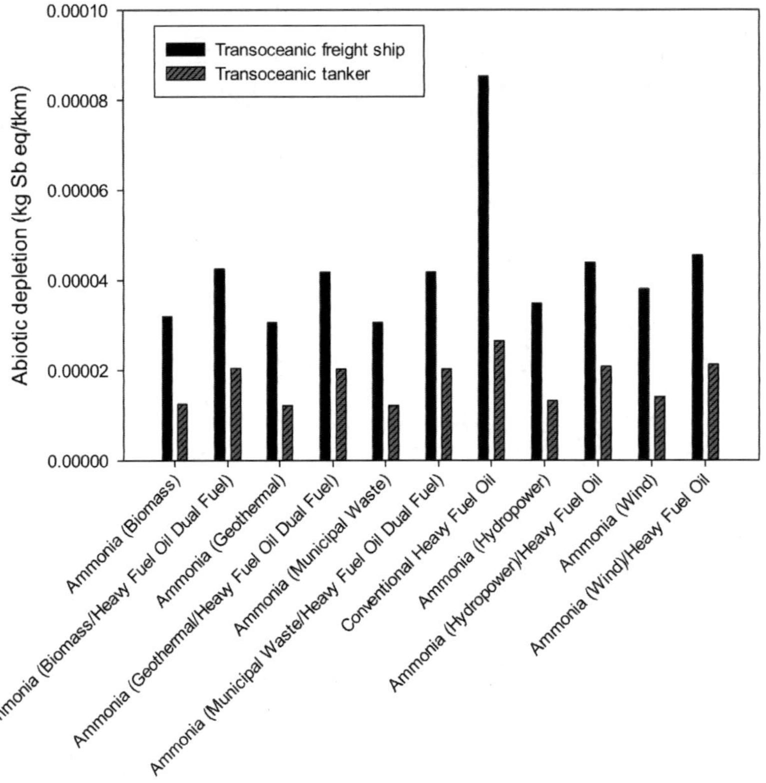

Figure 69 Abiotic depletion values of transoceanic tanker and transoceanic freight ship per tonne kilometer for ammonia and conventional heavy fuel oil

The abiotic depletion impact is calculated to be highest conventional heavy fuel oil followed ammonia (wind)/heavy fuel oil driven freight ships as it is shown in Fig. 69. Abiotic sources are natural sources counting energy sources, such as hard coal and crude oil, which are evaluated as non-living. This is because of fossil fuels are major basis of energy and feed resource, it shows the huge intake of hard coal and lignite (in total 64%) for tonne-kilometer travel of ammonia (wind) driven transoceanic freight ship as illustrated in Fig. 70.

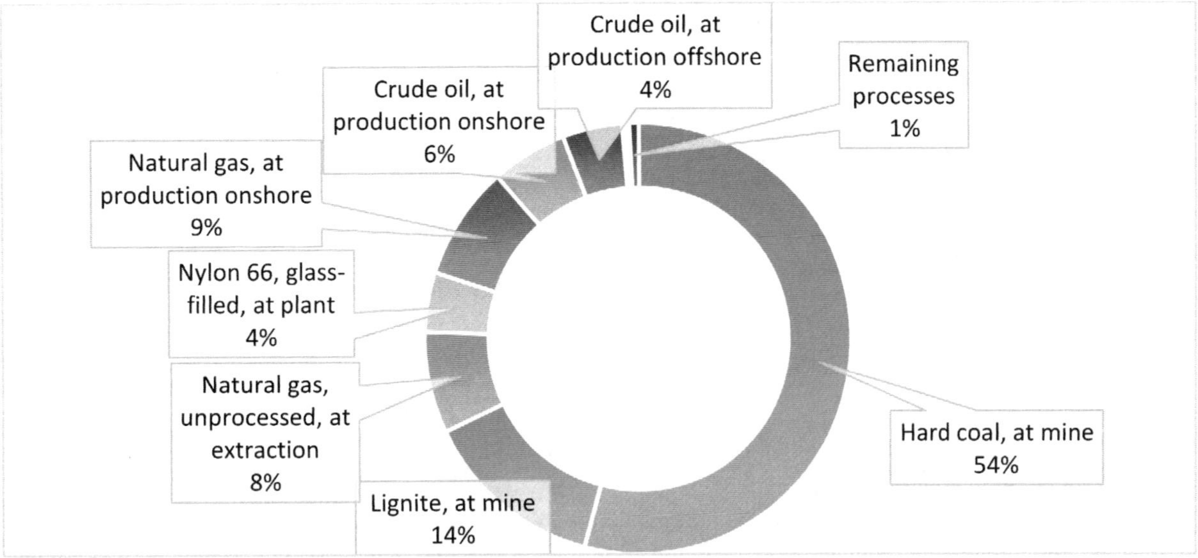

Figure 70 Process contributions to abiotic depletion values of transoceanic freight ship driven by ammonia from wind energy

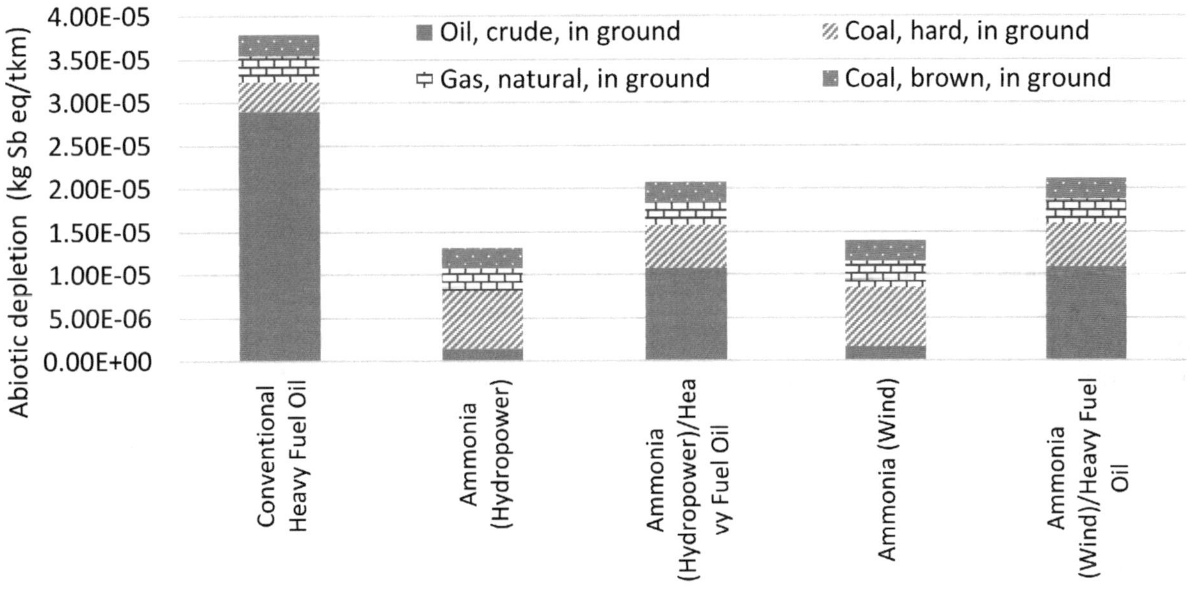

Figure 71 Abiotic substances causing depletion of abiotic resources for transoceanic tanker driven by various fuels

The reason of hard coal and lignite consumption is the electricity US mix usage in air separation plant for nitrogen production of ammonia synthesis process. Nitrogen production constitutes about 40% of overall abiotic depletion whereas hydrogen production from wind electricity is responsible for 16.1% of total abiotic depletion. Note that the operation and maintenance of the port has also a high share corresponding to 29.7% of total where it is similarly originated from electricity mix production. The causes of abiotic depletion values are comparatively shown in Fig. 71. For conventional heavy fuel oil driven transoceanic tanker, crude oil is the main depleted source similar to dual fuel options.

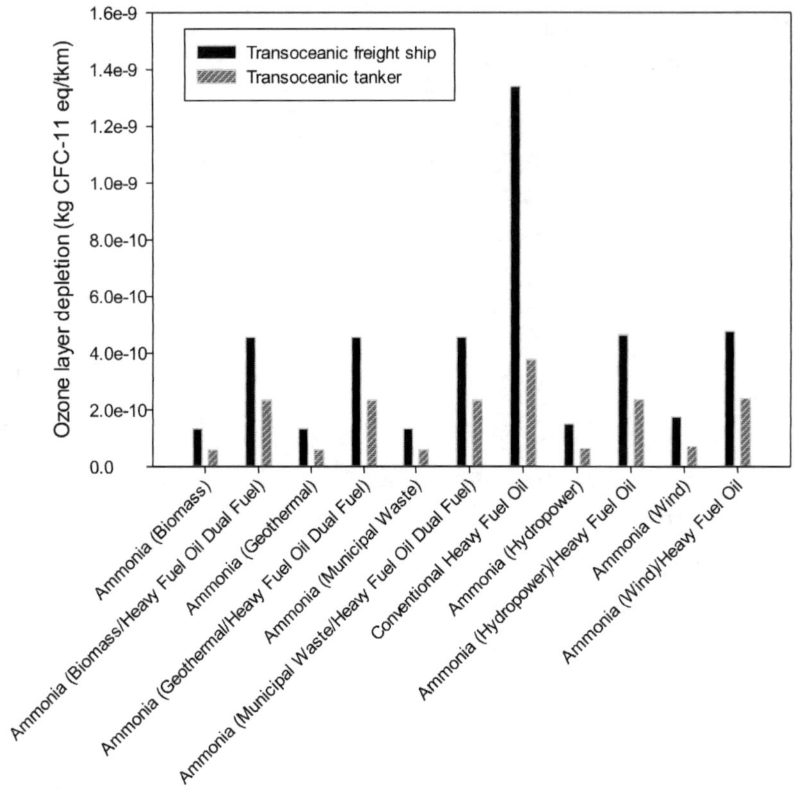

Figure 72 . Stratospheric ozone layer depletion values of transoceanic tanker and transoceanic freight ship per tonne kilometer for ammonia and conventional heavy fuel oil

Fig. 72 presents the life cycle kg CFC-11 eq. emissions of the transoceanic tanker and freight ship with different fuels per tonne-kilometer travelled. It is quite high for heavy fuel oil and dual fuel options while it is considerably less for hydrogen and ammonia fuels. Particularly, hydropower options have the lowest environmental impact in terms of ozone layer depletion.

5. Concluding Remarks

As non-carbon clean fuels for maritime ship engines, ammonia and hydrogen, yield considerably lower global warming impact during operation. The highest GHG values after sole heavy fuel oil are 0.0079 kg CO_2 eq/tkm and 0.0036 kg CO_2 eq/tkm for ammonia (wind)/heavy fuel oil combination, respectively for transoceanic freight ship and tanker whereas it is about 0.011 kg CO_2 eq/tkm for currently used heavy fuel oil freight ship. Using ammonia as dual fuel in the marine engines can decrease total greenhouse gas emissions up to 34.5% per tonne-kilometer. For ammonia (hydropower)/heavy fuel oil driven tanker, total GHG emissions are caused by mainly operation of tanker corresponding to about 64% whereas maintenance and operation of port has a share of 31%. This demonstrates that if clean fuels are even partially replaced with current hydrocarbon derived fuels, total GHG emissions in maritime transportation can be lowered significantly. By development and full utilization of renewable energy based ammonia and hydrogen fuels, GHG emissions during operation of the transoceanic tankers can be even zeroed.

CHAPTER 6: AMMONIA IN AVIATION

Petroleum based fuels have combination of accessibility, ease of handling, energy content, performance, and price because of being a mature product. Therefore, these type of fuels are heavily used by the transportation sector including air, road, and sea. However, limited nature, nonhomogeneous source distribution, changing prices, and end use related emissions of fossil fuels have driven most of the industries such as air transportation to search for other alternatives. The global aviation industry produces around 2% of all human induced carbon dioxide (CO_2) emissions. Specifically, aviation is responsible for 12% of CO_2 emissions from all transports sources whereas it is 74% for road transport [62]. Around 80% of aviation CO_2 emissions are originated from flights of over 1,500 km, for which there is no practical alternative mode of transport. The national and international flights overall the world produced 770 million tons of CO_2 in 2015 [62]. Presently, aviation industry consumes fuels derived from fossil fuels, mostly in specialized forms of petroleum based fuels. Such fuels tend to have higher qualities compared to some conventional applications, such as heating and transportation. Keeping in mind that aircraft fuels have additional requirements compared to land and sea transportation counterparts, such as resistance to extreme temperature changes. Hydrogen and ammonia, have zero or very little emissions when produced using water and renewable energy sources. Possibly, hydrogen, ammonia, methanol and ethanol can eliminate aviation industry's reliance on limited fossil fuel sources with fluctuating prices. Particularly, hydrogen and ammonia can additionally decrease aviation industry's impact on the greenhouse gas emissions considerably.

Table 20 illustrates the specific energy, energy density and density features of various alternative aviation fuels. Liquid hydrogen has the highest specific energy although currently utilized Jet A fuel has the highest energy density. According to the aviation fuel standards, a few of these fuels are not currently suitable for aircraft usage, however, by advancing technologies in engines and combustion, the aformentioned fuels are likely to be alternatives in the aviation industry.

Table 20 Specifications of some alternative aviation fuels

Fuel	Specific Energy (MJ/kg)	Density at 15°C (g/m^3)	Energy Density (MJ/L)
Kerosene (Jet A/Jet A-1)	43.2	0.808	34.9
Liquid Hydrogen	120	0.071	8.4
Liquid Methane	50	0.424	21.2
Methanol	19.9	0.796	15.9
Ethanol	27.2	0.794	21.6
Biodiesel (typical)	38.9	0.87	33.9
Liquid Ammonia	18.6	0.73	13.6

Source: Refs. [63–65]

Fig. 73 plots the mass of fuel essential to provide a definite quantity of energy in comparison with the volume of the equal fuel to give the equal quantity of energy. The quantity of energy, 100 MJ, brings a mutual base for assessment. Inferior values are favoured on both axes.

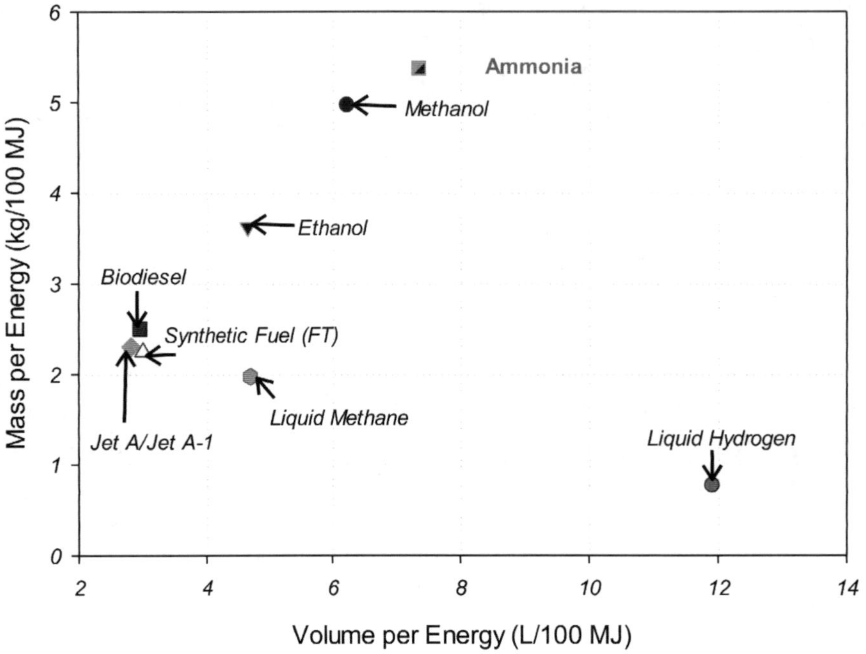

Figure 73 Comparisons of fuel mass and volume per unit energy (data from [66])

Hydrogen has the uppermost gravimetric energy content, however the low density of the liquid yields a very low volumetric energy content. Likewise for liquid methane, the high gravimetric energy content is offset by low density. The gravimetric energy content of the alcohols and biodiesel reveals their oxygen content. Biodiesels are nearer to conventional jet fuel since of the fact that oxygen signifies a smaller fraction of the fuel mass. FT synthetic fuels have a little higher gravimetric energy content than conventional jet fuel, but similarly slightly lower volumetric energy content [66].

In this regard, a life cycle assessment study of an average aircraft transport using various alternative aviation fuels to determine the relative environmental impact of each life cycle phase is presented. Using LCA methodology, the overall life cycle emissions of an aircraft running on various aviation fuels are calculated from well-to-wake. The steps considered in the analyses covers: (i) production, operation and maintenance of the aircraft, (ii) construction, maintenance and disposal of the airport, (iii) production, transportation and utilization of the aviation fuel. The environmental impact categories taken into account in this study are human toxicity, global warming, land use, depletion of abiotic resources and stratospheric ozone depletion.

1. Methodology

The evaluation of the life cycle of alternative fuels include emissions of the fuel cycle. This includes the extraction and transportation of raw materials from the well field or mine to the simple production, processing of these materials in fuel, transport and distribution of fuel in the tank on the plane and finally burn the fuel in the aircraft. The stages of this lifecycle analyzes well to wake as shown schematically in Fig. 74. For this analysis, we account for all the stages in the life cycle of aviation fuels, including feedstock recovery and transportation, fuel production and transportation, and fuel consumption in an aircraft. The exploration and recovery activities from the well to fuel production and the subsequent transportation to the pump constitute the well-to-pump (WTP) stage. The combustion of fuel during aircraft operation constitutes the pump-to-wake (PTWa) stage. These two stages combined comprise the well-to-wake (WTWa) fuel cycle as illustrated in Fig. 74.

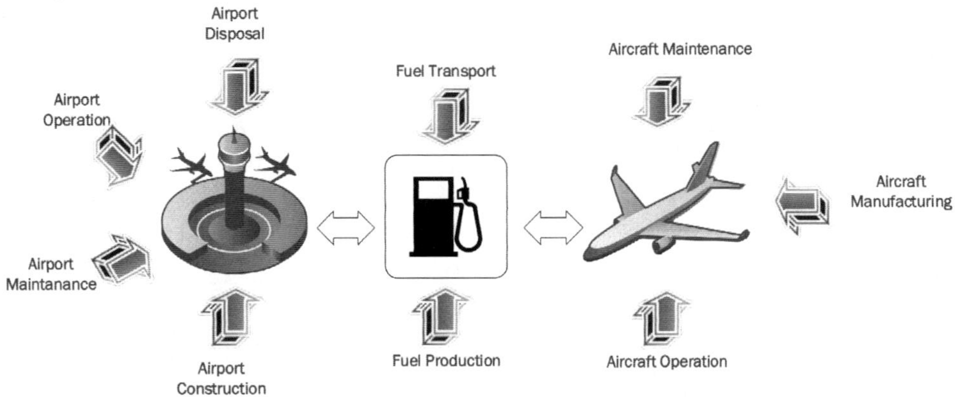

Figure 74 Complete well-to-wake life cycle phases used in the study

A ton kilometer by aviation is defined as the unit of measure of goods transport which represents the transport of one ton by an aircraft over one kilometer. To allocate energy use and emissions of a trip to a passenger, it is assumed that 100 kg of the payload represents an average weight of a passenger and luggage. This is equivalent to a mass allocation of energy and emissions between the passengers and cargo onboard the aircraft. A simple conversion of per-kg results to per-passenger results can be performed using the fixed weight of a passenger plus luggage. This allocation methodology works for all aircraft classes regardless of the split between passengers and cargo payloads onboard the aircraft. It is important that multiplying the passenger-based functional unit by the total passenger-distance flown will not provide the total emissions for a flight because this method will not account for the emissions that were allocated to the cargo in the aircraft. To obtain the total emissions from a flight, it should multiply the payload-based functional unit by the total payload-distance flown.

Environmental impacts are measured by selected environmental impact categories. Estimates of their potential for reducing greenhouse gases instead, compared with results generated for aviation kerosene using data from the results to observe aviation biofuel impacts relative to fossil fuels. It is assumed that sufficient operational fuel properties are modeled with acceptable characteristics of alternative fuels. Hence, the adaptation of the infrastructure of the aircrafts for other fuel combustion purposes are not considered in this study. All of the selected fuels are combusted in the same type of aircraft. Therefore, the aircraft manufacturing and airport operation/maintenance phases are identical for the selected cases. The LCA is performed using SimaPro LCA software in combination with the Ecoinvent database. Ecoinvent contains industrial life cycle inventory data on energy supply, resource extraction, material supply, chemicals, metals, agriculture, waste management services, and transport services. Hence, the inventory data for airport construction/maintenance/operation and aircraft manufacturing/maintenance phases are taken from SimaPro 7.3 software database whereas the other phases are generated based on the previous literature and work. CML 2001 is the selected method to calculate the environmental impact of the inventory data.

2. Description of the Processes

The LCA is divided into three main phases namely; manufacturing of the aircraft, construction and maintenance of the airport and operation of the aircraft.

Aircraft manufacturing

Material consumption values denote the production of one average aircraft in the referring class where the net weight of the aircraft is about 234 tonnes. Interventions for production processes of vehicles are derived from specific exchanges per seat. The aircraft is assumed to have about 367 seats. Inventory data represent 16 European production sites of Airbus company where they are located in Germany, France, Spain and the UK [60]. The inventory data of aircraft include processes of material, energy and water use in vehicle manufacturing. Rail and road transport of materials are considered together with VOC emissions. However, plant infrastructure is not included. Energy consumption for deionized water is included in the electricity value. Material expenditures are not accounted for. Transport of vehicle parts between the different sites is excluded.

Construction and maintenance of airport

In this step, maintenance, construction and disposal of airports are considered. The inventory data for construction and maintenance of the airport include material consumption and energy expenditures related with the construction sealed area such as aircraft parking, runways, etc. at airports. Furthermore, expenditures for buildings are considered. The data include expenditures due to refurbishment and demolition. The airport is chosen as a major one in Zurich, Switzerland and has the following specifications: Foundation layer is 40 cm gravel and a life span of 100 years is considered. Concrete floor is 22 cm with a concrete reinforcement of 1.8 kg steel/m^2 and a lifespan of 30 years is considered. The distance for transport is 25 km. In order to facilitate a first estimate, the following assumptions are made:

- 70% of the built up area is occupied with building halls.
- 30% are multi storey buildings with an average of 5 floors.
- The height between the floors is 2.7 m.

Operation of the aircraft

The fuel consumption and airborne emissions are especially significant for this process. For the airborne emissions, the location of emissions are separated into three main categories; low population density area, stratosphere and unspecified region. The inventory data include consumption of selected fuel, which is jet fuel kerosene in conventional applications, and direct emissions to air as gaseous emissions, particulate emissions, emissions of heavy metals. The results show the average fuel consumption and emission data caused by an average flight. The allocation between intercontinental (94.6%) and intra-European (5.4%) transport performance is made on basis of the yearly transport performance of aircraft departing from Swiss airports. Data refers to average transport conditions of aircraft departure from Swiss airports.

The aviation transportation segment necessitates fuels with great density of energy and hence it is primarily dependent on liquid hydrocarbon fuels. Alternative aircraft fuels need to carry around definite abilities such as good cold flow possessions, thermal steadiness and low freezing point. The fuel needs to be well matched for the current scheme of the aircraft engine. Sustainable aircraft fuels should satisfy low carbon emission in the complete life cycles. The energy crops used as the production source should not encounter the food manufacture and ecosystem and also do not damage the atmosphere and do not cause deforestation.

As alternative fuels, the following options are taken into account in the operation of the aircraft. The production method of the alternative fuels are also given here

- Hydrogen: 95% hydrocarbon cracking and 5% salt brine electrolysis as conventional methods, hydropower as renewable resource
- Methanol: Steam reforming process of methane
- Ethanol: Direct hydration of ethylene
- Ammonia: Heavy fuel oil and natural gas reforming as conventional methods, hydropower as renewable resource
- Liquefied natural gas: Liquefaction of natural gas in a liquefaction plant
- Jet Fuel (Kerosene): Production from crude oil

Kerosene

In the production of kerosene, all procedures on the refinery site discounting the emissions from combustion services, comprising waste water treatment, process emissions and direct releases to rivers are accounted for. Explanation of all streams of materials and energy because of the throughput of 1 kg crude oil in the refinery. The multioutput process crude oil in refinery delivers the co-products petrol, bitumen, diesel, light fuel oil, heavy fuel oil, kerosene, naphtha, propane/ butane, refinery gas, secondary sulphur and electricity. The influences of handling are allocated to the different products. Main pointers like energy use were projected based on a survey in European refineries. Inventory represents the distribution of petroleum product to the final consumer including all necessary transports. Transportation of product from the refinery to the end user is the full cycle. Operation of storage tanks and petrol stations are also considered. Emissions from evaporation and treatment of effluents are taken into account.

Ethanol

Though technical aspects are available to decrease the dangers throughout usual aircraft operations, important costs would be related with retrofitting present aircraft to admit on ethanol. Furthermore, the usage of ethanol or butanol would yield decreased operative competences and a lower energy efficiency of aircraft processes. The specific energy and energy density of ethanol are each about 40% lesser than the energy density of conventional jet fuel [67]. The production of ethanol in this study is considered from direct hydration of ethylene including materials, energy uses, infrastructure and emissions. The inventory is modelled with data from plants in Europe and India.

LNG

Liquefied natural gas is natural gas that is refrigerated to the liquid phase at around -162.1°C. Natural gas involves typically of methane, with minor quantities of ethane, propane, and butane, along with other hydrocarbons. LNG is utilized as an intermediate for natural gas transportation for over 50 years where the usage continues to grow. Natural gas is one of the greatest plentiful energy sources with considerable new unconventional resources, such as coal bed methane and shale gas, being established around the world. Additionally, methane hydrates can deliver more energy, in the form of methane, than is included in all other fossil fuels. Methane hydrates are methane particles covered in ice and are possibly accessible about the world. Presently, the extraction of methane hydrates is problematic because of the position of deposits. Moreover, treatment of the ice needs to be completed with attention so that the methane is not emitted into the air. The key benefit of LNG comparative to conventional jet fuel is the possible magnitude of the reserves. The chief drawback is the storing condition in terms of allocating with a fuel that not only has lower energy density than conventional jet fuel, but that is also cryogenic [67]. LNG could not be well-suited with current aircraft and it would also necessitate substantial expansion of the present worldwide distribution network as all of the airports that service LNG-fueled aircraft would need this fuel. Nevertheless, it is also probable that Fischer-Tropsch synthesis of natural gas to a synthetic fuel would be both financially and ecologically superior to LNG [67]. The carbon dioxide emission can be reduced about 25% by the usage of LNG fuel. Engine scheme is a trouble that has to be taken into account for the commercialization of the LNG fuels. The combustion of LNG fuel releases methane, which is one of the main GHG. The dataset in this study defines the liquefaction of natural gas in a liquefaction plant.

Liquid hydrogen

Liquid hydrogen is considered as alternative jet fuel in the last decades. Liquid hydrogen generated more energy per mass in comparison to conventional aviation fuel, however obliges great storage volume. The combustion of liquid hydrogen fuel origins low emission of GHG in comparison to petroleum based jet fuels. There are likewise a few drawbacks of liquid hydrogen. The generation cost, creation of char and tar as by-products are besides to be faced. The other problematic issue related with the usage of liquid hydrogen is that when mixed with air, it can burn in low concentration which will reason safety complications and the storage of hydrogen as liquid is challenging because it needs low temperature. The emission of moderately high volume of water vapor is a problem related with hydrogen aircrafts [67]. For the hydrogen generation in this study, it is assumed that 95% of hydrogen is produced from cracking of fossil fuels, the remaining 5% is from electrolysis of salt brine. The data represent from raw material extraction until delivery at plant for all processes. To compare the production of hydrogen from renewable raw materials, the methods of electrolysis of water from wind, hydropower, geothermal and solar energy are also evaluated. In renewable cases, the hydrogen production is conducted using electrolyzer consuming 53 kWh per kg of hydrogen.

Liquid ammonia

In terms of conventional resources, naphtha, heavy fuel oil, coal, natural gas coke oven gas and refinery gas can be used as feedstock in ammonia production. Natural gas is the primary feedstock used for producing ammonia worldwide. Hence in the current study, it is assumed that 85% of ammonia is produced via steam methane reforming and 15% is produced via partial oxidation of heavy fuel oil. The values represent mostly present state of the art technology used in European ammonia production plants. In order to compare the production of ammonia based on renewable resources, hydraulic, wind, solar and geothermal energy paths are considered in the analysis. For renewables based ammonia production, the hydrogen is initially produced from electrolysis as performed in previous step of hydrogen production and then it is combined with nitrogen in a Haber-Bosch ammonia plant. Manufacturing process starting with heavy fuel oil and natural gas, air and electricity is considered including the auxiliaries, energy, transportation, infrastructure and land use, as well as wastes and emissions into air and water. Transport of the raw materials, auxiliaries and wastes is include. Carbon dioxide is the by-product generated. Transient or unstable operations are not considered, but the production is assumed to be during stable operation conditions. Emissions to air are considered as emanating in a high population density area. Emissions into water are assumed to be emitted into rivers.

Methanol

The production of methanol also offers an important market for the use of otherwise flared natural gas. Modern natural gas-to-methanol plants are characterized by methanol selectivity above 99% and energy efficiencies higher than 70%. The raw materials, processing energy, estimate on catalyst use, and emissions to air and water from process, plant infrastructure are included. The process describes the production of methanol from natural gas via steam reforming process to obtain syngas for the production of methanol. There is no CO_2 use and hydrogen is assumed as burned in the furnace. Raw materials, average transportation, emissions to air from tank storage, estimation for storage infrastructure are included for the distribution part where 40% of the methanol is assumed to be transported from overseas. The subject matter particulate emissions cover exhaust- and abrasions emissions. Data for distances from production places to Switzerland are estimated with actual plant capacities. Similar overseas distances assumed. Additional 13% is assumed to be from Norway including tanker transport. Other production is within Europe continent.

Maintenance and operation of the airport

Operation of aircraft infrastructure contains the heat and electricity consumption for aircraft maintenance and buildings in the airport. A high amount of the heat is consumed at the airport is produced on-site via natural gas and oil. Additionally, consumption of water, waste of water and disposal of the wastes are considered. The inventory includes energy expenditures and airport energy infrastructure. Transport services on airport site are included. The use and emissions of de-icing materials are taken into account in addition to land transformation and occupation. The data represent a major airport in Europe. The use of de-icing materials is calculated as the geometric mean of 3 years.

Transportation via aircraft

In this step, all the previous processes are included for determining the overall life cycle results per tkm (tonne-kilometer) by considering the operation of aircraft, production of aircraft, construction and land use of airport, operation, maintenance and disposal of airport. Therefore, inventory refers to the entire transport life cycle.

Table 21 Fuel and energy requirements of various fueled aircrafts

Fuel	Energy Consumption (MJ/tkm)	Energy Consumption (MJ/km)	Fuel Consumption (kg/tkm)
Kerosene (Jet Fuel A)	9.35	343.5	0.21
Hydrogen	8.62	316.5	0.07
Ammonia	9.53	350	0.51
Methanol	9.80	360	0.49
Ethanol	9.80	360	0.33
Natural Gas	12.53	460	0.25

Airport infrastructure expenditures and environmental interventions are accounted for using a weighted average of Intra-European and intercontinental freight transport performance at the airport. Aircraft manufacturing is allocated based on the total life span of an aircraft and its transport performance with a value of 23.5 tonne per aircraft unit. For the manufacturing of aircrafts modern production technologies are taken into account. The fuel and energy requirements of various fueled aircrafts are listed in Table 21 which are taken into account in the analyses.

3. Results and Discussion

The full life cycle of an aircraft running on the conventional jet fuel and various alternative fuels, including hydrogen, are assessed in terms of environmental impact and performance in SimaPro LCA software based on the inventory data. The results presented here are given on per tonne-km basis which represents a wider approach for all aircraft classes regardless of the split between passengers and cargo payloads onboard the aircraft. The analyzed impact categories are: Human toxicity which is related with toxic substances on the human environment and given in terms of 1,4-dichlorobenzene DB equivalents, global warming which represents the greenhouse gases to air associated with the climate change and given in kg CO_2 eq. (equivalents) unit, depletion of abiotic resources which is related to extraction of minerals/fossils and given in terms of kg Sb Antimony equivalents, stratospheric ozone depletion is the ozone depletion potential of various gasses in kg CFC-11 equivalent unit, land use which is the extraction of raw materials, production processes, agricultural land, area of industrial territory, landfill sites, incineration plant area, transport, use processes and given in terms of m^2a. The land use refers to the total arrangements, activities and inputs undertaken in a certain land cover type. The term land use is also used in the sense of the social and economic purposes for which land is managed. In the CML2001 the life cycle impact assessment method, competition is measured as occupied area*time (m^2a) where a represents the annual (year) [60].

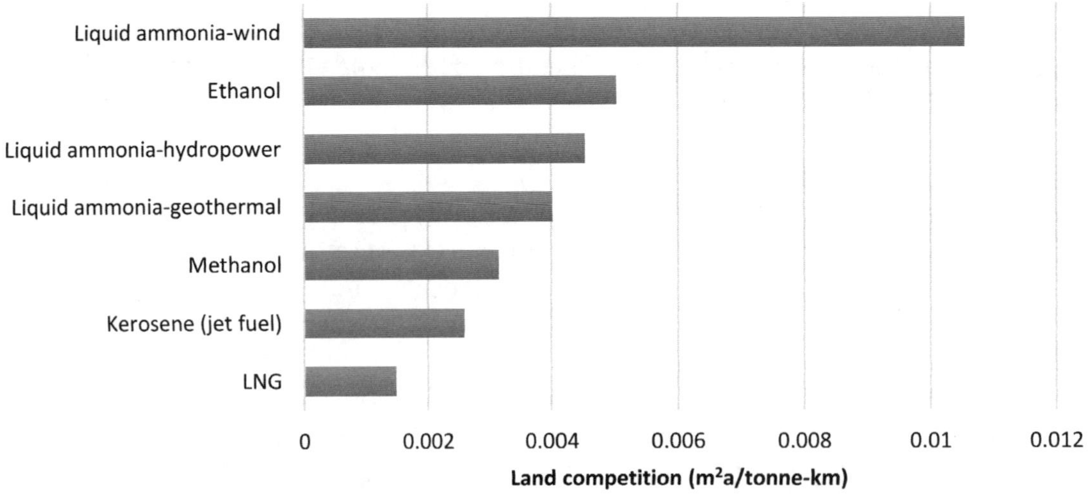

Figure 75 Land use in full life cycle of various fueled aircrafts per travelled tonne-km

The feedstock generation is the process which can cause land employment. Land usage per entity of fuel is lowermost for the LNG case as shown in Fig. 75 due to its higher overall fuel heating value. However land use intensities for jet fuel and methanol pathways are relatively similar and lower than liquid ammonia when compared with the wider collection of alternative fuel pathways. The annual land use intensities liquid ammonia (geothermal based) and LNG in the study are found to be 0.004 m^2/tonne-km and 0.0014 m^2/ tonne-km, respectively. For LNG, the land occupation and human toxicity potentials are lower than for kerosene. Liquid ammonia from geothermal energy has comparable land use values with methanol and ethanol fueled aircrafts.

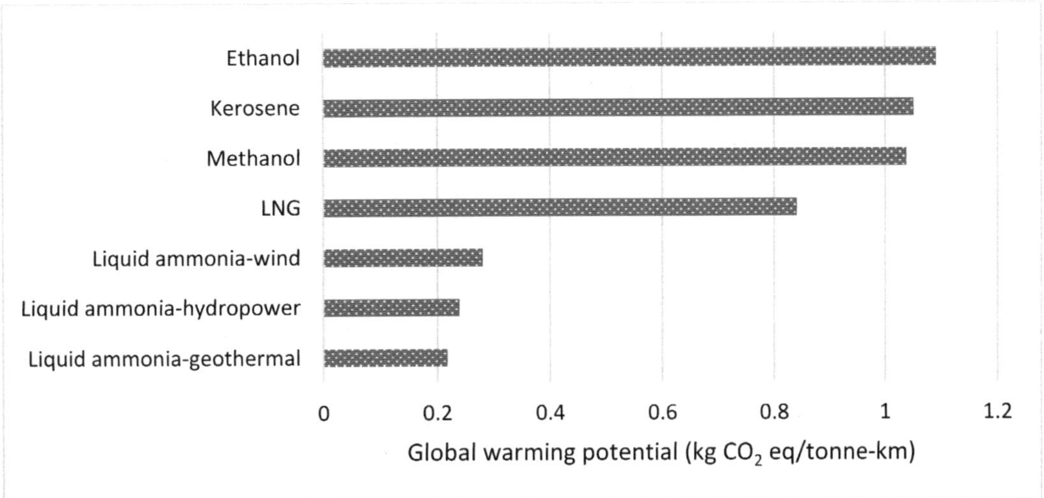

Figure 76 Global warming potential of various fueled aircrafts per travelled tonne-km

Though CO_2 is the furthermost significant GHG and is the major emission from the whole life cycle, quantifying the total amount of greenhouse gases produced is the key to examining the gross warming potential (GWP) of the different systems as shown in Fig. 76. The GWP of the cases is a mixture of CO_2, CH_4 and N_2O emissions. Ethanol fueled aircraft has the highest GWP among others. Liquid ammonia has the lowest values among all other fuels. Using renewable resources such as wind, geothermal and hydropower, the total GHG emissions are considerably lower for liquid ammonia. The overall GWP value of ethanol is the highest corresponding to 1.09 kg CO_2 eq/tonne-km since it comes from ethylene which is a hydrocarbon. Renewable energy usage in the fuel production processes decreases the overall GHG emissions. The hydropower based ammonia fueled aircraft releases about 0.24 kg CO_2 eq per tonne-km. This value goes down to 0.21 for kg CO_2 eq/tonne-km for ammonia fueled aircraft in case the renewable source is geothermal energy.

Ozone depletion potential of several gasses are specified in kg CFC-11 equivalent/kg emission where the time span is infinity. Fig. 77 presents the life cycle kg CFC-11 eq. emissions of the aircrafts with different fuels per kilometer travelled. It is quite high for methanol and kerosene while it is considerably less for liquid ammonia and LNG. Furthermore, utilization of renewable based ammonia is better than kerosene, methanol and ethanol in terms of ozone layer depletion.

In contrast to other categories, LNG has quite higher value in comparison with methanol as illustrated in Fig. 78. This is caused by the exploration and extraction process of natural gas. Similarly, natural gas dependent fuel such as methanol has high abiotic depletion values. Due to higher fuel flow rate for ammonia, it yields about 0.01 kg Sb eq/tonne-km. However, renewable based ammonia driven aircrafts can significantly decrease the depletion of abiotic resources down to 0.0014 kg Sb eq/tonne-km which corresponds to about 10% of conventional steam methane reforming based ammonia fueled aircraft.

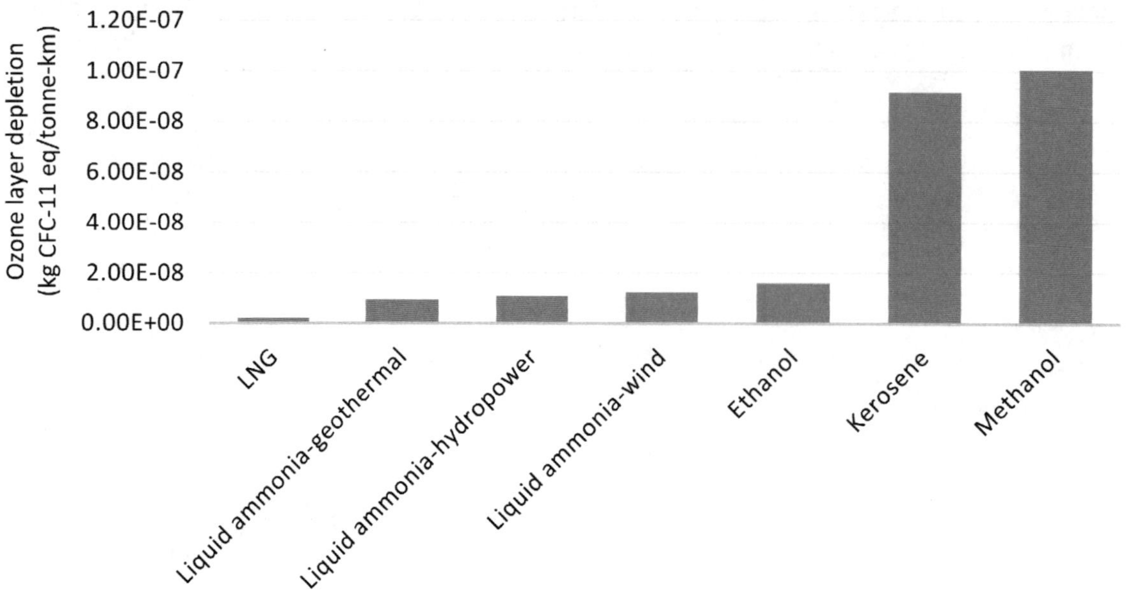

Figure 77 Ozone layer depletion values of various fueled aircrafts per travelled tonne-km

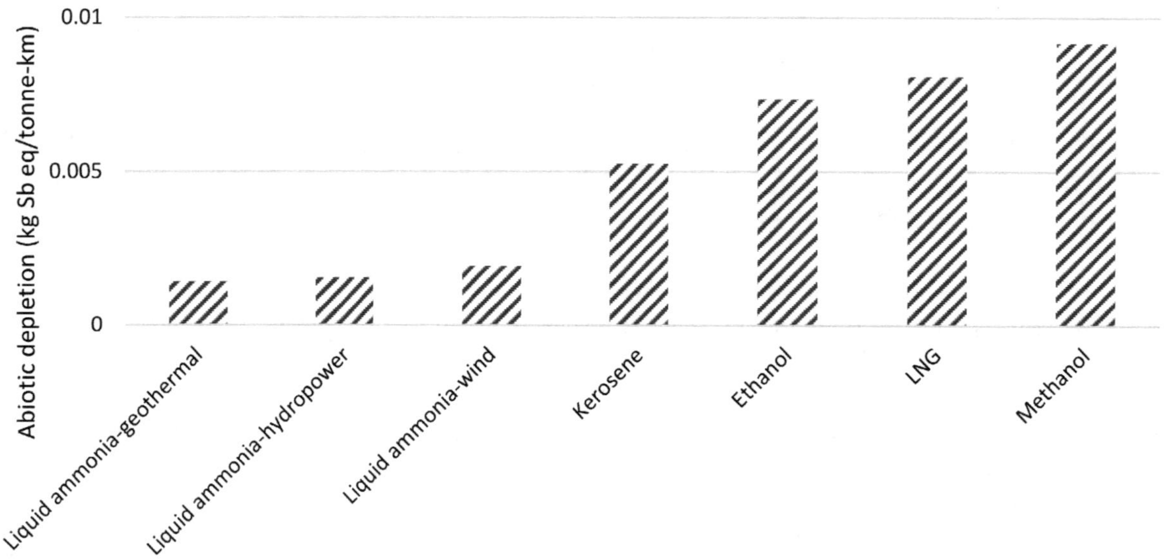

Figure 78 Abiotic depletion values of various fueled aircrafts per travelled tonne-km

Among all alternative aviation fuels, the kerosene jet fuel, methanol and ethanol human toxicity values are greater as shown in Fig. 79. On contrary, LNG and liquid ammonia have significantly lower toxic impacts on humans. Geothermal based ammonia fueled aircraft has the lowest human toxicity value among others whereas methanol has the highest value. It is attributed to operation of aircraft with a share of 93%. Natural gas burned in the furnace has a portion of 3% where it is also utilized in the methanol production process. Copper and ferrochromium are used for steel and building materials during the construction of the airport. Ammonia fuel from geothermal resource has comparable toxicity value corresponding to 0.04 kg 1,4- DB eq/tonne-km with LNG corresponding to 0.03 kg 1,4- DB eq/tonne-km.

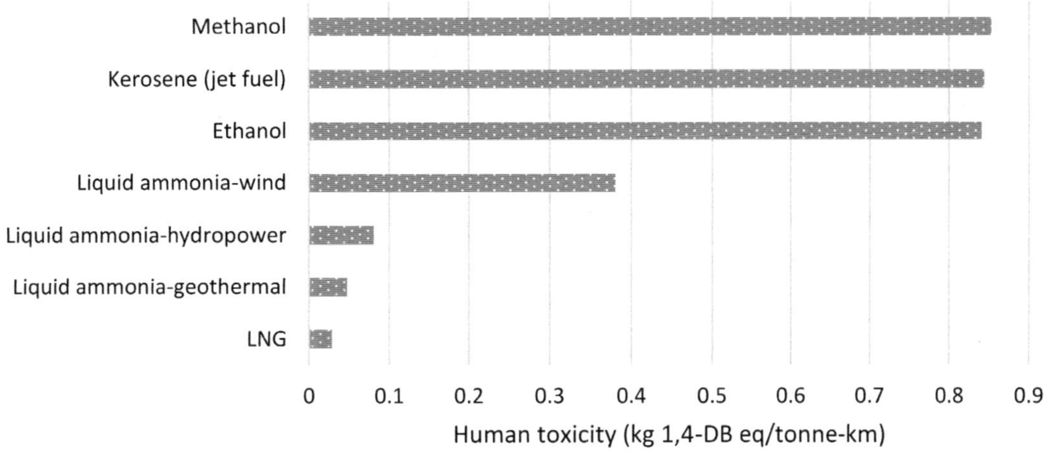

Figure 79 Human toxicity values of various fueled aircrafts per travelled tonne-km

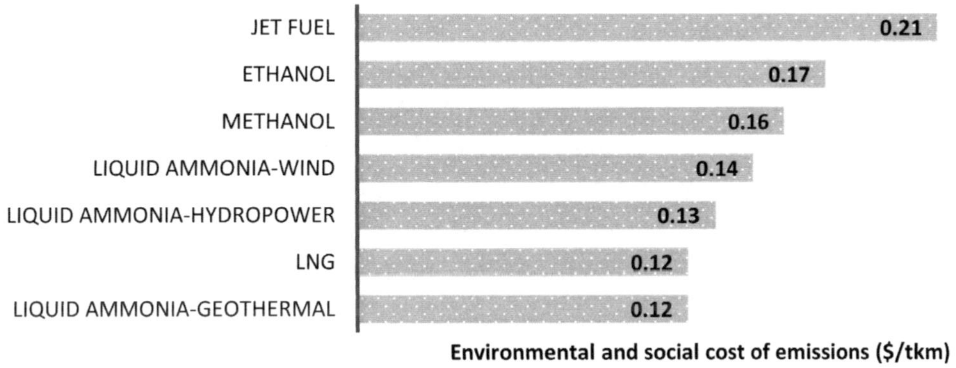

Figure 80 Total environmental and social cost of emissions for various fueled aircrafts from conventional and renewable resources

The marginal external price of a unit of GHG emissions is identified as environmental and social costs of emissions. In the literature, these values are estimated by using an integrated assessment (IAM) framework. IAM framework employs: a reference socio-economic situation, a model of the connection between emissions and temperature variation, and a link between the temperature variation and financial damages. Fig. 80 shows the calculated environmental and social cost of emissions. Since, they are associated with the amount of various type of emissions, kerosene jet fuel and fossil fuel based ammonia represent higher costs.

The environmental and social costs methodology is essential, for the reason that it gives a unique perspective on the economic impact of emissions on environment and human health. Environmental and social costs of HC, CO, NO_x, PM, and CO_2 emissions of various fueled aircrafts are evaluated in terms of USD/tonne-km as shown in Fig. 81 and Table 22.

Table 22 Environmental and social cost of emissions for complete life cycle of various fueled aircrafts from conventional resources

		Environmental and social cost of emissions (USD/tonne-km)						
Substance	Environmental and social impact (USD/kg)	Methanol	kerosene	liquid ammonia-hydropower	liquid ammonia-wind	liquid ammonia-geothermal	methanol	LNG
Particulates, < 2.5 um	229.2	0.0122	0.011	0.018	0.025	0.012	0.0942	0.0629
Carbon monoxide	4.16	0.0281	0.005	0.035	0.036	0.034	0.02794	0.0316
Nitrogen oxides	24.8	0.0240	0.113	0.055	0.056	0.053	0.02386	0.0121
Hydrocarbon	8.27	0.0314	0.008	0.007	0.009	0.007	0.02315	0.0124
Carbon dioxide	0.07	0.074	0.073	0.015	0.018	0.014	0.07086	0.057

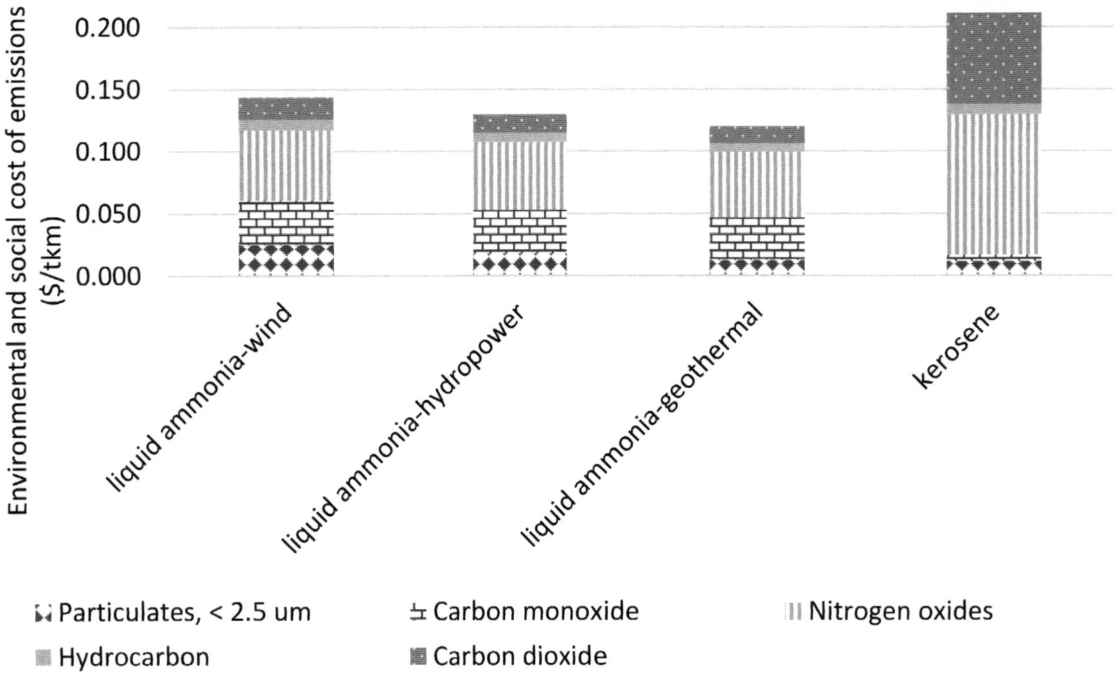

Figure 81 Comparison of total environmental and social cost of emissions for renewable based ammonia driven aircrafts and kerosene driven aircraft

However, as illustrated in Fig. 80, renewable based ammonia yields lower environmental and social cost of emissions in comparison with kerosene. Specifically, Fig. 81 shows renewable only based routes for ammonia in comparison with kerosene. It is noted that the total environmental and social costs for renewable based ammonia fueled aircrafts are considerably lower than conventional kerosene jet fuel.

Fig. 82 and Table 16 illustrate the cost of flight for a 5600 km distance in which alternative fuels are used for the aircrafts. Although LNG represented better environmental performance, the cost of aircraft operation in terms of fuel is the highest for ethanol and LNG.

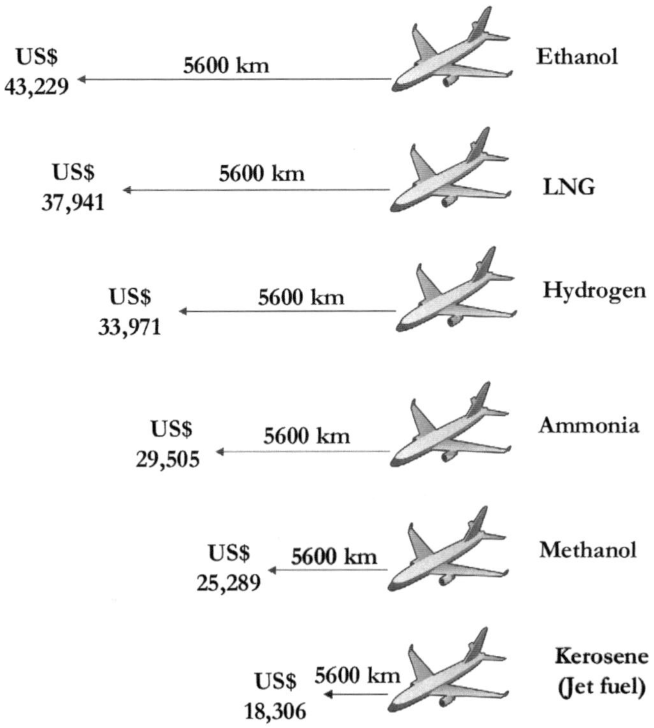

Figure 82 Comparison of fuel costs during the operation of aircrafts for the given range

In the cost calculations, current conventional based routes are taken into account for comparison purposes. Hydrogen fueled aircraft has lower cost compared to these two alternatives. Since the production of kerosene from crude oil is a more mature technology, it represents the lowest cost among all. Liquid ammonia is also low cost alternative compared to hydrogen as seen in Table 23. Hydrogen has the highest cost per unit mass, however, having higher heating value makes hydrogen a cost effective solution. Liquid ammonia here is considered from steam methane reforming, hence developing renewable technologies will enable lower cost of operation for liquid ammonia driven aircrafts.

Table 23 Average fuel consumption rates and fuel costs for selected alternative fuels

Fuel	Fuel Consumption (kg/km)	Fuel Consumption (kg/tonne-km)	Fuel Cost (USD/kg)
Kerosene (Jet fuel)	7.99	0.217	0.409
Methanol	18.06	0.492	0.250
Ammonia	18.82	0.512	0.280
Hydrogen	2.64	0.071	2.300
LNG	9.46	0.257	0.716
Ethanol	12.47	0.339	0.619

Source: Refs. [60,61]

The performances of the aircrafts are also compared in terms of energy and exergy efficiencies in Fig. 83. In the calculations, two different efficiencies are taken into account namely: production efficiency and combustion efficiency. The production efficiency represent the process from raw material to final product as fuel. The combustion efficiency is the process of utilization in the aircraft. Although, there can be storage and transportation losses, the overall efficiency can be calculated based on these two main efficiencies by neglecting the storage and transportation efficiencies. Here, conventional method (SMR) is taken into account for hydrogen and ammonia rather than renewable based options.

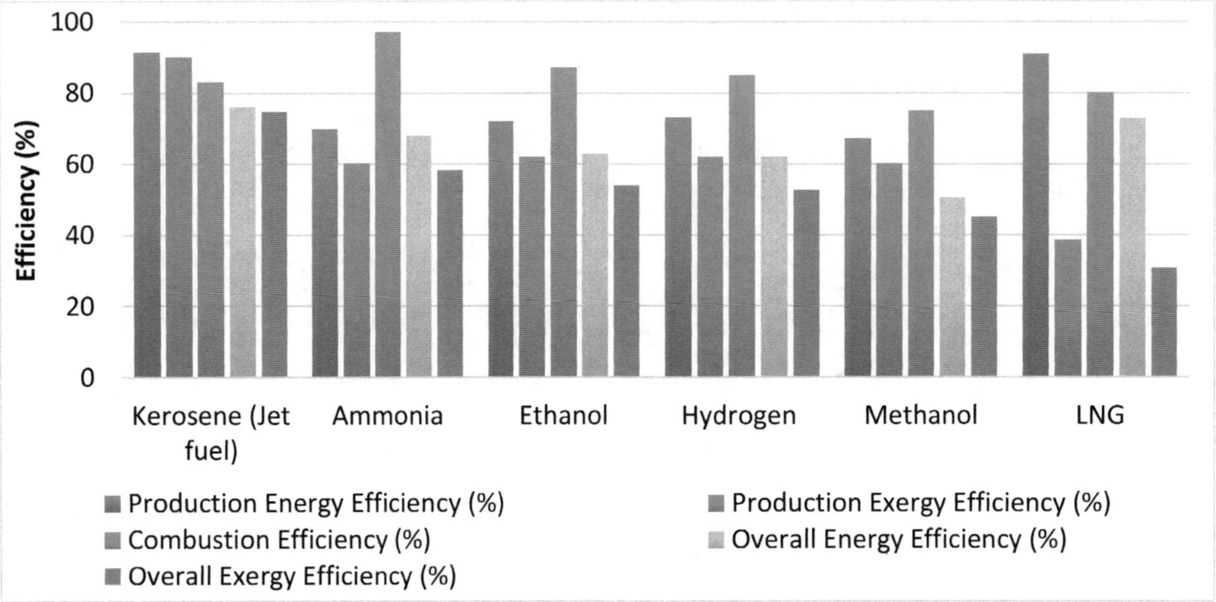

Figure 83 Comparison of energy and exergy efficiencies of various fueled aircrafts

The efficiency values are taken from the open literature [11,68,69] and calculated accordingly. Kerosene has the highest production energy and exergy efficiencies yielding the highest overall efficiency. Liquefaction of natural gas in an energy intensive process causing lower exergetic efficiency, hence lower overall efficiency for utilization in aircrafts. Ammonia has high combustion efficiency which results in the second highest overall exergy efficiency among the other alternative fuels. Overall efficiencies of methanol are lower than other fuels. It is noted that the selected alternative fuels may represent higher efficiencies but also higher environmental impacts depending on the production process and technology.

4. Concluding Remarks

In this study, a well-to-wake approach is used in order to determine the overall life cycle emissions of an aircraft running on various conventional and alternative fuels. Operation of the aircraft, construction, maintenance and operation of airport, manufacturing of the aircraft, production and utilization of fuel are considered as a complete LCA cycle. Although there are modifications required to achieve the aviation fuel qualifications for such alternative fuels, the long term viability and environmental sustainability promote them attractive solutions for the future of aviation industry. The following conclusion remarks can be derived from this study:

- Alternative aviation fuels including ammonia and LNG are more environmentally friendly options than kerosene.
- Renewable sources based ammonia routes represent the most preferable option in terms of environmental impact.

- Global warming potential of LNG (0.84 kg CO_2 per tonne-km) and methanol (1.03 kg CO_2 per tonne-km) driven aircrafts are lower than currently used kerosene based jet fuels.
- Renewable resources based ammonia production considerably lowers the environmental impacts corresponding to 0.23 kg CO_2 per tonne-km for hydropower route.
- Operation of the aircraft has almost equivalent share (40.7%) with operation and maintenance of the airport (44.6%) in total GHG emissions. Hence, the energy supply of airport facilities are also critical when complete life cycle is evaluated.
- The environmental cost analyses reveal that nitrogen oxides are the highest contributor followed by carbon dioxides.
- The cost of flight is currently lower for kerosene jet fuels however by developing technologies the cost of flight for ammonia can compete with conventional jet fuels.

CHAPTER 7:AMMONIA IN ROAD TRANSPORTATION

Current vehicles can be converted to run on ammonia by minor modifications. The general requirements for the conversion of a diesel vehicle engine to dual fuel ammonia engine can be written as follows:

- Installation of separate low pressure ammonia injectors
- Replacement of copper and brass parts with stainless steel parts.
- Specific arrangement to reduce the injected flow rate of diesel
- Replacement of gaskets, nuts and bolts interacting with ammonia with stainless steel.
- Installation of control unit for ammonia injection
- Installation of temperature and pressure measurement units at intake manifold, engine cooling fluid, ammonia flow regulator and exhaust.
- In case of on-board hydrogen production, a decomposition unit which reduces total diesel consumption to increase the ammonia combustion performance.
- Installation of non-corrosive fuel ammonia tank

Ammonia can be used for multiple purposes in the vehicle. It is a fuel, refrigerant and reduction agent. The patent developed by UOIT proposes the utilization of ammonia in these major areas. The requirements for patent US8272353 B2 "Apparatus for using ammonia as a sustainable fuel, refrigerant and NOx reduction agent" [70] can be summarized as follows:

1. *Configuration: ammonia-fueled system for vehicular power and cooling generation*

- Adaptation of fuel storage tank to store ammonia
- A heat exchanger operably connected to the fuel tank
- A decomposition and separation unit operably connected to the heat exchangers and having a hydrogen conduit and a nitrogen conduit,
- The decomposition and separation unit is adapted to separate the heated ammonia into hydrogen and nitrogen and stream them into the hydrogen and nitrogen conduits
- an internal combustion engine operably connected to the hydrogen conduit.

2. *Configuration: ammonia-fueled hybrid system for propulsion, power, heating and air-conditioning*

- a thermally insulated fuel tank adapted to receive ammonia;
- a condenser operably connected to the fuel tank;
- an evaporator operably connected to the fuel tank,
- the condenser and the linear piston;
- a selective catalytic reductor operably connected to the fuel tank and operably connected to the generator,
- the selective catalytic reductor is adapted to combine ammonia with nitrogen to reduce the production of nitrogen oxides.

3. *Configuration: ammonia based fuel-cell system with combined power, heating and refrigeration*

- the fuel tank adapted to receive ammonia;
- a heater operably connected to the fuel tank;
- a fuel-cell operably connected to the heater;

- a turbocharger operably connected to the fuel-cell;
- an electrical drive generator operably connected to the turbocharger.

4. *Configuration: ammonia based cooling system*
- a thermally insulated tank adapted to receive ammonia;
- a heat exchanger; and a throttling valve operably connected between the tank and the heat exchanger;
- adjusting the throttling valve adjusts the evaporation temperature of the ammonia in the tank.

5. *Configuration: a fueling system*
- a decomposition and separation unit adapted to receive ammonia, delivering hydrogen and nitrogen as separate streams;
- a nitrogen expanding turbine operably connected to a heat recovery heat exchanger.

6. *Configuration: an ammonia fueled internal combustion engine*
- a thermally insulated fuel tank adapted to store ammonia;
- a mechanical refrigeration unit having a sub-cooler coil immersed in ammonia in the fuel tank;
- an internal combustion engine operably connected to the fuel tank and the mechanical refrigeration unit.

Conversion process to ammonia vehicle is quite similar with compressed natural gas and propane vehicles. If current fuel tank size of a vehicle is used for ammonia only, the range would be about 50-60% of gasoline only vehicle. The estimated conversion costs of a conventional diesel/gasoline passenger vehicle to various alternative fueled vehicles are given below.

- Propane Conversion Kit: $600-$1300
- Compressed Natural Gas Conversion Kit: $4000-$8000
- Ammonia Conversion Kit: $5000-$10,000

1. Life Cycle Assessment of Vehicles

A characteristic life cycle of a vehicle technology can be categorized into two main steps, namely fuel cycle and vehicle cycle. In the fuel cycle, the processes beginning from the feedstock production to fuel utilization in the vehicle are considered as shown in Fig 84. In the vehicle cycle, utilization of fuel is considered. The results presented here are given on per km basis based on the fuel consumption rates given in Table 24.

Figure 84 Complete life cycle of vehicles including fuel/vehicle cycle.

Table 24 Energy consumptions per km for the selected vehicles

Fuel		Fuel/Energy Consumption	Unit
Gasoline		0.0649108	kg/km
Diesel		0.0551536	kg/km
M90	Methanol (90%)	0.1180535	kg/km
	Gasoline (10%)	0.0060664	kg/km
Hydrogen		0.0195508	kg/km
Ammonia		0.0926600	kg/km
EV		0.2167432	kWh/km
HEV	Electric (50%)	0.1083716	kWh/km
	Gasoline (50%)	0.0324554	kg/km
CNG		0.0603914	kg/km
LPG		0.0576296	kg/km

Source: [71]

The specific conditions for the selected vehicles are presented herein:

- Gasoline: All processes on the refinery site excluding the emissions from combustion facilities, including waste water treatment, process emissions and direct discharges to rivers are accounted for. The inventory data also includes the distribution of petroleum product to the final consumer including all necessary transports. Transportation of product from the refinery to the end user is considered together with operation of storage tanks and petrol stations. Emissions from evaporation and treatment of effluents are accounted for. Particulate emissions cover exhaust- and abrasions emissions.
- Diesel: Diesel is evaluated as low-Sulphur at regional storage with an estimation for the total conversion of refinery production to low-Sulphur diesel. An additional energy use (6% of energy use for diesel production in the refinery) has been estimated. The other processes are similar to gasoline. Particulate emissions cover exhaust- and abrasions emissions.
- CNG: Natural gas with a production mix at service station is taken into account. The inventory data contains electricity necessities of a natural gas service station together with emissions from losses. The data set represents service stations with high (92%), medium (6%) and low (2%) initial pressure. VOC emissions are obtained from gas losses and contents of natural gas. Particulate emissions cover exhaust- and abrasions emissions.
- Hydrogen: Hydrogen is produced during cracking of hydrocarbons. It includes combined data for all processes from raw material extraction until delivery at plant. The output fractions from an oil refinery are composite combinations of mainly unreactive saturated hydrocarbons. The first processing step in converting such elements into feedstock suitable for the petrochemical industries is cracking. Particulate emissions cover exhaust- and abrasions emissions. In order to have comparable results where hydrogen comes from non-fossil fuels such as solar PV and nuclear, they are also taken into account in the analyses by applying water electrolysis route. The electrolyzer is assumed to consume about 53 kWh electricity for one kg of hydrogen production.
- Ammonia: Ammonia synthesis process is Haber-Bosch which is the most common method in the world. Ammonia production requires nitrogen and hydrogen. In this study, hydrogen is assumed to be from hydrocarbon cracking as explained in the previous paragraph. In the life cycle assessment of nitrogen production, electricity for process, cooling water, waste heat and infrastructure for air separation plant are included. Particulate emissions cover exhaust- and

abrasions emissions. In order to have comparable results where ammonia comes from non-fossil fuels such as solar PV and nuclear, they are also taken into account in the analyses. The generated hydrogen from electrolyzers are used for ammonia synthesis plant.

- EV: Electricity consumption is included. Particulate emissions comprise exhaust and abrasions emissions. Heavy metal emissions to soil and water caused by tire abrasion are accounted for. In the electricity usage process, electricity production mix, the transmission network and direct SF6-emissions to air are included. In order to present a renewable based scenario for electric vehicles, a mixture of renewables for energy requirement during the operation are also evaluated consisting of 25% biomass, 25% solar PV, 25% wind power and 25% hydropower.
- HEV: Hybrid car is assumed to be 50% electric and 50% gasoline with ICE. Electricity and gasoline consumptions are included. Particulate emissions comprise exhaust and abrasions emissions. Heavy metal emissions to soil and water caused by tire abrasion are accounted for. For the hybrid vehicle's electricity, a mixture of renewables for energy requirement during the operation are also evaluated consisting of 25% biomass, 25% solar PV, 25% wind power and 25% hydropower.
- Methanol: The selected fuel M90 consists of 90% methanol and 10% gasoline. The raw materials, processing energy, estimate on catalyst use, and emissions to air and water from process, plant infrastructure are included. The process describes the production of methanol from natural gas via steam reforming process to obtain syngas for the production of methanol. There is no CO_2 use and hydrogen is assumed as burned in the furnace. Raw materials, average transportation, emissions to air from tank storage, estimation for storage infrastructure are included for the distribution part where 40% of the methanol is assumed to be transported from overseas. Particulate emissions cover exhaust- and abrasions emissions.
- LPG: All processes on the refinery site excluding the emissions from combustion facilities, including waste water treatment, process emissions and direct discharges to rivers are considered. All flows of materials and energy due to the throughput of 1 kg crude oil in the refinery is accounted for. Refinery data include desalting, distillation (vacuum and atmospheric), and hydro treating operations. Particulate emissions cover exhaust- and abrasions emissions.

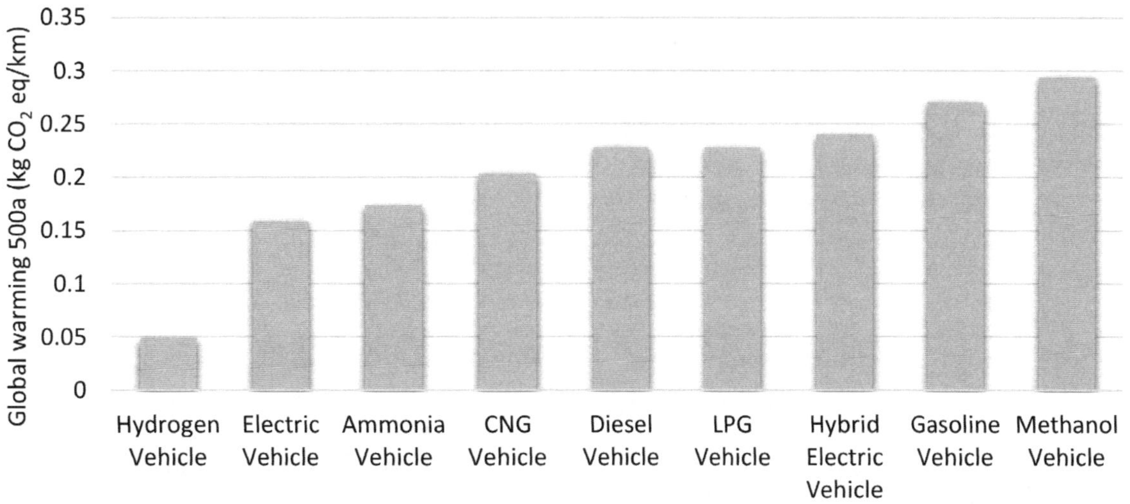

Figure 85 Life cycle comparison of global warming results for various vehicles

The global warming potentials of assessed vehicles are comparatively shown in Fig. 85. The lowest GHG emissions are observed in hydrogen, electric and ammonia vehicles corresponding to 0.049 kg CO_2 eq/km, 0.15 kg CO_2 eq/km and 0.17 kg CO_2 eq/km, respectively. Hydrogen consumption is quite lower than ammonia consumption in the passenger car because of higher energy density. It is an expectable result that EVs also yield lower global warming potential, however production pathway of electricity has a key role in GHG emissions. If electricity production can be realized by renewable sources such as solar, biomass, hydropower and wind energy, total emissions would decrease for both EVs and HEVs.

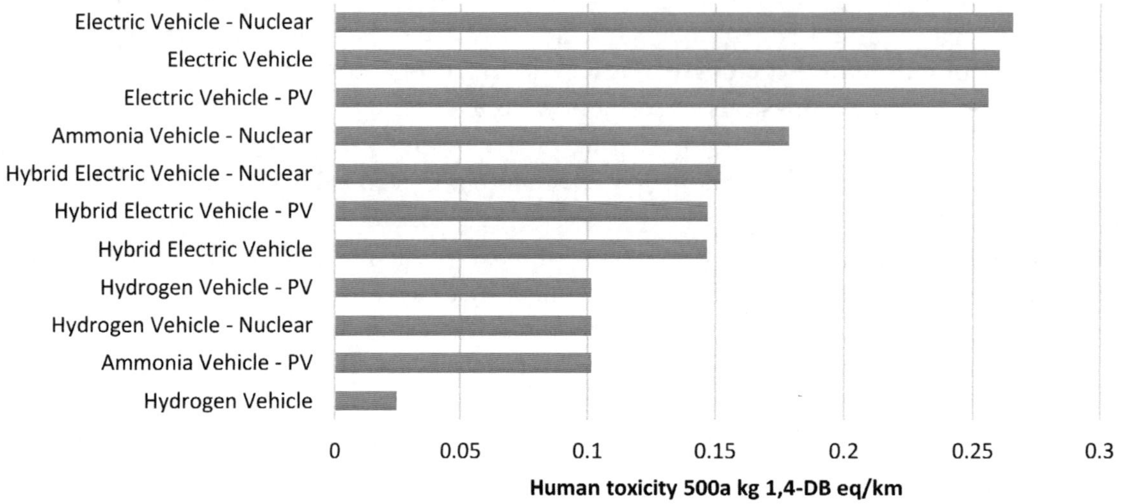

Figure 86 Life cycle comparison of human toxicity results for various vehicles from nuclear energy and solar PV routes

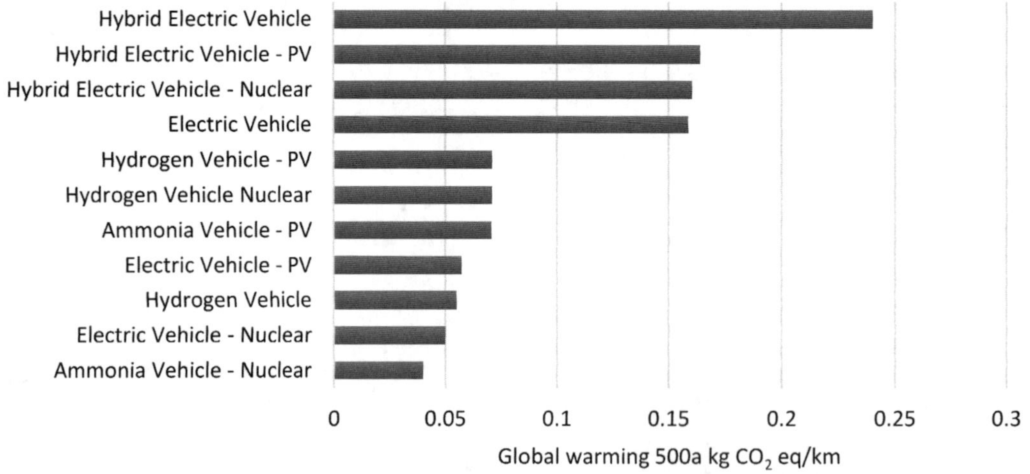

Figure 87 Life cycle comparison of global warming results for various vehicles from nuclear energy and solar PV routes

The electricity from nuclear (25%), biomass (25%), hydropower (25%) and solar PV (25%) are equally used only for the operation processes of the EVs and HEVs vehicles as renewable mix. Utilization of renewable electricity for EVs, HEVs, ammonia and hydrogen vehicles are comparatively shown for various impact categories. For human toxicity category, using solar and nuclear energy does not cause significant reductions for EVs and HEVs as shown in Fig. 86. However, for global warming and abiotic depletion categories, both solar energy and nuclear energy routes lower the environmental impacts more than 50%.

Ammonia driven vehicle where the ammonia is produced from nuclear electrolysis method yield the lowest GHG emissions corresponding to about 0.04 kg CO_2 eq. per km as shown in Fig. 87. Hydrogen and EVs (from nuclear and PV) have quite similar greenhouse gas emissions in the range of 0.049-0.057 kg CO_2 eq. per km. In terms of abiotic depletion values, hydrogen vehicle still yields the lowest value whereas nuclear routes for ammonia and EVs further decrease the abiotic depletion impact as shown in Fig. 88.

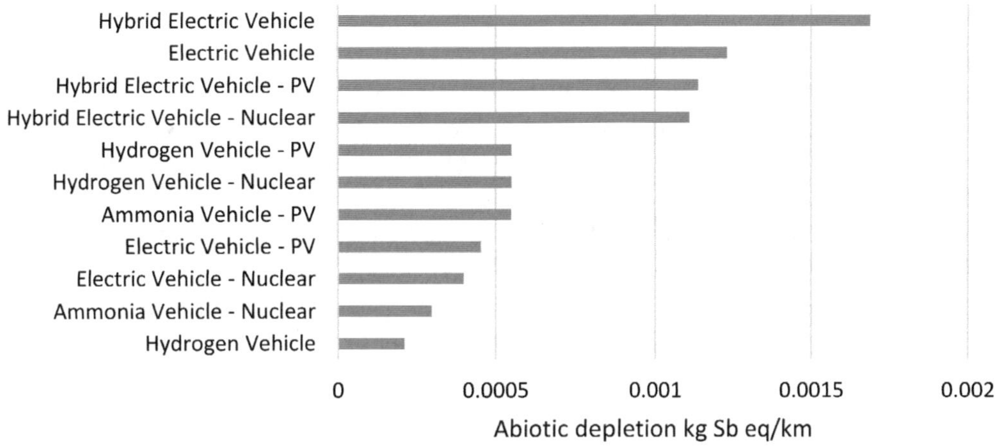

Figure 88 Life cycle comparison abiotic depletion for various vehicles from nuclear energy and solar PV routes

The results show that hydrogen and ammonia vehicles are the most environmentally benign ones in most of the environmental impact categories. Ammonia as a sustainable and clean fuel has lowest global warming potential after EVs and yield lower ozone layer depletion values than EVs. However, in case renewables are used both for ammonia vehicles and EVs, ammonia can suggest lower environmental impacts. Although EVs do not emit direct CO_2 during operation, the production and disposal processes of batteries bring some consequences which harm the environment in terms of acidification, eutrophication and human toxicity.

CHAPTER 8: ON-BOARD AMMONIA UTILIZATION

In this section, a comparative life cycle assessment of a hydrogen fueled car with on-board ammonia decomposition is conducted. Ammonia is used as the fuel source of hydrogen which is stored in anhydrous ammonia storage tank. Ammonia is decomposed into hydrogen and nitrogen on-board to fuel internal combustion engine of the car. Hydrogen production via ammonia decomposition is an endothermic process which is also a reversible chemical reaction. The storage and transportation of ammonia is similar to liquefied petroleum gas (LPG) which is commonly used for the current automobile industry. Ammonia can be produced by extracting nitrogen from air and hydrogen from water and combining them with the help of any type of energy source. The theoretical thermal efficiency under adiabatic conditions for a thermocatalytic reaction is about 85% as relative to the energy of the released hydrogen. If no other source of energy is available, at least 15% of the available hydrogen energy content will be sufficient to supply the necessary heat of hydrogen production reaction. In the present study, the life cycle environmental assessment of a hydrogen fueled car using on-board ammonia decomposition is carried out in which ammonia is synthesized via (i) steam methane reforming (SMR) , (ii) solar PV (iii) wind energy. SMR based ammonia production is the mostly used option in the world as of year 2016. In the analyses, manufacturing process starting with heavy fuel oil, natural gas, air and electricity is considered together with auxiliaries, energy, transportation, infrastructure and land use, as well as wastes and emissions into air and water. In addition, transport of the raw materials, auxiliaries and wastes is also included. Using the LCA methodology, the environmental impacts of using a hydrogen fueled internal combustion engine vehicle is comparatively identified and quantified from cradle to grave.

1. Life Cycle Assessment of On-Board Ammonia Cracking Vehicles

Ammonia can be decomposed thermo-catalytically to achieve hydrogen according to the following endothermic reaction [72]:

$$\frac{2}{3}NH_3 + 30.1\frac{kJ}{mol}H_2 \rightarrow H_2 + \frac{1}{3}N_2 \qquad (7)$$

Here, the obligatory enthalpy signifies 10.6% of HHV or 12.5% of the lower heating value (LHV) of the generated hydrogen. The ammonia decomposition reaction does not need catalysis to be performed at high temperatures for example over 1,000 K; though, at inferior temperatures, the reaction rate is too little for practical applications such as hydrogen generation for energy conversion. At 400°C, the equilibrium conversion of NH_3 is very high at 99.1% [72] and at about 430°C, almost all ammonia is converted to hydrogen at equilibrium, below atmospheric pressure circumstances [72]. A schematic diagram of on-board ammonia usage is shown in Fig. 89.

There is a big array of catalysts appropriate to ammonia decomposition (e.g., Fe, Ni, Pt, Ir, Pd, and Rh), nonetheless ruthenium (Ru) seems to be the finest one when reinforced with carbon nanotubes, making hydrogen at additional than 60 kW equal power per kilogram of catalyst [72]. Over ruthenium catalysts, at temperatures lower than about 300°C, recombination of nitrogen atoms is rate limiting, while at temperatures higher than 550°C, the cleavage of ammonia's N–H bond is rate limiting. Though, the activation energy is greater at low temperature (180 kJ/mol) and inferior at higher temperatures (21 kJ/mol). The finest temperature range for ammonia decomposition over ruthenium catalysts may be 350°C to 525°C, which proposes that flue gasses from hydrogen ICEs, other hot exhausts from burning procedures, or electrochemical power conversion in high-temperature fuel cells can be used to drive ammonia decomposition [72,73].

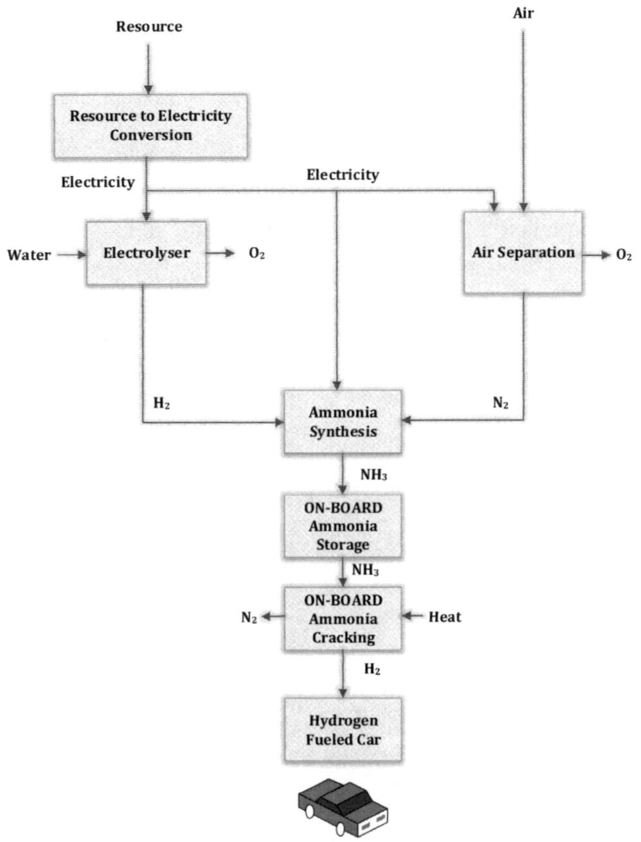

Figure 89 The complete process describing the production, storage and decomposition of ammonia for hydrogen driven vehicle

Here, the following hydrogen production routes required for ammonia synthesis are considered in LCA analyses:

- Steam methane reforming (SMR)
- Solar PV assisted water electrolysis
- Wind energy assisted water electrolysis

In LCA, ammonia synthesis process is Haber-Bosch. Catalyst is the ruthenium based catalyst for on-board ammonia cracking.

The following LCA phases are accounted for:

- manufacturing of the vehicle
- operation of the vehicle (including the fuel production)
- maintenance of the vehicle
- disposal of the vehicle

Considering all these processes, the environmental impact categories are defined and results are obtained as shown in Fig. 90 to Fig. 93 for ammonia cracking hydrogen vehicle. Here, ammonia is decomposed at elevated temperature thermo-catalytically. The abiotic depletion results of on-board ammonia cracking vehicles from various resources and gasoline vehicle are shown in Fig. 90. Gasoline vehicle has higher values than renewable based routes for ammonia cracking.

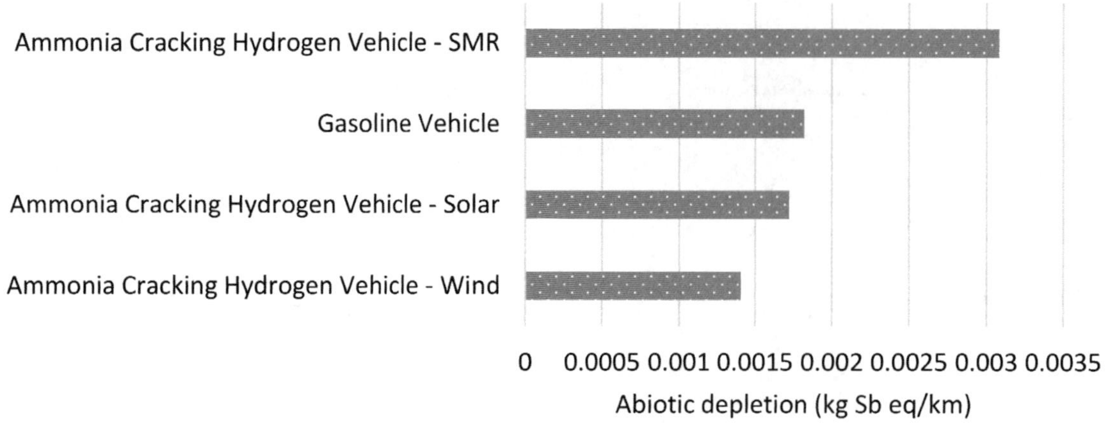

Figure 90 Abiotic depletion results of on-board ammonia cracking vehicles from various resources and gasoline vehicle

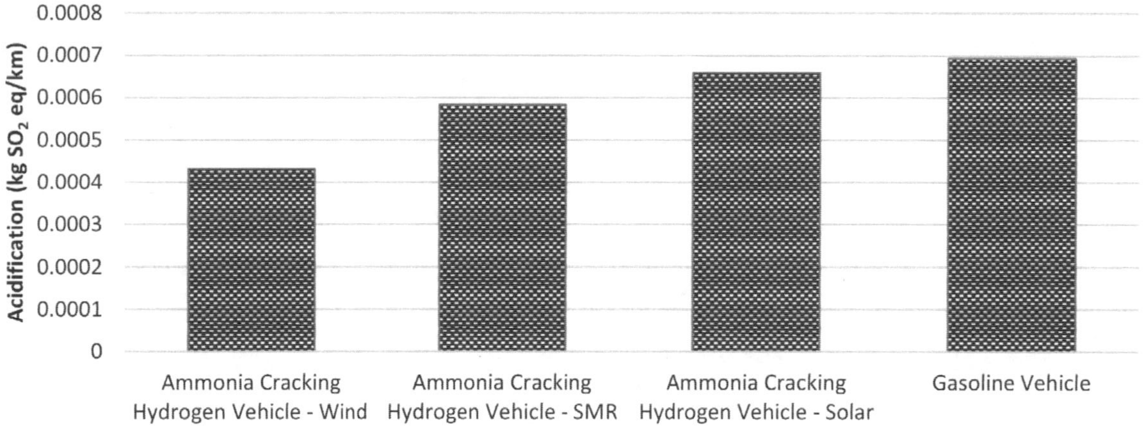

Figure 91 Acidification depletion results of on-board ammonia cracking vehicles from various resources and gasoline vehicle

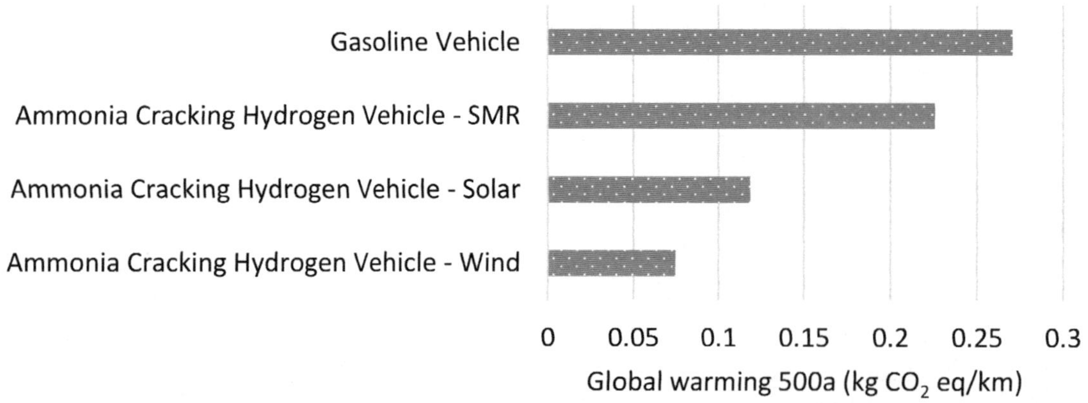

Figure 92 Global warming results of on-board ammonia cracking vehicles from various resources and gasoline vehicle

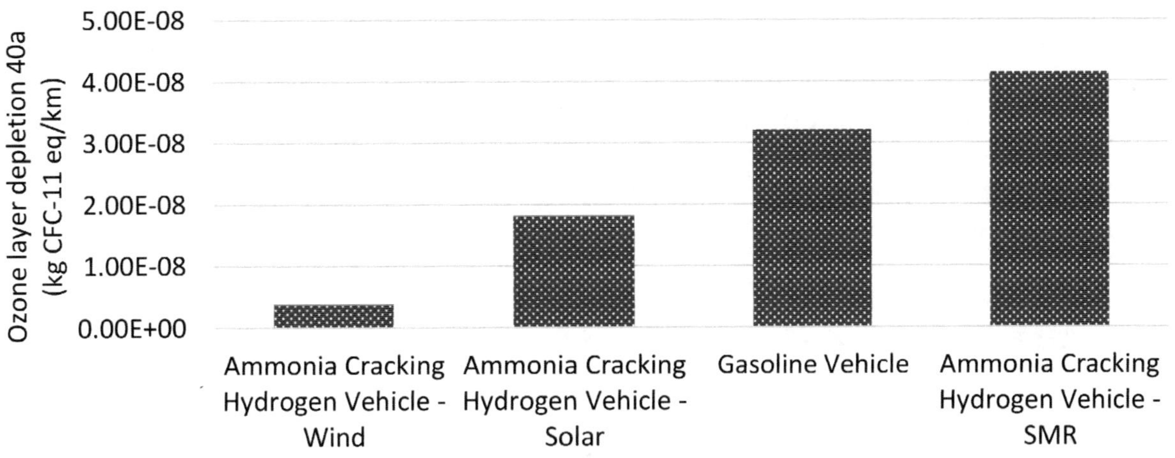

Figure 93 Ozone layer depletion results of on-board ammonia cracking vehicles from various resources and gasoline vehicle

The acidification depletion results of on-board ammonia cracking vehicles from various resources and gasoline vehicle are shown in Fig. 91. Gasoline vehicle has the highest acidification value in comparison with other ammonia cracking vehicles. Global warming potentials of ammonia cracking vehicles are quite better than conventional gasoline vehicles as shown in Fig. 92. Although ammonia is produced from SMR, the total GHG emissions are considerably lower. If ammonia is produced using wind energy and cracked on-board in the vehicle, the GHG emissions decrease down to 0.075 kg per every km considering the complete cycle. The ozone layer depletion results of on-board ammonia cracking vehicles from various resources and gasoline vehicle are shown in Fig. 93 where gasoline vehicle is higher than wind and solar based ammonia cracking vehicles.

2. Ammonia Decomposition

Ammonia decomposition (cracking) is simply the reverse of the synthesis reaction. NH_3 (g) → 1/2 N_2 (g) + 3/2 H_2 (g) $\Delta H = +46$ kJ/mol Note that the reaction is endothermic. The temperature required for efficient cracking depends on the catalyst. There are a wide variety of materials that have been found to be effective, but some require temperatures above 1000°C. Others have high conversion efficiency at temperatures in the range of 650-700°C. As these temperatures are well above PEM fuel cell operating temperatures, some of the fuel or, perhaps, the fuel cell purge gas, would need to be burned to maintain an efficient reaction [74].

The theoretical adiabatic efficiency for the thermocatalytic reaction is about 85% relative to the energy (LHV) of the released hydrogen. If no other energy source were available, at least 15% of the available hydrogen energy content would have to be burned to supply the heat of reaction. Of course, additional energy would be required to overcome thermal losses in the cracking reactor. Since the reaction occurs at high temperature, this heat would likely come from the combustion of ammonia and/or hydrogen in onboard storage applications. In the forecourt, one would have the option of using other energy sources (electricity, natural gas, etc.) to supply the heat of reaction.

3. On-Board Ammonia

Table 25 shows the equilibrium conversion of ammonia, calculated using HSC10, as a function of temperature at reactor pressures of 1 and 10 bar. Actual decomposition conditions used in real-world operations involve tradeoffs between the costs of operations at higher temperatures and cost of removing unconverted ammonia. Since proton electrolyte membrane (PEM) fuel cells require ammonia concentrations below 0.1 ppm, significant purification will be necessary. Even at 900° C, the equilibrium calculations indicate 1500 ppm unconverted NH_3 at 10 bar.

Table 25 Ammonia conversion rates at equilibrium

Temperature (°C)	Unconverted Ammonia at 10 bar (%)	Unconverted Ammonia at 1 bar (%)
400	7.91	0.88
500	2.55	0.26
600	1	0.1
700	0.47	0.047
800	0.25	0.025
900	0.15	0.015

As the reactor temperature is increased, other problems emerge. In particular, it becomes more difficult for the reactor materials, such as the catalyst, its supports and the reactor container, to sustain exposure to this environment. Exposure of the container materials to stress corrosion cracking (SCC) and high temperature hydrogen embrittlement, for example, would have to be considered in the materials choices, particularly since the unit would be subjected to temperature swings from ambient temperatures (when not used) to the highest operating temperature. Furthermore, just from a strength and stability point of view, metal alloy choices become more limited (and generally more expensive) for high temperature applications. Although a stainless steel alloy might be used below 500°C, a Ni based or other alloy would be required at higher temperatures. A higher operating temperature would perhaps also require more insulation, increasing the weight and volume of the reactor. It could also increase the time required for the reactor to start producing hydrogen from a cold start, an important parameter for on-board vehicular applications [74].

A potential design for an ammonia cracking reactor is schematically shown in Fig. 94. Liquid ammonia would be pumped from a storage tank through a heat exchanger to capture waste heat from the hot gases exiting the cracking reactor. The preheated gases would then go through a furnace or catalytic combustor to heat them to the temperatures necessary for the reaction. The stream exiting the reaction would go to a separation system which would be optimized to produce a very pure stream of hydrogen while still leaving sufficient hydrogen with the nitrogen and unreacted ammonia to provide heat for the endothermic cracking reaction. This waste stream would be combusted to supply heat for the reaction and to remove unreacted ammonia.

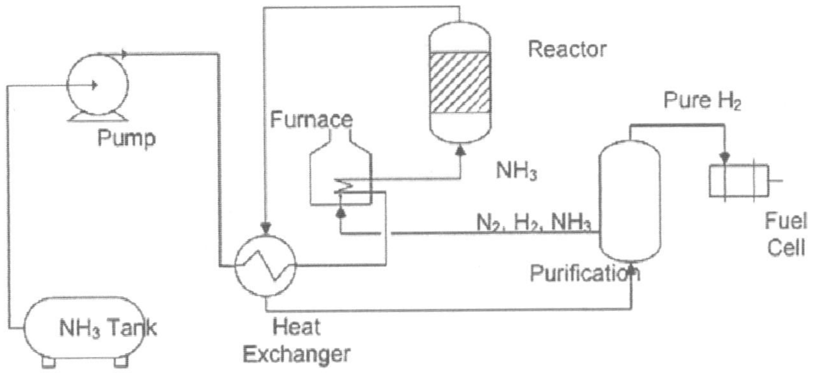

Figure 94 Ammonia fuel processing system (adapted from [74])

For onboard storage applications, the reactor would have to function over a very extensive dynamic range and with very fast response time to supply a fuel cell or ICE under all drive conditions. It would also have to be sized based on its ability to supply enough hydrogen for full power operation. For instance, the input hydrogen flow rate at full power for a 100 kW fuel cell (assuming a full power efficiency of about 45%) would be 2 g/s, or about 1,350 standard L/min of hydrogen. At 65% overall conversion efficiency, the cracking reactor would be consuming about 1.75 liters of liquid ammonia per minute and would need to process about 23 standard liters of ammonia per second [74]. Additional requirements on the cracking reactor component for on-board use would be to eliminate even trace amounts of ammonia in the outlet stream and to filter out the nitrogen. The ammonia removed from the outlet stream must be recycled into the inlet stream of the tank or used to supplement the inlet stream for source heat, since venting even small amounts of ammonia is not an option. Further details of ammonia cracking systems can be found in the literature, where a number of papers discuss the performance of different catalysts and other aspects of ammonia cracking. The system above is but one of many potential designs for an integrated reactor for ammonia cracking and purification. It does, however, serve to illustrate the complexity of practical reactor systems. An onboard system would likely be very different from a system designed for use in a forecourt application, but the basic components of reactor, heat recovery, waste stream recycling and highly effective purification will be vital parts of any successful system [74].

Advances in catalyst, high temperature materials, and separations will be necessary to produce an integrated reactor/separator that meets all the requirements. Metkemeijer et al. [75] compared the indirect use of both methanol (CH_3OH) and ammonia (NH_3) for the production of hydrogen to use in a fuel cell. Their findings extensively showed that the decomposition of ammonia was more attractive than methanol reforming from both thermodynamic and economic perspectives. They implied that the maximum specific energy produced from ammonia decomposition is 2.68 kWh_e/kg_f which is approximately 229% more than the 1.17 kWh_e/kg_f obtained from the reforming of methanol. They also assumed that both ammonia and methanol are both synthesized out of methane (CH_4) with a specific energy requirement of 8.05 $kWh/kg_{CH4}$for both cases. The researchers deduce that the overall efficiency with respect to the lower heating value of methane of the indirect ammonia and methanol approaches are 33.3% and 22.9%, respectively. According to Yin et al. [76] ruthenium catalyst supported on carbon nanotubes and promoted by potassium hydroxide was found to be the most effective catalyst for the thermal decomposition of ammonia. The researchers had discussed the kinetics of ammonia decomposition and had discovered that for temperatures less than 650.0 K the combinative desorption of nitrogen atoms was determined to be a rate limiting step. Chellappa et al. [77] examined the high temperature decomposition of ammonia at near atmospheric pressures over a nickel catalyst. The researchers were prompted to conduct this study to further understand the kinetic mechanisms of ammonia decomposition and to verify the existence of a transitional region where the Tymken-Pyzhev mechanism changes to a Tamaru mechanism. Roy et al. [78] had extensively reviewed the catalytic decomposition over alumina supported nickel catalyst at temperatures ranging from 400°C – 600°C with two catalysts; 10.0% and 15.0% nickel content, respectively. The researchers had found that the partial pressure of hydrogen reduced the ammonia decomposition rate over the entire temperature range and the presence of nitrogen had negligible effect on ammonia decomposition. The comparison of various chemical reactions for thermal H_2 production are shown in Table 26.

Table 26 Various chemical reactions for thermal H_2 production.

Reaction	Chemical Equation	Standard Enthalpy of Reaction (kJ/mol)	Operating Temperatures (K)
Ammonia Decomposition	$2NH_3 \rightarrow 3H_2 + N_2$	92.8	673.0 – 873.0
Methanol Reforming	$CH_3OH + H_2O \rightarrow CO_2 + 3H_2$	131.70	400.0 – 570.0
Steam Methane Reforming	$CH_4 + 2H_2O \rightarrow CO_2 + 4H_2$	164.70	973.0 – 1173.0

Source: [79]

Anhydrous ammonia has very high gravimetric (about 17 wt.% or 5.8 kWh/kg) and volumetric (about 0.105 kg/liter or 3.6 kWh/L at 25° C) hydrogen energy densities and, hence, looks appealing for use as a potential hydrogen carrier for onboard storage. Also, compared to reforming hydrocarbon fuels, the dissociation, or cracking, of hydrogen from ammonia may require a less complex process. The byproduct of the hydrogen dissociation process, nitrogen, can be vented from the vehicle with no adverse environmental effects so there is no need for a recycling process, as in some other chemical hydride systems. However, ammonia has specific chemical and physical properties which require attention in the design and engineering of an onboard storage system. These issues are: high coefficient of thermal expansion, high vapor pressure at ambient conditions, propensity for reacting with water, reactivity with container materials and high toxicity of the vapor if released into the air. These will be discussed below. Importantly, PEM fuel cells cannot tolerate ammonia, even at very low ppm levels and, hence, an onboard storage system based on ammonia would require a cracking reactor with essentially no pass-through of undissociated ammonia and/or a very effective filtration system.

The onboard tank must also sustain a modest overpressure. At 60° C, the vapor pressure of anhydrous ammonia is about 365 psig (about 25 bar). With a safety factor of approximately 2 to 2.2, the tank should sustain a maximum pressure of about 800-900 psi. In order to maintain the high gravimetric density, lightweight tank fabrication would be desirable. Currently, fixed site ammonia tanks are often constructed of steel or ductile iron. Many polymers are compatible with ammonia, so it is likely that composite tanks or lightweight aluminum tanks with polymer liners could be used for onboard storage of ammonia. Filling the onboard tank from a forecourt source will require a demountable, leak-proof coupling. This is mainly due to the toxicity of ammonia, but also because of the high affinity of ammonia for water resulting in the formation of ammonium hydroxide. Humidity from the air leaking into the refueling tube or coupling would, over time, lead to corrosion of the piping and tank by the hydroxide. The byproduct of the ammonia dissociation process, nitrogen, can be vented from the vehicle with no adverse environmental effects so there is no need for a recycling process, as in some other chemical hydride systems, Hydrogen production via ammonia decomposition is an endothermic process which is also a reversible chemical reaction. At 400°C, the equilibrium conversion of NH_3 is very high 99.1% [76]. Fe, Ni, Pt, Ir, Pd, Rh are a few of the possible catalysts whereas Ru and Ir appears to be more appropriate power per kg of catalyst. The waste heat could be supplied from sources such as: gas turbine, SOFC, or diesel power plants. All of these power plants have exhausts at temperatures of or greater than 600°C and would be able to meet the thermal energy requirement of an NH_3 decomposition and separation reactor. For this scenario to be realized, an exhaust gas to reactor heat exchanger must be designed and built to facilitate efficient heat transfer to the reactor. A membrane reactor system to produce hydrogen by the thermal decomposition of ammonia could also take place in either large scale marine or rail applications. This is envisioned to be a process where a fraction of the hydrogen energy produced would be used to decompose additional ammonia for hydrogen. A major point of interest pertaining to this scenario would be the very high energy ratio of the ammonia decomposition reaction. Thermodynamically, the ammonia decomposition reaction only requires approximately 13.0% - 15.0% of the hydrogen's calorific value.

4. On-Board Ammonia Electrolysis

Ammonia stored in the tank of the vehicle can be electrolyzed to generate hydrogen and use in the fuel cell or engines. Ammonia electrolysis is a method by which ammonia is electrochemically oxidized on a suitable electrocatalyst to nitrogen and hydrogen. Based on thermodynamics, the energy required to produce hydrogen by ammonia electrolysis is 97% lower than that required by electrolyzing water [80].

The reactions for an ammonia electrolytic cell comprise of ammonia electro oxidation at the anode and alkaline water reduction at the cathode as shown in the following equations [80]:

Anode: $2NH_3 + 6OH^- \rightarrow N_2 + 6H_2O + 6e^-$, $E_0 = -0.770$ V vs SHE (8)

Cathode: $6H_2O + 6e^- \rightarrow 3H_2 + 6OH^-$, $E_0 = -0.828$ V vs SHE (9)

Overall: $2NH_3 \rightarrow N_2 + 3H_2$, $E_0 = 0.058$ V (10)

Therefore, thermodynamic values are much in favor of the production of hydrogen coupled to the oxidation of ammonia compared to hydrogen production by electrolysis of water, for which the theoretical cell voltage is 1.223V. The advantage of this process is its ease of integration with renewable energy (electricity) sources. Because the energy consumption is low, the cell could operate with renewable energy.

The theoretical energy consumption during ammonia electrolysis (assuming that there are not kinetics limitations for the reaction to take place at the thermodynamics conditions) can be calculated from the standard potential of the cell and is equal to 1.55Wh/g H_2 while the electrolysis of water requires at least 33 Wh/g H_2 at standard conditions; this means that theoretically the electrolysis of ammonia consumes 95% lower energy than a water electrolyzer [80]. The commercialization of the technology depends on the development of effective electrodes for the electro-oxidation of ammonia. The development of superior electrolytes and electrocatalysts for ammonia electrolysis are vital for the improvement of the electrolysis efficiency. Vitse et al. [81] stated that hydrogen produced from the electrolysis of ammonia costs 0.89 US$ /kg of H_2 opposed to 7.10 US$/kg of H_2 from water electrolysis. These numbers were based on an ammonia cost of 275 US$/ton and a solar energy cost of 0.214 US$/kWh. Thermodynamically for 1 g of H_2, ammonia electrolysis consumes 1.55 Wh. For this same gram of H_2, a PEMFC, which is the reverse reaction of water electrolysis, generates 33 Wh. After sending 1.55 Wh back to the ammonia electrolysis cell (AEC) from the PEMFC, making the system self-sustaining, there is potential for a net energy of 31.45 Wh that can be used to recharge the batteries used for system start-up, to power a motor, or for any other applications. However, PEM fuel cells have efficiencies that range from 50 to 70% [82,83]. Additionally, ammonia is converted to hydrogen with 100% Faradaic efficiency, but kinetic problems creating large ammonia oxidation overpotentials exist. Boggs and Botte [84] integrated proton exchange membrane fuel cell with ammonia electrolyte cell (AEC), the electricity generated from the PEM fuel cell is utilized to feed the AEC with electricity. The study shows that using 203 L of aqueous ammonia will allow a hybrid fuel cell vehicle to travel 483 km before demanding another ammonia refill.

5. Case Study for Ammonia Electrolysis Vehicle

In this section, a case study is employed to show the practicability of the ammonia electrolysis vehicle for a 500 km driving range. Here, we follow the same procedure for AEC calculations as explained in the study by Boggs and Botte [84].The storage system costs, gravimetric, and volumetric capacities of an ammonia electrolysis vehicle are performed. A lightweight hydrogen fuel cell vehicle is considered. The driving range of the vehicle is taken as 500 km. A lightweight vehicle can consume about 5.3 km/L which is 2.5 less efficient than a hydrogen fuel cell vehicle. A gallon of gasoline is equivalent (gge) to a kg of hydrogen. Hence, the energy required for internal combustion engine (ICE) is found as 12.5 gal × 33.3 kWh/gal= 416.25 kWh [84].

Using this value, the amount of hydrogen is calculated as follows:

$$m_{H_2} = 416.25\ kWh/(2.5 \times 33\ kWh/kg) = 5\ kg \tag{11}$$

In order to travel about 500 km with this vehicle, about 5 kg of hydrogen is required. On average, if the vehicle is driven with the speed of 80 km/h, it will require refilling in every 6 hours. The nominal fuel cell power required to move a lightweight fuel cell vehicle is:

$$\dot{W}_{PEMFC} = \frac{416.25\ kWh}{(6h \times 2.5)} = 27.75\ kW \tag{12}$$

The generated electricity from the fuel cell is both used for vehicle propulsion and ammonia electrolysis. Therefore, the PEM fuel cell is required to be higher power rate. Ammonia electrolysis consumes about 60% of the energy. The increase of 60% in fuel cell cost, weight, and volume is considered as part the storage system.

To find the total PEM fuel cell power, the following calculation is used:

$$\dot{W}_{TOTAL,PEMFC} = \frac{27.75\ kW}{(1-0.6)} = 69.38\ kW \tag{13}$$

For the storage of the ammonia on-board, the ammonia storage vessel, Teflon tubing, centrifugal pump, start-up hydrogen drum, compressor, and controller are mostly common and commercially available devices. Based on the market, the estimated cost of ammonia storage unit is taken as US$ 3,633 (inflated to 2017) based on the study by Boggs and Botte [84]. The cost, weight, and volume of the ammonia electrolysis unit and PEM fuel cell modules are affected by many factors. Since there is an increase in power required from the PEM fuel cell, an increase in the amount of hydrogen is required between refueling.

$$m_{H_2} = \frac{69.375\ kW \times 6h}{(33\frac{kWh}{kg} \times 0.6)} = 21.2\ kg \tag{14}$$

In order to find the current required for ammonia electrolysis, under the assumption of 100% Faradaic efficiency, the Faraday law can be used:

$$I = 3500\frac{kg}{h} \times \frac{6e^-}{mol} \times \frac{26.8\frac{Ah}{e^-}}{3 \times 2\frac{g}{mol}} = 93,800\ A \quad (15)$$

Here, we assume a current density of 0.15 A/cm^2 in the cell and a corresponding cell area of 100 cm^2. This implies that about 15 A is needed for one cell. Hence, a total of about 6254 cells would be required to obtain 27.2 kW net energy for the motor.

The total catalyst (ruthenium is selected) required for the ammonia electrolysis can be calculated as follows [84]:

$$m_{catalyst} = 15\frac{mg}{cm^2} \times 2 \times 6254\ cells \times 100\ cm^2 = 18.762\ kg \quad (16)$$

Here, it is multiplied by 2 because of anode and cathode. Due to the expense of noble metals, the ammonia electrolysis cell cost is entirely dependent on the loading and catalyst costs.

We assume a ruthenium catalyst to be used in the ammonia electrolysis cell (AEC). The loading of the electrodes is kept at 15 mg/cm^2. In the open literature on catalysts for this reaction, loadings of up to 51 mg/cm^2 have been reported. The cost of ruthenium is US$1286 per kg in the market [85].

Therefore, the cost of ammonia electrolysis can be approximated as

$$Cost_{AEC,storage} = 18.762\ kg \times 1286\frac{US\$}{kg} = US\$\ 24,128 \quad (17)$$

About 41.6 kW of the PEM fuel cell is required for storage calculations. Using the 35 US$/kW for fuel cells which was reported by DOE Boggs and Botte [84]:

$$Cost_{PEM,storage} = 35\frac{US\$}{kW} \times 41.6\ kW = US\$\ 1456 \quad (18)$$

Counting the PEM fuel cell and ammonia electrolysis cell, the total system storage cost is found to be US$ 25,584. The DOE technical storage targets are summarized in the following Table 27. Capacities are defined as the usable quantity of hydrogen deliverable to the power plant divided by the total mass/volume of the complete storage system, including all stored hydrogen, media, reactants, and system components. Tank designs that are conformable and have the ability to be efficiently packaged onboard vehicles may be beneficial even if they do not meet the full volumetric capacity targets. Capacities must be met at end of service life. Hydrogen cost is independent of pathway and is defined as the untaxed cost of hydrogen produced, delivered, and dispensed to the vehicle.

Table 27 Technical targets for onboard hydrogen storage for light-duty vehicles

STORAGE PARAMETER	UNITS	2020	ULTIMATE
System Gravimetric Capacity			
Usable, specific-energy from H_2 (net useful energy/max system mass)	kWh/kg	1.8	2.5
	kg H_2/kg system	0.06	0.075
System Volumetric Capacity			
Usable energy density from H_2 (net useful energy/max system volume)	kWh/L	1.3	2.3
	kg H_2/L system	0.04	0.07
Storage System Cost			
System cost	$/kWh net	10	8
	$/kg H_2 stored	333	266
Fuel cost	$/gge at pump	2–4	2–4

Source: [86]

$$Cost_{per\ kg} = \frac{US\$\ 25{,}584}{21.2\ kg\ H_2} = 1206.8\frac{US\$}{kg} \quad (19)$$

$$Cost_{per\ kWh} = \frac{US\$\ 25{,}584}{\frac{33kWh}{kg} \times 60\% \times 21.2\ kg} = 60.95\frac{US\$}{kWh} \quad (20)$$

The ammonia storage vessel, Teflon tubing, start-up drum, compressor, and process controller are estimated to weigh about 26.5 kg based on commercially available products. The weight of fuel, storage part of the fuel cell, and AEC are calculated.

The required ammonia mass can be found using molecular weight and percentage of hydrogen in ammonia:

$$m_{NH_3} = \frac{21.2\ kg\ H_2 \times 1\ kg\ NH_3}{0.177\ kg\ H_2} = 119.8\ kg \quad (21)$$

According to Satyapal et al. [87], the target power density for PEM fuel cell is 2000 W/kg. As a result, the storage part of the fuel cell would weigh about $\frac{41.6\ kW}{2\frac{kW}{kg}} = 20.8\ kg$. It is assumed that the ammonia electrolysis unit is twice as heavy as the fuel cell that means 69.4 kg.

$$m_{AEC} = \frac{2 \times 69.4\ kW}{2\frac{kW}{kg}} = 69.4\ kg \quad (22)$$

The total gravimetric capacity is determined using the total estimated storage system weight, which is 249.9 kg.

$$Capacity_{Gravimetric\ (per\ kg\ system)} = \frac{21.2\ kg}{249.9\ kg} = 0.0.084 \quad (23)$$

$$Capacity_{Gravimetric\ \left(\frac{kWh}{kg}\right)} = 21.2\ kg \times 33\frac{kWh}{kg} \times \frac{60\%}{249.9\ kg} = 1.68\frac{kWh}{kg} \quad (24)$$

Similarly, the volumes of the tubing, start-up drum, compressor, and process control are estimated to occupy about 100 L. Using the fact that 119.8 kg of ammonia is required and considering a density of 682 kg/m^3, the storage vessel volume is 174 L. The targeted power/volume density for PEM fuel cell is 2000 W/L according to the DOE. Based on the 41.6 kW of fuel cell power that is used for storage, the storage part of the PEM fuel cell requires 20.8 L. Using the 2 times relation for ammonia : PEM fuel cell for volume and weight, the AEC occupies:

$$V_{AEC} = \frac{2 \times 69.4\ kW}{2\frac{kW}{L}} = 69.4\ L \quad (25)$$

Total storage system volume required is calculated to be 364.2 L. The volumetric storage parameters are as follows:

$$Capacity_{Volumetric\left(kg\frac{H_2}{L\ system}\right)} = \frac{21.2\ kg}{364.2\ L} = 0.058\frac{kg}{L} \quad (26)$$

$$Capacity_{Volumetric\ \left(\frac{kWh}{L}\right)} = 21.2\ kg \times 33\frac{kWh}{kg} \times \frac{60\%}{364.2\ L} = 1.15\frac{kWh}{L} \quad (27)$$

6. Thermodynamic Analyses of On-Board Ammonia Electrolysis

The theoretical electrolysis voltage of liquid NH_3 at any temperature is evaluated from Nernst's equation as follows [88]:

$$E_{r,AEC} = -\frac{\Delta G}{3\ F} + \frac{R\ T_{operation}}{3\ F}\ln\left(P_{N_2}^{0.5} P_{H_2}^{1.5}\right) \quad (28)$$

where, P_{H_2} is the partial pressure of H_2, and P_{N_2} is the partial pressure of N_2.

To obtain the required practical cell voltage, it is necessary to add all the cell losses (activation, concentration, ohmic overpotentials) to reversible cell voltage as below:

$$E_{actual} = E_{open} + E_{act} + E_{ohm} + E_{conc} \quad (29)$$

The ohmic overpotential is identified as

$E_{ohm,i} = \rho_i \, \delta_i \, J$ (30)

where ρ is the material resistivity, i represents the component and δ is the element thickness. The ohmic overpotential for each component of the cell can be calculated and added.

The activation polarization is given in terms of current density and exchange current density by Butler-Volmer equation:

$$J = J_0 \left\{ \exp\left(\frac{\alpha_a n F E_{act,a}}{RT}\right) - \exp\left(-\frac{\alpha_c n F E_{act,c}}{RT}\right) \right\} \quad (31)$$

where J_0 is the exchange current density, α_a and α_c are the electron transfer coefficients for anode and cathode, respectively, n is number of transferred electrons, and $E_{act,a}$ and $E_{act,c}$ are the activation overpotentials related with anode, and cathode respectively.

The total activation overpotential is the summation of each electrode:

$$E_{act} = E_{act,c} + E_{act,a} \quad (32)$$

The concentration polarizations are derived for each electrode:

$$E_{conc,c} = -\frac{RT}{nF} \ln\left(1 - \frac{J}{J_{L,c}}\right) \quad (33)$$

$$E_{conc,a} = -\frac{RT}{nF} \ln\left(1 - \frac{J}{J_{L,a}}\right) \quad (34)$$

Here, $J_{L,a}$ and $J_{L,c}$ are the limiting current densities for anode and cathode, respectively.

The values for different parameters used in the analyses of AEC unit are shown in Table 28.

Table 28 Data used in the analyses for ammonia electrolysis

Parameter	Range or value
Current density	0.15 A/cm^2
Exchange current density	0.037 mA/cm^2
Cell operating temperature	30°C
Cell operating pressure	10 bar
No of cells	6254
Anode thickness	0.0020 cm
Cathode thickness	0.0020 cm
Electrolyte thickness	0.0040 cm
Liquefied ammonia temperature	20°C
Liquefied ammonia pressure	10 bar

Source: [80,89]

The energy and exergy efficiencies of the ammonia electrolysis cell can be defined as follows:

$$\eta_{en,AEC} = \frac{LHV_{Hydrogen} \, \dot{m}_{H_2}}{LHV_{ammonia} \, \dot{m}_{NH_3} + J \, E_{actual} \, A_{cell}} \quad (35)$$

$$\eta_{ex,AEC} = \frac{ex_{Hydrogen} \, \dot{m}_{H_2}}{ex_{ammonia} \, \dot{m}_{NH_3} + J \, E_{actual} \, A_{cell}} \quad (36)$$

where ex represents the total exergy counting chemical and physical components.

The open cell voltage is found to be 0.078 V whereas the actual cell voltage counting the overpotentials is calculated to be 0.29 V. In total, 6254 cells are used to obtain the required power of 27.2 kW for the ammonia electrolysis. Under these conditions, the exergy destruction rate 28.93 kW. The inlet mass flow rate of ammonia is 220.8 g/min which yields 39.2 g/min hydrogen. The energy and exergy efficiencies of the AEC are found to be 82.03% and 79.51%, respectively as shown in Fig. 95.

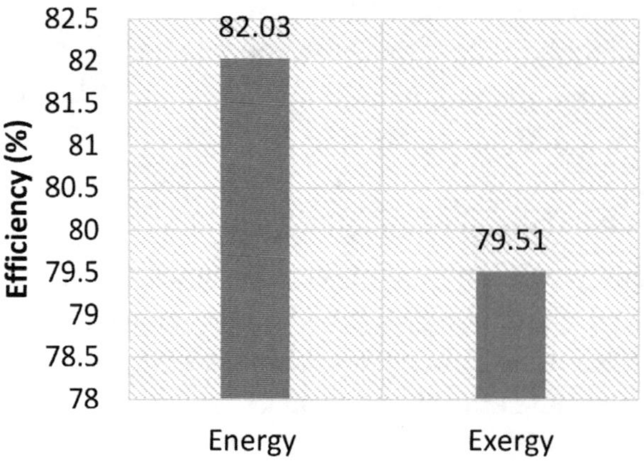

Figure 95 Energy and exergy efficiencies of the ammonia electrolysis cell unit

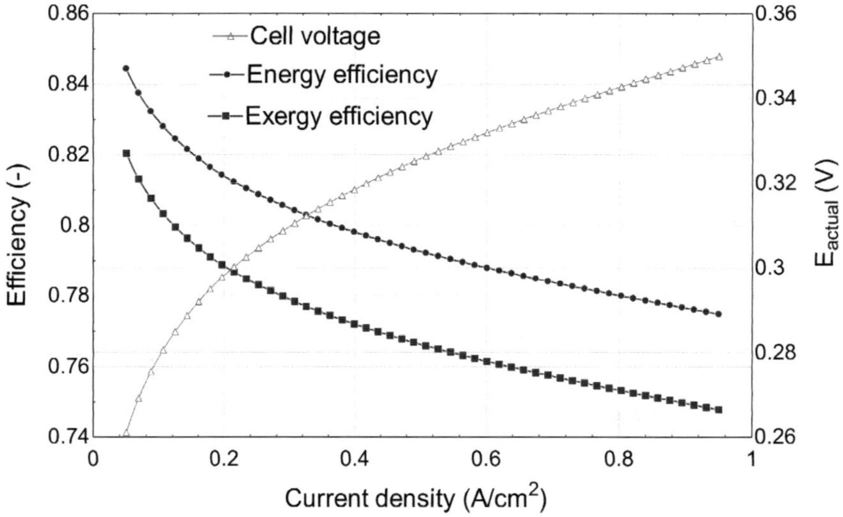

Figure 96 Effect of changing AEC current density on electrolysis voltage and overall AEC energy and exergy efficiencies

The overpotentials and actual cell voltage in the AEC are listed in Table 29.

Table 29 The overpotentials and actual cell voltage in the AEC

Activation overpotential (V)	0.2102
Concentration overpotential (V)	0.0008567
Ohmic overpotential (V)	0.001134
Actual cell potential (V)	0.2902

In Fig. 96, the effect of changing the AEC current density on the AEC electrolysis voltage, energy and exergy efficiencies of the AEC unit are illustrated. The electrolysis voltage required to separate the ammonia increased from 0.26 V to 0.35 V due to the increase of the AEC current density from 0.05 A/cm^2 to 0.95 A/cm^2. However, the same range of change of the AEC current density leads to a drop in energy and exergy efficiency of the AEC unit from 84.44% to 77.49% and from 82.04% to 74.79% respectively.

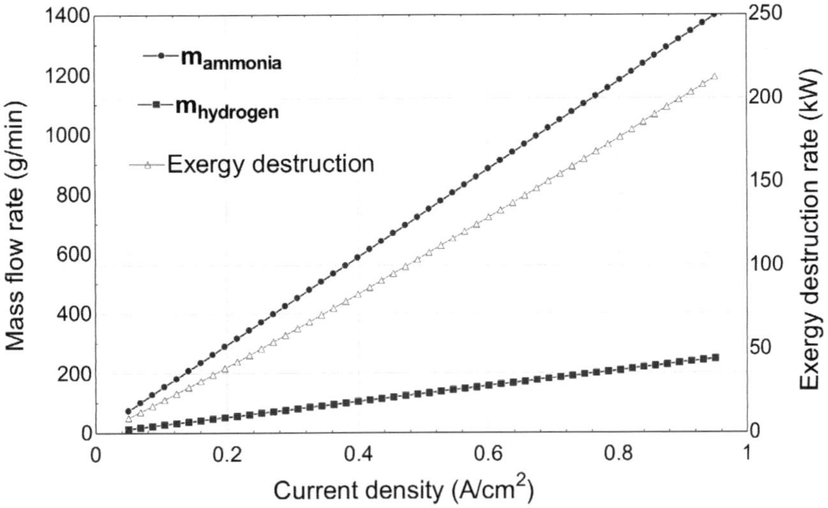

Figure 97 Effect of changing AEC current density on AEC exergy destruction, ammonia amount required for electrolysis and hydrogen produced by AEC

The increase in the electrolysis voltage and the decrease in the energy and exergy efficiency of the AEC unit can be interpreted by the growth in the AEC voltage losses. As it increase directly proportional to the increase in the current density, which leads to an increase in the exergy destruction rate and eventually a drop in the AEC efficiency. Fig. 97 shows the relation between applied current density and yielding mass flow rate of the chemicals. When the current density is 4 A/cm^2, then the required ammonia flow rate is about 600 g/min to produce about 100 g/min hydrogen. The exergy destruction also increases by rising current density. The effects of changing AEC operating temperature on exergy destruction and overall AEC energy and exergy efficiencies are shown in Fig. 98. Although operating at higher temperature decreases the exergy destruction in the AEC, the efficiencies slightly lowers at higher temperatures.

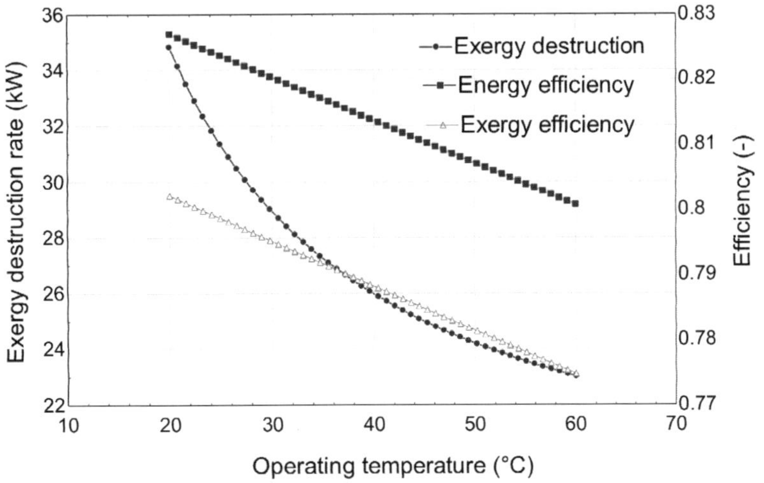

Figure 98 Effect of changing AEC operating temperature on exergy destruction and overall AEC energy and exergy efficiencies

Operating pressure is also important for the system performance as shown in Fig. 99. At higher pressure, the required potential increases causing a higher power requirements in the AEC for hydrogen production from ammonia. In Fig. 100, the effects of ambient temperature are illustrated. At higher ambient temperatures, the total exergy destruction rate is higher.

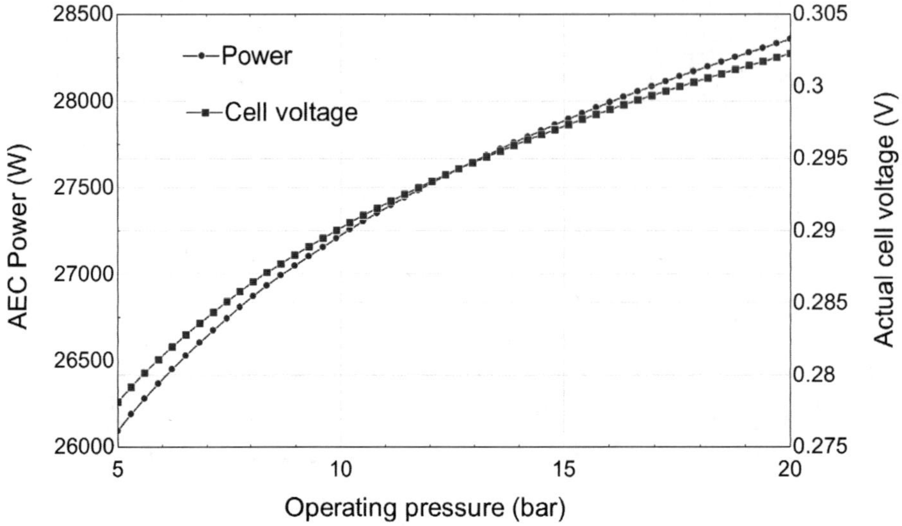

Figure 99 Effect of changing AEC operating pressure on AEC cell voltage and required power

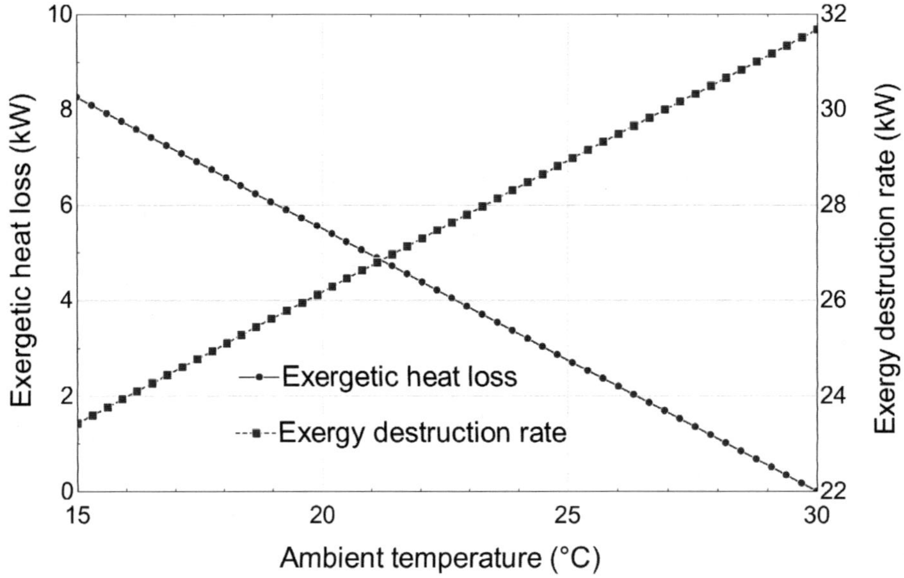

Figure 100 The effects of ambient temperature on the total exergy destruction rate and heat loss of the AEC

One of the key parameters is the mass flow rate of ammonia entering the AEC unit as shown in Fig. 101. Increasing the mass flow rate causes a decrease in efficiency which is mainly caused by the definition of efficiency.

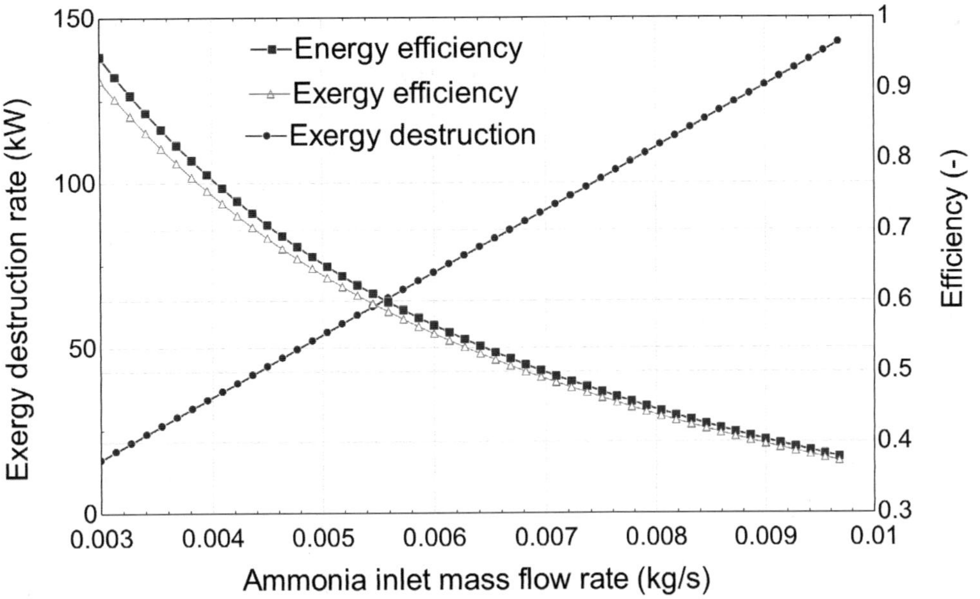

Figure 101 The effects of ammonia inlet mass flow rate on exergy destruction and efficiencies of AEC

CHAPTER 9:ECONOMIC ANALYSES OF SOLAR ENERGY BASED AMMONIA PRODUCTION

In this section, the exergoeconomic analyses of the experimental systems are performed. The purchased costs of the experimental systems in this study are presented in the following tables. The experimental systems are divided into three main sub-systems;

- Photoelectrochemical hydrogen production reactor
- Electrochemical ammonia production reactor
- Integrated system comprising of solar light concentrator and splitter, PV cell and support mechanism.

The purchased costs of the materials used in the PEC reactor are listed in Table 30.

Table 30 The cost of materials used in the PEC hydrogen production reactor

Material	Quantity	Unit Price ($)	Total Price ($)
Reactor casing (HDPE)	2	$130	$260
Stainless steel electrodes	2	$200	$400
Nafion membrane	1	$2,000	$2,000
Chemicals for electrodeposition	3	$150	$450
Washers, bolt nuts	50	$1	$50
Acrylic or polycarbonate reactor window	1	$100	$100
Piping (plastic)	4	$25	$100
Rubber gasket	6	$10	$60
Machining (for casing and electrodes)	1	$450	$450
Others (adhesive, silicon etc.)	1	$120	$120
TOTAL			$3,990

The reactor casing is chosen as HDPE for the reactor because of the advantages explained in the experimental apparatus. The machining is easier and requires low cost. For higher solar concentration ratios, the temperature levels on the PEC reactor body may rise more than material specification. It is similar for the viewing panel of the reactor which is made of acrylic.

Table 31 The cost of materials used in the electrochemical ammonia production reactor

Material	Price ($)
Nickel wiring	$40
Nickel electrodes	$160
Reactor casing alumina crucible (Alumina Al_2O_3)	$140
Reactor lids (Stainless steel 316 Alloy)	$120
Bolt Nuts and Washers	$55
Reactor tubes (Alumina Al_2O_3)	$160
Piping (plastic)	$50
Heating tape	$110
Gaskets	$25
Others (adhesive, insulation etc.)	$65
TOTAL	$925

Therefore, the temperature levels on the PEC reactor surface need to be checked before deciding the materials selection. In total, the cost of the PEC reactor in the experimental setup is calculated to be 3990$. The purchased costs of the materials used in the electrochemical ammonia production reactor are listed in Table 31. The ammonia reactor casing has 500 mL capacity. The tubes for the gas inlet and outlet the reactor are also made of Alumina (Al_2O_3) which is non-corrosive. After the gasses exit the reactor, plastic pipes are used. The heating tape used in the experiments are for sustaining the reaction temperature. That is an additional equipment to the reactor construction. The total cost of ammonia reactor is found to be 925$. The PEC hydrogen production reactor is used under concentrated and split spectrum. Therefore, the solar concentrator, dielectric mirrors and PVs are included in the integrated system costs as shown in Table 32. These two sub-systems for hydrogen and ammonia production are integrated in the experimental setup which yield the total system capital cost.

Table 32 The cost of materials used in the integrated system for PEC hydrogen based electrochemical ammonia production system

Photovoltaic cell, multicrystalline silicon	$80
Fresnel lens	$75
Dielectric mirrors (6 in total) (Borosilicate glass)	$672
PEC hydrogen production reactor	$3,990
Ammonia production reactor	$925
Support structure (wood)	$45
Metal support mechanism including nuts and bolts	$75
TOTAL	$5,862

The support mechanism used in the integrated system consists of wood and metal parts. The highest cost is for the PEC hydrogen production reactor which corresponds to about 68% of total cost as show in Fig. 102.

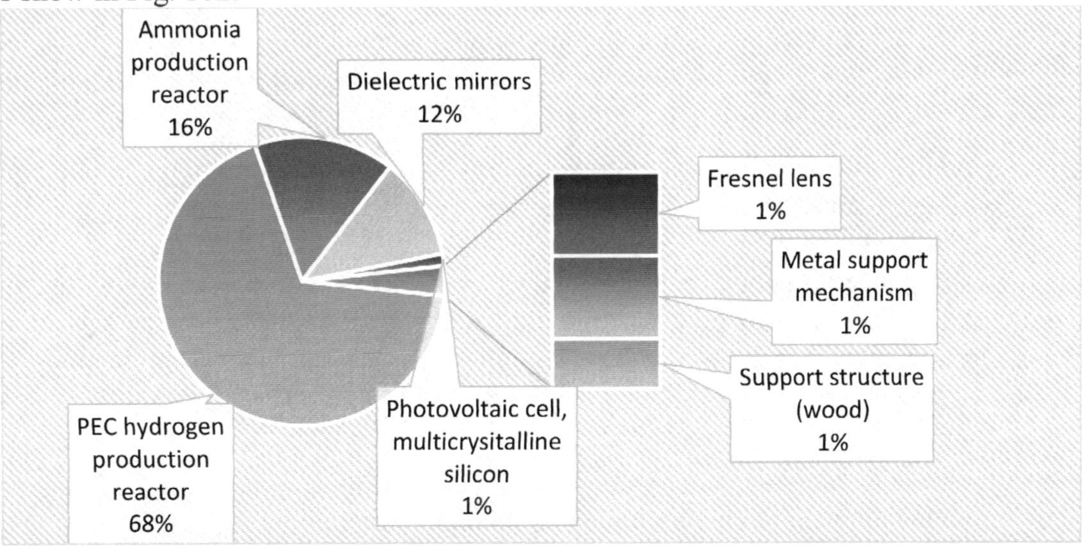

Figure 102 Cost breakdown of the integrated system for hydrogen and ammonia production

The exergoeconomic analysis requires that a specific cost is put on the exergy streams in an exergy balance on a component. On top of putting costs on the exergy streams, capital and running costs are taken into account in order to get a complete cost analysis. The exergoeconomic analyses are performed for the experimental integrated system. The capital costs are taken from the experimental setup costs as listed above. Exergy cost for the streams in any cost rate balance is given as [90,91]

$$\dot{C} = c\,\dot{Ex} \tag{37}$$

Here, c is in given in \$/kWh and $\dot{Ex}$ is given in W. The capital costs of the components is given as $\dot{Z}$ in \$/h.

Typical cost rate balance for a component is given below [92,93]:

$$\sum \dot{C}_{in} + \dot{W}\,c_{in} + \dot{Z} = \sum \dot{C}_{out} + \dot{W}\,c_{out} \tag{38}$$

CRF refers to capital recovery factor and depends on the interest rate and equipment life time, and is determined here as follows:

$$CRF = \frac{i\,(1+i)^n}{(1+i)^n - 1} \tag{39}$$

Here, i denotes the interest rate and n the total operating period of the system in years. Total costs for each of the components in the system are needed in \$/h in order to use them in cost rate balance equations. The capital cost and the operating and maintenance costs are added. The total costs are then divided by the number of hours in a year to get a cost in \$/h. Operating and maintenance costs are assumed to be a ratio of the capital costs as

$$OM = CC\ OM_{ratio} \tag{40}$$

where OM_{ratio} depends on the type of application and material.

The capital costs of the equipment are calculated based on the experimental setup costs as explained in the previous tables. The total cost balance is written as follows:

$$TCC = CRF\,(CC + OM) \tag{41}$$

The annual investment cost rate of any component, $\dot{Z}$ is calculated for the components of the experimental integrated system. It is the summation of the annual capital investment cost rate and the annual O&M cost rate and defined as follows:

$$\dot{Z} = \frac{TCC}{t_{operation}} \tag{42}$$

where $t_{operation}$ is the total operational hours in a year.

The cost rate of exergy destruction for each component is expressed as [92,93]

$$\dot{C}_D = c\,\dot{Ex}_d \tag{43}$$

Summation of additional cost caused by exergy destruction, $\dot{C}_D$ and final capital and operating cost rate $\dot{Z}$ gives a critical parameter named as total cost rate $\dot{C}_D + \dot{Z}$:

$$\dot{C}_{total} = \dot{C}_D + \dot{Z} \tag{44}$$

Total cost rates of the system consists of the total investment cost and cost of exergy destruction. In general, the smaller the sum of this parameter, it means that the component is more cost effective. Therefore, this parameters is taken as optimization function in the optimization analyses.

The exergoeconomic factor, which is a measure of system effectiveness in terms of cost, obtained through exergoeconomic analysis is given as

$$f = \frac{\dot{Z}}{\dot{Z} + \dot{C}_D} \tag{45}$$

The exergoeconomic variables $\dot{Z}$ and $\dot{C}_D$ provide the significance of component in the system optimization, whereas the variable f exergoeconomic factor is a relative measure of the component cost effectiveness.

The obtained results in the exergoeconomic analysis of the streams for each of the sub-systems in the experimental setup are presented in results and discussion chapter. The following financial parameters shown in Table 33 are used in the exergoeconomic analysis.

Table 33 The financial and operation parameters used in the exergoeconomic analyses

Parameter	Value
Interest rate	7%
Lifetime of all components	10 years
Calculated capital recovery factor	0.1424
Calculated hydrogen cost	3.24 \$/kg
Calculated ammonia cost	0.73 \$/kg
Cost of electricity	0.06 \$/kWh
Cost of thermal energy	0.02 \$/kWh
O&M percentage of capital cost	2.2%
System annual operation hours	2500 hours

The main findings of the exergoeconomic assessment is based on stream exergy rates and corresponding exergy destruction ratios. Thus, exergy destruction rates of the system components is illustrated in Fig. 103. In the Fresnel lens and dielectric mirror, only light interactions occur. Therefore, the exergy destruction rates are quite higher than other components. In addition, inlet irradiance is about 946 W/m^2 and it is concentrated about 6 to 10 times. The concentration and light splitting processes destruct more exergy than PV and PEC processes.

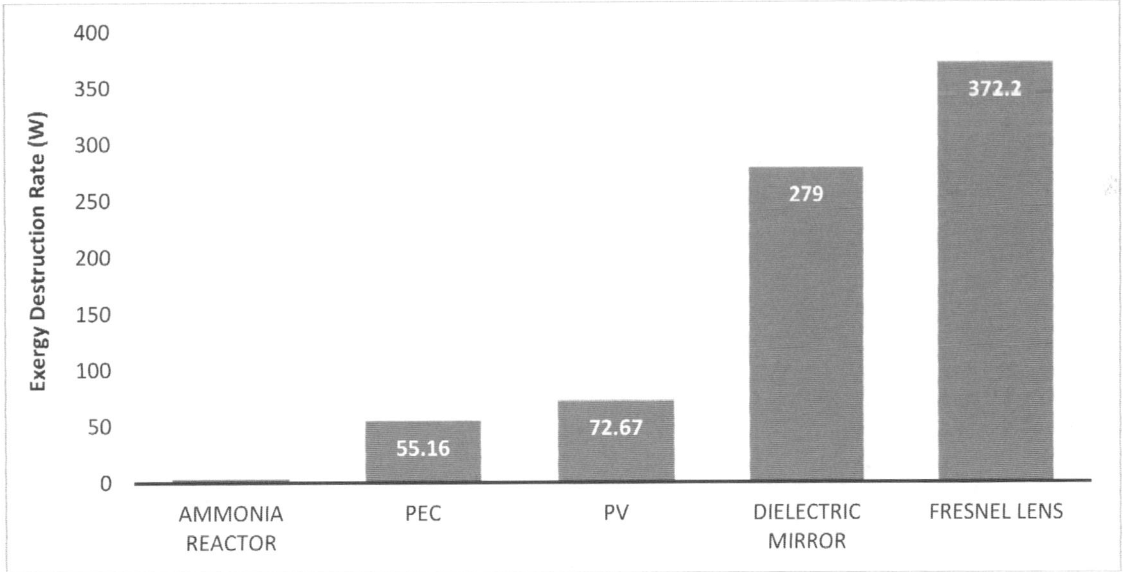

Figure 103 The exergy destruction rates of the integrated system components

The cost rates and costs of exergy destructions for each component are tabulated in Table 34. The highest capital cost is observed in PEC reactor because of high purchased cost and electricity input. Secondly, ammonia reactor has highest cost rate as shown in Fig. 104. These two reactors are the only electricity consuming devices resulting in a larger cost rates. Furthermore, since PV generates electricity, the total cost rate is quite lower than other components.

Table 34 The exergoeconomic results of the components in the integrated system

Component	Cost Rate of Exergy Destruction - $\dot{C}_D$ (\$/h)	Exergoeconomic Factor - f (%)	Total Cost Rate - $\dot{C}_{total}$ (\$/h)	Annual Investment Cost Rate - $\dot{Z}$ (\$/h)
Ammonia reactor	0.07295	42.46	0.1268	0.05384
Fresnel Lens	0.004365	50	0.008731	0.004365
Dielectric Mirror	0.03911	50	0.07823	0.03911
PEC	0.1596	59.27	0.3918	0.2322
PV	0.003902	54.4	0.008559	0.004656
TOTAL	0.2799	54.42	0.6141	0.3342

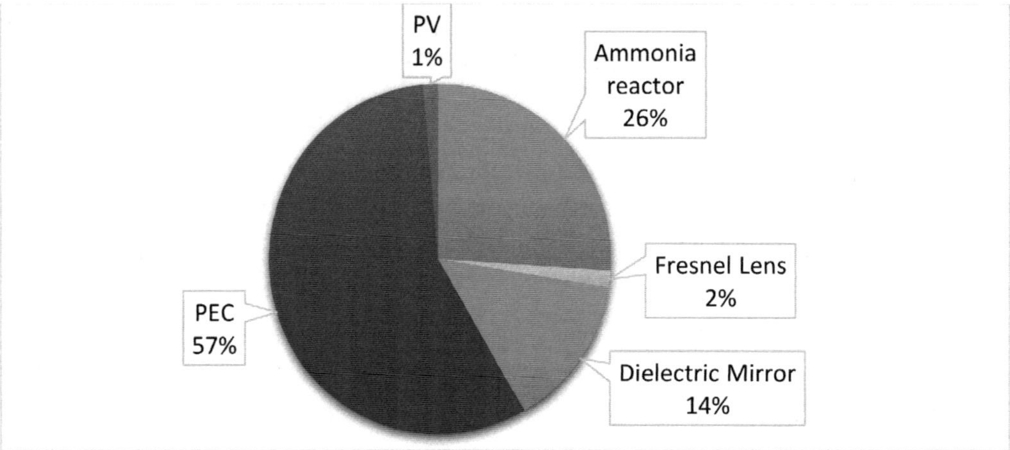

Figure 104 The cost rate of exergy destruction in each component of the integrated system

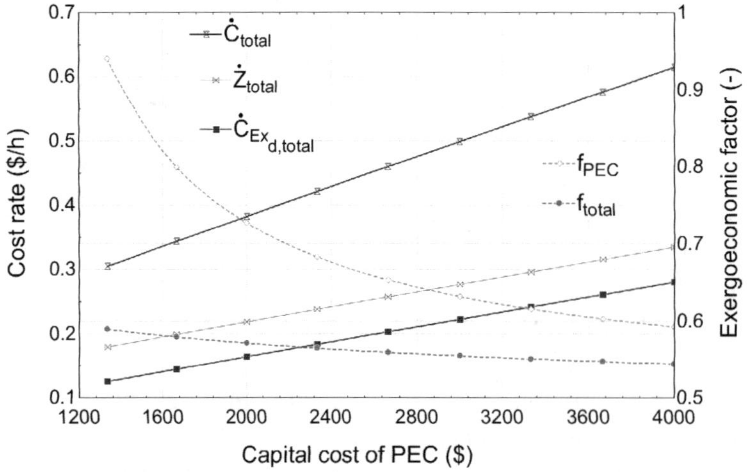

Figure 105 The effects of PEC reactor capital cost on the system cost rates and exergoeconomic factors

Since the PEC reactor capital cost is the highest contributor to the system cost, a parametric study is conducted to investigate the effect on the total cost rates as shown in Fig 105. In case the PEC reactor can be built in a more cost effective way corresponding to about 2000$, the total exergy destruction cost rate decreases to 0.1641 $/h whereas total exergoeconomic factor increases to 57.1%. Also, the exergoeconomic factor of PEC reactor component increases to 72.7 % from 54.41%.

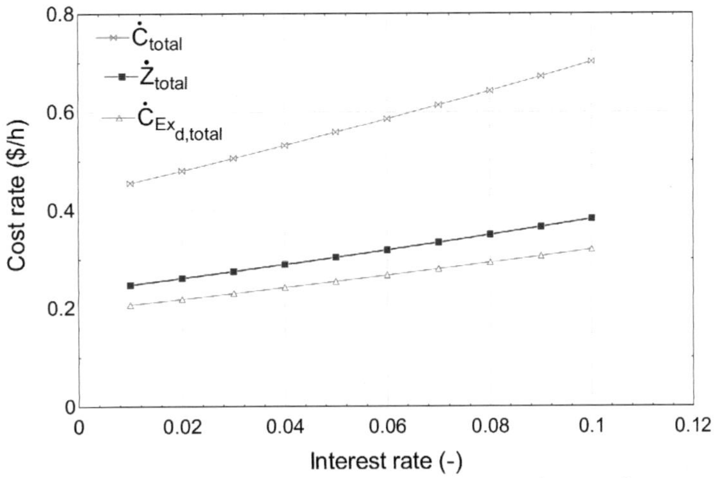

Figure 106 The effects of increasing interest rate on the total system cost rates

Fig 106 clearly shows that interest rate has a negative impact on the system cost rates. Although the total cost rate of the system is 0.4801 $/h at 2% interest rate, it rises to 0.702 $/h at 10% interest. The lifetime of the system and components has also important role in the total cost rate as shown in Fig. 107. Each component can have different lifetime periods. For example, the PEC electrodes may need to be replaced in two years whereas the solar concentrator may have up to ten years operation. In the base case, the system lifetime is taken to be 10 years for the experimental system that is about 0.6141 $/h total cost rate. However, in case the lifetime can be increased up to 40 years, the total cost rate can be decreased down to 0.3233 $/h.

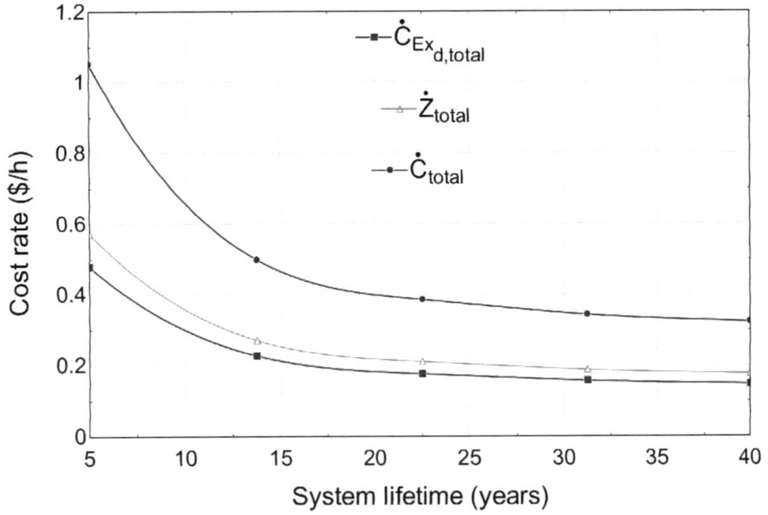

Figure 107 The effects of system total lifetime on the system cost rates

Annual operation time of the systems are also critical for the solar energy applications. The intermittency of the solar energy decreases the total annual operation time. The availability of the sunshine depends on the season and location. This is also named as capacity factor. In general, the capacity factor of the solar energy applications range between 10-30% corresponding to about 876 to 2628 hours annually. In the base case of the system, the operation time is set to 2500 hours. However, if the operation time diminishes to 1000 hours, the total cost rate rises to about 1 $/h as shown in Fig. 108.

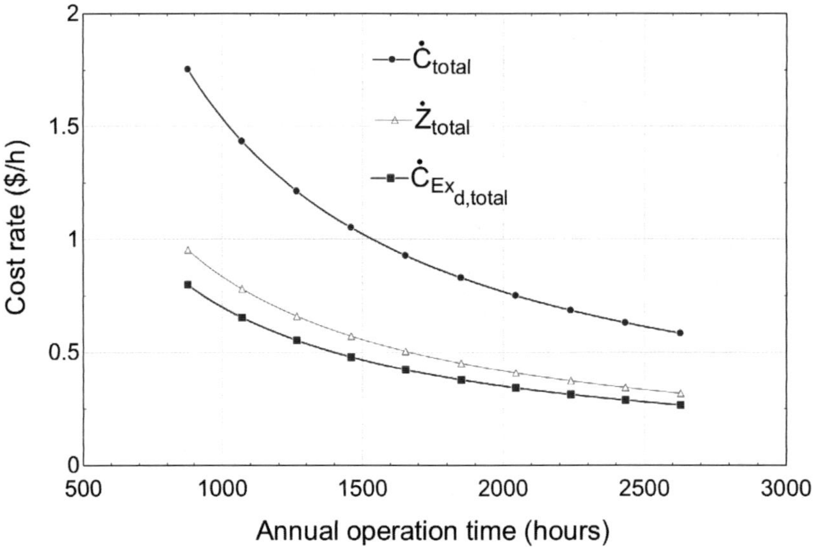

Figure 108 The effects of system total lifetime on the system cost rates

CHAPTER 10: SCALE-UP ANALYSES FOR SOLAR ENERGY BASED AMMONIA PRODUCTION

In case of larger production scales, the mass manufacturing of these equipment will be considerably lower. In order to analyze the cost of hydrogen and ammonia at larger production capacities such as 1000 kg/day, the scale up analyses are conducted as explained in the following paragraphs. PEC systems use solar photons to generate a voltage in an electrolysis cell sufficient to electrolyze water, producing H_2 and O_2 gases. For the economic analyses of the photoelectrochemical hydrogen production system, the Hydrogen Analysis (H2A) production model [94] is used which is developed by U.S. DOE Hydrogen & Fuel Cells Program. The H2A Production Model analyzes the technical and economic aspects of central and forecourt hydrogen production systems. Using a standard discounted cash flow rate of return methodology, it determines the minimum hydrogen levelized cost, including a specified after-tax internal rate of return from the production technology. The employed scenario models a PEC solar concentrator system using reflectors to focus the solar flux with a concentration ratio of 10 intensity ratio onto multi-junction PEC cell receivers immersed in an electrolyte reservoir and pressurized to 300 psi. The PEC cells are in electrical contact with a small electrolyte reservoir and produce oxygen gas on the anode side and hydrogen gas on the cathode side. The start-up year of the plant is taken as 2020. The Chemical Engineering Plant Cost Index (CEPCI) is used to adjust the capital cost of the H_2 Production facility from the basis year to the current year. The Consumer Price Inflator (CPI) is used to deflate all dollars from the current year to the Reference Year. The available model is quite similar to the designed and tested concentrated PEC system except for the solar light splitting part. Hence, the solar splitting mechanism is not considered in the scaled-up cost assessment. A solar tracking system is employed to make best use of direct radiation capture. Solar concentrators, which can use reflectors or lenses to concentrate the solar energy, considerably lessen the cost influence of the PV component of the system, but add the costs of the concentrators and directing systems. For the concentrator PEC system, the water reservoir and the H_2 and O_2 collected are pressurized by the inlet water pump at relatively low added cost. Pressurization to 300 psi avoids the need for a separate compressor, minimizes water vapor loss by the reactor, and reduces O_2 gas bubble size, which minimizes potential bubble scattering of incident photons at the anode face.

The H2A Costing Model [94] delivers an organized layout to enter factors which impact cash inflows and outflows associated with the construction and operation of a hydrogen production plant. The system practices a solar concentrator reflector to focus solar direct radiation onto the PEC cell. A PEC concentrator system can possibly use a concentration ratio of 10-50 suns; nonetheless, since the experimentally tested system uses about a concentration ratio of 6 to 10, the scale-up analyses are considered for 10 suns. Plant control arrangements perform many duties including local and remote monitoring, alarming and controlling of plant equipment and functions. The model comprises the control and instrumentation mechanisms including the functionality and safety. In the scaled up analyses for hydrogen production, the capacity factor and plant outputs are listed in Table 35. The main financial parameters used in the cost analyses are shown in Table 36. Industrial electricity prices are taken in the calculations as $0.06/kWh [95]. The overall solar-to-hydrogen conversion efficiency of the concentrated PEC hydrogen production system is taken as 16% which is the expected efficiency by 2020 as mentioned in the literature review section. It is assumed that the average solar irradiance is 6.55 $kWh/m^2/day$.

Table 35 The capacity and hydrogen production plant output

Operating Capacity Factor (%)	85.0%
Plant Design Capacity (kg of H_2/day)	1,000
Plant Output (kg/day)	850
Plant Output (kg/year)	310,250

Table 36 The financial input parameters used to calculate the unit hydrogen production cost

Reference year	2009
Assumed start-up year	2020
Basis year	2009
Length of Construction Period (years)	2
% of Capital Spent in 1st Year of Construction	20%
% of Capital Spent in 2nd Year of Construction	80%
Start-up Time (years)	0.4
Plant life (years)	40
Analysis period (years)	40
Depreciation Schedule Length (years)	20
Decommissioning costs (% of depreciable capital investment)	10%
Salvage value (% of total capital investment)	5%
Inflation rate (%)	1.1%

Source: [94]

Engineering & design and up-Front permitting costs are assumed to be 7.5% of the direct capital cost whereas process contingency cost is assumed to be 10% of direct capital cost. Furthermore, the land cost in Ontario, Canada is taken as $6,500 per acre for rural area. The costs are expressed in U.S dollar. The total required land area is calculated based on the solar-to-hydrogen efficiency and light absorption efficiencies. The basis year costs (2009) are inflated to 2017 dollars using inflation tool [96].

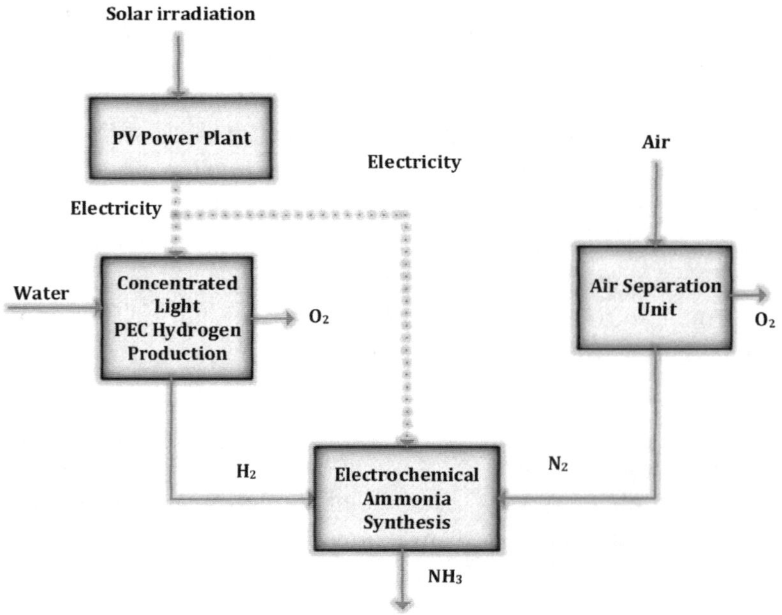

Figure 109 Illustration of large scale electrochemical ammonia production plant

The cost of production for the ammonia facility is of principal importance: a great portion of the complete costs will come from acquiring electricity, presumed to be the utility cost. The cost of production is the summation of waste disposal, labor costs, utilities, general expenses, raw materials, taxes, maintenance expenses as well as other minor costs. Cryogenic air separation methods are frequently used in medium to large scale facilities to yield nitrogen, oxygen, and argon as gases or liquid products. Cryogenic air separation is generally favored technology for generating very high purity oxygen and nitrogen. The plants producing only nitrogen are less complex and need less power to function than an oxygen-only plant making the same amount of product. Producing these products in liquid form necessitates additional apparatus and more power required per unit of delivered product. The average cost of ammonia production from the electrolysis-based systems are approximately 20-40% of hydrogen production cost as previously given in [97] for various ammonia production methods such as PV electrolysis. 17.8% of ammonia is hydrogen in weight, and around 3% of ammonia production cost comes from air separation based nitrogen production [98]. The schematic diagram of the large scale electrochemical ammonia production plant using photoelectrochemical hydrogen is depicted in Fig 109. In Haber-Bosch ammonia synthesis plants, there are two main compressors to compress the feed gases into the Haber-Bosch reactor. These compressors consume most of the electricity in the synthesis loop. The synthesis loop requires about 5.5% of the total power requirements in an electrolyzer based Haber-Bosch plant [98]. On the other hand, in the tested electrochemical ammonia synthesis, the gases are in atmospheric pressure in the reaction. Hence, compression costs are eliminated. However, currently electrochemical ammonia synthesis requires more electricity than Haber-Bosch process per unit kg of product because of the strong chemical bonds. Considering these conditions, a conversion factor is calculated for ammonia cost determination based on the hydrogen cost. Therefore, ammonia production costs are calculated as the 26% of hydrogen production cost. The prices are production prices for hydrogen and ammonia which means that the transportation to the final end user or storage processes are not included. The economic analyses results scale-up study are presented in this section. The plant is scaled-up to 1000 kg/day. The capital costs, operation/maintenance costs etc. are considered in the analyses. Table 37 shows the total capital costs of the hydrogen production plant with cost breakdown of indirect and direct capital costs. The total capital cost of the concentrated light PEC hydrogen production plant is calculated as $6,428,852.03. Total Variable Operating Costs per year is estimated as $4,800 because of unpredicted Replacement Capital Cost. The fixed operating costs are tabulated in Table 38.

Table 37 The direct and indirect depreciable capital costs

Indirect Depreciable Capital Costs	2017$
Site Preparation ($)	$98,690.81
Engineering & design ($)	$370,089.69
Project contingency ($)	$493,452.92
Up-Front Permitting Costs ($)	$370,089.69
Total Depreciable Capital Costs (Including direct capital costs)	$6,266,851.18
Cost of land ($/acre)	$7,256.69
Land required (acres)	22.32
Land Cost ($)	$162,001.32
Total Non-Depreciable Capital Costs	$162,000.85
Total Capital Costs	$6,428,852.03

Table 38 The fixed operating costs of the PEC hydrogen production plant

Fixed Operating Costs	**2017$**
Burdened labor cost, including overhead ($/man-hour)	$45.20
Labor cost ($/year)	$24,444.16
General and administrative expense ($/year)	$4,888.38
Licensing, Permits and Fees ($/year)	$589.11
Property tax and insurance rate (% of total capital investment per year) (In Ontario)	1.45%
Property taxes and insurance ($/year)	$93,218.22
Production Maintenance and Repairs ($/year)	$19,973.88
Total Fixed Operating Costs	$143,114.50

Production maintenance and repair costs are predicted to 4% of the overall cost excluding the replacement parts which are mainly electrodes and lens. The state, federal taxes and after-tax real rate of return are taken as 6.0%, 35.0% and 1%, respectively in the cost analysis. The after-tax real rate of return affects the cost considerably which would increase the cost of hydrogen. The after-tax real rate of return is the actual financial benefit of an investment after accounting for inflation and taxes. The calculations include the replacement of the PEC reactor electrodes every two years and replacement of the PEC cells including the Fresnel lens solar concentrator every ten years as tabulated for 40 year operation in Table 39. The other components are allocated for 20 years.

Table 39 The cost of material replacements of the system components

Operations Year	Total Yearly Replacement Costs (2017$)	Replacement
1	$35,061	
2	$632,966	Replacement of PEC electrodes every 2 years.
3	$35,061	
4	$632,966	Replacement of PEC electrodes every 2 years.
5	$35,061	
6	$632,966	Replacement of PEC electrodes every 2 years.
7	$35,061	
8	$632,966	Replacement of PEC electrodes every 2 years.
9	$35,061	
10	$4,997,679	Replacement of PEC electrodes every 2 years and replacement of PEC cell including solar concentrator, windows and sealing every 10 years plus installation
11	$35,061	
12	$632,966	Replacement of PEC electrodes every 2 years.
13	$35,061	
14	$632,966	Replacement of PEC electrodes every 2 years.
15	$35,061	
16	$632,966	Replacement of PEC electrodes every 2 years.
17	$35,061	
18	$632,966	Replacement of PEC electrodes every 2 years.

19	$35,061	
20	$4,997,679	Replacement of PEC electrodes every 2 years and replacement of PEC cell including solar concentrator, windows and sealing every 10 years plus installation
21	$35,061	
22	$632,966	Replacement of PEC electrodes every 2 years.
23	$35,061	
24	$632,966	Replacement of PEC electrodes every 2 years.
25	$35,061	
26	$632,966	Replacement of PEC electrodes every 2 years.
27	$35,061	
28	$632,966	Replacement of PEC electrodes every 2 years.
29	$35,061	
30	$4,997,679	Replacement of PEC electrodes every 2 years and replacement of PEC cell including solar concentrator, windows and sealing every 10 years plus installation
31	$35,061	
32	$632,966	Replacement of PEC electrodes every 2 years.
33	$35,061	
34	$632,966	Replacement of PEC electrodes every 2 years.
35	$35,061	
36	$632,966	Replacement of PEC electrodes every 2 years.
37	$35,061	
38	$632,966	Replacement of PEC electrodes every 2 years.
39	$35,061	
40	$35,061	

The capital cost of the system components including the control unit, PEC cell, pumps and other components are shown in Table 40.

Table 40 The direct capital costs of the components in 1000 kg/day PEC concentrated light hydrogen production plant

Major components/systems	**Installed Costs (2017$)**
PEC cell body, concentrating and containment system	$ 3,139,192
PEC Electrodes	$ 506,321
Make-up Water Pump	$ 237
Manifold Piping	$ 18,062
Collection Piping	$ 4,475
Column Collection Piping	$ 2,111
Final Collection Piping	$ 481
Condenser	$ 7,924
Manifold Piping (diameter)	$ 18,062
Collection Piping (diameter)	$ 4,475
Column Collection Piping (diameter)	$ 2,111
Final Collection Piping (diameter)	$ 481
PLC	$ 3,349

Control Room building	$ 19,567
Control Room Wiring Panel	$ 3,349
Computer and Monitor	$ 1,675
LabVIEW Software	$ 4,799
Water Level Controllers	$ 78,902
Pressure Sensors	$ 6,933
Hydrogen Area Sensors	$ 152,725
Hydrogen Flow Meter	$ 6,140
Instrument Wiring	$ 453
Power Wiring	$ 227
Conduit	$ 6,771
Piping Installation	$ 8,870
Reactor Foundation & Erection	$ 556,953
Reactor feed install	$ 71
Gas processing Subassembly install	$ 2,377
Control System Install	$ 85,467
TOTAL DIRECT CAPITAL COST	$ 4,675,618

Here, the PEC cell reactor body cost is taken as $145.23/m^2 including the concentrating and containment system. In addition, the PEC electrodes are taken as $234.24/m^2 unit cost. For a 1 ton/day hydrogen production plant, an overall solar capturing area of 21,615 m^2 and electrode area of about 2162 m^2 are required. The highest cost is the PEC cell with the concentrators which is followed by the PEC electrodes.

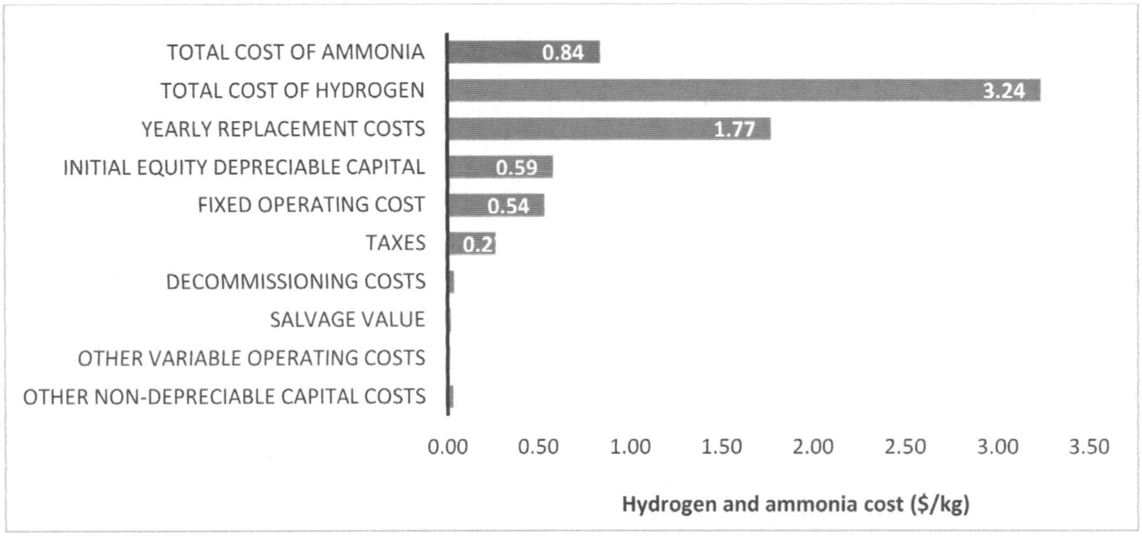

Figure 110 The calculated cost of hydrogen and ammonia with contributing factors for a 1000 kg/day concentrated PEC hydrogen production plant

The hydrogen and ammonia costs calculated based on the model are shown in Fig. 110. Moreover, the breakdown of the hydrogen cost are illustrated. Yearly replacement costs are about 1.77 $/kg hydrogen. This implies that if the durability and stability of the PEC electrodes and solar concentrators can be improved, the unit cost of hydrogen would decrease considerably. The hydrogen cost per kg is calculated to be 3.24 \$/kg. The cost of ammonia is found to be 0.84 \$/kg. The fixed operation and maintenance costs represent about 14% of the overall hydrogen cost whereas the capital cost of the plant has about 84% share as shown in Fig. 111

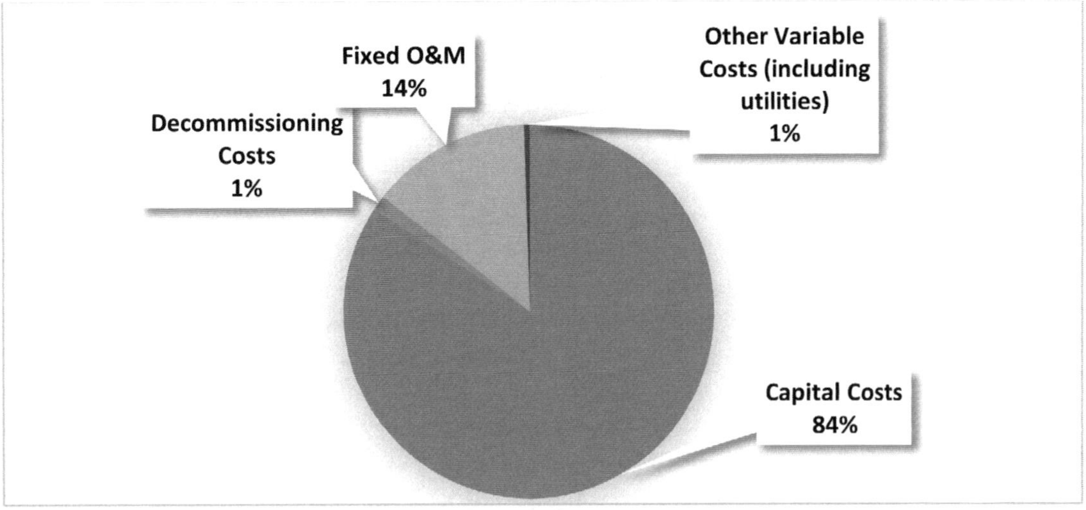

Figure 111 The cost breakdown of hydrogen production plant

The cost of hydrogen is expected to decrease with higher concentration ratio. The bars within the tornado chart in Fig. 112 show the range of minimum hydrogen cost values obtained by entering a base value for each specified variable, a reducing value, and an increasing value while holding all other variables constant at their base values.

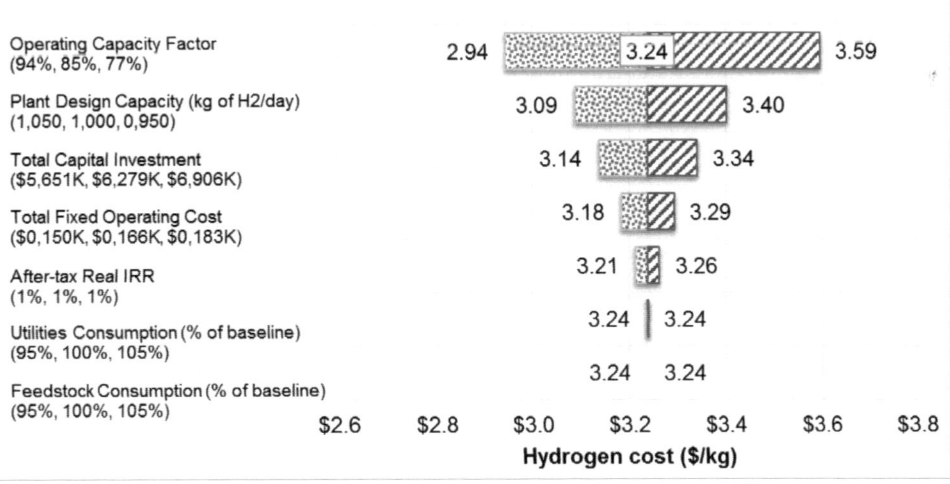

Figure 112 The sensitivity of the hydrogen cost based on different parameters

The operating capacity is chosen as 85% in the base case. As shown in Fig. 113, when operating capacity is increased to 94%, the cost of hydrogen can decrease down to 2.94 \$/kg.

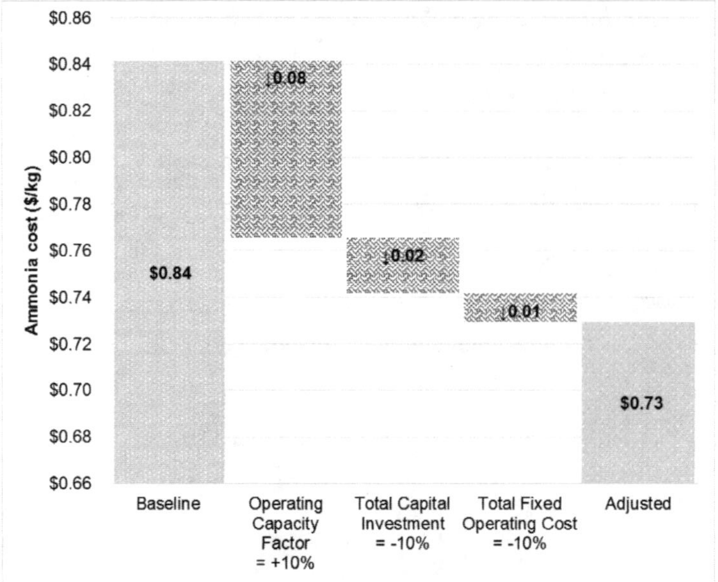

Figure 113 Waterfall diagram for ammonia cost considering better plant operating capacity and lower capital, operating costs

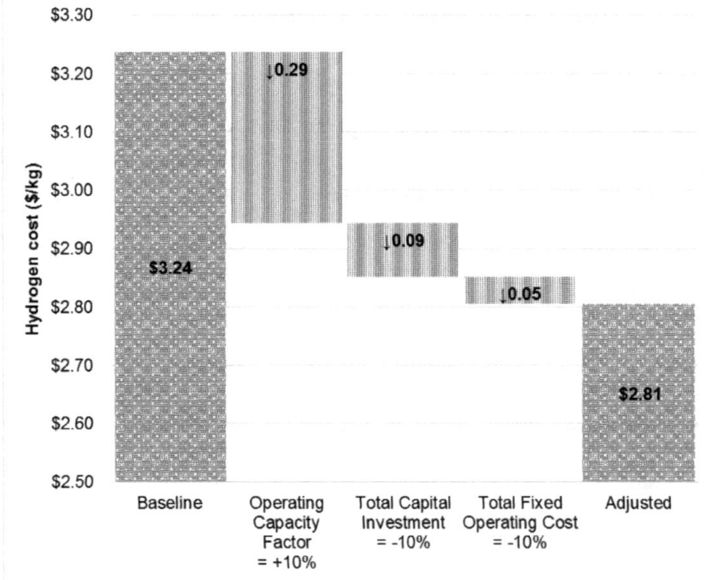

Figure 114 Waterfall diagram for hydrogen cost considering better plant operating capacity and lower capital, operating costs

On the other hand, if there is 10% lower capital investment, the cost of hydrogen can drop by 0.1 $/kg. The plant capacity factor is increased 10%, the capital investment and fixed operating costs are lowered 10% and utilities consumption is decreased by 5%. Under these conditions, the adjusted hydrogen cost can decrease down to 2.82 $/kg and ammonia cost in this case is found to be 0.73$/kg. The waterfall diagram of the hydrogen cost by increasing the operating conditions is shown in Fig. 114 where multiple improvements are considered.

7. Case Study for Ontario

In this section, the total cost of ammonia under different production pathways is calculated. The hydrogen production costs for wind and hydropower routes are calculated using H2A Central

Hydrogen Production Model, Version 3.1 of U.S. DOE [94] whereas the cost for photoelectrochemical hydrogen is previously calculated in scale-up analyses. The cost of electricity considered in the calculations is shown in Table 41. The land transportation is assumed to be for 800 km where 0.025 US$/kg is considered by truck transportation to the port in Quebec. The overseas transport is realized by the ocean tankers similar to LPG tankers in which the unit cost is taken as 0.1097 US$/kg of liquid ammonia.

Table 41 Cost of electricity considered in the analyses

Method	Cost of Electricity (US$/kWh)
Hydropower	0.0275
Wind	0.045
SMR	0.05
Solar	0.06

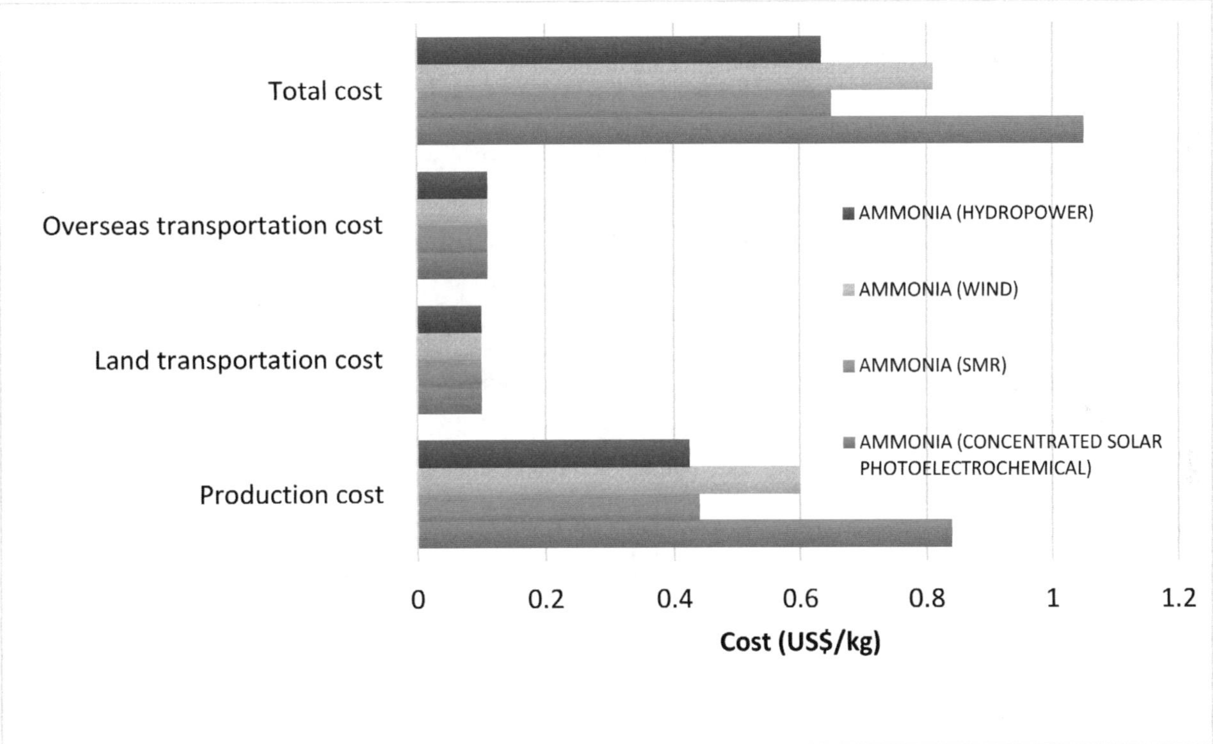

Figure 115 Comparison of ammonia costs using different production routes in Ontario

The required hydrogen for ammonia is produced using electrolysis of water except for the SMR method. The low-cost hydropower yields slightly lower total cost compared to SMR method as shown in Fig. 115. Since the solar energy route employs electrochemical ammonia synthesis rather than Haber-Bosch plant, the production cost is higher than other methods. The electrochemical ammonia synthesis route is in development phase and the costs will decrease by improving technology. However, Haber-Bosch process is a mature and saturated technology in which cost reduction is less likely to happen.

CHAPTER 11: CONCLUDING REMARKS

Ammonia is a carbon-free fuel suitable for use in transportation sector. It has a well-established production and distribution infrastructure, and has zero global warming potential during operation. In addition to its attractive qualities as a fuel, ammonia is widely used as a NOx reducing agent for combustion exhaust gases using selective catalytic reduction (SCR), and its capacity as a refrigerant can be applied to recover and further utilize engine heat that would otherwise be lost. Renewable ammonia (NH_3), which is a carbon-free fuel, refrigerant and working fluid; and storage media of hydrogen, are unique solutions to Canada's energy and environmental challenges. Renewable ammonia and hydrogen can serve almost all economic sectors, ranging from transportation to residential, industrial to commercial, public to utility, and agricultural to chemical. Ammonia as a potential fuel for vehicles or power generation can also be economically produced using conventional hydrocarbons in a cleaner manner by implying current technologies and developments. Ammonia can be produced using fossil fuels, or any renewable energy source using heat and/or electricity. Ammonia and hydrogen applications have been developed for sectors including but not limited to: transportation, industrial, commercial, utility, agricultural, and the chemical industries.

High-efficiency ammonia/urea plants using natural gas and other hydrocarbon feed stocks can be built beside natural gas power plants and oil sand extraction sites. Utilizing the waste, low-grade heat and excess oxygen results in a significant reduction of costs and emissions. The 200 million metric tonnes of ammonia produced globally each year comes from combining nitrogen from air with the hydrogen in coal and natural gas. The carbon in these hydrocarbons is usually emitted as carbon dioxide (CO_2), but about 40 per cent of the ammonia produced is combined with CO_2 sequestered in the a well-known Haber-Bosch synthesis process to make urea.

Dissociation of hydrocarbons such as methane and oil sand bitumen into hydrogen which can be then converted to ammonia is a promising option for oil sand and natural gas reserves in Western Canada and stranded gas reserves in Newfoundland and Labrador. Ammonia has significant environmental advantages. Even if ammonia is produced from hydrocarbons, it has similar greenhouse gas emissions with a solar energy based route.

In terms of environmental sustainability, ammonia can be produced using either fossil fuels, or any renewable energy source, using heat and/or electricity, which allows for evolution of ammonia production methods and technologies in parallel with sustainable development. Ammonia as a sustainable fuel can be used in all types of combustion engines, gas turbines, burners with only small modifications and directly in fuel cells which is a very significant advantage compared to another type of fuels. Reducing the total greenhouse gas emissions from marine transportation is possible by using ammonia which is carbon-free fuel. When ammonia is produced using renewable energy such as wind (versus conventional unleaded gasoline), a further 30 per cent greenhouse gas reduction is possible. When compared to propane, greenhouse gas emissions decrease about 18 per cent. An ammonia-driven passenger vehicle emits less greenhouse gas than compressed natural gas (CNG), liquefied petroleum gas (LPG), diesel, gasoline, and even hybrid electric vehicles. They can be utilized for maritime ship engines directly as supplementary fuels or individual fuels. Ammonia fueled ships yield considerably lower global warming impact during operation. Ammonia as a sustainable and clean fuel in road vehicles yield also the lowest global warming potential after electric and hydrogen vehicles. As a result, ammonia usage in the communities for transportation sector will bring significant cost and environmental benefits together with public satisfaction.

REFERENCES

1. Yara Fertilizer Industry Handbook. 2017.
2. Weekly Fertilizer Review [Internet]. Ammonia prices. [cited 2017 Jan 7]. Available from: http://www.farmfutures.com/story-weekly-fertilizer-review-0-30765
3. U.S., Mexico to source 50% of electricity from clean energy by 2025 - Politics - CBC News [Internet]. [cited 2017 Mar 24]. Available from: http://www.cbc.ca/news/politics/three-amigos-summit-climate-policies-1.3654166
4. Muradov NZ, Veziroğlu TN. From hydrocarbon to hydrogen–carbon to hydrogen economy. Int J Hydrogen Energy. 2005;30(3):225–37.
5. Muradov N. Hydrogen via methane decomposition: an application for decarbonization of fossil fuels. Int J Hydrogen Energy. 2001;26(11):1165–75.
6. Muradov NZ. How to produce hydrogen from fossil fuels without CO2 emission. Int J Hydrogen Energy. 1993;18(3):211–5.
7. Brohi EA. Ammonia as fuel for internal combustion engines? An evaluation of the feasibility of using nitrogen-based fuels in ICE. CHALMERS UNIVERSITY OF TECHNOLOGY; 2014.
8. Iki N, Kurata O, Matsunuma T, Inoue T, Suzuki M, Tsujimura T, et al. Micro Gas Turbine Firing Kerosene and Ammonia. In: ASME Turbo Expo 2015: Turbine Technical Conference and Exposition. American Society of Mechanical Engineers; 2015. p. V008T23A023-V008T23A023.
9. Haputhanthri SO, Maxwell TT, Fleming J, Austin C. Ammonia and Gasoline Fuel Blends for Spark Ignited Internal Combustion Engines. J Energy Resour Technol. 2015;137(6):62201.
10. Nozari H, Karabeyoğlu A. Numerical study of combustion characteristics of ammonia as a renewable fuel and establishment of reduced reaction mechanisms. Fuel. 2015;159:223–33.
11. Reiter AJ, Kong S-C. Combustion and emissions characteristics of compression-ignition engine using dual ammonia-diesel fuel. Fuel. 2011;90(1):87–97.
12. Ryu K, Zacharakis-Jutz GE, Kong S-C. Performance enhancement of ammonia-fueled engine by using dissociation catalyst for hydrogen generation. Int J Hydrogen Energy. 2014;39(5):2390–8.
13. Zamfirescu C, Dincer I. Using ammonia as a sustainable fuel. J Power Sources. 2008;185(1):459–65.
14. Zamfirescu C, Dincer I. Ammonia as a green fuel and hydrogen source for vehicular applications. Fuel Process Technol. 2009;90(5):729–37.
15. Hinnemann B, Nørskov JK. Catalysis by Enzymes: The Biological Ammonia Synthesis. Top Catal. 2006 Mar;37(1):55–70.
16. Lan R, Irvine JTS, Tao S. Synthesis of ammonia directly from air and water at ambient temperature and pressure. Sci Rep. 2013;3:1145.
17. Li F-F, Licht S. Advances in Understanding the Mechanism and Improved Stability of the Synthesis of Ammonia from Air and Water in Hydroxide Suspensions of Nanoscale Fe2O3. Inorg Chem. 2014;53(19):10042–4.
18. Licht S, Cui B, Wang B, Li F-F, Lau J, Liu S. Ammonia synthesis by N2 and steam electrolysis in molten hydroxide suspensions of nanoscale Fe2O3. Science (80-). 2014;345(6197):637–40.
19. Paschkewitz TM. Ammonia Production at Ambient Temperature and Pressure: An

Electrochemical and Biological Approach. PhD (Doctor of Philosophy) thesis. University of Iowa; 2012.

20. Skodra A, Stoukides M. Electrocatalytic synthesis of ammonia from steam and nitrogen at atmospheric pressure. Solid State Ionics. 2009;180(23–25):1332–6.
21. Kordali V, Kyriacou G, Lambrou C. Electrochemical synthesis of ammonia at atmospheric pressure and low temperature in a solid polymer electrolyte cell. Chem Commun. 2000;(17):1673–4.
22. Ganley JC. An intermediate-temperature direct ammonia fuel cell with a molten alkaline hydroxide electrolyte. J Power Sources. 2008;178(1):44–7.
23. Serizawa N, Miyashiro H, Takei K, Ikezumi T, Nishikiori T, Ito Y. Dissolution Behavior of Ammonia Electrosynthesized in Molten LiCl–KCl–CsCl System. J Electrochem Soc. 2012;159(4):E87–91.
24. Amar I, Lan R, Petit CG, Tao S. Solid-state electrochemical synthesis of ammonia: a review. J Solid State Electrochem. 2011;15(9):1845–60.
25. Murakami T, Nohira T, Ogata YH, Ito Y. Electrolytic Ammonia Synthesis in Molten Salts under Atmospheric Pressure Using Methane as a Hydrogen Source. Electrochem Solid-State Lett. 2005;8(4):D12–4.
26. Yiokari CG, Pitselis GE, Polydoros DG, Katsaounis AD, Vayenas CG. High-Pressure Electrochemical Promotion of Ammonia Synthesis over an Industrial Iron Catalyst. J Phys Chem A. 2000;104(46):10600–2.
27. Marnellos G, Stoukides M. Ammonia Synthesis at Atmospheric Pressure. Science (80-). 1998;282(5386):98–100.
28. Zhang Z, Zhong Z, Liu R. Cathode catalysis performance of SmBaCuMO5+δ (M=Fe, Co, Ni) in ammonia synthesis. J Rare Earths. 2010;28(4):556–9.
29. Wang WB, Cao XB, Gao WJ, Zhang F, Wang HT, Ma GL. Ammonia synthesis at atmospheric pressure using a reactor with thin solid electrolyte BaCe0.85Y0.15O3−α membrane. J Memb Sci. 2010;360(1–2):397–403.
30. Marnellos G, Zisekas S, Stoukides M. Synthesis of Ammonia at Atmospheric Pressure with the Use of Solid State Proton Conductors. J Catal. 2000;193(1):80–7.
31. Goto T, Ito Y. Electrochemical reduction of nitrogen gas in a molten chloride system. Electrochim Acta. 1998;43(21–22):3379–84.
32. Kugler K, Luhn M, Schramm JA, Rahimi K, Wessling M. Galvanic deposition of Rh and Ru on randomly structured Ti felts for the electrochemical NH3 synthesis. Phys Chem Chem Phys. 2015;17(5):3768–82.
33. Yun DS, Joo JH, Yu JH, Yoon HC, Kim J-N, Yoo C-Y. Electrochemical ammonia synthesis from steam and nitrogen using proton conducting yttrium doped barium zirconate electrolyte with silver, platinum, and lanthanum strontium cobalt ferrite electrocatalyst. J Power Sources. 2015;284:245–51.
34. Xu G, Liu R, Wang J. Electrochemical synthesis of ammonia using a cell with a Nafion membrane and SmFe0. 7Cu0. 3− xNixO3 (x= 0− 0.3) cathode at atmospheric pressure and lower temperature. Sci China Ser B Chem. 2009;52(8):1171–5.
35. Garagounis I, Kyriakou V, Skodra A, Vasileiou E, Stoukides M. Electrochemical Synthesis of Ammonia in Solid Electrolyte Cells. Front Energy Res. 2014;2.
36. Skodras G, Kaldis S, Topis S, Koutsonikolas D, Grammelis P, Sakellaropoulos G. NH3 decomposition and simultaneous H 2 separation with a commercial Pd-Cu-Ag/V membrane. In: Proceedings of the Second International Green Energy Conference June.

2006. p. 25–9.
37. Norby T. Solid-state protonic conductors: principles, properties, progress and prospects. Solid State Ionics. 1999;125(1–4):1–11.
38. Iwahara H, Esaka T, Uchida H, Maeda N. Proton conduction in sintered oxides and its application to steam electrolysis for hydrogen production. Solid State Ionics. 1981;3:359–63.
39. Zhai Y, Ye C, Xiao J, Dai L. A microwave-induced solution-polymerization synthesis of doped LaGaO3 powders. J Power Sources. 2006;163(1):316–22.
40. Nowick AS, Du Y, Liang KC. Some factors that determine proton conductivity in nonstoichiometric complex perovskites. Solid State Ionics. 1999;125(1–4):303–11.
41. Giddey S, Badwal SPS, Kulkarni A. Review of electrochemical ammonia production technologies and materials. Int J Hydrogen Energy. 2013;38(34):14576–94.
42. Kyriakou V, Garagounis I, Vasileiou E, Vourros A, Stoukides M. Progress in the Electrochemical Synthesis of Ammonia. Catal Today.
43. Shipman MA, Symes MD. Recent progress towards the electrosynthesis of ammonia from sustainable resources. Catal Today.
44. Muradov NZ, Veziroğlu TN. "Green" path from fossil-based to hydrogen economy: An overview of carbon-neutral technologies. Int J Hydrogen Energy. 2008;33(23):6804–39.
45. Transportation of liquefied natural gas - Energy Education [Internet]. [cited 2017 Mar 23]. Available from: http://energyeducation.ca/encyclopedia/Transportation_of_liquefied_natural_gas
46. Jefferies LLC. LNG & LPG Shipping Fundamentals. 2013.
47. WHOLESALE GAS PRICE SURVEY 2016 EDITION. 2016.
48. U.S. Department of Energy. Potential Roles of Ammonia in a Hydrogen Economy.
49. Resources / Nitrogen / Production Cost Profile and Operational Capability [Internet]. [cited 2017 Mar 25]. Available from: http://www.potashcorp.com/overview/resources/nitrogen/production-cost-profile-operational-capability#
50. Nutrients / Nitrogen / Overview / Nitrogen Production Cash Costs [Internet]. [cited 2017 Mar 25]. Available from: http://www.potashcorp.com/overview/nutrients/nitrogen/overview/nitrogen-production-cash-costs
51. Advantages / Nitrogen / US Midwest Delivered Ammonia Cost [Internet]. [cited 2017 Mar 25]. Available from: http://www.potashcorp.com/overview/advantages/nitrogen/us-midwest-delivered-ammonia-cost
52. Alternative Fuels Data Center– Fuel Properties Comparison.
53. Barton R. Estimation of Costs of Heavy Vehicle Use Per Vehicle-Kilometre. 2006.
54. Johnson AP, LO CM, FARJO SF. Microwave-assisted bitumen extraction with vacuum-assisted sediment filtration. Google Patents; 2016.
55. Bosisio RG, Cambon JL, Chavarie C, Klvana D. Exprimental Result on the Hesting of Athabasca Tar Sand Samples with Microwave Power. J Microw Power. 1977 Jan 17;12(4):301–7.
56. Pringle FG. Microwave-based recovery of hydrocarbons and fossil fuels. Google Patents; 2009.
57. Renouf G, Scoular RJ, Soveran D. Treating Heavy Slop Oil With Variable Frequency Microwaves. In: Canadian International Petroleum Conference. Petroleum Society of

Canada; 2003.
58. Process for the recovery of shale oil, heavy oil, kerogen or tar from their natural sources. 1981.
59. Mutyala S, Fairbridge C, Paré JRJ, Bélanger JMR, Ng S, Hawkins R. Microwave applications to oil sands and petroleum: A review. Fuel Process Technol. 2010;91(2):127–35.
60. Consultants P. SimaPro Life Cycle Analysis version 7.2 (software).
61. GREET 2016. IL 60439-4844 USA: Argonne National Laboratory; 2016.
62. Facts and FIGURES - Air Transport Action Group (ATAG) [Internet]. [cited 2017 Mar 25]. Available from: http://www.atag.org/facts-and-figures.html
63. Koroneos C, Dompros A, Roumbas G, Moussiopoulos N. Advantages of the use of hydrogen fuel as compared to kerosene. Resour Conserv Recycl. 2005;44(2):99–113.
64. Contreras A, Yiğit S, Özay K, Veziroğlu TN. Hydrogen as aviation fuel: A comparison with hydrocarbon fuels. Int J Hydrogen Energy. 1997;22(10–11):1053–60.
65. A Full Fuel-Cycle Analysis of Energy and Emissions Impacts of Transportation Fuels Produced from Natural Gas.
66. Alternative Jet Fuels, Addendum 1 to Aviation Fuels Technical Review (FTR-3/A1).
67. Hileman JI, Stratton RW. Alternative jet fuel feasibility. Transp Policy. 2014;34:52–62.
68. Turner D, Xu H, Cracknell RF, Natarajan V, Chen X. Combustion performance of bio-ethanol at various blend ratios in a gasoline direct injection engine. Fuel. 2011;90(5):1999–2006.
69. Kong S-C. Combustion Efficiency and Exhaust Emissions of Ammonia Combustion in Diesel Engines. In: 2008 Annual NH3 Fuel Conference. MINNEAPOLIS: NH3 Fuel Association; 2008. p. SEPTEMBER 29-30.
70. Dincer I, Zamfirescu C. Apparatus for using ammonia as a sustainable fuel, refrigerant and NOx reduction agent. Google Patents; 2012.
71. GREET 2015. IL 60439-4844 USA: Argonne National Laboratory; 2015.
72. Appl M. Ammonia: principles and industrial practice. Vch Verlagsgesellschaft Mbh; 1999.
73. Dincer I, Zamfirescu C. Ammonia as a Potential Substance. In: Sustainable Energy Systems and Applications. Boston, MA: Springer US; 2011. p. 203–32.
74. Thomas G, Parks G. Potential roles of ammonia in a hydrogen economy: a study of issues related to the use ammonia for on-board vehicular hydrogen storage. US Dep Energy. 2006;23.
75. Metkemeijer R, Achard P. Ammonia as a feedstock for a hydrogen fuel cell; reformer and fuel cell behaviour. J Power Sources. 1994;49(1):271–82.
76. Yin SF, Xu BQ, Zhou XP, Au CT. A mini-review on ammonia decomposition catalysts for on-site generation of hydrogen for fuel cell applications. Appl Catal A Gen. 2004;277(1):1–9.
77. Chellappa AS, Fischer CM, Thomson WJ. Ammonia decomposition kinetics over Ni-Pt/Al 2 O 3 for PEM fuel cell applications. Appl Catal A Gen. 2002;227(1):231–40.
78. Roy SK, Ray N, Mukheriee D. Kinetics and mechanism of ammonia decomposition over alumina supported nickel catalysts. Plan Dev Div. 1975;41:48.
79. Chorkendorff I, Niemantsverdriet JW. Concepts of modern catalysis and kinetics. John Wiley & Sons; 2006.
80. Palaniappan R. Improving the efficiency of ammonia electrolysis for hydrogen production. Ohio University; 2013.

81. Vitse F, Cooper M, Botte GG. On the use of ammonia electrolysis for hydrogen production. J Power Sources. 2005;142(1):18–26.
82. Barbir F. PEM fuel cells : theory and practice. Academic Press; 2013. 518 p.
83. Barbir F. PEM electrolysis for production of hydrogen from renewable energy sources. Sol Energy. 2005;78(5):661–9.
84. Boggs BK, Botte GG. On-board hydrogen storage and production: An application of ammonia electrolysis. J Power Sources. 2009 Jul 15;192(2):573–81.
85. 1 Month Ruthenium Prices and Ruthenium Price Charts - InvestmentMine [Internet]. [cited 2017 Mar 22]. Available from: http://www.infomine.com/investment/metal-prices/ruthenium/1-month/
86. DOE Technical Targets for Hydrogen Storage Systems for Material Handling Equipment | Department of Energy [Internet]. [cited 2017 Mar 22]. Available from: https://energy.gov/eere/fuelcells/doe-technical-targets-hydrogen-storage-systems-material-handling-equipment
87. Satyapal S, Petrovic J, Read C, Thomas G, Ordaz G. The U.S. Department of Energy's National Hydrogen Storage Project: Progress towards meeting hydrogen-powered vehicle requirements. Catal Today. 2007 Feb 28;120(3–4):246–56.
88. Dong B-X, Tian H, Wu Y-C, Bu F-Y, Liu W-L, Teng Y-L, et al. Improved electrolysis of liquid ammonia for hydrogen generation via ammonium salt electrolyte and Pt/Rh/Ir electrocatalysts. Int J Hydrogen Energy. 2016 Sep;41(33):14507–18.
89. Mulder FM. Electrolytic cell for the production of ammonia. Google Patents; 2016.
90. Tsatsaronis G. Thermoeconomic analysis and optimization of energy systems. Prog Energy Combust Sci. 1993;19(3):227–57.
91. Dincer I, Zamfirescu C. Sustainable hydrogen production. 2016;494.
92. Bejan A, Tsatsaronis G (George), Moran MJ. Thermal design and optimization. Wiley; 1996. 542 p.
93. Lazzaretto A, Tsatsaronis G. SPECO: A systematic and general methodology for calculating efficiencies and costs in thermal systems. Energy. 2006 Jul;31(8–9):1257–89.
94. DOE Hydrogen and Fuel Cells Program: DOE H2A Production Analysis [Internet]. H2A Central Hydrogen Production Model, Version 3.1. [cited 2017 Jan 7]. Available from: https://www.hydrogen.energy.gov/h2a_production.html
95. Comparison of Electricity Prices in Major North American Cities | Publications | Hydro-Québec [Internet]. [cited 2017 Jan 7]. Available from: http://www.hydroquebec.com/publications/en/corporate-documents/comparaison-electricity-prices.html
96. Inflation Calculator | Find US Dollar's Value from 1913-2017 [Internet]. [cited 2017 Mar 3]. Available from: http://www.usinflationcalculator.com/
97. Bartels J. A feasibility study of implementing an Ammonia Economy. Graduate Theses and Dissertations. Iowa State University; 2008.
98. Morgan E. Techno-Economic Feasibility Study of Ammonia Plants Powered by Offshore Wind. Dissertations. University of Massachusetts - Amherst; 2013.

484565 Multiscale Characterization and CFD Simulation of W/O Emulsions

Abstract:

MULTISCALE CHARACTERIZATION AND CFD SIMULATION OF W/O EMULSIONS

Juan Pablo Gallo-Molina, Nicolás Ratkovich, Óscar Álvarez.

Process and Product Design Group (GDPP).

Department of Chemical Engineering, Universidad de los Andes, Bogotá, Colombia.

Emulsions are a type of metastable colloids composed by two or more immiscible liquids. Both oil-in-water (O/W) and water-in-oil (W/O) emulsions are widely used in a variety of applications, such as cosmetics, drug delivery, food, etc. Although there exist theoretical foundations which offer insights into these systems, industry practices often focus on formulation and process; commonly using empirical methods. In this work, a multiscale approximation was used for the study of W/O emulsions. This approach allows for the analysis of interrelationships among macroscopic, microscopic, process and formulation variables. Additionally, the emulsions were modelled with Computational Fluid Dynamics (CFD), which permitted a better understanding of the role process variables plays. Using the incorporated energy during the emulsification process as a transversal factor, it was possible to establish relationships between rheological measurements (i.e. the macroscopic response) and droplet size distribution observations (i.e. the microscopic variable). Additionally, stability measurements were made and differences among impeller types used during preparation were established.

In accordance to previous literature (Pradilla et al. 2015), average droplet size was found to be inversely proportional to the dispersed phase concentration, while the opposite behavior was observed in incorporated energy. This was explained by the increments in the elasticity of the interphase associated with increases in the dispersed phase concentration (Langevin, 2000). Results also showed that an increase in the incorporated energy leads to larger values of the elastic modulus in the linear viscoelastic region. That is: more elasticity at the interphase and more solid-like behavior in the product. Furthermore, larger amounts of incorporated energy tended to generate more stable emulsions with lower average droplet diameters. This was attributed to the lower sensitivities to gravitational effects and the reduction in Ostwald ripening associated with smaller droplets in the system (Leal-Calderon et al., 2007).

Emulsions were prepared using four impeller types: propeller, straight paddles turbine, 45° pitched blade turbine and Rushton turbine. Sizable differences in the response variables were observed among the mentioned types. For instance, significantly different average mean droplet diameters were observed in function of the impeller type, which led to a tendency towards different degrees of stability. This phenomenon occurs because different impeller geometries add different amounts of energy and make different quantities of shear available for the emulsification process (Torres & Zamora, 2002). In turn, the

divergences in the incorporated energies generated generally slight differences in the values of the elastic modulus.

CFD simulations were developed using commercial software STAR-CCM+ v11.04 (Siemens). The modelling was limited to emulsions prepared with the propeller type impeller and only the homogenization stage (steady state) was considered. An Eulerian approach was preferred and no turbulence model was implemented, for the reason that flow regimes were laminar under all conditions. The Morris and Boulay model was used for describing the rheological behavior of the systems, while the S-gamma approximation was chosen for describing the drop size distribution as well as breakage and coalescence. There was good agreement between experimental data and CFD results for relative viscosity and incorporated energy and it was possible to observe three dimensional profiles of relevant variables. Gradients in droplet diameter, relative viscosity and dispersed phase volume fraction were found to exist in the mixing tank, which indicated that the chosen impeller and tip velocity don't provide optimal mixing, while generating effects in the final macroscopic and microscopic properties of the product.

REFERENCES

Langevin, D. (2000). Influence of interfacial rheology on foam and emulsion properties. *Advances in Colloid and Interface Science*, *88*(1-2), 209–222. https://doi.org/10.1016/S0001-8686(00)00045-2-

Leal-Calderon, F., Schmitt, V., & Bibette, J. (2007). *Emulsion science,* (Second Edi). New York: Springer. https://doi.org/10.1007/978-0-387-39683-5

Pradilla, D., Vargas, W., & Alvarez, O. (2015). The application of a multi-scale approach to the manufacture of concentrated and highly concentrated emulsions. *Chemical Engineering Research and Design*, *95*(February), 162–172. https://doi.org/10.1016/j.cherd.2014.10.016

Torres, L. G., & Zamora, E. R. (2002). Preparation and power consumption of surfactant-fuel oil-water emulsions using axial, radial, and mixed flow impellers. *Fuel*, *81*(17), 2289–2302. https://doi.org/10.1016/S0016-2361(02)00137-0

497413 Procafd: A Tool for Generating Sustainable Hybrid Process Flowsheets

Abstract:

The chemical industry is always looking for innovative process designs, which are both efficient and sustainable. To generate an optimal flowsheet, what is needed is a systematic way to identify the types of tasks-operations that need to be performed, the corresponding design of the operation-equipment, their configuration, mass-energy flows etc. Due to the fact that process synthesis problems are by nature combinatorial and open ended, a number of different solution approaches have been proposed. However, the solution for any synthesis-design problem is dependent on the search space of alternatives and the process performance criteria which in most cases are influenced by economic factors. This work focuses on the development of a computer aided tool (ProCAFD), capable of enumerating the entire feasible search space, analyze and determine the most sustainable process. ProCAFD is based on decomposition based method developed by Tula et al. (2015). On a higher level the workflow of ProCAFD can be divided into three stages: 1) Synthesis stage â€“ Here a hybrid approach based on group contribution is used to generate and rank the alternatives. In this stage, chemical process flowsheets are synthesized in the same way as atoms or groups of atoms are synthesized to form molecules in computer aided molecular design (CAMD) techniques. The building blocks in flowsheet synthesis are called process-groups, which represent a single or set of unit operations that are selected by employing a thermodynamic insights based method. These building blocks are then combined using connectivity rules to generate all the feasible flowsheet alternatives. The main advantage of representing the flowsheet with process-groups is that, the performance of the entire process can be evaluated from the contributions of the individual process-groups towards the selected flowsheet property (for example, energy consumed). Available flowsheet property models include energy consumption, carbon footprint, product recovery, product purity etc. In this way, the entire list of feasible chemical process flowsheets is quickly generated, screened and selected. 2) Analysis stage â€“ In this stage, comprehensive investigation consisting of sustainability, life cycle and economic analysis is performed to identify areas of improvement for the generated process alternative. The identified areas or process hotspots are translated into design targets that, if satisfied will guarantee process improvement. 3) Innovation stage â€“ In this stage different innovative design strategies are used to match the set design targets. Different design strategies include simultaneous process optimization and heat integration where determination of optimal operating conditions and design parameters is solved together with heat integration involving minimizing utility consumption. One other strategy ProCAFD uses is a generalized method based on Tula et al. (2016) to replace distillation columns in the selected design with a hybrid distillation-membrane system to reduce the operational cost substantially. Other strategies include process intensification, where the phenomena based intensification method developed by Babi et al. (2015) is employed to generate intensified processes.

In this work, the software implementation of the developed methodology will be highlighted through several examples involving chemical and biochemical processes. In all the cases studied, the developed software could generate innovative and more sustainable solutions very quickly.

References:

A. K. Tula, B. Befort, N. Garg, K. V. Camarda, R. Gani, Sustainable Process Design & Analysis of Hybrid Separations, Computers & Chemical Engineering (2016). http://dx.doi.org/10.1016/j.compchemeng.2016.11.031.

A. K. Tula, M. R. Eden, R. Gani, 2015, Process synthesis, design and analysis using a process-group contribution method, Computers & Chemical Engineering, 81, 245-259.

D. K. Babi, J. Holtbruegge, P. Lutze, A. Gorak, J.M. Woodley, R. Gani, 2015, Sustainable Process Synthesis-Intensification, Computers & Chemical Engineering, 81, 218-44.

501345 Bio-Ionic Liquid Functionalzed Biomaterial

Abstract:

Conventional methods to engineer electroconductive hydrogels (ECHs) through the incorporation of conductive nanomaterials and polymers exhibit major technical limitations. These are mainly associated with the cytotoxicity, as well as poor solubility, processability, and biodegradability of their components. Here, we describe the engineering of a new class of ECHs through the functionalization of non-conductive polymers with a conductive choline-based bio-ionic liquid (Bio-IL). Bio-IL conjugated hydrogels exhibited a wide range of highly tunable physical properties, remarkable in vitro and in vivo biocompatibility, and high electrical conductivity without the need for additional conductive components.

500277 Effect of Electrical Stimulation on Nerve Cells As a Function of Hydrogel Stiffness and Electrical Conductivity with a Custom Designed Device

Abstract:

Electrical stimulation of nerve cells has shown promising results related to cell migration, alignment and neurite outgrowth, which can be used to promote and guide axon regeneration across an injury. Polymer scaffolds are also used to direct and support nerve regeneration. The goal of this study was to design a device for electrically stimulating nerve cells seeded on a polymeric substrate and test the combined effect of substrate properties and electrical stimulation on cell alignment as a means to achieve directional cell growth.

Here, we designed a custom device for electrically stimulating neural cells seeded on a hydrogel. Polyacrylamide (PA) hydrogels were prepared by first mixing bis-acrylamide and acrylamide in various ratios for gels of varying stiffness. Then, 0.1% w/v Irgacure 2959 was added and the solution was degassed for 60 min. Hydrogels were formed by exposure to 365 nm UV light for 10 min. 4-arm polyethylene glycol acrylate (4-arm PEG-Ac)-carbon nanotube (CNT) gels (7.5% and 20% w/v in PEG and 0.05% and 0.1% w/v in CNTs) were prepared. PEG-dithiol was used as a crosslinker (thiol and acrylate were kept at 1:1 molar ratio). Bovine serum albumin (0.25% w/v) was used as a surfactant to aid CNT dispersion. PC12 cells were seeded on top of the gels at a density of 5Ã—10^4 cells/cm^2 for 24 h, stimulated with a direct current at 0.07 mA and voltage of 6 V/cm for 2 h and imaged at 24 h post-stimulation. MATLAB was used to simulate electrical current density, direction, and distribution through the hydrogels.

We first determined the effect of electrode geometry and positioning on electrical current density and distribution through the hydrogel. We determined that rectangular shape electrodes spaced 26 mm apart could uniformly distribute current density within the hydrogel. For the electrodes, we used biocompatible and cost-effective Ni-Cr alloys. Neural-like PC12 cells were seeded on polyacrylamide hydrogels of varying stiffness (2-70 kPa) as well as resistivity (0.08 to 1 Î©m) and stimulated for 2 h. Cells exhibited alignment along the applied current direction at 24 h post-stimulation, where the extent of alignment was dependent on hydrogel stiffness and resistivity. The number of PC12 cells aligned to the direction of applied current at 60 ± 10 was most frequent on soft hydrogels ranging from 2-10 kPa with higher conductivity. We also studied the alignment of PC12 cells on PEG-CNT nanocomposite gels of various stiffness and resistivity. Cells showed more uniform alignment on soft substrates with higher CNT concentration (i.e. higher conductivity).

In this study, we designed a custom device to electrically stimulate cells seeded on a hydrogel substrate. We optimized electrode geometry and spacing to achieve a uniform current density and distribution through the substrate. We applied the device towards the stimulation of neural-like PC12 cells on hydrogels of varying stiffness and electro-conductivity. We showed that cells aligned along the direction of the applied current at 24 h post-stimulation and that cell alignment was inversely proportional to substrate stiffness and proportional to substrate conductivity.

492512 IVF Modeling, Optimal Control, and Customized Drug Treatment: Results of First Clinical Trial

Abstract:

In-vitro fertilization (IVF) is the most common technique in assisted reproductive technology and most cases the last resort for infertility treatment. It has four basic stages: superovulation, egg retrieval, insemination/fertilization and embryo transfer. Superovulation is a drug induced method to enable multiple ovulation per menstrual cycle. The success of IVF majorly depends upon successful superovulation, defined by the number and uniform high quality of eggs retrieved in a cycle. In the earlier part of our research, we have created theoretical framework for customized modeling of superovulation for each patient depending on first two days of cycle data. The models then can be used to customize optimal drug treatment (hormonal doses each day) for each patient. Optimal control theory is used to find customized optimal drug treatment. In this paper, we have carried out the first clinical double blind trial with customized doses predicted by the theory and doses specified by the IVF specialist doctor. We present case reports of six patients in this paper.

501616 Tissue Patterning By Spatially Defined Addressable Microfluidic Delivery of Differentiated Growth Factors

Abstract:

Now that tissue engineering is targeting organs of high complexity, the design and manufacturing of scaffold should take into account all of these factors: (1) Active Microfluidic Pores for Delivery of Nutrients, Oxygen and Chemical Signals Throughout the Scaffold, (2) Transparent Material for Microscopy Observation, Understanding Tissue Growth in Artificial Environments, (3) "Addressable" Design for Localized Chemical Delivery and/or Sampling, (4) Cell Control, Tissue Patterning, Monitoring Tissue Development. Therefore, an 'addressable' microfluidics platform will be modified to deliver predetermined amounts of growth-factors to specified XY locations within the chip. And continuous real-time feedback for differentiation-guiding will be accomplished by coupling valve automation with robotized microscopy, computer vision and transport modeling of the chemo-stimuli release within the device. The deliverables will be: A) an experimental platform capable of precise chemo-stimuli delivery to single cells, and B) a computational framework capable of adaptive tissue patterning. Since this platform is lacking in previous works, we aim to provide a proof-of-concept of an adaptive control of cellular differentiation via an addressable release platform. The platform is a microfluidic device consisting of four layers of polydimethylsiloxane (PDMS) elastomer which is connected to an automatic pumping system controlled by Matlab code. The first layer locating on top of the chip acts as a valve which helps switching the flow of media in the second layer between through and by-passing delivery holes in the third layer. These holes are used to deliver differentiation factors to a cell chamber located at the bottom of the chip. Using this platform, we assessed the ability to adaptively control the tissue formation of mouse embryonic fibroblasts by platelet-derived growth factor-BB (PDGF-BB) gradients. Cell behavior is monitored in real time using an automated microscope while the translocation of the cells is purposely navigated to various locations in a migration chamber in an addressable manner through an adjustable chemoattractant delivery.

492295 Multiscale Modeling of Drug Transport through Human Skin Stratum Corneum

Abstract:

Skin is the largest organ of the human body. Owing to the high surface area, it provides a convenient route for administration of drugs. Thus, transdermal delivery provides a potential alternative to other regular modes of administration of drugs. Understanding the mechanism of drug permeation through the skin, and, dermal uptake, is beneficial for the development of effective trans-dermal delivery formulations. The current standard in both pharma and cosmetic industries is to conduct extensive *in-vivo/in-vitro* experiments to arrive at the optimized formulations. This is expensive and also highly time consuming In view of this, a realistic *in-silico* molecular model of human skin coupled with accurate transport models, which could supplement/replace some of the elaborate *in-vivo/in-vitro* tests with *in-silico* tests, are required.

Most of the SC's molecular level models reported in the literature comprise of cholesterol and phospholipids only, which is far from the reality. Also, in the previous studies, permeation data from the experiments was fitted to the model to predict the diffusion coefficient and partition coefficient. The fitted diffusion coefficient is an approximate one because it doesn't account for the heterogeneity of the SC lipids. In this study, we have implemented a multiscale modelling framework to obtain the release profile of three drugs namely Caffeine, Fentanyl and Naphthol through skin SC. We report for the first time diffusion of drugs through a realistic skin molecular model comprised of ceramides, cholesterol and free fatty acid. The diffusion coefficients of drugs in the SC lipid matrix were determined from multiple constrained molecular dynamics simulations. The calculated diffusion coefficients were then used in the macroscopic models to predict the release profiles of drugs through the SC. The obtained release profiles were in good agreement with the available experimental data. Our simulations show that the partition coefficient exhibits more effect on the release profiles. The reported multiscale modelling framework would provide insight into the delivery mechanisms of the drugs through the skin, and shall act as a guiding tool in performing targeted experiments to come up with a suitable delivery system.

References

1. Elias, P. M. Epidermal Lipids, Barrier Dunction, and Desquamation. J. Invest. Dermatol. 1983,80.
2. Pirot, F.; Kalia, Y.; Stinchcomb, A.; Keating, G.; Bunge, A.; Guy, R. Characterization Of The Permeability Barrier Of Human Skin In Vivo. Proceedings of the National Academy of Sciences 1997, 94, 1562-1567.
3. Kushner, J., Deen, W., Blankschtein, D., & Langer, R. (2007). First-principles, structure-based transdermal transport model to evaluate lipid partition and diffusion coefficients of hydrophobic permeants solely from stratum corneum permeation experiments. Journal of pharmaceutical sciences, 96(12), 3236-3251.
4. Rim, Jee E., Peter M. Pinsky, and William W. van Osdol. "Multiscale modeling framework of transdermal drug delivery." Annals of biomedical engineering 37.6 (2009): 1217-1229.
5. Dias, M., et al. "Topical delivery of caffeine from some commercial formulations." International journal of pharmaceutics 182.1 (1999): 41-47.
6. Hansen, Steffi, et al. "In-silico model of skin penetration based on experimentally determined input parameters. Part I: Experimental determination of partition and diffusion coefficients." European Journal of Pharmaceutics and Biopharmaceutics 68.2 (2008): 352-367.
7. Gupta, Rakesh, D. B. Sridhar, and Beena Rai. "Molecular Dynamics Simulation Study of Permeation of Molecules through Skin Lipid Bilayer." The Journal of Physical Chemistry B 120.34 (2016): 8987-8996.
8. Gupta, Rakesh, Balarama Sridhar Dwadasi, and Beena Rai. "Molecular Dynamics Simulation of Skin Lipids: Effect of Ceramide Chain Lengths on Bilayer Properties." The Journal of Physical Chemistry B 120.49 (2016): 12536-12546

492612 Integrated Design of Sulfur Host Materials to Enhanced the Performance of Li-Sulfur Batteries

Abstract:

Lithiumâ€"sulfur batteries (LSBs) are considered as a promising candidate for next generation electrical energy storage systems, because they can deliver a significantly higher theoretical energy density of 2600 Wh/kg compared to lithium ion batteries (LIBs), and offer intriguing structural advantages, such as very high specific theoretical capacity (1675 mAh g^{-1}), low cost, and nontoxicity. Despite of multiple advantages of LSBs, they still suffer from some significant challenges, including low electrical conductivity of sulfur, fast capacity fading at high discharge rate, and low Coulombic efficiency due to the shuttle effect from polysulfides ions. The insulating nature of sulfur, the dissolution of intermediate polysulfides and final Li_2S products affecting the utilization rate of active sulfur, lead to low specific capacity. Particularly, the big stone for LSBs is the redox reactions of polysulfides at both the cathode and anode surfaces, caused by the dissolution of the reaction intermediates in the organic electrolyte, which consume a large proportion of the capacity. In order to overcome the aforementioned issues we have designated various metal/metal oxide-carbo hybrid sulfur host materials that have the ability to efficiently restrained polysulfides via both structural restriction and chemical encapsulations effect. The developed sulfur host have enhanced the performance of LSBs in term of capacity, rate capability and cyclability.

References:

1. *Sarish Rehman*, Shaojun Guo, & Yanglong Hou; Rational Design of Si/SiO_2@Hierarchical Porous Carbon Spheres as Efficient Polysulfide Reservoirs for High-Performance Liâ€"S Battery; ***Adv. Mater***. 2016, *28,* 3167â€"3172
2. *Sarish Rehman*, Tianyu Tang, Zeeshan Ali, Xiaoxiao Huang and Yanglong Hou; Integrated Design of MnO_2@Carbon Hollow Nanoboxes to Synergistically Encapsulate Polysulfides for Empowering Lithium Sulfur Batteries, ***Small***, 2017 DOI: 10.1002/smll.201700087.

493247 Porous and Chemically Functional Polymeric Hydrogel Microspheres for Improved Biomacromolecular Conjugation

Abstract:

Porous and chemically functional polymeric hydrogel microspheres for improved biomacromolecular conjugation

Eric Liu, Sukwon Jung, Chang-Hyung Choi[2] and Hyunmin Yi[1]

[1] Department of Chemical and Biological Engineering, Tufts University

4 Colby St. Medford, MA 02155 USA

hyunmin.yi@tufts.edu, (TEL) 617-627-2195

[2] Cosmetic Science and Technology Division, Daegu Haany University

1 Haanydaero, Gyeongsan-si, Gyeongsangbuk-Do, Korea (Zip code 712-715)

changhyung.choi@gmail.com (TEL) 82 053 819 7721

Suspension arrays of biofunctionalized polymeric hydrogel microparticles hold significant potential for rapid and reliable biosensing and biodiagnostics, yet there exist several critical hurdles in the existing fabrication technologies. We demonstrate a robust and tunable micromolding method to fabricate chemically functional poly(acrylamide-*co*-acrylic acid) (p(AAm-*co*-AA)) hydrogel microspheres with uniform dimensions and controlled porous network structures. The inherently batch processing-based nature of our micromolding process offers several distinct advantages over microfluidic fabrication processes; for example accommodating a variety of fabrication parameters including monomer systems with slow polymerization rates, without the need for surfactants, delicate flow control, or viscosity tuning. The p(AAm-*co*-AA) microspheres with abundant carboxylate functional groups are fabricated via surface tension-induced droplet formation in patterned poly(dimethylsulfoxide) (PDMS) molds and photo-induced radical polymerization. To demonstrate chemical functionality, we enlisted a rapid EDC/NHS (1-ethyl-3-(3-dimethylaminopropyl)carbodiimide (EDC) and N-hydroxysuccinimide (NHS)) chemistry for fluorescent labeling of the microspheres with a small molecule dye fluorescein glycine amide. Epifluorescence imaging and image analysis illustrate the uniform incorporation of carboxylate groups within the microspheres and rapid kinetics without diffusion limitation. Furthermore, confocal imaging of the conjugation of the p(AAm-*co*-AA) microspheres with fluorescent protein R-phycoerythrin (M.W. 240 kDa, D_h ~11nm) demonstrate the highly porous nature of the microspheres as well

as the utility of the microspheres and the EDC/NHS scheme for facile biomacromolecular conjugation and rapid biosensing. Finally, we also report our recent results on thorough investigation on the effects of two potent conjugation handles, namely AA and a potent aminopolysaccharide chitosan, on the pH-dependent conjugation efficiency, mesh size and biomacromolecular conjugation kinetics of the hydrogel microspheres fabricated via a rapid microcapillary device technology. Here, we utilized simple size-based multiplexing to further demonstrate orthogonal one-pot conjugation of multiple proteins on microspheres with either AA or chitosan conjugation handles. Combined, these results illustrate significant potential for our fabrication-conjugation strategy in the development of biofunctionalized polymeric hydrogel microspheres toward rapid biosensing, bioprocess monitoring, and biodiagnostics.

493883 Engineered CRISPR/Cas9 System for Multiplex Genome Engineering of Industrial Yeast Strains

Abstract:

Compared with laboratory yeast strains, industrial yeast strains have the advantages of higher productivity and robustness under harsh industrial conditions. However, industrial yeast strains are polyploid and the presence of multiple copies of chromosomes makes their genome engineering rather challenging. The clustered regularly interspaced short palindromic repeats (CRISPR) system has been adopted for efficient and multiplex genome engineering in nearly all kingdoms of life by introducing a double strand break followed by DNA repair machinery. Although successful in the engineering of laboratory yeast strains, the application of CRISPR/Cas9 system in polyploid industrial yeast strain engineering is less explored. During CRISPR/Cas9 based genome editing process, the double strand break can be repaired by the provided DNA donor (edited yeast strains) or the other copies of the chromosome (wild-type yeast strains), leading to very low genome editing efficiency. We developed an efficient CRISPR/Cas9 system for industrial yeast strain engineering by introducing double strand breaks to all copies of the chromosome simultaneously, which can only be repaired using the provided DNA donors. Using plasmids with increased copy numbers for SgRNA expression, we could knock out four genes in a diploid industrial yeast strain (Ethanol Red, 8 alleles in total) and in a triploid industrial yeast strain (ATCC4124, 12 alleles in total) in a single step with 100% efficiency. The metabolic engineering application was demonstrated by the construction of efficient xylose-fermenting and lactate-producing industrial yeast strains, in which *ALD6*, *PHO13*, *LEU2*, and *URA3* were disrupted in a single step followed by the introduction of a xylose utilization pathway (*XR-XDH-XKS*) with *URA3* auxotrophic marker and a lactate dehydrogenase gene (*LDH*) with *LEU2* auxotrophic marker. The engineered CRISPR/Cas9 system can be used for the development of industrial yeast strains based microbial cell factories.

494235 Accelerating Build and Test of Microbial Libraries Via Integration of Synthetic Biology, Robotic Automation and Mass Spectrometry

Abstract:

Whereas synthetic biology initially focuses on modular assembly of standardized 'function' units for predicable phenotypes, the importance of standardized 'physical' processes in constructing biological systems has been increasingly recognized. Here we demonstrated synergistic integration of both genetic design and workflow/instrument development are indispensable to accelerate the design-build-test-learn cycle in microbial engineering.

For the first example of optimizing *Saccharomyces cerevisiae* as a microbial cell factory, an automated engineering workflow was established using an integrated robotic system (Illinois Biological Foundry for Advanced Bioengineering, iBioFAB). On the genetic aspect, as robotic automation excels at iterative execution of defined processes, it is desirable to use the same procedure to introduce different genetic modifications. Therefore, we chose to apply *trans-acting* elements for genetic modulation so that mutations can be created independent of the target chromosomal sequences using standardized vectors. Repetitive genomic sequences were selected as vectors because scalable accumulation of combinatorial mutations can be achieved during repeated integration of *trans-acting* donors. On the instrument aspect, as the yeast engineering workflow includes several process modules including heat-shock transformation, cell cultivation, on-line spectrophotometric monitoring of cell growth, and sample preparation for off-line analyses, a flexible framework is needed to complete all the distinct tasks on a single platform for full automation. To achieve such flexibility, we enumerate all common unit operations in microbial engineering, including liquid handling, centrifugation, temperature control, and so on, and match them with component instruments in iBioFAB. Then any process module in the yeast workflow can be converted into a custom sequence of unit operations, which is then executed using robotic arms to transfer samples among component instruments. An integrated computational framework is also developed to orchestrate physical operation and data acquisition. With the optimized genetic constructs and process parameters, yeast strain libraries were iteratively created and subjected to serial transfer under different HAc stresses. After three automated engineering cycles in ~1 month, a mutant strain was isolated harboring 26 *trans-acting* elements from a genome-scale modulation collection for both overexpression and knockdown mutations in yeast. The mutant strain can ferment 2% glucose in the presence of 1.1% (v/v) HAc within 4 days, the highest HAc resistance ever reported in yeast to the best of our knowledge.

In the second example of engineering natural product analogs in *Escherichia coli*, a structure-based screening method was developed called optically-guided matrix-assisted laser desorption/ionization mass spectrometry (MALDI-MS). On the instrument aspect, MALDI-MS provides label-free characterization of target molecules with high specificity, but individual samples need to be processed and deposited into defined arrays on MALDI target plates at low throughput. To overcome this limitation, we integrated machine vision for automated mass spectra acquisition at randomly distributed colonies on agar. However, such instrumental design creates additional constrains on the genetic aspect for direct MS analysis of colonies. For a peptidic antibiotic plantazolicin, pathway refactoring was performed to enhance production so that sufficient analytes can be accumulated in a single colony. For another peptidic antibiotic haloduracin, inducible self-lysis of the host cells was programmed for extracellular release of target analytes produced in cytosol. To acquire high-resolution mass spectra only obtainable currently at a low molecular range (<6,000 Da), protease expression was compartmentalized in periplasm so that proteolytic processing cannot be initiated until cell lysis, as the protease activity will interfere with haloduracin biosynthesis.

Together, we found continuing crosstalk between synthetic biology and instrument/workflow development is critical to establish high-throughput technologies in microbial engineering. In addition to rapid isolation of mutants with desirable traits, accelerating build and test helps to collect large data sets of a target biological system. Development of computational tools for data processing will be the future focus to complete the design-build-test-learn cycle.

494550 Cell-Free Synthetic Biology: An Emerging Strategy to Revolutionize the Biomedical Industry

Abstract:

Advances in gene editing technology (e.g., CRISPR) and sequencing technology have endowed the biologist with unprecedented power to program cells at will. The ability of synthetic biology to engineer biological functions hold great promise for applications ranging from life science to biofuel research. For the most part, synthetic biology is still tied to the living cell. One major advantage of using the living cell is its self-reproduction. However, the daunting complexity of living cells and the barriers of cell membrane make engineering difficult, and therefore make synthetic biology face four insurmountable challenges: hard to standardize, unwieldy complexity, incompatibility, and variability. From the standpoint of synthetic biology, it is most desirable for these problems to be overcome using a systematic and standardized set of solutions.

To address these challenges, we adopt a cell-free synthetic biology strategy. Cell-free synthetic biology is emerging as a powerful interdisciplinary approach aimed to harness and expand the capabilities of natural biological systems without using intact cells. Our research lies on developing and applying cell-free systems and synthetic biology technologies to engineer DNAs, RNAs, proteins, pathways, and networks for addressing most daunting challenges in human health. To this end, our current research interests center on three distinct yet related areas, including development of fully automated medical diagnostics and therapeutics, non-natural modification of biopharmaceutical macromolecules, and assembly of bio-inspired molecular machines.

498551 Encapsulation, Protection and Programmed Release of Active Ingredients from Silicone Gel Particles for Topical Applications

Abstract:

New active pharmaceutical ingredients (APIs) are developed each year to address major challenges in skin care (e.g., for treating psoriasis, eczema and various carcinomas). However, many of these APIs fail to impact current dermatological practices due to the instability, poor solubility and/or narrow therapeutic windows of the APIs, leading to limited therapeutic benefits, or in some cases, deleterious side effects. While some have looked to microencapsulation to overcome these challenges, current technologies fail to meet all of the key demand drivers: increased stability of API, controlled release of the API, low cost, size control of the delivery platform, high payloads, regulatory compliance, and marketability. To address these needs, we have developed a highly scalable process to synthesize programmable silicone particles via sol-gel chemistry that enables direct control over the release of APIs. Our process allows us to control the size (i.e., 0.15 to 15 Î¼m) and stiffness of the particles (600 MPa to 2.50 GPa), while maintaining a narrow size distribution (i.e., C.V. <15%), even when scaling the reaction up to gallons at a time, which is critical for manufacturing and standardization. We have shown that the particles can encapsulate retinol (a proof-of-concept molecule used in the skin care industry that is highly sensitive to breakdown from light and oxygen) with efficiencies up to 85%, that our particles can protect retinol from degradation 6x longer than unencapsulated retinol, and that our particles can slowly release retinol over several hours, thus leading to less irritation (23% less irritation on human skin compared to a predicate technology). Importantly, we will discuss how modifying the composition of the particles (i.e., crosslink density) through regulating the silane monomers used in the reaction imparts the ability to tune the release rate of APIs over two orders of magnitude. These findings reveal that these particles can encapsulate, protect and controllably release retinol and a variety of hydrophobic APIs limited by narrow therapeutic windows. We will show that this technology can be used to encapsulate prescription drugs (i.e., ingenol mebutate, brand name Picato®, which is for treating actinic keratosis) with efficiencies up to 70%. We will also show that these particles have been used to encapsulate fluorescent dyes, antioxidants and pesticides, establishing this technology as a true platform technology. The methodology for particle synthesis and encapsulation is currently patent-pending and serves as the basis of Encapsio LLC, which is a startup company that spun out of Duke University in 2016.

492856 Long-Term Adaptive Evolution of Amberless *Escherichia coli* strains Reveals Selective Mutations in Translation Machinery

Abstract:

Advances in genome engineering and DNA synthesis technologies have facilitated large-scale recoding of bacterial genomes for objectives that include non-standard amino acid incorporation, multi-virus resistance, and biocontainment. An example is the recoded *Escherichia coli* strain C321.ΔA, in which all instances of UAG ("amber") stop codons were substituted for UAA and release factor 1 (RF1) was deleted. Unfortunately, C321.ΔA exhibits a significant growth defect, which is particularly severe in industrially relevant glucose minimal media (Glu-MM), where its growth rate is nearly half that of its parental strain (ECNR2). To improve strain fitness and concurrently shed light on recoded genome evolution, we performed adaptive evolution of C321.ΔA strains for more than 1,000 generations each in 14 independent replicate populations in Glu-MM. Average growth rates of evolved recoded populations significantly exceed the growth rates of the ancestral C321.ΔA and ECNR2 strains in Glu-MM and in some lineages approach the growth rate of ECNR2 in LB media. We used whole genome sequencing of two clones each from population to determine statistically significant mutations and subsequently reconstructed them in ancestral strains using multiplex automatable genome engineering (MAGE). Overall, we observe that recoding introduced a burden to *E. coli* cellular translation machinery under a wide range of industrially relevant defined media, but that point mutations in release factor 2 (RF2) known to provide increased activity on UAA codons are selected for and recover much of the fitness. Furthermore, we observe that natural selection exhibits a variety of mechanisms to correct the most detrimental off-target mutations introduced during engineering of the recoded strain, including the use of premature termination codons in essential genes. Our findings demonstrate that laboratory evolution can be applied after engineering of recoded genomes to streamline fitness recovery compared to application of additional rounds of genome engineering that may introduce further unintended mutations.

499567 Award Session: Laser-Activated Sealants for Skin Tissue Repair

Abstract:

Sutures and staples are ubiquitous in surgery, but present their own complications and cause local trauma to tissue during application. Glues, adhesives and sealants have been investigated as means for approximating wounds, but many suffer from toxicity and poor strength. Photothermal tissue sealing and repair, facilitated by near-infrared laser-responsive nanoglues, has emerged as a potential alternative to existing tissue apposition methods. Here, we demonstrate the photothermal sealing of lacerted skin and other ruptured tissues using laser activation of novel nanoglues in which, gold nanorods (GNRs) are embedded throughout a polypeptide matrix. Critical factors including laser power density, exposure time, and laser type (continuous wave or pulsed wave) were investigated in order to determine optimal treatment parameters that led to the highest recovery in sealing strength. Laser-activated sealing of full thickness lacerations in skin of live mice resulted recovery of tissue mechanical properties and integrity that were comparable to that seen with commercial sutures. Local inflammation caused by photothermal sealing approach was minimal and was compared to that seen with commercial sutures. Furthermore, laser-activated sealing was also employed to reinforce suture lines in skin, as a combination approach that can provide significant benefit for rapid sealing and healing of skin lacerations. Our results demonstrate that polypeptide-GNR nanoglues may have applications in photothermal sealing and repair of ruptured tissues and may emerge as attractive alternatives to sutures.

490528 Non-Natural Redox Cofactor-Wired Metabolic Circuits

Abstract:

Substance transformation and energy transfer are two tightly-connected aspects of cellular metabolism. While tremendous progresses have been achieved in regulating substance transformation, our capacity remains limited to control energy transfer. To facilitate an energy transfer process, it requires three basic components, energy supplying module (ESM), energy carrier (EC) and energy utilizing module (EUM). As ESM and EUM are usually linked to substance transformation, the process can be considered as an EC-wired metabolic circuit. Because natural EC, such as the reduced form of NAD(P), are shared by many ESM and EUM, it is intrinsically challenging to transfer energy selectively. We devised non-natural redox cofactor-wired subsystem to better control energy transfer. We used phosphite dehydrogenase as ESM, nicotinamide cytosine dinucleotide (NCD) and NAD as EC, and malic enzyme and lactate dehydrogenase as EUM. We engineered these ESM and EUM to use NCD. We showed that chemical energy in phosphite could be converted into charged EC and then selectively drive NCD– or NAD–dependent EUM. When such systems operated in engineered *Escherichia coli*, it reversed the flux direction of malic enzyme-catalyzed reaction and led to improved malate production by up to 38%. This synthetic system generated NCDH from phosphite to drive the metabolic pathway, which is essentially different from other known approaches linked to endogenous substance transformation reactions. Currently we are engineering other redox enzymes to use non-natural redox cofactors and constructing more synthetic pathways in different hosts.

Representative references

1. Wang L, Ji D, Liu Y, et al. Synthetic cofactor-linked metabolic circuits for selective energy transfer. *ACS Catal.* **2017**, *7*, 1977–1983.
2. Ji D, Wang L, Liu W, et al. Synthesis of NAD analogs to develop bioorthogonal redox system. *Sci. Chin. Chem.* **2013**, *56*, 296–300.
3. Ji D, Wang L, Hou S, et al. Creation of bioorthogonal redox systems depending on nicotinamide flucytosine dinucleotide. *J. Am. Chem. Soc.* **2011**, 133, 20857–20862.

491340 Interplay Between Dopant and Oxygen Vacancy in a TiO_2 Support Enhances the Oxygen Reduction Reaction

Abstract:

The activity and durability of the catalysts used for the oxygen reduction reaction (ORR) are crucial to the commercialization of polymer electrolyte membrane fuel cells (PEMFCs). Pt is still the most common catalytic material for the PEMFCs. Many factors, such as the catalyst's composition, size, shape, support, and adsorbates, are related to the catalyst's performance; such matters have been extensively studied over the past several decades in pursuit of high ORR activity, long-term stability, and reduced Pt consumption. Titanium oxide-based materials compared to traditional carbon supports have been seen as promising supports for nanosized catalysts in electrochemical applications because of their low cost, nontoxicity, and high stability in acidic and oxidative environments[1,2]. In this presentation, dual doping of anatase TiO_2, used as a Pt catalyst support, both augments resistance against the carbon corrosion that commonly occurs in oxygen reduction reaction (ORR) Pt/C catalysts and promotes the generation of oxygen vacancies that allow better electron transfer from the nanosupport to Pt, thereby facilitating the oxygen dissociation reaction. The effects of the oxygen vacancies within the doped TiO_2 nanosupport on ORR activity and stability are investigated both experimentally and by density functional theory analysis. The mass activity of Pt supported doped anatase TiO_2 is shown to be 9.1 times higher than that of a commercial standard Pt/C catalyst after hydrogen reduction[3]. The oxide-supported nanocatalysts also show improved stability against Pt sintering under during cycling, because of strong metal–support interactions (SMSI).

References

1. VTT Ho, BJ Hwang et al., "Nanostructure $Ti_{0.7}Mo_{0.3}O2$ Support Enhances Electron Transfer to Pt: Used as High Performance Catalyst for Oxygen Reduction Reaction", 2011, *J. Amer. Chem. Soc., 133,* 11716–11724.
2. CJ Pan, BJ Hwang et al., "Tuning/exploiting Strong Metal-Support Interaction (SMSI) in Heterogeneous Catalysis", **2017**, *J. Taiwan Inst. Chem. Eng.*, DOI.org/10.1016/j.jtice.2017.02.012.
3. MC Tsai, BJ Hwang et al., "Interplay between Molybdenum Dopant and Oxygen Vacancies in a TiO_2 Support Enhances the Oxygen Reduction Reaction", **2016**, *ACS Catalysis*, 6 (10), 6551–6559.

491458 The Use of Gas Pressure Profiles to Enhance Blending in Conical Hoppers and Cone-in-Cone Blenders

Abstract:

In traditional mass flow theory the velocity in hoppers and bins depends in part on the wall friction angle in the hopper and the internal friction angle of the solid. The radial stress theory pioneered by Dr. Jerry Johanson can be used to compute the expected solids velocity profiles in conical hoppers and plane flow hoppers, as well as hoppers that have internal structures such as the cone-in-cone. However, stress gradients, gas pressure gradients, and changes in the flow direction relative to the direction of gravity all induce flow patterns that are different from the traditional radial flow patterns which generally exist in normal gravity flow. The focus of this paper is to relate these external forces to the expected velocity in bins and hoppers, and then relate the velocity profiles to residence time distribution, combining the Danckwertsâ€™ theory with a modified version of the radial stress theory to quantify continuous blending in process equipment. We will present work that highlights a parametric study of various mass flow devices used as blending devices including the cone-in-cone, the Diamondback® hopper, traditional conical hopper, and offset conical hoppers. The parameters studied will include wall friction angle, gas pressure gradient, stress gradient, hopper geometry and gravitational direction. Each of these mass flow devices will have, associated with it, a characteristic residence time distribution that can be used with incoming concentration distributions to blend a bulk solid material. The question to be answered is: what, friction angles, gas pressure gradients, solid stress gradients and hopper configurations will optimize the continuous blending from these mass flow devices?

493421 Alkylation of Isobutane and Butene Using Mixed Acid As Catalyst

Abstract:

Improving isobutane solubility in H2SO4 is an effective method to enhance alkylation of isobutane and butene so as to improve the oil product quality. This research focused on alkylation of isobutane and butene with mixed acid includes sulfuric acid (H2SO4) and trifluoroacetic acid (TFA). With the mixing of TFA in H2SO4, selectivity of C8 was obviously improved due to the improved solubility of isobutane in H2SO4 and the decreased acidity. Different mixing amounts of TFA were studied, showing that 20 wt% is optimal and selectivity of alkane C8 could be improved from nearly 36.8% to 75.6% under given conditions. And the effects of stirring speed, reaction temperature, acid to hydrocarbon volume ratio (H/C), isobutane to butene molar ratio (I/O) and reaction time were also studied carefully and respectively. The mixed acid is potential to be applied in large scale industrial processes to improve the quality of alkylation.

495392 Strategies for Improving Active Chemistry, Mitigation of Coke Formation and Sustaining Selectivity to Benzene in the Catalytic Aromatization of Methane

Abstract:

The heterogeneous catalytic conversion of methane, CH_4, to a mixture of benzene and hydrogen, by the non-oxidative methane aromatization reaction (6 CH_4(g) --> C_6H_6(g) + 9H_2(g)) constitutes a direct, one step process that upgrades methane to added-value products. Since this process does not require reagents, it can be performed at the gas source at lower cost. Our group is carrying out a fundamental and systematic study of this catalytic reaction with the goal to answer some fundamental questions, which can mitigate the technological challenges of this process.

Zeolite-supported molybdenum oxide catalysts have so far been the most widely studied catalysts for the methane aromatization reaction. Rapid catalyst deactivation and low conversions are disadvantages that impede commercialization of the methane aromatization process. It is agreed that Mo carbide species, originated by exposure of the Mo oxide species to the reactant methane, are responsible for methane activation, however, the nature and amount of carbide phases formed in the induction period of the reaction are not controlled.

We have explored pretreatment conditions that are necessary to purposefully prepare zeolite-supported metal carbide phases with the best catalytic properties and relate the reaction and deactivation pathways to the structure of specific carbide phases.

495460 Preparation of the Graphite Phase Carbonic Nitrogen(g-C_3N_4) for Photocatalytically Reducing CO_2

Abstract:

The graphite phase carbon nitride (g-C_3N_4) semiconductors with different morphologies and photocatalytic performance were prepared by polycondensation method, using urea and melamine as precursors respectively. The catalytic performance of the g-C_3N_4 was measured by the reduction of CO_2 in methanol under light irradiation. X-ray diffraction (XRD)、Scanning electron microscope (SEM)、Transmission electron microscope (TEM)、Uv-vis spectrophotometry (Uv-vis)、Fourier transform infrared spectroscopy (FTIR)、Specific surface area measurements (BET) and Fluorescence spectrum (PL)were applied to analyze the influence of different precursors to the photocatalytic activity of g-C_3N_4. Results showed that g-C_3N_4 using urea as precursor had better photocatalytic activity for CO_2 reduction and the formation rate of methyl formate (MF) in methanol reached up to 512.8μmol/gcat·h, nevertheless that of the g-C_3N_4 made from melamine was only 257.3 μmol/gcat·h.

495607 The Synthesization of SAPO-11 and Its Catalytic Performance for the Alkylation of Naphthalene

Abstract:

The SAPO-11 zeolite is a member of the silico-aluminophosphate (SAPO-n) family which was first synthesized by Union Carbide Corporation. It has novel pore structures and exhibits milder acidity due to the presence of phosphorus. The pore size of SAPO-11 was more than 0.60nm, which can sieve effectively the products of the methylation of naphthalene. In this study, SAPO-11 zeolites were successfully synthesized by using three different templates (diethylamine (DEA), di-n-propylamine (DPA) and di-iso-propylamine (DIPA)) under hydrothermal conditions. The structures and acidities of SAPO-11 were characterized by means of XRD, BET and NH_3-TPD. XRD results indicated that the directing effect of different templates for AEL structure decreased in the order of DPA > DEA > DIPA. N_2 adsorption-desorption results showed that SAPO-11(DPA,1.2) possesses higher surface area and larger pore volume than other samples, which corresponds to the higher crystallinity. This is in agreement with the results of the XRD.

The alkylation of naphthalene (NAPH) with methanol (ME) on the SAPO-11 zeolite was investigated. 1,3,5-trimethylbenzene (TMB) was used as solvent. The effects of the reactant mol ratio, the reaction temperature and the weight hourly space velocity (WHSV) on the conversion of NAPH and the selectivity to 2,6-dimethylnaphthalene (2,6-DMN) was studied. The results showed that the optimum reaction conditions are as follows: 0.1 MPa, 425 , n(NAPH):n(ME):n(TMB)=1:5:3.5, WHSV 0.06 h^{-1}.

497604 Role of Active Sites in the CO_2 and Steam Gasification of Model Rdf Char

Abstract:

Refuse Derived Fuel (RDF) is a refined form of Municipal Solid Waste (MSW) derived by following certain pre-processing steps to remove fractions like glass and heavy metals from MSW. For this study, a model RDF composition was established based on the EPA data and pyrolysis results of individual components in different categories, e.g., plastics, paper and food waste. The chars were prepared in a thermogravimetric analyzer (TGA) by heating at a rate of 20 K/min up to 800 °C. Gasification of pyrolyzed feed was conducted at 800 °C using different mixtures of CO_2 and steam as the gasification reagents.

Active sites in the char correspond to different entities depending on the char constitution and pore structure. For chars with negligible inorganics, the concentration of active sites is a function of the surface area. However, in the presence of certain inorganic elements, active sites correspond to the inorganic-bound-carbon. It is well established that alkali and alkaline earth metals are more catalytically active than the other inorganics. Potassium present in the orange peel char and calcium present in the paper components were the dominant alkali and alkaline earth metals in the model RDF char, respectively. Several studies were conducted to determine the role of potassium and calcium bound carbon active sites in CO_2 and H_2O gasification.

The chemical phases of potassium and calcium species present in the model RDF char were investigated by weight – change studies in different gas mediums. Change in the chemical form of each of these inorganics was studied at different temperatures during the various stages of gasification. The activity of these inorganic-rich active sites in CO_2 and H_2O gasification was compared and analyzed. Both the type and amount of inorganics had an effect on the gasification reactivity in the two gasifying mediums. Nevertheless, potassium had a considerably greater catalytic activity than calcium. Reaction mechanisms were established to explain char gasification in both CO_2 and H_2O. Co-gasification studies with CO_2 and H_2O were conducted at different partial pressures to analyze the affinity of the active sites towards the two gasification agents. In addition to the effect of inorganics, steam gasification was influenced by hydrogen inhibition. Gasification studies were conducted at higher temperatures of 900 °C and 1000 °C to counteract the inhibition caused by hydrogen. Essentially, the role of inorganic-rich active sites was found to have a great effect on the reactivity of the model RDF char in both CO_2 and H_2O gasification which can be extended to other potassium-rich or calcium-rich char gasification systems.

498631 Computer Generated Microkinetic Mechanisms: Applications for Catalytic Combustion of Methane on Pt

Abstract:

Many processes for methane activation are performed at high temperatures and pressures, such as catalytic partial oxidation, [1] catalytic combustion, [2] and oxidative coupling.[3] Under these conditions, the heterogeneous surface kinetics can be coupled with homogeneous gas-phase chemistry. Creating a homogeneous/heterogeneous coupled mechanism that is based upon elementary reactions and is thermodynamically consistent is a major challenge in modeling and analysis of chemical reactors under industrial conditions. In most instances, either the surface or the gas-phase chemistry is treated in an ad hoc manner. Important pathways may be neglected, and the mechanism my violate fundamental principles of thermodynamics.

In this talk, we present a new tool for generating microkinetic mechanisms in heterogeneous catalysis, RMG-Cat.[4] RMG-Cat previously has been applied successfully to methane dry reforming on Nickel. In the present work, RMG-Cat was expanded to include methane oxidation on Platinum as well. Additionally, the new RMG-Cat can seamlessly handle both gas-phase and surface chemistry within a self-consistent, unified framework.

To demonstrate the new functionality of RMG-Cat, we look at recent high-temperature experimental data for the catalytic combustion of methane on a Pt gauze.2 RMG-Cat successfully generated a mechanism that includes bond fission and abstraction reactions on the catalyst surface, radical-driven chemistry in the gas phase, and adsorption/desorption kinetics for the coupling between the two regimes. The resulting mechanism ensures thermodynamic consistency by providing the rate coefficient for each reaction in a single direction and then computing the equilibrium constant from temperature-dependent free energies for gas-phase and surface intermediates.

[1] Korup, O., Goldsmith, C. F., Weinberg, G., Geske, M., Kandemir, ., Schlogl, R., Horn, R. "Catalytic partial oxidation of methane on platinum investigated by spatial reactor profiles, spatially resolved spectroscopy, and microkinetic modeling" Journal of Catalysis (2013) 297, 1-16

[2] Scwartz, H., Dong, Y., Horn, R. "Catalytic Methane Combustion on a Pt Gauze: Laser-Induced Fluorescence Spectroscopy, Species Profiles, and Simulations" Chem. Eng. Technol. (2016) 39, 2011-209

[3] Karakaya, C., Kee, R. J. "Progress in the direct catalytic conversion of methane to fuels and chemicals" Prog. Energy Comb. Sci (2016) 55, 60-97

[4] Goldsmith, C. F., West, R. H. "Automatic Generation of Microkinetic Mechanisms for Heterogeneous Catalysis" J. Phys. Chem. C. http://doi.org/10.1021/acs.jpcc.7b02133

499986 Optimizing Acid-Stable Metal-Oxides for Oxygen Evolution Reaction

Abstract:

Oxygen Evolution Reaction (OER) is the most important reaction for electro-chemical proton production from water. It occurs exclusively at high anodic potentials (> 1.23 eV vs SHE) in water solution when metal ellectrode is subject to corrosion and passivation by derived metal-oxides. In alkaline media, a number of stable and active first-row 3d-transition metal-oxides is known for this process. In acid media, only the oxides of noble metals (PtO_2 and IrO_2) are suffeciently active and stable at these conditions and no earth-abundant alternatives exits.

In our computational work, we explore how common and stable passivation oxides such as TiO_2, MoO_3, WO_3, Cr_2O_3, SnO_2 and PbO_2 can be utilized as OER catalyts in acidic media. We explore the number of strategies such as doping with OER active elements, formation of stable binary or ternary alloys from opposite sites of OER volcano and oxygen substition by other ligands.

500746 An Improved Catalyst Deactivation Protocol on Commercial FCC Catalysts for Higher Conversion of Residual Feedstock

Abstract:

It is often a challenging task to select fluid catalytic cracking (FCC) catalysts from various manufactures as it directly related to huge investment and expensive operating process. To achieve this task, a proper test protocol is mandatory to replicate the products yield pattern as close as to the respective commercial FCC units [1-5]. Few standard protocols are available in open literature and however development or improvement on existing deactivation protocols is deemed mandatory for high conversion of residual feedstock and improved yield pattern of the products for both commercial and environmental view point. This work demonstrates the improved deactivation protocol on commercial FCC catalysts for high conversion of residue feedstock and enhanced yield on products selectivity over standard deactivation method. Fresh catalyst samples from 4 different manufacturers were subjected to CMD (Cyclic Metal Deactivation) to achieve the targeted metal contamination by mixing known concentration of non-aqueous metal naphthenates (Ni, V) with TAKREER atmospheric residue. The improved deactivation study was carried out for 45 cycles under identical conditions, whereas each cycle consists of cracking, stripping and regeneration. In the improved deactivation method, cracking was carried out at 650 °C and both stripping and regeneration was carried out at 805 °C and these catalysts are subjected to hydrothermal treatment in presence of 98 % steam and 2% N_2 at 780 °C for 20 h at the end of each cycle. The same fresh catalysts were also prepared by standard deactivation protocol. All the prepared catalysts were well characterized by XRD, Nitrogen sorption analysis, TEM and acidity measurements. The catalytic evaluation studies were carried out in a fluidized bed (ACE R+MM, KT Inc, USA) unit using TAKREER atmospheric residue feedstock for the catalysts deactivated by standard method, improved method and Refinery E-cat. It was found that the catalysts prepared by improved deactivation protocol offers 4 to 7% higher conversion than standard method deactivation catalysts and E-cat for a given Cat-to-Oil ratio. In addition, this improved method deactivated catalysts exhibited better yield patterns on product selectivities over standard method and E-cat samples. The major products obtained are Propylene, C_4 Olefins, LPG, Gasoline, LCO and Coke. A thorough catalytic evaluation performance study was carried out to optimize higher conversion and better product selectivities.

Keywords: Catalyst deactivation, cracking, stripping, CMD, FCC

References

1) A.C. Psarras, E.F. Iliopoulou, K. Kostaras, A.A. Lappas, C. Pouwels, Micropor. Mesopor. Mater. **2009**, 120, 141-146.

2) A.C. Psarras, E.F. Iliopoulou, L. Nalbandian, A.A. Lappas, C. Pouwels, Catal.Today **2007**, 127(1-4), 44-53.

3) P. Oâ€™Connor, J.P.J. Verlaan, S.J. Yanik, Catal. Today **1998**, 43 305-313.

4) J.E. Otterstedt, S.B. Gevert, S.G. Jaras, P.G. Menon, Appl. Catal., **1986**, 22, 159-179.

5) H.S. Cerqueira, G. Caeiro, L. Costa, F. Ramôa Ribeiro, J. Mol. Catal. A: Chem., **2008**, 292, 1-13.

500989 Computational Design of Near Surface Alloyed Oxide for Water Splitting

Abstract:

Near-surface alloy (NSA) of bimetallic or multimetallic system has been intensively studied for many catalytic reactions. A close analogy can also be drawn to the oxide materials. The most important scientist question we are going to address is how the surface catalytic properties, *i.e.* binding energy of surface species, reaction barrier, are correlated with the composition and ordering. To make this challenge manageable, we will focus on the effect of varying B-site metal. Two sets of calculations will be performed for 1) oxygen evolution reaction (OER) activity and 2) oxygen vacancy formation energy. Our calculations show that surface reactivity of NSA oxide can be widely tuned by the dopant and its ordering in the material. Our study leads to a new design scheme of efficient oxide catalyst. The correlation between the ionic structure and electronic structure serves as the guideline of the followed research.

484528 Photocatalytical Degradation of Congo Red (CR) Dye By Nano Titanium Dioxide Coated Glass Bead Under UV Light

Abstract:

Titanium dioxide (TiO_2) synthesized on nano-scale was coated firmly on soda lime glass beads used as the catalyst support. The application of the sol-gel method resulted in the formation of fine and nano-sized TiO_2 (n-TiO_2) particles uniformly dispersed and anchored on the soda lime glass beads by dry mixing method. A multi-column immobilized glass reactor was designed to evaluate the performance of the n-TiO_2 coated glass bead by subjecting Congo-red (CR) dye to photo-catalytic degradation reactor under U.V. light. The effect of calcinations temperature on the TiO_2 Anatase and Rutile phase transformation; preparation of Glass surface (etching) and effect of glass support on the catalyst performance; the CR dye concentration used and pH of solution; the dye degradation in dark condition and the impact of U.V light exposure; the duration of U.V. light irradiation and its impact on the dye degradation; the catalyst (n-TiO_2) loading used with the catalyst regeneration were some of the parameters studied to evaluate and optimize the performance of the n-TiO_2 coated glass bead. It was observed that increased calcination temperature resulted in the phase transformation of anatase to rutile phase with complete rutile phase transformation at 800° C calcination temperature. The U.V. spectroscopy studies signified that photo-catalytic degradation is optimized in acidic zone as lower pH favored the formation of more soluble CR ions. A reduction in the n-TiO_2 desecration was observed with profound catalyst regeneration cycle up to five times for the same catalyst used due to the employed glass bead support. The U.V. light has optimized the CR dye degradation with 66% photo-catalytic dye degradation in just 10 minutes of the U.V. exposure. The TiO_2 spherical structure and nano-sized particles formation was confirmed by SEM analysis and recorded in the range of 130 to 270 nm. The elemental analysis was evaluated by EDX analysis and XRD analysis detected no characterization peaks other than TiO_2, signifying the synthesis of pure TiO_2 in the presented study. The surface area was measured using BET technique and was found 26.18 and 5.32 m^2/g for Titania samples prepared at 600° C and 800° C calcinations temperature respectively. U.V. spectrophotometer analysis appraised the CR dye degradation.

500003 Highly Efficient Photocatalytic Degradation ofÂ Organic Pollutants By TiO $_2$Â -PDMS Composite Sponge

Abstract:

In this study, a composite of a polydimethylsiloxane (PDMS) sponge with TiO_2 has been successfully prepared using the very simple sugar-template method. PDMS sponge is a suitable substrate for selective absorption of organic pollutants from water and can be elastically deformed into any shape, and can repeatedly be compressed in air or liquids without collapsing enabling excellent recyclability. Additionally, TiO_2-PDMS composite sponge shows nice photocatalytic activity thus promoting potential in environmental applications. The photocatalytic degradation of Rhodamine B (RhB) dye under the solar light in the presence of composite sponge was examined to assess their efficiency towards degradation and removal of organic pollutants. The proposed composite is shown to first remove a significant amount of the organic dyes (Rh B) from solution by absorption (70% removal) even when incubated in the dark. Upon exposure to solar light, most of the remaining dyes in solution was photo-catalytically degraded on the surface of sponge (resulting in total 90% removal). More interestingly, these composite sponges can be recycled by exposing them to the solar light in the presence of clean water promoting the degradation of the dye absorbed in the sponge. The degraded products can be mostly removed by washing and squeezing the composite sponge. TiO_2-PDMS composite sponge exhibited enhanced performance towards degradation and removal of organic pollutants, e.g., RhB, in comparison with TiO_2 nanocatalyst(<100nm) alone. The enhanced performance is attributed to the combined effect of strong absorption of organic molecules and the photocatalytic degradation. Application of TiO_2 sponge for photodegradation also eliminates the need for separating catalyst from solution. The excellent performance and the recyclability of the PDMS sponge@TiO_2 suggest its promising applications in water treatment and environmental remediation.

490560 Kinetics of the Water Gas Shift over a Cu-Based Catalyst for Pyrolysis Vapor Upgrading

Abstract:

Kinetics of the Water Gas Shift over a Cu-based Catalyst for Pyrolysis Vapor Upgrading

Ross Houston[a], Nicole Labbé[b] and Nourredine Abdoulmoumine[a,b]

[a]Biosystems Engineering and Soil Science Department, University of Tennessee, Knoxville, TN 37996

[b]Center for Renewable Carbon, University of Tennessee, Knoxville, TN 37996

Abstract

Biomass pyrolysis vapors have limited applications due chiefly to its high oxygen content. While many approaches are explored to reduce oxygen content in these vapors, hydrodeoxygenation has emerged as the most effective method to minimize carbon loss by removing oxygen as water. The goal of this study is to investigate hydrogen production for deoxygenation via the water gas shift reaction using carbon monoxide (CO), one of the major non-condensable gases of pyrolysis and steam, a by-product of hydrodeoxygenation.

Prior to the catalytic experiments, the Cu-based catalyst was characterized by physisorption for surface area and pore volume and elemental composition. Water gas shift experiments were carried out over the size reduced catalyst (0.425-0.600 mm) on a bench-scale catalytic packed bed reactor (PBR) in a CO-lean environment (70 vol. % water, 20 vol. % He, and 10 vol. % CO). The experimental temperatures varied from 150°C to 400°C at 50°C interval and three different weight hourly space velocities (1221, 2036 and 6107 cm^3/g-min). A full-factorial experiment with three replicates was developed, and the conversion of CO was analyzed using a gas chromatograph (GC) equipped with a thermal conductive detector (TCD).

The surface area and pore volume of the catalyst were 62.174 m^2/g and 0.218 cm^3/g, respectively and its elemental composition is as follows: 50% copper oxide, 25% zinc oxide, <25% aluminum oxide, and a balance of graphite. Preliminary results show that CO conversion increases with increasing temperature and catalyst weight to maximum CO conversion of 85-90 % at 300°C and 350°C.

483129 Superwetting Electrodes for Gas-Involved Electrocatalysis

Abstract:

Superwetting electrodes for Gas-involved Electrocatalysis

Wenwen Xu, Zhiyi Lu, Yingjie Li, Xiaoming Sun*

State Key Laboratory of Chemical Resource Engineering, Beijing University of Chemical Technology, Beijing 100029, China

ABSTRACT

Electrochemical gas-involved reactions, including gas-evolution reaction and gas-consumption reaction, are essential parts in current energy conversion processes and industries. Although the exploration of the highly active catalysts has been very mature, less attention was paid on the gas management during the gas-involved reactions. Inspired from bio-inspired materials, scientists find that bio-mimicked electrodes with superwetting property will influence the gas transportation process during the electrochemical reactions. Our group fortunately found that the interface behavior of electrode could be tuned by surface architecture construction, for example, transferring from aerophobic to superaerophobic by engineering a series of superwetting micro-/nanostructured electrodes, eg. MoS_2. Cu nanoarray and Pt pine-like films [1-4]; transferring from aerophibic to superaerophibic by poly(tetrafluoroethylene)(PTFE) modifying, eg. CoNCNT@CFP.[5] For gas-evolution reaction, constructing nanostructured superaerophobic electrodes is effective to improve the performance by enlarging the bubble contact angle and reducing the bubble adhesion force with the surface of the electrode, thus insuring smooth leaving of the gas products. As to the gas consumption reactions, the superaerophibic electrodes are able to improve the performance by providing an unblocked gas diffusion pathway and a smooth electron transport. Therefore, construction of superwetting surface (auperserophobic for gas evolution reaction and superaerophibic for gas consumption reaction) can boost the performances of the electrodes by managing the surface bubbles.

References

[1] Z. Lu, W. Zhu, X. Yu, H. Zhang, Y. Li, X. Sun, X. Wang, H. Wang, J. Wang, J. Luo, X. Lei, L. Jiang. *Adv. Mater.* **26**(2014) 2683-2687.

[2] Y. Li, H. Zhang, T. Xu, Z. Lu, X. Wu, P. Wan, X. Sun, L. Jiang. *Adv. Funct. Mater.* **25**(2015) 1737-1744.

[3] Z. Lu,M. Sun, T. Xu, Y. Li, W. Xu, Z. Chang, Y. Ding, X. Sun, L. Jiang. *Adv. Mater.* **27**(2015) 2361-2366.

[4] X. Liu, Z. Chang, L. Luo, T. Xu, X. Lei, J. Liu, X. Sun. *Chem. Mater.*, **26**(2014) 1889-1895.

[5] Z. Lu, W. Xu, J. Ma, Y. Li, X. Sun, L. Jiang. *Adv. Mater*, **28**(2016)155-7161.

Short BIOGRAPHY

Prof. Xiaoming Sun

State Key Laboratory of Chemical Resource Engineering, Beijing University of Chemical Technology

Email: sunxm@mail.buct.edu.cn

Professor Xiaoming Sun gained his B.S. degree and Ph.D. in Department of Chemistry, Tsinghua University in 2000 and 2005, respectively. After postdoctoral work at Stanford University, he joined State Key Laboratory of Chemical Resource Engineering, Beijing University of Chemical Technology at 2008. His main research interests focus on wetting regulation of electrode interface for electrocatalysis, separation and assembly of inorganic nanostructures, synthesis and separation of carbon nanomaterials and their composites, and structure control and opto-/electro-property investigations of oxide nanoarrays. Recently, Prof. Sun has made original and innovative contributions to the discovery and exploration of gas-involved behaviour of superwetting electrodes.

Professor Xiaoming Sun makes great contributions to nanomaterial and nanotechnology for his outstanding creativity and productivity. He has authored and co-authored 68 journal articles (eg. ***J. Am. Chem. Soc., Angew. Chem. Int. Ed., Adv. Mater., ACS Nano.***). His entire publications have been cited for over 14717 times [an updated report from google scholars: https://scholar.google.com/citations?user=_G0EETMAAAAJ&hl=zh-CN&oi=ao]. The H-index of his publications is 49 per SCI data base.

492076 Anodic Aluminum Oxide Supported Cu-Zn Catalyst for Steam Reforming of Methanol

Abstract:

Anodic Aluminum Oxide Supported Cu-Zn Catalyst for Steam Reforming of Methanol

Jung Hyeon Kim, Young Shin Jang, Dong Hyun Kim*

Department of Chemical Engineering, Kyungpook National University, Daegu, Korea

*E-mail: dhkim@knu.ac.kr

Recently, hydrogen production from steam reforming of methanol (SRM) has been proposed for supplying hydrogen to fuel cells[1]. For this purpose, metal monolith reactors have been studied since such reactors have a high thermal conductivity, good mechanical strength and low pressure drop[2]. However, the reactors suffered from peeling of the catalyst layer due to the weak bonding between the metal substrate and the catalyst layer. The catalyst layer prepared on the porous support grown on aluminum metal surface has been shown to be resistant to peeling and as active as conventional particle catalysts [3].

In this study, an anodic aluminum oxide (AAO) support was prepared on an Al plate by anodizing in oxalic acid electrolyte solution. The AAO was treated with warm oxalic acid solution to widen the pores and then treated with hot water to increase the surface area. Thus prepared support was immersed into a Cu and Zn nitrate solution, dried and then calcined. These impregnation steps were repeated to increase the loading of the active components.

It was observed that the time of immersion of AAO support into the nitrate solution (5 min or 1 h) had little effect on the performance of the catalyst, but the activity increased noticeably with repetition of the impregnation up to 4 times. A direct relation between the Cu metal area and the catalytic activity was observed. Also in this study, the reaction rate of methanol steam reforming over the AAO supported catalyst was also obtained for design of the reformer.

REFERENCES

[1] Daniel R. Palo, Robert A. Dagle, Jamie D. Holladay, Chem. Rev., 107 (2007) 3992.

[2] Achim K. Heibel, Charles M. Sorensen, Ind. Eng. Chem. Res., 43 (2004) 4602.

[3] E.L. Reddy, H.C. Lee, D.H. Kim, Int. J. Hydrogen energy, 40 (2015) 2509.

496139 Mixing and Interaction of Two Reactive Droplets in a Powder Bed

Abstract:

Among all 3D printing technologies, binder-jetting is famous for printing multiple materials simultaneously with multi-nozzles, and thus is convenient to integrate several elements in a single manufacturing process. Standard binder-jetting procedure prints glue inks on powder layer to bind powder particles together. By printing epoxy inks with curing agents (or so-called AB glues) from separate nozzles simultaneously on powder, the printed objects can yield in high mechanical strength after curing. Therefore, the distribution of the two inks in powder and the reaction extent are critical to the printing resolution and mechanical properties.

Because it is technically difficult to observe fluid flows in powder experimentally in the binder-jetting process, a computational fluid dynamics (CFD) approach is used in this study to explore capillary flows in between micron-size particles. A CFD model is developed to describe the mixing and reaction of two chemical components in the capillary flow inside the powder bed using a commercial software, Flow-3D. Two droplets, one carrying expoxy ink and the other containing curing agents, fall onto and penetrate into a powder bed of micron-size particles. After droplet impact and spread over the particles, flows driven by capillary force causes the liquid gradually penetrate into the void space in the powder bed. The capillary flow of the two adjacent droplets leads to mixing and reaction between the two chemical species carried by separate droplets. The degree of mixing and reaction extent of the AB glue will be optimized by alternating droplet velocity, the relative positions of droplets, and the hydrophilicity and sizes of powder particles. In summary, this study will provide a general guideline to optimize the mixing of inks in powders, and can be further apply to improve the printing quality of printed objects from binder-jetting.

498540 Removal of Color By Eletrocoagulation Method - Preliminary Results in Textile Dyes

Abstract:

This work is aimed to the study of the electrocoagulation process (EC) applied to the removal of two characteristic dyes from the textile industry: Danix Marinho S-2G 200% (Disperse, Dye A), Remazol Brilliant Blue Reactive (Reactive, Dye B). A lab-scale electro-coagulation reactor consisting of a set of six equidistant and parallel aluminum plates within a 2,5 L container was used whose performance change according to two important variables: effluent initial pH (Dye A: 5-7; B: 7-9) and electric current (6-8 A). To obtain a significant effect in the process, it was also evaluated by Response Surface Methodology (RSM), with a 2^2 factorial design. The response variable to evaluate the efficiency of the process was the percentage of color removal. An important influence of the initial pH on the two dyes was observed, showing that in the ranges of study, higher initial values achieve to higher color removal results in both dyes. In relation to the intensity of current it was evidenced that the highest value, the best as color removal. Resulting in the best reactor performance was achieved at 8 A and 7 initial pH for the dye A, while 8 A and 9 initial pH, was achieved for the dye B. Under these conditions, more than 99% of color was removed, for both dyes. The EC treatment to study Danix Marinho S-2G 200% and Remazol Brilliant Blue Reactive showed favorable responses. Thus, indicating clearly that the method of electrocoagulation is very promising alternative for textile industrial applications.

499998 A Win-Win Strategy for Chemical Plant Shutdown: Integrating Economic and Environmental Objectives

Abstract:

Ground-level ozone is a pervasive air pollutant, which can be potentially aggravated by flaring emissions from olefin plant shutdown operations. Although, flaring is crucial to chemical plant safety, the excessive flaring, for instance the intensive flaring from plant shutdown operations, emits huge amounts of ozone pollution precursors (VOCs and NOx). This will cause tremendous industrial material loss and environmental damage; meanwhile result in tremendous economic loss and potential adverse impact on the regional air quality. Thus, the plant shutdown flare emissions should be minimized at any possibility. In this paper, plant-wide dynamic simulations for flare emissions with regional air-quality modeling are coupled together to quantify the air-quality impact and roughly evaluate the cost savings. It proposed a systematic methodology on both economy and air-quality conscious study for shutdown operations of an olefin plant. Through studies of original and optimal cases, it shows that the original plant shutdown operation procedure has a serious effect on the regional air quality and tremendous loss of raw material; however, the optimal shutdown operation procedure would significantly reduce such ozone impact together with the saving of tremendous raw material. This study can provide valuable and quantitative supports for decision makings involving multiple stakeholders, including environmental agencies, regional plants, and local communities.

486219 Three Dimensional Photovoltaic Microyarns with Efficient Optoelectronic Performance and Enhanced Exciton-Hole Pair Separation

Abstract:

The solar energy conversion efficiency of flat dye-sensitized solar cells (DSC) could reach up to 12.1% using platinized transparent conductive oxide (TCO) positive electrodes/support materials. Highly interaligned carbon nanotubes (CNTs) demonstrate excellent mechanical, electronic, and catalytic properties, which exhibit high surface area with enhanced optoelectronic and electronic performance. A yarn-shaped and high efficiency DSC has been developed using coaxial, inter-alighned, ultrastrong and highly conductive CNT yarns (CNY) in both electrtodes with solid state electrolyte. The hydrothermally crystalized TiO_2 film along with a thin (50-100nm) underlayer interfaced with the high surface area CNY has been grafted around the working electrode (WE), which has been found capable in wide range of structural flexibility (30-330^0) without changing photovoltaic characteristics. The maximum photoconvesion efficiency achieved is 6.15%, which is independent of incedent ilumination and cell shape or posision. Due to the yarn shape, the CNY cells can easily be woven to engineering textile, which may be used in wide field of practical application, such as portable/embedable intrisic photosensor, 3D photovoltaics, clothing with integrated photovoltaic yarns *etc*. This innovative solid state, 3D and low cost CNY based photovolatic yarn opens a new era of high efficiency 3D flexible solar cells.

Kinetic Study of Thermal Degradation of 2-Amino-2-Methyl-1-Propanol to Cyclic 4,4-Dimethyl-1,3-Oxazolidin-2-One

Naser S. Matin, Jesse Thompson, Femke M. Onneweer, and Kunlei Liu
Center for Applied Energy Research, University of Kentucky, 2540 Research Park Drive, Lexington, KY 40511, United States

Abstract: Through initial rate data, the apparent (power law-PL) and mechanistic rate equations for the thermal degradation of 2-amino-2-methyl-1-propanol (AMP) to 4,4-Dimethyl-1,3-oxazolidin-2-one (DMOZD) as function of amine and CO_2 concentration in the solution were extracted. The reaction kinetics parameters including activation energies and pre-exponential factors for both approaches were defined.[1,2]

Experimental Conditions: The rate experiments were carried out at 120, 135 and 150 °C, in the aqueous solutions of 1.12 - 3.36 M, AMP and CO_2 loadings of 0.17- 0.7, mol_{CO2}/mol_{AMP}.

Method of Analysis: IC and HPLC/MS have been used for AMP and DMOZD, analysis, respectively.

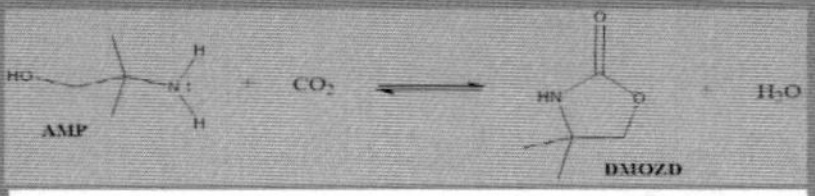

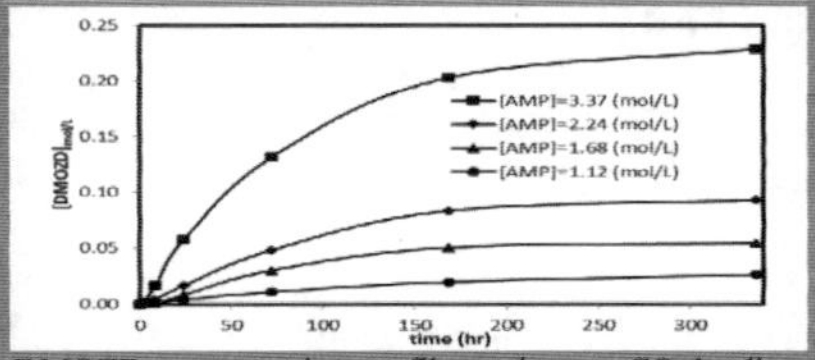

DMOZD concentration profile vs time, at CO_2 loading of around 0.7 (±0.04) mol_{CO2}/mol_{AMP}, and 135° C.

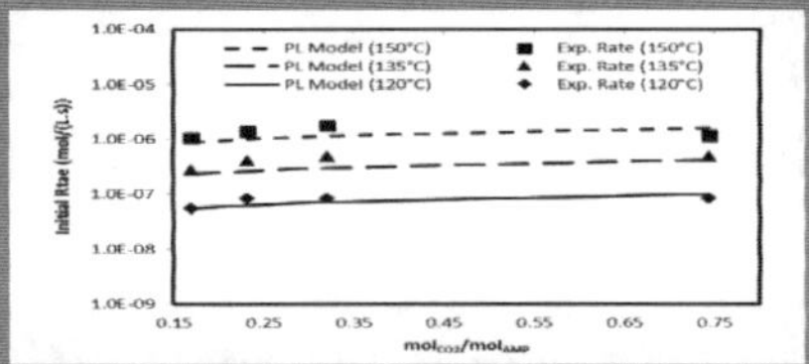

The power law (PL) model prediction compared to experimental reaction rate data, the initial AMP concentration is 3.36 mol/L.

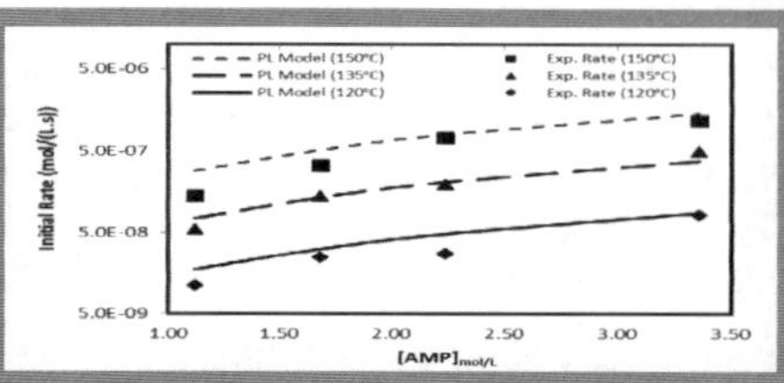

The power law (PL) model prediction compared to experimental reaction rate data, CO_2 loading around 0.7(± 0.04) mol_{CO2}/mol_{AMP}.

$$Rate\ (mol.L^{-1}.s^{-1}) = 1.69\times10^{9}e^{-129.12/RT}[AMP]^{1.18}[CO_2]^{0.45}$$

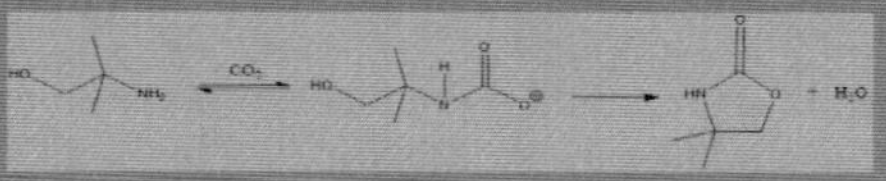

Mechanism (I)

$(n)AMP + (m)CO_2 \overset{K_1}{\leftrightarrow} carb.$

$carb. \overset{K_2}{\rightarrow} DMOZD$

Mechanism (II)

$(n)AMP + (m)CO_2 \overset{K_1}{\leftrightarrow} carb.$

$carb. \overset{K_2}{\leftrightarrow} carb\text{-}prime$

$carb\text{-}prime \overset{K_3}{\rightarrow} DMOZD$

Mechanism (III)

$AMP+CO_2 \overset{K_1}{\leftrightarrow} zwitterion$

$zwitterion + AMP \overset{k_2}{\rightarrow} carbamate$

$carbamate \overset{k_3}{\rightarrow} DMOZD$

$$Rate_{Mech_{(I)}} = \frac{k_1k_2}{k_{-1}+k_2}[AMP]^n[CO_2]^m$$

	Mechanism I (w. adjust. react. order)
$E_{a1}(kJ.mol^{-1})$	109.03 (± 25.10)
$E_{a-1}(kJ.mol^{-1})$	94.99 (± 24.65)
$E_{a2}(kJ.mol^{-1})$	104.40 (± 23.11)
$k_{01}(lit.mol^{-1}.s^{-1})$	1.63×10^8 ($\pm0.47\times10^8$)
$k_{0-1}(s^{-1})$	1.11×10^8 ($\pm0.19\times10^8$)
$k_{02}(s^{-1})$	6.86×10^7 ($\pm2.78\times10^7$)
m	0.35 (±0.03)
n	1.4 (±0.14)

Results: Power law and mechanistic approach both present good prediction from experimental data. While mechanisms (II) and (III) have reasonable estimations, mechanism (I) with adjustable reaction order, has best prediction from experimental data.

However, results from mechanism (III) provide the activation energy of 23.91 (kJ/mol) for the forward reaction of AMP with CO_2 to form a zwitterion, and 24.47 (kJ/mol) for the reverse reaction, i.e. zwitterion conversion to AMP and CO_2, both values are fairly consistent with literature reported values.

According to the mechanism (III) zwitterion shows higher tendency to decay to its predecessor reactants, $k_1/k_{-1} \cong 0.03$.

Conclusion: The temperature dependent reaction rate constant for the power law rate equation varies from 16×10^{-9} $(mol/L)^{(1-a-b)}.s^{-1}$ at 120 °C to 26×10^{-8} $(mol/L)^{(1-a-b)}.s^{-1}$ at 150 °C. In the case of $k_2 << k_{-1}$, the reaction rate model extracted from mechanism (I), can simply reduce to power law equation. Considering different mechanisms and non-integer reaction orders for reactants demonstrates that the actual reaction mechanism may be more complex.

Acknowledgments: The Carbon Management Research Group (CMRG) members, including Duke Energy, Electric Power Research Institute (EPRI), Kentucky Department of Energy Development and Independence (KY-DEDI), and Louisville Gas & Electric (LG&E) and Kentucky Utilities (KU) for their financial support.

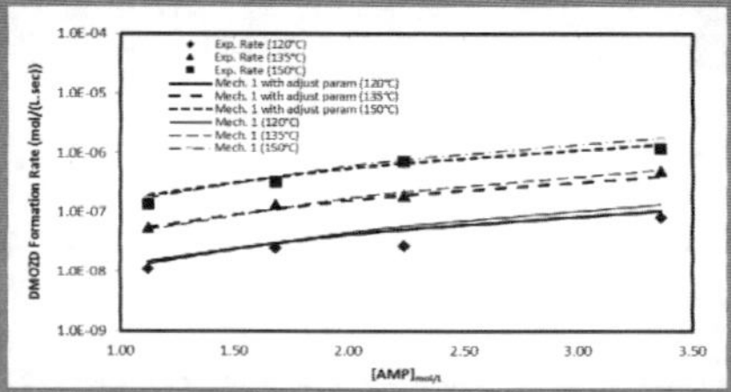

Mechanism (I) prediction compared to experimental data, CO_2 loading around 0.7(± 0.04) mol_{CO2}/mol_{AMP}.

(1) Matin, et al., Ind. Eng. Chem. Res., 55 (2016), 9586.
(2) Matin, et al., Ind. Eng. Chem. Res., 56 (2017), 9437.

Center for Applied Energy Research

caer.uky.edu

1

501497 Energy Integrated Natural Gas Liquid Recovery Process By Introducing Vapor Recompressed Internally Driven Reboiler

Abstract:

Abstract

Processing of Natural gas, liquid recovery is a complex industrial process designed to clean the raw natural gas by separating the impurities and various non-methane hydrocarbons, which are value added chemicals need to separate out to fulfil the dry natural gas as pipeline quality to satisfy the end user requirement for their process. This step has to be carried out at an offshore plant location, where the space and sources of energy are limited which plays a vital role in the process economics. In this study considered NGL recovery in energy integrated configuration to improve the energy efficiency and process economics. Simulated the NGL recovery from pre-treated Natura gas by considering a propane C3MR cycle as a refrigerant cycle separated out of the propane, LPG and C5+ in two column distillation units represented in Aspen Hysys environment and considered this case study as a base case to analyze the explored different energy integrated configuration for NGL recovery. Explored the different energy integrated schemes to enhance the energy efficiency of the NGL recovery with compared to the conventional NGL recovery. Performed the different case studies and analyses to select the best case study based on energy efficiency and economics.

Specific report for discussion paper "Developing a modern renewable fuel standard for gasoline in Ontario"

Ammonia (NH_3) as a Potential Transportation Solution for Ontario

University of Ontario Institute of Technology

Prof. Dr. Ibrahim Dincer

Yusuf Bicer

Hydrofuel Inc.

Greg Vezina (Chairman and CEO)

Frank Raso

February 21, 2017

INTRODUCTION

Ammonia as a Carbon-Free Fuel for Use in the Transportation Sector

The action plan lays out the specific commitments Ontario is making to meet its 15% overall greenhouse gas emissions reduction target by 2020. Emissions from total passenger transportation (cars, trucks, bus, rail and domestic aviation) have grown almost 15% since 1990, to 36 million tonnes of CO_2e, approximately 66% of Ontario's 2014 transportation emissions. This growth was driven by an increase in vehicle-kilometres travelled as well as a shift in the composition of the fleet from cars to sport-utility vehicles, pick-ups and minivans where the specific contributions of the vehicle types are shown in Fig 1.

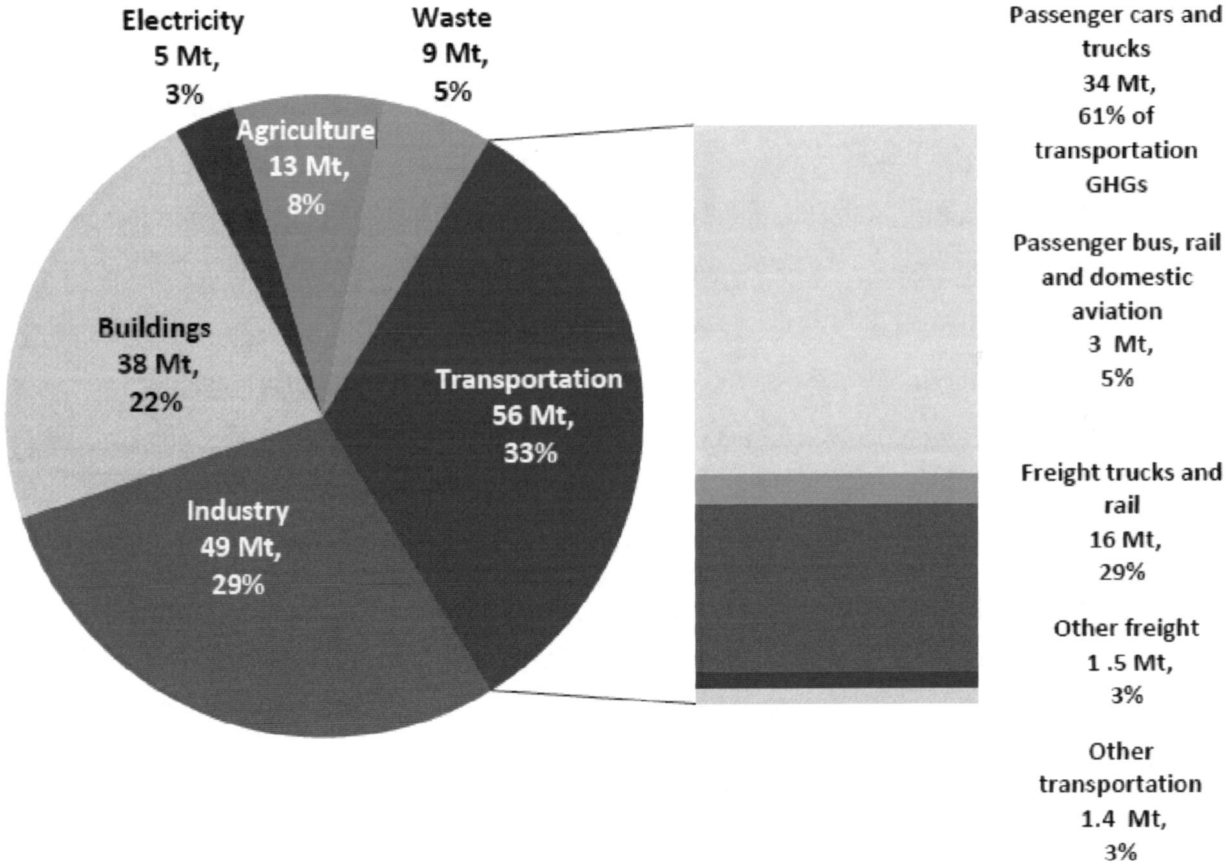

Fig. 1. Ontario's greenhouse gas emissions in 2014 (data from Ref. 1).

Here, it is important to note that under Environment and Climate Change Canada's economic sector categorization, most off-road transportation emissions are allocated to their host economic sectors and consequently are not included under transportation in Fig. 1. For example, emissions from diesel combustion in farm equipment are categorized under Agriculture. Therefore, in reality GHG emissions caused by the transportation vehicles are higher than 33% in Ontario.

Additionally, emissions from total freight transportation (trucks, rail and other) have increased more drastically over the period, rising 85% since 1990, to almost 18 million tonnes of CO_2e (approximately one third of Ontario's current transportation emissions). This was driven by a significant increase in the use of diesel-fuelled heavy-duty trucks, with additional kilometres travelled offsetting improvements in efficiency [1].

This brief report tries to address the following points:

1. Targets and blending requirements:
    a. Ontario's has existing content requirements for ethanol in gasoline. What minimum level of ethanol blending and GHG performance would help support the objectives of the RFS?
    b. Given Ontario's GHG reduction targets for 2030 and 2050, what factors should be considered in setting RFS targets post-2020?
2. Flexibility mechanisms:
    a. Should activities to lower the carbon intensity of other conventional transportation fuels be eligible for compliance purposes?
    b. Should investments in low-carbon transportation projects also be eligible for compliance purposes? If yes, what types of projects?
3. Assessing lifecycle emissions
    a. Should an RFS consider impacts from indirect land-use changes (ILUC),7 even though science in this area continues to evolve? If so, how?
4. Transparency:
    a. What measures can be taken to increase transparency and support business decision making under an RFS (e.g. an information registry, bulletins, guidance material)?
5. Others:
    a. What other considerations should be included in the discussion?

AMMONIA FACTS

Ammonia is one of the largest synthesized industrial chemical in the world having over 200 million tonne per production per year.

Ammonia (NH_3):

- consists of one nitrogen atom from air separation and three hydrogen atoms from any conventional or renewable resources.
- is the second largest synthesized industrial chemical in the world.
- is a significant hydrogen carrier and transportation fuel that does not contain any carbon atoms and has a high hydrogen ratio.
- contains about 48% more hydrogen by volume than liquefied hydrogen.
- does not emit direct greenhouse gas emission during utilization
- can be used as solid and/or liquid for many purposes.
- can be stored and transported under relatively lower pressures.
- can be produced from various type of resources ranging from oil sands to renewables.
- is a suitable fuel to be transferred using steel pipelines with minor modifications which are currently used for natural gas and oil.
- can be used in all types of combustion engines, gas turbines, burners as a sustainable fuel with only small modifications and directly in fuel cells which is a very important advantage compared to other type of fuels.
- brings a non-centralized power generation via fuel cells, stationary generators, furnaces/boilers and enables smart grid applications.
- can be used as a refrigerant for cooling in the car.

WHAT ARE THE USES OF AMMONIA?

Ammonia is considered a possible working fluid for thermodynamic cycles, working for refrigeration, heating, power or any mixture of those can be coupled with internal combustion engines, and using exhaust gasses to drive automotive absorption refrigeration system.

Ammonia has been recognized and employed as a leading refrigerant in the industrial sector due to its outstanding thermal properties, zero ozone depletion and global warming potential (GWP). Ammonia has the highest refrigerating effect per unit mass compared to all the refrigerants being used including the halocarbons. The remarkable advantages of ammonia over R-134a could be lower overall operating costs of ammonia systems, the flexibility in meeting complex and several refrigeration needs, and lower initial costs for numerous applications. Ammonia has better heat transfer properties than most of chemical refrigerants and consequently allow for the use of equipment with a smaller heat transfer area. Thereby plant construction cost will be lower. But as these properties also benefit the thermodynamic efficiency in the system, it also reduces the operating costs of the system. In many countries the cost of ammonia per mass is considerably lower than the cost of HFCs. This kind of advantage is even multiplied by the fact that ammonia has a lower density in liquid phase. Modern ammonia systems are fully contained closed-loop systems with fully integrated controls, which regulate pressures throughout the system. Ammonia is used as refrigerant highly in the refrigeration structures of food industry like dairies, ice creams plants, frozen food production plants, cold storage warehouses, processors of fish, poultry and meat and a number of other uses.

It is also stimulating to note that NH_3 is a reduction agent for the NOx typically current in combustion releases. The reaction of NOx with ammonia over catalysts produces only steam and nitrogen. An average car needs only approximately 30 ml of NH_3 per 100 km to neutralize any NOx emissions. If the vehicles run with NH_3 as fuel, this amount is unimportant with respect to the fuel tank volume.

Ammonia is used as fertilizer in the agriculture. It is also converted into urea by reacting with CO_2. The majority of growth in ammonia usage is expected to be for industrial uses and the production of fertilizer products.

It is also worth to examine the option to cool the engine with ammonia that can act as a refrigerant while it is heated to the temperature at which it is fed to the power producer (ICE or fuel cell). Optionally, the cooling outcome of ammonia, i.e., its high latent heat of evaporation, may be used to harvest some air conditioning onboard.

IS AMMONIA REALLY A FUEL?

Ammonia as a sustainable fuel can be used in all types of combustion engines, gas turbines, and burners with only small modifications and directly in fuel cells. Ammonia was initially used as a fuel for buses in Belgium in 1940s [2]. Many studies have already been performed and many applications have been implemented so far. A prototype unit for combustion which enabled liquid kerosene and gaseous ammonia to be fed, and ammonia was combusted in a gas-turbine unit. Further studies have been performed by various researchers which have proven the practicality of using ammonia as fuel [3-10]. Numerical studies of combustion characteristics of ammonia as a renewable fuel have been conducted. Ammonia can also be used a fuel blending option for current gasoline and diesel engines. Combustion and emissions characteristics of compression-ignition engine using dual ammonia-diesel fuel have been performed. Performance enhancement of ammonia-fueled engine by using dissociation catalyst has been studied. These are just a few examples to show the current progress in the ammonia utilization options in transportation applications.

IS AMMONIA A SUITABLE FOR TRANSPORTATION SECTOR?

The storage and delivery infrastructure of ammonia is similar to liquefied petroleum gas (LPG) process. Under medium pressures (5-15 bar), both of the substances are in liquid form which brings the significant advantage because of storage benefits. Today, vehicles running with propane are mostly accepted and used by the public since their on-board storage is possible and it is a good example for ammonia fueled vehicle opportunities since the storage and risk characteristics of both substances are similar to each other. An ammonia pipeline from the Gulf of Mexico to Minnesota and with divisions to Ohio and Texas has served the ammonia industry for many years. It indicates that there is a working ammonia pipeline transportation which can be spread overall the world. The potential of ammonia usage in many applications will be dependent on the availability of ammonia in the cities. Ammonia is a suitable substance to be transferred using steel pipelines with minor modifications which are currently used for natural gas and oil. In this way, the problem of availability of ammonia can be eliminated.

HOW CAN AMMONIA BE USED IN TRANSPORTATION?

Ammonia has significant potential as an alternative fuel to further the sustainable development of transportation sector. A few of the following alternatives are shown in Fig. 2 for direct ammonia usage in transportation applications.

Currently, the majority of the locomotive fleet is made up of diesel-electric locomotives, operating with either two-stroke or four-stroke prime mover diesel engines that is coupled to an electric generator. Application of ammonia fuel for internal combustion engine (ICE) with the alternative locomotive configurations direct feed, or a combination of direct feed and decomposition subcategory options will bring more sustainable solutions. Additionally, fuel cell driven vehicles and locomotives may contribute to solve the associated matters of urban air superiority and national energy security influencing the rail and transportation sector.

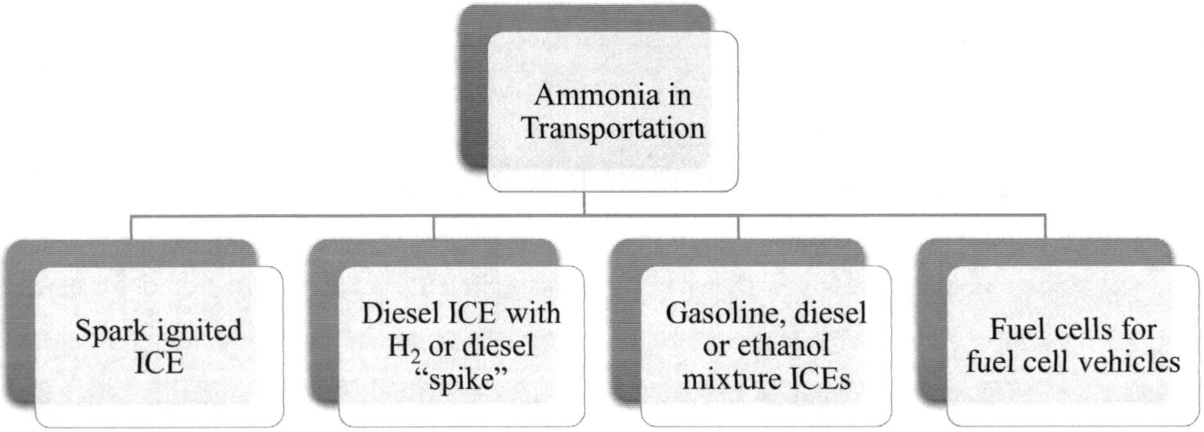

Fig. 2. Ammonia utilization options in transportation sector

IS AMMONIA A CLEAN FUEL?

Compared to gasoline vehicles, ammonia-fueled vehicles do not produce direct CO_2 emission during operation. Since ammonia produces mainly water and nitrogen on combustion, replacing a part of conventional fuel with ammonia will have a large effect in reducing carbon dioxide emissions.

WHAT FACTORS SHOULD BE CONSIDERED IN SETTING RFS TARGETS POST-2020?

- Environmental impact (including the indirect land-use changes)
- Cost
- Availability of clean production routes (e.g. solar, hydropower, wind)
- Blending properties (e.g. combustion characteristics)
- Fuel production life cycle emissions
- Fuel storage and transport infrastructure
- Integrability
- Multi-purpose usage (e.g. power, cooling, heating)

WHICH ENVIRONMENTAL IMPACTS SHOULD BE CONSIDERED?

A life cycle is the set of phases of a product or service system, from the extraction of natural resources to last removal. Overall environmental impact of any process is not complete if only operation is considered, all the life steps from resource extraction to disposal during the lifetime of a product or process should be considered. The selection of future vehicle options can strongly depend on the emission characteristics. As the world struggles with greenhouse gas emission reduction policies, global warming potential is the main characteristics to compare the total CO_2 equivalent emission from the alternative vehicles. Abiotic depletion, human toxicity, ozone layer depletion appear to play an important role for decision of using clean transportation vehicles because there are vast amount of road vehicles in the cities which can cause severe side effects. Moreover, when considering alternative fuels, issues such as land use, fertilizer use, water for irrigation, waste products etc. are necessary points to be addressed. Therefore, indirect land-use changes (ILUC) should be also considered. Indirect land-use changes can also have important social and environmental impacts which can include biodiversity, water quality, food prices and supply, community and cultural stability. Assessing the indirect land-use changes is a knowns as challenging topic. Some methods of quantifying indirect land-use changes can be listed as follows:

- Implementing empirical calculations based on previously experienced indirect land-use changes
- Developing life cycle analyses methodology with lower uncertainty ranges
- Developing integrated models combining the life cycle, sustainability, efficiency, social cost etc.

The following environmental impact categories represent higher significance in life cycle assessment approach, hence suggested to be included in decision making processes:

- Global warming potential is the main characteristics to be compare the total CO_2 equivalent emission from any source.
- Abiotic resources are natural resources including energy resources. Since fossil fuels resources are declining gradually, abiotic depletion potential is also a significant category.
- Human toxicity may play an important role for decision of using alternative fuels.
- Acidification potential is for acidifying substances which causes a wide range of impacts on soil, groundwater, surface water, organisms, ecosystems and materials.
- Marine aquatic eco-toxicity refers to impacts of toxic substances on marine aquatic ecosystems which is more important for maritime transportation sector.
- Land occupation/land use refers to the total arrangements, activities and inputs undertaken in a certain land cover type. The term land use is also used in the sense of the social and economic purposes for which land is managed.

HOW MUCH GREENHOUSE GAS CAN I SAVE IF I DRIVE AN AMMONIA CAR?

Considering a complete life cycle counting the production, transport and usage of the fuel, a diesel driven car can emit greenhouse gas emissions of about 220 g per km. Ammonia driven car can decrease this number down to about 70 g per km if it is produced from solar energy and about 150 g per km if it is produced from hydrocarbon cracking.

IS AMMONIA A COST EFFECTIVE FUEL?

The illustrative cost comparison of various fueled vehicles is shown in Fig. 3 and 4. Considering the current market prices of the fuels, ammonia is the lowest cost fuel corresponding to about 3.1 US$ in a 100 km driving range. This shows that ammonia is a promising transportation fuel in terms of cost. There is an advantage of by-product refrigeration which reduces the costs and maintenance during vehicle operation. Some additional advantages of ammonia are commercial availability and viability, global distribution network and easy handling experience. Ammonia is a cost effective fuel per unit energy stored onboard compared to methanol, CNG, hydrogen, gasoline and LPG as shown in Fig. 3.

In Table 1, the exact fuel energy per mass is given regarding fuel's higher heating value. The volumetric energy of the fuel is found by multiplying the HHV with the density value listed in the third column. Ammonia's HHV is around half of the one of gasoline, and its density is also inferior. Therefore liquid NH_3 stores 2.5 fewer energy per unit capacity than gasoline. If the NH_3 is stored in the form of hexaamminemagnesium chloride to remove the hazard related to its toxicity, the energetic cost to pay for discharging ammonia reduces its HHV. Among alternative fuels, ammonia yields the lowest cost per energy basis. Therefore, it is important to note that low-carbon transportation projects should be eligible for compliance purposes.

Table 1. Comparison of ammonia with other fuels

Fuel/storage	Pressure (bar)	Density (kg m^{-3})	HHV (MJ kg^{-1})	HHV per Volume (GJ m^{-1})	Energy per Volume (GJ m^{-1})	Cost per Mass (US$ kg^{-1})	Cost Per Volume (US$ m^{-1})	Cost per Energy (US$ GJ^{-1})
Gasoline, C_8H_{18}/liquid	1	736	46.7	34.4	34.4	1.03	754.86	21.97
CNG, CH_4/integrated storage	250	188	42.5	10.4	7.8	0.91	170.60	21.29
LPG, C_3H_8/pressurized tank	14	388	48.9	19	11.7	1.06	413.66	21.74
Methanol, CH_3OH/liquid	1	786	14.3	11.2	9.6	0.41	317.80	28.31
Hydrogen, H_2/metal hydrides	14	25	142	3.6	3	3.02	75.49	21.29

Ammonia, NH_3/pressurized tank	10	603	22.5	13.6	11.9	0.23	136.63	10.04
Ammonia, NH_3/metal amines	1	610	17.1	10.4	8.5	0.23	138.14	13.21

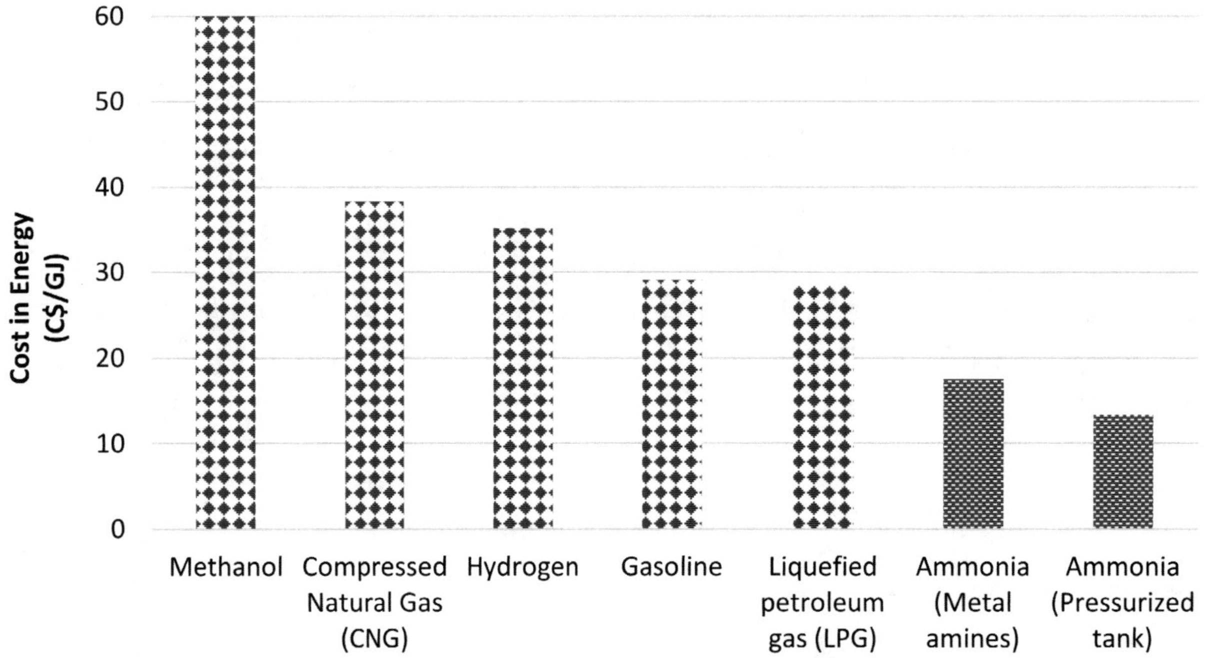

Fig. 3. Comparison of various vehicle fuels in terms of energy cost per gigajoule

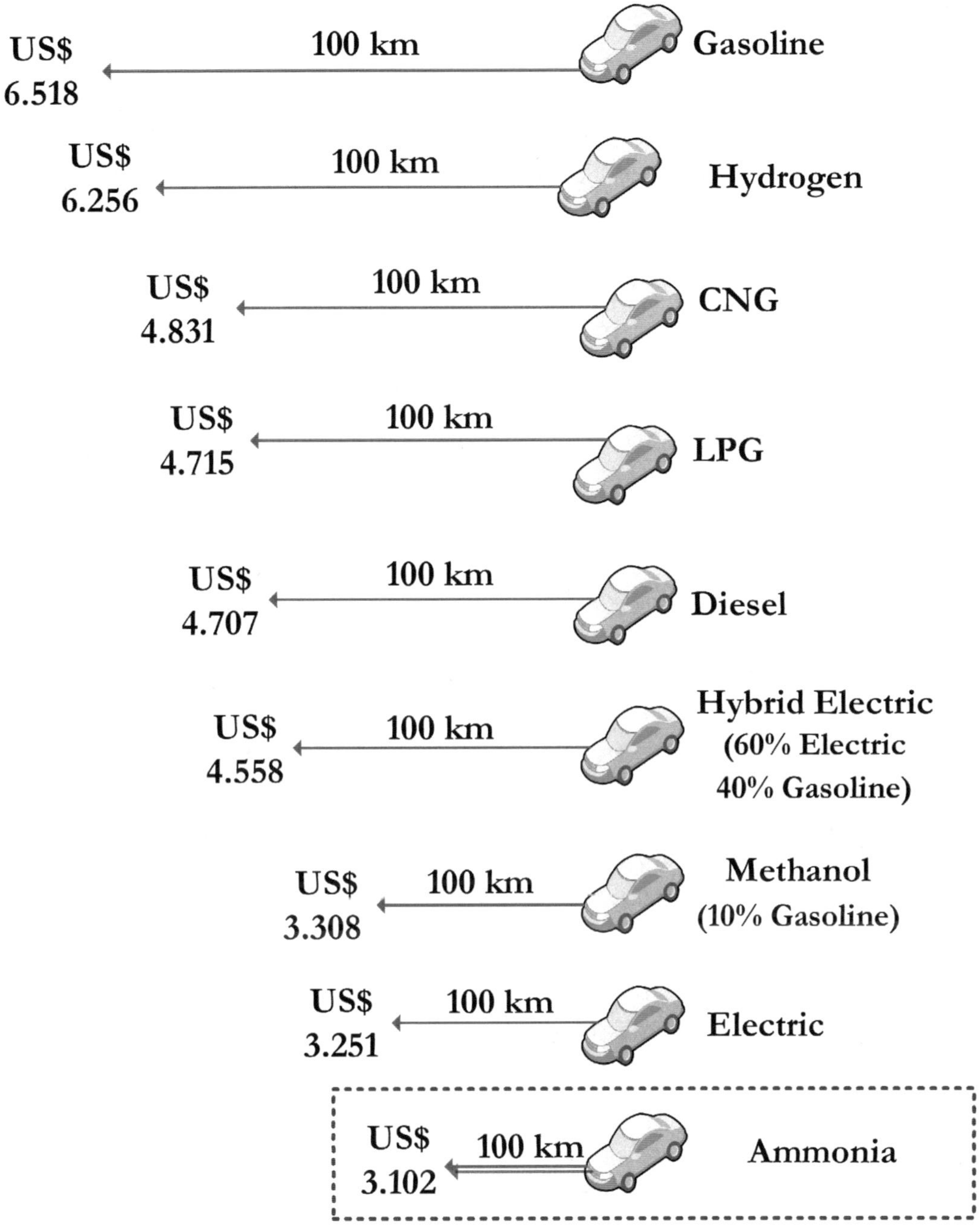

Fig. 4. Comparison of driving cost for various fueled vehicles

WHAT IS THE PROCESS OF AMMONIA PRODUCTION?

A most common ammonia synthesis technique is recognized as Haber-Bosch process in the world. In this process, nitrogen is supplied through air separation process. Hydrogen is mainly supplied using steam methane reforming or coal gasification. However the source of hydrogen can be renewable resources. The Haber-Bosch is an exothermic process that combines hydrogen and nitrogen in 3:1 ratio to produce ammonia. The reaction is facilitated by catalyst and the optimal temperature range is 450-600°C.

Alternative production pathways are also available and under investigation including electrochemical and biological routes. These routes can easily be integrated to renewable energy sources for cleaner production. The electrochemical process can be carried out under ambient conditions or at higher temperatures depending on the type of the electrolyte material used. There are various electrochemical pathways such as molten salt, polymer membrane, liquid electrolyte etc. are intensively being researched at the moment [11-16].

The electrochemical process can be carried out under ambient conditions or at higher temperatures depending on the type of the electrolyte material used. For high temperature electrolytic routes of ammonia production, the use of waste heat from thermal or nuclear power plants or heat from renewable energy sources like solar would make the overall process more environmentally friendly.

One of the key advantages of ammonia is to be a storage medium. Renewable energy generation does not often match electrical demand which causes a requirement of storage. Green ammonia can be manufactured from surplus renewable sources, which would reduce the amount of electricity exported to neighboring jurisdictions at a negative cost.

WHAT IS THE SOURCE OF AMMONIA AND IS IT CLEANER THAN OTHER FUELS?

In terms of conventional resources, naphtha, heavy fuel oil, coal, natural gas coke oven gas and refinery gas can be used as feedstock in ammonia production. Natural gas is the primary feedstock used for producing ammonia in worldwide corresponding to about 72%. However, renewable resources can easily be integrated for ammonia production. In this way, decentralized ammonia production can be realized which further decreases the delivery cost of the fuel. Many studies have been performed to investigate the ammonia production routes and their environmental impacts. Here, some of them are briefly shown.

The production of the different fuels is compared in terms of abiotic depletion of sources as shown in Fig. 5. Ammonia fuel has the lowest abiotic depletion value compared to others although the production process may be fossil fuel based. There are multiple pathways for ammonia production. Ammonia is cleaner when produced from renewable resources. Fig. 6 shows the comparison of ozone layer depletion values for various transportation fuels. Ammonia has lowest ozone layer depletion even if it is produced from steam methane reforming and partial oxidation of heavy oil.

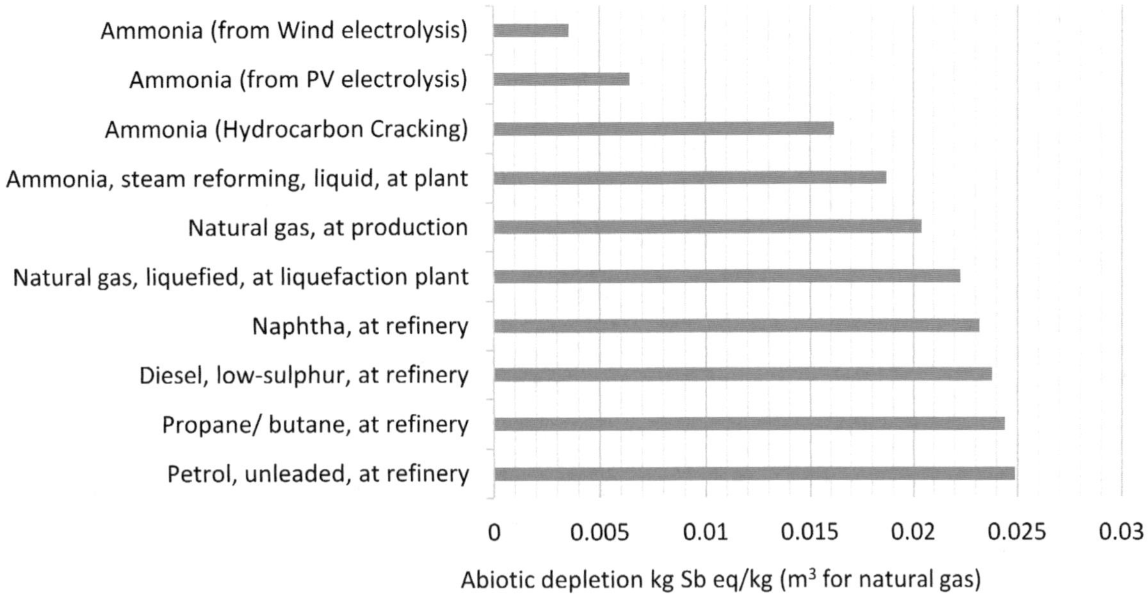

Fig. 5. Abiotic depletion values during production of various fuels

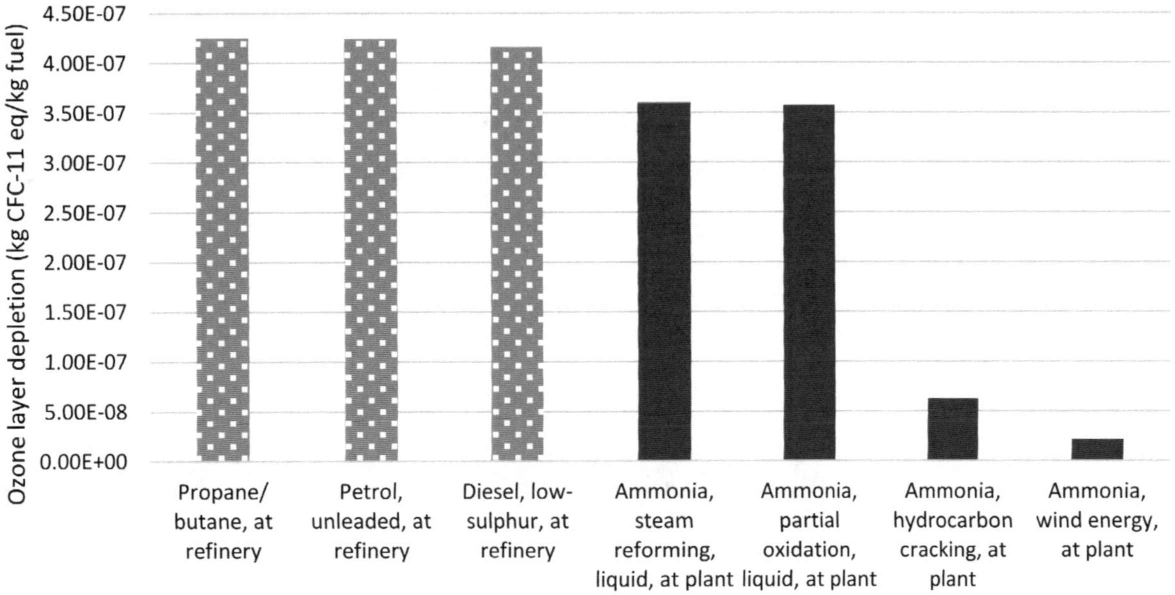

Fig. 6. Ozone layer depletion during productions of various fuels

Fig. 7 compares the total greenhouse gas emissions during production of 1 MJ energy from various resources including gasoline, LPG, diesel, natural gas and ammonia. Production of 1 MJ energy from ammonia has lower emissions than gasoline, LPG, diesel, oil and natural gas.

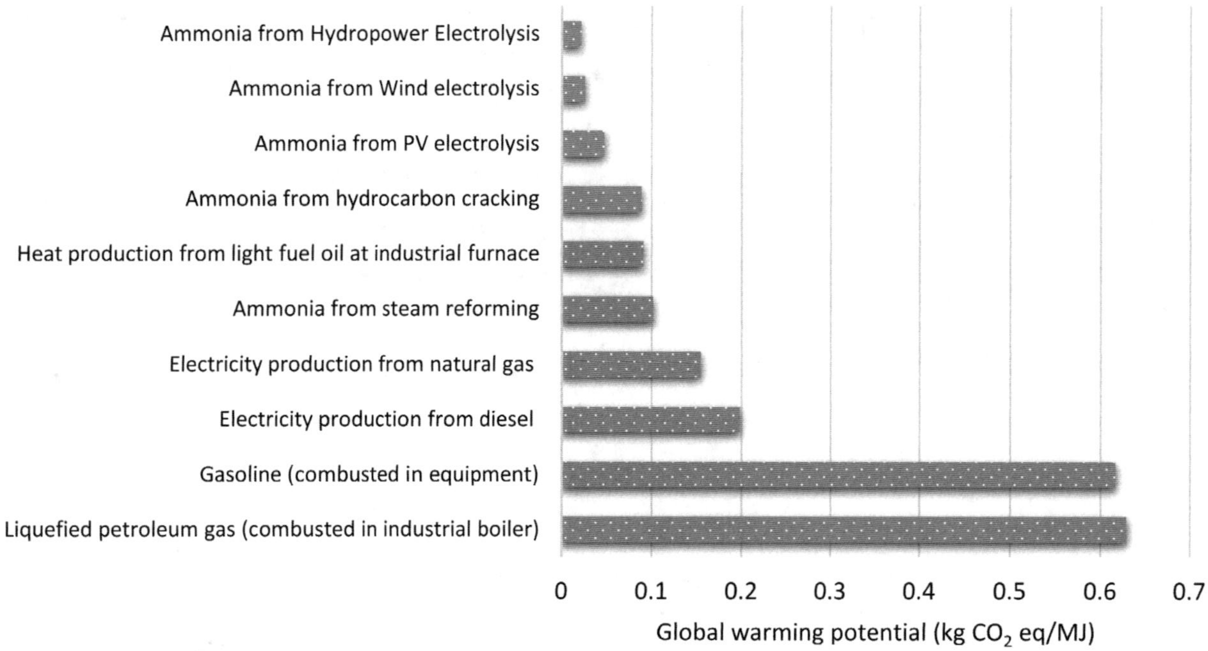

Fig. 7. Comparison of global warming potential of 1 MJ energy production from various resources

Giving priority for complete conversion from fossil fuel based fuels to carbon-free fuels will bring short term and long term solutions to combat global warming. Therefore, the activities to lower the carbon intensity of conventional transportation fuels be eligible for compliance purposes

AMMONIA IN ROAD TRANSPORTATION

Assessing the life cycle emissions from vehicles is a powerful method for transportation sector. The usage of fossil fuel based electricity decreases the attractiveness of electric (EV) and hybrid electric (HEV) vehicles. Henceforth, noteworthy attention should be paid to the power generation technologies and their CO_2 intensity, used to supply electricity to EVs or HEVs. The

GHG emissions of EVs and HEVs are therefore dependent on the CO_2 intensity of the energy mix and differs based on the countries. A characteristic life cycle of a vehicle technology can be categorized into two main steps, namely fuel cycle and vehicle cycle. In the fuel cycle, the processes beginning from the feedstock production to fuel utilization in the vehicle are considered. In the vehicle cycle, utilization of fuel is considered. Among the selected categories, global warming, abiotic depletion and human toxicity results carry more significant decision parameters for road vehicles. The results presented here are given on per km basis based on the fuel consumption rates given in Table 2.

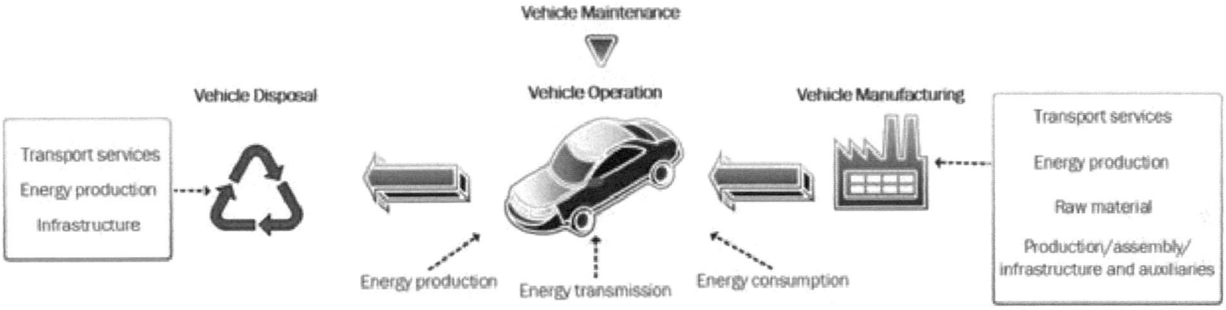

Fig. 8. Complete life cycle of vehicles including fuel/vehicle cycle.

Table 2. Energy consumptions per km for the selected vehicles

Fuel		Fuel/Energy Consumption	Unit
Gasoline		0.0649108	kg/km
Diesel		0.0551536	kg/km
M90	Methanol	0.1180535	kg/km
	Gasoline	0.0060664	kg/km
Hydrogen		0.0195508	kg/km
Ammonia		0.0926600	kg/km
EV		0.2167432	kWh/km
HEV	Electric	0.1083716	kWh/km
	Gasoline	0.0324554	kg/km
CNG		0.0603914	kg/km
LPG		0.057629687	kg/km

The specific conditions for the selected vehicles are presented herein:

- Gasoline: All processes on the refinery site excluding the emissions from combustion facilities, including waste water treatment, process emissions and direct discharges to rivers are accounted for. The inventory data also includes the distribution of petroleum product to the final consumer including all necessary transports. Transportation of product from the refinery to the end user is considered together with operation of storage tanks and petrol stations. Emissions from evaporation and treatment of effluents are accounted for. Particulate emissions cover exhaust- and abrasions emissions.
- Diesel: Diesel is evaluated as low-Sulphur at regional storage with an estimation for the total conversion of refinery production to low-Sulphur diesel. An additional energy use (6% of energy use for diesel production in the refinery) has been estimated. The other processes are similar to gasoline. Particulate emissions cover exhaust- and abrasions emissions.
- CNG: Natural gas with a production mix at service station is taken into account. The inventory data contains electricity necessities of a natural gas service station together with emissions from losses. The data set represents service stations with high (92%), medium (6%) and low (2%) initial pressure. VOC emissions are obtained from gas losses and contents of natural gas. Particulate emissions cover exhaust- and abrasions emissions.
- Hydrogen: Hydrogen is produced during cracking of hydrocarbons. It includes combined data for all processes from raw material extraction until delivery at plant. The output fractions from an oil refinery are composite combinations of mainly unreactive saturated hydrocarbons. The first processing step in converting such elements into feedstock suitable for the petrochemical industries is cracking. Essentially a cracker achieves two tasks in (i) rising the complexity of the feed mixture into a smaller number of low molecular mass hydrocarbons and (ii) presenting unsaturation into the hydrocarbons to enable more reactivity. The raw hydrocarbon input from the refinery is fed to the heater unit where the temperature is increased. The forming reaction products vary based on the composition of the input, the temperature of the heater and the residence time. The cracker operator selects temperature and residence time to enhance product mix from a supplied input. Cracker feeds can be naphtha from oil refining or natural gas or a mixture of both. After exiting the heater, the hydrocarbon gas is cooled to prevent extra reactions. After that, it is sent to the separation phase where the individual hydrocarbons are separated from one another by fractional distillation. Particulate emissions cover exhaust- and abrasions emissions. In order to have comparable results where hydrogen

comes from non-fossil fuels such as solar PV and nuclear, they are also taken into account in the analyses by applying water electrolysis route. The electrolyzer is assumed to consume about 53 kWh electricity for one kg of hydrogen production.

- Ammonia: Ammonia synthesis process is Haber-Bosch which is the most common method in the world. Ammonia production requires nitrogen and hydrogen. In this study, hydrogen is assumed to be from hydrocarbon cracking as explained in the previous paragraph. Cryogenic air separation is mostly used method for massive amount of nitrogen production. In the life cycle assessment of nitrogen production, electricity for process, cooling water, waste heat and infrastructure for air separation plant are included. Haber-Bosch process is an exothermic method that combines hydrogen and nitrogen in 3:1 ratio to produce ammonia. The reaction is facilitated by catalyst (iron-oxide based) and the optimal temperature range is 450-600°C. Particulate emissions cover exhaust- and abrasions emissions. In order to have comparable results where ammonia comes from non-fossil fuels such as solar PV and nuclear, they are also taken into account in the analyses. The generated hydrogen from electrolyzers are used for ammonia synthesis plant.
- EV: Electricity consumption is included. Particulate emissions comprise exhaust and abrasions emissions. Heavy metal emissions to soil and water caused by tire abrasion are accounted for. In the electricity usage process, electricity production mix, the transmission network and direct SF6-emissions to air are included. In order to present a renewable based scenario for electric vehicles, a mixture of renewables for energy requirement during the operation are also evaluated consisting of 25% biomass, 25% solar PV, 25% wind power and 25% hydropower.
- HEV: Hybrid car is assumed to be 50% electric and 50% gasoline with ICE. Electricity and gasoline consumptions are included. Particulate emissions comprise exhaust and abrasions emissions. Heavy metal emissions to soil and water caused by tire abrasion are accounted for. For the hybrid vehicle's electricity, a mixture of renewables for energy requirement during the operation are also evaluated consisting of 25% biomass, 25% solar PV, 25% wind power and 25% hydropower.
- Methanol: The selected fuel M90 consists of 90% methanol and 10% gasoline. The raw materials, processing energy, estimate on catalyst use, and emissions to air and water from process, plant infrastructure are included. The process describes the production of methanol

from natural gas via steam reforming process to obtain syngas for the production of methanol. There is no CO_2 use and hydrogen is assumed as burned in the furnace. Raw materials, average transportation, emissions to air from tank storage, estimation for storage infrastructure are included for the distribution part where 40% of the methanol is assumed to be transported from overseas. Particulate emissions cover exhaust- and abrasions emissions.

- LPG: All processes on the refinery site excluding the emissions from combustion facilities, including waste water treatment, process emissions and direct discharges to rivers are considered. All flows of materials and energy due to the throughput of 1 kg crude oil in the refinery is accounted for. Refinery data include desalting, distillation (vacuum and atmospheric), and hydro treating operations. Particulate emissions cover exhaust- and abrasions emissions.

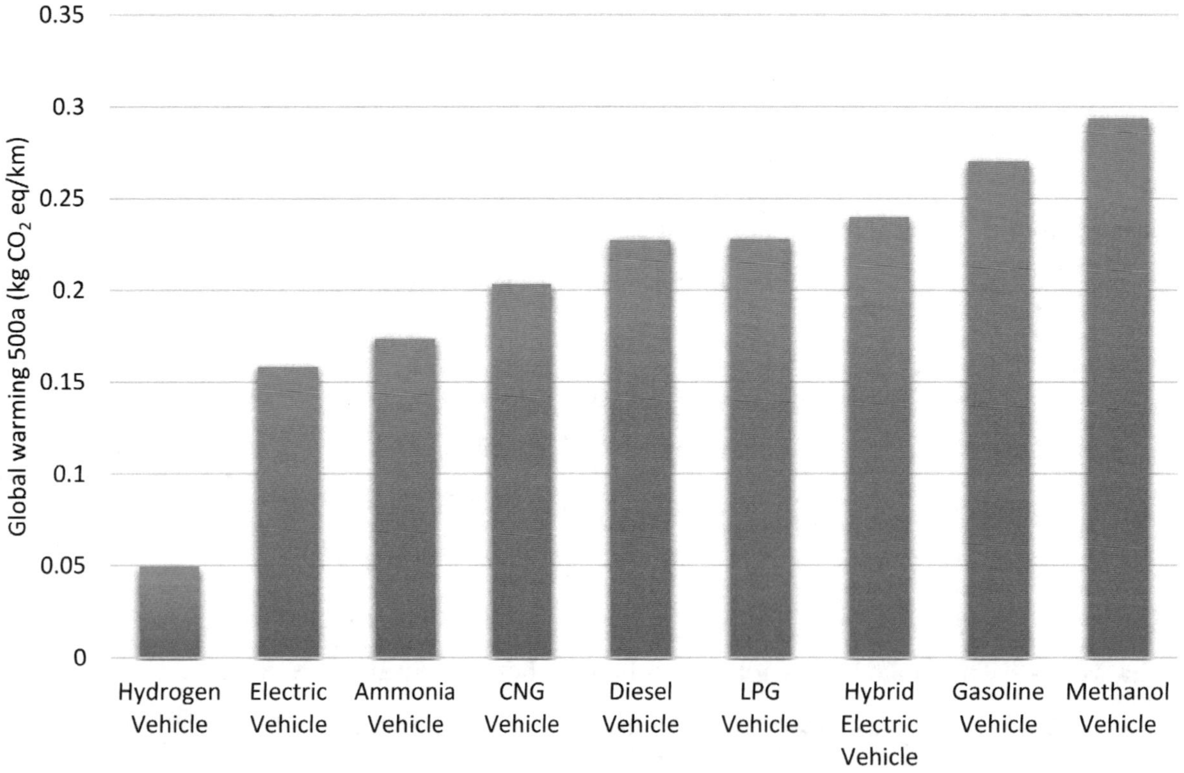

Fig. 9. Life cycle comparison of global warming results for various vehicles

The global warming potentials of assessed vehicles are comparatively shown in Fig. 9. The lowest GHG emissions are observed in hydrogen, electric and ammonia vehicles corresponding to 0.049 kg CO_2 eq/km, 0.15 kg CO_2 eq/km and 0.17 kg CO_2 eq/km, respectively. Hydrogen consumption is quite lower than ammonia consumption in the passenger car because of higher energy density. It is an expectable result that EVs also yield lower global warming potential, however production pathway of electricity has a key role in GHG emissions. If electricity production can be realized by renewable sources such as solar, biomass, hydropower and wind energy, total emissions would decrease for both EVs and HEVs.

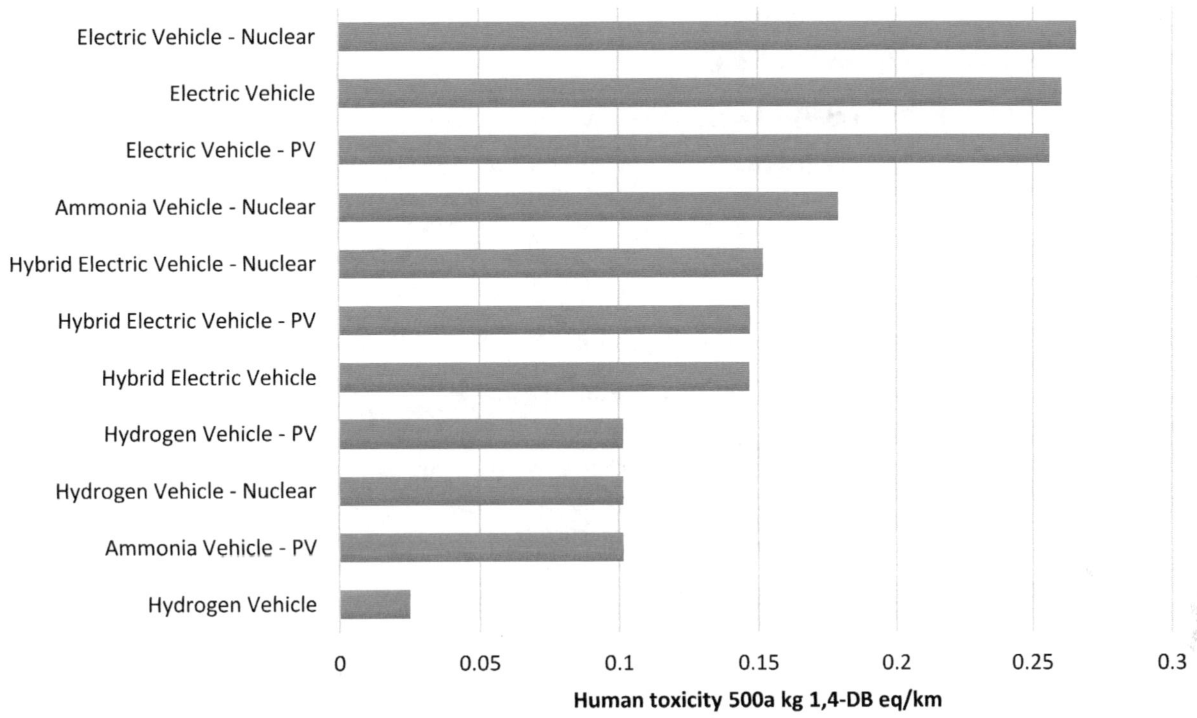

Fig. 10. Life cycle comparison of human toxicity results for various vehicles from nuclear energy and solar PV routes

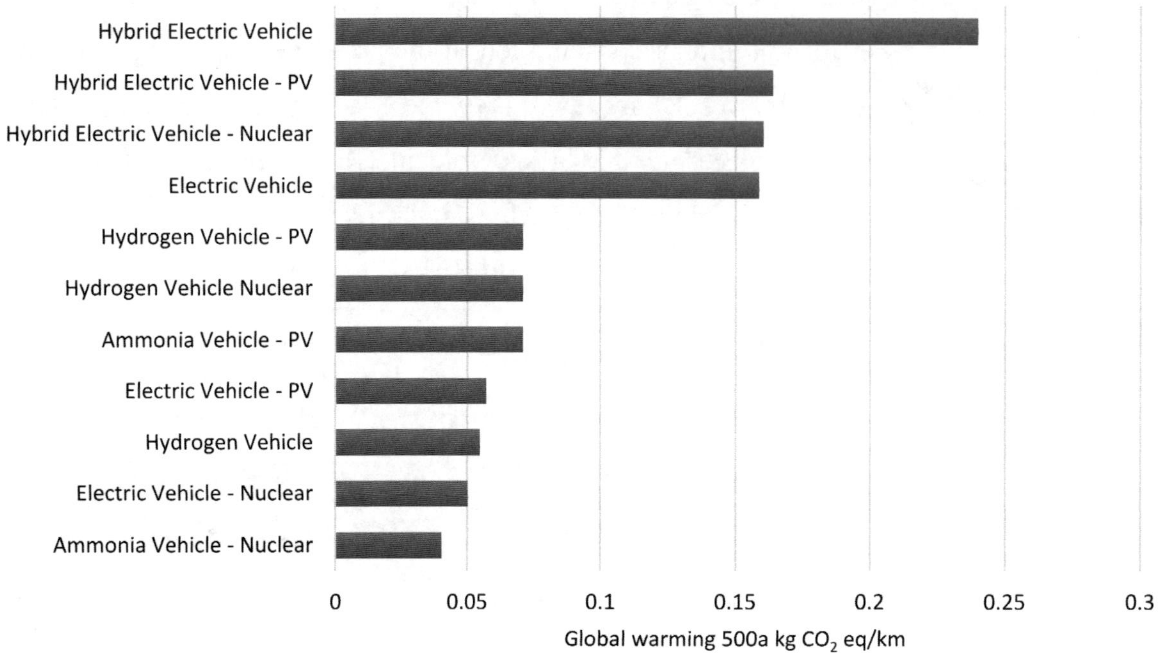

Fig. 11. Life cycle comparison of global warming results for various vehicles from nuclear energy and solar PV routes

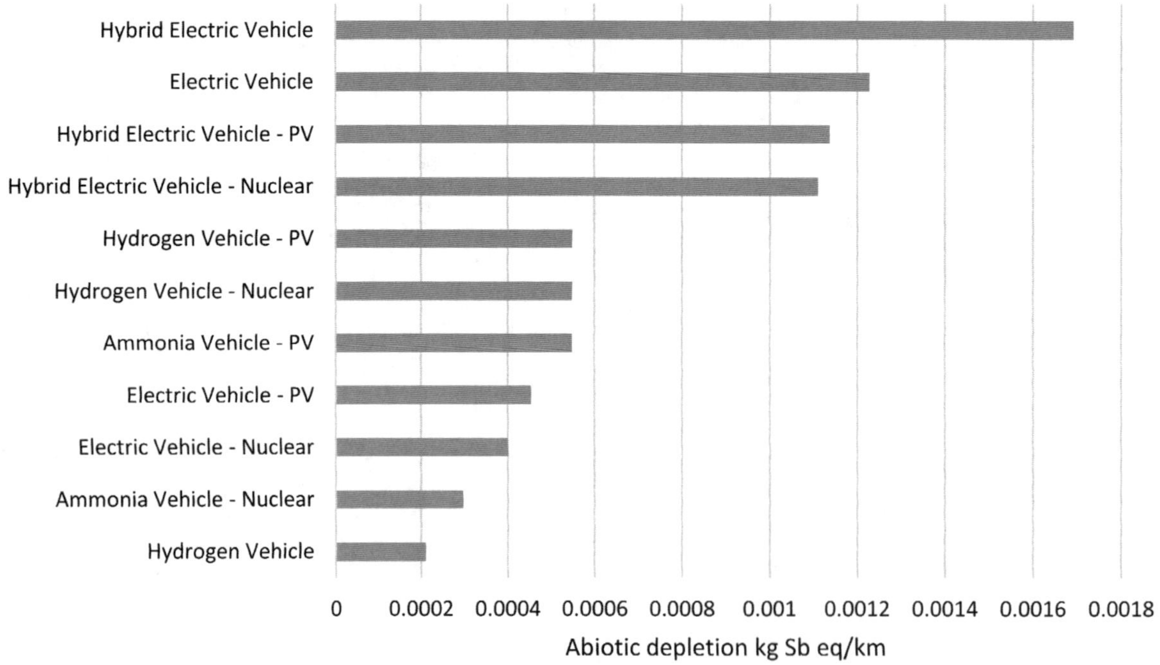

Fig. 12. Life cycle comparison abiotic depletion for various vehicles from nuclear energy and solar PV routes

The electricity from nuclear (25%), biomass (25%), hydropower (25%) and solar PV (25%) are equally used only for the operation processes of the EVs and HEVs vehicles as renewable mix. Utilization of renewable electricity for EVs, HEVs, ammonia and hydrogen vehicles are comparatively shown for various impact categories. For human toxicity category, using solar and nuclear energy does not cause significant reductions for EVs and HEVs as shown in Fig. 10. However, for global warming and abiotic depletion categories, both solar energy and nuclear energy routes lower the environmental impacts more than 50%. Ammonia driven vehicle where the ammonia is produced from nuclear electrolysis method yield the lowest GHG emissions corresponding to about 0.04 kg CO_2 eq. per km as shown in Fig. 11. Hydrogen and EVs (from nuclear and PV) have quite similar greenhouse gas emissions in the range of 0.049-0.057 kg CO_2 eq. per km. In terms of abiotic depletion values, hydrogen vehicle still yields the lowest value whereas nuclear routes for ammonia and EVs further decrease the abiotic depletion impact as shown in Fig. 12.

The results show that hydrogen and ammonia vehicles are the most environmentally benign ones in most of the environmental impact categories. Ammonia as a sustainable and clean fuel has lowest global warming potential after EVs and yield lower ozone layer depletion values than EVs. However, in case renewables are used both for ammonia vehicles and EVs, ammonia can suggest lower environmental impacts. Although EVs do not emit direct CO_2 during operation, the production and disposal processes of batteries bring some consequences which harm the environment in terms of acidification, eutrophication and human toxicity.

AMMONIA IN MARITIME TRANSPORTATION

Decreasing the global warming potential caused by current transportation technologies and fuels can be reduced significantly by replacing alternative clean fuels. Sea transportation vehicles mostly use heavy fuel oil or diesel fuel for power generation. Ocean tankers and freight ships require massive amount of energy for operation.

Being a sustainable energy carrier that can be generated from any primary energy source, ammonia can subsidize to a broadening of maritime fuel resources and may offer the long term option of being generated from renewable resources.

A brief life cycle consideration of maritime vehicles, ocean tanker and freight ship, is shown here where the tankers are driven with hydrogen and ammonia instead of heavy fuel oils in the power engines. Additionally, dual fuel options, heavy fuel oil and hydrogen/ammonia, are investigated.

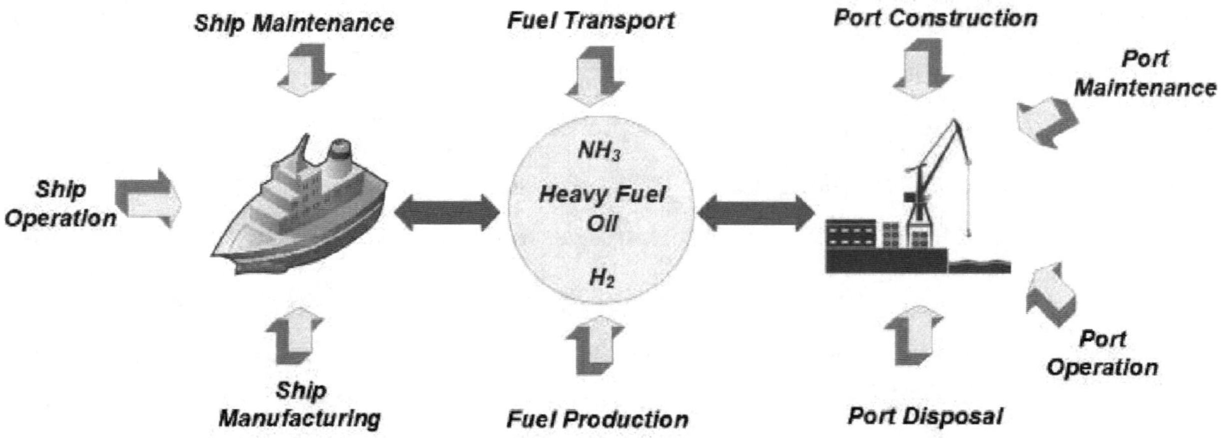

Fig. 13. Life cycle steps of maritime transportation

A comparative life cycle assessment of transoceanic fright ship and transoceanic tanker, is performed to examine the effects of clean fuel driven maritime vehicles on the environment. The complete transport life cycle is evaluated in the life cycle analyses comprising of manufacture of tanker/freight ship; operation of tanker/freight ship; construction and land use of seaport; operation, maintenance and disposal of seaport; production and transportation of hydrogen and ammonia. A tonne kilometer by shipping is defined as unit of measure of goods transport which represent the transport of one tonne by a vessel over one kilometer.

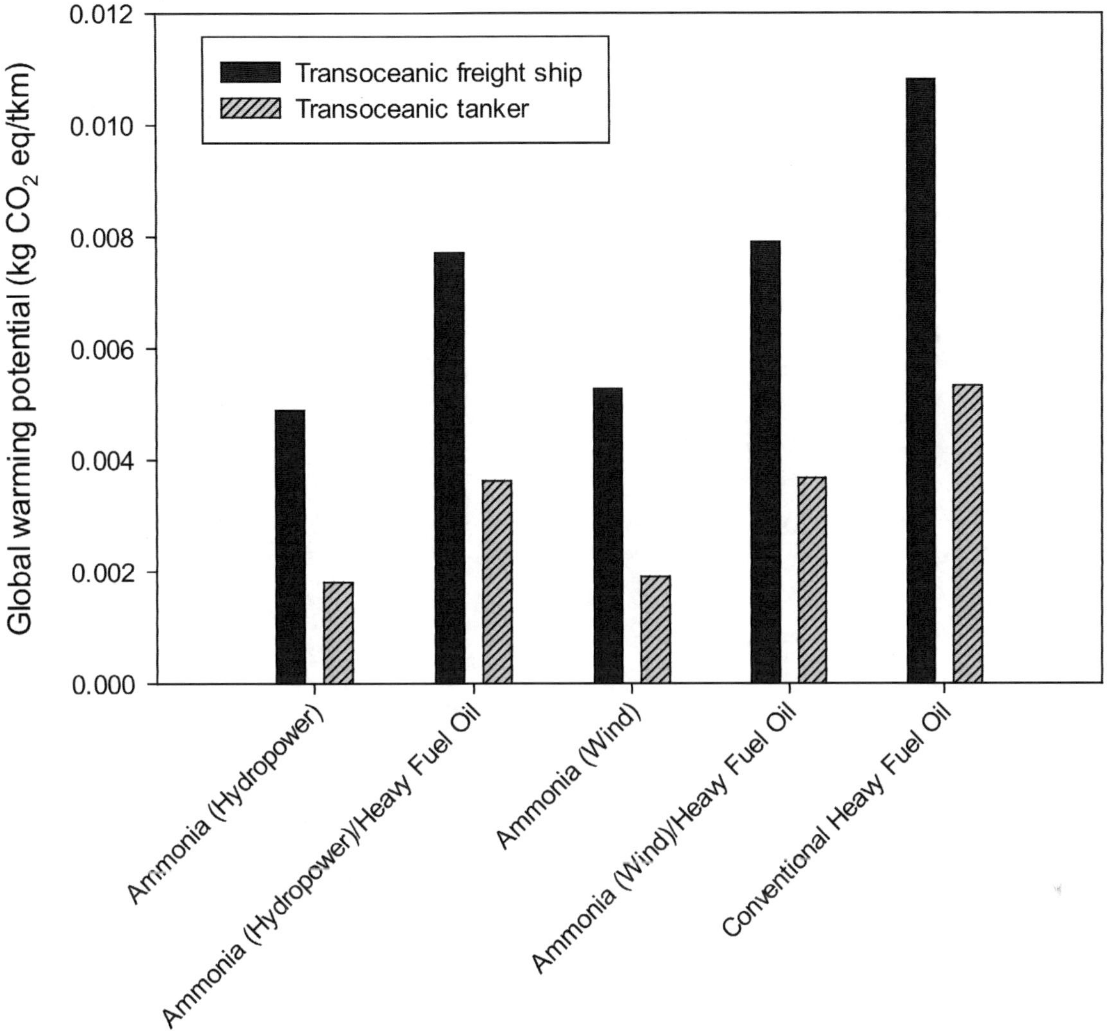

Fig. 14. Global warming potential of transoceanic tanker and transoceanic freight ship per tonne kilometer for clean fuels and conventional heavy fuel oil

The heavy fuel oil combustion significantly contributes to many of the environmental impact. Using ammonia as dual fuel in the marine engines can decrease total greenhouse gas emissions up to 34.5% per tkm. Similarly, sole ammonia driven transoceanic tanker releases about 0.0018 kg CO_2 eq/tkm greenhouse gas compared to 0.0055 kg CO_2 eq/tkm for sole heavy fuel oil tanker as shown in Fig. 14.

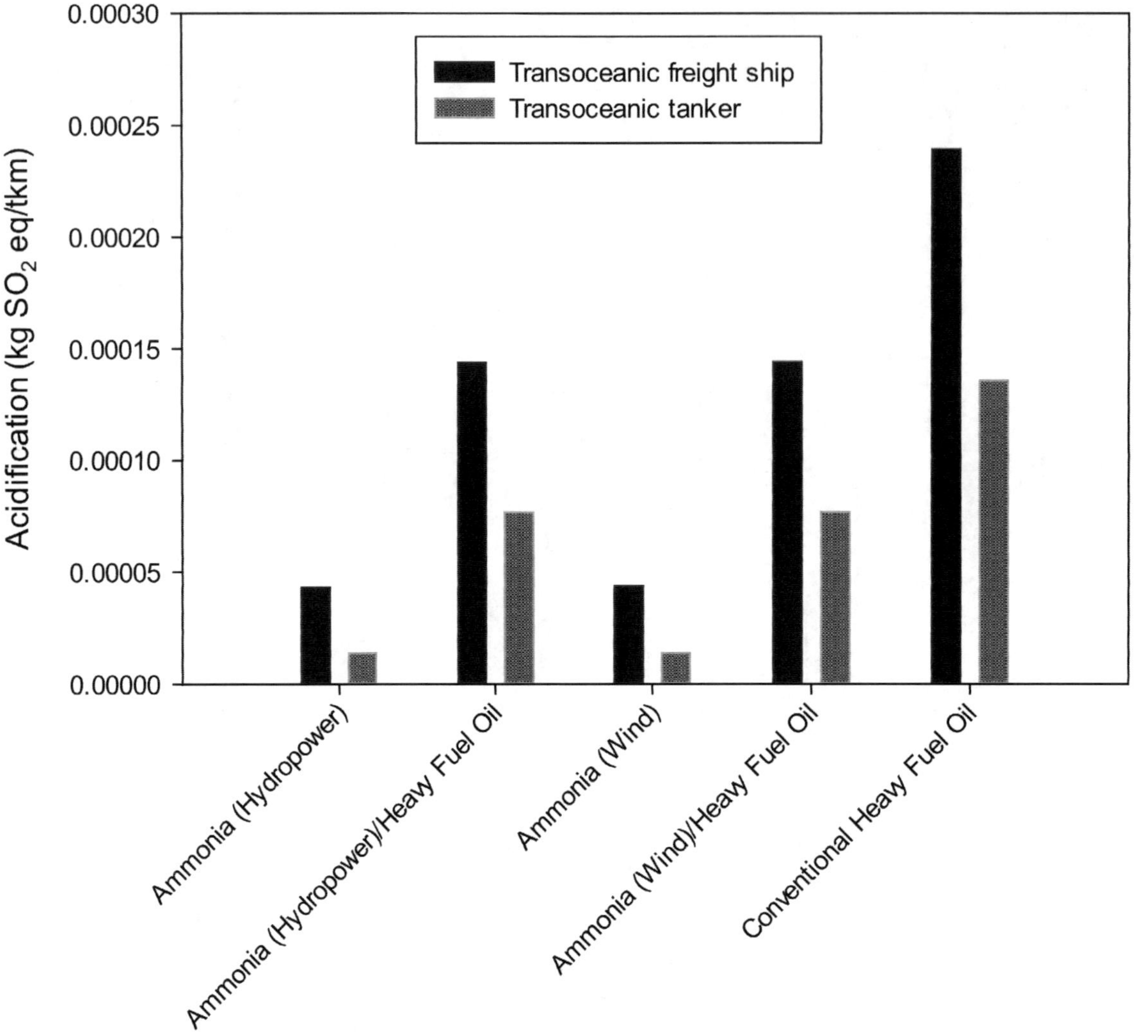

Fig. 15. Acidification values of transoceanic tanker and transoceanic freight ship per tonne kilometer for clean fuels and conventional heavy fuel oil

The acidification values of heavy fuel oil driven transoceanic tanker and freight ship are mainly caused by SO_2 and NO_x emissions (Fig. 15) which corresponds to more than 90% of overall acidification value. The source of SO_2 emission is predominantly the operation of tanker and freight ship (96.8%). This is caused by the sulfur content of the heavy fuel oil hence it is mostly eliminated if ammonia is used.

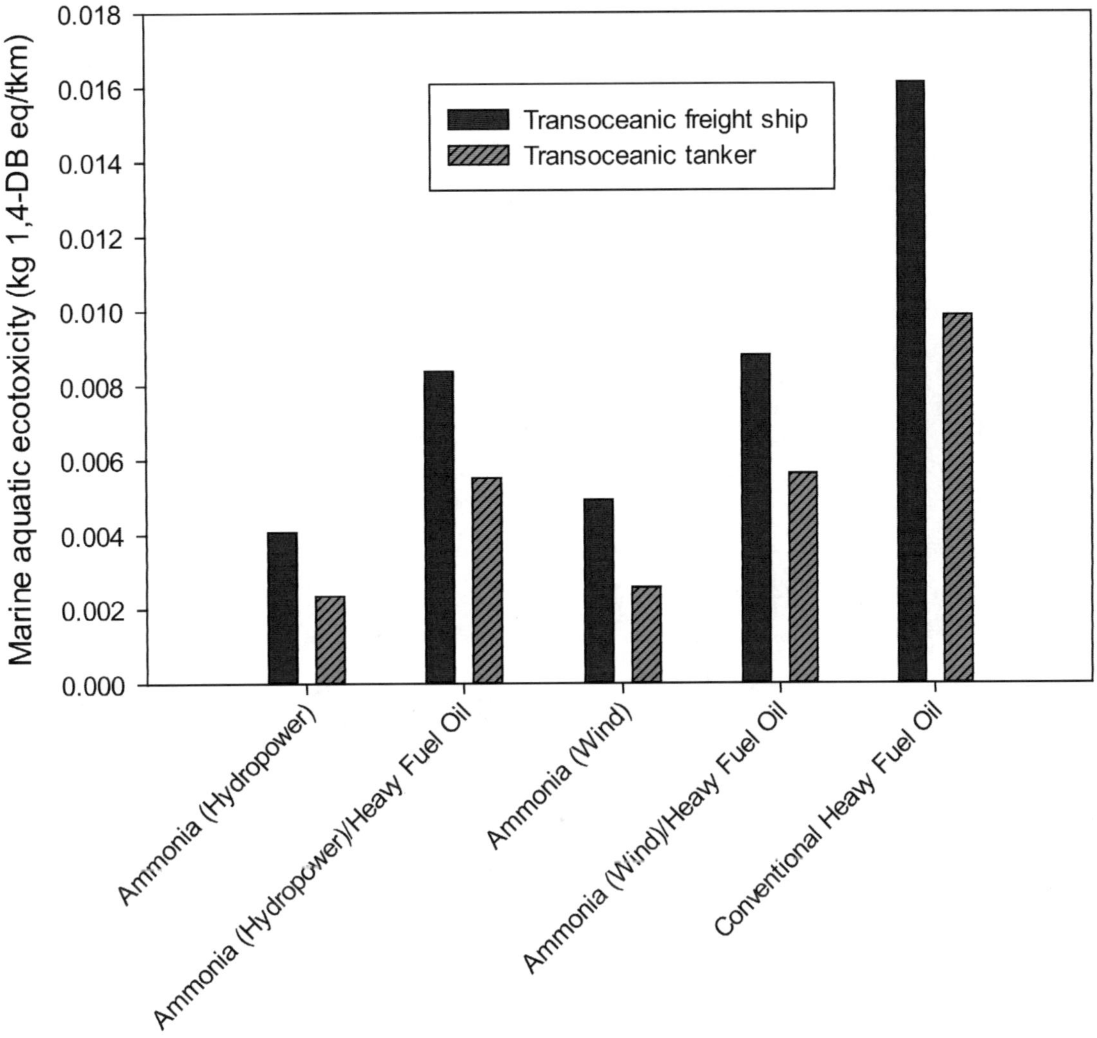

Fig. 16. Marine aquatic ecotoxicity values of transoceanic tanker and transoceanic freight ship per tonne kilometer for clean fuels and conventional heavy fuel oil

Using ammonia (wind energy) in transoceanic tanker as dual fuel with heavy fuel oil lowers the ecotoxicity level about 47% as shown in Fig. 16. For conventional heavy fuel oil driven transoceanic tanker, crude oil is the main depleted source similar to dual fuel options as shown in Fig. 17. For ammonia, the depletion is mainly caused by port operation process rather than tanker/ship operation. As non-carbon clean fuels for maritime ship engines, ammonia yields considerably lower global warming impact during operation. This demonstrates that if clean fuels are even partially replaced with current hydrocarbon derived fuels, total GHG emissions in maritime transportation can be lowered significantly.

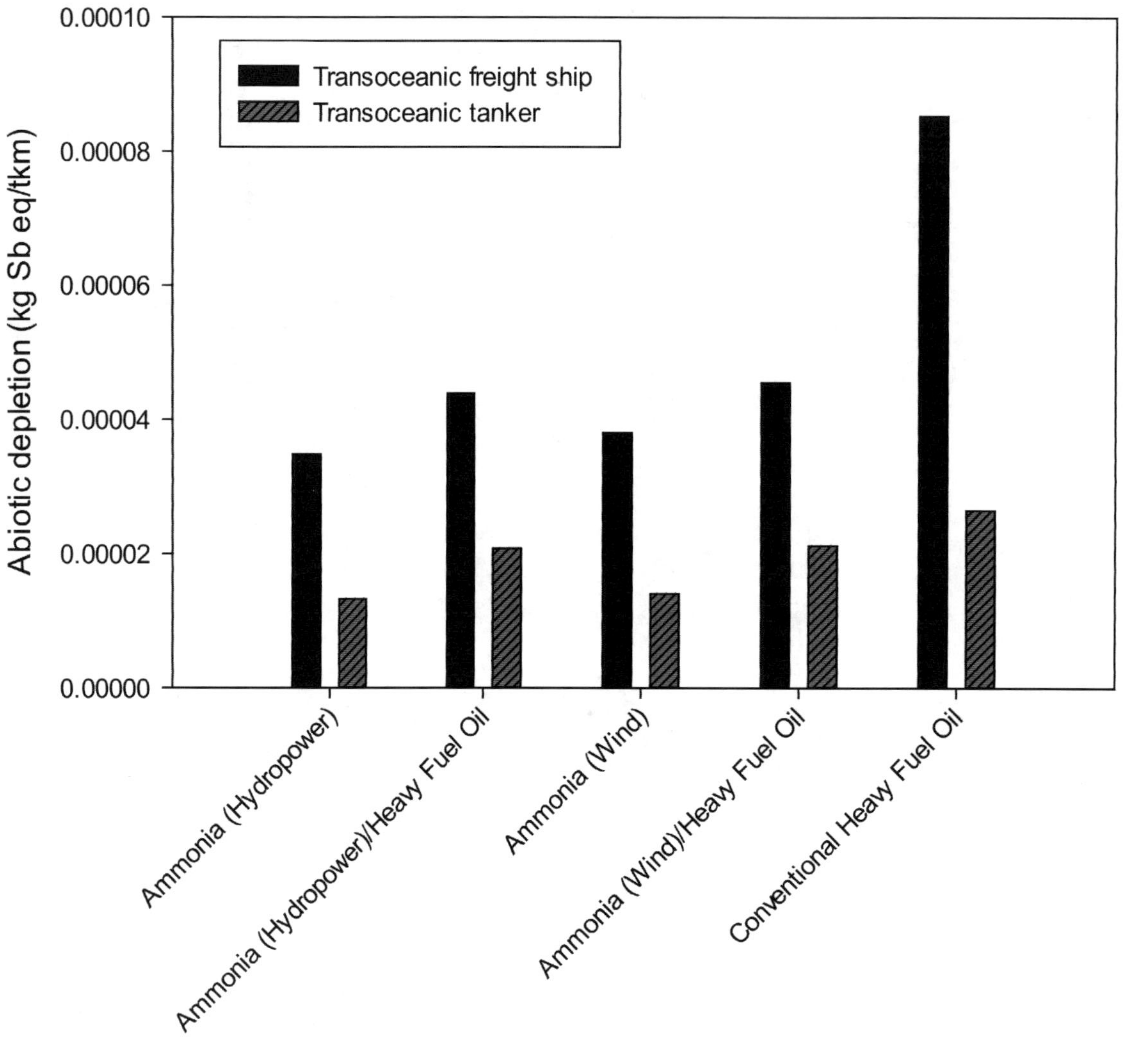

Fig. 17. Abiotic depletion values of transoceanic tanker and transoceanic freight ship per tonne kilometer for clean fuels and conventional heavy fuel oil

AMMONIA IN AVIATION

Petroleum based fuels have combination of accessibility, ease of handling, energy content, performance, and price because of being a mature product. Therefore, these type of fuels are heavily used by the transportation sector including air, road, and sea. However, limited nature, nonhomogeneous source distribution, changing prices, and end use related emissions of fossil fuels have driven most of the industries such as air transportation to search for other alternatives. Aviation is responsible for 12% of CO_2 emissions from all transports sources. The national and international flights overall the world produced 770 million tons of CO_2 in 2015 [17].

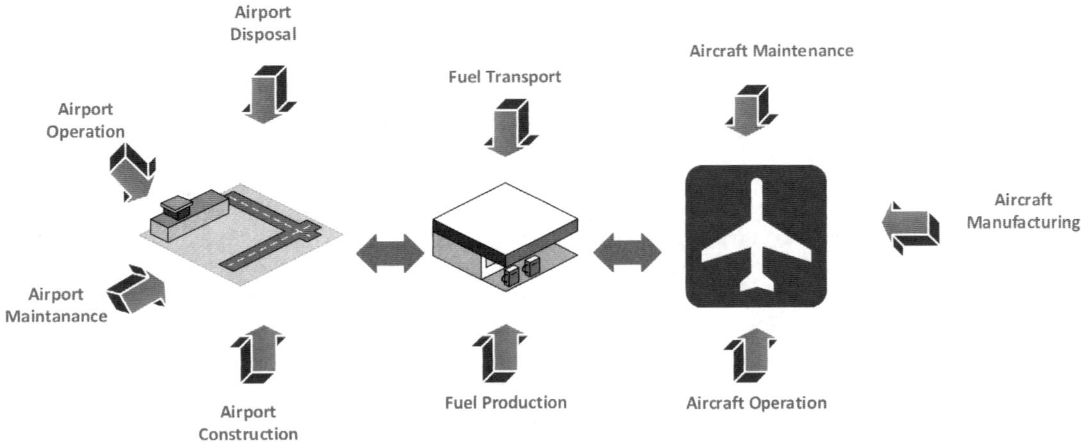

Fig. 18. Illustration of complete well-to-wake life cycle phases used in the LCA study

The overall life cycle emissions of an aircraft running on various aviation fuels are calculated from well-to-wake. The life cycle phases included in the analyses (shown in Fig. 18) are as follows: (i) production, operation and maintenance of the aircraft, (ii) construction, maintenance and disposal of the airport, (iii) production, transportation and utilization of the aviation fuel in the aircraft.

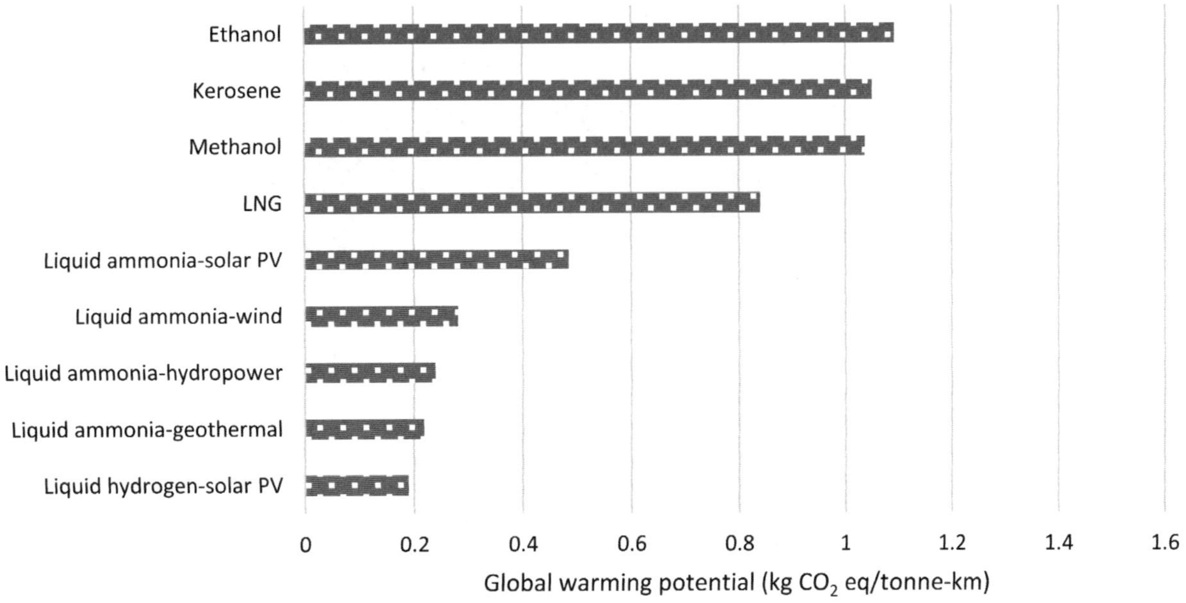

Fig. 19. Global warming potential of various fueled aircrafts per travelled tonne-km

The overall GWP value of ethanol is the highest corresponding to 1.09 kg CO_2 eq/tonne-km since it comes from ethylene which is a hydrocarbon as shown in Fig. 19. Renewable energy usage in the fuel production processes decreases the overall GHG emissions. The hydropower based ammonia fueled aircraft releases about 0.24 kg CO_2 eq per tonne-km. This value goes down to 0.21 for kg CO_2 eq/tonne-km for ammonia fueled aircraft and 0.18 kg CO_2 eq/tonne-km for hydrogen fueled aircraft in case the renewable source is solar energy. Although there are several processes, operation of the aircraft is the second largest contributor corresponding to 34%. Operation and maintenance of the airport is the primary responsible for GWP corresponding to 48.9% in total where it is distributed into sub-processes such as natural gas burning in the furnace (22%), light fuel oil burning in the furnace (5%), lignite burning in the power plant (7%) and hard coal burning (8%)

Renewable based ammonia driven aircrafts can significantly decrease the depletion of abiotic resources down to 0.0014 kg Sb eq/tonne-km which corresponds to about 10% of conventional steam methane reforming based ammonia fueled aircraft as shown in Fig. 20.

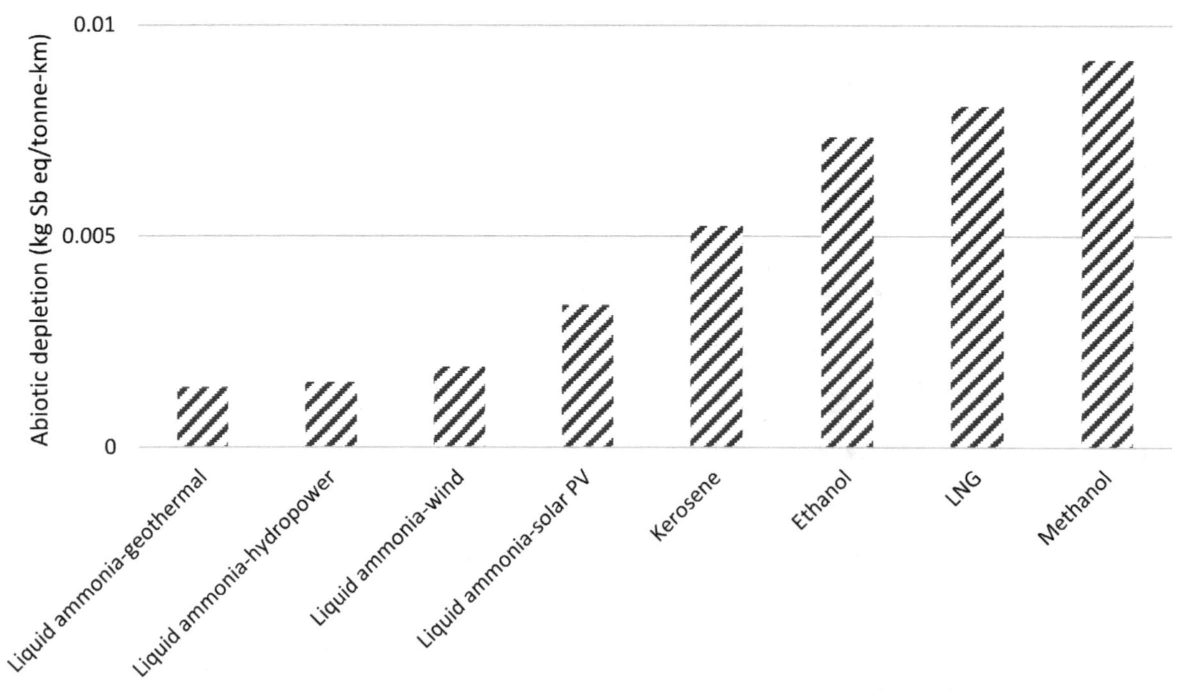

Fig. 20. Abiotic depletion values of various fueled aircrafts per travelled tonne-km

It is common knowledge that of HC, CO, NO_x, PM, and CO_2 which are named as GHG emissions cause environmental damages and adverse effect on human health. The marginal external price of a unit of these emissions is identified as environmental and social costs of emissions. Environmental and social costs of HC, CO, NO_x, PM, and CO_2 emissions of various fueled aircrafts are evaluated in terms of USD/tonne-km as shown in Fig. 21. renewable based ammonia yields lower environmental and social cost of emissions in comparison with kerosene.

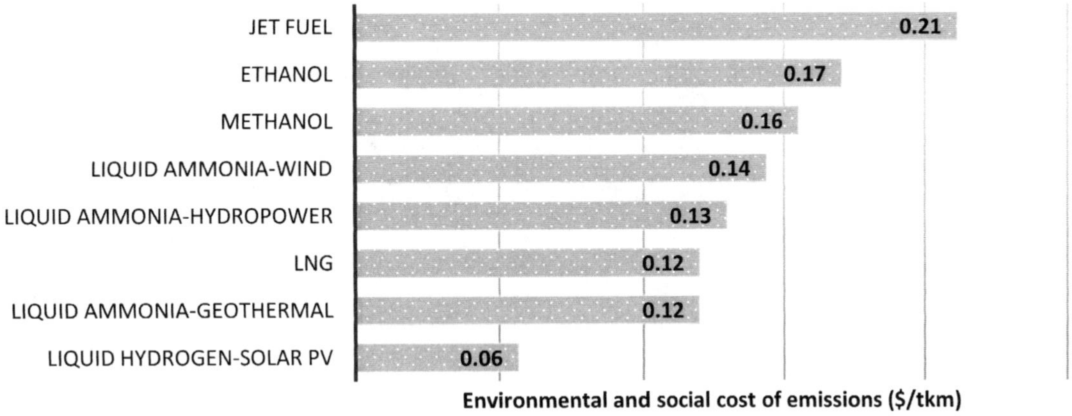

Fig. 21. Total environmental and social cost of emissions for various fueled aircrafts from conventional and renewable resources

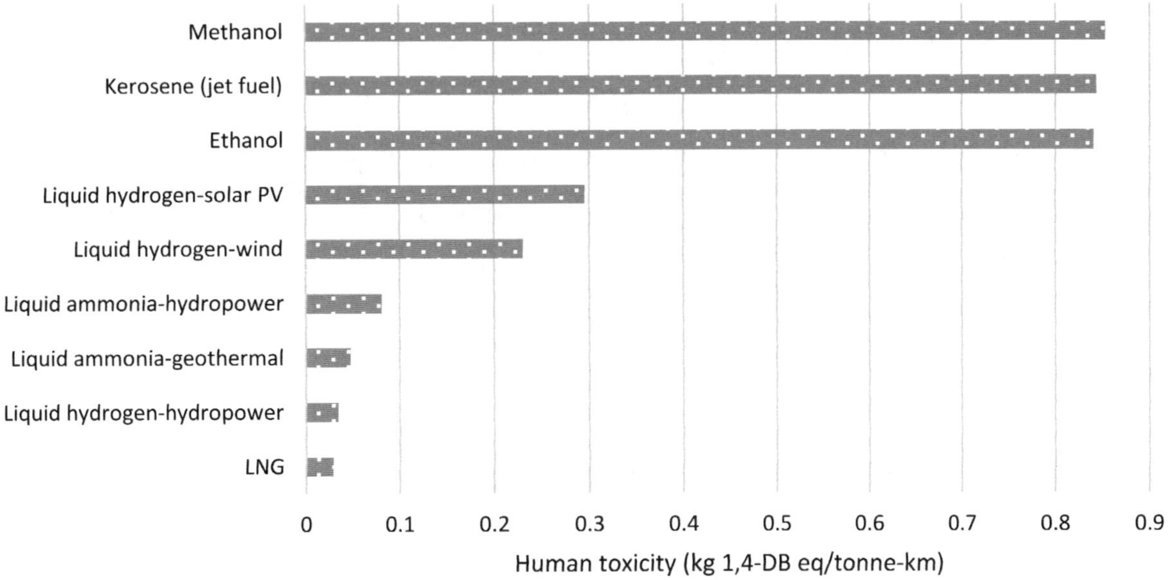

Fig. 22. Human toxicity potential of various fueled aircrafts per travelled tonne-km

Among all alternative aviation fuels, the kerosene jet fuel, methanol and ethanol human toxicity values are greater. Ammonia fuel from hydropower resource has comparable toxicity value corresponding to 0.08 kg 1,4- DB eq/tonne-km with LNG corresponding to 0.03 kg 1,4-DB eq/tonne-km as shown in Fig. 22.

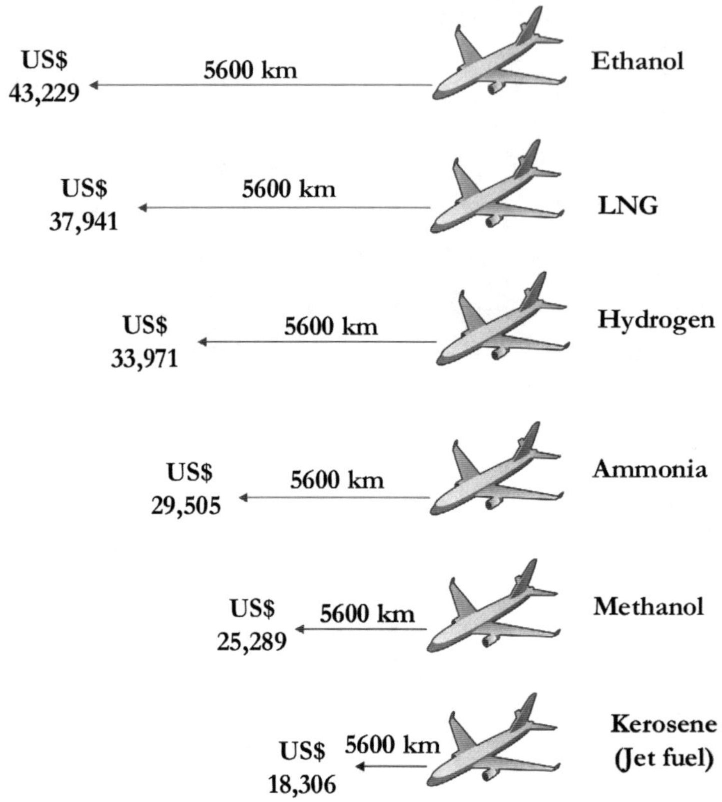

Fig. 23. Comparison of fuel costs during the operation of aircrafts for the given range

Table 3. Average fuel consumption rates and fuel costs for selected alternative fuels

Fuel	Fuel Consumption (kg/km)	Fuel Consumption (kg/tonne-km)	Fuel Cost (USD/kg)
Kerosene (Jet fuel)	7.99	0.217666815	0.409
Methanol	18.06	0.492185866	0.250
Ammonia	18.82	0.512730362	0.280
Hydrogen	2.64	0.071866485	2.300
LNG	9.46	0.257732767	0.716
Ethanol	12.47	0.339891348	0.619

Fig. 23 illustrates the cost of flight for a 5600 km distance in which alternative fuels are used for the aircrafts when the fuel costs and consumption rates in Table 3 are considered. Although LNG represented better environmental performance, the cost of aircraft operation in terms of fuel is the highest for ethanol and LNG. In the calculations, current conventional based routes are taken into account for comparison purposes. Hydrogen fueled aircraft has lower cost compared to these two alternatives. Since the production of kerosene from crude oil is a more mature technology, it represents the lowest cost among all. Liquid ammonia is also low cost alternative compared to hydrogen.

CONCLUDING REMARKS

Ammonia is a carbon-free fuel suitable for use in transportation sector. It has a well-established production and distribution infrastructure, and has zero global warming potential during operation. In addition to its attractive qualities as a fuel, ammonia is widely used as a NOx reducing agent for combustion exhaust gases using selective catalytic reduction (SCR), and its capacity as a refrigerant can be applied to recover and further utilize engine heat that would otherwise be lost. In terms of environmental sustainability, ammonia can be produced using either fossil fuels, or any renewable energy source, using heat and/or electricity, which allows for evolution of ammonia production methods and technologies in parallel with sustainable development. Ammonia as a sustainable fuel can be used in all types of combustion engines, gas turbines, burners with only small modifications and directly in fuel cells which is a very significant advantage compared to another type of fuels. Reducing the total greenhouse gas emissions from marine transportation is possible by using ammonia which is carbon-free fuel. They can be utilized for maritime ship engines directly as supplementary fuels or individual fuels. Ammonia fueled ships yield considerably lower global warming impact during operation. Ammonia as a sustainable and clean fuel in road vehicles yield also the lowest global warming potential after electric and hydrogen vehicles. As a result, ammonia usage in the communities for transportation sector will bring significant cost and environmental benefits together with public satisfaction.

References

1. Environment and Climate Change Canada (2016), National Inventory Report 1990-2014: Greenhouse Gas Sources and Sinks in Canada.
2. Kroch E.: Ammonia-a fuel for motorbuses, Journal of the Institute of Petroleum, 1945, pp 213-223
3. Iki N, Kurata O, Matsunuma T, Inoue T, Suzuki M, Tsujimura T, et al., editors. Micro Gas Turbine Firing Kerosene and Ammonia. ASME Turbo Expo 2015: Turbine Technical Conference and Exposition; 2015: American Society of Mechanical Engineers.
4. Haputhanthri SO, Maxwell TT, Fleming J, Austin C. Ammonia and Gasoline Fuel Blends for Spark Ignited Internal Combustion Engines. Journal of Energy Resources Technology. 2015;137:062201-.
5. Nozari H, Karabeyoğlu A. Numerical study of combustion characteristics of ammonia as a renewable fuel and establishment of reduced reaction mechanisms. Fuel. 2015;159:223-33.
6. Miller AR, Ammonia Fuel For Rail Transportation, Vehicle Projects LLC, 2006 Annual NH3 Fuel Conference, OCTOBER 9-10, 2006 Denver West Marriott, Golden, CO
7. Reiter AJ, Kong S-C. Combustion and emissions characteristics of compression-ignition engine using dual ammonia-diesel fuel. Fuel. 2011;90(1):87-97.
8. Ryu K, Zacharakis-Jutz GE, Kong S-C. Performance enhancement of ammonia-fueled engine by using dissociation catalyst for hydrogen generation. International Journal of Hydrogen Energy. 2014;39(5):2390-8.
9. Zamfirescu C, Dincer I. Ammonia as a green fuel and hydrogen source for vehicular applications. Fuel Processing Technology. 2009;90(5):729-37.
10. Zamfirescu C, Dincer I. Using ammonia as a sustainable fuel. Journal of Power Sources. 2008;185(1):459-65.
11. Hinnemann B, Nørskov JK. Catalysis by Enzymes: The Biological Ammonia Synthesis. Topics in Catalysis. 2006;37(1):55-70.
12. Lan R, Irvine JTS, Tao S. Synthesis of ammonia directly from air and water at ambient temperature and pressure. Scientific Reports. 2013;3:1145.

13. Li F-F, Licht S. Advances in Understanding the Mechanism and Improved Stability of the Synthesis of Ammonia from Air and Water in Hydroxide Suspensions of Nanoscale Fe2O3. Inorganic Chemistry. 2014;53(19):10042-4.
14. Licht S, Cui B, Wang B, Li F-F, Lau J, Liu S. Ammonia synthesis by N2 and steam electrolysis in molten hydroxide suspensions of nanoscale Fe2O3. Science. 2014;345(6197):637-40.
15. Paschkewitz TM. Ammonia Production at Ambient Temperature and Pressure: An Electrochemical and Biological Approach. PhD (Doctor of Philosophy) thesis. University of Iowa; 2012
16. Skodra A, Stoukides M. Electrocatalytic synthesis of ammonia from steam and nitrogen at atmospheric pressure. Solid State Ionics. 2009;180(23-25):1332-6.
17. Facts and Figures, Air Transport Action Group, May 2016, http://www.atag.org/facts-and-figures.html

Transparency in Renewable Fuels

Ontario's Cap & Trade Program states:

The RFS is a technology-neutral policy and a technology-neutral approach lets the alternatives compete on their merits.

In order for renewable fuel technology to compete on its own merits and for the market to decide which technology has the best merits, energy consumers must ultimately participate in selection of renewable fuels.

Ontario's Cap & Trade program has the following participants:

Mandatory participants	Facilities and natural gas distributors with emissions of 25,000 tonnes or more of greenhouse gas emissions per year are required by law to participate in the program. Additionally, fuel suppliers that sell more than 200 litres of fuel per year and electricity importers must also participate in the program.
Voluntary participants	Facilities generating more than 10,000 tonnes but less than 25,000 tonnes of emissions may choose to opt in to the program. These companies will be subject to the same rules as mandatory participants.
Market participants	Companies that do not have emissions to report and therefore do not have a compliance obligation can also participate in the auction. Market participants can include individuals, not-for-profit organizations and companies without compliance obligations.

For the market to decide the winners in the Cap & Trade program, each energy consumer must also be a Cap & Trade participant.

This could be administered through the income tax system, with each family or business entity being issued its own baseline carbon credit. At income tax time, each energy or fuel supplier would provide each consumer a carbon statement that the energy consumer would include in his income tax return. Depending upon each tax entity's carbon consumption, the net carbon credits would translate into additional carbon tax payments or refunds.

TRANSPARENCY VIA A FLOW-THROUGH CARBON TAX

From an energy consumer's perspective, it is impractical to determine his own individual emissions. However, it is simple to determine their expected emissions by the amount of fuel consumed. Therefore, a more practical and transparent way of putting a cost on carbon emissions would be through a flow-through carbon taxation program similar to the GST/HST system. The carbon tax would replace the fuel excise tax and would be applicable to ALL consumable energy and fuel.

This taxation system would start at the producer level. For energy produced in Ontario, the carbon tax would replace the fuel excise tax. The carbon tax would be based on the carbon content of the fuel sourced from fossil sources or of the amount of fossil-carbon consumed to convert a fuel to a consumable form of energy. The carbon tax would then flow through from the producer to the distributor to the consumer.

For a technology-neutral carbon policy, there must be a completely level playing field and there is more to the life cycle environmental impact of the fuel than the amount of carbon dioxide generated from the its combustion. There are nine categories of life cycle environmental impacts caused by the production and utilization of energy that must be considered for a completely level playing field:

Environmental Impact	**Description**
Abiotic Depletion	Abiotic resources are natural resources including energy resources. Since fossil fuels resources are declining gradually, abiotic depletion potential is also a significant category. It is expressed in kg antimony (Sb) equivalents.

Environmental Impact	Description
Acidification	Acidification potential is for acidifying substances which causes a wide range of impacts on soil, groundwater, surface water, organisms, ecosystems and materials. kg SO2 equivalents is used to express the acidification potential.
Eutrophication	Eutrophication is the impact of excessive levels of macro-nutrients in the environment which is mainly caused by disposal processes. It is stated as kg PO_4 equivalents.
Global Warming	Global warming potential is the main characteristics to be compare the total CO2 equivalent emission from any source. The greenhouse gasses to air are related to the climate change. Methane, carbon monoxide etc. are in this category.
Human Toxicity	Human toxicity may play an important role for decision of using alternative fuels. It indicates the toxicity level and presented in 1,4-dichlorobenzene ($C_6H_4Cl_2$) equivalent.
Ozone Layer Depletion	Ozone layer depletion increases the amount of UVB that reaches the Earth's surface. Laboratory and epidemiological studies demonstrate that UVB causes non-melanoma skin cancer and plays a major role in malignant melanoma development. Ozone depletion potential of several gasses is specified in kg CFC-11 equivalent.
Terrestrial Ecotoxicity	Terrestrial ecotoxicity refers to the potential for biological, chemical or physical stressors to affect terrestrial ecosystems. Such stressors might occur in the natural environment at densities, concentrations or levels high enough to disrupt the natural biochemistry, physiology, behavior and interactions of the living organisms that comprise terrestrial ecosystems. The results are stated as 1,4-dichlorobenzene equivalents per kg emission.

Environmental Impact	Description
Marine Aquatic Eco-Toxicity	Marine aquatic eco-toxicity refers to impacts of toxic substances on marine aquatic ecosystems which is more important for maritime transportation sector. It considers each substance emitted to the air, water or/and soil. The unit of this factor is kg of 1,4-dichlorobenzene equivalents (1,4-DB eq.) per kg of emission.
Land Occupation/Land Use	Land occupation/land use refers to the total arrangements, activities and inputs undertaken in a certain land cover type. The term land use is also used in the sense of the social and economic purposes for which land is managed. It is mainly expressed as m2a which implies area usage in a year.

Focusing on carbon dioxide alone precludes any consideration of the global warming potential of other greenhouse gas emissions. From the US EPA [1]:

Greenhouse Gas	Global Warming Potential Over 100 years	Notes
CO2	1 by definition	Regardless of the time period used, CO2 is the reference gas. CO2 remains in the climate system for a very long time: CO2 emissions cause increases in atmospheric concentrations of CO2 that will last thousands of years.

Greenhouse Gas	Global Warming Potential Over 100 years	Notes
Methane (CH_4)	28–36	CH_4 emitted today lasts about a decade on average, which is much less time than CO_2. But CH_4 also absorbs much more energy than CO_2. The net effect of the shorter lifetime and higher energy absorption is reflected in the GWP. The CH_4 GWP also accounts for some indirect effects, such as the fact that CH_4 is a precursor to ozone, and ozone is itself a GHG. (Learn why EPA's U.S. Inventory of Greenhouse Gas Emissions and Sinks uses a different value.)
Nitrous Oxide (N_2O)	265–298	N_2O emitted today remains in the atmosphere for more than 100 years, on average.
Chlorofluorocarbons (CFCs), Hydrofluorocarbons (HFCs), Hydrochlorofluorocarbons (HCFCs), Perfluorocarbons (PFCs), Sulfur Hexafluoride (SF_6)	Thousands or Tens of Thousands	Sometimes called high-GWP gases because, for a given amount of mass, they trap substantially more heat than CO_2.

The carbon tax would be designed to improve upon the reduction of greenhouse gas emissions by including other detrimental effects on the environment. Rather than basing the tax rate solely on the carbon content of the fuel, global warming would be one of the nine environmental impact categories.

Conceivably, some fuels could have a positive effect on a particular type of pollution. In this case, such a process should be given a negative environmental impact factor to reduce its tax rate. An environmental impact multiplier would then be applied to the baseline rate that takes into account the nine environmental impact categories. Each environmental impact category would have an assessment scale of -10 to 10. The environmental impact multiplier would then be based on the sum of the nine assessments.

Ultimately, only electricity or fuels with a zero or negative environmental impact assessment would have a zero carbon tax rate. The cost of the inputs of varying environmental impacts used to generate the each source of electricity would be included in the cost of electricity. This would ensure that that the lowest-cost electricity is produced with the lowest environmental impact.

MANUFACTURING PRODUCTS FROM FUELS

If part of the fuel was then converted to a product, the manufacturer of that carbon-containing product would then get a credit for the carbon that has been prevented from entering the atmosphere.

For example, some biofuel processes can take carbon out of the atmosphere and can permanently sequester it into the ground as biochar. This type of permanent sequestration can result in Terra Preta [2], which has enhanced soil fertility. This process would get a credit for the carbon content of the biochar, less whatever biofuel was consumed to produce the biochar.

For another example, if a manufacturer had a process to produce anhydrous ammonia and elemental carbon from natural gas, the manufacturer would pay the full carbon tax on the natural gas used in its process. It would then get a credit for the elemental carbon, less whatever natural gas was used for process heat. The elemental carbon would carry the maximum carbon tax but the consumer of the elemental carbon would get the full carbon credit if that elemental carbon would be used to manufacture a carbon product, such as graphene. Since anhydrous ammonia has own environmental impact, its carbon tax would be different than that of the natural gas.

TAX ADMINISTRATION

A carbon tax program would be easier to administer at a consumer level than a Cap & Trade program. Businesses are already familiar with the GST/HST system and would not require significantly more bureaucracy to administer. This would also be transparent because the carbon tax portion of the energy cost could be separately broken out in the price as well as the methodology used to calculate the pollution multiplier factors. Consumers would therefore be able to determine the most effective way of reducing their energy costs.

In addition, having a carbon tax would also create growth in the economy with businesses providing ways of reducing energy costs through new equipment sales or retrofitting existing equipment.

This carbon tax should be revenue-neutral and overseen by the Environmental Commissioner of Ontario. The entire carbon tax revenue stream should be collected in its own fund and not go into the general revenue fund. The carbon tax revenue should be returned to the citizens of Ontario in the form of an annual dividend at income tax time, less the cost of administering the program.

Since the carbon tax would replace the fuel excise tax, the government may take some money from the carbon tax fund equivalent to the tax that would have been collected from the fuel excise tax. This is necessary to keep the carbon-tax revenue-neutral. The equivalent revenue removed from the carbon tax fund should be used to fund public transit and/or climate change programs including research and innovation.

CARBON TAX VS CAP & TRADE

The website *carbontax.org* makes an excellent case for why a carbon tax is superior in terms of transparency to cap & trade for reducing carbon emissions:

CAP-AND-TRADE'S INHERENT DEFECTS [3]

A tax on carbon emissions isn't the only way to "put a price on carbon" and provide incentives to reduce use of high-carbon fuels. A **carbon cap-and-trade system** is an alternative approach supported by some prominent politicians, corporations and mainstream environmental groups.

Cap-and-trade was the structure embodied in the Waxman-Markey climate bill that passed the House in 2009 but died in the Senate. And cap-and-trade is the cornerstone of the European Union's "Emissions Trading Scheme" (ETS).

Cap-and-trade systems can be effective under certain conditions. The U.S. *sulfur dioxide* cap-and-trade system instituted in the early 1990s efficiently reduced acid rain emissions from power plants. However, the scale of a *carbon* trading system — it would be up to 100 times larger than that for sulfur — combined with the lack of "technical fixes" for filtering or capturing CO2, rules out sulfur cap-and-trade as a model for carbon. Moreover, evidence from the EU's ETS suggests that price volatility and gaming by market participants have undermined the effectiveness of this complex, opaque indirect method of pricing carbon pollution.

The Carbon Tax Center along with most economists regard a carbon tax as vastly superior and preferable to a carbon cap-and-trade system. Here's why:

- Carbon taxes lend predictability to energy prices, whereas cap-and-trade systems exacerbate the price volatility that historically has discouraged investments in carbon-reducing energy efficiency and carbon-replacing renewable energy.
- Carbon taxes are transparent and understandable, making them more likely to elicit public support than an opaque and difficult to understand cap-and-trade system. The co-author of the U.S. Senate cap-and-trade bill, Sen. John Kerry (now Secretary of State) told a reporter in 2009, "I don't know what 'cap and trade' means. I don't think the average American does."
- Carbon taxes can be implemented more quickly than complex permit-based cap-and-trade systems.
- Carbon taxes aren't easily manipulable by special interests, whereas the complexity of cap-and-trade leaves it rife for exploitation by the financial industry.
- Carbon tax revenues can be more or less guaranteed and integrated into state or federal fiscal policy, owing to their predictability, whereas the price-volatility of cap-and-trade precludes its being counted on as a revenue source.

- Carbon taxes are replicable across borders, since the price "metric" embodied in a carbon tax is far more universal than the quantity-reduction metric underlying cap-and-trade.
- Perversely, cap-and-trade discourages voluntary/individual carbon reductions, since those cause a lowering of prices of emission permits which undercuts low-carbon investments; carbon taxes are free of this unintended negative consequence.

Politically, cap-and-trade has functioned as a "safe harbor" for politicians who grasp the need for pricing carbon emissions but cling to the need to "hide the price" to appease interest groups and/or ordinary citizens. But the point of carbon emissions pricing is to raise the price of emitting carbon. Better to make the price explicit, via a tax, and protect households by making the tax revenue-neutral.

CARBON TAX IN CANADA

On December 9, 2016, Prime Minister Justin Trudeau made the most important energy and environmental policy announcement in half a century with his pledge to create a national clean fuel standard "based on life cycle analysis".

The flow-through carbon tax we propose is consistent with the federal government's **National Clean Fuel Standard** policies [4]:

> OBJECTIVE
>
> The overall objective of a clean fuel standard would be to achieve annual reductions of 30 megatonnes (Mt) of GHG emissions by 2030. This reduction will provide a significant contribution towards achieving Canada's commitment of 30 percent emissions reduction below 2005 levels, by 2030. This reduction is like removing over 7-million vehicles from the roads for a year.
>
> BROAD COVERAGE
>
> A clean fuel standard will encourage the use of cleaner fuels in many sectors of the economy, including the fuels we use in transportation, in our homes and buildings, and the fuels that power our industries. It would address a broad suite of fuels, which could

include liquid fuels (e.g., gasoline, diesel, and heavy-fuel oil), gaseous fuels (e.g., natural gas and propane), and solid fuels (e.g., petroleum coke).

PERFORMANCE-BASED APPROACH

The clean fuel standard would set requirements to reduce the lifecycle carbon intensities of fuels supplied in a given year, based on lifecycle analysis. By contrast to renewable-fuel mandates, this approach would not prescribe the particular low-carbon fuel or technology that must be used; instead, it would focus on emissions reduction. The clean fuel standard would result in decreased emissions while minimizing compliance costs.

This approach would foster the deployment of a broad range of lower-carbon fuels and alternative technologies such as electricity, biogas, hydrogen, and renewable fuels.

MEASURING CARBON INTENSITY BY LIFECYCLE ANALYSIS

Carbon-intensity standards will be set to reduce the lifecycle carbon intensities of all fuels supplied, based on lifecycle analysis. The standards could be set at the facility level, at a sector-wide average, or set on some other basis. Carbon intensity is the measure of how much carbon is emitted into the atmosphere relative to the amount of energy in the fuel consumed.

FLEXIBILITY

The standard will be designed to provide maximum flexibility to fuel suppliers, and it may include provisions to take into account regional differences, similar to those that currently exist under the Renewable Fuels Regulations.

CARBON PRICING AND CLEAN FUEL STANDARD

The clean fuel standard would complement carbon pricing by ensuring consumers have access to a suite of lower-carbon fuels. In doing so, it will also drive down emissions and contribute to a clean-growth economy. By specifically addressing the carbon footprint of fuels, the clean fuel standard will drive innovation and create jobs and opportunities across a number of sectors, including transportation, agriculture, and clean-technology sectors.

The National Clean Fuel Standard policy mandates a broader consideration than simply transportation fuels because it covers ALL fuel uses, which represents about 80% of Canadian energy use rather than just transportation fuels as Ontario's review does. Indeed, Ontario is mandated by the Federal policy to expand their policy review to cover all fuels and to submit them and a reply to the proposed new Federal Policy.

Although Ontario’s Renewable Fuel Standard is applicable to transportation fuels, the Standard’s mandate should be expanded to include 100% of energy production and utilization, in order not only to be compliant with the National Clean Fuel Standard policy, but to exceed it by including the 20% of energy that is electricity that Ottawa left out.

THE GREEN SHIFT – STEPHAN DION

At the federal level, a carbon tax was proposed by Stephan Dion with his Green Shift program [5]. Dion proposed a carbon tax based on the greenhouse gas emissions. It would start out at $10/tonne, rising to $40/tonne within 4 years and be revenue neutral through of tax credits offsetting the revenue. While not specific about the implementation, it appeared that Dion intended to replace the federal excise tax on fuel and replace it with carbon tax applied at the wholesale level. With gasoline being taxed at the equivalent of $42/tonne of carbon dioxide, this carbon tax would not have resulted in any price increases to gasoline.

BRITISH COLUMBIA

Similar to Dion’s Green Shift plan, British Columbia introduced a $10/tonne carbon tax on Carbon Dioxide Equivalent (CO_2e) emissions [6], which would rise to $30/tonne by 2012. BC’s carbon tax was revenue neutral by reducing corporate and income taxes subject to annual legislative planning.

ALBERTA

Alberta enacted the *Specified Gas Emitters Regulation* in 2007 [7]. This carbon tax requires a $15/tonne contribution be made to the "Climate Change and Emissions Management Fund"

(CCEMF) by large carbon emitter companies (ie, emitting more than 100,000 tonnes/year of greenhouse gas) to either reduce their CO2 emissions per barrel by 12 percent, or buy an offset in Alberta to apply against their total emissions. In 2016, the contribution was raised to $40/tonne for those large emitters. The intent of tax was to get oil companies and coal-fired generators to reduce their greenhouse gas emissions.

Starting 2017, Alberta will have a similar carbon tax scheme as British Columbia, which will be applied to the entire economy. All businesses and residents will pay a carbon tax based upon the carbon dioxide equivalent emissions, including the burning of wood and biofuels. The initial tax will be $20 per tonne, increasing to $30 per tonne in 2018, and increase thereafter by the rate of inflation plus 2%.

QUEBEC

Quebec enacted Canada's first carbon tax in October of 2007. It introduced a $0.008/litre tax on diesel fuel, which the equivalent of $3.00/tonne equivalent CO2e emissions [8]. This is not a revenue-neutral program because the revenue collected is used to fund energy-efficiency programs including public transit. Quebec's program also does not appear to be transparent because the carbon tax is not broken out as component of Quebec's fuel taxes.

CARBON TAX IN THE USA

There is no carbon tax under consideration at the federal level in the USA. Several states have considered implementing carbon tax programs but no proposals have yet been passed into law. In addition, some municipal level jurisdictions implemented carbon taxes:

- Boulder, Colorado, November 2006
- Bay Area Air Quality Management District, which covers nine counties in the San Francisco Bay Area, May 2008
- Montgomery County, Maryland, May 2010

CARBON TAX IN OTHER COUNTRIES

The following countries have the some sort of carbon tax in place [7]. All appear to base their carbon pricing solely on carbon dioxide emissions.

- Costa Rica
- Denmark
- Finland
- France
- Germany
- Netherlands
- Norway
- Republic of Ireland
- Sweden
- Switzerland
- United Kingdom

Other Considerations

ENERGY STORAGE THROUGH VEHICLE BATTERIES

For Ontario to maximize the consumption of the renewable energy generated in the province and minimize the amount of surplus power exported to neighbouring jurisdictions, Ontario must have the ability to store renewable energy. One such means of storing electricity is through battery technology and every electric vehicle has a battery.

Most vehicles are used for commuting and are therefore parked most of the time either at work or at home. If sufficiently large enough, Ontario's entire fleet of electric vehicles could therefore be used to help balance Ontario's electricity supply when they are plugged-in so that every vehicle owner would be an IESO market participant. This could be accomplished with smart chargers that charge vehicles when there is a surplus and draw power when there is a demand. The vehicle's charge controller would ensure that the vehicle battery would not be drawn down below the necessary charge level for the homeward commute. The spread between the market price to charge the battery at low demand and the price to return electricity at high demand could help with the public's acceptance of electric vehicles.

Besides vehicle owners, IESO should allow each home or business owner to become an IESO market participant. Already, there is home battery technology available to store electricity for the home owner. In addition, other technology is being developed to store electricity (such as solid state ammonia synthesis and ammonia fuel cells) as home appliances, which would help to transform Ontario's grid from central generation to distributed generation. This would facilitate vehicle-to-grid (V2G) technology as well.

EMISSION COMPLIANCE

Greenhouse gas emissions are not limited to combustion. Many fuels themselves are greenhouse gases and fugitive emissions prior to combustion can also have significant adverse effects on global warming. It is crucial that the Ministry of Environment conduct regular inspections of fuel systems to ensure that system operators do not allow any leakage. There must be vigilance about fugitive emissions from the wellhead to the tailpipe.

Methane, for example, is emitted during the production and transport of coal, natural gas, and oil. Methane emissions also result from livestock and other agricultural practices and by the decay of organic waste in municipal solid waste landfills. According to the US EPA, 11% of greenhouse gas emissions in 2014 were from methane compared with 81% for carbon dioxide [9]. Since methane has a global warming potential of 28-36, the relatively smaller methane emission fraction had 3.8 to 4.9 times worse effect on global warming than carbon dioxide.

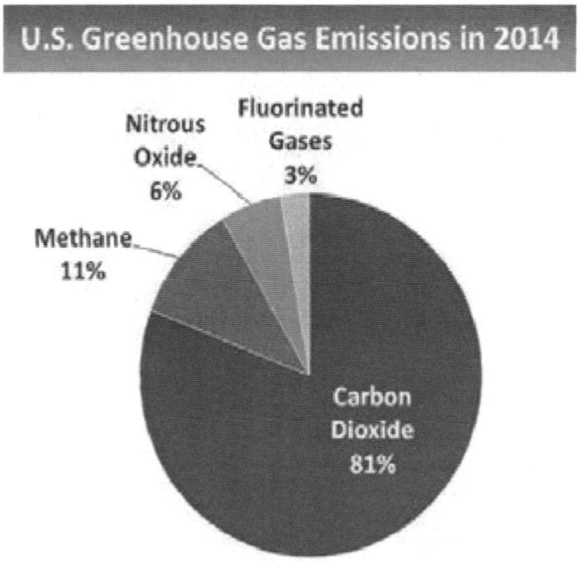

Figure 24 – U.S. Greenhouse Gas Emissions [9]

Natural gas-fueled vehicles have methane emissions both before and after combustion. In the case of LNG vehicles, as the cryogenic fuel warms, some of the methane leaks into the atmosphere as the safety relief valves blow off excess pressure. Both compression and spark ignition engines will have "methane slip" through the exhaust from incomplete natural gas combustion. Both of these sources of methane emissions must be mitigated and these vehicles should be regularly tested for emission compliance.

Figure 25 – LNG Truck Venting Excess Methane Pressure [10]

DEVELOPMENTS IN NH3 FUEL TECHNOLOGY

There has been much research that shows that ammonia can be used to fuel the most polluting forms of transport including marine, truck, and aircraft at less than half the cost and emissions of conventional fuels, even when the ammonia is made from natural gas or coal using appropriate technologies.

There have been some huge developments regarding green ammonia production, conversion of carbon in hydrocarbons into urea and use of NH3 as a fuel, for energy storage, especially in the past few months, including Mr. Gates Breakthrough Energy moving to develop technologies to produce "Zero emission ammonia", many of which actually exist commercially today.

Commercially available technologies have proven that energy and heating and cooling costs can reduced by more than 50% when ammonia is used in the mix, and there are opportunities to use a combination of fossil fuels and electricity in much more profitable ways. There are patented engines that provide CHPC (combined heat, power and cooling) at over 60%, at a fraction of the cost of any other technologies.

Existing natural gas infrastructure and power plants can use ammonia and where NH3 plants are built beside gas plants that burn NH3 with the vast amounts of pure oxygen available results in increases in efficiency as much as 50% with similar reductions in costs and virtually zero carbon, sulfur or nitrogen oxide emitted, and nothing but nitrogen (N2) and water (H2O) created as a result.

A joint submission between *Hydrofuel Inc.* and the *University of Ontario Institute of Technology's (UOIT)* submission to the *Canadian Environmental Assessment Agency (CEAA)* for the Pacific NorthWest LNG Project concluded [11]:

> "For the reasons stated in the above document and supported by Dr Ibrahim Dincer's group at UOIT (comments and MITACS report attached), Hydrofuel Inc. believes that the Pacific NorthWest LNG Project should not proceed and that the alternative of transporting energy in the form of anhydrous ammonia should be considered instead."

This submission proposes we use ammonia instead, an idea that has gained worldwide academic and public and private sector support as you will see from the details in the 13 points below and

the attachments.
Hydrofuel Inc. has completed several projects with UOIT including one that will be ready by the end of March 2017, which will provide an analysis of the economics of converting gas into NH3 and urea for transport to worldwide markets instead of using LNG. We already know that NH3 can be burned and utilized at much higher efficiencies in engines and turbines with lower emissions, adding a price for carbon changes the numbers and adding the life cycle costs will make the economics of do so even more compelling.

In late September, 2016, Hydrofuel Inc. released the *UOIT Green NH3 and Key Life cycle energy research* [12] at the *13th NH3 Fuel Association conference at UCLA* [13], where SIEMENS (supported by Innovate UK. Collaborators include the University of Oxford, Cardiff University and the Science & Technology Facilities Council) announced their "Green" Ammonia technology, pilot plant project, and strategy that *"Green" ammonia is the key to meeting the twin challenges of the 21st century.* [14]

There is worldwide development of dozens of NH3 production and utilization technologies being reported now, and in addition to the *NH3 Fuel Association* website [15] a new *Ammonia Energy* Global Information Portal website [16]

UOIT RESEARCH

- Key Life-Cycle Numbers for NH3, Fossil Fuel and Green Energy production and utilization in agriculture, energy and utilities, and transportation systems [17]
- Comparative Life Cycle Assessment of Various Ammonia Production Methods [18]
- The most recent UOIT/Hydrofuel Inc. peer-reviewed paper published free to read by Springer as part of their Nature SharedIt and formally in *Environmental Management* on "Impact Assessment and Environmental Evaluation of Various Ammonia Production Processes". The previous paper a "Comparative life cycle assessment of various ammonia production methods" was published in the Journal of Cleaner Power on November 1, 2016. [19]
- Other research Hydrofuel has done with UOIT may be found in the Latest News section of the Hydrofuel web site. [20]

EUROPEAN POWER TO AMMONIA CONFERENCE

1st European Power to Ammonia® Conference announced for May 18-19, 2017 - 1st European Conference on Sustainable Ammonia Solutions aims to gather thought societies, industries and academics, including well-known experts, developers and scientists to present the latest research results, present achievements, application fields and business prospects in energy solutions. This conference welcomes all researchers, industrialists, scientists as well as student and corporate delegates to participate and to have a great experience. [21]

JAPANESE AMMONIA USAGE

Japan has been using ammonia to fuel coal power plants reducing emissions. (Using similar technologies to this used in China for forty years.) Japanese utilities team on CO2-reducing tech for coal plants - Coal-ammonia fuel mix seen lowering emissions by at least 20%. [22]

DR. STEVE WITTRIG, US CLEAN AIR TASK FORCE

The development of the ammonia economy will depend on successful deployment of a range of new technologies in the context of the existing ammonia industry and the world's many existing energy markets. This report *Ammonia Fuel: Opportunities, Markets, Issues* [23] provides a framework for:

- Supply/demand and status of the global ammonia industry.
- Plausible response of growth and prices in response to a giant new market in fuels.
- New incremental applications which may lead to 'disruptive innovation' and rapid takeoff for ammonia fuels.
- Barriers to takeoff.

HYPER HUB PATENT

An improved system of hardware and controls, known as a Hyper Hub, that absorbs electric power from any source, including hydropower, wind, solar, and other renewable energy resources, chemically stores the power in hydrogen-dense anhydrous ammonia, then reshapes the stored energy to the power grid with zero emissions by using anhydrous ammonia to fuel diesel-type, spark-ignited internal combustion, combustion turbine, fuel cell or other electric power generators, and for other purposes. [24]

OTHER DEVELOPMENTS

1. On November 10, 2016, *Ammonia Energy* announced a 2018 pilot plant in Japan will demonstrate a new way to produce ammonia at industrial-scale, with a low carbon footprint. [25]

2. On November 28, 2016, the Trudeau Government announced its new life cycle Clean Fuels Standards policy being formulated for February 2017. [26]

3. On December 2, 2016, *Ammonia Energy* announced *Nuon*'s Power-to-Ammonia update, and the first European ammonia fuel conference in 2017. [27]

4. On December 6, 2016, the Massachusetts Institute of Technology (MIT) released their *PLAN B - FOSSIL FUELS without CO2 emissions*. Using clean-tech and ammonia makes fossil fuels clean and much more profitable. The stored chemical potential in fossil fuels from thermonuclear derived solar energy may be utilized by less conventional means without producing carbon dioxide. Chemical pathways to produce hydrogen or ammonia from hydrocarbons without co-production of carbon dioxide are possible in new process configurations. Such processes may be more cost effective than other options and more readily implemented. [28]

5. Phoenix Energy, a developer of biomass gasification plants based in San Francisco, shared some interesting data today during a Webinar on "The Future of Renewable Methane in Today's Regulatory & Policy Environment." The company operates two plants in California that together produce 5.5 tonnes/day of byproduct carbon in the form of biochar. They are currently selling all of the biochar they can produce into agricultural markets at a price of $1,700 per tonne. [29]

- *Ammonia as a fuel is a real interesting target, where it would allow you to make the transfer of that chemical potential to something that fits in well to our infrastructure ... It has the same heat of combustion as methanol, so it has plenty of power per weight. When you burn it though, you only make nitrogen and water. There's ways to get around the issues with NOx and stuff, so this is really a nice fuel. And the wonderful thing about ammonia, if you think about it, if you made it at a price for fuel, it would do something else that's really important for the poor, which is to lower the price of food ...*
- *All of the automobile companies have looked at ammonia as a possible fuel, and it's one that is certainly worthy of consideration if one wants to think about what we can do about a liquid fuel that fits into our own infrastructure that can be used more or less in our world as we know it.*

6. On December 14, 2106, *Ammonia Energy* announced the US DOE ARPA-E's REFUEL program awarded $35 Million (US) in funding to a total of 16 projects for carbon-neutral liquid fuels (of these 16 projects, 13 were focused on carbon free ammonia (none for hydrogen not associated with using ammonia), of the three others that went to carbon based energy, two were for Dimethyl ether, one looked at ethanol). [30]

7. On December 14, 2106, *Ammonia Energy* announced the Bill Gates' *Breakthrough Energy Coalition* [30] is starting work on viable grid scale and world class Green NH3 technologies after concluding that one of its initial Technical Quests is to make "Zero-GHG Ammonia Production" a reality. [31]

8. On December 16, 2016, the Trudeau government announced it has delayed the notification date of NRCan's decision on clean energy innovation funding proposals [32]. The date has been changed from late fall 2016 to winter (mid to late February) 2017. As Greg Vezina points out in the upcoming Ottawa Life Magazine article mentioned below, it appears the recent developments in support of ammonia and moving away from bio-fuels and other carbon based technologies combined with the government's new Life Cycle Clean Fuels Policy has caused a major rethink of how they will spend the $billions in proposed funding. All of this is very good news for ammonia energy proponents.

9. On January 27, 2017, *Ammonia Energy* announced more US DOE funding research into sustainable ammonia synthesis. These *Sustainable Ammonia Synthesis projects*, announced in August 2016 and administered by the Office of Basic Energy Sciences, aim "to investigate some of the outstanding scientific questions in the synthesis of ammonia (NH_3) from nitrogen (N_2) using processes that do not generate greenhouse gases." There are two substantial differences between these awards and the other recent DOE funding for sustainable ammonia technologies, announced in December 2016 and administered through ARPA-E's "REFUEL" program. [33]

10. Below are links to a number of articles we have written or co-written, and one that has been written about our work.

    A. Four articles co-written for Ottawa Life Magazine.
        i. One of Four: An Alternate View on Pipelines — Transport Ammonia not Crude [34]
        ii. Two of four: Green Ammonia [35]
        iii. Three of four: Canada's Ammonia Energy Option [36]
        iv. Four of Four: Canada's Life Cycle Clean Fuels Policy Game Changer [37]

    B. Most of the information about ammonia as a renewable fuel is not covered by the mainstream media in Canada with the exception of Ottawa Life Magazine and the Toronto Sun Editor Lorrie Goldstein who has let Greg Vezina write many SUN editorial page articles [38] and who has written about my NH3 work, most recently in:

    "High Cost of Ignorance" a Toronto Sun Editorial page article on Oct. 23, 2016. [39]

11. Research on Cost comparisons of pipelines, distribution and marine transport for natural gas and ammonia. [40]

12. Included below are four studies Hydrofuel Inc. have already done with UOIT on point, however, as mentioned above, we are completing our third *MITACS* research report in February, 2017, which will be much more extensive.

   A. Ammonia as a Potential Solution for Alberta (Submitted to Alberta Premier) [41]
   B. Natural Gas to Ammonia as a Potential Solution for British Columbia (Submitted to BC Premier) [42]
   C. *MITACS Report 1* November 2015 - "Comparative assessment of NH3 production and utilization in transportation systems" [43]
   D. *MITACS Report 2* June 2016 - "Comparative assessment of NH3 production and utilization in agriculture, energy and utilities, and transportation systems" [44]

After reviewing the links and attachments, you will see that there is very solid scientific and economic research and numerous practical applications that support using ammonia to green hydrocarbons and store renewable energy already commercially available and many more emerging almost daily.

Transparency & Other Considerations References

[1] "Understanding Global Warming Potentials," 23 02 2017. [Online]. Available: https://www.epa.gov/ghgemissions/understanding-global-warming-potentials.

[2] "Terra preta," 23 02 2017. [Online]. Available: https://en.wikipedia.org/wiki/Terra_preta.

[3] Cap-and-Trade's Inherent Defects, "Cap-and-Trade's Inherent Defects," 24 02 2017. [Online]. Available: https://www.carbontax.org/cap-and-trade-problems/.

[4] "Government of Canada to work with provinces, territories, and stakeholders to develop a clean fuel standard," 25 11 2016. [Online]. Available: http://news.gc.ca/web/article-en.do?nid=1160579.

[5] S. Dion, "the Green Shift," 23 02 2017. [Online]. Available: https://www.poltext.org/sites/poltext.org/files/plateformes/ca2008lib_plt_eng._05012009_111617.pdf.

[6] "British Columbia carbon tax," 23 02 2017. [Online]. Available: https://en.wikipedia.org/wiki/British_Columbia_carbon_tax.

[7] Wikipedia, "Carbon Tax," 23 02 2017. [Online]. Available: https://en.wikipedia.org/wiki/Carbon_tax.

[8] "Quebec's Green Future: The Lowest-Cost Route to Greenhouse Gas Reductions," 01 10 2009. [Online]. Available: https://www.cdhowe.org/sites/default/files/attachments/research_papers/mixed//backgrounder_118_English.pdf.

[9] "Overview of Greenhouse Gases," 23 02 2017. [Online]. Available: https://www.epa.gov/ghgemissions/overview-greenhouse-gases.

[10] Ultra Blaze, "Venting LNG Tanks," YouTube, 02 11 2015. [Online]. Available: https://youtu.be/vT4nG-Z3zdE.

[11] Hydrofuel Inc. & UOIT, "Pacific NorthWest LNG Project," 16 03 2016. [Online]. Available: http://www.ceaa-acee.gc.ca/050/documents/p80032/108890E.pdf. [Accessed 08 03 2017].

[12] Hydrofuel Inc & UOIT, "Key Life-Cycle Numbers for NH3 and other fuels," 16 09 2016. [Online]. Available: http://www.marketwired.com/press-release/new-evidence-on-green-ammonia-energy-solutions-be-presented-los-angeles-conference-2159195.htm. [Accessed 08 03 2017].

[13] NH3 Fuel Association, "NH3 FUEL CONFERENCE 2016: SCHEDULE," [Online]. Available: https://nh3fuelassociation.org/events-conferences/conference2016/#schedule. [Accessed 08 03

2017].

[14] Siemens PLC, ""Green" ammonia is the key to meeting the twin challenges of the 21st century.," [Online]. Available: https://www.siemens.co.uk/en/insights/potential-of-green-ammonia-as-fertiliser-and-electricity-storage.htm. [Accessed 08 03 2017].

[15] NH3 Fuel Association, "NH3 Fuel Association," [Online]. Available: http://nh3fuelassociation.org/. [Accessed 08 03 2017].

[16] Ammonia Energy, "The global information portal for ammonia as an energy vector," [Online]. Available: http://www.ammoniaenergy.org/. [Accessed 08 03 2017].

[17] Hydrofuel Inc. & UOIT, "Key Life-Cycle Numbers for NH3, Fossil Fuel and Green Energy production and utilization in agriculture, energy and utilities, and transportation systems," [Online]. Available: https://www.academia.edu/28681084/Key_Life-Cycle_Numbers_for_NH3_Fossil_Fuel_and_Green_Energy_production_and_utilization_in_agriculture_energy_and_utilities_and_transportation_systems. [Accessed 08 03 2017].

[18] Hydrofuel Inc. & UOIT, "Comparative Life Cycle Assessment of Various Ammonia Production Methods," [Online]. Available: https://www.academia.edu/19786635/Comparative_Life_Cycle_Assessment_of_Various_Ammonia_Production_Methods. [Accessed 08 03 2017].

[19] Hydrofuel Inc. & UOIT, "Impact Assessment and Environmental Evaluation of Various Ammonia Production Processes," [Online]. Available: http://link.springer.com/article/10.1007%2Fs00267-017-0831-6. [Accessed 08 03 2017].

[20] Hydrofuel Inc., "Hydrofuel Inc. Latest News," [Online]. Available: http://nh3fuel.com/index.php?option=com_content&task=category&sectionid=1&id=1&Itemid=36. [Accessed 08 03 2017].

[21] Proton Ventures, "NH3 Event - The Biggest Ammonia Event in Europe," [Online]. Available: http://nh3event.com/. [Accessed 08 03 2017].

[22] Nikkei Asian Review, "Japanese utilities team on CO2-reducing tech for coal plants," 02 03 2017. [Online]. Available: http://asia.nikkei.com/Business/Deals/Japanese-utilities-team-on-CO2-reducing-tech-for-coal-plants?n_cid=NARAN1507. [Accessed 08 03 2017].

[23] D. S. Wittrig, "Clean Air Task Force - Ammonia Fuel: Opportunities, Markets, Issues," [Online]. Available: https://arpa-e.energy.gov/sites/default/files/Wittrig_Ammonia_TransportationFuels_Workshop.pdf. [Accessed 08 03 2017].

[24] J. S. Robertson, "Energy conversion system," Google Patents, 26 09 2013. [Online]. Available: https://www.google.com/patents/US20130252120. [Accessed 08 03 2017].

[25] T. Brown, "Low-carbon ammonia synthesis: Japan's 'Energy Carriers'," Ammonia Industry, 10 11 2016. [Online]. Available: http://www.ammoniaenergy.org/low-carbon-ammonia-synthesis-japans-energy-carriers/. [Accessed 08 03 2017].

[26] Canadian Manufacturing, "Feds pledge to create national clean fuel standard in latest piece of climate plan," 28 11 2016. [Online]. Available: http://www.canadianmanufacturing.com/environment-and-safety/feds-pledge-to-create-national-clean-fuel-standard-in-latest-piece-of-climate-plan-179601/. [Accessed 08 03 2017].

[27] T. Brown, "Nuon's Power-to-Ammonia update, and the first European ammonia fuel conference in 2017," Ammonia Industry, 02 12 2016. [Online]. Available: http://www.ammoniaenergy.org/nuons-power-to-ammonia-update-and-the-first-european-ammonia-fuel-conference-in-2017/. [Accessed 08 03 2017].

[28] MIT Energy Initiative, "Plan B: Fossil fuels without CO2," 06 12 2016. [Online]. Available: http://energy.mit.edu/news/plan-b-fossil-fuels-without-co2/. [Accessed 08 03 2017].

[29] S. Crolius, "Methane to Ammonia via Pyrolysis," Ammonia Energy, 26 01 2017. [Online]. Available: http://www.ammoniaenergy.org/methane-to-ammonia-via-pyrolysis/. [Accessed 08 03 2017].

[30] T. Brown, "ARPA-E funding for renewable ammonia synthesis technologies," Ammonia Industry, 22 12 2016. [Online]. Available: https://ammoniaindustry.com/arpa-e-funding-for-renewable-ammonia-synthesis-technologies/. [Accessed 08 03 2017].

[31] T. Brown, "Breakthrough Energy Coalition targets carbon-free ammonia," Ammonia Industry, 14 12 2016. [Online]. Available: https://ammoniaindustry.com/breakthrough-energy-coalition-targets-carbon-free-ammonia/. [Accessed 08 03 2017].

[32] Natural Resources Canada, "The Energy Innovation Program : Frequently Asked Questions," 20 10 2016. [Online]. Available: http://www.nrcan.gc.ca/energy/science/programs-funding/18384#li-15. [Accessed 08 03 2017].

[33] T. Brown, "US DOE funding research into sustainable ammonia synthesis," Ammonia Industry, 27 01 2017. [Online]. Available: https://ammoniaindustry.com/us-doe-funding-research-into-sustainable-ammonia-synthesis/. [Accessed 08 03 2017].

[34] G. Vezina and F. Raso, "An Alternate View on Pipelines — Transport Ammonia not Crude," Ottawa Life Magazine, 11 07 2016. [Online]. Available: http://www.ottawalife.com/article/an-alternate-view-on-pipelines-transport-ammonia-not-crude. [Accessed 08 03 2017].

[35] G. Vezina and I. Dincer, "Green Ammonia," Ottawa Life Magazine, 16 09 2016. [Online]. Available: http://www.ottawalife.com/article/green-ammonia. [Accessed 08 03 2017].

[36] G. Vezina and S. Wittrig, "Canada's Ammonia Energy Option," Ottawa Life Magazine, 15 11 2016. [Online]. Available: http://www.ottawalife.com/article/canada-s-ammonia-energy-option?c=9. [Accessed 08 03 2017].

[37] G. Vezina and F. Raso, "Canada's Life Cycle Clean Fuels Policy Game Changer," Ottawa Life Magazine, 23 01 2017. [Online]. Available: http://www.ottawalife.com/article/canada-s-life-cycle-clean-fuels-policy-game-changer. [Accessed 08 03 2017].

[38] G. Vezina, "Editorial Article Author - Greg Vezina," Toronto Sun, [Online]. Available: http://www.torontosun.com/search?cx=016362519718727753442:ubvjwz0hpe0&cof=FORID:11&ie=UTF-8&q=%22Greg+Vezina%22&siteurl=www.torontosun.com/&ref=&ss=1514j2292196j2 [Accessed 08 03 2017].

[39] L. Goldstein, "High cost of ignorance," Toronto Sun, 22 10 2016. [Online]. Available: http://www.torontosun.com/2016/10/21/high-cost-of-ignorance. [Accessed 08 03 2017].

[40] NH3 Fuel Association, "Comparisons - The Logical Path Forward," [Online]. Available: https://nh3fuelassociation.org/comparisons/. [Accessed 08 03 2017].

[41] Hydrofuel Inc. & UOIT, "Ammonia as a Potential Solution for Alberta," 07 08 2016. [Online]. Available: http://nh3fuel.com/images/documents/2016-08-07%20-%20Ammonia%20Production%20for%20Alberta.pdf. [Accessed 08 03 2017].

[42] Hydrofuel Inc. & UOIT, "Natural Gas to Ammonia as a Potential Solution for British Columbia," 09 08 2016. [Online]. Available: http://nh3fuel.com/images/documents/2016-08-09%20-%20Ammonia%20Instead%20of%20Natural%20Gas%20for%20BC.pdf. [Accessed 08 03 2017].

[43] Hydrofuel Inc. & UOIT, "Comparative assessment of NH3 production and utilization in transportation systems for Ontario," 23 11 2015. [Online]. Available: http://nh3fuel.com/images/documents/2015-11-23%20-%20MITACS-Final-Report-P1%20.pdf. [Accessed 08 03 2017].

[44] Hydrofuel Inc. & UOIT, "Comparative assessment of NH3 production and utilization in agriculture, energy and utilities, and transportation systems," 17 06 2016. [Online]. Available: http://nh3fuel.com/images/documents/2016-06-17%20-%20MITACS-Final%20Report-P2.pdf. [Accessed 08 03 2017].

490296 Block-Copolymer Derived Nanoporous Carbon Membranes for High Throughput Gas Separation

Abstract:

Ultrathin, gas-selective membranes are highly desired to reduce the total area for a given membrane operation, reducing capital and operating costs, as well as production time and complexity for a potential scale-up. Moreover, such membranes are expected to increase the overall energy-efficiency.

Nanoporous carbon films can be obtained by pyrolysis of block-copolymer thin films. However, for their application in high permeance membrane, a crack-free synthesis on a macroporous support is not trivial. Herein, a gas-selective nanoporous carbon (NPC) membrane is reported[1] which displayed a gigantic hydrogen permeance (5.9 million GPU) that is several orders of magnitude higher than the membranes in the literature (barring the bilayer graphene membrane reported by Celebi et al.).[2] A moderate H_2/CO_2 ideal selectivity of 4.3 was obtained.

A defect-free, ultra-high flux nanoporous carbon film was synthesized by optimizing the synthesis and transfer protocols. SEM and TEM images reveal the crack-free mesoporous structure with an average pore size of 20 ± 16 nm. Permeances of He and CH4 were tested at 25 °C and at 100 °C, whereas H2, N2 and CO2 were tested in the temperature range of 25-300 °C. NPC exhibited an extremely high permeance (above 1 million GPU) for all the gases. The gas permeance scaled inversely to square root of molecular weight in the temperature range of 25-250 °C, confirming Knudsen transport regime attributed to the mesoporous structure of the NPC film.

The NPC film was stable up to a transmembrane pressure of 4 bar. XPS data confirmed that NPC film primarily consisted of C-C bonded structure (284.8 eV), with only a small contribution of C-O bonds (286.7 eV), which typically arise from the C-O-C and C-OH groups on the surface of the film.

The ultrahigh hydrogen permeance (5.9 GPU), a moderate gas selectivity (H_2/CO_2 ideal selectivity of 4.3) and a high temperature stability (250 °C) make NPC film highly attractive in separation of gases.

References

1. Dakhchoune M. et. al, Patent Application, PCT/EP2017/057684

2. Celebi, K. et. al, Science. 344, 289-292 (2014)

ACHIEVEMENTS OF HIGH CAPACITY AND LOW ENERGY CONSUMPTION WITH AMMONIA CONVERTER REPLACEMENT

Alvina Elysia Dharmawangsa[1], *Ahmad Mardiani*[1]
PT. Pupuk Kalimantan Timur, Bontang, Indonesia

Abstract

As the biggest Urea fertilizer manufacturer in Indonesia, Pupuk Kaltim currently has production capacity of 3.43 million tons per year Urea, 2.74 million tons per year Ammonia, and 350 thousand tons per year NPK. PKT-3 is one of the Ammonia and Urea plant of Pupuk Kaltim with licensing process from Haldor Topsoe for Ammonia plant and Stamicarbon for Urea plant. The production capacity of PKT-3 is 1,000 tons per day Ammonia and 1,725 tons per day Urea.

At February – March 2015, PKT-3 changed its Ammonia Converter from S-200 type to S-300 type after 26 years of operation. The improvement was expected to give greater production with less energy consumption. It was expected to reach 1,100 tons per day of Ammonia production with lower energy consumption.

The comparison of operation data before and after the improvement shows that the increase of production is gained from higher conversion in Ammonia Converter. The energy consumption decreases as the result of lower synthesis loop pressure thus decrease the import steam. The increase of conversion also decreases the ratio of process fuel and result less energy consumption for Ammonia plant.

Keywords: Ammonia converter, production capacity, less energy consumption

Introduction

Pupuk Kaltim is a subsidiary of PT Pupuk Indonesia (Persero). In 2014, PT Pupuk Kaltim took over the operation of PT Kaltim Pasifik Amoniak that currently has production capacity 3.43 million tons of Urea per year, 2.74 million tons Ammonia per year, and 350 thousand tons NPK per year. Pupuk Kaltim also produces organic fertilizer with 45 thousand tons per year capacity. The company was officially established on December 7, 1977 and located in Bontang, East Kalimantan.

PKT-3 is a 1,000 tons/day capacity of Ammonia plant and 1,725 tons/day capacity of Urea plant. PKT-3 applies the Haldor Topsoe process for Ammonia and Stamicarbon process for Urea. In February-March 2015, PKT-3 carried out Turn Around with the main point of repair was replacement of Ammonia Basket Converter from S-200 type to S-300 type. The replacement was made after 26 years of operation. During Turn Around, there were also the replacements of catalyst in Primary Reformer, High Temperature Shift Converter, and Low Temperature Shift Converter. The target of Turn Around was to obtain greater production with less energy consumption.

Ammonia Process

In general, the process of Ammonia generated from the catalytic reaction between Hydrogen (H_2) and Nitrogen (N_2) with a ratio of $H_2 : N_2 = 3 : 1$ by equation [1)] :

$$N_{2(g)} + 3H_{2(g)} \leftrightarrow 2NH_{3(g)} \qquad \Delta H = -92.200 \text{ J/mol (JM. Smith V, 1971)}$$

Ammonia synthesis reaction above is an exothermic equilibrium reaction. Hydrogen is obtained from the reaction of hydrocarbon (natural gas) with steam (reforming) and Nitrogen is obtained from air. Reforming reaction of natural gas with steam is as follows:

$$C_nH_m + 2H_2O \rightarrow C_{n-1}H_{m-2} + 3H_2 + CO_2$$
$$CH_4 + H_2O \leftrightarrow CO + 3H_2$$
$$CO + H_2O \leftrightarrow CO_2 + H_2$$

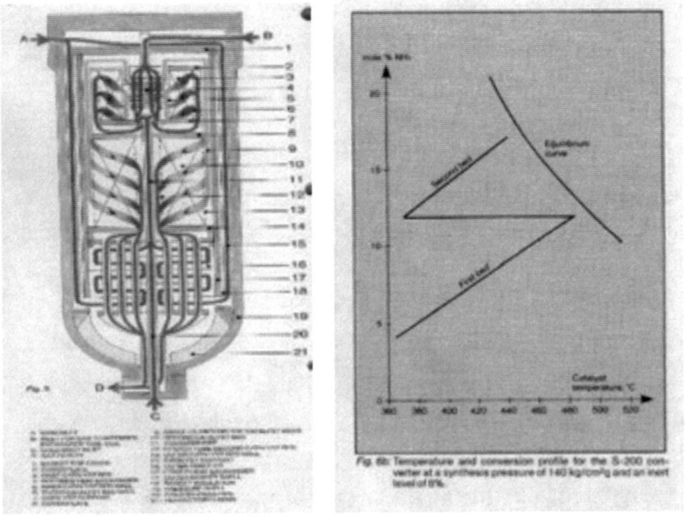

Fig. 1. S-200 type Ammonia Converter

Quality of liquid Ammonia produced at 99,9 %w and 0,1 %w of impurities. While the quality of CO2 (byproduct) at 99,9 %vol and impurities such as H_2, N_2, CH_4, CO, Ar at maximum 0,1 %vol. The process of Ammonia production with natural gas as feedstock can be divided into 5 stages as follows:

1. Feedstock purification
2. Synthesis gas preparation
3. Synthesis gas purification
4. Ammonia synthesis
5. Refrigeration

Synthesis gas with ratio of $H_2/N_2 = 3$ is compressed to reaction pressure by *Synthesis Gas Compressor* and reacted in *Ammonia Converter* with Fe catalyst, in accordance to the following reaction:

$$3H_2 + N_2 \leftrightarrow 2\ NH_3 \qquad \Delta H_{298} = -\ 21.900 \text{ kcal/kgmol}$$

Ammonia forming reaction is *reversible* and only around 25% N_2 and H_2 converted to Ammonia. After Ammonia product separation, residual unconverted gas is recirculated back to *Ammonia Converter.*

S-200 type *Ammonia Converter* is a radial flow reactor with 2 beds of catalyst and 1 interbed heat exchanger. While S-300 type Ammonia Converter has 3 beds of catalyst with 2 interbeds heat exchanger.

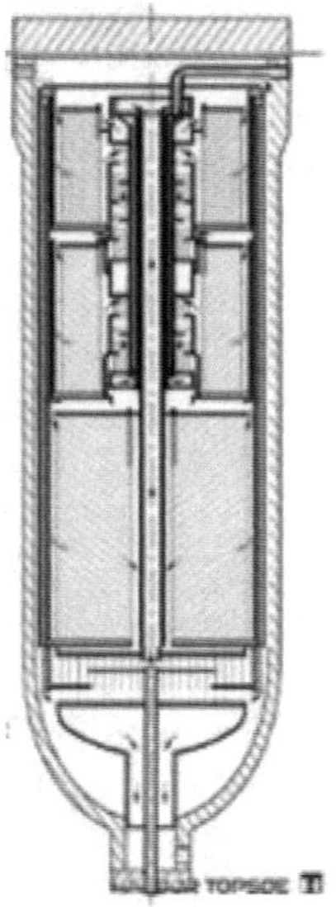
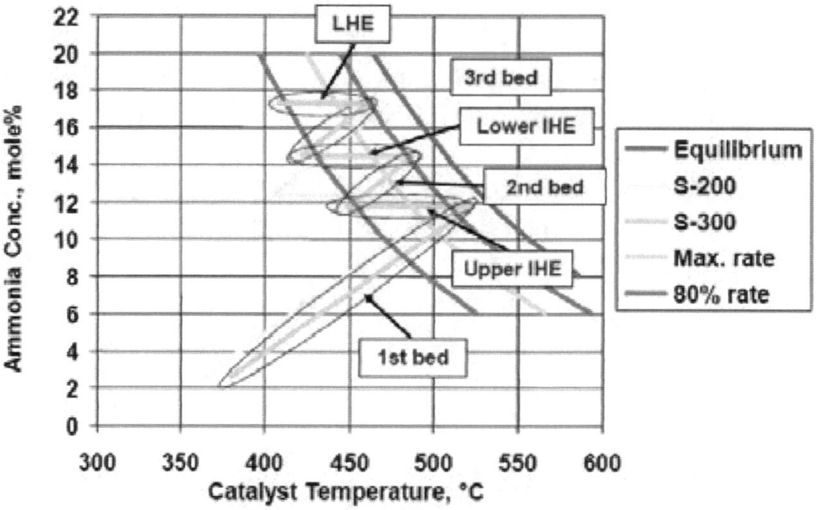

Fig. 2. S-300 type Ammonia Converter

The advantages of S-300 type Ammonia Converter compared to S-200 type are [2)]:

1. Higher conversion efficiency
2. Lower synthesis loop pressure
3. Lower energy consumption

Ammonia Converter Replacement

Interbed HE lifting

Ammonia Converter unloading activity was started at February 20, 2015 at 23:30. The catalyst was initially cooled to 30 °C by Syngas and Purging Nitrogen circulation. It took 159.5 hours for unloading catalyst at first bed and 215.25 hours for second bed of PKT-3 S-200 type Ammonia Converter. The total time required for unloading the catalyst was 374.75 hours. The unloading scheme is shown at Fig. 3.

First bed lifting

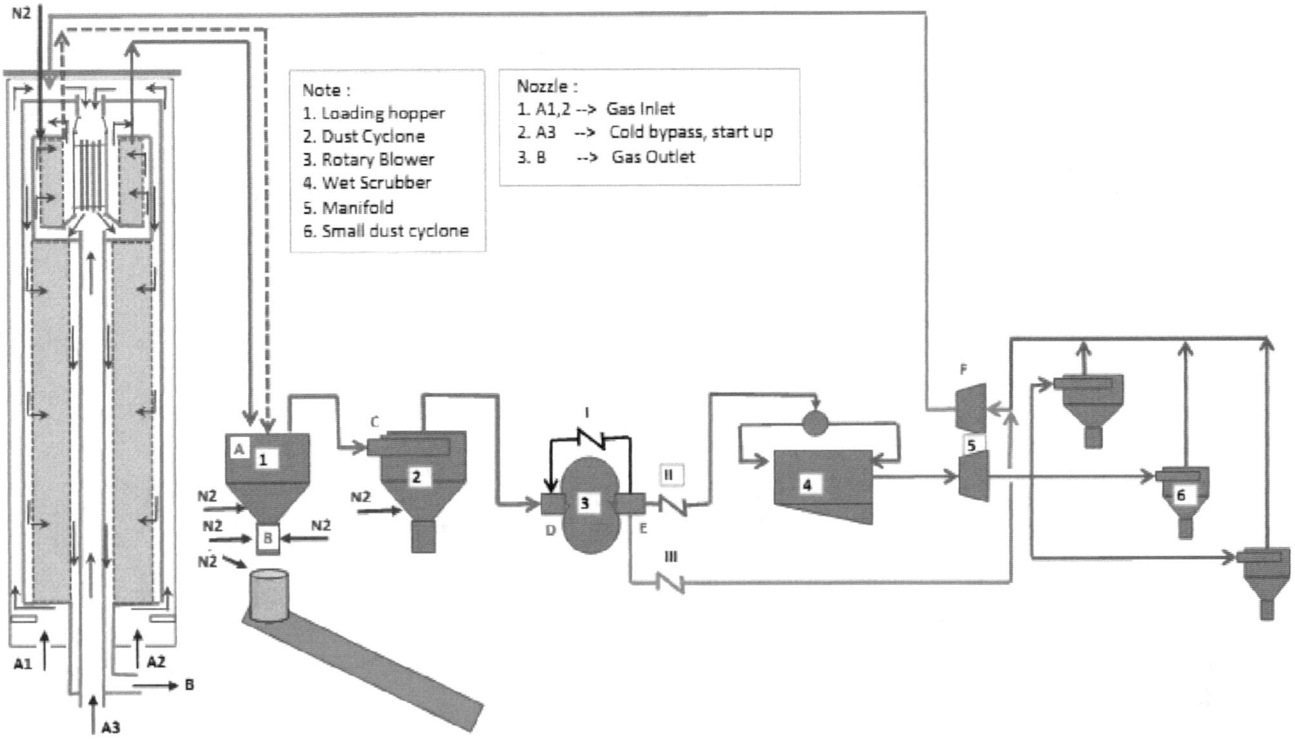

Fig. 3. Unloading Scheme of PKT-3 S-200 type Ammonia Converter

As for loading, it tooks 41.5 hours for third bed, 6.5 hours for second bed, and 10 hours for first bed of new PKT-3 S-300 type Ammonia Converter. The loading scheme is shown at Fig. 4.

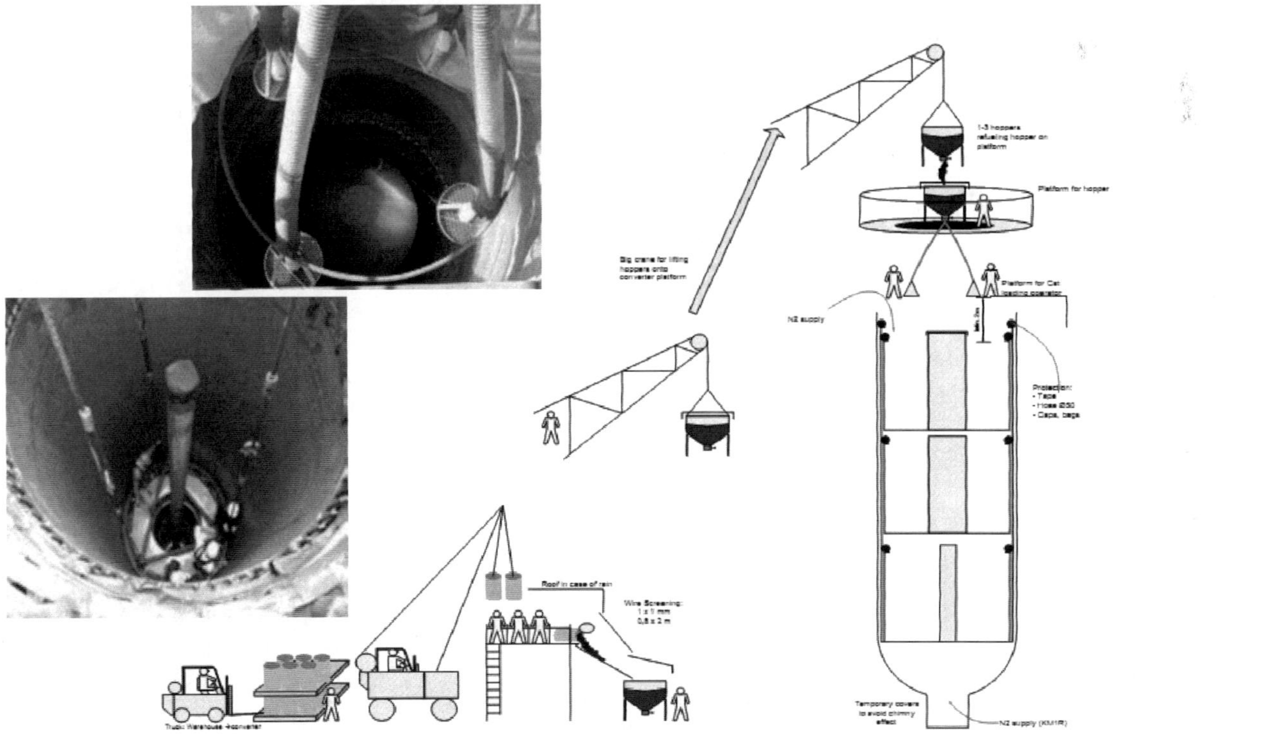

Fig. 4. Loading Scheme of PKT-3 New S-300 type Ammonia Converter

Result Analysis and Discussion

The plant operational data shows improvement in performance between PKT-3 performance before and after Turn Around. Two improvements in PKT-3 plant performance which is caused by upgrading the S-300 type Ammonia Converter are the decrease of synthesis loop pressure and the increase of Ammonia conversion at the Ammonia Converter. These improvements give the most impact on reducing PKT-3 energy consumption per ton of Ammonia after Turn Around.

The synthesis loop pressure decreases from 132.8 kg/cm^{2}G to 124,8 kg/cm^{2}G due to new S-300 type Ammonia Converter. As the synthesis loop pressure decreases, the load of Synthesis Gas Compressor becomes lower and decreases the steam consumption. Thus, steam import will also decrease as well as the overall energy consumption.

Table 1. PKT-3 Energy Consumption Difference

	Unit	Difference After – Before Turn Around	Percentage (%)
Natural Gas Process	mmbtu/ton	- 0.178	50
Natural Gas Fuel	mmbtu/ton	0.081	- 23
Net Steam Import	mmbtu/ton	- 0.248	70
H_2 Import from HRU	mmbtu/ton	- 0.009	3
Ammonia Product	ton/day	22	
Energy Consumption	mmbtu/ton	- 0.354	100

By replacing the catalyst and upgrading the S-300 type Ammonia converter, the conversion at Ammonia Converter reaction increases which has impact in the increase of Ammonia production. When the Ammonia product increased, the consumption ratio of Natural Gas Process will decrease. Thus, it also gives impact on reducing overall energy consumption per ton of Ammonia production.

After upgrading S-300 type Ammonia Converter, the Ammonia content at the outlet stream of Ammonia Converter gets higher than before. It shows the higher conversion efficiency and better performance of the Ammonia Converter. The comparison of Ammonia content at the outlet stream of Ammonia Converter before and after Turn Around is shown at Fig. 5.

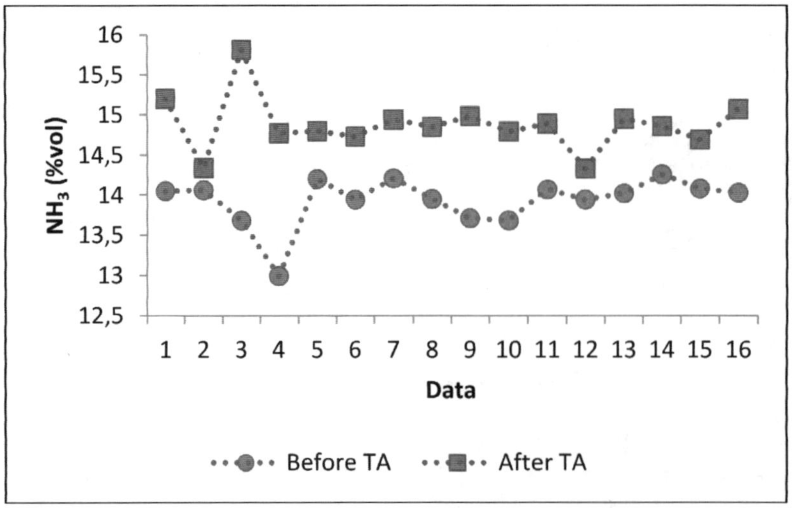

Fig. 5. Comparison of Ammonia content at the outlet stream of Ammonia Converter

Conclusion

After 26 years of operation, S-200 type Ammonia Converter at PKT-3 was replaced and upgraded to S-300 type Ammonia Converter during Turn Around at 2015. The purposes of this upgrading are to get higher conversion efficiency, lower synthesis loop pressure, and lower energy consumption.

On the same operating condition and gas rate, PKT-3 energy consumption per ton of Ammonia production after Turn Around is lower than before. The upgrading of PKT-3 S-300 type Ammonia Converter gives better performance for the plant. Two improvements in PKT-3 plant performance which is caused by upgrading the S-300 type Ammonia Converter are the decrease of synthesis loop pressure and the increase of Ammonia conversion at the Ammonia Converter. These improvements give the most impact on reducing PKT-3 energy consumption per ton of Ammonia.

The synthesis loop pressure decreases from 132.8 kg/cm^2G to 124,8 kg/cm^2G so the load of Synthesis Gas Compressor becomes lower and decreases the steam consumption. Thus, steam import will also decrease as well as the overall energy consumption.

The conversion at Ammonia Converter reaction increases which has impact in the increase of Ammonia production and reduce the consumption ratio of Natural Gas Process. Thus, it also gives impact on reducing overall energy consumption per ton of Ammonia production.

References

1. Mardiani A., dkk (2012), "Pengenalan Proses Bisnis Bagi Calon Karyawan Pupuk Kaltim", Revisi-1, *PT. Pupuk Kalimantan Timur.*
2. Abbas Naqvi Y. (2014), *"*Ammonia Basket Revamp*"*, *Haldor Topsoe.*

502971 Transition Metal Halides for Solid State Ammonia Storage: The CoX_2-NH_3Â System (X=Cl-I)

Abstract:

Transition Metal Halides for Solid State Ammonia Storage; the CoX_2-NH_3 System (X=Cl-I)"

J.Alnawmasi, D.H.Gregory,*

* *School of Chemistry, University of Glasgow, G12 8QQ, Glasgow. Tel: 0141 330 8128, duncan.gregory@glasgow.ac.uk*

Keywords: ammonia, storage, cobalt, halides, reversible, solid state.

Abstract: Ammonia has attracted interest as a promising alternative fuel for internal combustion engines due to the fact that ammonia does not release carbon dioxide during combustion but equally NH_3 could be used directly with fuel cells.[1] Storing ammonia in the form of metal ammines is the most notable way to overcome the challenges that are related to the use of NH_3 in liquid form namely the toxicity and corrosive nature of ammonia.[2] Ammine complexes of light transition metal halides such as (Co, Ni) are very promising candidates because they exhibit high ammonia contents and they are relatively stable at room temperature.We have found that the CoX_2 (X = Cl, Br, I) systems are interesting candidates and especially cobalt chloride as it possesses a NH_3 capacity of 44 wt. % and desorption is completed by 330 °C. We are currently investigating the chemical and structural properties of the Co halides during both ammonia uptake (at ambient conditions) and ammonia release (upon heating). The ammoniated metal halides were studied by powder X-ray diffraction, FTIR and Raman spectroscopy. The crystal structure of the hexammine has been determined and the products of the first and second decomposition steps were confirmed as the respective diammine and monoammine. The thermal properties of the hexammine cobalt halides were studied by TPD(TG-DTA-MS) measurements. These measurements were intended to investigate their deammoniation properties and the products of decomposition. In addition, the cycling behaviour and the stability of the ammoniated cobalt chloride in air were also investigated (figure1).

Figure1. *The cycling behaviour of $CoCl_2.6NH_3$.*

References:

[1] Steele, R. B. Chemtech 1999, (August), 28.

[2] C. H. Christensen, R. Z. Sørensen, T. Johannessen, U. J. Quaade, K. Honkala, T. D. Elmøe, R. Køhler, J. K. Nørskov, J. Mater. Chem., 2005, 15, 4106–4108.

503273 Adaptive Test Bed for Anhydrous Ammonia-Based Energy Systems

Abstract:

Proof-of-concept requires bench-top testing of material coupons and working fluids. Engineering development, including hardware design and fabrication, system integration, and scale-up testing can be even more difficult. The authors have designed a platformâ€•or test bedâ€•for developmental testing of subsystems and components of anhydrous ammonia-based energy systems.

We refer to our concept as a *composite ammonia flow battery* (CAFB). It consists of the following subsystems: wind turbines, power conditioning, ammonia synthesis unit, ammonia storage, system data acquisition and control (SCADA), and output. If the end product is electricity, the output subsystem can be either a fuel cell or a motor-generator; no additional subsystem is required if the system output is to be liquid fuel or fertilizer. Numerous options exist for the synthesizerâ€•for example Haber-Bosch, molten salt, polymer-electrolyte membrane, solid oxide, non-thermal plasma reactor, or others. For long-term testing an air separation unit and electrolyzer could be incorporated in the synthesis subsystem; but short-term tests could operate on simulated external power and bottled gases. The storage subsystem could be refrigerated; but because NH_3 is readily stored with no losses at ambient temperature and 8-10 atm, the latter is more attractive for energy systems. The output subsystem could include a splitter to reduce NH_3 to H_2 and N_2 and the H_2 fed to a hydrogen fuel cell. However, recent progress toward developing a direct ammonia fuel cell (DAFC) indicates that the splitter may become unnecessary.

We have developed an algorithm for predicting full or partial system efficiency from measured performance of the subsystems and components; and we are configuring the whole system so that it can be packaged in a CONEX container for transportation to a field test site (e.g., a wind farm).

Clean Fuel Standard Discussion Paper

Ammonia (NH_3) as a Carbon-Free Fuel

University of Ontario Institute of Technology

Prof. Dr. Ibrahim Dincer

Yusuf Bicer

Hydrofuel Inc.

Greg Vezina (Chairman and CEO)

Frank Raso

April 21, 2017

Ammonia as a Carbon-Free Fuel for Use in Transportation and Power Generation

The action plan lays out the specific commitments Canada is making from Ontario's point of view to meet its 15% overall greenhouse gas emissions reduction target by 2020. Emissions from total passenger transportation (cars, trucks, bus, rail and domestic aviation) have grown almost 15% since 1990, to 36 million tonnes of CO_2e, approximately 66% of Ontario's 2014 transportation emissions. This growth was driven by an increase in vehicle-kilometres travelled as well as a shift in the composition of the fleet from cars to sport-utility vehicles, pick-ups and minivans where the specific contributions of the vehicle types are shown in Fig 1.

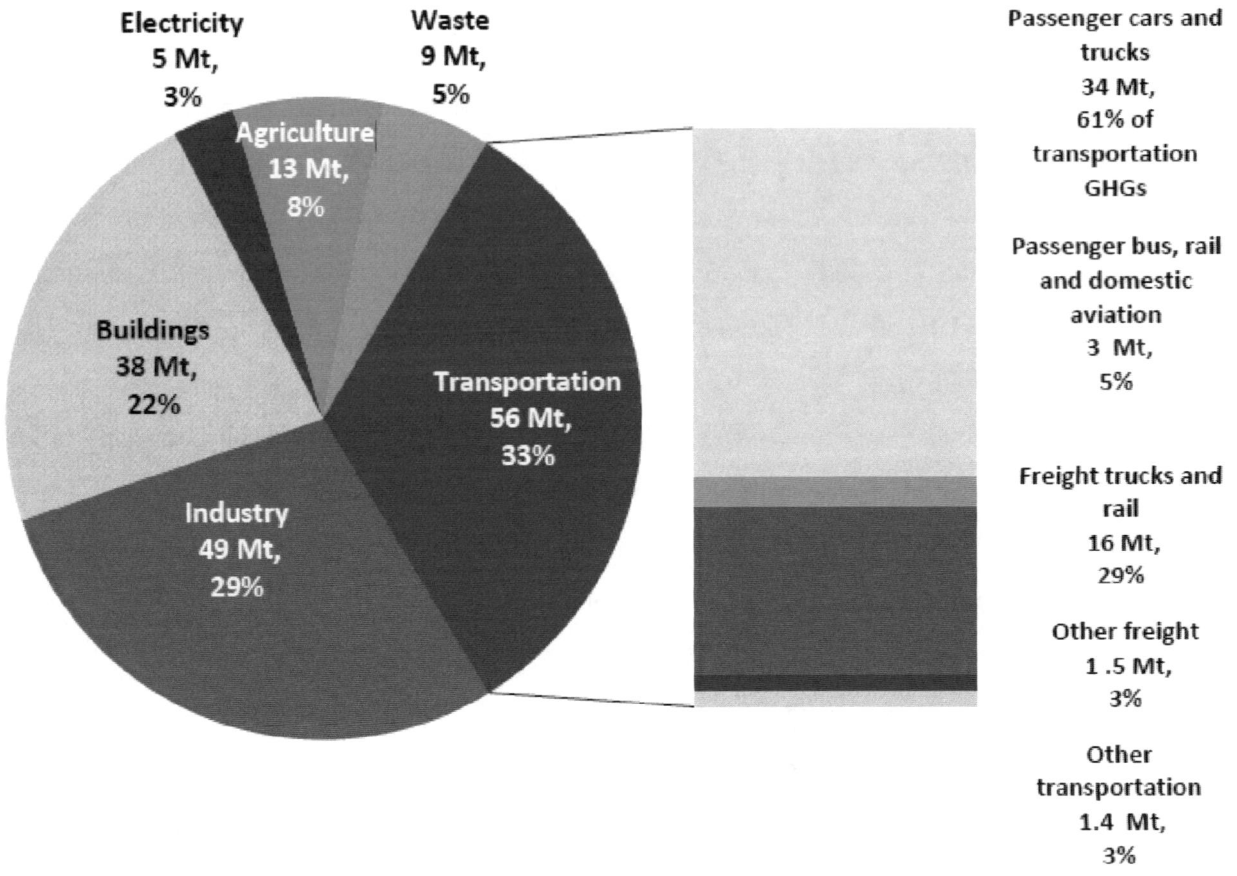

Fig. 1. Ontario's greenhouse gas emissions in 2014 (data from Ref. 1).

Here, it is important to note that under Environment and Climate Change Canada's economic sector categorization, most off-road transportation emissions are allocated to their host economic sectors and consequently are not included under transportation in Fig. 1. For example, emissions from diesel combustion in farm equipment are categorized under Agriculture. Therefore, in reality GHG emissions caused by the transportation vehicles are higher than 33% in Ontario.

Additionally, emissions from total freight transportation (trucks, rail and other) have increased more drastically over the period, rising 85% since 1990, to almost 18 million tonnes of CO_2e (approximately one third of Ontario's current transportation emissions). This was driven by a significant increase in the use of diesel-fuelled heavy-duty trucks, with additional kilometres travelled offsetting improvements in efficiency [1].

This brief report tries to address the following points:

1. Targets and blending requirements:
    a. Ontario's has existing content requirements for ethanol in gasoline. What minimum level of ethanol blending and GHG performance would help support the objectives of the RFS?
    b. Given Ontario's GHG reduction targets for 2030 and 2050, what factors should be considered in setting RFS targets post-2020?
2. Flexibility mechanisms:
    a. Should activities to lower the carbon intensity of other conventional transportation fuels be eligible for compliance purposes?
    b. Should investments in low-carbon transportation projects also be eligible for compliance purposes? If yes, what types of projects?
3. Assessing lifecycle emissions
    a. Should an RFS consider impacts from indirect land-use changes (ILUC),7 even though science in this area continues to evolve? If so, how?
4. Transparency:
    a. What measures can be taken to increase transparency and support business decision making under an RFS (e.g. an information registry, bulletins, guidance material)?
5. Others:
    a. What other considerations should be included in the discussion?

AMMONIA FACTS

Ammonia is one of the largest synthesized industrial chemical in the world having over 200 million tonne per production per year.

Ammonia (NH_3):

- consists of one nitrogen atom from air separation and three hydrogen atoms from any conventional or renewable resources.
- is the second largest synthesized industrial chemical in the world.
- is a significant hydrogen carrier and transportation fuel that does not contain any carbon atoms and has a high hydrogen ratio.
- contains about 48% more hydrogen by volume than liquefied hydrogen.
- does not emit direct greenhouse gas emission during utilization
- can be used as solid and/or liquid for many purposes.
- can be stored and transported under relatively lower pressures.
- can be produced from various type of resources ranging from oil sands to renewables.
- is a suitable fuel to be transferred using steel pipelines with minor modifications which are currently used for natural gas and oil.
- can be used in all types of combustion engines, gas turbines, burners as a sustainable fuel with only small modifications and directly in fuel cells which is a very important advantage compared to other type of fuels.
- brings a non-centralized power generation via fuel cells, stationary generators, furnaces/boilers and enables smart grid applications.
- can be used as a refrigerant for cooling in the car.

WHAT ARE THE USES OF AMMONIA?

Ammonia is considered a possible working fluid for thermodynamic cycles, working for refrigeration, heating, power or any mixture of those can be coupled with internal combustion engines, and using exhaust gasses to drive automotive absorption refrigeration system.

Ammonia has been recognized and employed as a leading refrigerant in the industrial sector due to its outstanding thermal properties, zero ozone depletion and global warming potential (GWP). Ammonia has the highest refrigerating effect per unit mass compared to all the refrigerants being used including the halocarbons. The remarkable advantages of ammonia over R-134a could be lower overall operating costs of ammonia systems, the flexibility in meeting complex and several refrigeration needs, and lower initial costs for numerous applications. Ammonia has better heat transfer properties than most of chemical refrigerants and consequently allow for the use of equipment with a smaller heat transfer area. Thereby plant construction cost will be lower. But as these properties also benefit the thermodynamic efficiency in the system, it also reduces the operating costs of the system. In many countries the cost of ammonia per mass is considerably lower than the cost of HFCs. This kind of advantage is even multiplied by the fact that ammonia has a lower density in liquid phase. Modern ammonia systems are fully contained closed-loop systems with fully integrated controls, which regulate pressures throughout the system. Ammonia is used as refrigerant highly in the refrigeration structures of food industry like dairies, ice creams plants, frozen food production plants, cold storage warehouses, processors of fish, poultry and meat and a number of other uses.

It is also stimulating to note that NH_3 is a reduction agent for the NOx typically current in combustion releases. The reaction of NOx with ammonia over catalysts produces only steam and nitrogen. An average car needs only approximately 30 ml of NH_3 per 100 km to neutralize any NOx emissions. If the vehicles run with NH_3 as fuel, this amount is unimportant with respect to the fuel tank volume.

Ammonia is used as fertilizer in the agriculture. It is also converted into urea by reacting with CO_2. The majority of growth in ammonia usage is expected to be for industrial uses and the production of fertilizer products.

It is also worth to examine the option to cool the engine with ammonia that can act as a refrigerant while it is heated to the temperature at which it is fed to the power producer (ICE or fuel cell). Optionally, the cooling outcome of ammonia, i.e., its high latent heat of evaporation, may be used to harvest some air conditioning onboard.

IS AMMONIA REALLY A FUEL?

Ammonia as a sustainable fuel can be used in all types of combustion engines, gas turbines, and burners with only small modifications and directly in fuel cells. Ammonia was initially used as a fuel for buses in Belgium in 1940s [2]. Many studies have already been performed and many applications have been implemented so far. A prototype unit for combustion which enabled liquid kerosene and gaseous ammonia to be fed, and ammonia was combusted in a gas-turbine unit. Further studies have been performed by various researchers which have proven the practicality of using ammonia as fuel [3, 4, 5, 6, 7, 8, 9, 10]. Numerical studies of combustion characteristics of ammonia as a renewable fuel have been conducted. Ammonia can also be used a fuel blending option for current gasoline and diesel engines. Combustion and emissions characteristics of compression-ignition engine using dual ammonia-diesel fuel have been performed. Performance enhancement of ammonia-fueled engine by using dissociation catalyst has been studied. These are just a few examples to show the current progress in the ammonia utilization options in transportation applications.

IS AMMONIA A SUITABLE FOR TRANSPORTATION SECTOR?

The storage and delivery infrastructure of ammonia is similar to liquefied petroleum gas (LPG) process. Under medium pressures (5-15 bar), both of the substances are in liquid form which brings the significant advantage because of storage benefits. Today, vehicles running with propane are mostly accepted and used by the public since their on-board storage is possible and it is a good example for ammonia fueled vehicle opportunities since the storage and risk characteristics of both substances are similar to each other. An ammonia pipeline from the Gulf of Mexico to Minnesota and with divisions to Ohio and Texas has served the ammonia industry for many years. It indicates that there is a working ammonia pipeline transportation which can be spread overall the world. The potential of ammonia usage in many applications will be dependent on the availability of ammonia in the cities. Ammonia is a suitable substance to be transferred using steel pipelines with minor modifications which are currently used for natural gas and oil. In this way, the problem of availability of ammonia can be eliminated.

HOW CAN AMMONIA BE USED IN TRANSPORTATION?

Ammonia has significant potential as an alternative fuel to further the sustainable development of transportation sector. A few of the following alternatives are shown in Fig. 2 for direct ammonia usage in transportation applications.

Currently, the majority of the locomotive fleet is made up of diesel-electric locomotives, operating with either two-stroke or four-stroke prime mover diesel engines that is coupled to an electric generator. Application of ammonia fuel for internal combustion engine (ICE) with the alternative locomotive configurations direct feed, or a combination of direct feed and decomposition subcategory options will bring more sustainable solutions. Additionally, fuel cell driven vehicles and locomotives may contribute to solve the associated matters of urban air superiority and national energy security influencing the rail and transportation sector.

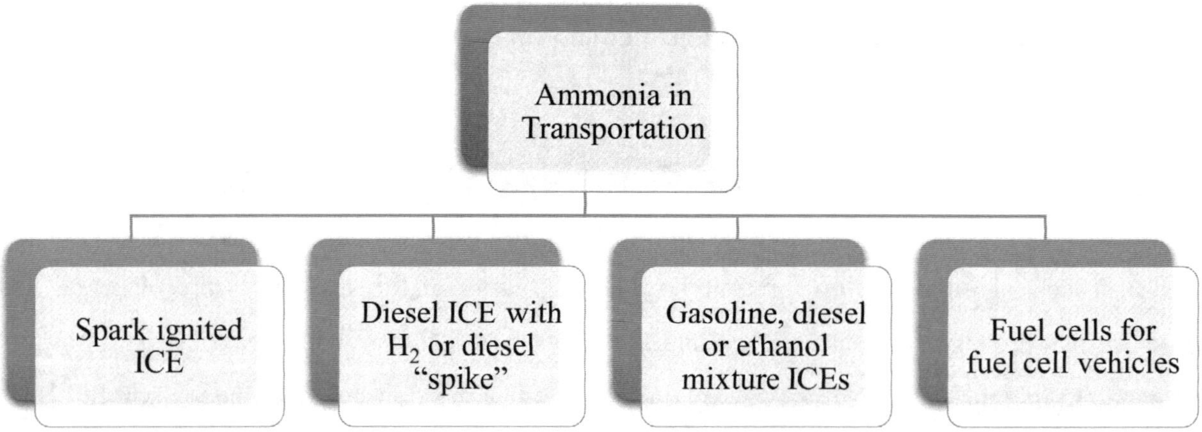

Fig. 2. Ammonia utilization options in transportation sector

IS AMMONIA A CLEAN FUEL?

Compared to gasoline vehicles, ammonia-fueled vehicles do not produce direct CO_2 emission during operation. Since ammonia produces mainly water and nitrogen on combustion, replacing a part of conventional fuel with ammonia will have a large effect in reducing carbon dioxide emissions.

WHAT FACTORS SHOULD BE CONSIDERED IN SETTING RFS TARGETS POST-2020?

- Environmental impact (including the indirect land-use changes)
- Cost
- Availability of clean production routes (e.g. solar, hydropower, wind)
- Blending properties (e.g. combustion characteristics)
- Fuel production life cycle emissions
- Fuel storage and transport infrastructure
- Integrability
- Multi-purpose usage (e.g. power, cooling, heating)

WHICH ENVIRONMENTAL IMPACTS SHOULD BE CONSIDERED?

A life cycle is the set of phases of a product or service system, from the extraction of natural resources to last removal. Overall environmental impact of any process is not complete if only operation is considered, all the life steps from resource extraction to disposal during the lifetime of a product or process should be considered. The selection of future vehicle options can strongly depend on the emission characteristics. As the world struggles with greenhouse gas emission reduction policies, global warming potential is the main characteristics to compare the total CO_2 equivalent emission from the alternative vehicles. Abiotic depletion, human toxicity, ozone layer depletion appear to play an important role for decision of using clean transportation vehicles because there are vast amount of road vehicles in the cities which can cause severe side effects. Moreover, when considering alternative fuels, issues such as land use, fertilizer use, water for irrigation, waste products etc. are necessary points to be addressed. Therefore, indirect land-use changes (ILUC) should be also considered. Indirect land-use changes can also have important social and environmental impacts which can include biodiversity, water quality, food prices and supply, community and cultural stability. Assessing the indirect land-use changes is a knowns as challenging topic. Some methods of quantifying indirect land-use changes can be listed as follows:

- Implementing empirical calculations based on previously experienced indirect land-use changes
- Developing life cycle analyses methodology with lower uncertainty ranges
- Developing integrated models combining the life cycle, sustainability, efficiency, social cost etc.

The following environmental impact categories represent higher significance in life cycle assessment approach, hence suggested to be included in decision making processes:

- Global warming potential is the main characteristics to be compare the total CO_2 equivalent emission from any source.
- Abiotic resources are natural resources including energy resources. Since fossil fuels resources are declining gradually, abiotic depletion potential is also a significant category.
- Human toxicity may play an important role for decision of using alternative fuels.
- Acidification potential is for acidifying substances which causes a wide range of impacts on soil, groundwater, surface water, organisms, ecosystems and materials.
- Marine aquatic eco-toxicity refers to impacts of toxic substances on marine aquatic ecosystems which is more important for maritime transportation sector.
- Land occupation/land use refers to the total arrangements, activities and inputs undertaken in a certain land cover type. The term land use is also used in the sense of the social and economic purposes for which land is managed.

HOW MUCH GREENHOUSE GAS CAN I SAVE IF I DRIVE AN AMMONIA CAR?

Considering a complete life cycle counting the production, transport and usage of the fuel, a diesel driven car can emit greenhouse gas emissions of about 220 g per km. Ammonia driven car can decrease this number down to about 70 g per km if it is produced from solar energy and about 150 g per km if it is produced from hydrocarbon cracking.

IS AMMONIA A COST EFFECTIVE FUEL?

The illustrative cost comparison of various fueled vehicles is shown in Fig. 3 and 4. Considering the current market prices of the fuels, ammonia is the lowest cost fuel corresponding to about 3.1 US$ in a 100 km driving range. This shows that ammonia is a promising transportation fuel in terms of cost. There is an advantage of by-product refrigeration which reduces the costs and maintenance during vehicle operation. Some additional advantages of ammonia are commercial availability and viability, global distribution network and easy handling experience. Ammonia is a cost effective fuel per unit energy stored onboard compared to methanol, CNG, hydrogen, gasoline and LPG as shown in Fig. 3.

In Table 1, the exact fuel energy per mass is given regarding fuel's higher heating value. The volumetric energy of the fuel is found by multiplying the HHV with the density value listed in the third column. Ammonia's HHV is around half of the one of gasoline, and its density is also inferior. Therefore liquid NH_3 stores 2.5 fewer energy per unit capacity than gasoline. If the NH_3 is stored in the form of hexaamminemagnesium chloride to remove the hazard related to its toxicity, the energetic cost to pay for discharging ammonia reduces its HHV. Among alternative fuels, ammonia yields the lowest cost per energy basis. Therefore, it is important to note that low-carbon transportation projects should be eligible for compliance purposes.

Table 1. Comparison of ammonia with other fuels

Fuel/storage	Pressure (bar)	Density (kg m^{-3})	HHV (MJ kg^{-1})	HHV per Volume (GJ m^{-1})	Energy per Volume (GJ m^{-1})	Cost per Mass (US\$ kg^{-1})	Cost Per Volume (US\$ m^{-1})	Cost per Energy (US\$ GJ^{-1})
Gasoline, C_8H_{18}/liquid	1	736	46.7	34.4	34.4	1.03	754.86	21.97
CNG, CH_4/integrated storage	250	188	42.5	10.4	7.8	0.91	170.60	21.29
LPG, C_3H_8/pressurized tank	14	388	48.9	19	11.7	1.06	413.66	21.74
Methanol, CH_3OH/liquid	1	786	14.3	11.2	9.6	0.41	317.80	28.31
Hydrogen, H_2/metal	14	25	142	3.6	3	3.02	75.49	21.29

hydrides								
Ammonia, NH_3/pressurized tank	10	603	22.5	13.6	11.9	0.23	136.63	10.04
Ammonia, NH_3/metal amines	1	610	17.1	10.4	8.5	0.23	138.14	13.21

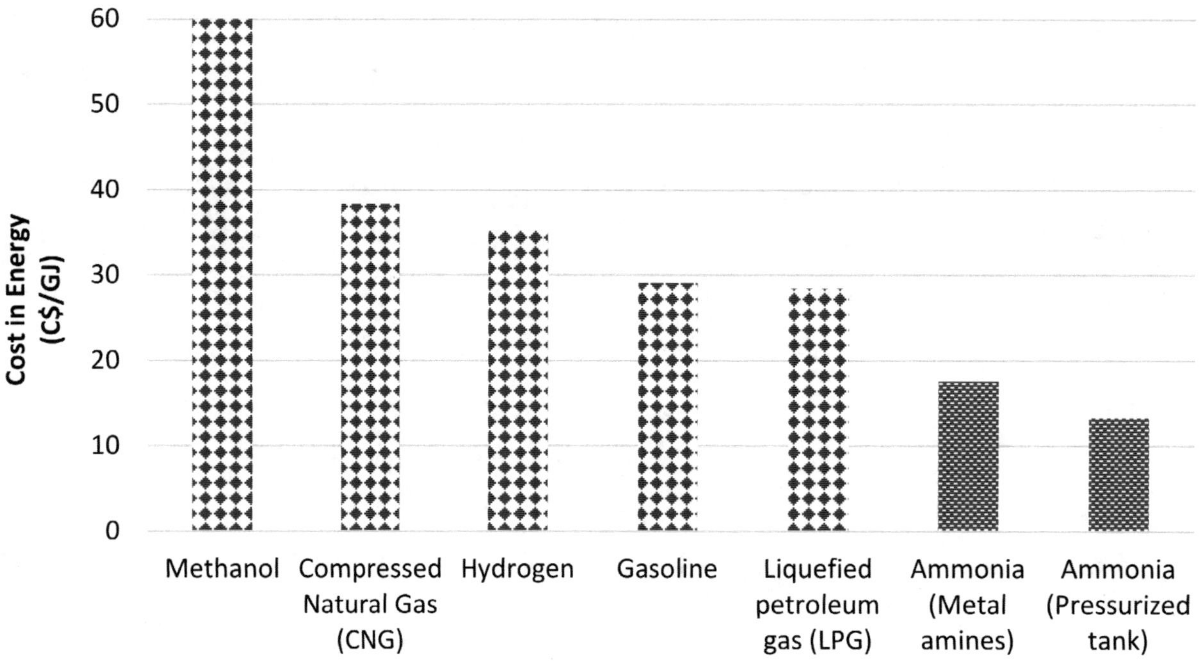

Fig. 3. Comparison of various vehicle fuels in terms of energy cost per gigajoule

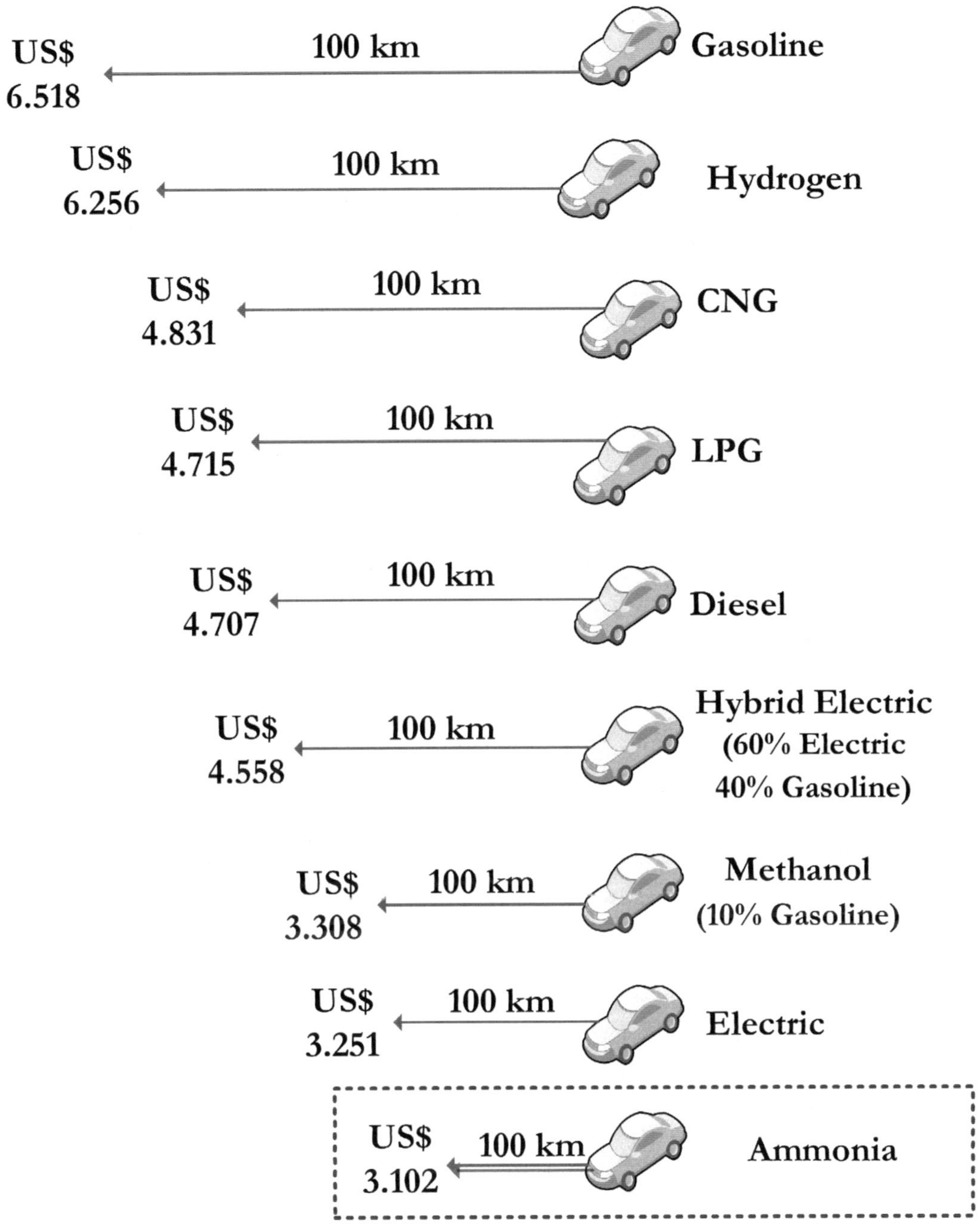

Fig. 4. Comparison of driving cost for various fueled vehicles

WHAT IS THE PROCESS OF AMMONIA PRODUCTION?

A most common ammonia synthesis technique is recognized as Haber-Bosch process in the world. In this process, nitrogen is supplied through air separation process. Hydrogen is mainly supplied using steam methane reforming or coal gasification. However the source of hydrogen can be renewable resources. The Haber-Bosch is an exothermic process that combines hydrogen and nitrogen in 3:1 ratio to produce ammonia. The reaction is facilitated by catalyst and the optimal temperature range is 450-600°C.

Alternative production pathways are also available and under investigation including electrochemical and biological routes. These routes can easily be integrated to renewable energy sources for cleaner production. The electrochemical process can be carried out under ambient conditions or at higher temperatures depending on the type of the electrolyte material used. There are various electrochemical pathways such as molten salt, polymer membrane, liquid electrolyte etc. are intensively being researched at the moment [11, 12, 13, 14, 15, 16].

The electrochemical process can be carried out under ambient conditions or at higher temperatures depending on the type of the electrolyte material used. For high temperature electrolytic routes of ammonia production, the use of waste heat from thermal or nuclear power plants or heat from renewable energy sources like solar would make the overall process more environmentally friendly.

One of the key advantages of ammonia is to be a storage medium. Renewable energy generation does not often match electrical demand which causes a requirement of storage. Green ammonia can be manufactured from surplus renewable sources, which would reduce the amount of electricity exported to neighboring jurisdictions at a negative cost.

WHAT IS THE SOURCE OF AMMONIA AND IS IT CLEANER THAN OTHER FUELS?

In terms of conventional resources, naphtha, heavy fuel oil, coal, natural gas coke oven gas and refinery gas can be used as feedstock in ammonia production. Natural gas is the primary feedstock used for producing ammonia in worldwide corresponding to about 72%. However, renewable resources can easily be integrated for ammonia production. In this way, decentralized ammonia production can be realized which further decreases the delivery cost of the fuel. Many studies have been performed to investigate the ammonia production routes and their environmental impacts. Here, some of them are briefly shown.

The production of the different fuels is compared in terms of abiotic depletion of sources as shown in Fig. 5. Ammonia fuel has the lowest abiotic depletion value compared to others although the production process may be fossil fuel based. There are multiple pathways for ammonia production. Ammonia is cleaner when produced from renewable resources. Fig. 6 shows the comparison of ozone layer depletion values for various transportation fuels. Ammonia has lowest ozone layer depletion even if it is produced from steam methane reforming and partial oxidation of heavy oil.

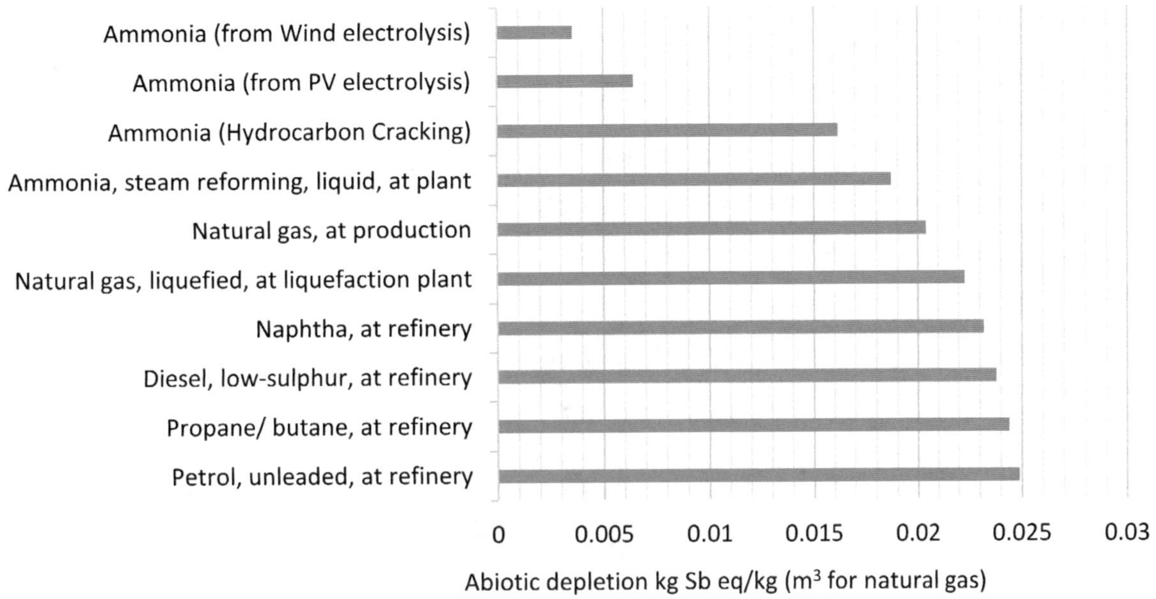

Fig. 5. Abiotic depletion values during production of various fuels

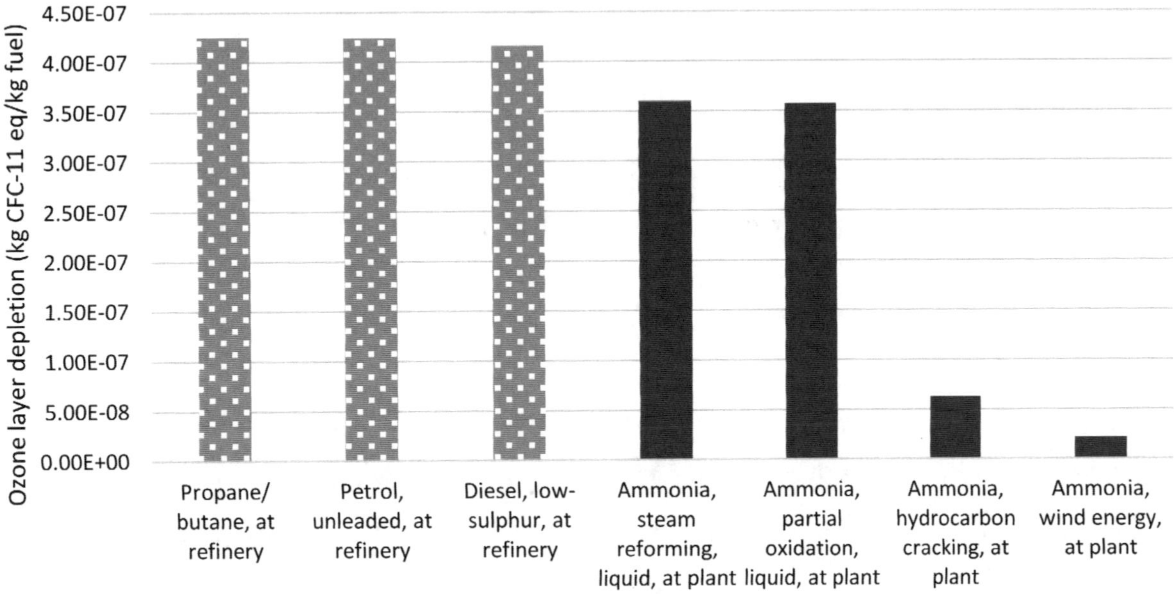

Fig. 6. Ozone layer depletion during productions of various fuels

Fig. 7 compares the total greenhouse gas emissions during production of 1 MJ energy from various resources including gasoline, LPG, diesel, natural gas and ammonia. Production of 1 MJ energy from ammonia has lower emissions than gasoline, LPG, diesel, oil and natural gas.

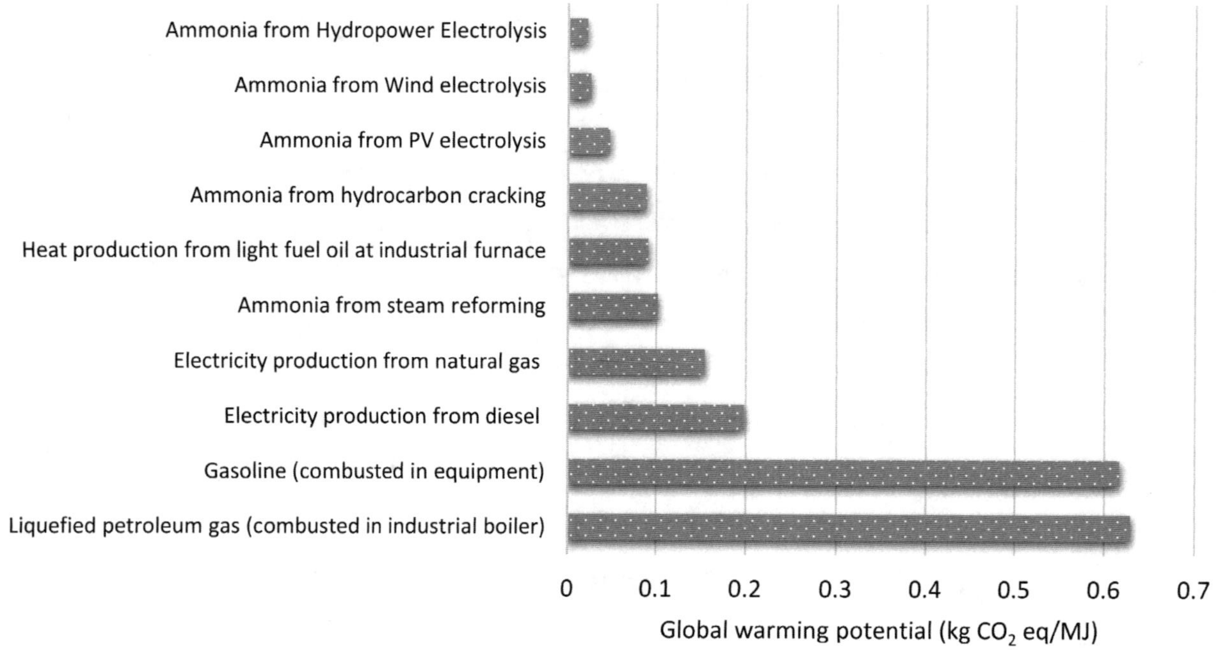

Fig. 7. Comparison of global warming potential of 1 MJ energy production from various resources

Giving priority for complete conversion from fossil fuel based fuels to carbon-free fuels will bring short term and long term solutions to combat global warming. Therefore, the activities to lower the carbon intensity of conventional transportation fuels be eligible for compliance purposes

AMMONIA IN ROAD TRANSPORTATION

Assessing the life cycle emissions from vehicles is a powerful method for transportation sector. The usage of fossil fuel based electricity decreases the attractiveness of electric (EV) and hybrid electric (HEV) vehicles. Henceforth, noteworthy attention should be paid to the power generation technologies and their CO_2 intensity, used to supply electricity to EVs or HEVs. The

GHG emissions of EVs and HEVs are therefore dependent on the CO_2 intensity of the energy mix and differs based on the countries. A characteristic life cycle of a vehicle technology can be categorized into two main steps, namely fuel cycle and vehicle cycle. In the fuel cycle, the processes beginning from the feedstock production to fuel utilization in the vehicle are considered. In the vehicle cycle, utilization of fuel is considered. Among the selected categories, global warming, abiotic depletion and human toxicity results carry more significant decision parameters for road vehicles. The results presented here are given on per km basis based on the fuel consumption rates given in Table 2.

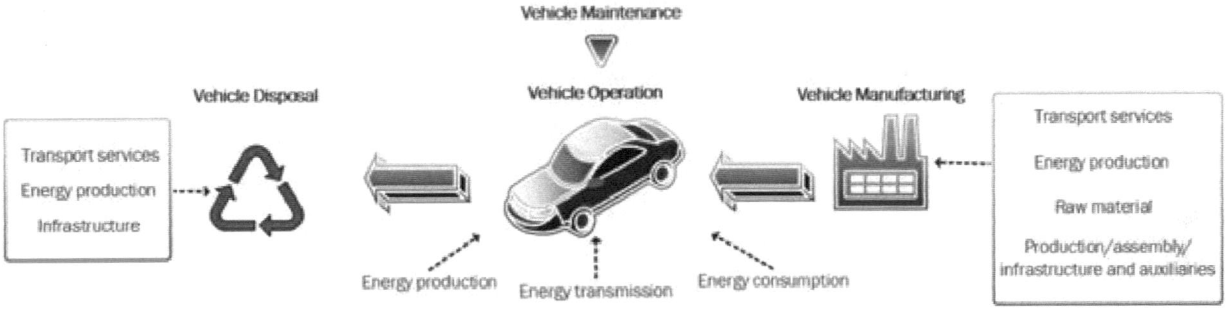

Fig. 8. Complete life cycle of vehicles including fuel/vehicle cycle.

Table 2. Energy consumptions per km for the selected vehicles

Fuel		Fuel/Energy Consumption	Unit
Gasoline		0.0649108	kg/km
Diesel		0.0551536	kg/km
M90	Methanol	0.1180535	kg/km
	Gasoline	0.0060664	kg/km
Hydrogen		0.0195508	kg/km
Ammonia		0.0926600	kg/km
EV		0.2167432	kWh/km
HEV	Electric	0.1083716	kWh/km
	Gasoline	0.0324554	kg/km
CNG		0.0603914	kg/km
LPG		0.057629687	kg/km

The specific conditions for the selected vehicles are presented herein:

- Gasoline: All processes on the refinery site excluding the emissions from combustion facilities, including waste water treatment, process emissions and direct discharges to rivers are accounted for. The inventory data also includes the distribution of petroleum product to the final consumer including all necessary transports. Transportation of product from the refinery to the end user is considered together with operation of storage tanks and petrol stations. Emissions from evaporation and treatment of effluents are accounted for. Particulate emissions cover exhaust- and abrasions emissions.
- Diesel: Diesel is evaluated as low-Sulphur at regional storage with an estimation for the total conversion of refinery production to low-Sulphur diesel. An additional energy use (6% of energy use for diesel production in the refinery) has been estimated. The other processes are similar to gasoline. Particulate emissions cover exhaust- and abrasions emissions.
- CNG: Natural gas with a production mix at service station is taken into account. The inventory data contains electricity necessities of a natural gas service station together with emissions from losses. The data set represents service stations with high (92%), medium (6%) and low (2%) initial pressure. VOC emissions are obtained from gas losses and contents of natural gas. Particulate emissions cover exhaust- and abrasions emissions.
- Hydrogen: Hydrogen is produced during cracking of hydrocarbons. It includes combined data for all processes from raw material extraction until delivery at plant. The output fractions from an oil refinery are composite combinations of mainly unreactive saturated hydrocarbons. The first processing step in converting such elements into feedstock suitable for the petrochemical industries is cracking. Essentially a cracker achieves two tasks in (i) rising the complexity of the feed mixture into a smaller number of low molecular mass hydrocarbons and (ii) presenting unsaturation into the hydrocarbons to enable more reactivity. The raw hydrocarbon input from the refinery is fed to the heater unit where the temperature is increased. The forming reaction products vary based on the composition of the input, the temperature of the heater and the residence time. The cracker operator selects temperature and residence time to enhance product mix from a supplied input. Cracker feeds can be naphtha from oil refining or natural gas or a mixture of both. After exiting the heater, the hydrocarbon gas is cooled to prevent extra reactions. After that, it is sent to the separation phase where the individual hydrocarbons are separated from one another by fractional distillation. Particulate emissions cover exhaust- and abrasions emissions. In order to have comparable results where hydrogen

comes from non-fossil fuels such as solar PV and nuclear, they are also taken into account in the analyses by applying water electrolysis route. The electrolyzer is assumed to consume about 53 kWh electricity for one kg of hydrogen production.

- Ammonia: Ammonia synthesis process is Haber-Bosch which is the most common method in the world. Ammonia production requires nitrogen and hydrogen. In this study, hydrogen is assumed to be from hydrocarbon cracking as explained in the previous paragraph. Cryogenic air separation is mostly used method for massive amount of nitrogen production. In the life cycle assessment of nitrogen production, electricity for process, cooling water, waste heat and infrastructure for air separation plant are included. Haber-Bosch process is an exothermic method that combines hydrogen and nitrogen in 3:1 ratio to produce ammonia. The reaction is facilitated by catalyst (iron-oxide based) and the optimal temperature range is 450-600°C. Particulate emissions cover exhaust- and abrasions emissions. In order to have comparable results where ammonia comes from non-fossil fuels such as solar PV and nuclear, they are also taken into account in the analyses. The generated hydrogen from electrolyzers are used for ammonia synthesis plant.
- EV: Electricity consumption is included. Particulate emissions comprise exhaust and abrasions emissions. Heavy metal emissions to soil and water caused by tire abrasion are accounted for. In the electricity usage process, electricity production mix, the transmission network and direct SF6-emissions to air are included. In order to present a renewable based scenario for electric vehicles, a mixture of renewables for energy requirement during the operation are also evaluated consisting of 25% biomass, 25% solar PV, 25% wind power and 25% hydropower.
- HEV: Hybrid car is assumed to be 50% electric and 50% gasoline with ICE. Electricity and gasoline consumptions are included. Particulate emissions comprise exhaust and abrasions emissions. Heavy metal emissions to soil and water caused by tire abrasion are accounted for. For the hybrid vehicle's electricity, a mixture of renewables for energy requirement during the operation are also evaluated consisting of 25% biomass, 25% solar PV, 25% wind power and 25% hydropower.
- Methanol: The selected fuel M90 consists of 90% methanol and 10% gasoline. The raw materials, processing energy, estimate on catalyst use, and emissions to air and water from process, plant infrastructure are included. The process describes the production of methanol

from natural gas via steam reforming process to obtain syngas for the production of methanol. There is no CO_2 use and hydrogen is assumed as burned in the furnace. Raw materials, average transportation, emissions to air from tank storage, estimation for storage infrastructure are included for the distribution part where 40% of the methanol is assumed to be transported from overseas. Particulate emissions cover exhaust- and abrasions emissions.

- LPG: All processes on the refinery site excluding the emissions from combustion facilities, including waste water treatment, process emissions and direct discharges to rivers are considered. All flows of materials and energy due to the throughput of 1 kg crude oil in the refinery is accounted for. Refinery data include desalting, distillation (vacuum and atmospheric), and hydro treating operations. Particulate emissions cover exhaust- and abrasions emissions.

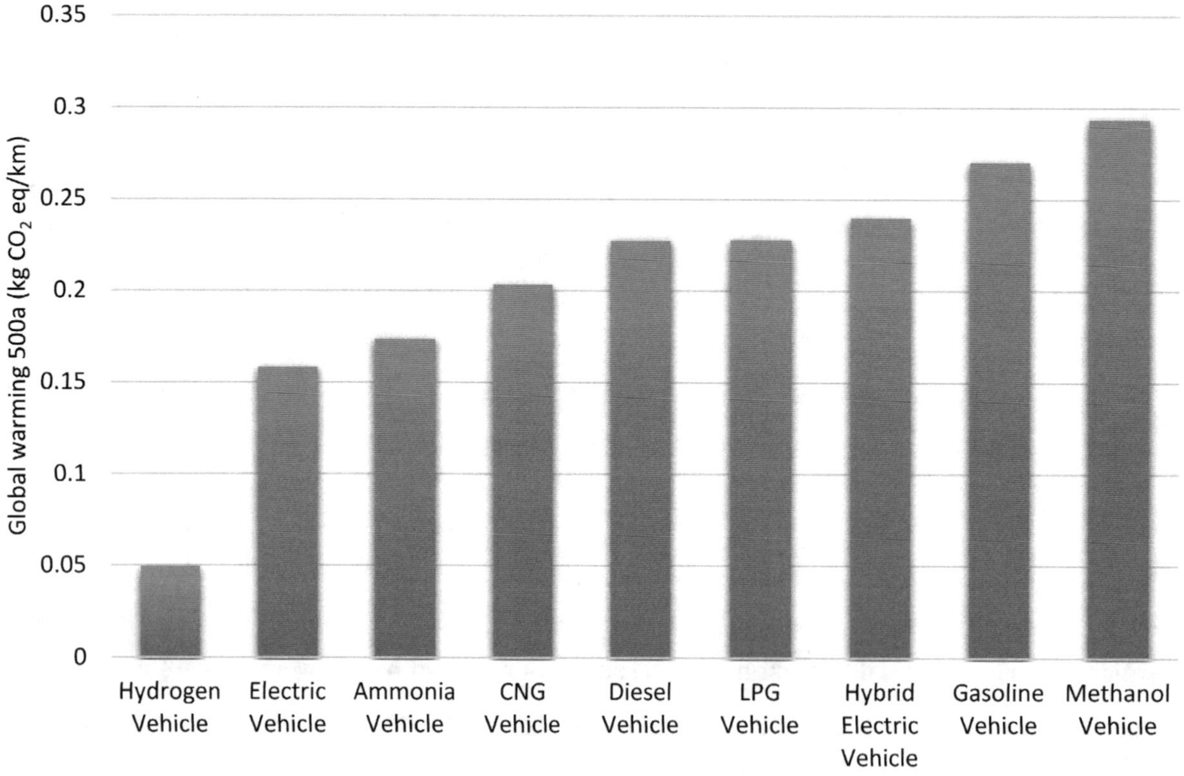

Fig. 9. Life cycle comparison of global warming results for various vehicles

The global warming potentials of assessed vehicles are comparatively shown in Fig. 9. The lowest GHG emissions are observed in hydrogen, electric and ammonia vehicles corresponding to 0.049 kg CO_2 eq/km, 0.15 kg CO_2 eq/km and 0.17 kg CO_2 eq/km, respectively. Hydrogen consumption is quite lower than ammonia consumption in the passenger car because of higher energy density. It is an expectable result that EVs also yield lower global warming potential, however production pathway of electricity has a key role in GHG emissions. If electricity production can be realized by renewable sources such as solar, biomass, hydropower and wind energy, total emissions would decrease for both EVs and HEVs.

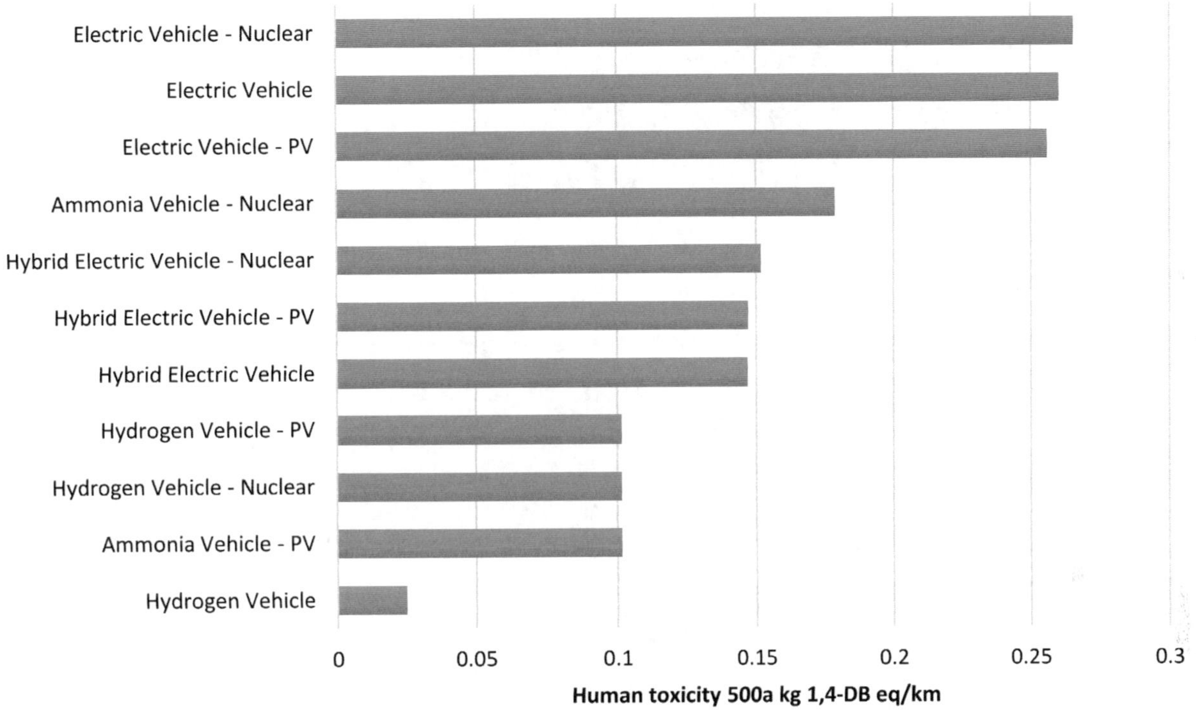

Fig. 10. Life cycle comparison of human toxicity results for various vehicles from nuclear energy and solar PV routes

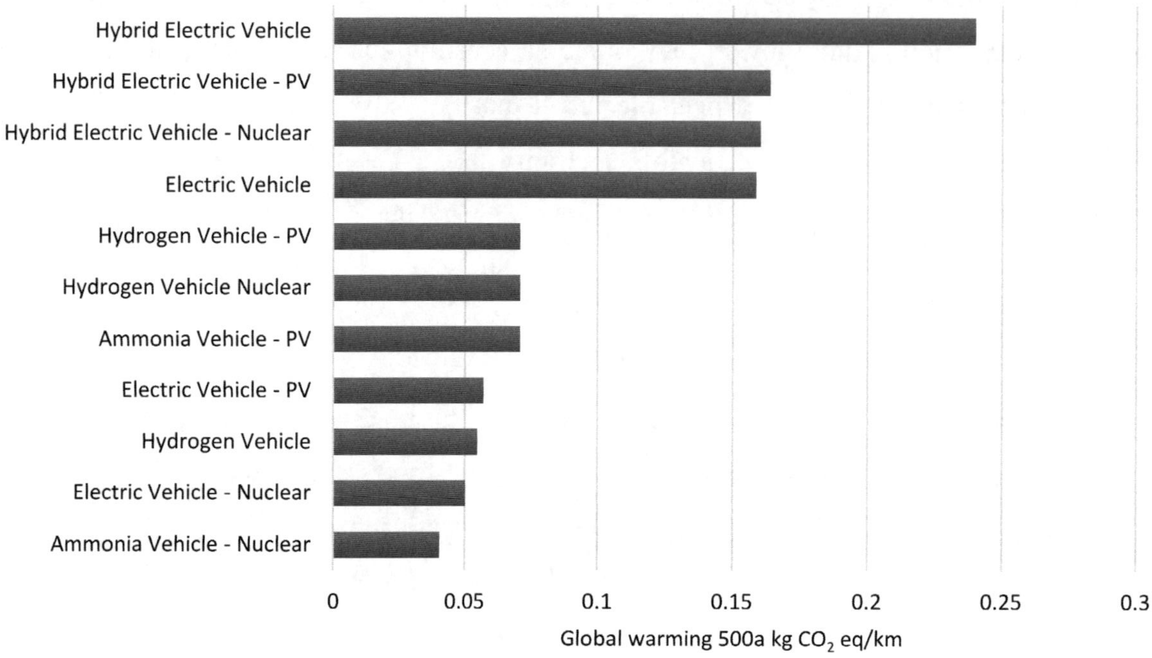

Fig. 11. Life cycle comparison of global warming results for various vehicles from nuclear energy and solar PV routes

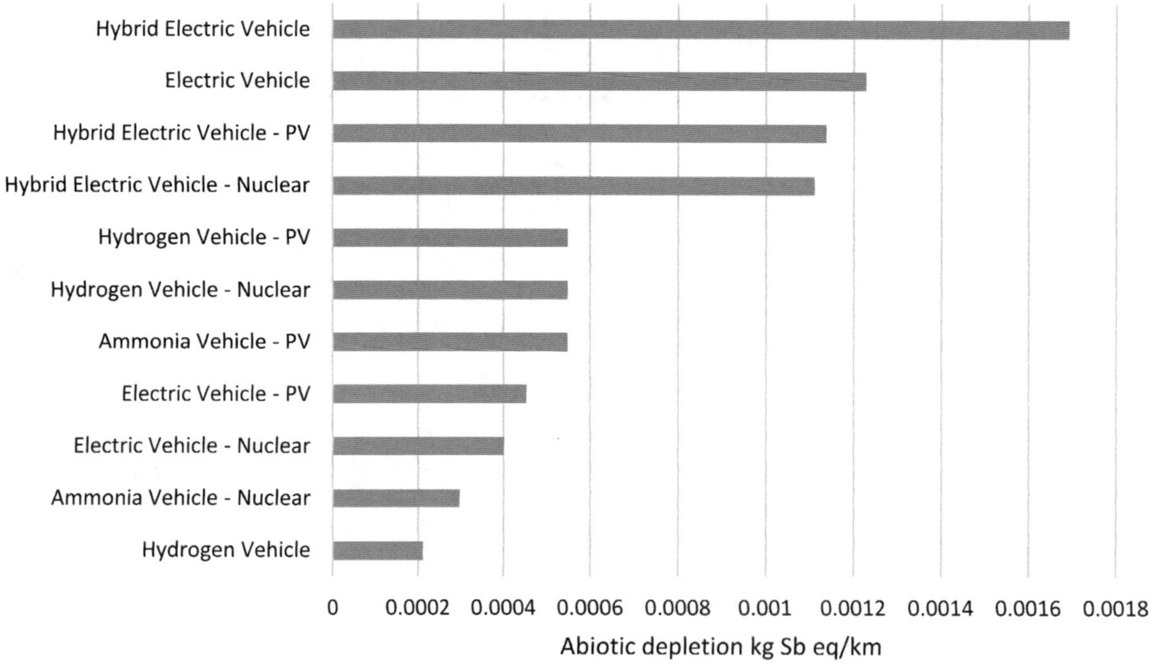

Fig. 12. Life cycle comparison abiotic depletion for various vehicles from nuclear energy and solar PV routes

The electricity from nuclear (25%), biomass (25%), hydropower (25%) and solar PV (25%) are equally used only for the operation processes of the EVs and HEVs vehicles as renewable mix. Utilization of renewable electricity for EVs, HEVs, ammonia and hydrogen vehicles are comparatively shown for various impact categories. For human toxicity category, using solar and nuclear energy does not cause significant reductions for EVs and HEVs as shown in Fig. 10. However, for global warming and abiotic depletion categories, both solar energy and nuclear energy routes lower the environmental impacts more than 50%. Ammonia driven vehicle where the ammonia is produced from nuclear electrolysis method yield the lowest GHG emissions corresponding to about 0.04 kg CO_2 eq. per km as shown in Fig. 11. Hydrogen and EVs (from nuclear and PV) have quite similar greenhouse gas emissions in the range of 0.049-0.057 kg CO_2 eq. per km. In terms of abiotic depletion values, hydrogen vehicle still yields the lowest value whereas nuclear routes for ammonia and EVs further decrease the abiotic depletion impact as shown in Fig. 12.

The results show that hydrogen and ammonia vehicles are the most environmentally benign ones in most of the environmental impact categories. Ammonia as a sustainable and clean fuel has lowest global warming potential after EVs and yield lower ozone layer depletion values than EVs. However, in case renewables are used both for ammonia vehicles and EVs, ammonia can suggest lower environmental impacts. Although EVs do not emit direct CO_2 during operation, the production and disposal processes of batteries bring some consequences which harm the environment in terms of acidification, eutrophication and human toxicity.

AMMONIA IN MARITIME TRANSPORTATION

Decreasing the global warming potential caused by current transportation technologies and fuels can be reduced significantly by replacing alternative clean fuels. Sea transportation vehicles mostly use heavy fuel oil or diesel fuel for power generation. Ocean tankers and freight ships require massive amount of energy for operation.

Being a sustainable energy carrier that can be generated from any primary energy source, ammonia can subsidize to a broadening of maritime fuel resources and may offer the long term option of being generated from renewable resources.

A brief life cycle consideration of maritime vehicles, ocean tanker and freight ship, is shown here where the tankers are driven with hydrogen and ammonia instead of heavy fuel oils in the power engines. Additionally, dual fuel options, heavy fuel oil and hydrogen/ammonia, are investigated.

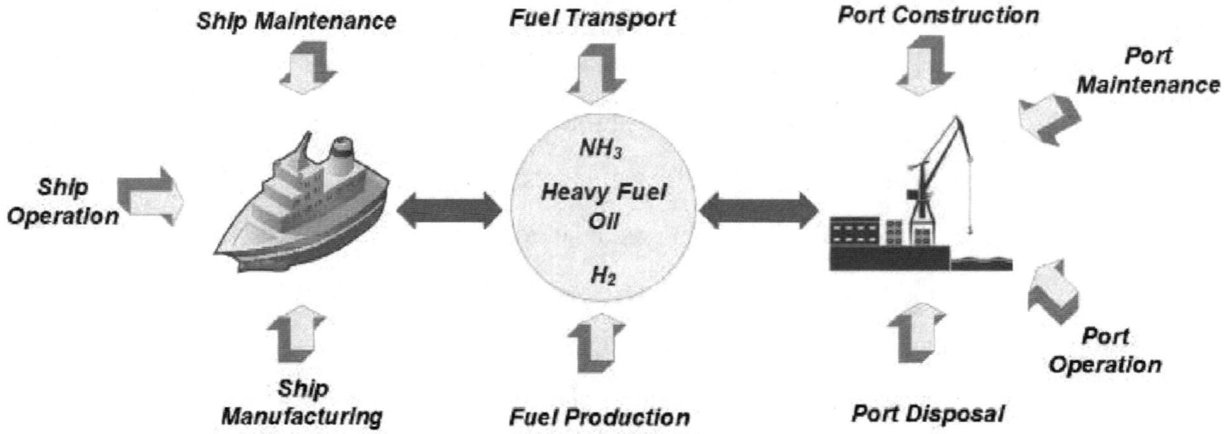

Fig. 13. Life cycle steps of maritime transportation

A comparative life cycle assessment of transoceanic fright ship and transoceanic tanker, is performed to examine the effects of clean fuel driven maritime vehicles on the environment. The complete transport life cycle is evaluated in the life cycle analyses comprising of manufacture of tanker/freight ship; operation of tanker/freight ship; construction and land use of seaport; operation, maintenance and disposal of seaport; production and transportation of hydrogen and ammonia. A tonne kilometer by shipping is defined as unit of measure of goods transport which represent the transport of one tonne by a vessel over one kilometer.

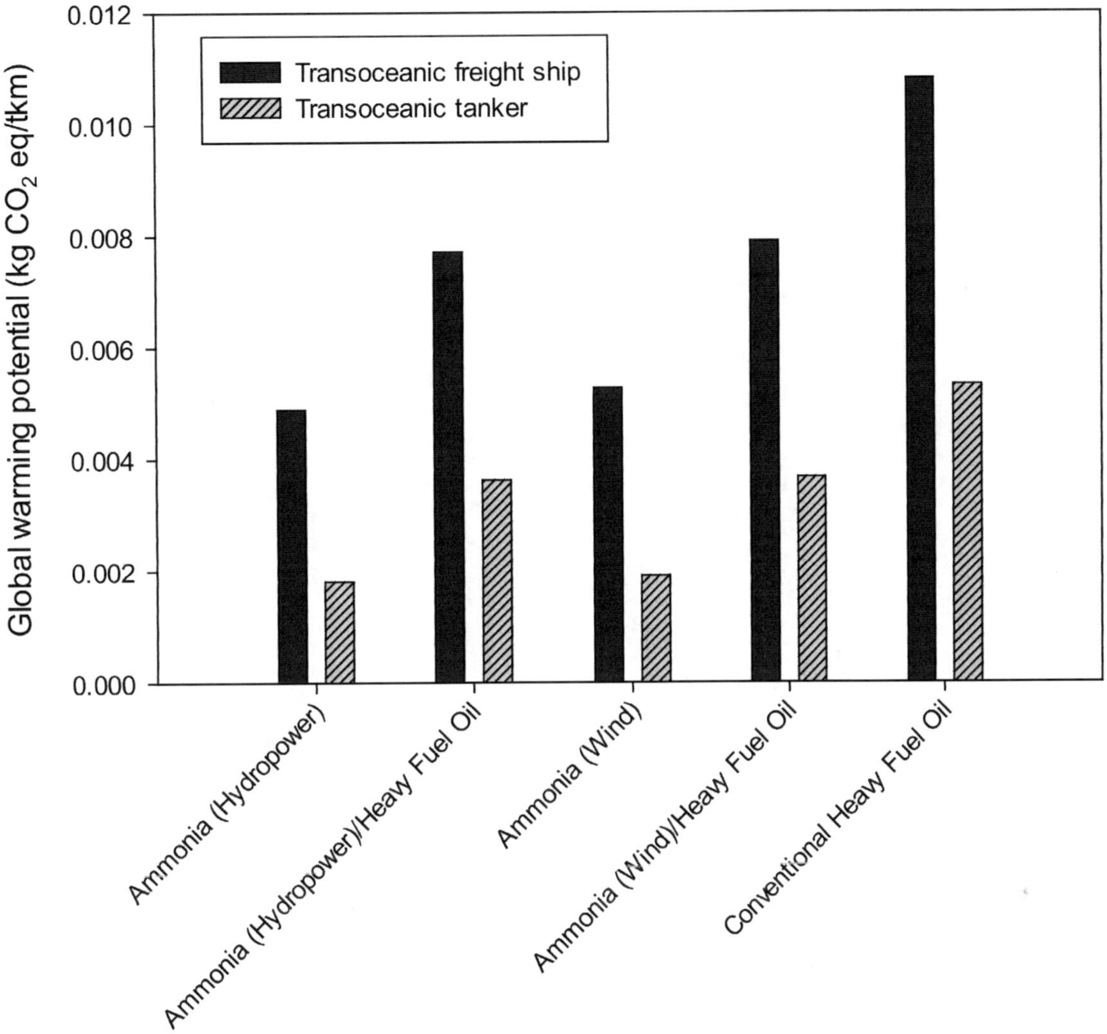

Fig. 14. Global warming potential of transoceanic tanker and transoceanic freight ship per tonne kilometer for clean fuels and conventional heavy fuel oil

The heavy fuel oil combustion significantly contributes to many of the environmental impact. Using ammonia as dual fuel in the marine engines can decrease total greenhouse gas emissions up to 34.5% per tkm. Similarly, sole ammonia driven transoceanic tanker releases about 0.0018 kg CO_2 eq/tkm greenhouse gas compared to 0.0055 kg CO_2 eq/tkm for sole heavy fuel oil tanker as shown in Fig. 14.

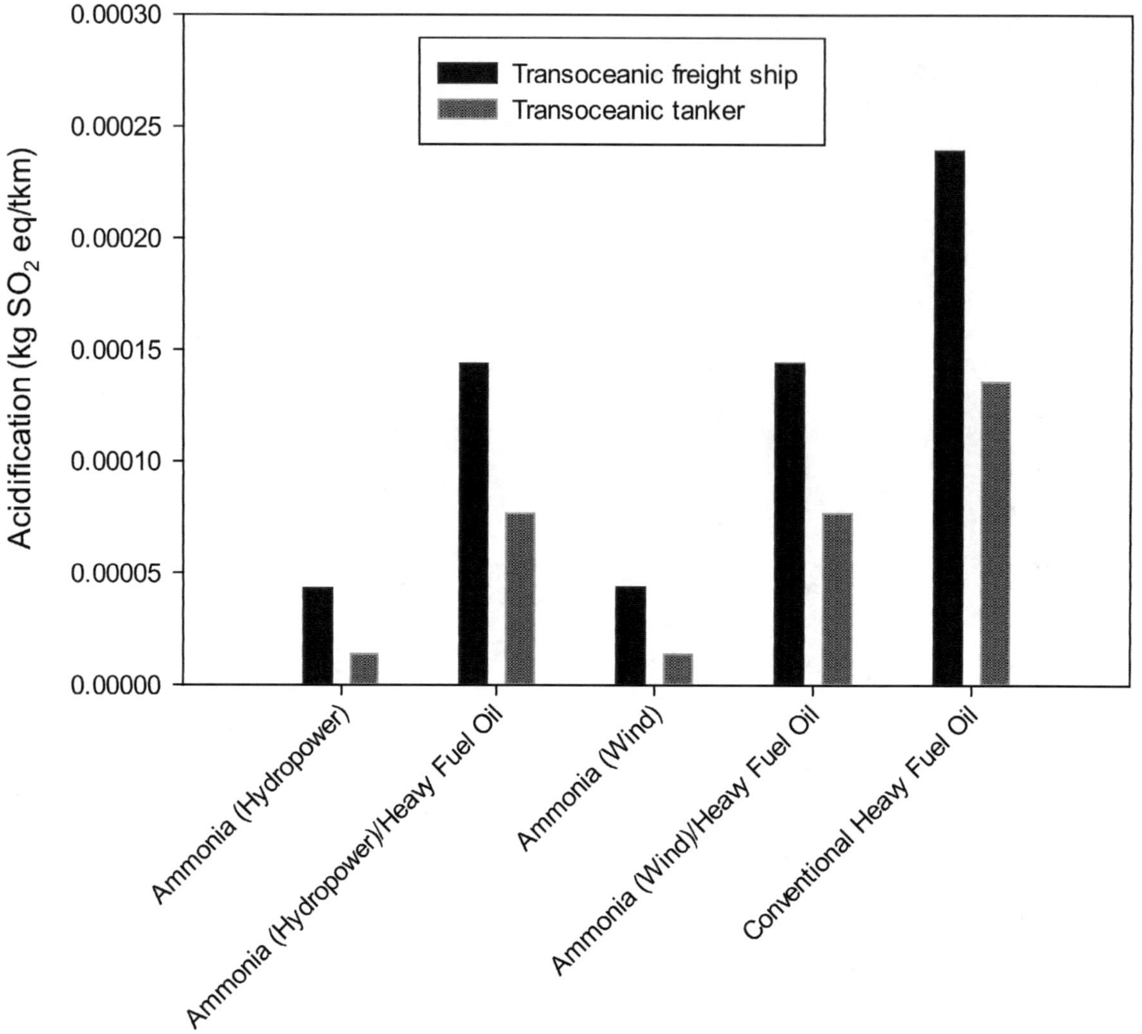

Fig. 15. Acidification values of transoceanic tanker and transoceanic freight ship per tonne kilometer for clean fuels and conventional heavy fuel oil

The acidification values of heavy fuel oil driven transoceanic tanker and freight ship are mainly caused by SO_2 and NO_x emissions (Fig. 15) which corresponds to more than 90% of overall acidification value. The source of SO_2 emission is predominantly the operation of tanker and freight ship (96.8%). This is caused by the sulfur content of the heavy fuel oil hence it is mostly eliminated if ammonia is used.

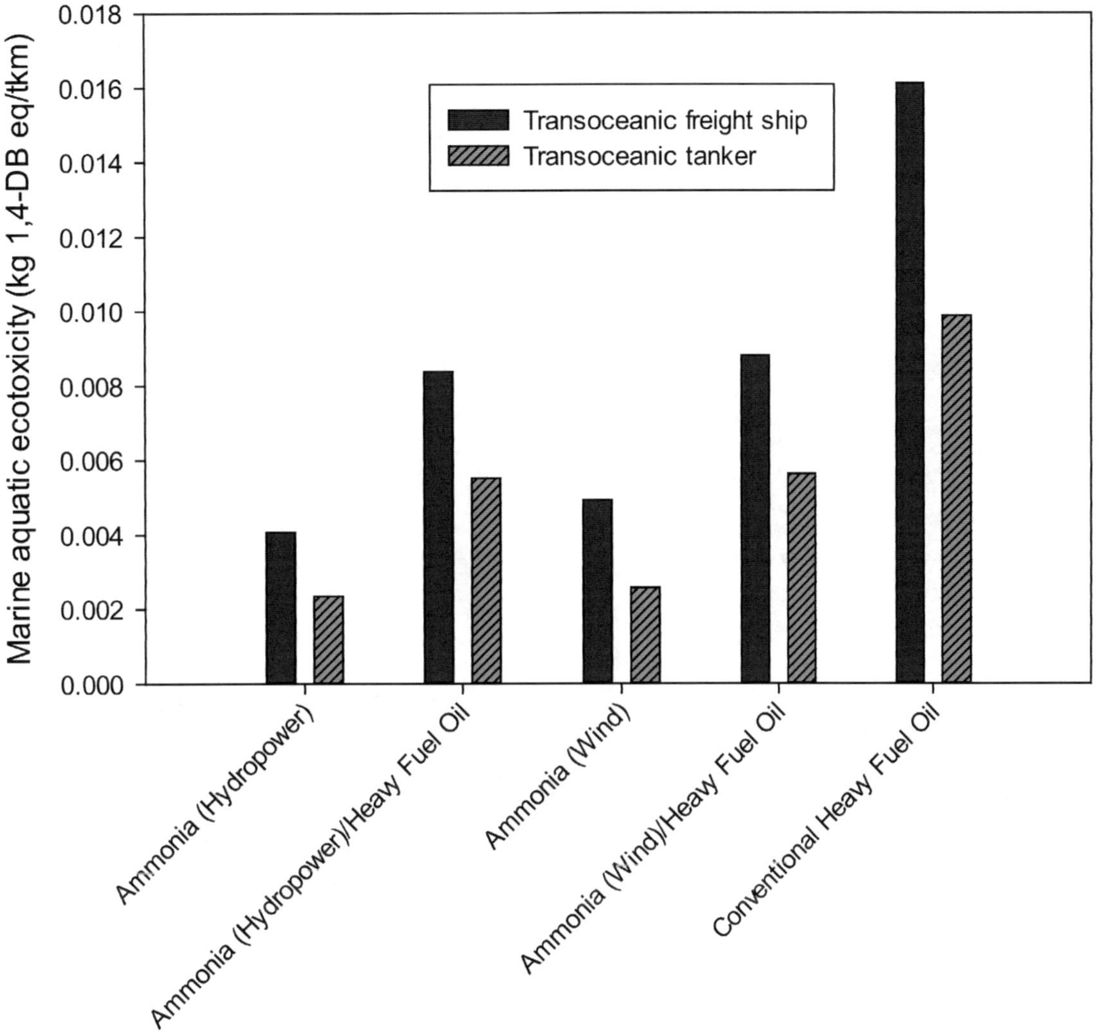

Fig. 16. Marine aquatic ecotoxicity values of transoceanic tanker and transoceanic freight ship per tonne kilometer for clean fuels and conventional heavy fuel oil

Using ammonia (wind energy) in transoceanic tanker as dual fuel with heavy fuel oil lowers the ecotoxicity level about 47% as shown in Fig. 16. For conventional heavy fuel oil driven transoceanic tanker, crude oil is the main depleted source similar to dual fuel options as shown in Fig. 17. For ammonia, the depletion is mainly caused by port operation process rather than tanker/ship operation. As non-carbon clean fuels for maritime ship engines, ammonia yields considerably lower global warming impact during operation. This demonstrates that if clean fuels are even partially replaced with current hydrocarbon derived fuels, total GHG emissions in maritime transportation can be lowered significantly.

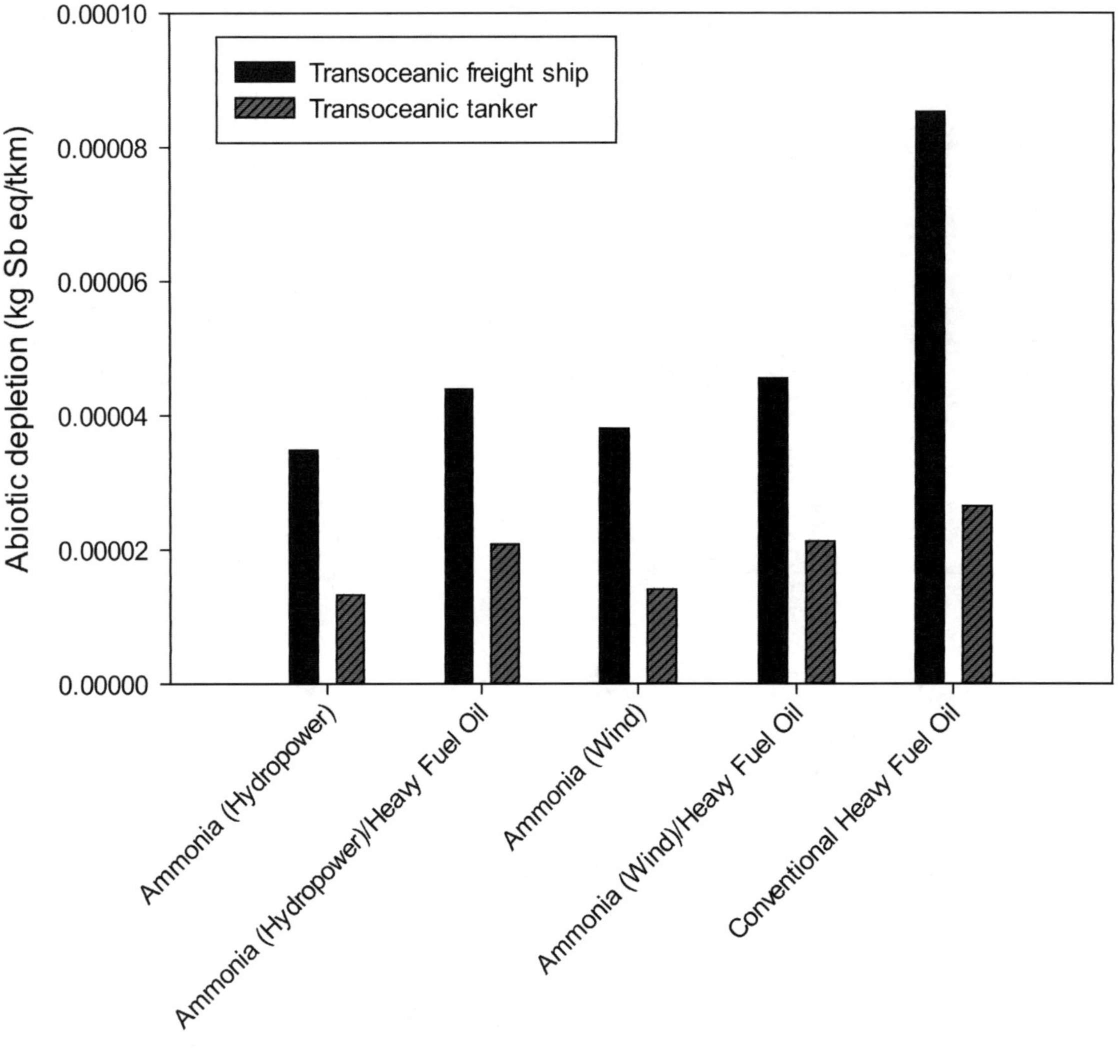

Fig. 17. Abiotic depletion values of transoceanic tanker and transoceanic freight ship per tonne kilometer for clean fuels and conventional heavy fuel oil

AMMONIA IN AVIATION

Petroleum based fuels have combination of accessibility, ease of handling, energy content, performance, and price because of being a mature product. Therefore, these type of fuels are heavily used by the transportation sector including air, road, and sea. However, limited nature, nonhomogeneous source distribution, changing prices, and end use related emissions of fossil fuels have driven most of the industries such as air transportation to search for other alternatives. Aviation is responsible for 12% of CO_2 emissions from all transports sources. The national and international flights overall the world produced 770 million tons of CO_2 in 2015 [17]

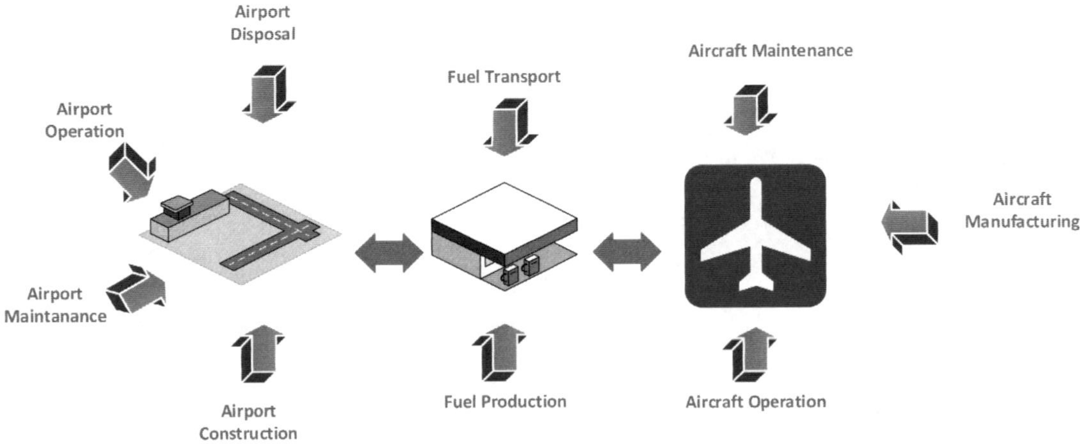

Fig. 18. Illustration of complete well-to-wake life cycle phases used in the LCA study

The overall life cycle emissions of an aircraft running on various aviation fuels are calculated from well-to-wake. The life cycle phases included in the analyses (shown in Fig. 18) are as follows: (i) production, operation and maintenance of the aircraft, (ii) construction, maintenance and disposal of the airport, (iii) production, transportation and utilization of the aviation fuel in the aircraft.

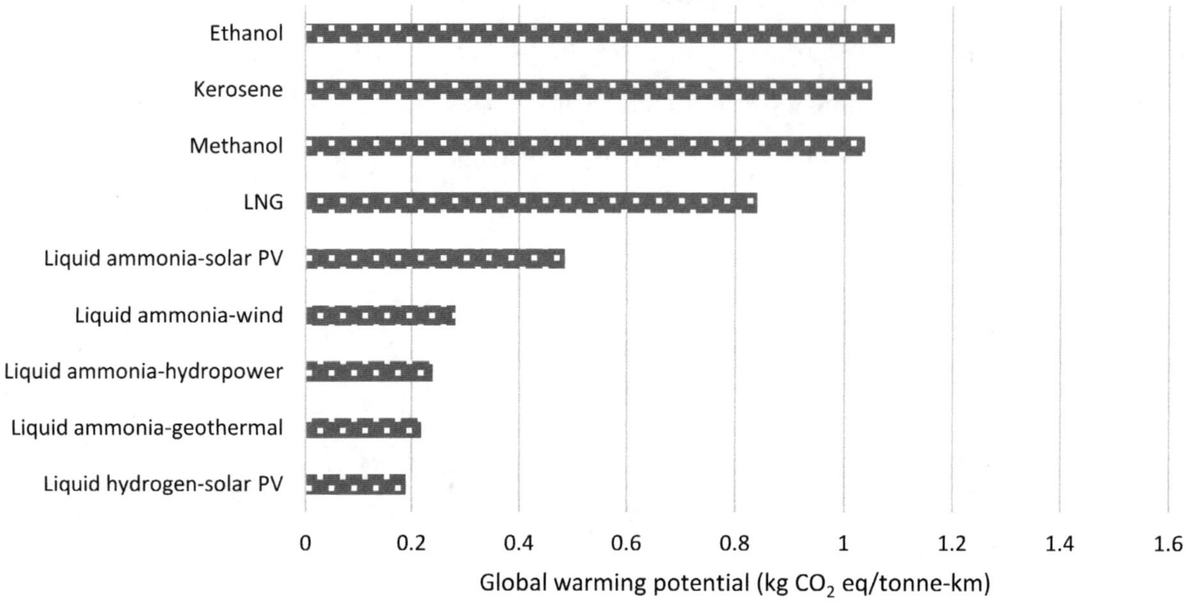

Fig. 19. Global warming potential of various fueled aircrafts per travelled tonne-km

The overall GWP value of ethanol is the highest corresponding to 1.09 kg CO_2 eq/tonne-km since it comes from ethylene which is a hydrocarbon as shown in Fig. 19. Renewable energy usage in the fuel production processes decreases the overall GHG emissions. The hydropower based ammonia fueled aircraft releases about 0.24 kg CO_2 eq per tonne-km. This value goes down to 0.21 for kg CO_2 eq/tonne-km for ammonia fueled aircraft and 0.18 kg CO_2 eq/tonne-km for hydrogen fueled aircraft in case the renewable source is solar energy. Although there are several processes, operation of the aircraft is the second largest contributor corresponding to 34%. Operation and maintenance of the airport is the primary responsible for GWP corresponding to 48.9% in total where it is distributed into sub-processes such as natural gas burning in the furnace (22%), light fuel oil burning in the furnace (5%), lignite burning in the power plant (7%) and hard coal burning (8%)

Renewable based ammonia driven aircrafts can significantly decrease the depletion of abiotic resources down to 0.0014 kg Sb eq/tonne-km which corresponds to about 10% of conventional steam methane reforming based ammonia fueled aircraft as shown in Fig. 20.

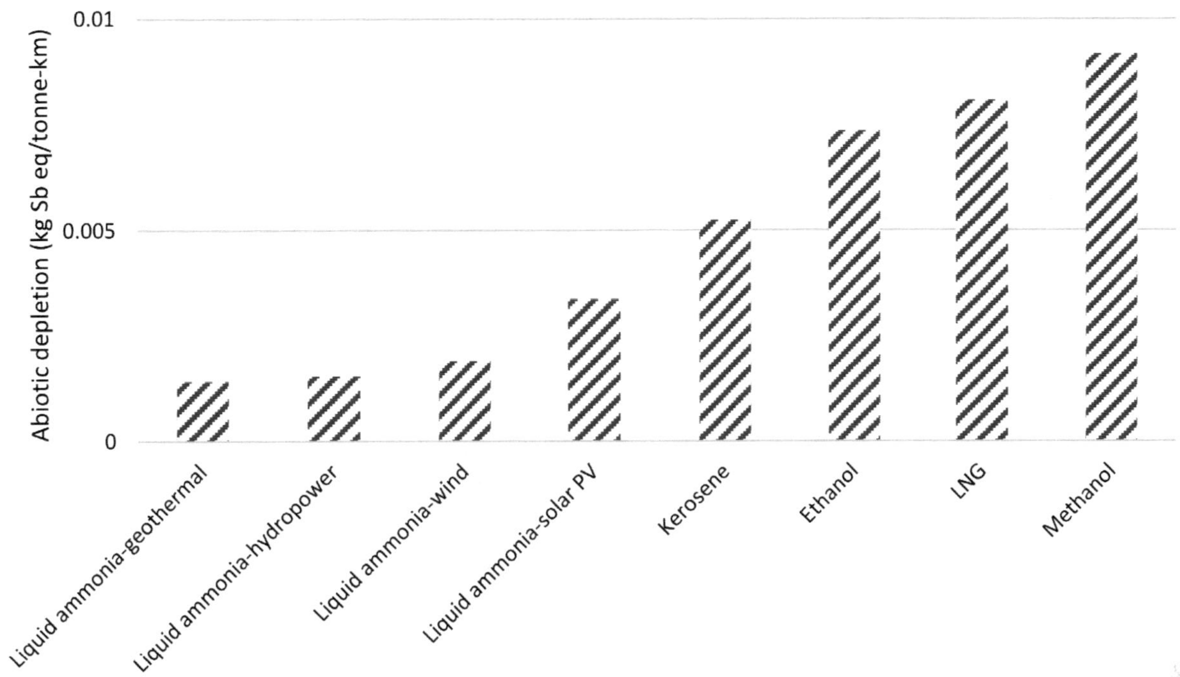

Fig. 20. Abiotic depletion values of various fueled aircrafts per travelled tonne-km

It is common knowledge that of HC, CO, NO_x, PM, and CO_2 which are named as GHG emissions cause environmental damages and adverse effect on human health. The marginal external pricc of a unit of these emissions is identified as environmental and social costs of emissions. Environmental and social costs of HC, CO, NO_x, PM, and CO_2 emissions of various fueled aircrafts are evaluated in terms of USD/tonne-km as shown in Fig. 21. renewable based ammonia yields lower environmental and social cost of emissions in comparison with kerosene.

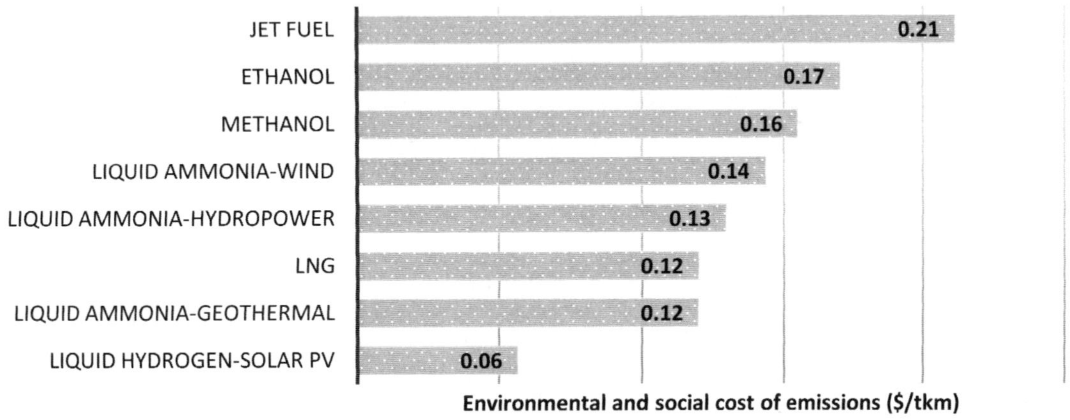

Fig. 21. Total environmental and social cost of emissions for various fueled aircrafts from conventional and renewable resources

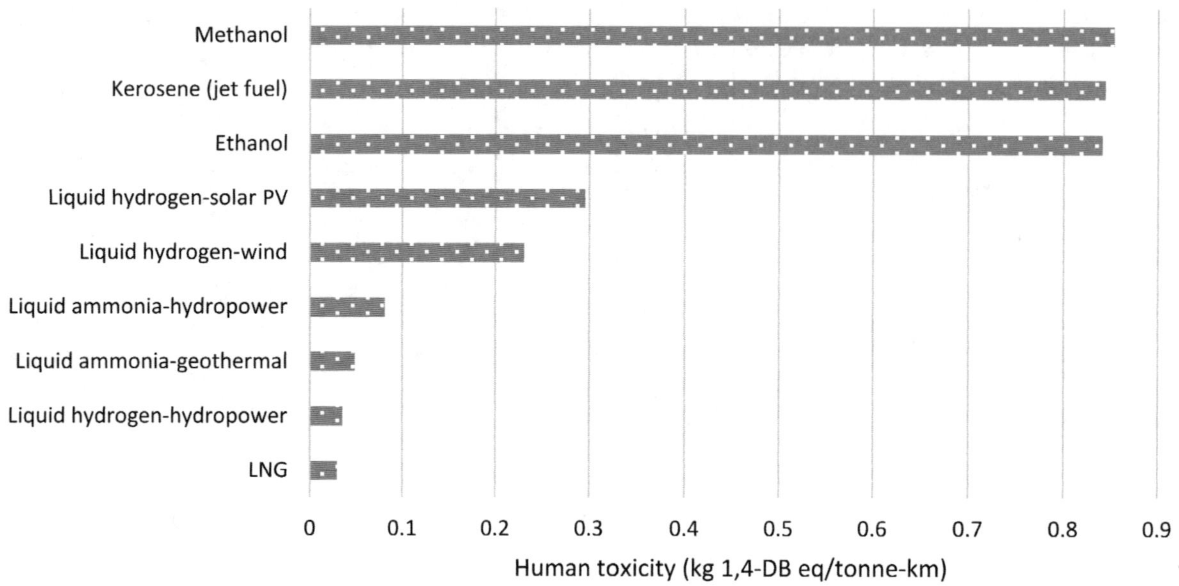

Fig. 22. Human toxicity potential of various fueled aircrafts per travelled tonne-km

Among all alternative aviation fuels, the kerosene jet fuel, methanol and ethanol human toxicity values are greater. Ammonia fuel from hydropower resource has comparable toxicity value corresponding to 0.08 kg 1,4- DB eq/tonne-km with LNG corresponding to 0.03 kg 1,4-DB eq/tonne-km as shown in Fig. 22.

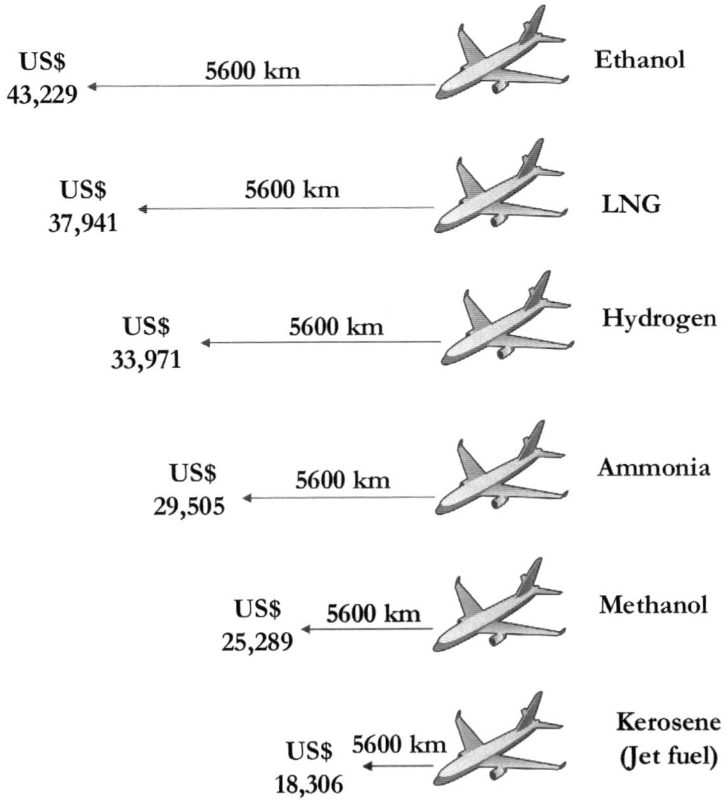

Fig. 23. Comparison of fuel costs during the operation of aircrafts for the given range

Table 3. Average fuel consumption rates and fuel costs for selected alternative fuels

Fuel	Fuel Consumption (kg/km)	Fuel Consumption (kg/tonne-km)	Fuel Cost (USD/kg)
Kerosene (Jet fuel)	7.99	0.217666815	0.409
Methanol	18.06	0.492185866	0.250
Ammonia	18.82	0.512730362	0.280
Hydrogen	2.64	0.071866485	2.300
LNG	9.46	0.257732767	0.716
Ethanol	12.47	0.339891348	0.619

Fig. 23 illustrates the cost of flight for a 5600 km distance in which alternative fuels are used for the aircrafts when the fuel costs and consumption rates in Table 3 are considered. Although LNG represented better environmental performance, the cost of aircraft operation in terms of fuel is the highest for ethanol and LNG. In the calculations, current conventional based routes are taken into account for comparison purposes. Hydrogen fueled aircraft has lower cost compared to these two alternatives. Since the production of kerosene from crude oil is a more mature technology, it represents the lowest cost among all. Liquid ammonia is also low cost alternative compared to hydrogen.

Transparency in Renewable Fuels

Ontario's Cap & Trade Program states:

The Renewable Fuel Standard is a technology-neutral policy and a technology-neutral approach lets the alternatives compete on their merits.

In order for renewable fuel technology to compete on its own merits and for the market to decide which technology has the best merits, energy consumers must ultimately participate in selection of renewable fuels.

Ontario's Cap & Trade program has the following participants:

Mandatory participants	Facilities and natural gas distributors with emissions of 25,000 tonnes or more of greenhouse gas emissions per year are required by law to participate in the program. Additionally, fuel suppliers that sell more than 200 litres of fuel per year and electricity importers must also participate in the program.
Voluntary participants	Facilities generating more than 10,000 tonnes but less than 25,000 tonnes of emissions may choose to opt in to the program. These companies will be subject to the same rules as mandatory participants.
Market participants	Companies that do not have emissions to report and therefore do not have a compliance obligation can also participate in the auction. Market participants can include individuals, not-for-profit organizations and companies without compliance obligations.

For the market to decide the winners in the Cap & Trade program, each energy consumer must also be a Cap & Trade participant.

This could be administered through the income tax system, with each family or business entity being issued its own baseline carbon credit. At income tax time, each energy or fuel supplier would provide each consumer a carbon statement that the energy consumer would include in his income tax return. Depending upon each tax entity's carbon consumption, the net carbon credits would translate into additional carbon tax payments or refunds.

TRANSPARENCY VIA A FLOW-THROUGH CARBON TAX

From an energy consumer's perspective, it is impractical to determine his own individual emissions. However, it is simple to determine their expected emissions by the amount of fuel consumed. Therefore, a more practical and transparent way of putting a cost on carbon emissions would be through a flow-through carbon taxation program similar to the GST/HST system. The carbon tax would replace the fuel excise tax and would be applicable to ALL consumable energy and fuel.

This taxation system would start at the producer level. For energy produced in Ontario, the carbon tax would replace the fuel excise tax. The carbon tax would be based on the carbon content of the fuel sourced from fossil sources or of the amount of fossil-carbon consumed to convert a fuel to a consumable form of energy. The carbon tax would then flow through from the producer to the distributor to the consumer.

For a technology-neutral carbon policy, there must be a completely level playing field and there is more to the life cycle environmental impact of the fuel than the amount of carbon dioxide generated from the its combustion. There are nine categories of life cycle environmental impacts caused by the production and utilization of energy that must be considered for a completely level playing field:

Environmental Impact	**Description**
Abiotic Depletion	Abiotic resources are natural resources including energy resources. Since fossil fuels resources are declining gradually, abiotic depletion potential is also a significant category. It is expressed in kg antimony (Sb) equivalents.

Environmental Impact	**Description**
Acidification	Acidification potential is for acidifying substances which causes a wide range of impacts on soil, groundwater, surface water, organisms, ecosystems and materials. kg SO2 equivalents is used to express the acidification potential.
Eutrophication	Eutrophication is the impact of excessive levels of macro-nutrients in the environment which is mainly caused by disposal processes. It is stated as kg PO_4 equivalents.
Global Warming	Global warming potential is the main characteristics to be compare the total CO2 equivalent emission from any source. The greenhouse gasses to air are related to the climate change. Methane, carbon monoxide etc. are in this category.
Human Toxicity	Human toxicity may play an important role for decision of using alternative fuels. It indicates the toxicity level and presented in 1,4-dichlorobenzene ($C_6H_4Cl_2$) equivalent.
Ozone Layer Depletion	Ozone layer depletion increases the amount of UVB that reaches the Earth's surface. Laboratory and epidemiological studies demonstrate that UVB causes non-melanoma skin cancer and plays a major role in malignant melanoma development. Ozone depletion potential of several gasses is specified in kg CFC-11 equivalent.
Terrestrial Ecotoxicity	Terrestrial ecotoxicity refers to the potential for biological, chemical or physical stressors to affect terrestrial ecosystems. Such stressors might occur in the natural environment at densities, concentrations or levels high enough to disrupt the natural biochemistry, physiology, behavior and interactions of the living organisms that comprise terrestrial ecosystems. The results are stated as 1,4-dichlorobenzene equivalents per kg emission.

Environmental Impact	Description
Marine Aquatic Eco-Toxicity	Marine aquatic eco-toxicity refers to impacts of toxic substances on marine aquatic ecosystems which is more important for maritime transportation sector. It considers each substance emitted to the air, water or/and soil. The unit of this factor is kg of 1,4-dichlorobenzene equivalents (1,4-DB eq.) per kg of emission.
Land Occupation/Land Use	Land occupation/land use refers to the total arrangements, activities and inputs undertaken in a certain land cover type. The term land use is also used in the sense of the social and economic purposes for which land is managed. It is mainly expressed as m2a which implies area usage in a year.

Focusing on carbon dioxide alone precludes any consideration of the global warming potential of other greenhouse gas emissions. From the US EPA [18]:

Greenhouse Gas	Global Warming Potential Over 100 years	Notes
CO2	1 by definition	Regardless of the time period used, CO2 is the reference gas. CO2 remains in the climate system for a very long time: CO2 emissions cause increases in atmospheric concentrations of CO2 that will last thousands of years.

Greenhouse Gas	Global Warming Potential Over 100 years	Notes
Methane (CH_4)	28–36	CH_4 emitted today lasts about a decade on average, which is much less time than CO_2. But CH_4 also absorbs much more energy than CO_2. The net effect of the shorter lifetime and higher energy absorption is reflected in the GWP. The CH_4 GWP also accounts for some indirect effects, such as the fact that CH_4 is a precursor to ozone, and ozone is itself a GHG. (Learn why EPA's U.S. Inventory of Greenhouse Gas Emissions and Sinks uses a different value.)
Nitrous Oxide (N_2O)	265–298	N_2O emitted today remains in the atmosphere for more than 100 years, on average.
Chlorofluorocarbons (CFCs), Hydrofluorocarbons (HFCs), Hydrochlorofluorocarbons (HCFCs), Perfluorocarbons (PFCs), Sulfur Hexafluoride (SF_6)	Thousands or Tens of Thousands	Sometimes called high-GWP gases because, for a given amount of mass, they trap substantially more heat than CO_2.

The carbon tax would be designed to improve upon the reduction of greenhouse gas emissions by including other detrimental effects on the environment. Rather than basing the tax rate solely on the carbon content of the fuel, global warming would be one of the nine environmental impact categories.

Conceivably, some fuels could have a positive effect on a particular type of pollution. In this case, such a process should be given a negative environmental impact factor to reduce its tax rate. An environmental impact multiplier would then be applied to the baseline rate that takes into account the nine environmental impact categories. Each environmental impact category would have an assessment scale of -10 to 10. The environmental impact multiplier would then be based on the sum of the nine assessments.

Ultimately, only electricity or fuels with a zero or negative environmental impact assessment would have a zero carbon tax rate. The cost of the inputs of varying environmental impacts used to generate the each source of electricity would be included in the cost of electricity. This would ensure that that the lowest-cost electricity is produced with the lowest environmental impact.

MANUFACTURING PRODUCTS FROM FUELS

If part of the fuel was then converted to a product, the manufacturer of that carbon-containing product would then get a credit for the carbon that has been prevented from entering the atmosphere.

For example, some biofuel processes can take carbon out of the atmosphere and can permanently sequester it into the ground as biochar. This type of permanent sequestration can result in Terra Preta [19], which has enhanced soil fertility. This process would get a credit for the carbon content of the biochar, less whatever biofuel was consumed to produce the biochar.

For another example, if a manufacturer had a process to produce anhydrous ammonia and elemental carbon from natural gas, the manufacturer would pay the full carbon tax on the natural gas used in its process. It would then get a credit for the elemental carbon, less whatever natural gas was used for process heat. The elemental carbon would carry the maximum carbon tax but the consumer of the elemental carbon would get the full carbon credit if that elemental carbon would be used to manufacture a carbon product, such as graphene. Since anhydrous ammonia has own environmental impact, its carbon tax would be different than that of the natural gas.

TAX ADMINISTRATION

A carbon tax program would be easier to administer at a consumer level than a Cap & Trade program. Businesses are already familiar with the GST/HST system and would not require significantly more bureaucracy to administer. This would also be transparent because the carbon tax portion of the energy cost could be separately broken out in the price as well as the methodology used to calculate the pollution multiplier factors. Consumers would therefore be able to determine the most effective way of reducing their energy costs.

In addition, having a carbon tax would also create growth in the economy with businesses providing ways of reducing energy costs through new equipment sales or retrofitting existing equipment.

This carbon tax should be revenue-neutral and overseen by the Environmental Commissioner of Ontario. The entire carbon tax revenue stream should be collected in its own fund and not go into the general revenue fund. The carbon tax revenue should be returned to the citizens of Ontario in the form of an annual dividend at income tax time, less the cost of administering the program.

Since the carbon tax would replace the fuel excise tax, the government may take some money from the carbon tax fund equivalent to the tax that would have been collected from the fuel excise tax. This is necessary to keep the carbon-tax revenue-neutral. The equivalent revenue removed from the carbon tax fund should be used to fund public transit and/or climate change programs including research and innovation.

CARBON TAX VS CAP & TRADE

The website *carbontax.org* makes an excellent case for why a carbon tax is superior in terms of transparency to cap & trade for reducing carbon emissions:

CAP-AND-TRADE'S INHERENT DEFECTS [20]

A tax on carbon emissions isn't the only way to "put a price on carbon" and provide incentives to reduce use of high-carbon fuels. A **carbon cap-and-trade system** is an alternative approach supported by some prominent politicians, corporations and mainstream environmental groups. Cap-and-trade was the structure embodied in the Waxman-Markey climate bill that passed the House in 2009 but died in the Senate. And cap-and-trade is the cornerstone of the European Union's "Emissions Trading Scheme" (ETS).

Cap-and-trade systems can be effective under certain conditions. The U.S. *sulfur dioxide* cap-and-trade system instituted in the early 1990s efficiently reduced acid rain emissions from power plants. However, the scale of a *carbon* trading system — it would be up to 100 times larger than that for sulfur — combined with the lack of "technical fixes" for filtering or capturing CO2, rules out sulfur cap-and-trade as a model for carbon. Moreover, evidence from the EU's ETS suggests that price volatility and gaming by market participants have undermined the effectiveness of this complex, opaque indirect method of pricing carbon pollution.

The Carbon Tax Center along with most economists regard a carbon tax as vastly superior and preferable to a carbon cap-and-trade system. Here's why:

- Carbon taxes lend predictability to energy prices, whereas cap-and-trade systems exacerbate the price volatility that historically has discouraged investments in carbon-reducing energy efficiency and carbon-replacing renewable energy.
- Carbon taxes are transparent and understandable, making them more likely to elicit public support than an opaque and difficult to understand cap-and-trade system. The co-author of the U.S. Senate cap-and-trade bill, Sen. John Kerry (now Secretary of State) told a reporter in 2009, "I don't know what 'cap and trade' means. I don't think the average American does."
- Carbon taxes can be implemented more quickly than complex permit-based cap-and-trade systems.
- Carbon taxes aren't easily manipulable by special interests, whereas the complexity of cap-and-trade leaves it rife for exploitation by the financial industry.

- Carbon tax revenues can be more or less guaranteed and integrated into state or federal fiscal policy, owing to their predictability, whereas the price-volatility of cap-and-trade precludes its being counted on as a revenue source.
- Carbon taxes are replicable across borders, since the price "metric" embodied in a carbon tax is far more universal than the quantity-reduction metric underlying cap-and-trade.
- Perversely, cap-and-trade discourages voluntary/individual carbon reductions, since those cause a lowering of prices of emission permits which undercuts low-carbon investments; carbon taxes are free of this unintended negative consequence.

Politically, cap-and-trade has functioned as a "safe harbor" for politicians who grasp the need for pricing carbon emissions but cling to the need to "hide the price" to appease interest groups and/or ordinary citizens. But the point of carbon emissions pricing is to raise the price of emitting carbon. Better to make the price explicit, via a tax, and protect households by making the tax revenue-neutral.

CARBON TAX IN CANADA

On December 9, 2016, Prime Minister Justin Trudeau made the most important energy and environmental policy announcement in half a century with his pledge to create a national clean fuel standard "based on life cycle analysis".

The flow-through carbon tax we propose is consistent with the federal government's **National Clean Fuel Standard** policies [21]:

> OBJECTIVE
> The overall objective of a clean fuel standard would be to achieve annual reductions of 30 megatonnes (Mt) of GHG emissions by 2030. This reduction will provide a significant contribution towards achieving Canada's commitment of 30 percent emissions reduction below 2005 levels, by 2030. This reduction is like removing over 7-million vehicles from the roads for a year.

BROAD COVERAGE

A clean fuel standard will encourage the use of cleaner fuels in many sectors of the economy, including the fuels we use in transportation, in our homes and buildings, and the fuels that power our industries. It would address a broad suite of fuels, which could include liquid fuels (e.g., gasoline, diesel, and heavy-fuel oil), gaseous fuels (e.g., natural gas and propane), and solid fuels (e.g., petroleum coke).

PERFORMANCE-BASED APPROACH

The clean fuel standard would set requirements to reduce the lifecycle carbon intensities of fuels supplied in a given year, based on lifecycle analysis. By contrast to renewable-fuel mandates, this approach would not prescribe the particular low-carbon fuel or technology that must be used; instead, it would focus on emissions reduction. The clean fuel standard would result in decreased emissions while minimizing compliance costs.

This approach would foster the deployment of a broad range of lower-carbon fuels and alternative technologies such as electricity, biogas, hydrogen, and renewable fuels.

MEASURING CARBON INTENSITY BY LIFECYCLE ANALYSIS

Carbon-intensity standards will be set to reduce the lifecycle carbon intensities of all fuels supplied, based on lifecycle analysis. The standards could be set at the facility level, at a sector-wide average, or set on some other basis. Carbon intensity is the measure of how much carbon is emitted into the atmosphere relative to the amount of energy in the fuel consumed.

FLEXIBILITY

The standard will be designed to provide maximum flexibility to fuel suppliers, and it may include provisions to take into account regional differences, similar to those that currently exist under the Renewable Fuels Regulations.

CARBON PRICING AND CLEAN FUEL STANDARD

The clean fuel standard would complement carbon pricing by ensuring consumers have access to a suite of lower-carbon fuels. In doing so, it will also drive down emissions and contribute to a clean-growth economy. By specifically addressing the carbon footprint of fuels, the clean fuel standard will drive innovation and create jobs and opportunities across a number of sectors, including transportation, agriculture, and clean-technology sectors.

The National Clean Fuel Standard policy mandates a broader consideration than simply transportation fuels because it covers ALL fuel uses, which represents about 80% of Canadian energy use rather than just transportation fuels as Ontario's review does. Indeed, Ontario is mandated by the Federal policy to expand their policy review to cover all fuels and to submit them and a reply to the proposed new Federal Policy.

Although Ontario's Renewable Fuel Standard is applicable to transportation fuels, the Standard's mandate should be expanded to include 100% of energy production and utilization, in order not only to be compliant with the National Clean Fuel Standard policy, but to exceed it by including the 20% of energy that is electricity that Ottawa left out.

THE GREEN SHIFT – STEPHAN DION

At the federal level, a carbon tax was proposed by Stephan Dion with his Green Shift program [22]. Dion proposed a carbon tax based on the greenhouse gas emissions. It would start out at $10/tonne, rising to $40/tonne within 4 years and be revenue neutral through of tax credits offsetting the revenue. While not specific about the implementation, it appeared that Dion intended to replace the federal excise tax on fuel and replace it with carbon tax applied at the wholesale level. With gasoline being taxed at the equivalent of $42/tonne of carbon dioxide, this carbon tax would not have resulted in any price increases to gasoline.

BRITISH COLUMBIA

Similar to Dion's Green Shift plan, British Columbia introduced a $10/tonne carbon tax on Carbon Dioxide Equivalent (CO2e) emissions [23], which would rise to $30/tonne by 2012.

BC's carbon tax was revenue neutral by reducing corporate and income taxes subject to annual legislative planning.

ALBERTA

Alberta enacted the *Specified Gas Emitters Regulation* in 2007 [24]. This carbon tax requires a $15/tonne contribution be made to the "Climate Change and Emissions Management Fund" (CCEMF) by large carbon emitter companies (ie, emitting more than 100,000 tonnes/year of greenhouse gas) to either reduce their CO2 emissions per barrel by 12 percent, or buy an offset in Alberta to apply against their total emissions. In 2016, the contribution was raised to $40/tonne for those large emitters. The intent of tax was to get oil companies and coal-fired generators to reduce their greenhouse gas emissions.

Starting 2017, Alberta will have a similar carbon tax scheme as British Columbia, which will be applied to the entire economy. All businesses and residents will pay a carbon tax based upon the carbon dioxide equivalent emissions, including the burning of wood and biofuels. The initial tax will be $20 per tonne, increasing to $30 per tonne in 2018, and increase thereafter by the rate of inflation plus 2%.

QUEBEC

Quebec enacted Canada's first carbon tax in October of 2007. It introduced a $0.008/litre tax on diesel fuel, which the equivalent of $3.00/tonne equivalent CO2e emissions [25]. This is not a revenue-neutral program because the revenue collected is used to fund energy-efficiency programs including public transit. Quebec's program also does not appear to be transparent because the carbon tax is not broken out as component of Quebec's fuel taxes.

CARBON TAX IN THE USA

There is no carbon tax under consideration at the federal level in the USA. Several states have considered implementing carbon tax programs but no proposals have yet been passed into law. In addition, some municipal level jurisdictions implemented carbon taxes:

- Boulder, Colorado, November 2006
- Bay Area Air Quality Management District, which covers nine counties in the San Francisco Bay Area, May 2008
- Montgomery County, Maryland, May 2010

CARBON TAX IN OTHER COUNTRIES

The following countries have the some sort of carbon tax in place [24]. All appear to base their carbon pricing solely on carbon dioxide emissions.

- Costa Rica
- Denmark
- Finland
- France
- Germany
- Netherlands
- Norway
- Republic of Ireland
- Sweden
- Switzerland
- United Kingdom

Other Considerations

ENERGY STORAGE THROUGH VEHICLE BATTERIES

For Ontario to maximize the consumption of the renewable energy generated in the province and minimize the amount of surplus power exported to neighbouring jurisdictions, Ontario must have the ability to store renewable energy. One such means of storing electricity is through battery technology and every electric vehicle has a battery.

Most vehicles are used for commuting and are therefore parked most of the time either at work or at home. If sufficiently large enough, Ontario's entire fleet of electric vehicles could therefore be used to help balance Ontario's electricity supply when they are plugged-in so that every vehicle owner would be an IESO market participant. This could be accomplished with smart chargers that charge vehicles when there is a surplus and draw power when there is a demand. The vehicle's charge controller would ensure that the vehicle battery would not be drawn down below the necessary charge level for the homeward commute. The spread between the market price to charge the battery at low demand and the price to return electricity at high demand could help with the public's acceptance of electric vehicles.

Besides vehicle owners, IESO should allow each home or business owner to become an IESO market participant. Already, there is home battery technology available to store electricity for the home owner. In addition, other technology is being developed to store electricity (such as solid state ammonia synthesis and ammonia fuel cells) as home appliances, which would help to transform Ontario's grid from central generation to distributed generation. This would facilitate vehicle-to-grid (V2G) technology as well.

EMISSION COMPLIANCE

Greenhouse gas emissions are not limited to combustion. Many fuels themselves are greenhouse gases and fugitive emissions prior to combustion can also have significant adverse effects on global warming. It is crucial that the Ministry of Environment conduct regular inspections of fuel systems to ensure that system operators do not allow any leakage. There must be vigilance about fugitive emissions from the wellhead to the tailpipe.

Methane, for example, is emitted during the production and transport of coal, natural gas, and oil. Methane emissions also result from livestock and other agricultural practices and by the decay of organic waste in municipal solid waste landfills. According to the US EPA, 11% of greenhouse gas emissions in 2014 were from methane compared with 81% for carbon dioxide [26]. Since methane has a global warming potential of 28-36, the relatively smaller methane emission fraction had 3.8 to 4.9 times worse effect on global warming than carbon dioxide.

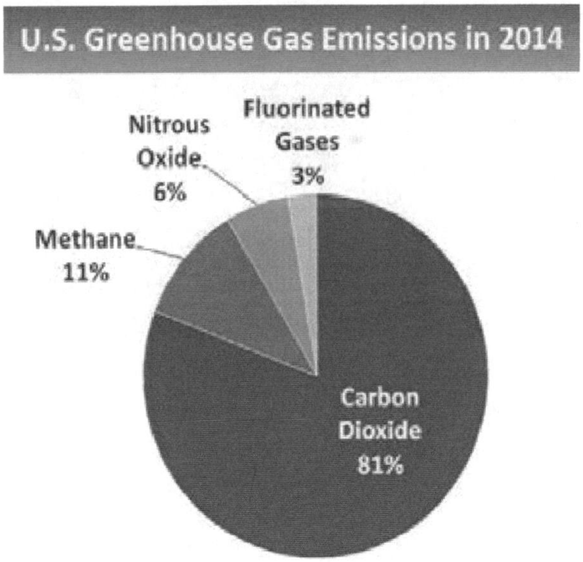

Figure 24 – U.S. Greenhouse Gas Emissions [26]

Natural gas-fueled vehicles have methane emissions both before and after combustion. In the case of LNG vehicles, as the cryogenic fuel warms, some of the methane leaks into the atmosphere as the safety relief valves blow off excess pressure. Both compression and spark ignition engines will have "methane slip" through the exhaust from incomplete natural gas combustion. Both of these sources of methane emissions must be mitigated and these vehicles should be regularly tested for emission compliance.

Figure 25 – LNG Truck Venting Excess Methane Pressure [27]

DEVELOPMENTS IN NH3 FUEL TECHNOLOGY

There has been much research that shows that ammonia can be used to fuel the most polluting forms of transport including marine, truck, and aircraft at less than half the cost and emissions of conventional fuels, even when the ammonia is made from natural gas or coal using appropriate technologies.

There have been some huge developments regarding green ammonia production, conversion of carbon in hydrocarbons into urea and use of NH3 as a fuel, for energy storage, especially in the past few months, including Mr. Gates Breakthrough Energy moving to develop technologies to produce "Zero emission ammonia", many of which actually exist commercially today.

Commercially available technologies have proven that energy and heating and cooling costs can reduced by more than 50% when ammonia is used in the mix, and there are opportunities to use a combination of fossil fuels and electricity in much more profitable ways. There are patented engines that provide CHPC (combined heat, power and cooling) at over 60%, at a fraction of the cost of any other technologies.

Existing natural gas infrastructure and power plants can use ammonia and where NH3 plants are built beside gas plants that burn NH3 with the vast amounts of pure oxygen available results in increases in efficiency as much as 50% with similar reductions in costs and virtually zero carbon, sulfur or nitrogen oxide emitted, and nothing but nitrogen (N2) and water (H2O) created as a result.

A joint submission between *Hydrofuel Inc.* and the *University of Ontario Institute of Technology's (UOIT)* submission to the *Canadian Environmental Assessment Agency (CEAA)* for the Pacific NorthWest LNG Project concluded [28]:

> "For the reasons stated in the above document and supported by Dr Ibrahim Dincer's group at UOIT (comments and MITACS report attached), Hydrofuel Inc. believes that the Pacific NorthWest LNG Project should not proceed and that the alternative of transporting energy in the form of anhydrous ammonia should be considered instead."

This submission proposes we use ammonia instead, an idea that has gained worldwide academic and public and private sector support as you will see from the details in the 13 points below and the attachments.

Hydrofuel Inc. has completed several projects with UOIT including one that will be ready by the end of March 2017, which will provide an analysis of the economics of converting gas into NH3 and urea for transport to worldwide markets instead of using LNG. We already know that NH3 can be burned and utilized at much higher efficiencies in engines and turbines with lower emissions, adding a price for carbon changes the numbers and adding the life cycle costs will make the economics of do so even more compelling.

In late September, 2016, Hydrofuel Inc. released the *UOIT Green NH3 and Key Life cycle energy research* [29] at the *13th NH3 Fuel Association conference at UCLA* [30], where SIEMENS (supported by Innovate UK. Collaborators include the University of Oxford, Cardiff University and the Science & Technology Facilities Council) announced their "Green" Ammonia technology, pilot plant project, and strategy that *"Green" ammonia is the key to meeting the twin challenges of the 21st century*. [31]

There is worldwide development of dozens of NH3 production and utilization technologies being reported now, and in addition to the *NH3 Fuel Association* website [32] a new *Ammonia Energy* Global Information Portal website [33]

UOIT RESEARCH

- Key Life-Cycle Numbers for NH3, Fossil Fuel and Green Energy production and utilization in agriculture, energy and utilities, and transportation systems [34]
- Comparative Life Cycle Assessment of Various Ammonia Production Methods [35]
- The most recent UOIT/Hydrofuel Inc. peer-reviewed paper published free to read by Springer as part of their Nature SharedIt and formally in *Environmental Management* on "Impact Assessment and Environmental Evaluation of Various Ammonia Production Processes" [36]. The previous paper a "Comparative life cycle assessment of various ammonia production methods" was published in the Journal of Cleaner Power on November 1, 2016 [35] .

- Other research Hydrofuel has done with UOIT may be found in the Latest News section of the Hydrofuel web site. [37]

EUROPEAN POWER TO AMMONIA CONFERENCE

1st European Power to Ammonia® Conference announced for May 18-19, 2017 - 1st European Conference on Sustainable Ammonia Solutions aims to gather thought societies, industries and academics, including well-known experts, developers and scientists to present the latest research results, present achievements, application fields and business prospects in energy solutions. This conference welcomes all researchers, industrialists, scientists as well as student and corporate delegates to participate and to have a great experience. [38]

JAPANESE AMMONIA USAGE

Japan has been using ammonia to fuel coal power plants reducing emissions. (Using similar technologies to this used in China for forty years.) Japanese utilities team on CO2-reducing tech for coal plants - Coal-ammonia fuel mix seen lowering emissions by at least 20%. [39]

DR. STEVE WITTRIG, US CLEAN AIR TASK FORCE

The development of the ammonia economy will depend on successful deployment of a range of new technologies in the context of the existing ammonia industry and the world's many existing energy markets. This report *Ammonia Fuel: Opportunities, Markets, Issues* [40] provides a framework for:

- Supply/demand and status of the global ammonia industry.
- Plausible response of growth and prices in response to a giant new market in fuels.
- New incremental applications which may lead to 'disruptive innovation' and rapid takeoff for ammonia fuels.
- Barriers to takeoff.

HYPER HUB PATENT

An improved system of hardware and controls, known as a Hyper Hub, that absorbs electric power from any source, including hydropower, wind, solar, and other renewable energy resources, chemically stores the power in hydrogen-dense anhydrous ammonia, then reshapes the stored energy to the power grid with zero emissions by using anhydrous ammonia to fuel diesel-type, spark-ignited internal combustion, combustion turbine, fuel cell or other electric power generators, and for other purposes. [41]

OTHER DEVELOPMENTS

1. On November 10, 2016, *Ammonia Energy* announced a 2018 pilot plant in Japan will demonstrate a new way to produce ammonia at industrial-scale, with a low carbon footprint. [42]

2. On November 28, 2016, the Trudeau Government announced its new life cycle Clean Fuels Standards policy being formulated for February 2017. [43]

3. On December 2, 2016, *Ammonia Energy* announced *Nuon*'s Power-to-Ammonia update, and the first European ammonia fuel conference in 2017. [44]

4. On December 6, 2016, the Massachusetts Institute of Technology (MIT) released their *PLAN B - FOSSIL FUELS without CO2 emissions*. Using clean-tech and ammonia makes fossil fuels clean and much more profitable. The stored chemical potential in fossil fuels from thermonuclear derived solar energy may be utilized by less conventional means without producing carbon dioxide. Chemical pathways to produce hydrogen or ammonia from hydrocarbons without co-production of carbon dioxide are possible in new process configurations. Such processes may be more cost effective than other options and more readily implemented. [45]

5. Phoenix Energy, a developer of biomass gasification plants based in San Francisco, shared some interesting data today during a Webinar on "The Future of Renewable Methane in Today's Regulatory & Policy Environment." The company operates two plants in California

that together produce 5.5 tonnes/day of byproduct carbon in the form of biochar. They are currently selling all of the biochar they can produce into agricultural markets at a price of $1,700 per tonne. [46] Unlike what many uninformed politicians, environmentalists and special interests believe or are lead to, Pheonix Energy said that their data proved:

- *Ammonia as a fuel is a real interesting target, where it would allow you to make the transfer of that chemical potential to something that fits in well to our infrastructure ... It has the same heat of combustion as methanol, so it has plenty of power per weight. When you burn it though, you only make nitrogen and water. There's ways to get around the issues with NOx and stuff, so this is really a nice fuel. And the wonderful thing about ammonia, if you think about it, if you made it at a price for fuel, it would do something else that's really important for the poor, which is to lower the price of food ...*
- *All of the automobile companies have looked at ammonia as a possible fuel, and it's one that is certainly worthy of consideration if one wants to think about what we can do about a liquid fuel that fits into our own infrastructure that can be used more or less in our world as we know it.*

6. On December 14, 2106, *Ammonia Energy* announced the US DOE ARPA-E's REFUEL program awarded $35 Million (US) in funding to a total of 16 projects for carbon-neutral liquid fuels (of these 16 projects, 13 were focused on carbon free ammonia (none for hydrogen not associated with using ammonia), of the three others that went to carbon based energy, two were for Dimethyl ether, one looked at ethanol). [47]

7. On December 14, 2106, *Ammonia Energy* announced the Bill Gates' *Breakthrough Energy Coalition* [48] is starting work on viable grid scale and world class Green NH3 technologies after concluding that one of its initial Technical Quests is to make "Zero-GHG Ammonia Production" a reality. [49]

8. On December 16, 2016, the Trudeau government announced it has delayed the notification date of NRCan's decision on clean energy innovation funding proposals [50]. The date has been changed from late fall 2016 to winter (mid to late February) 2017. As Greg Vezina points out in the upcoming Ottawa Life Magazine article mentioned below, it appears the

recent developments in support of ammonia and moving away from bio-fuels and other carbon based technologies combined with the government's new Life Cycle Clean Fuels Policy has caused a major rethink of how they will spend the $billions in proposed funding. All of this is very good news for ammonia energy proponents.

9. On January 27, 2017, *Ammonia Energy* announced more US DOE funding research into sustainable ammonia synthesis. These *Sustainable Ammonia Synthesis projects*, announced in August 2016 and administered by the Office of Basic Energy Sciences, aim "to investigate some of the outstanding scientific questions in the synthesis of ammonia (NH_3) from nitrogen (N_2) using processes that do not generate greenhouse gases." There are two substantial differences between these awards and the other recent DOE funding for sustainable ammonia technologies, announced in December 2016 and administered through ARPA-E's "REFUEL" program. [51]

10. On February 2, 2017, *Ammonia Energy* announced studies in both Canada and Italy including our work and a "Comparative studies of ammonia production, combining renewable hydrogen with Haber-Bosch." [52] The Japan Science and Technology Agency (JST) announced [53] that another consortium – this one led by Hiroshima University and including Showa Denko, Taiyo Nichi Company, Toyota Industries and the National Institute of Advanced Industrial Science and Technology – had succeeded in developing "viable technology to produce high-purity hydrogen [from an] ammonia hydrogen station."

11. On February 3, 2017, *Ammonia Energy* announced seven International R&D projects [54] on sustainable ammonia synthesis technologies, representing research coming out of eight countries spread across four continents. These projects span the breadth of next-generation ammonia synthesis research, from nanotechnology and electrocatalysis to plasmas and ionic liquids.

12. On February 9, 2017, *Ammonia Energy* announced the German Aerospace Center (DLR) at the Institute of Thermodynamic Engineering is developing a method for electrochemical ammonia synthesis [55] at ambient conditions.

13. On February 16, 2017, *Ammonia Energy* announced the Institute of Fluid Science at Tohoku University in Sendai, Japan, has developed the world's first technology for direct combustion of ammonia in a gas turbine [56], and a start-up company in Denmark is commercializing technology to generate low-cost, high-purity hydrogen from ammonia, for use in fuel cells [57].

14. On March 3, 2017, *Ammonia Energy* announced this new study [58], published in the January 2017 AIChE Journal by researchers from RWTH Aachen University and JARA-ENERGY, which examines ammonia energy storage "for integrating intermittent renewables on the utility scale," and in Israel, researchers recently published a complementary Power-to-fuel-to-power [59] study that compared carbon-based and nitrogen-based synthetic fuels.

15. On March 6, 2017, *Ammonia Energy* announced Nikkei Asian Review has published two articles outlining plans by major engineering and power firms to build utility-scale demonstrations using ammonia as a fuel for electricity generation [60] in ammonia-coal and ammonia-natural gas applications to reduce costs and emissions.

16. On March 16, 2017, *Ammonia Energy* announced a recent The Japan Times opinion piece that predicts a "revolutionary disruption coming to the energy sector," and suggests that using ammonia for energy storage will prove to be "a game-changer [61] at least on the scale of the shale oil and gas revolution."

17. On March 23, 2017, *Ammonia Energy* announced the shipping industry is beginning to evaluate ammonia as a potential "bunker fuel," [62] a carbon-free alternative to the heavy fuel oil (HFO) used in maritime transport.

18. The market for ammonia/diesel dual fuel engines in the marine industry is now quantified for 2016-2020. [63]

    PART 10: Market trends
    - Ammonia gas as substitute for natural gas in dual fuel engines
    - Increased OEM collaboration to study dual fuel engine technology for automobiles
    - LNG: Maritime green fuel of the future

19. On March 30, 2017, *Ammonia Energy* announced Gifu University in Japan's breakthrough in technology for generating hydrogen from ammonia [64], and the Institute for Sustainable Process Technology's Power to Ammonia feasibility study [65] that represents a major step toward commercializing renewable ammonia and which examines the "value chains and business cases to produce CO2-free ammonia," analysing the potential for commercial deployment at three companies with existing sites in The Netherlands

20. On April 6, 2017, *Ammonia Energy* announced the 2017 NH3 Fuel Conference that will be staged on November 1, 2017, in conjunction with the American Institute of Chemical Engineers (AIChE) conference in Minneapolis, MN, from October 29 to November 3. The two organizations will co-sponsor a topical conference on ammonia energy called "NH3 Energy+: Enabling Optimized, Sustainable Energy and Agriculture" [66]. This year's annual AIChE meeting is expected to draw more than 8,000 attendees.

21. On April 20, 2017, *Ammonia Energy* announced the publication of "Auto-ignition of a carbon-free aqueous ammonia/ammonium nitrate monofuel: a thermal and barometric analysis." The authors clearly see nitrogen-based fuels as a key element of a sustainable energy future. [67]

22. On April 21, 2017, *Ammonia Energy* announced an important new voice joined the chorus of support for renewable ammonia and its potential use as an energy vector – the International Energy Agency (IEA). Cédric Philibert, Senior Energy Analyst at the IEA, identifies a major problem with the hydrogen economy: hydrogen is currently made from fossil fuels. But his argument for producing hydrogen from renewable energy leads almost inevitably to ammonia: "In some not-too-distant future, ammonia could be used on its own as a carbon-free fuel or as an energy carrier to store and transport energy conveniently" [68].

23. York University Disaster and Emergency Management degree reports:
    - "Anhydrous Ammonia to Reduce Critical Infrastructure Risk" [69]
    - "Can the Smog in China be Reduced or Eliminated Using Ammonia?" [70].

24. Below are links to a number of articles we have written or co-written, and one that has been written about our work.

A. Four articles co-written for Ottawa Life Magazine.

i. One of Four: An Alternate View on Pipelines — Transport Ammonia not Crude [71]

ii. Two of four: Green Ammonia [72]

iii. Three of four: Canada's Ammonia Energy Option [73]

iv. Four of Four: Canada's Life Cycle Clean Fuels Policy Game Changer [74]

B. Most of the information about ammonia as a renewable fuel is not covered by the mainstream media in Canada with the exception of Ottawa Life Magazine and the Toronto Sun Editor Lorrie Goldstein who has let Greg Vezina write many SUN editorial page articles [75] and who has written about my NH3 work, most recently in:

"High Cost of Ignorance" a Toronto Sun Editorial page article on Oct. 23, 2016. [76]

25. Research on Cost comparisons of pipelines, distribution and marine transport for natural gas and ammonia [77]. Included are <u>five</u> studies Hydrofuel Inc. have already done with UOIT on point, however, as mentioned above, we have completed our third *MITACS* research report in February, 2017, which is much more extensive. MITACS Report 3 will be released at the US DOE ARPA-E REFUEL [78] program launch April 25-26, 2017, in Houston Texas, at which Greg Vezina will speak about "Life cycle analysis of carbon neutral liquid fuels" on the first day and will participate in the panel discussion: *Investment in chemical energy storage/VC perspective* on the second day. [79].

A. Ammonia as a Potential Solution for Alberta (Submitted to Alberta Premier) [80]

B. Natural Gas to Ammonia as a Potential Solution for British Columbia (Submitted to BC Premier) [81]

C. *MITACS Report 1* November 2015 - "Comparative assessment of NH3 production and utilization in transportation systems" [82]

D. *MITACS Report 2* June 2016 - "Comparative assessment of NH3 production and utilization in agriculture, energy and utilities, and transportation systems" [83]

E. *MITACS Report 3* March 2017 - " Comprehensive Evaluation of NH3 Production and Utilization Options for Clean Energy Applications" [84]

The three MITACS Reports prove beyond a doubt the viability of the technologies UOIT and Hydrofuel Inc. have developed and demonstrated to date and the others we have under development.

Hydrofuel Inc. and UOIT submitted a joint report "Ammonia (NH3) as a Potential Transportation Solution for Ontario" [85] in response to the Ontario government's discussion paper "Developing a modern renewable fuel standard for gasoline in Ontario" [86] for which submissions closed on March 12, 2017.

While Ontario's review was limited to transportation fuels our comments were expanded to cover all fuel uses in accordance with the provisions outlined in the Federal government's new "Life Cycle Clean Fuels Policy" [87] review discussion paper released in late February, 2017, for which comments close April 25, 2017. The new federal standards will cover all fuels, which represents 80% of energy use, unlike Ontario's which only cover about one third of fuel use emissions, but which will none the less have to be expanded to comply with the new federal standards regardless.

Hydrofuel Inc. intends to continue our work with UOIT to complete six more projects that have many expected benefits:

- Work Package 1: to develop a system for clean ammonia production using fossil fuel, renewable and waste energy as feedstocks
- Work Package 2: to develop efficient and effective ammonia fuel cells
- Work Package 3: to improve and optimize ammonia decomposer for on-board hydrogen production and utilization in vehicles
- Work Package 4: to develop and implement the production of commercially ammonia kits for converting diesel generators to ammonia power generators

- Work Package 5: to improve and optimize ammonia heat engine for power generation and for heating and cooling applications
- Work Package 6: to study feasibility and life cycle assessments on using ammonia in various systems and applications, expanding our most recent work

These projects will help:

- to implement novel technologies for production and utilization of energy more cleanly and efficiently.
- to achieve significant emissions reductions, while maintaining Canada's economic advantage and its ability to create jobs for Canadians.
- to reduce and eliminate air pollutants from energy production and utilization.
- to contribute to the transition to a low-carbon economy and help reduce the economic, environmental, and health risks associated with air pollution and climate change.
- to significantly reduce energy use at the community level and increase its renewable content through the innovative integration of renewable energy sources for heating & cooling & electricity production and energy storage & management
- to reduce the use of diesel fuel at industrial sites through innovative alternative fuel applications
- to implement highly efficient heating and cooling technologies for micro-combined heat and power generation

For example, the retrofit technology for diesel/ammonia multifuels conversions kits for diesel engines and generators we have developed could be used in Ontario successfully in many applications according to an analysis in our Summary of the Proposed Project "Connection of Remote First Nation Communities in Northwest Ontario" [88] . The report proposes to use ammonia initially shipped in and a decade later produced locally using renewable resources verses plans for a $1.5 billion 1,500 km hydro transmission line based on a forty year business case to provide interruptible power backed up by... wait for it... diesel power.

Conclusions of UOIT's Diesel Generator Alternatives Report:

- As the global warming and energy issues become important topics to be considered for the future of remote communities, alternative solutions for power and heat generation need to be investigated.
- Currently, diesel generators running on diesel are utilized for Northwestern Ontario remote communities. This is one of the most expensive and environmentally damaging option.
- First Nations request and propose a 1,500 km power transmission line to the remote communities instead of diesel generators which is a high cost investment.
- The significance of distributed power generation is emphasized in many studies throughout the world. Hence, producing power and heat via stand-alone facilities are being encouraged by the governments and decision makers.
- As an alternative and sustainable fuel, ammonia, can be utilized in the diesel generators by minor modifications. Ammonia has no greenhouse gas emission during utilization in the diesel generators. Hence, it is the most environmentally benign fuel among other alternatives.
- Transportation and storage of ammonia is already available and well-known since it is the second largest produced chemical in the world. This implies that instead of diesel, ammonia can easily be transported and stored.
- Ammonia can be produced on-site using renewable energy resources such as wind, solar and hydropower which are already available in these remote communities. This brings minimum transportation cost. However, diesel needs to be transported for long distances.
- The operation of diesel engines with fuel ammonia has the lowest cost based on the current market prices. Considering the technology development of ammonia production, the cost of ammonia will continue to decrease which will bring additional reductions in total cost.
- Production and utilization of ammonia in diesel generators have significantly lower environmental impacts in terms of climate change and global warming.
- As a result, ammonia usage in remote communities for power and heat production will bring significant cost and environmental benefits together with public satisfaction.

We have details about and hyperlinks to much of the Hydrofuel Inc. - UOIT and other available research on the Hydrofuel Inc. website [89]. There is very solid scientific and economic research and numerous practical applications that support using ammonia to green hydrocarbons and store renewable energy already commercially available and many more emerging almost daily.

NH3 is the Obvious Carbon-Free Fuel and Energy Currency

If these technologies and opportunities are so good, why doesn't anyone know about them and why aren't the major companies in these sectors developing these technologies?

The following three examples directly on point explain the answer as succinctly as is possible under these circumstances, where there actually is a mountain of research and a century and a half of history involved to digest to quickly come up to speed on this subject matter, an experience which might otherwise be compared to drinking out of a fire hydrant.

EXAMPLE 1

In 1984, the National Research Council of Canada ("NRC") commissioned a report by International consulting firm Chemetics International Co. (Toronto), a division of C-I-L Inc. ("Chemetics") called, "*Ammonia Vehicle Fuel - Assessment of the technical, economic and safety problems associated with ammonia use as a ground transportation fuel.*" [90] Among the conclusions and recommendations were:

> 6.0 CONCLUSIONS AND RECOMMENDATIONS
>
> This report development started from a rather negative preconceived feeling as to the potential of ammonia as a vehicle fuel. The study has indicated a very real potential from a purely technical point of view and some positive features for safety and environmental concerns. If the economics of capital costs continue to improve and the high costs and scarcity of hydrocarbons become more acute, ammonia will become a viable option either as a carrier for hydrogen fuel or possibly as a direct combustion fuel. It is conceivable that the conditions in remote areas or in developing countries where there could be available, renewable resources to generate electricity and hence, ammonia.
>
> 6.1 CONCLUSIONS

The Following listing is a generalized summarization of the conclusion that has been formed:

1. There are no technical or practical engineering problems that are not very logically solvable for the general usage of ammonia as an ICE fuel.

2. The general safety and public acceptance of ammonia as an ICE fuel should not be difficult to achieve.

3. There are no known ecological or environmental problems with ammonia as an ICE fuel and in fact indications are that significant improvements could be achieved over hydrocarbon fuels.

4. The replacement of hydrocarbon by electricity as the energy source and raw material for ammonia-fuelled ICE is technically feasible.

6.2 RECOMMENDATIONS

As the concerns for hydrocarbon conservation and environmental damage caused are on the increase, all viable options must be looked at in depth. The alternatives are many and varied and the ramifications are astronomical. It follows then that a knowledgeable planned evolution to new and probably numerous options must be set up.

No funding was provided for any ammonia fuel and energy related research in Canada for decades that followed until the Harper government funded limited research into NH3 fuel cells, renewable energy storage in ammonia and clean diesel locomotive conversion to ammonia in the GTA in the latter part of its mandate.

Example 2

We should have moved in this direction way back in 1967 when Dr. Leon Green first proposed this idea in "Energy needs versus environmental pollution: A reconciliation?" [91].

His proposal was supported by several subsequent studies [92] including in the 2012 extensive IEEE Report by Cal Poly Tech professor William Ahlgren "The Dual-Fuel Strategy: An Energy

Transition Plan," [93] the subsequent report "Planning for a hundred fold increase in ammonia production," [94] and the recently added "Prologue to the Dual Fuel Plan," by Jim Esch, IEEE Contributing Editor. [95]

In the Introduction to the *Dual Fuel Strategy* Dr. Ahlgren wrote that many of the ideas presented have a long history.

> "As early as 1967 Leon Green, Jr., writing in Science magazine [1], formulated a concept for the large-scale use of ammonia as fuel. He observed: The long-term consequences of the greenhouse effect due to CO2 buildup in the atmosphere are of serious concern. . . . To remove the offending elements (carbon and sulfur) from the fuel prior to combustion is a much more efficient and less expensive procedure than trying to clean up the combustion products. ... Outlined below is a concept for energy generation in which the fossil fuels are not burned directly but serve as raw materials for the synthesis of a clean fuel. ... This clean fuel is ammonia. ... In commercial high-tonnage production of ammonia, natural gas is used as raw material for steam reforming to generate hydrogen for the synthesis reaction. In the course of this process sulfur is removed and recovered in elemental form, and CO2 is scrubbed from the stream and may be recovered for sale or use. Although current practice is to discharge this CO2 to the atmosphere, the point is that the CO2 is under control and can be condensed or caused to react so that the carbon is tied up in some useful form. ... large amounts of CO2 are recovered per unit of ammonia produced, and the commercial value of this CO2 will have a major bearing on the economic attractiveness of the concept.
>
> Green's concept, published nearly half a century ago, is stunning in its prescience. Why has it received so little attention? Over three decades ago the U.S. Department of Energy conducted a comprehensive study of liquefied gaseous fuels, including LNG, LPG, hydrogen, and ammonia. The latter alternatives were studied because of the extreme explosion hazard associated with LNG and LPG. Bomelberg and McNaughton coauthored the report on ammonia [2], published in 1980. After a careful comparison of hydrogen and ammonia, they wrote:

> It is not understandable why hydrogen as a future fuel is widely promoted, whereas ammonia is presently not considered at all. The most likely explanation appears to be that the potential use of ammonia as a substitute fuel is just too unknown, even within the technical community.

In 2012, the potential use of ammonia as fuel remains just too unknown. Why is this? Perhaps it is because the well-known hazards of ammonia cause its use as fuel to be dismissed out-of-hand. There is also a widely held (but erroneous) notion that nitrogen-based fuels must necessarily produce excessive NOx in their exhaust. These misconceptions must be dispelled; one objective of the present essay is to do so. At the same time, the validity of these concerns must be acknowledged. Ammonia is a hazardous substance, and NOx is found in the exhaust of ammonia combustion processes. We argue that ammonia can nevertheless be safely used as fuel if it is not required to serve all purposes. We outline a plan, the dual-fuel strategy that supplements ammonia with a complementary substance, methanol, well known as an alternative fuel. Ammonia and methanol each has strength that compensates the other's weakness: ammonia is carbon-free, but has high relative toxicity;2 methanol has low relative toxicity, but contains carbon. Their use together yields a good (not perfect) solution for liquid renewable fuel, which we assert to be sine qua non for a post-petroleum global energy system. Further, we note that ammonia and methanol are but one example of such a dual-fuel pair. Other, better pairs might be found, and must be sought. What we reject is hydrogen. Hydrogen is disqualified because it is a gas. In our view, the use of hydrogen as an energy vector has been thoroughly explored over a period of half a century, and has failed to come to fruition. It is now time to move on.

A purpose of this essay is to contribute to a discussion initiated in these Proceedings by Bossel [3] and continued by Abbott [4], [5]. Bossel points out the deficiencies of the hydrogen economy, and advocates an electron economy in which batteries play a central role and fuel is unnecessary. Abbott insists that fuel is essential and that the hydrogen economy is the only realistic alternative to business as done now. We agree with Bossel that hydrogen is deficient, and with Abbott that renewable fuel is indispensible. We suggest that liquid fuels, ammonia and methanol, answer many of the objections raised by Bossel, while acknowledging Abbott's observation that renewable fuel is an essential component of any future energy strategy.

This essay is an unapologetically rhetorical work. Our purpose is to convince as many readers as possible to adopt the dual-fuel strategy and to adjust their research priorities accordingly. What the reader will find is a mixture of scientific fact, reports on demonstrated technology, speculation about what might be possible, and future-scenario narratives. This mixture is appropriate to the grand scope of our topic. The astute reader will have no difficulty identifying the character of any given statement or passage, and treating it accordingly. Further, we approach the problem at hand how best to put the global energy system on a sustainable course as one that cannot be compartmentalized. Issues of energy, environment, and economics are inextricably intertwined yet are traditionally classified in distinct intellectual categories. These three categories cannot be separated; all must be addressed simultaneously. Commercial and financial issues in particular are touched on in this essay. It is essential to do so."

Example 3

There have been a huge number of developments in technologies and applications for the production and utilization of ammonia as a fuel in the last few years and an industrial and academic stampede to demonstrate pilot projects and commercial applications.

While the history of using ammonia [96] to power engines and in solar applications dates back to the mid-1860s, long before the invention of the internal combustion engine, the reality is it has been used to power trolleys and locomotives, cars, trucks, buses, engines, generators, and even jet aircraft. Most recently it has powered hybrid electric buses in China, fuel cells in remote cell towers in Africa, farm tractors and irrigation pumps in the USA, and our retrofit conversion kits to run diesel engines and generators are tested and ready for commercial demonstration.

CONCLUDING REMARKS

Ammonia is a carbon-free fuel suitable for use in transportation sector. It has a well-established production and distribution infrastructure, and has zero global warming potential during operation. In addition to its attractive qualities as a fuel, ammonia is widely used as a NOx reducing agent for combustion exhaust gases using selective catalytic reduction (SCR), and its capacity as a refrigerant can be applied to recover and further utilize engine heat that would otherwise be lost. In terms of environmental sustainability, ammonia can be produced using either fossil fuels, or any renewable energy source, using heat and/or electricity, which allows for evolution of ammonia production methods and technologies in parallel with sustainable development. Ammonia as a sustainable fuel can be used in all types of combustion engines, gas turbines, burners with only small modifications and directly in fuel cells which is a very significant advantage compared to another type of fuels. Reducing the total greenhouse gas emissions from marine transportation is possible by using ammonia which is carbon-free fuel. They can be utilized for maritime ship engines directly as supplementary fuels or individual fuels. Ammonia fueled ships yield considerably lower global warming impact during operation. Ammonia as a sustainable and clean fuel in road vehicles yield also the lowest global warming potential after electric and hydrogen vehicles. As a result, ammonia usage in the communities for transportation sector will bring significant cost and environmental benefits together with public satisfaction.

You can store twice as much energy in ammonia as a liquid refrigerated to -32°F or under pressure at 250 psi as opposed to storing or transporting natural gas at much greater pressures of 5,000 psi or much lower temperatures, especially LNG at −162°C., and that the profits from using our vast natural gas reserves to make ammonia, urea (with any income from carbon black) would be five to ten times what we can get transporting LNG half way around the world, or even across the country. In terms of energy density, liquid ammonia contains 4.32 kWh per liter, which is 70% more than liquid hydrogen (2.54 kWh per liter at cryogenic temperature) or almost three times more than compressed hydrogen (1.55 kWh per liter at 700 bar).

Although stationary applications are superior and on road use may not be the most appropriate initial use of ammonia as a fuel in comparison to other fuels in terms of driving range, a 16-gallon fuel tank of ammonia provides a driving range of 470 miles, almost twice the range of the same volume of liquid hydrogen (259 miles) and three times the range of the same volume of compressed hydrogen (158 miles).

Some of the findings of this research are astounding and such knowledge is very valuable in corporate planning over the short to long term. As you can clearly see when today tens of thousands of scientists, engineers and experts in energy, clean tech and environmental technologies in the public and private sector and academia have quickly learned of the huge existing commercial and potential future benefits related to the production and utilization of ammonia as a fuel, energy currency, for power generation and multi-purpose uses including for refrigerated transport and combined heat, power and cooling applications in commercial or residential buildings.

Most importantly, research shows the real benefits of using *Anhydrous Ammonia to Reduce Critical Infrastructure Risk* and low cost infrastructure re-purposing applications that will delay or negate the requirements to replace existing infrastructure due to incompatibility with emerging technologies, as has been clearly demonstrated with the difficulties integrating renewable energy into the power grid.

We put together this report with all the details on the state of the art of ammonia production and utilization as a fuel, energy currency and for CO2 reduction and conversions into urea, using fossil fuel, nuclear and renewable energy and waste as the feedstocks. Hydrofuel Inc. takes great pleasure to finally get to the point where may get the fair chance to prove the viability of our technologies and business plans in an emerging global market place with a level playing field.

References

[1] Environment and Climate Change Canada, "National Inventory Report 1990-2014: Greenhouse Gas Sources and Sinks in Canada," 2016.

[2] E. Kroch, "Ammonia-a fuel for motorbuses," *Journal of the Institute of Petroleum,* pp. 213-223, 1945.

[3] "ASME Turbo Expo 2015: Turbine Technical Conference and Exposition," 2015.

[4] S. Haputhanthri, T. Maxwell, J. Fleming and C. Austin, "Ammonia and Gasoline Fuel Blends for Spark Ignited Internal Combustion Engines," *Journal of Energy Resources Technology,* vol. 137, no. 6, pp. 062201-, 2015.

[5] H. Nozari and A. Karabeyoğlu, "Numerical study of combustion characteristics of ammonia as a renewable fuel and establishment of reduced reaction mechanisms," *Fuel,* vol. 159, pp. 223-33, 2015.

[6] A. Miller, "Ammonia Fuel For Rail Transportation," in *2006 Annual NH3 Fuel Conference*, Denver West Marriott, Golden, CO, OCTOBER 9-10, 2006.

[7] A. Reiter and S.-C. Kong, "Combustion and emissions characteristics of compression-ignition engine using dual ammonia-diesel fuel," *Fuel,* vol. 90, no. 1, pp. 87-97, 2011.

[8] K. Ryu, G. Zacharakis-Jutz and S.-C. Kong, "Performance enhancement of ammonia-fueled engine by using dissociation catalyst for hydrogen generation," *International Journal of Hydrogen Energy,* vol. 39, no. 5, pp. 2390-8, 2014.

[9] C. Zamfirescu and I. Dincer, "Ammonia as a green fuel and hydrogen source for vehicular applications," *Fuel Processing Technology,* vol. 90, no. 5, pp. 729-37, 2009.

[10] C. Zamfirescu and I. Dincer, "Using ammonia as a sustainable fuel," *Journal of Power Sources,* vol. 185, no. 1, pp. 459-65, 2008.

[11] B. Hinnemann and J. Nørskov, "Catalysis by Enzymes: The Biological Ammonia Synthesis," *Topics in Catalysis,* vol. 37, no. 1, pp. 55-70, 2006.

[12] R. Lan, J. Irvine and S. Tao, "Synthesis of ammonia directly from air and water at ambient temperature and pressure," *Scientific Reports,* vol. 3, no. 1145, 2013.

[13] F.-F. Li and S. Licht, "Advances in Understanding the Mechanism and Improved Stability of the Synthesis of Ammonia from Air and Water in Hydroxide Suspensions of Nanoscale Fe2O3," *Inorganic Chemistry,* vol. 53, no. 19, pp. 10042-4, 2014.

[14] S. Licht, B. Cui, B. Wang, F.-F. Li, J. Lau and S. Liu, "Ammonia synthesis by N2 and steam electrolysis

in molten hydroxide suspensions of nanoscale Fe2O3," *Science,* vol. 345, no. 6197, pp. 637-40, 2014.

[15] T. Paschkewitz, *Ammonia Production at Ambient Temperature and Pressure: An Electrochemical and Biological Approach.,* University of Iowa, 2012.

[16] A. Skodra and M. Stoukides, "Electrocatalytic synthesis of ammonia from steam and nitrogen at atmospheric pressure," *Solid State Ionics,* vol. 180, no. 23-25, pp. 1332-6, 2009.

[17] "Facts and Figures," Air Transport Action Group, May 2016. [Online]. Available: http://www.atag.org/facts-and-figures.html.

[18] "Understanding Global Warming Potentials," 23 02 2017. [Online]. Available: https://www.epa.gov/ghgemissions/understanding-global-warming-potentials.

[19] "Terra preta," 23 02 2017. [Online]. Available: https://en.wikipedia.org/wiki/Terra_preta.

[20] Cap-and-Trade's Inherent Defects, "Cap-and-Trade's Inherent Defects," 24 02 2017. [Online]. Available: https://www.carbontax.org/cap-and-trade-problems/.

[21] "Government of Canada to work with provinces, territories, and stakeholders to develop a clean fuel standard," 25 11 2016. [Online]. Available: http://news.gc.ca/web/article-en.do?nid=1160579.

[22] S. Dion, "the Green Shift," 23 02 2017. [Online]. Available: https://www.poltext.org/sites/poltext.org/files/plateformes/ca2008lib_plt_eng._05012009_111617.pdf.

[23] "British Columbia carbon tax," 23 02 2017. [Online]. Available: https://en.wikipedia.org/wiki/British_Columbia_carbon_tax.

[24] Wikipedia, "Carbon Tax," 23 02 2017. [Online]. Available: https://en.wikipedia.org/wiki/Carbon_tax.

[25] "Quebec's Green Future: The Lowest-Cost Route to Greenhouse Gas Reductions," 01 10 2009. [Online]. Available: https://www.cdhowe.org/sites/default/files/attachments/research_papers/mixed//backgrounder_118_English.pdf.

[26] "Overview of Greenhouse Gases," 23 02 2017. [Online]. Available: https://www.epa.gov/ghgemissions/overview-greenhouse-gases.

[27] Ultra Blaze, "Venting LNG Tanks," YouTube, 02 11 2015. [Online]. Available: https://youtu.be/vT4nG-Z3zdE.

[28] Hydrofuel Inc. & UOIT, "Pacific NorthWest LNG Project," 16 03 2016. [Online]. Available: http://www.ceaa-acee.gc.ca/050/documents/p80032/108890E.pdf. [Accessed 08 03 2017].

[29] Hydrofuel Inc & UOIT, "Key Life-Cycle Numbers for NH3 and other fuels," 16 09 2016. [Online]. Available: http://www.marketwired.com/press-release/new-evidence-on-green-ammonia-energy-solutions-be-presented-los-angeles-conference-2159195.htm. [Accessed 08 03 2017].

[30] NH3 Fuel Association, "NH3 FUEL CONFERENCE 2016: SCHEDULE," [Online]. Available: https://nh3fuelassociation.org/events-conferences/conference2016/#schedule. [Accessed 08 03 2017].

[31] Siemens PLC, ""Green" ammonia is the key to meeting the twin challenges of the 21st century.," [Online]. Available: https://www.siemens.co.uk/en/insights/potential-of-green-ammonia-as-fertiliser-and-electricity-storage.htm. [Accessed 08 03 2017].

[32] NH3 Fuel Association, "NH3 Fuel Association," [Online]. Available: http://nh3fuelassociation.org/. [Accessed 08 03 2017].

[33] Ammonia Energy, "The global information portal for ammonia as an energy vector," [Online]. Available: http://www.ammoniaenergy.org/. [Accessed 08 03 2017].

[34] Hydrofuel Inc. & UOIT, "Key Life-Cycle Numbers for NH3, Fossil Fuel and Green Energy production and utilization in agriculture, energy and utilities, and transportation systems," [Online]. Available: https://www.academia.edu/28681084/Key_Life-Cycle_Numbers_for_NH3_Fossil_Fuel_and_Green_Energy_production_and_utilization_in_agriculture_energy_and_utilities_and_transportation_systems. [Accessed 08 03 2017].

[35] Hydrofuel Inc. & UOIT, "Comparative Life Cycle Assessment of Various Ammonia Production Methods," 01 11 2016. [Online]. Available: https://www.academia.edu/19786635/Comparative_Life_Cycle_Assessment_of_Various_Ammonia_Production_Methods. [Accessed 08 03 2017].

[36] Y. Bicer, I. Dincer, G. Vezina and F. Raso, "Impact Assessment and Environmental Evaluation of Various Ammonia Production Processes," Hydrofuel Inc. & UOIT, 14 02 2017. [Online]. Available: http://www.readcube.com/articles/10.1007/s00267-017-0831-6?author_access_token=MYnucYZr5t7vIs0I-yGtZ_e4RwlQNchNByi7wbcMAY4AiStKfv68gRrCJvRKLH6G6FVerSnoUyRDjpAoU_jbEHe-5vxAQg92U9ydQMSO45LFpCYfUdW7efZDV-G7XbrDCJmgMFYc22Ca2CJ_zSZ-8A%3D%3D. [Accessed 21 04 2017].

[37] Hydrofuel Inc., "Hydrofuel Inc. Latest News," [Online]. Available: http://nh3fuel.com/index.php?option=com_content&task=category&sectionid=1&id=1&Itemid=36. [Accessed 08 03 2017].

[38] Proton Ventures, "NH3 Event - The Biggest Ammonia Event in Europe," [Online]. Available: http://nh3event.com/. [Accessed 08 03 2017].

[39] Nikkei Asian Review, "Japanese utilities team on CO2-reducing tech for coal plants," 02 03 2017. [Online]. Available: http://asia.nikkei.com/Business/Deals/Japanese-utilities-team-on-CO2-

reducing-tech-for-coal-plants?n_cid=NARAN1507. [Accessed 08 03 2017].

[40] D. S. Wittrig, "Clean Air Task Force - Ammonia Fuel: Opportunities, Markets, Issues," [Online]. Available: https://arpa-e.energy.gov/sites/default/files/Wittrig_Ammonia_TransportationFuels_Workshop.pdf. [Accessed 08 03 2017].

[41] J. S. Robertson, "Energy conversion system," Google Patents, 26 09 2013. [Online]. Available: https://www.google.com/patents/US20130252120. [Accessed 08 03 2017].

[42] T. Brown, "Low-carbon ammonia synthesis: Japan's 'Energy Carriers'," Ammonia Industry, 10 11 2016. [Online]. Available: http://www.ammoniaenergy.org/low-carbon-ammonia-synthesis-japans-energy-carriers/. [Accessed 08 03 2017].

[43] Canadian Manufacturing, "Feds pledge to create national clean fuel standard in latest piece of climate plan," 28 11 2016. [Online]. Available: http://www.canadianmanufacturing.com/environment-and-safety/feds-pledge-to-create-national-clean-fuel-standard-in-latest-piece-of-climate-plan-179601/. [Accessed 08 03 2017].

[44] T. Brown, "Nuon's Power-to-Ammonia update, and the first European ammonia fuel conference in 2017," Ammonia Industry, 02 12 2016. [Online]. Available: http://www.ammoniaenergy.org/nuons-power-to-ammonia-update-and-the-first-european-ammonia-fuel-conference-in-2017/. [Accessed 08 03 2017].

[45] MIT Energy Initiative, "Plan B: Fossil fuels without CO2," 06 12 2016. [Online]. Available: http://energy.mit.edu/news/plan-b-fossil-fuels-without-co2/. [Accessed 08 03 2017].

[46] S. Crolius, "Methane to Ammonia via Pyrolysis," Ammonia Energy, 26 01 2017. [Online]. Available: http://www.ammoniaenergy.org/methane-to-ammonia-via-pyrolysis/. [Accessed 08 03 2017].

[47] T. Brown, "ARPA-E funding for renewable ammonia synthesis technologies," Ammonia Industry, 22 12 2016. [Online]. Available: https://ammoniaindustry.com/arpa-e-funding-for-renewable-ammonia-synthesis-technologies/. [Accessed 08 03 2017].

[48] Breakthrough Energy, "Breakthrough Energy Coalition," [Online]. Available: http://www.b-t.energy/coalition/. [Accessed 08 03 2017].

[49] T. Brown, "Breakthrough Energy Coalition targets carbon-free ammonia," Ammonia Industry, 14 12 2016. [Online]. Available: https://ammoniaindustry.com/breakthrough-energy-coalition-targets-carbon-free-ammonia/. [Accessed 08 03 2017].

[50] Natural Resources Canada, "The Energy Innovation Program : Frequently Asked Questions," 20 10 2016. [Online]. Available: http://www.nrcan.gc.ca/energy/science/programs-funding/18384#li-15. [Accessed 08 03 2017].

[51] T. Brown, "US DOE funding research into sustainable ammonia synthesis," Ammonia Industry, 27 01 2017. [Online]. Available: https://ammoniaindustry.com/us-doe-funding-research-into-

sustainable-ammonia-synthesis/. [Accessed 08 03 2017].

[52] T. Brown, "Comparative studies of ammonia production, combining renewable hydrogen with Haber-Bosch," Ammonia Energy, 02 02 2017. [Online]. Available: http://www.ammoniaenergy.org/comparative-studies-of-ammonia-production-combining-renewable-hydrogen-with-haber-bosch/. [Accessed 21 04 2017].

[53] Hiroshima University Showa Denko KK Taiyo Nichi Co., Ltd. Science and Technology Promotion Organization (JST), "Manufacture hydrogen fuel for fuel cell vehicles from ammonia," 19 07 2016. [Online]. Available: http://www.jst.go.jp/pr/announce/20160719-2/#YOUGO3. [Accessed 21 04 2017].

[54] T. Brown, "International R&D on sustainable ammonia synthesis technologies," Ammonia Industry, 03 02 2017. [Online]. Available: https://ammoniaindustry.com/international-rd-on-sustainable-ammonia-synthesis-technologies/. [Accessed 21 04 2017].

[55] T. Brown, "On the Ground in Germany," Ammonia Energy, 09 02 2017. [Online]. Available: http://www.ammoniaenergy.org/renewable-ammonia-in-germany/. [Accessed 21 04 2017].

[56] T. Brown, "Ammonia-Fueled Gas Turbine Power Generation," Ammonia Energy, 16 02 2017. [Online]. Available: http://www.ammoniaenergy.org/ammonia-fueled-gas-turbine-power-generation/. [Accessed 21 04 2017].

[57] T. Brown, "Ammonia cracking to high-purity hydrogen, for PEM fuel cells in Denmark," Ammonia Energy, 16 02 2017. [Online]. Available: http://www.ammoniaenergy.org/ammonia-cracking-to-high-purity-hydrogen-for-pem-fuel-cells-in-denmark/. [Accessed 21 04 2017].

[58] G. Wang, A. Mitsos and W. Marquardt, "Conceptual design of ammonia-based energy storage system: System design and time-invariant performance," *AIChE J,* vol. 63, no. 5, p. 1620–1637, 2017.

[59] T. Brown, "Ammonia for energy storage: economic and technical analysis," Ammonia Industry, 02 03 2017. [Online]. Available: https://ammoniaindustry.com/ammonia-for-energy-storage-economic-and-technical-analysis/. [Accessed 21 04 2017].

[60] T. Brown, "Industrial demonstrations of ammonia fuel in Japan," Ammonia Industry, 06 03 2017. [Online]. Available: https://ammoniaindustry.com/industrial-demonstrations-of-ammonia-fuel-in-japan/. [Accessed 21 04 2017].

[61] T. Brown, "Ammonia for energy storage: a "revolutionary disruption"," Ammonia Energy, 16 03 2017. [Online]. Available: http://www.ammoniaenergy.org/ammonia-for-energy-storage-a-revolutionary-disruption/. [Accessed 21 04 2017].

[62] T. Brown, "Bunker Ammonia: carbon-free liquid fuel for ships," Ammonia Energy, 23 03 2017. [Online]. Available: http://www.ammoniaenergy.org/bunker-ammonia-carbon-free-liquid-fuel-for-ships/. [Accessed 21 04 2017].

[63] "Global Dual Fuel Engine Market 2016-2020," Technavio - Discover Market Opportunities, 07 2016. [Online]. Available: https://www.technavio.com/report/global-tools-and-components-dual-fuel-engine-market. [Accessed 21 04 2017].

[64] S. Crolius, "New Technology for Generating Hydrogen from Ammonia," Ammonia Energy, 30 03 2017. [Online]. Available: http://www.ammoniaenergy.org/new-technology-for-generating-hydrogen-from-ammonia/. [Accessed 21 04 201].

[65] T. Brown, "Power to Ammonia feasibility study," Ammonia Energy, 30 03 2017. [Online]. Available: http://www.ammoniaenergy.org/power-to-ammonia-feasibility-study/. [Accessed 21 04 2017].

[66] S. Crolius, "NH Energy+: Topical Conference at AIChE Annual Meeting," Ammonia Energy, 06 04 2017. [Online]. Available: http://www.ammoniaenergy.org/nh-energy-topical-conference-at-aiche-annual-meeting/. [Accessed 21 04 2017].

[67] S. Crolius, "Ammonia – and Other Nitrogen-Based Fuels," Ammonia Energy, 20 04 2017. [Online]. Available: http://www.ammoniaenergy.org/ammonia-and-other-nitrogen-based-fuels/. [Accessed 21 04 2017].

[68] T. Brown, "IEA calls for renewable hydrogen and carbon-free ammonia," Ammonia Energy, 21 04 2017. [Online]. Available: http://www.ammoniaenergy.org/iea-calls-for-renewable-hydrogen-and-carbon-free-ammonia/. [Accessed 21 04 2017].

[69] A. Vezina, "Anhydrous Ammonia to Reduce Critical Infrastructure Risk," York University, [Online]. Available: http://www.nh3fuel.com/images/documents/A-Vezina%20-%20Anhydrous%20Ammonia%20to%20Reduce%20Critical%20Infrastructure%20Risk.pdf. [Accessed 21 04 2017].

[70] A. Vezina, "Can the Smog in China be Reduced or Eliminated Using Ammonia?," York University, [Online]. Available: http://www.nh3fuel.com/images/documents/A-Vezina%20-%20China%20Ammonia%20Research%20Final.pdf. [Accessed 21 04 2017].

[71] G. Vezina and F. Raso, "An Alternate View on Pipelines — Transport Ammonia not Crude," Ottawa Life Magazine, 11 07 2016. [Online]. Available: http://www.ottawalife.com/article/an-alternate-view-on-pipelines-transport-ammonia-not-crude. [Accessed 08 03 2017].

[72] G. Vezina and I. Dincer, "Green Ammonia," Ottawa Life Magazine, 16 09 2016. [Online]. Available: http://www.ottawalife.com/article/green-ammonia. [Accessed 08 03 2017].

[73] G. Vezina and S. Wittrig, "Canada's Ammonia Energy Option," Ottawa Life Magazine, 15 11 2016. [Online]. Available: http://www.ottawalife.com/article/canada-s-ammonia-energy-option?c=9. [Accessed 08 03 2017].

[74] G. Vezina and F. Raso, "Canada's Life Cycle Clean Fuels Policy Game Changer," Ottawa Life Magazine, 23 01 2017. [Online]. Available: http://www.ottawalife.com/article/canada-s-life-cycle-clean-fuels-policy-game-changer. [Accessed 08 03 2017].

[75] G. Vezina, "Editorial Article Author - Greg Vezina," Toronto Sun, [Online]. Available: http://www.torontosun.com/search?cx=016362519718727753442:ubvjwz0hpe0&cof=FORID:11&ie=UTF-8&q=%22Greg+Vezina%22&siteurl=www.torontosun.com/&ref=&ss=1514j2292196j2. [Accessed 08 03 2017].

[76] L. Goldstein, "High cost of ignorance," Toronto Sun, 22 10 2016. [Online]. Available: http://www.torontosun.com/2016/10/21/high-cost-of-ignorance. [Accessed 08 03 2017].

[77] NH3 Fuel Association, "Comparisons - The Logical Path Forward," [Online]. Available: https://nh3fuelassociation.org/comparisons/. [Accessed 08 03 2017].

[78] "REFUEL: Renewable Energy to Fuels Through Utilization of Energy-Dense Liquids," U.S. Department of Energy , 15 12 2016. [Online]. Available: https://arpa-e.energy.gov/?q=arpa-e-programs/refuel. [Accessed 21 04 2017].

[79] "ARPA-E REFUEL Kickoff Meeting, Houston, TX 2017," US Department of Energy, 16 03 2017. [Online]. Available: https://custom.cvent.com/AA09371532024F85AB8D6B671E9321D5/files/1d76b74231dd4941832fd4489b3ce6fc.docx. [Accessed 21 04 2017].

[80] Hydrofuel Inc. & UOIT, "Ammonia as a Potential Solution for Alberta," 07 08 2016. [Online]. Available: http://nh3fuel.com/images/documents/2016-08-07%20-%20Ammonia%20Production%20for%20Alberta.pdf. [Accessed 08 03 2017].

[81] Hydrofuel Inc. & UOIT, "Natural Gas to Ammonia as a Potential Solution for British Columbia," 09 08 2016. [Online]. Available: http://nh3fuel.com/images/documents/2016-08-09%20-%20Ammonia%20Instead%20of%20Natural%20Gas%20for%20BC.pdf. [Accessed 08 03 2017].

[82] Hydrofuel Inc. & UOIT, "Comparative assessment of NH3 production and utilization in transportation systems for Ontario," 23 11 2015. [Online]. Available: http://www.nh3fuel.com/images/documents/2015-11-23%20-%20MITACS-Final-Report-P1.pdf. [Accessed 08 03 2017].

[83] Hydrofuel Inc. & UOIT, "Comparative assessment of NH3 production and utilization in agriculture, energy and utilities, and transportation systems," 17 06 2016. [Online]. Available: http://nh3fuel.com/images/documents/2016-06-17%20-%20MITACS-Final%20Report-P2.pdf. [Accessed 08 03 2017].

[84] I. Dincer and Y. Bicer, "Comprehensive Evaluation of NH," Hydrofuel Inc. & UOIT, 25 03 2017. [Online]. Available: http://www.nh3fuel.com/images/documents/2017-03-25%20-%20MITACS-Final-Report-P3-IT08015.pdf. [Accessed 21 04 2017].

[85] I. Dincer, Y. Bicer, G. Vezina and F. Raso, "Ammonia (NH3) as a Potential Transportation Solution for Ontario," Hydrofuel & UOIT, 10 03 2017. [Online]. Available: http://www.nh3fuel.com/images/documents/2017-03-10%20-%20Ontario-RFS-Discussion-Paper.pdf. [Accessed 21 04 2017].

[86] Ontario, Environmental Registry, "Discussion Paper: Developing a Modern Renewable Fuel Standard For Gasoline in Ontario," 11 01 2017. [Online]. Available: https://www.ebr.gov.on.ca/ERS-WEB-External/displaynoticecontent.do?noticeId=MTI4OTAz&statusId=MTk0OTQ4. [Accessed 21 04 2017].

[87] "Clean Fuel Standard: Discussion Paper," Environment and Climiate Change Canada, 06 03 2017. [Online]. Available: http://ec.gc.ca/lcpe-cepa/default.asp?lang=En&xml=D7C913BB-13D0-42AF-9BC7-FBC1580C2F4B. [Accessed 21 04 2017].

[88] Hydrofuel Inc. & UOIT, "Summary of the Proposed Project “Connection of Remote First Nation Communities in Northwest Ontario”," [Online]. Available: http://nh3fuel.com/images/documents/Diesel%20Generator%20Alternatives.pdf. [Accessed 21 04 2017].

[89] Hydrofuel Inc., "Hydrofuel Inc.," [Online]. Available: http://nh3fuel.com/. [Accessed 21 04 2017].

[90] R. Knowlton and H. Wright, "Ammonia Vehicle Fuel - Assessment of the technical, economic and safety problems associated with ammonia use as a ground transportation fuel," Chemetics International Company, 04 1984. [Online]. Available: http://www.nh3fuel.com/images/documents/1984-04%20-%20Chemetics%20NCR%20CAN.pdf. [Accessed 21 04 2017].

[91] L. J. Green, "Energy Needs versus Environmental Pollution: A Reconciliation?," *American Association for the Advancement of Science,* vol. 156, no. 3781, pp. 1448-1450, 16 06 1967.

[92] W. Ahlgren, "William Ahlgren comment to AB1257 Natural Gas Act Report," California Polytechnic State University, 04 05 2015. [Online]. Available: http://docketpublic.energy.ca.gov/PublicDocuments/15-IEPR-04/TN204490_20150504T125450_William_Ahlgren_comment_to_AB1257_Natural_Gas_Act_Report.pdf. [Accessed 21 04 2017].

[93] W. Ahlgren, "The Dual-Fuel Strategy: An Energy Transition Plan," Proceedings of the IEEE, 11 2012. [Online]. Available: http://www.nh3fuel.com/images/documents/IEEE%20-%20Dual%20Fuel%20Strategy.pdf. [Accessed 21 04 2017].

[94] W. Ahlgren, "Planning for Hundred-Fold Increase," California Polytechnic State University, San Luis Obispo, 2013. [Online]. Available: http://www.nh3fuel.com/images/documents/Planning%20for%20Hundred-Fold%20Increase.pdf. [Accessed 21 04 2017].

[95] J. Esch, "Prolog to "The Dual-Fuel Strategy: An Energy Transition Plan'," IEEE, 16 10 2012. [Online]. Available: http://ieeexplore.ieee.org/document/6331607/. [Accessed 21 04 2017].

[96] "What are some surprising details about using ammonia as a fuel?," Hydrofuel Inc., [Online]. Available: http://nh3fuel.com/index.php?option=com_content&task=view&id=37&Itemid=37. Accessed 21 04 2017].

498765 Economic Analysis of Ammonia Production Using Renewable Energy

Abstract:

Economic Analysis of Ammonia Production Using Renewable Energy

Douglas G. Tiffany

Bioproducts & Biosystems Engineering

University of Minnesota

Ammonia production has been fundamental to worldwide agricultural productivity growth for over a century with the biggest impacts occurring between 1957 and 1967 in the U.S. This presentation will consider the economic merits of producing this critical input, of which 80% is used in agriculture, at smaller scales in a distributed fashion and from renewable energy versus the prevailing large-scale, centralized production facilities using fossil energy and close to abundant natural gas supplies.

Installed facility cost and production data of a small-scale, renewable ammonia plant offer opportunities to analyze and compare the competitive position of ammonia production at sites close to ammonia markets with wind energy or other renewable sources. Production economics will be presented based upon the pilot plant on site and in operation for three years. In addition, scaled production will be explored to identify scale economies that may be attractive for operational and financial considerations. Segregated analyses of stages of ammonia production will be presented for hydrogen from water hydrolysis, nitrogen from pressure swing absorption and ammonia synthesis based on the Haber-Bosch process using iron catalysts. Absorbent technologies have been developed by this research team that offer potential capital savings due to reduced pressure and smaller operational costs, requiring less energy. Logistical advantages of distributed production include smaller transportation costs and reduced Greenhouse Gas (GHG) emissions. Potential state and federal policy incentives to reward distributed renewable production of this key agricultural and industrial input will be discussed.

POSTER: Group: Liaison Functions Abstract # 503381
Session: Poster Session: General Topics on Chemical Engineering II
Time: Wednesday, November 1, 2017 - 3:15pm-4:45pm

(5853) (585f) Ammonia Renewable Energy Systems at Continental Scale: Alternative to Electricity for Transmission, Storage, and Integration for Deep Decarbonization of the World's Largest Industry

We must soon "run the world on renewables"? but cannot, and should not try to, accomplish this entirely with electricity transmission. New, abundant, low-cost, unconventional natural gas supplies are finite; burning adds CO2 to Earth's atmosphere. Humanity's goal must be nothing less than: Transforming the world's largest industry from ~ 80% fossil to ~ 100 renewable, CO2-emission-free, energy sources as quickly as we prudently and profitably can.

Anhydrous Ammonia, NH3, as a fuel, may have a vital role in humanity's global Grand Challenge: Transforming the world's largest industry, from ~ 80% fossil to ~ 100% renewable, CO2-emission-free (CEF), renewable energy (RE) sources, as quickly as we prudently and profitably can.

Attempting this via electricity systems, alone, will be technically and economically suboptimal, suffering delays while citizens oppose building many of the new electricity transmission lines necessary to gather, transmit, and deliver the very large quantity of energy required to supply all humanity's energy, for all purposes, from diverse sources. Hydrogen and Ammonia are the only C-free fuels.

Because NH3 is a carbon-free fuel, nearly 18% hydrogen by weight, with slightly over half the energy density of gasoline by volume, it is potentially the foundation of a complete, integrated, optimized energy system -- rivaling the electricity "Grid" in magnitude and importance in meeting this Challenge. Iowa State University, recognizing this potential, has hosted thirteen annual Ammonia Fuel Conferences, which include NH3 as a renewable energy (RE) transmission and storage medium, as well as a transportation and distributed generation fuel.
Most of our RE is today generated as electricity from wind, solar, and hydro -- which vary in output at time scales of seconds to seasons. We therefore need low-cost energy storage, for annual-scale firming storage.

A safe, reliable, proven ammonia delivery and storage infrastructure already exists in the US. Approximately 3,000 miles of carbon-steel ammonia pipeline is in service in Americaï¿½s agricultural heartland, mainly in the Corn Belt. Almost a hundred large terminals for refrigerated ammonia storage are distributed along the pipeline. Barges, trains, and trucks round out the delivery system, which supplies the ammonia from the terminal to the farmer when he needs it for the growing season. The state of Iowa, alone, has over 800 retail outlets where farmers buy "anhydrous" or "nitrogen" , the vernacular for ammonia fertilizer, NH3.

We now store liquid NH3 in "atmospheric" steel on-grade tanks, of typical 30,000 ton capacity, which is ~ 200,000 MWh as the chemical energy in NH3. Capital expenditute (capex) is ~ $ 0.10 / kWh, far below the likely cost of electricity battery storage, assuming continued progress in battery technology and capex and O&M costs.

We should now carefully consider using pipeline networks, rather than the electricity grid, for solving the three salient technical problems of renewable energy (RE) at lower cost:

1. Transmission: from diverse, stranded, remote, rich RE resources
2. Storage: intermittent RE becomes annually firm and dispatchable
3. Integration: with conventional, extant energy, for firm quality supply

Electricity systems are not well suited for gathering, transmission, storage, and delivery of time-varying solar, wind, and other renewable energy resources. The renewable energy industry should consider emulating the natural gas and ammonia industries, because:

1. Gaseous hydrogen (GH2) fuel may be stored in large, deep, solution-mined salt caverns for ~ $ 0.20 / kWh capital cost;

2. Liquid anhydrous ammonia (NH_3) fuel may be stored in large, refrigerated, steel, "atmospheric" surface tanks for ~ $0.10 / kWh capital cost

We need to supply all energy, not just electricity, from diverse renewable energy (RE) resources, both distributed and centralized, where the worldâ??s richest RE resources â?? of large geographic extent and high intensity â?? are stranded: far from end-users with inadequate or nonexistent gathering and transmission systems to deliver the energy. Electricity energy storage cannot affordably firm large, intermittent renewables at annual scale, while carbon-free gaseous hydrogen (GH2) and liquid anhydrous ammonia (NH3) fuels can.

Thus, we need to now investigate and plan for a diversity of complementary RE transmission and storage systems: media, fuels, and strategies. GH2 and NH3 are especially attractive. Transmission pipelines for both have multi-GW capacity over hundreds of km and provide valuable inherent storage. Capital costs per GW-km capacity, and transmission losses and costs per GWh-km, are comparable.

But we lack a process by which to synthesize NH3 directly from RE-source electricity, water, and air. Electrolysis plus Haber-Bosch (EHB) is too costly, in capex and energy conversion losses, especially at the low capacity factor (CF) inflicted on any downstream process by the low CF of the RE sources: typically 30 - 45 %.
This novel synthesis process is chemical engineering's Grand Challenge.

Like hydrogen, ammonia can burn directly in spark-ignited internal combustion engines (ICE), and in combustion turbines (CT), and may also be fed directly to medium temperature solid oxide, proton-conducting ceramic, and molten-salt direct-ammonia fuel cells, with only nitrogen and water vapor as combustion products. Like hydrogen, ammonia is lighter than air and is not a greenhouse gas.

All of ammonia's energy is derived from its hydrogen content; it can be easily reformed to hydrogen and nitrogen, with N2 returned to its source, Earthï¿½s atmosphere. NH3 has the highest hydrogen content by volume of any liquid fuel, including gasoline, liquefied natural gas (LNG), liquefied petroleum gas (LPG, propane), ethanol, and even liquid hydrogen. Liquid anhydrous ammonia, NH3, has more atoms of hydrogen per liter than liquid hydrogen. This ability of NH3 to store hydrogen very compactly at ambient temperature and moderate pressure is a key advantage for NH3 over GH2.

Chemical engineering now needs to collaborate with the electricity industry's many facets -- including utilities, USDOE national laboratories, Electric Power Research Institute (EPRI), Edison Electric Institute, and many non-profit organizations (UCS, NRDC, RFF, WRI, et al) -- to model the total global future energy system, including GH2 and NH3 as complete, integrated, optimized energy systems for gathering from diverse sources, transmission, low-cost storage, distribution, and end-use of RE-CEF energy.

Therefore, we should now:

- Design and optimize complete RE systems, at local and continental scales, from sunlight, wind, and water resources to dispatchable, delivered energy services:
  - Generation, Gathering, Firming storage, End use
  - Conversion, Transmission, Combined-heat-and-power (CHP)

- Render annually-firm and dispatchable RE supplied, via very low capital cost storage, less than $US 1.00 / kWh:
  - Gaseous Hydrogen (GH2) in large salt caverns, where geology is available
  - Liquid Ammonia (NH3) in carbon steel surface tanks
  - Interconnected via continental underground pipelines, adding storage
  - Lower cost than any contemplated â??electricityâ?? storage components

- Design and build pilot plants for both GH2 and NH3 RE systems, by which to:
  - Discover and demonstrate scaleable technical proof-of-concept and economics

- Explore optimum system topology for sources, components, and end-uses
- Motivate private-public collaboratives to conceive RPFs and RFQs for the plants

To achieve this goal, we must overcome these obstacles:

- Earth's richest RE is stranded, far from markets with no transmission
- We cannot do this entirely via electricity, and should not try to do so; "Smart Grid"⍰ is primarily demand side management (DSM); it adds no inherent or physical new capacity , and only slight effective new capacity.

Now, we need to simultaneously pursue the several RD & D projects and pathways below, because we:

a. Must deeply decarbonize humanity's entire energy system quickly, prudently, and profitably, optimizing our investment strategy, avoiding overinvestment in electricity;
b. Don't know whether future RE-CEF energy will be most conveniently and economically harvested as electricity, as from wind and PV, or as GH2 directly from photo-, thermo-, or biochemical processes, splitting water molecules in both "central" and "distributed" sites;
c. Need both R&D and pilot plants by which to discover and demonstrate the technical and economic advantages -- if any -- of GH2 and NH3 as complete RE-CEF energy systems vis-a-vis electricity systems, at local, regional, continental scales, for all energy services;
d. Must prevent suboptimal investments in enlarging and "Smartening" the electricity Grid, insuring that "electrifying transportation" includes GH2 and NH3 as on-board, CEF fuels.
e. Must consider the EMP threat: NH3 infrastructures could be built with simpler controls than the electricity Grid's, for better resilience as we face more probable threats from aerial nuclear explosions attacking our electricity grid and other infrastructures.

1. Complete RE-CEF energy systems. Conceive, design, build, and demonstrate complete RE-CEF energy systems optimized for capturing energy, in all ways, from incident photons and moving air and water molecules, from which to deliver an affordable, annually-firm, always-dispatchable supply of distributed energy services, for all purposes, to everyone on Earth. GH2 and NH3 systems may be technically and economically superior to electricity systems at regional to continental scales. Attempts to upgrade, enlarge, and "smarten" the electricity Grid must be evaluated in this context, to avoid suboptimal investments in the Grid: http://energy.gov/eere/articles/revolutionnow-rewind-modernizing-grid-ac... powered-future Collaborators in this major work should include: NREL Energy Systems Integration, with LBL and other DOE labs, EPRI, GTI, IGU, EEI, other industry groups, solar, wind, other energy OEM's, Oil & Gas industry, universities, and NGO's.

2. Wind and solar industries need to think "beyond electricity". "Transmission" now means to them only electric wires; "storage" generally means batteries. By 2050, hydrogen and ammonia fuels may be a bigger market for CEF energy than the electricity grid. http://leightyfoundation.org/w/wp-content/uploads/WP16-A-1.pdf OEM's and project developers shall embrace GH2 and NH3 systems.

3. Pilot plants: wind and solar dedicated to Hydrogen and Ammonia fuel production. Discover and demonstrate reduction in cost of wind-source and solar-source H2 and NH3 fuels from plants with no Grid connection, simplified in design without transformers, underground wire, electronics, delivering only "wild DC" to electrolyzers and NH3 synth reactors instead of Grid-quality electricity.

Wind-to-H2 example: http://leightyfoundation.org/w/wp-content/uploads/WP16-B.pdf

4. Embrace Anhydrous Ammonia, NH3, as "the other Hydrogen". ARPA-E's 2016 "REFUEL" FOA includes NH3 because of its high Hydrogen volumetric energy density, ease of transportation and handling, and low-cost energy storage. It is a C-free fuel for the ICE, CT, and direct ammonia fuel cell (DAFC); we need more R&D to improve energy conversion efficiency, in both NH3 synthesis and energy recovery, and to prevent NOx. Pipelining and storage in low-cost carbon steel, at ~10 bar and in large, refrigerated "atmospheric" storage, are well known. Siemens has embraced "Green Ammonia": http://www.siemens.co.uk/en/insights/potential-of-green-ammonia-as-ferti... Japan includes Ammonia in its

"Strategic Innovation Promotion" (SIP) program, "Energy Carriers": http://www.jst.go.jp/sip/pdf/SIP_energycarriers2015_en.pdf Fig. 6.

5. Dedicated "distributed" GH2 and NH3 fuel plants. With no connection to, nor energy delivery to, the electricity grid, wind and solar plants are simplified, eliminating field transformers, power wiring, substations necessary to deliver grid-quality electricity. Turbine and PV output is "wild DC" close-coupled to electrolysis stacks and / or to NH3 synthesis reactors. Capex and O&M cost savings may pay for electrolysis and NH3 plants and for some energy conversion losses. These tradeoffs must be engineered, field-tested, and quantified at GW scale.

The "smarter"? grid may be more vulnerable to cyberattack. Adding storage, control, and quality adjunct devices to the electricity grid, to accommodate very high renewables content, may be technically and economically inferior to GH2 and NH3 RE systems. Thus, we need to look beyond "smart grid" , expanding our concept of "transmission" , to synergistically and simultaneously solve the transmission, firming storage, and RE integration "balancing"? problems now severely constraining our progress toward "running the world on renewables" ?.

At GW scale, renewable-source electricity from diverse sources can be converted to hydrogen and byproduct oxygen, and/or to NH3 fuels, both processes producing copious by- product oxygen, and pipelined underground to load centers for use as vehicle fuel and combined-heat-and-power generation on the wholesale or retail side of the customers' meters.

This enables very low cost energy storage: less than $ 1.00 / kWh capital cost:

- Gaseous Hydrogen (GH2) in large, deep, solution-mined salt caverns, where the salt geology is available: Gulf of Mexico coast
- Liquid Anhydrous Ammonia (NH3) in large, refrigerated, "atmospheric", carbon steel surface tanks, extant in the Corn Belt
- Interconnected via continental underground pipelines, adding "free" storage by packing the GH2 pipelines (not liquid NH3 lines)
- At lower cost than any contemplated "electricity"? storage technology, components, or systems

Pipelined GH2 and NH3 fuels free those wind and solar PV plants, which would be dedicated to delivering all their captured RE as GH2 and NH3 fuels to pipelines, from the capital and O&M costs of generating and delivering grid-quality AC or DC electricity: the required complex generators and power electronics, field transformers, cables and substations, transmission lines.

The internal combustion engine (ICE), combustion turbine (CT), and direct ammonia fuel cell (DAFC) operate very efficiently on GH2 and NH3 fuels. USA has extensive extant NH3 pipeline and tank storage infrastructure serving the N-fertilizer industry.

Humanity's goal is to eventually "Run the World on Renewables" plus perhaps some nuclear. Therefore, we design alternatives to, and adjuncts to, the electricity grid:

- Convert all RE at sources to Gaseous Hydrogen (GH2) or Ammonia (NH3) fuels
- Deliver these C-free fuels via underground pipelines for transport and CHP

491297 Optimal Design Strategy of an Aerated Stirred Tank Reactor Using Computational Fluid Dynamics and Bayesian Multi-Objective Optimization Combined Method

Abstract:

Optimal Design Strategy of an Aerated Stirred Tank Reactor Using Computational Fluid Dynamics and Bayesian Multi-Objective Optimization Combined Method

Seongeon Park, Minjun Kim, Jonggeol Na, Jinjoo An and Chonghun Han*

School of Chemical & Biological Engineering, Seoul National University, Seoul, South Korea

ABSTRACT

Optimal reactor design is still a poorly understood problem compared to process optimization problems, although the performance of a reactor greatly influences the efficiency of whole process. This is partly due to the inability of numerical reactor models to predict the non-ideal mixing behaviors. Nowadays, computational fluid dynamics (CFD) has emerged as a solution to overcome such limitations by reflecting the complex 3D hydrodynamics inside a reactor.

However, the optimization problem involving CFD simulation is difficult to solve, because the computational cost for runs of simulation is very expensive, and gradient information which may guide the parameter search is unavailable. Bayesian optimization (BO), a recently popularized gradient-free black-box global optimization algorithm, is a very promising approach for such problems. BO builds a surrogate model of the objective function based on the data points already explored, and exploits information from the model to decide where to search next.

In this research, a strategy to find the optimal design of an aerated stirred tank reactor with two Rushton turbine impellers using CFD and BO combined method is proposed. A multi-objective extension of BO was employed in order to maximize the effective gas holdup and minimize the power consumption simultaneously. Six optimization variables to define the geometrical structure of the reactor were chosen: the aspect ratio of the tank, the impeller diameter to the tank diameter ratio, the clearance of each impeller to the tank height ratio, the gas sparger diameter to impeller diameter ratio and the sparger clearance to the tank height ratio. The CFD reactor model works as a black-box function of which the inputs are design variables and the outputs are objective function values. The reactor structure of Alves (2002) was chosen to be a base case and its experimental data were used to validate the CFD reactor model.

As a result, the algorithm found the best solutions, which comprise a Pareto front, within 50 iterations. Furthermore, the whole optimal design process could be automated without any need of human decision practices.

REFERENCE

Alves, S. S., et al., "Experimental and modelling study of gas dispersion in a double turbine stirred tank," *Chemical Engineering Science,* **57**(3), pp. 487-496 (2002)

497313 Chemical Product Design Using a Novel Computer-Aided Model-Based Tool

Abstract:

In chemical product design, the application of model-based methodologies can help to design/improve products so as to reach the market faster by reducing costly and time-consuming experiments at the early stages of design [1]. That is, experiments are only performed during the last stage as a verification step. Since product candidates involve thousands of mixtures/blends of chemicals that need to be evaluated, a huge amount of data on physico-chemical properties of chemicals and their mixtures and/or models that can reliably predict them are needed for tailor-made design of products. This is a challenging task requiring data acquisition, data testing, model development, model-based design method development, etc, that needs to be integrated within a computer-aided framework so that chemicals based products can be designed, analyzed and verified in a fast, efficient and systematic manner [2].

In this work, the development of a systematic model-based framework for product design and evaluation, implemented in a new product design simulator called ProCAPD is presented. The chemical product simulator works in the same way as a chemical process simulator, but has additional features. Like a process simulator, the chemical product simulator is able to perform calculations on the specified product. Instead of the process flowsheet, the product molecular structure (in the case of a single molecule) or a mixture of molecules is given as input. Analogous to the process conditions, the product application conditions are specified and similar to the mass and energy balance calculations in the process simulator, the properties-functions of the product are calculated. Like the contents of the process simulator, the product simulator needs a database of chemicals and properties, a library of models, numerical routines to solve mathematical problems as well as various calculation options. Also, like the process simulator, the product simulator comes with a user-interface to describe the problems and to obtain the simulation results.

In order to make the chemical product simulator versatile with a wide range of applications, it includes a suite of a built-in knowledge base that has a suit of databases containing properties of different classes of chemicals (lipids, solvents, aroma, polymers, cosmetics, etc.) and blended products (gasoline, jet-fuels, lubricants, hair spray, etc.). It can calculate 55 pure component properties (such as critical properties, acentric factor and solubility parameter) and 10 functional properties (such as vapor pressure, liquid viscosity, and surface tension) and includes databases covering more than 24,000 compounds. Also, it has models for phase equilibria predictions such as UNIQUAC, UNIFAC, NRTL and PC-SAFT. It has special model parameters and data for lipids and ionic liquids. Transport related properties and new interaction parameters for a UNIFAC model for lipids systems have also been incorporated. These features together allow the design of a very wide range chemical-based products. The large amount of data, models and calculation-algorithms are managed through a specially developed ontology. All these capabilities are based on the prototype tool developed by Kalakul et al. (2017) [3].

The presentation will highlight the software architecture, the implemented computer aided methods-tools and the scope-significance will be illustrated through new chemical product design-evaluation applications.

Â

References

[1] M. Hill, 2009. Chemical Product Engineering â€“ The third paradigm, Computer & Chemical Engineering, 33(5), 947-953.

[2] R. Gani, 2004, Chemical Product Design: Challenges & Opportunities, Computers and Chemical Engineering, 28, 2441-2457.

[3] S. Kalakul, S. Cignitti, L. Zhang, R. Gani, 2017, Chapter 3 â€“ VPPD-Lab: The Chemical Product Simulator, Tools For Chemical Product Design, Computer Aided Chemical Engineering, 39, 61-94

499728 Accelerated Process Innovation through Hybrid Computational Modeling

Abstract:

Process innovation in the chemical, food, pharma and consumer good industries has become increasingly dependent on modeling. High fidelity computational models are being regularly used to evaluate novel process concepts and designs. Such models though suffer from two key drawbacks: first, they tend to be computationally very intensive and second, the numerical techniques involved in their solution tend to be complex and require significant domain expertise. Taken together, these two challenges have hampered the widespread adoption of advanced computational models by the larger engineering community.

With desktop computers becoming increasingly powerful of late, the first challenge has been met with some degree of success. The second, however, is a more fundamental challenge. Process development engineers are not always neccessarily well versed in advanced numerical computing. Nevertheless, they are very well versed in building process/engineering models that are based on engineering fundamentals as applied to their specific processes. Thus the second challenge noted above can be overcome by bridging the gap between the two paradigms - computational modeling and process/engineering modeling.

Hybrid computational models which combine the best aspects of computational modeling paradigms (CFD, DEM, LBM etc.) with traditional process/engineering modeling ideas present a viable means to bridge the gap mentioned above. CFD- and/or DEM-based "compartmental" or zonal models are a good example of this class of models. These models use flow information from CFD, DEM models of process equipment to partition the equipment into "zones" (each zone is uniform with respect to some physical quantity characterizing turbulence, flow, mixing etc.). The equipment is represented as a network of zones with each zone exchanging flow information with its neighbours. Each zone can now be modeled as an idealized flow vessel to perform calculations pertinent to the multiphysics of interest (reactions, heat and mass transfer, phase transfer etc.).

In this talk, we show how such zonal models can be used to describe multiscale, multiphysics engineering problems of different levels of complexity. We also describe ZoneIT, a software package that streamlines the hybrid zonal modeling workflow described above. Using ZoneIT, engineers can easily and readily implement hybrid zonal modeling to analyze their processes. We present case studies where ZoneIT has been successfully employed to analyze processes in the food, pharma and chemicals industries.

499076 Single- and Multi-Objective Optimizations Using Parallelized Process Simulators

Abstract:

Equation-oriented (EO) optimization methods are difficult to apply to rigorous thermodynamic methods, discontinuous variables, or systems where the number of equations is not constant. Despite these challenges, EO methods are still quite commonly used for process optimization, due in part to a perception that EO methods are faster than process simulation-based methods.
We have found that by integrating parallel processing techniques into modern process simulation and optimization, we can achieve rapid results comparable to the speed of more traditional EO modeling methods. We can attain these speeds even while using first-principle models, without risking validity of the model through too many simplifying assumptions. In this presentation, we will consider several process examples to explore the advantages of using simulators for optimization projects.

493659 Liquid-Liquid Extraction in Stratified Flow in a Wavy-Wall Microchannel

Abstract:

A theoretical investigation of liquid-liquid extraction in stratified flows in a wavy-wall microchannel is carried out in this work. Several studies in the past have focused on heat transfer augmentation in single phase and two-phase fluid flow in a wavy-wall microchannel. This work focuses on the improvement in extraction in a wavy-wall microchannel. This is studied for two flow configurations i.e. flow (i) parallel to corrugations and (ii) perpendicular to the corrugations. A semi-analytical solution approach is used in which the velocity profile is obtained using a domain perturbation method. The species transport equations are solved using method of lines (MOL) for flow parallel to the corrugations whereas for flow perpendicular to the corrugations, an operator splitting technique is used. CFD simulation are also carried out considering the exact boundary using *Fluent 16.1* to verify the approximate solution obtained by the perturbation method of the mathematical model. The improvement in liquid-liquid extraction in a wavy-wall microchannel is analysed by comparing the performance with the extraction found in a smooth wall microchannel at same flow-rates. Our study shows that a significant improvement in extraction is possible in a wavy-wall microchannel for flow perpendicular to the corrugations.

503400 Correlating Molecular Details to Emergent Phenomena for Colloidal Dispersions

Abstract:

To gain a better understanding of emergent phenomena at macroscopic scales, we need to rigorously understand relevant physicochemical feature at smaller scales, which would immediately bring a challenge on â€œconnecting scalesâ€. This is especially true if one would like to control emergent phenomena. For example, understanding slurry rheology (one of the important emergent phenomena for colloidal dispersions) is critical to control flows and transport of colloidal dispersions in various engineering applications such as nuclear waste processing at the Hanford site. Slurry rheology indeed results from a delicate balance between repulsive and attractive forces between particles, inter-correlating to details at both molecular and aggregate scales (e.g., crystallinity of solid particles and microstructure of aggregates respectively). Therefore, one should build self-consistent frameworks via a combination of experiments, simulations, and theories to implement correlations between molecular details, particle forces, aggregate structures, and slurry rheology. In this talk, I am going to briefly overview our approach and recent studies/on-going research activities to address such challenge.

503401 Interfacial Dynamics of Ionic Liquids under Nanoconfinement

Abstract:

The Ionic liquids (ILs) are a relatively new class of liquids that have been used or considered for use in various important fields, such as energy storage, lubrication, catalysis, etc. While many of those applications often require the ILs near surfaces/interfaces under confined geometry at nano- and/or micro-length scales, the structure-property relationships of ILs and their interfacial responses under confinement are not well understood. This talk is primarily aimed at addressing this knowledge gap. Here, we present normal force and rheological measurements of two types of ionic liquids, [C4mim$^+$][BF4$^-$] and [C8mim$^+$][BF4$^-$], confined in gaps comparable to their molecular dimensions. Forces between two mica surfaces across ILs showed distinctive differences at different degrees of nanoconfinement. In the long range (~10 nm to 50 nm), monotonic repulsive force showed exponential decay with increasing gap distance. The fitting of Poisson-Boltzmann equation in this region resulted in a shorter Debye screening length for [C4mim$^+$][BF4$^-$] compared to [C8mim$^+$][BF4$^-$], implying the species with smaller cation has a higher degree of effective ion pair dissociation. In the short range (< 10 nm), oscillatory features were observed in the force profiles for both species, indicating strategized assembly. By considering and comparing the dimension of ion species and hard-wall distances in the force profiles, we propose a possible bilayer formation in the more amphiphilic pair, i.e. [C8mim$^+$][BF4$^-$] , which is absent in [C4mim$^+$][BF4$^-$]. Dynamic rheological properties such as viscosity and storage/loss moduli were characterized in the short range (for the first time to our knowledge) with different numbers of ion or ion-pair layers across the interface. Viscosity of both ionic liquid species increased by four orders of magnitude from bulk values as the degree of confinement progressed from the long range into the short range oscillatory region, implying the breakdown of continuum behavior. In the long range (bulk) region, both ionic liquid species demonstrated Newtonian liquid behavior. A transition from liquid to solid-like behavior was observed in the oscillatory region, marked by a substantially higher storage modulus in comparison to the loss modulus. We anticipate the findings obtained from this unique experimental approach ultimately serve as new fundamental insights for rational design of ILs with desired interfacial properties in use of fabricating superior functional materials and devices.

503396 Computational Materials Design for Developing High Performance Solid Oxide Fuel Cell Electrodes

Abstract:

Solid oxide fuel cells (SOFCs) operated at high temperature have received much interests as prospective device due to their high efficiency and fuel flexibility. There are, however, several critical issues for enhancing the SOFC performance. First, the rate of reduction, oxidation or transport of oxygen drastically deteriorates at reduced temperature. Second, severe degradation observed in electrode affects long-term stability in negative ways. To resolve these problems, it is essential to rationally develop the materials of SOFC components. Unfortunately, it is not easy to completely achieve it by depending only on conventional experimental methods. In this presentation, we introduce computational approaches to design SOFC electrode materials mainly based on density functional theory (DFT). Our results provide mechanistic information required to understand surface reactions and discover the promising potential materials. Specifically, enhanced electrochemical performance through both the metallic nanoparticle exsolution and the strain-driven chemical stabilization of perovskite electrode surfaces will be discussed. We believe that our computational study will play an important role in improving SOFC performance by guiding or complementing the relevant experiments.

503397 Systems Biotechnology for Understanding and Designing Microbial, Plant and Mammalian Cell Factories

Abstract:

The goals of process systems engineering are to better understand complex real systems and to effectively achieve desirable properties of the systems by exploiting and integrating statistical analysis, modeling, control and optimization techniques. Such systems approaches have been successfully applied to a variety of chemical processes. Recently, special attention has been paid to Systems Biotechnology to elucidate complex biological systems and improve/design their characteristics and properties towards biotechnological applications. The central task of this data- and model-driven research is to comprehensively collect, manage and analyze the global cellular information, i.e., high-throughput omics data, and to generate predictive computational models of the intricate biological processes, i.e., metabolic, signaling and regulatory networks. Thus, better understanding the cellular physiology, regulation and metabolism at the systems level and subsequently designing strategies for achieving desirable states using various systems analysis techniques are a prime target of this research. To this end, we have established core platform technologies and innovative methodological frameworks, developed computational models describing various microbial, plant and mammalian cells, and combined them systematically to characterize and engineer such cell factories towards the enhanced cellular properties. In this talk, several applications of the platform will be presented and current challenges and future direction in the field will be discussed.

503398 Integration of Iterative Learning Control and Model Predictive Control for Point-to-Point Tracking Problem

Abstract:

Iterative learning control (ILC) is an effective control technique for handling repetitive, cyclic, iterative or batch processes which repeat a same task on a finite time interval. ILC was originally introduced for manipulating robots and has been applied to many industrial processes. However, conventional ILC cannot reject real-time disturbances because control actions are calculated based on the tracking error of the previous batch.

Much research has been conducted to incorporate feedback controller into ILC. Among them, the ILC method combined with model predictive control (MPC), referred to as iterative learning model predictive control (ILMPC), has attracted much attention due to their applicability to constrained multivariable control problems in the process industry. The main purpose of the ILMPC method is to track an entire reference trajectory at every time step of the operation. However, the entire reference trajectory is not always relevant for many practical applications such as the robotic â €œpick and placeâ€ task, crane control, rapid thermal process, and chemical batch reactor. The errors at certain time points are only relevant, and regulating the system at the time points of interest is the point-to-point (PTP) tracking problem.

For real-time feedback control of PTP tracking problem, we propose a PTP ILC method combined with MPC, called point-to-point iterative learning model predictive control (PTP ILMPC). Unlike general tracking problems, PTP tracking problems have output errors at desired time points only. Thus, conventional feedback controller cannot be used for PTP tracking problems because it requires output errors at every time step as the real-time feedback. The proposed PTP ILMPC uses state variables instead of output errors of each time step as feedback signals and minimizes the output errors at the specified time points in the future time window. Thus, it can be applied using the desired reference points without an arbitrary reference trajectory that passes through the desired reference points. The proposed method can seamlessly handle tracking, quality control, and constraints handling problems with a single controller. This study also presents the nominal and robust stability analyses of the proposed method.

503399 First-Principles Modeling of Redox Potential of Organic Materials for Lithium-Ion Batteries

Abstract:

In this talk, we present the first-principles density functional theory modeling of redox properties of various organic materials as positive electrode materials for lithium-ion batteries. First, from the density functional theory (DFT) calculations to understand the binding behaviors of Li atoms interacting with the quinone derivatives, we find that the Li atoms prefer to bind with the carbonyl groups in the molecules. Next, we observe that the redox properties of organic materials can be harnessed by designing their chemical structures. In particular, introducing electron-withdrawing functional groups is beneficial to enhance the redox potential. In addition, we present the change in the redox properties of organic materials as a function of the number of bound Li atoms, which would be related to the change of redox characteristics during the discharging process. In general, the redox potential is decreased with increasing the number of bound Li atoms, meaning that the charge capacity is limited by the number of carbonyl groups. However, it should be noted that the functionalization of organic materials with carboxylic acids can improve their charge capacity. Through this study, it is highlighted that the cathodic activity of organic materials during the discharging process relies strongly on the solvation effect as well as on the number of carbonyl groups available for further Li binding.

AUTHOR INDEX

AUTHOR INDEX

AUTHOR INDEX

AUTHOR INDEX

AUTHOR INDEX